Advanced Polymers for Biomedical Applications

Advanced Polymers for Biomedical Applications

Editor

Luis García-Fernández

MDPI • Basel • Beijing • Wuhan • Barcelona • Belgrade • Manchester • Tokyo • Cluj • Tianjin

Editor
Luis García-Fernández
Grupo de Biomateriales
Instituto de Ciencia y
Tecnología de Polímeros
(ICTP), CSIC
Madrid
Spain

Editorial Office
MDPI
St. Alban-Anlage 66
4052 Basel, Switzerland

This is a reprint of articles from the Special Issue published online in the open access journal *Polymers* (ISSN 2073-4360) (available at: www.mdpi.com/journal/polymers/special_issues/Adv_Polym_Biomed_Appl).

For citation purposes, cite each article independently as indicated on the article page online and as indicated below:

LastName, A.A.; LastName, B.B.; LastName, C.C. Article Title. *Journal Name* **Year**, *Volume Number*, Page Range.

ISBN 978-3-0365-4614-8 (Hbk)
ISBN 978-3-0365-4613-1 (PDF)

© 2022 by the authors. Articles in this book are Open Access and distributed under the Creative Commons Attribution (CC BY) license, which allows users to download, copy and build upon published articles, as long as the author and publisher are properly credited, which ensures maximum dissemination and a wider impact of our publications.

The book as a whole is distributed by MDPI under the terms and conditions of the Creative Commons license CC BY-NC-ND.

Contents

About the Editor ix

Preface to "Advanced Polymers for Biomedical Applications" xi

María Puertas-Bartolomé, Ana Mora-Boza and Luis García-Fernández
Emerging Biofabrication Techniques: A Review on Natural Polymers for Biomedical Applications
Reprinted from: *Polymers* **2021**, *13*, 1209, doi:10.3390/polym13081209 1

Nibedita Saha, Nabanita Saha, Tomas Sáha, Ebru Toksoy Öner, Urška Vrabič Brodnjak and Heinz Redl et al.
Polymer Based Bioadhesive Biomaterials for Medical Application—A Perspective of Redefining Healthcare System Management
Reprinted from: *Polymers* **2020**, *12*, 3015, doi:10.3390/polym12123015 27

Julian A. Serna, Laura Rueda-Gensini, Daniela N. Céspedes-Valenzuela, Javier Cifuentes, Juan C. Cruz and Carolina Muñoz-Camargo
Recent Advances on Stimuli-Responsive Hydrogels Based on Tissue-Derived ECMs and Their Components: Towards Improving Functionality for Tissue Engineering and Controlled Drug Delivery
Reprinted from: *Polymers* **2021**, *13*, 3263, doi:10.3390/polym13193263 47

Zhiyuan Zhao, Tong Wu, Yu Cui, Rui Zhao, Qi Wan and Rui Xu
Design and Fabrication of Nanofibrous Dura Mater with Antifibrosis and Neuroprotection Effects on SH-SY5Y Cells
Reprinted from: *Polymers* **2022**, *14*, 1882, doi:10.3390/polym14091882 81

Suheda Yilmaz-Bayraktar, Katharina Foremny, Michaela Kreienmeyer, Athanasia Warnecke and Theodor Doll
Medical-Grade Silicone Rubber–Hydrogel-Composites for Modiolar Hugging Cochlear Implants
Reprinted from: *Polymers* **2022**, *14*, 1766, doi:10.3390/polym14091766 97

Xuhui Sun, Chengcheng Yu, Lin Zhang, Jingcao Cao, Emrullah Hakan Kaleli and Guoxin Xie
Tribological and Antibacterial Properties of Polyetheretherketone Composites with Black Phosphorus Nanosheets
Reprinted from: *Polymers* **2022**, *14*, 1242, doi:10.3390/polym14061242 117

Chih-Hao Chang, Chih-Hung Chang, Ya-Wen Yang, Hsuan-Yu Chen, Shu-Jyuan Yang and Wei-Cheng Yao et al.
Quaternized Amphiphilic Block Copolymers as Antimicrobial Agents
Reprinted from: *Polymers* **2022**, *14*, 250, doi:10.3390/polym14020250 131

Kazuma Sakura, Masao Sasai, Takayuki Mino and Hiroshi Uyama
Non-Woven Sheet Containing Gemcitabine: Controlled Release Complex for Pancreatic Cancer Treatment
Reprinted from: *Polymers* **2022**, *14*, 168, doi:10.3390/polym14010168 143

Pichapar O-chongpian, Mingkwan Na Takuathung, Chuda Chittasupho, Warintorn Ruksiriwanich, Tanpong Chaiwarit and Phornsawat Baipaywad et al.
Composite Nanocellulose Fibers-Based Hydrogels Loading Clindamycin HCl with Ca^{2+} and Citric Acid as Crosslinking Agents for Pharmaceutical Applications
Reprinted from: *Polymers* **2021**, *13*, 4423, doi:10.3390/polym13244423 157

Wei-Lun Qiu, Wei-Hung Hsu, Shu-Ming Tsao, Ai-Jung Tseng, Zhi-Hu Lin and Wei-Jyun Hua et al.
WSG, a Glucose-Rich Polysaccharide from *Ganoderma lucidum*, Combined with Cisplatin Potentiates Inhibition of Lung Cancer *In Vitro* and *In Vivo*
Reprinted from: *Polymers* **2021**, *13*, 4353, doi:10.3390/polym13244353 171

Carmen M. González-Henríquez, Fernando E. Rodríguez-Umanzor, Matías N. Alegría-Gómez, Claudio A. Terraza-Inostroza, Enrique Martínez-Campos and Raquel Cue-López et al.
Wrinkling on Stimuli-Responsive Functional Polymer Surfaces as a Promising Strategy for the Preparation of Effective Antibacterial/Antibiofouling Surfaces
Reprinted from: *Polymers* **2021**, *13*, 4262, doi:10.3390/polym13234262 183

Huidong Wei, James S. Wolffsohn, Otavio Gomes de Oliveira and Leon N. Davies
Characterisation and Modelling of an Artificial Lens Capsule Mimicking Accommodation of Human Eyes
Reprinted from: *Polymers* **2021**, *13*, 3916, doi:10.3390/polym13223916 207

Chuda Chittasupho, Jakrapong Angklomklew, Thanu Thongnopkoon, Wongwit Senavongse, Pensak Jantrawut and Warintorn Ruksiriwanich
Biopolymer Hydrogel Scaffolds Containing Doxorubicin as A Localized Drug Delivery System for Inhibiting Lung Cancer Cell Proliferation
Reprinted from: *Polymers* **2021**, *13*, 3580, doi:10.3390/polym13203580 223

Younghyun Shin, Dajung Kim, Yiluo Hu, Yohan Kim, In Ki Hong and Moo Sung Kim et al.
pH-Responsive Succinoglycan-Carboxymethyl Cellulose Hydrogels with Highly Improved Mechanical Strength for Controlled Drug Delivery Systems
Reprinted from: *Polymers* **2021**, *13*, 3197, doi:10.3390/polym13183197 237

Xiukun Xue, Yanjuan Wu, Xiao Xu, Ben Xu, Zhaowei Chen and Tianduo Li
pH and Reduction Dual-Responsive Bi-Drugs Conjugated Dextran Assemblies for Combination Chemotherapy and In Vitro Evaluation
Reprinted from: *Polymers* **2021**, *13*, 1515, doi:10.3390/polym13091515 253

Felix Reisbeck, Alexander Ozimkovski, Mariam Cherri, Mathias Dimde, Elisa Quaas and Ehsan Mohammadifar et al.
Gram Scale Synthesis of Dual-Responsive Dendritic Polyglycerol Sulfate as Drug Delivery System
Reprinted from: *Polymers* **2021**, *13*, 982, doi:10.3390/polym13060982 271

Bruno Thorihara Tomoda, Murilo Santos Pacheco, Yasmin Broso Abranches, Juliane Viganó, Fabiana Perrechil and Mariana Agostini De Moraes
Assessing the Influence of Dyes Physico-Chemical Properties on Incorporation and Release Kinetics in Silk Fibroin Matrices
Reprinted from: *Polymers* **2021**, *13*, 798, doi:10.3390/polym13050798 285

Marion Gradwohl, Feng Chai, Julien Payen, Pierre Guerreschi, Philippe Marchetti and Nicolas Blanchemain
Effects of Two Melt Extrusion Based Additive Manufacturing Technologies and Common Sterilization Methods on the Properties of a Medical Grade PLGA Copolymer
Reprinted from: *Polymers* **2021**, *13*, 572, doi:10.3390/polym13040572 301

Pedro Guerrero, Tania Garrido, Itxaso Garcia-Orue, Edorta Santos-Vizcaino, Manoli Igartua and Rosa Maria Hernandez et al.
Characterization of Bio-Inspired Electro-Conductive Soy Protein Films
Reprinted from: *Polymers* **2021**, *13*, 416, doi:10.3390/polym13030416 315

Jorge A Roacho-Pérez, Kassandra O Rodríguez-Aguillón, Hugo L Gallardo-Blanco, María R Velazco-Campos, Karla V Sosa-Cruz and Perla E García-Casillas et al.
A Full Set of In Vitro Assays in Chitosan/Tween 80 Microspheres Loaded with Magnetite Nanoparticles
Reprinted from: *Polymers* **2021**, *13*, 400, doi:10.3390/polym13030400 331

Ana Mora-Boza, Elena López-Ruiz, María Luisa López-Donaire, Gema Jiménez, María Rosa Aguilar and Juan Antonio Marchal et al.
Evaluation of Glycerylphytate Crosslinked Semi- and Interpenetrated Polymer Membranes of Hyaluronic Acid and Chitosan for Tissue Engineering
Reprinted from: *Polymers* **2020**, *12*, 2661, doi:10.3390/polym12112661 347

Hafiz Muhammad Basit, Mohd Cairul Iqbal Mohd Amin, Shiow-Fern Ng, Haliza Katas, Shefaat Ullah Shah and Nauman Rahim Khan
Formulation and Evaluation of Microwave-Modified Chitosan-Curcumin Nanoparticles—A Promising Nanomaterials Platform for Skin Tissue Regeneration Applications Following Burn Wounds
Reprinted from: *Polymers* **2020**, *12*, 2608, doi:10.3390/polym12112608 365

María Puertas-Bartolomé, Małgorzata K. Włodarczyk-Biegun, Aránzazu del Campo, Blanca Vázquez-Lasa and Julio San Román
3D Printing of a Reactive Hydrogel Bio-Ink Using a Static Mixing Tool
Reprinted from: *Polymers* **2020**, *12*, 1986, doi:10.3390/polym12091986 385

Mar Fernández-Gutiérrez, Bárbara Pérez-Köhler, Selma Benito-Martínez, Francisca García-Moreno, Gemma Pascual and Luis García-Fernández et al.
Development of Biocomposite Polymeric Systems Loaded with Antibacterial Nanoparticles for the Coating of Polypropylene Biomaterials
Reprinted from: *Polymers* **2020**, *12*, 1829, doi:10.3390/polym12081829 403

About the Editor

Luis García-Fernández

Dr. Luis García Fernández obtained his Bachelor's Degree in Chemical Engineering at the University of Castilla la Mancha in 2004. At the beginning of 2005, he obtained an FPI grant to develop his doctoral thesis in the Biomaterials group of the Institute of Polymer Science and Technology (ICTP, CSIC). He obtained his European Ph.D. degree with the thesis "Synthesis and development of new antiangiogenic polymers". The thesis was qualified "cum laude"and awarded with the European Society of Biomaterials Doctoral Award by the European Society of Biomaterials (ESB). During his PhD, he worked at several well-recognized European Research Centers in Italy and Germany and participated in different projects in the field of polymers for drug delivery systems and tissue regeneration.

After his Ph.D., he spent several years at Max-Planck Institute for Polymer Research (MPIP) in Mainz (Germany) where he developed different research lines and collaborations in the field of polymers for biomaterials and their interaction with the cell environment.

In the year 2014, he returned to the Institute of Polymer Science and Technology, developing different innovative research lines based on tissue regeneration and new treatments with hydrogels based on natural polymers. During this period, he worked at the Center of Biological Research "Margarita Salas"(CIB-CSIC) to conduct bactericidal studies on different polymer surfaces. In the year 2018, he moved to the Centro de Investigaciones Biomédicas en Red (CIBER-BBN) where he continues with his research on tissue regeneration and the development of new technologies for biomedical applications.

Preface to "Advanced Polymers for Biomedical Applications"

Polymers are the largest and most versatile class of biomaterials, being extensively applied for therapeutic applications. From natural to synthetic polymers, the possibilities to design and modify their physical-chemical properties make these systems of great interest in a wide range of biomedical applications as diverse as drug delivery systems, organ-on-a-chip, diagnostics, tissue engineering, and so on.

In recent years, advances in the synthesis and modification of polymers and characterization techniques have allowed the design of novel biomaterials as well as the study of their biological behavior in vitro and in vivo.

The purpose of this reprint is to highlight recent achievements in the synthesis and modification of polymers for biomedical applications for final applications in the field of biomedicine.

Luis García-Fernández
Editor

Review

Emerging Biofabrication Techniques: A Review on Natural Polymers for Biomedical Applications

María Puertas-Bartolomé [1,2,*], Ana Mora-Boza [3,4,*] and Luis García-Fernández [4,5,*]

1. INM—Leibniz Institute for New Materials, Campus D2 2, 66123 Saarbrücken, Germany
2. Saarland University, 66123 Saarbrücken, Germany
3. Woodruff School of Mechanical Engineering and Petit Institute for Bioengineering and Bioscience, Georgia Institute of Technology, 315 Ferst Drive, 2310 IBB Building, Atlanta, GA 30332-0363, USA
4. Institute of Polymer Science and Technology (ICTP-CSIC), Juan de la Cierva 3, 28006 Madrid, Spain
5. Networking Biomedical Research Centre in Bioengineering, Biomaterials and Nanomedicine (CIBER-BBN), Monforte de Lemos 3-5, Pabellón 11, 28029 Madrid, Spain

* Correspondence: maria.puertas@leibniz-inm.de (M.P.-B.); aboza3@gatech.edu (A.M.-B.); luis.garcia@csic.es (L.G.-F.); Tel.: +49-(0)681-9300-214 (M.P.-B.); +1-404-894-2000 (A.M.-B.); +34-915622900 (L.G.-F.)

Abstract: Natural polymers have been widely used for biomedical applications in recent decades. They offer the advantages of resembling the extracellular matrix of native tissues and retaining biochemical cues and properties necessary to enhance their biocompatibility, so they usually improve the cellular attachment and behavior and avoid immunological reactions. Moreover, they offer a rapid degradability through natural enzymatic or chemical processes. However, natural polymers present poor mechanical strength, which frequently makes the manipulation processes difficult. Recent advances in biofabrication, 3D printing, microfluidics, and cell-electrospinning allow the manufacturing of complex natural polymer matrixes with biophysical and structural properties similar to those of the extracellular matrix. In addition, these techniques offer the possibility of incorporating different cell lines into the fabrication process, a revolutionary strategy broadly explored in recent years to produce cell-laden scaffolds that can better mimic the properties of functional tissues. In this review, the use of 3D printing, microfluidics, and electrospinning approaches has been extensively investigated for the biofabrication of naturally derived polymer scaffolds with encapsulated cells intended for biomedical applications (e.g., cell therapies, bone and dental grafts, cardiovascular or musculoskeletal tissue regeneration, and wound healing).

Keywords: biofabrication; microfluidics; electrospinning; 3D printing; electrospraying; natural polymers; cell encapsulation

1. Introduction

Polymeric biomaterials have been developed to provide an artificial matrix that can mimic the cell microenvironment. This artificial matrix needs to provide appropriate biophysical and structural properties (e.g., stiffness, roughness, topography, and alignment) as well as biochemical cues (e.g., signaling, growth factors, and proteins) in order to promote the native capacity of cells to adhere, migrate, proliferate, and differentiate towards the growth of new tissue [1].

Natural polymers extracted from biological systems such as plants, microorganisms, algae, or animals have been used for decades in the biomedical field. These materials retain the biochemical cues and properties necessary to improve their biocompatibility and present similar structures to the extracellular matrix (ECM) of native tissues [2–5]. Therefore, they usually present good cellular attachment, improve cellular behavior, and avoid immunological reactions, although in some cases, these properties are limited due to batch variability within production and purification processes. The most common natural polymers used in biomedical applications include polysaccharides (e.g., alginate [5–7], hyaluronic acid [3,8], and chitosan [9,10]), proteins (e.g., collagen [11], silk [12,13], gelatin [14–16], and fibrin [17]), and bacterial polyesters (e.g., bacterial cellulose [18]). However, the poor mechanical strength of natural polymers frequently makes the manipulation and biofabrication process difficult. For this reason, the use of derivatives or blends with different polymers are usually required to obtain appropriate mechanical properties for their use. An example is the modification of gelatin with methacrylamide to obtain a photopolymerizable biomaterial that can be used for 3D bioprinting and microfluidics [19–22].

Actual biomedical challenges require the use of complex polymer matrixes that can mimic the native ECM and regenerate the lost or damaged tissues [23–25]. Recent advances in biofabrication techniques allow the production of a polymer matrix with biophysical and structural properties similar to the ECM, and its combination with different cell lines is capable of proliferating and differentiating into the desired tissue. Moreover, the incorporation of different growth factors or other biomolecules can improve the migration, growth, and differentiation of the cells [3,26].

Currently, numerous research lines for polymer matrix biofabrication follow two different strategies for the incorporation of the cells: (i) cell implantation on the previously formed polymer matrix and (ii) fabrication of a polymer matrix with encapsulated cells.

The first strategy was used in the last decade, and it is restricted to the method of cell implantation. Normally, these systems do not present a good integration between cells and the polymer matrix, and their efficacy for tissue regeneration depends on the physical properties of the polymer matrix such as hydrophobicity, degradation rate, or stiffness [14,27]. Among the most used techniques, we can highlight layer-by-layer [28,29], melt molding [30], photolithography [31], and self-assembling [32].

The second strategy is the most investigated in recent years, since it allows the fabrication of advanced cell-laden structures with complex cellular microenvironments. Recently, some advanced techniques (i.e., microfluidics [33,34], electrospinning [10,35], and 3D printing [36,37]) allow the integration of cells directly into the polymer matrix with the adequate physical and biological properties to imitate the ECM of the desired tissue.

This review focuses on the biofabrication techniques of microfluidics, electrospinning, and 3D printing using natural polymers. These techniques have been recently explored to create polymer matrixes with embedded cells for biomedical applications, and they are in continuous evolution, as we are going to illustrate in the present review.

2. Microfluidics

Microfluidics has emerged as a powerful tool for the high throughput generation of monodisperse microgels [33,34]. Microgels are defined as 3D-crosslinked particles that provide a porous polymeric network and can recapitulate the cellular microenvironment (i.e., ECM), mimicking in vivo conditions and diffusion of nutrients and metabolic waste [38–40] (Figure 1A,B). Specifically, microgels are fabricated in microfluidic devices by the generation of polymer droplets (i.e., droplet-based microfluidics) through water/oil emulsions followed by physical or chemical crosslinking. The most frequently used geometry configurations to generate the droplets in the devices are T-junction, flow-focusing, and co-flowing (or capillary) laminar streams, which are illustrated in Figure 1C [41–44]. Microgels are especially attractive as cell carriers, because their large surface-to-volume ratio promotes efficient mass transport and enhances cell-matrix interactions, but it is important to notice that cell microencapsulation requires a polymer network that ensures cell viability during microgel preparation and adequate crosslinking chemistry to form a polymer network [45]. Microfluidics technology provides a tight control over microgel chemical properties and composition by easily tuning the flow rates and components in the microfluidic channels, being a versatile biofabrication platform, where different crosslinking strategies can be applied [41]. As mentioned, the microdroplets generated in the microfluidic devices should undergo physical (e.g., electrostatic interaction, thermal gelation, and hydrogen bond interaction) or chemical crosslinking (photopolymerization, Michael addition, and enzymatic reaction) to form solidified microgels [38]. Physical or chemical gelation will be chosen based on different factors like the type of polymer, the strategy for tissue encapsulation, as well as the final biomedical application. In addition, different crosslinking mechanisms can be combined to fulfil the desired features of the microgel systems [38,43,46,47], and these crosslinking processes can take place inside the microfluidic device (in situ crosslinking) or after microgel collecting [38]. In this review, most recent examples of interesting processes for the microfluidics generation of cell-laden microgels prepared using natural polymers and different crosslinking strategies are exposed (Table 1).

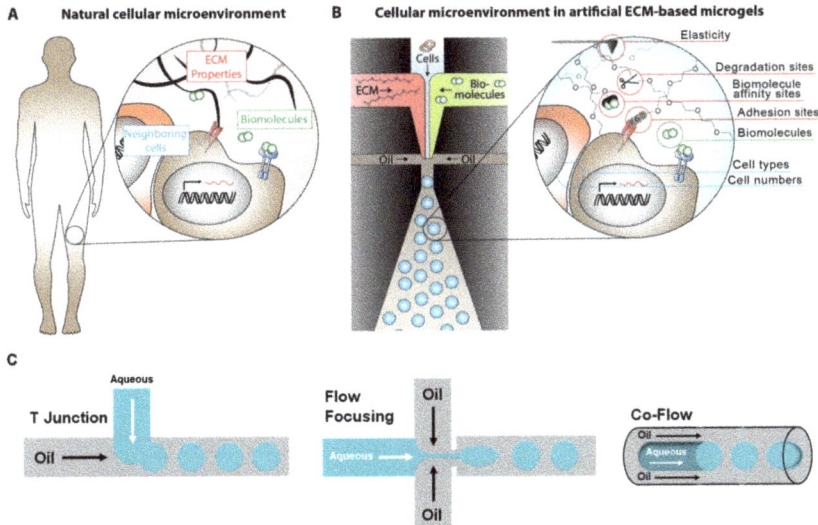

Figure 1. Recapitulating the natural cellular microenvironment in biomimetic microgels using droplet-based microfluidics. (A) The natural cellular microenvironment is composed of different cell types, ECM, and biomolecules such as growth factors. (B) Droplet-based microfluidics allows for versatile and high throughput generation of cell-laden microgels that can mimic the natural cellular environment. By mixing defined amounts of selected cells, ECM, and biomolecules, the microenvironment can be designed in a bottom-up approach with defined properties [34]. (C) Schematic illustration of different types of droplet generators, including T-Junction, flow-focusing, and co-flow (capillary) configurations. Adapted with permission from John Wiley and Sons Copyright®.

Table 1. Summary of the studies exposed in this review regarding microfluidics generation of cell-laden microgels.

Polymer	Microfluidics Approach	Crosslinking Strategy	Microgel Size Range	Additives	Cell Type	Ref.
Alginate	Flow-focusing	Ionic crosslinking (Calcium-EDTA)	10–50 µm	No	MSCs	[48]
Alginate	Flow-focusing	Ionic crosslinking (calcium)	20–50 µm	poly-D-lysine	bMSCs	[49]
Alginate	Flow-focusing	Ionic crosslinking (Calcium-EDTA)	≈140 µm	PNiPAM	HepG2	[33]
Alginate	Centrifugal microfluidics	Ionic crosslinking (calcium)	Tunable (also fibers)	No	HepG2	[50]
Alginate	Double emulsion (w/o/w) flow focusing	Ionic crosslinking (calcium)	≤200 µm	Collagen	Hepatocytes and endothelial cells	[11]
Acrylamide hyaluronic acid	Flow-focusing	Enzymatic reaction and photopolymerization	≈80 µm	No	Human dermal fibroblasts	[8]
Furylamine and tyramine hyaluronic acid	T-junction	Enzymatic crosslinking, Diels-Alder click chemistry, or a combination	≈250 µm	MAL-PEG-MAL	ATDC-5 cells	[51]
N-carboxylic chitosan	Asymmetric cross-section	Schiff base reaction	≈200 µm	Oxidized dextran	NIH-3T3 fibroblasts	[52]
Chitosan Lactate	Flow-focusing	Ionic crosslinking (G_1Phy and TPP)	100–130 µm	No	hMSCs	[45]
GelMA	Double flow-focusing	Photopolymerization	100–200 µm	No	macrophages	[20]
GelMA	Capillary	Photopolymerization	≈165 µm	no	bMSCs	[53]
GelMA	T-junction	Photopolymerization	300–1100 µm	PEGDA Poly(ethylene glycol)-fibrinogen	ECFCs breast cancer cells hiPSCs	[21]
GelNB	Capillary	Photopolymerization	300–600 µm	PEG-SH	bMSCs	[22]
Thiolated gelatin	T-junction	Thiol-Michael addition reaction	100–250 µm	Vinyl sulfonated hyaluronic acid	bMSCs	[16]
Dextran-tyramine	Flow-focusing	enzymatic crosslinking	120–200 µm	No	hMSCs	[54]
Dextran	Flow-focusing	Ionic crosslinking(calcium)	≈90 µm	PEG and Alginate	rat pancreatic islet	[55]
Methacrylated heparin	Flow-focusing	Michael addition	60–120 µm	PEG diacrylate monomers with 8-arm PEG-thiol	mESCs	[56]

2.1. Naturally Derived Polymer Used for the Preparation of Cell-Laden Microgels Using Microfluidics

2.1.1. Alginate

Alginate is a classic polymer used for the generation of microgels through microfluidics [38]. Typically, an aqueous alginate solution is emulsified in an oil phase and crosslinked ionically with bivalent ions such as Ca^{2+}, which can be found for example in $CaCl_2$ or $CaCO_3$. The ionic crosslinking process occurs immediately upon contact of alginate chains and Ca^{2+} ions [48]. Kumacheva and co-workers have reported the most representative works of alginate-based cell-laden microgels for the last years [43,46,57,58]. Alginate microgels can be prepared through an internal or external gelation approach [38]. In the internal crosslinking, also called in situ crosslinking methodology, alginate is exposed directly to the crosslinking agent, triggering gelation [59]. In the external approach, alginate droplets are firstly formed and then put into contact with the crosslinker solution [60]. This last approach provides a better control over the final morphology of the microgels [38].

Utech et al. [48] reported an interesting method for the fabrication of alginate microgels using a water-soluble calcium-ethylenediaminetetraacetic acid (calcium-EDTA) complex as a crosslinking agent. They were able to encapsulate individual MSCs with high cell viability due to the mild polymerization approach. Moreover, encapsulated MSCs grew and proliferated over two weeks. Encapsulation of MSCs in polymeric microgels is an excellent approach to improving cell persistence and immunomodulation [61]. Cell therapies based on MSCs are particularly interesting in ameliorating immune-related diseases and dysregulations, but they are limited due to short in vivo persistence [44,45,61,62]. Mao et al. [61] reported the encapsulation of MSCs in alginate-polylysine microgels using a microfluidic device. The encapsulated MSCs in their microgel formulation significantly increased their in vivo persistence after intravenous injection and responded to inflammatory cytokines, improving immunomodulatory effect of MSCs in a model of allogeneic transplantation.

Alginate has been also combined with synthetic polymers to fabricate microgels [33,63,64]. Chen et al. [33] synthetized a functional diblock copolymer, alginate-conjugated poly(N-isopropylacrylamide) (PNiPAM), to fabricate microgels in a flow-focusing device also using calcium-EDTA complex as a crosslinking agent. The permeability of the as obtained microgels could be modified by controlling temperature at low critical solution temperature (LCST), and the encapsulated human hepatocellular carcinoma cell (HepG2) showed high cell viability thanks to the mild conditions of the crosslinking process.

In recent years, other interesting microfluidic methodologies have been developed to overcome some limitations of conventional droplet-based microfluidics. This is the case for the work carried out by Cheng et al. [50], where an efficient centrifugal microfluidic system for controllable fabrication of simple structured alginate hydrogel beads and fibers was exposed. Among the advantages of centrifugal microfluidics, it is highlighted by the use of simple experiment facilities (i.e., a centrifuge) and the absence of an oil continuous phase and subsequent necessary washing steps. HepG2 cells were encapsulated in the developed alginate capsules and fibers, demonstrating the high validity of the method, and showing excellent potential for biomedical applications. Other studies have optimized a water-in-oil-in-water (W/O/W) double emulsion methodology to encapsulate cells. Chan et al. [11,35] developed a double emulsion platform to encapsulate rat hepatocytes and endothelial cells using a combination of alginate and collagen. The developed microgels provided an excellent physical support for the spheroids.

2.1.2. Hyaluronic Acid

Hyaluronic acid-based building blocks have been prepared by pseudo Michael addition crosslinking in a microfluidic device [8,38]. Sideris et al. [8] reported a microfluidics methodology for the fabrication of hyaluronic acid microgels that could self-assemble to form a biodegradable scaffold through two orthogonal chemistries. Human dermal fibroblasts were seeded after microgel preparation, demonstrating good cell spreading after two days of culture. Ma et al. [51] developed a new hyaluronic acid derivative, furylamine and tyramine hyaluronic acid, that can be crosslinked using enzymatic crosslinking, Diels-Alder click chemistry, or a combination of both methods. The versatility of this strategy provided control over crosslinking time and elasticity by simply switching the crosslinking strategy. After evaluating the mechanical properties, gelation time, microgel size, swelling, enzymatic degradation, and bioactivity of the obtained microgels, the group concluded that the microgels synthetized through the combination of both crosslinking methods were the most promising candidates for cell encapsulation and delivery, since the use of the strategies alone resulted in low elasticity and poor cell encapsulation performance.

2.1.3. Chitosan

Jang et al. [52] proposed a new in situ crosslinking methodology for the fabrication of microgels by merging two droplets of different viscosities in an asymmetric cross-junction microfluidic device. Thus, oxidized dextran (ODX) and N-carboxymethyl chitosan (N-CEC) were mixed to undergo an in situ crosslinking via a Schiff base reaction, resulting in microgel formation. This asymmetric cross-junction geometry was an interesting approach to overcoming the high surface tension of microdroplets of contrasting viscosities, which usually requires significant surfactant concentrations. In addition, the crosslinking methodology allowed the encapsulation of NIH-3T3 fibroblasts, which showed high viability after two days of culture, demonstrating the biocompatibility of the entire process. Mora-Boza et al. [45] reported the fabrication of hMSCs-laden microgels, applying also an in situ crosslinking approach for chitosan lactate (ChLA), a water-soluble chitosan derivative. The ionotropic gelation was based on a combination of glycerylphytate (G_1Phy) and tripolyphosphate (TPP) as ionic crosslinkers, obtaining polymeric microgels with homogeneous size distribution between 104 and 127 µm. The authors demonstrated that the presence of G_1Phy, which has been recognized as a potent antioxidant and bioactive compound [65], supported encapsulated hMSC viability over time and modulated hMSCs secretome at adverse conditions, resulting in an appealing cell delivery platform for hMSCs therapy applications.

2.1.4. Gelatin

Photocrosslinkable gelatin derivative, methacrylated gelatin (GelMA), has been widely applied for the preparation of cell-laden microgels through microfluidics [15,38]. Lee et al. [20] developed microtissues containing macrophages through flow-focusing microfluidics using GelMA as a macromer solution. The macrophages' viability was well maintained, and mechanical properties of the microgels could be controlled through GelMA concentration, which had a strong influence on the proliferation and polarization of the encapsulated cells. A similar methodology was applied by Weitz's group to encapsulate bone marrow derived MSCs. The authors demonstrated that the encapsulated cells migrated to the surface of the microgels after four weeks of culture, indicating their capacity to participate in regenerative processes. Moreover, they demonstrated in vivo osteogenic potential by increasing the percentage of calcium deposits and expression of bone-related proteins like BMP-2 [53].

Photocrosslinkable gelatin has also been applied in combination with other synthetic polymers like polyethyleneglycol (PEG) [21,22]. Seeto et al. [21] used a custom designed microfluidic device with a T-junction geometry that allowed the production of microgels with a wide range of diameters from 300 to 1100 µm. The group used a combination of poly(ethylene glycol) diacrylate (PEGDA), poly(ethylene glycol)-fibrinogen (PF), and GelMA, which underwent fast photocrosslinking using a full spectrum light source and Eosin Y as a light photoinitiator. High cellular densities of different cell lines, including horse endothelial colony forming cells (ECFCs), breast cancer cells, or human induced pluripotent stem cells (hiPSCs), were encapsulated in the microspheres, showing good cell distribution, high viability, and functional cellular activities. Forsythe's group combined PEG with another photocrosslinkable derivative of gelatin, gelatin norbornene (GelNB), to fabricate cell-laden microgels using visible light. The encapsulated hBMSCs in the GelNB-PEG microspheres demonstrated chondrogenesis properties when incubated with chondroinductive media, including significant upregulation of collagen-II expression in comparison to bulk hydrogels [22]. Cartilage repair properties have also been observed in the stem cell-laden microgels reported by Feng et al. [16]. In their work, thiolated gelatin and vinyl sulfonated hyaluronic acid were mixed in a microfluidic device to generate microgels through a thiol-Michael addition reaction. Encapsulated bMSCs showed excellent viability, proliferation, and chondrogenic properties. Furthermore, the in vivo experiments demonstrated that the cell-laden microgels were injectable and could self-assemble into cartilage-like structures, providing an effective method for cartilage tissue regeneration, since they were able to inhibit vascularization and hypertrophy.

2.1.5. Dextran

Dextran application in microfluidics technology has also been explored [38,54,55]. Henke et al. [54] developed very stable dextran-tyramine microgels through enzymatic crosslinking for hMSCs encapsulation, which demonstrated significantly higher cell viability in comparison to PEGDA and alginate microgels. Dextran-based microgels supported cells' metabolic activity and allowed cell analysis for 28 days of culture. Liu et al. [55] combined dextran with PEG to generate water-in-water droplets in a cross-flow microfluidic device. This strategy allowed avoiding the use of organic solvent and its subsequent removal. The droplets could be also used as templates for the fabrication of alginate microgels. Moreover, the platform was demonstrated to be a promising system for tissue engineering applications, since the encapsulated rat pancreatic islets maintained high viability and the function of insulin secretion after seven days of culture.

2.1.6. Heparin

Heparin has been combined with PEG to generate bioactive microgels via Michael addition to encapsulate and enhance the differentiation of mESCs [38,56]. Siltanen et al. [56] mixed heparin methacrylate and PEG diacrylate monomers with 8-arm PEG-thiol to fabricate bioactive microgels that provided a suitable environment for endodermal differentiation. The authors also incorporated growth factors FGF-2 and Nodal to evaluate the differentiation processes of the encapsulated cells, showing that 3D differentiation processes significantly upregulated the expression levels of endoderm markers.

2.2. Future Perspectives in Fabrication of Cell-Laden Microgels through Microfluidics

Microfluidics is a versatile technology to generate monodisperse cell-laden microgels, whose properties can be easily tuned if a sensible selection of biomaterials and crosslinking strategies is applied [38,47,66]. These microgels can be applied as building blocks that can self-assemble into mesoscale tissue structures and replicate structures of native tissues [8,67,68]. Nevertheless, some limitations must be overcome before clinical implementation of microfluidic microgels see a bright future [38]. One of the foremost concerns is related to the scalability of current microfluidic strategies. A higher and more robust mass production of cell-laden microgels is necessary to scale-up this technology and be able to obtain macroscale tissue assemblies that can be implemented in the clinic [38,67]. Therefore, new devices that can support large-scale production of microgels with complex geometries, such as core-shell morphology, are needed [38]. Another limitation of current cell-laden microgels is the lack of proper vascularization. Vascularization is essential for effective tissue implantation. Thus, a biomimetic tissue construct should contain essential elements like different cell lines, ECM components, and a vasculature network to maintain cellular interaction and normal tissue function [38,69–71]. Regarding this issue, many efforts are being made in recent years to develop devices and strategies to fabricate vasculature in the hydrogels or incorporate well-perfused vasculature networks through microfluidics systems [38].

3. Cell-Electrospinning (CE) and Bio-Electrospraying (BES)

Electrospinning is a well-known technology that allows the fabrication of micro/nanofiber scaffolds using different synthetic and natural polymers [72–75]. A typical electrospinning set up requires a nozzle tip, a high voltage supply, a pump to control flow rate, and a grounded collector. The process is based on the application of an electric field between the metallic syringe needle and the grounded collector, while the polymer solution is pumped out from the needle at a controlled rate. A conical shape called a "Taylor cone" is generated at the end of the nozzle. When the electrostatic forces within the cone are higher than the surface tension of the solution, the polymer generates a jet, and it is accelerated toward the collector plate, forming a randomly oriented nanofibers mat [75,76].

The concept of electrospinning was first introduced by Anton Formhals in the 1930s. Since then, it has been widely applied in numerous fields such as textiles, agriculture, filtration, sensors, and the biomedical area [24,77–83]. Specially, this technique has had a great impact on the area of tissue engineering, since it presents several advantages: the nanofiber mats can create complex structures that can simulate the native structure of the ECM, promoting the normal functions of cells; it has an easy manufacture and availability; the high porosity and high surface area due to the nano size of the fibers enhance cellular activities such as cell attachment proliferation and differentiation [84–89].

Electrospinning has been a great advancement in the context of biomedical and tissue engineering applications. However, this technique presents some limitations, i.e., the use of cytotoxic solvents, and poor cell infiltration and distribution, since the cell seeding and incubation take place after the substrate processing. In order to overcome these limitations, a new methodology was developed called cell-electrospinning (CE), which differs from the conventional electrospinning on the use of living cells. CE consists of the application of the electrospinning process to a polymer solution combined with living cells, generating electrospun fibers with embedded cells. Figure 2a shows the schematic setup of the CE technique [90,91]. Jayasinghe et al. introduced this methodology for the first time in 2006. In this study, the authors were able to encapsulate living astrocytoma (1321N1) cells into polydimethylsiloxane electrospun fibers using a coaxial methodology. Cell viability, metabolic activity, and cell proliferation were examined and proven to be maintained for six days [90,92]. After this innovative study, the concept of CE was extended to different cell lines (stem cells, osteoblasts, cardiac myocytes, or neuroblastoma) and materials (polyvinyl alcohol (PVA), alginate, or Matrigel) [6,91,93–97]. Today, CE presents a breakthrough in polymer scaffolds processing, and offers remarkable opportunities in the area of biomedicine.

On the other hand, electrospray is a technique analogous to electrospinning that can be performed using the same device. The main difference between both techniques relies on the jet of the polymer generated after the high voltage application. In this case, the resulting jet suffers continuous break-ups, and the aerosolization of the solution takes place, resulting in the production of polymeric nanoparticles [98,99]. The properties and size of the particles will depend on the material and processing parameters. Particularly the viscosity of the polymer solution is a crucial parameter that can act as a switch between electrospinning and electrospraying [100]. Electrospray has been broadly used in different fields such as sensors, food processing, and biomedical applications due to its simplicity and ability to process different polymers. Even though it was developed before ES, today the use of electrospray is less common than electrospinning for the processing of polymer solutions.

Similarly to CE, the technique bio-electrospray (BES) was developed for the preparation of nano/microgels encapsulating living cells (images of living structures fabricated using BES are shown in Figure 2b,d). Jayasinghe et al. were also pioneers in using this technique in 2006 [92]. In this work, the group processed Jurkat cells obtaining deposited droplets in the range of tens of micrometers. BES does not influence cell viability, which has been verified on cell lines such as sperm or stem cells [101–105]. This methodology has been proven to be a useful tool for cell encapsulation, the controlled deposition of cells on planar surfaces, drug delivery, and immunotherapy [102,106].

The successful application of these techniques requires that the viability and bifunctionality of the encapsulated cells be not negatively affected during the process. Several conditions such as the solution's viscosity, electric field applied, distance to the collector, or feed rate can influence the fiber size and shape, as well as the viability of the loaded cells. Therefore, controlling both material and processing parameters is essential to avoid stress damage of cells during the process, and therefore to provide high cell viability values of encapsulated cells. In this review, we extensively investigate the most important research studies carried out to fabricate cell-laden scaffolds using selected natural polymers and applying CE and BES technologies. It is expected that this review can serve as a reference tool and can give a better understanding of the methodologies for future research works.

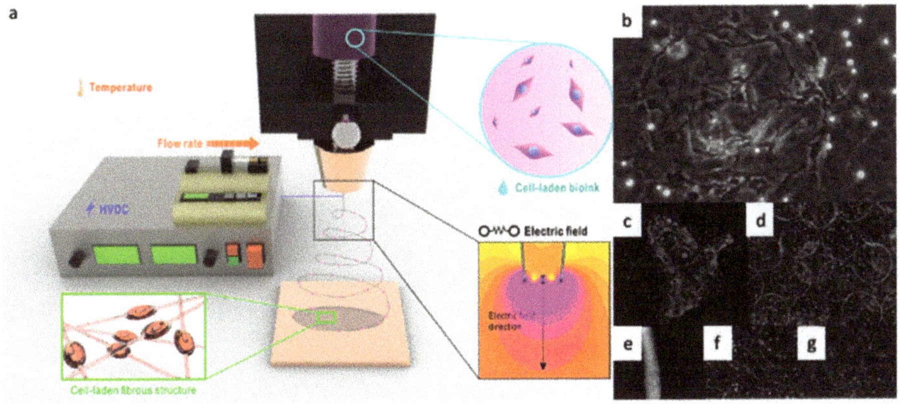

Figure 2. (**a**) Cell-electrospinning process with the basic components. Adapted from [79,91] with permission from MDPI. (**b–d**) Images of living structures fabricated using either BES or (**e–g**) CE. (**b**) Characteristic optical image illustrating a four cell culture system created in three dimensions (the image depicts cellular networks in three dimensions). (**c**) Confocal microscopy image of a three-dimensional culture prepared with the three major cell types of the myocardium (cardiac myocytes, endothelial cells, and fibroblasts). (**d**) Immobilized cells as composite living beads. (**e**) A vessel formed with cells embedded in the individual fibers. (**f,g**) The fiber configurations that could be altered from containing a single cell to a heterogeneous cell population. Adapted from [103] with permission from Wiley Materials.

3.1. Naturally Derived Polymers for CE and BES

Electrospinning and electrospraying techniques have shown great processability using both naturally and synthetically derived polymers. Synthetic polymers such as poly(dimethylsiloxane), polycaprolactone, or polylactic acid have been widely used for electrospinning and electrospray because of their versatility in the selection of the solvent that can provide adequate viscosity, conductivity, and surface tension of the polymer solution [91,106,107]. However, viability of encapsulated cells is highly affected by the use of the organic solvents commonly used, such as tetrahydrofuran, acetone, or chloroform [108]. Therefore, natural polymers such as alginate, collagen, and cellulose, compatible with non-toxic solvents, are frequently used for achieving biofabrication using living cells by cell-electrospinning and bio-electrospraying methodologies. Naturally derived polymers possess numerous advantages such as high cell affinity, low immunogenicity, or ECM biomimetic properties [109]. However, their low mechanical strength makes them challenging to process by electrospinning or electrospray [110]. These limitations can be overcome by blending natural polymers with synthetic polymers as well as modifying the processing methodology by employing a core-shell nozzle [109,111,112].

In this review, we will make an overview of the different biomaterials based on natural polymers and blends used for CE and BES techniques and the processing procedure used to obtain mechanically stable systems.

3.1.1. Alginate

Alginate is a low-cost biodegradable polymer biocompatible with numerous cell lines, but it also presents a poor mechanical strength. Alginate was used by Xie et al. for entrapment of living cells by BES technology. In this study, the droplet formation was analyzed to optimize the production of monodisperse cell-laden microcapsules with controllable size. The electrospray procedure was performed in dripping mode, and an additional ring electrode was used to improve stabilization. This modified set up allowed the successful encapsulation of Hep G2 cells into calcium alginate microbeads with narrow size distribution and more controlled conditions compared to conventional electrospray [113].

Several works have prepared blends of alginate with other polymers to increase its electrospinnability. For example, Yeo et al. described a cell-electrospun system based on a blend of alginate, poly(ethylene oxide) (PEO), and lecithin to encapsulate MG63 osteoblast cells for their application in bone regeneration [6]. The concentration of the polymers as well as the electric field applied were optimized to ensure adequate cell viability as well as mechanically stable nanofiber mats. As a result, the highest cell viability for encapsulated osteoblasts (around 80%) was obtained with 2×10^5 MG63 cells/mL, 2 wt% alginate, 2 wt% poly(ethylene oxide), and 0.7 wt% lecithin subjected to a 0.16 kV/mm electric field. Moreover, osteogenic differentiation of the cells was confirmed after 10 days of culture. Subsequently, hybrid scaffolds with high mechanical strength were prepared combining the cell-laden electrospun fibers and poly(e-caprolactone) microstructures prepared by 3D printing. It can be said that the cell-laden electrospun scaffolds enhanced the potential of 3D structures for bone regeneration, providing a high surface area and ECM-like structure.

Alginate/PEO blend was also used by Yeo et al. for encapsulating C2C12 myoblast cells by CE for skeletal muscle regeneration [114]. In this work, a high cell viability (around 90%) was obtained for encapsulated cells with an applied electric field of 0.075 kV/mm. It must be highlighted that alignment of the cell-laden fibers in the mat during the CE process allowed the achievement of highly aligned cells, which is proven to facilitate myogenic differentiation. Therefore, this study provides a new tool for achieving cell topographical cues by controlling fibers' orientation, which can be very advantageous, especially for muscle regeneration.

In order to go a step further in this direction, this group proposed a method to prepare a hierarchical platform with a topographical cue for co-culture of human umbilical vein endothelial cells (HUVECs) and C2C12 myoblasts cells [115]. An alginate/PEO blend was again used to develop aligned HUVECs-laden fibers by uniaxial CE, and cell viability at different electric fields was studied (schemes of native skeletal muscle structure and CE process, SEM and live/dead images, and quantitative analysis of orientation cell viability are presented in Figure 3a–e). Encapsulated HUVECs presented high cell viability (around 90%) at 10.5 kV of electrical field, homogeneous cell distribution, and efficient cell growth. The mat was combined with PCL/collagen struts prepared by 3D printing as a physical support. C2C12 cells were then seeded on the cell-laden fibers and co-cultured to facilitate myoblast regeneration. As a result, scaffolds containing HUVECs-laden electrospun fibers with a highly aligned topographical cue were able to enhance the myogenic-specific gene expressions.

3.1.2. Gelatin

Gelatin is a collagen derivative with great biodegradability and biocompatibility, but low mechanical strength, which limits its fiber-forming ability [14]. Nosoudi et al. have demonstrated the successful production of cell-laden nanofibers using a gelatin/pullulan blend [116]. In this work, the electrospinnability of gelatin is enhanced by the presence of pullulan that increases the tensile strength of the blend. An 8 kV voltage and a concentration of 5 mg/mL gelatin/pullulan were used during the process, and adipose-derived stem cells (ADSCs) encapsulated within the fibers presented a 90% viability. This work offers a new area to be studied, since the use of gelatin for CE has been restricted until now by its mechanical properties.

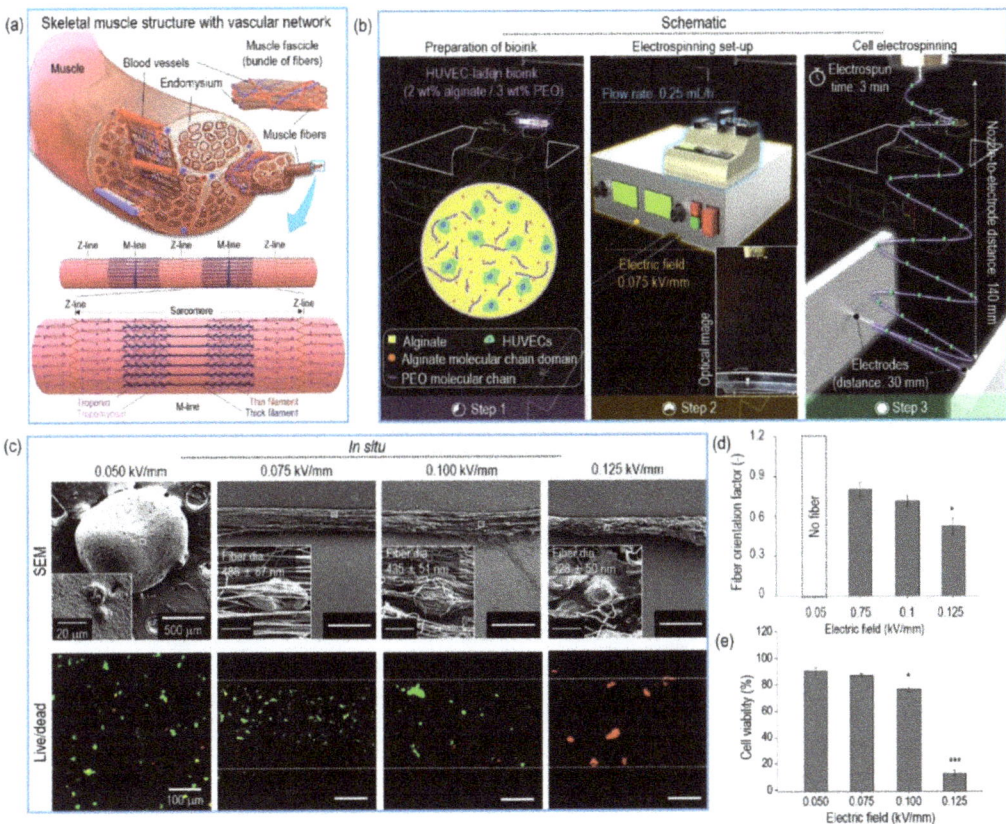

Figure 3. Schematics of (**a**) a native skeletal muscle structure with a vascular network and (**b**) the cell electrospinning process using human umbilical vein endothelial cells (HUVECs). (**c**) SEM and live/dead images of HUVECs-laden fibers fabricated using various electric fields. A quantitative analysis of (**d**) orientation factor of nanofibers and (**e**) cell viability where the analysis of variance (ANOVA) was used for the multiple comparisons and $p\ ^* < 0.05$, $p\ ^{**} < 0.01$, and $p\ ^{***} < 0.001$ indicate the statistical significance. Adapted from [115] with permission from Elsevier.

3.1.3. Fibrin

Fibrin matrix is formed by the polymerization of fibrinogen and thrombin in blood plasma. Due to good biocompatibility and fast biodegradability, it has been widely investigated for tissue engineering applications such as skin, cardiovascular, or musculoskeletal tissue regeneration. However, the mechanical properties of the fibrin matrix are very low. In a recent study by Guo et al., a fibrin matrix was used for cell encapsulation using CE technology [17]. C2C12s murine myoblasts were loaded as cellular aggregates (80–90 µm in diameter) into a fibrin/PEO polymer solution (schematic of CE process and cell suspension, cell-laden scaffold, and live/dead images are shown in Figure 4a–e). PEO was used to improve the mechanical properties of fibrin, as previously observed with alginate [6,114]. Electrospinning parameters were optimized to obtain homogeneous cells distribution inside the fibers and good proliferation after exposure to a 4.5 kV electric field and seven days of incubation. Moreover, myogenically induction provided elongated and multinucleated cells, demonstrating that encapsulated cells remained reactive to biological cues.

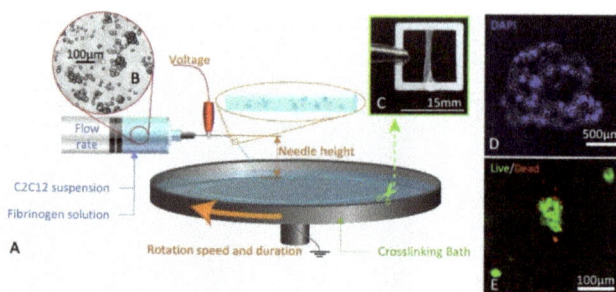

Figure 4. C2C12 can be electrospun into fibrin scaffolds. (**A**) Schematic of cell-laden wet-electrospinning setup identifying key parameters. (**B**) Bright field image of an aggregated cell suspension. (**C**) Cell-laden scaffold wrapped around ABS frame. (**D**) Cross-section of a cell-laden scaffold on Day 0 stained with DAPI (blue, nuclei). (**E**) High (20×) magnification of cell-laden microfiber bundles showing live (green) cells and dead (red) cells. For interpretation of the references to color in this figure caption, the reader is referred to the web version of this paper. Adapted from [17] with permission from Elsevier.

3.1.4. Collagen

Collagen is one of the most abundant proteins in mammals and a main protein of the ECM. It is highly biocompatible and relatively non-immunogenic. Matrigel™ is derived from extracts of Engelbreth-Holm-Swarm mouse tumors and consists principally of collagen type IV, entactin, perlecan (heparan sulfate proteoglycan), and laminin.

CE technology was applied to a Matrigel-rich collagen biopolymer to encapsulate primary cardiomyocytes within fibers for the first time [117]. In this case, the applied voltage was 230 V, which resulted in cell viability values of around 80%, similar to the controls. Immunofluorescence staining exposed that the integrity of the encapsulated cells was maintained after the CE process. Combination of CE methodology with this biopolymer system allowed creating 3D cardiac patches that were demonstrated to enhance the cardiac tissue regeneration.

Matrigel with a high concentration of laminin was used in another work by Sampson et al. to encapsulate N2A mouse neuroblastoma cells [95]. In this study, CE technology was compared to aerodynamically assisted bio-threading (AABT). Samples prepared by CE presented cell viability values from 60% to 85% until three days of incubation. In vivo evaluation in mice was performed, demonstrating a good biocompatibility of the electrospun samples and the CE technique, compared to the control by AABT.

3.2. Future Trends

CE and BES are emerging biomedical techniques with great capabilities for living cells' encapsulation into nano/microscale fibers. They allow the preparation of cell-laden scaffolds with high surface area and ECM-like structure using a simple methodology. Variation of both material's and processing parameters can be controlled, which has demonstrated to have a direct effect on cell viability. Embedded cells have been proven to present good cell viability values and proliferation, and are responsive to cell cues. Moreover, alignment of the fibers can guide the cells to grow in the fiber direction. Therefore, these technologies have been demonstrated not to have a negative effect on cells for optimized processing conditions. However, some challenges still need to be addressed. Preliminary in vivo studies tested on animal models have demonstrated a good biocompatibility of CE, but more research studies are necessary to assess the efficacy and true applicability for tissue regeneration. Improvement of the mechanical properties of the mats and the cell density is still required. In addition, restrictions to developing 3D structures must be solved. These limitations can be overcome by combining these techniques with other biofabrication methodologies such as 3D printing. In this way, more complex scaffolds that are able to better simulate the complexity of native tissues can be achieved and the range of potential biomedical applications can be expanded.

4. 3D Printing

3D printing is a technology with the capacity to create objects by adding materials layer by layer using computer aided design (CAD) software [109,118]. This technology converts an object into sliced horizontal cross-sections that can be printed layer by layer to sort out the complete object in 3D (Figure 5). This technology allows the preparation of complex scaffolds for tissue engineering in a fast and low-cost way without the use of other expensive techniques [119]. One of the main advantages is the capacity to prepare low-volume scaffolds with the appropriate geometry to use in tissue engineering, allowing huge advances in implant materials and personalized scaffolds.

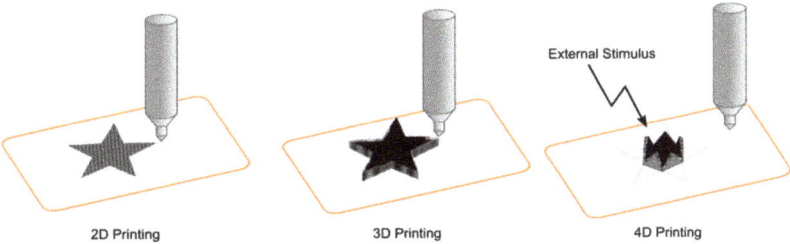

Figure 5. Differences between 2D, 3D, and 4D printing.

The development of new bioinks for 3D bioprinting has attracted attention in recent years. A bioink for biomedical applications can be defined as a formulation that can contain biologically active components and cells and is suitable for processing by an automated biofabrication technology [120]. The limitations of polymer-inspired bioink are toxicity, presence of toxic degradation products, and immune response between others. Recent studies have tried to develop new biocompatible bioinks and also polymer-free "bioink" consisting only of cells [121].

In recent years, 3D printing has been evolving into 4D printing. This technology is based on a shape transformation of the printed object in response to external stimulus, such as light, humidity, magnetic fields, enzymatic reactions, pH changes, or peptide detection [122,123] (Figure 5).

Therefore, functional 3D objects with the capacity to respond to biological conditions have been reported [123,124]. One of the main functions of 4D printing is the production of flexible-wearable biosensors with the capacity to detect small metabolites. In this sense, Nesaei et al. develop a bioink based on Prussian Blue and glucose oxidase enzyme solution to print two different microelectrodes that detect glucose in a concentration range between 100 and 1000 µM [125].

4D bioprinting also tries to include the use of cells to print living cellular structures with the capacity to evolve over time. The capacity to change the structure after receiving a stimulus could modify cell behavior and allow the formation of complex structures for tissue engineering [122]. For example, Kirillova et al. reported and advanced 4D bioprinting that allowed the fabrication of self-folding tubes based on hyaluronic acid and alginate. In this system, bone marrow stromal cells were encapsulated in a methacrylated alginate bioink and printed in different layers in combination with methacrylated hyaluronic acid layers. The system was crosslinked using a green light that is safe for the cells. Due to the difference in crosslinking degree between layers, the 3D bioprinted scaffolds have the capacity to fold forming tubes with the cells homogeneously distributed on the surface [126].

The technology used for 3D printing is an important factor to determine the resolution capacity, velocity, and cell viability. The most important technologies used in the biomedical field are inkjet, extrusion, laser-assisted, and stereolithography bioprinting (Figure 6) [127].

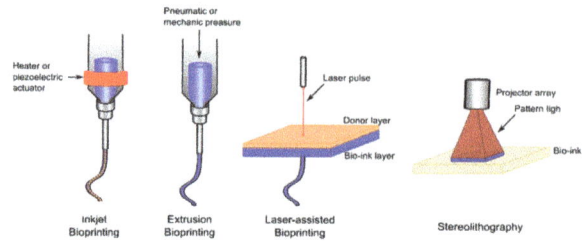

Figure 6. Schematic representation of bioprinting technologies.

Inkjet bioprinting is a droplet-based bioprinting system where the polymer solution in the chamber is extruded through a nozzle and droplets are generated on demand by the breaking of surface tension. The droplet can be generated using a thermal actuator, a piezoelectric actuator, or electrostatic forces. This technique only works with low viscosity liquids with low cell density [128].

Extrusion-based bioprinting is the most common and inexpensive technique. It is able to produce structures by staking multiple layers of a bioink by extrusion of a polymer solution through a micro-nozzle using continued pressure (pneumatic or mechanical) [129]. The properties of this technique are the capacity to deliver multiple cells and materials, i.e., high viscosity polymers with high cell densities, with a high cell viability.

Laser-induced forward transfer bioprinting consists of the deposition of a bioink layer that is in contact with a donor layer with the capacity to respond to a laser stimulation. During printing, a laser pulse is applied on the donor layer, and the bioink is propelled to the underneath substrate and immediately crosslinked [130]. This technique presents problems of cell viability due to the heating from the laser.

Finally, stereolithography uses light or laser to photolytically crosslink the bioinks layer by layer. This technique presents the highest resolution possible and high cell viability [128].

The reason for the increasing popularity of 3D bioprinting is the tremendous potential of the technique, which allows the production of tissues and other biological systems that mimic the in vivo tissue to repair. We are going to focus on the key points in 3D printing technology for biomedical applications: the development of new bioink and 3D printing for biomedical applications.

4.1. Recent Advances in Bioinks

Normally, the material used as bioink consists of natural polymers, cells, drugs, growth factors, and other materials that can be deposited in a controlled way. Bioinks should be non-toxic, easily printable, biocompatible, and biodegradable. We can define different families of natural polymers, as described in Table 2, that are commonly used for the preparation of bioinks.

Table 2. Most common natural polymers used for the preparation of bioinks.

	Compound	Advantages	Disadvantages	Bioprinting Technique	Ref.
Natural Polymers	Alginate	Low cytotoxicity, biodegradable, allow cell adhesion	Low mechanical properties	Extrusion	[19,130,131]
	Chitosan	Low cytotoxicity, biodegradable, antibacterial activity, allow cell adhesion	Low mechanical properties and depends on the origin and MW	Extrusion	[131,132]
	Gelatin	Lox cytotoxicity, improved cell adhesion, biodegradable	Poor mechanical properties and depends on the temperature. Low viscosity	Extrusion, Inkjet, Laser-assisted	[133–135]
	Hyaluronic acid	Similar to the ECM, biocompatible and biodegradable	Low mechanical strength and rapid degradation	Extrusion, Inkjet	[136–138]
	Collagen	Improved cell adhesion, good biocompatibility	Low mechanical strength and low viscosity	Extrusion, Inkjet, Laser-assisted	[139–141]
	Agarose	Good mechanical properties, biodegradable	Low cell adhesion	Extrusion	[142]
	Fibrin	Biocompatible, improved cell adhesion, non-cytotoxic	Low mechanical properties, rapid degradation	Extrusion, Inkjet	[143]

The main problems of the commonly used bioinks are their low cell affinity and their limited mechanical properties. Recent advances in the development of bioinks are focused on the synthesis of derivatives that improve the mechanical properties and cell affinity and that provide signals to promote cell growth, adhesion, and differentiation.

4.1.1. Alginate Based Bioinks

Alginate hydrogels have been extensively used as bioinks due to their good biocompatibility and similitude with the ECM. The easy way to use alginate as a bioink is by crosslinking with a solution of calcium chloride [144]. One of the main problems of alginates is their low printability and geometry accuracy due to their limited mechanical properties. To improve these properties, a variety of covalent crosslinking methods have been used. For example, Aldana et al. developed an alginate-based bioink with tunable mechanical properties using blends of alginate and gelatin methacrylamide (GelMA), obtaining a photopolymerizable biomaterial with different printability, accuracy, and mechanical and biological properties, depending on the ratio of alginate:GelMA [19].

A similar approximation was used by Soltan et al. The authors investigated the use of oxidized alginate (alginate dialdehyde, ADA) to obtain covalently crosslinked hydrogels with gelatin [145]. The mechanical properties of the hydrogel and therefore their printability and cell viability depend on the degree of oxidation and the ratio of ADA:gelatin. The authors printed different layers using two different cell types, human umbilical vein endothelial cells and rat Schwann cells, checking their viability over time.

In addition, the cell adhesion could be modified by the development of new blends. For example, the group of Boccaccini developed a hybrid hydrogel composed of alginate and keratin. This hydrogel promotes cell attachment, proliferation, spreading, and viability, being a good candidate for biomedical applications [146].

4.1.2. Chitosan Based Bioinks

Chitosan has been widely employed in tissue engineering and biomedical applications due to its biocompatibility, biodegradability, and antimicrobial activity [147–150]. Normally, chitosan is crosslinked using genipin or glutaraldehyde by a chemical crosslinking mechanism [151]. Recent advances have been focused on the development of new crosslinking agents to improve the printability and accuracy of chitosan systems and, at the same time, add biological properties to the system to promote cell adhesion, migration, or differentiation. For example, the group of Prof. San Roman developed an ionic crosslinker based on phytic acid (G_1Phy) for 3D printing [152]. This new crosslinker allowed the 3D printing of low concentrate chitosan/GelMA to obtain scaffolds with excellent mechanical and biological properties.

Another approximation to crosslink chitosan based bioinks for their use in laser-assisted bioprinting was developed by He et al. The authors provided a photocurable bioink based on the copolymerization of chitosan and acrylamide (AM) [153]. The capacity to use this bioink in laser-assisted bioprinting allows the preparation of complex 3D hydrogel scaffolds with high strength and good biocompatibility [154].

A different approach was presented by Puertas et al. in which the carboxymethyl derivative of chitosan was crosslinked with partially oxidized hyaluronic acid via Schiff base formation [155]. This study presented a novel bioprinting methodology based on a dual-syringe system with a static mixing tool that allowed the in situ crosslinking of the reactive hydrogel-based ink in the presence of living cells. This new approach allowed the use of low viscosity solutions while obtaining 3D printed scaffolds with good mechanical stability and proliferation of encapsulated cells.

4.1.3. Other Natural Polymer Based Bioinks

There are a great variety of natural polymers that are being used as bioinks. In the previous topics, we showed different examples using GelMA in combination with alginate or chitosan. Other natural polymers such as hyaluronic acid, fibrinogen, agarose, collagen, or silk have been used as bioinks. Due to their bad mechanical properties or low cell affinity, these natural polymers need to be modified, crosslinked, or blended to obtain adequate properties for their use as scaffolds. For example, Skardal et al. developed a bioink based on GelMA and methacrylated hyaluronic acid. This bioink allows the direct incorporation of cells due to the good biological properties [156]. This group also developed another bioink based on the combination of fibrin and collagen with stem cells [157]. This bioink was used to print a full-thickness skin as a carrier of stem cells for the treatment of wound healing.

4.1.4. Sacrificial Bioinks

Sacrificial bioinks are used to provide the necessary mechanical properties during the bioprinting step. After the scaffold is printed, the sacrificial bioink will be removed to create open spaces to allow cell adhesion and migration. Normally, water-soluble synthetic polymers are used as sacrificial bioinks due to their low adhesion to natural polymers.

Synthetic polymers such as pluronic, PVA, or PEG are commonly used as sacrificial bioinks in natural polymers scaffolds like in the research of Zou et al. Here, the authors used PVA as a sacrificial bioink to prepare a porous scaffold of alginate agarose. First, a support scaffold of PVA was printed, and then the empty space was filled with an alginate/agarose/HUVECs bioink. Once the scaffold was finished, PVA was removed, with cell media forming a porous structure (Figure 7).

Important progress has been made in recent years in the use of sacrificial polymers. Jian et al. developed a bioprinting method using two different inks for meniscal reconstruction [154]. The system consists of two nozzles; one of them prints PCL by high-temperature melt deposition, forming the principal construct that provided the physical properties, and the second nozzle uses a mix of GelMa, ECM, and chondrocytes, and it is deposited in the free space between PCLs. Once the scaffold is finished, PCL only provides the necessary mechanical properties to the scaffold in the first days and is degraded, forming a microchannel in the scaffold that allows the transport of nutrients to the cells.

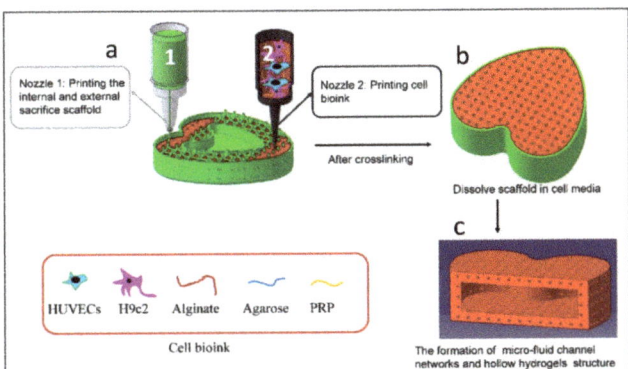

Figure 7. Flow diagram for the biofabrication of the large-size, hollow, and micro-fluid channel networks valentine-shaped heart. Reprinted from [158], with permission from Elsevier.

4.1.5. Evolution of the Bioinks

The continuous research into new materials and processes promotes the fast evolution of the bioinks. In recent years, potential candidates have been emerging quickly. Decellularized ECM, self-assembling peptides, cellular aggregates, or nanobiocomposites have emerged in recent years [159–162].

From the point of view of the polymer field, the incorporation of nanomaterials into a bioink in order to improve the stiffness, shear-thinning, degradation, or stability is interesting. One of the most important nanocomposites is nanocellulose. Nanocellulose is derived principally from bacteria [163], and its use, combined with other bioinks, integrates the common properties of the cellulose: high stiffness, modulus, hydrophilicity, and thermal stability. For example, Han et al. studied the effect of nanocellulose on alginate/gelatin bioinks [164]. Their result showed that the incorporation of a small amount of nanocellulose improved the printability, stability, and fidelity of the structure, but high amounts of nanocellulose promoted a negative impact on the elongation and compression yield.

4.2. 3D Printing for Biomedical Applications

Current research on 3D printing in biomedical applications can be classified into the following areas: (i) printing of bioactive and biodegradable scaffolds and (ii) directly printing tissues and organs.

4.2.1. Bioactive and Biodegradable Scaffolds

One of the main research fields in tissue engineering is the development of advanced scaffolds for tissue regeneration. In this case, the bioink will be a polymer system alone or a mixture incorporating cells, where the polymer systems play the role of the ECM. Compared with traditional scaffold-fabrication methods (salt-leaching, cryogels, or gas-foaming) that prepared simple-shapes supports with an inhomogeneity porosity, 3D printing can prepare complex structures, with the adequate shape to fill the defect and with an effective control of the porosity and microstructure. These systems require the presence of an interconnected porous network to allow cell growth and migration and flow transport of nutrients [120].

Bioactive scaffolds obtained by 3D printing can be divided into two families: scaffolds printed without cells (cell-free scaffolds) and scaffolds directly printed with cells (cell-loaded scaffolds).

Cell-free scaffolds are normally prepared from high water content polymer systems that present a high biocompatibility and a controlled biodegradation. The facility to prepared complex structures makes this kind of scaffold suitable for the reconstruction of complex tissues like osteochondral tissue. Osteochondral tissue is composed of different layers with different structures and compositions [25]. 3D printing is capable of producing scaffolds that simulate the structure of this tissue. For example, Gao et al. designed a multilayer system using GelMA with or without hydroxyapatite to obtain a scaffold that simulates the ECM in the osteochondral tissue [165]. The design of a proper bioink is a crucial point for the correct regeneration of the tissue. In this sense, Ma et al. developed a novel polymeric/ceramic bioink for customizing craniomaxillofacial bone reconstruction [166]. The bioink was based on a hyperelastic PEGylated urethane composited with microscale β-TCP. This bioink presented a high biocompatibility and osteoinductivity due to its biomimetic composition. In addition, it presented adequate mechanical properties for surgical manipulation and it had the capacity of osteoregeneration.

Cell-loaded scaffolds were developed due to the problems of seeding cells directly into 3D printed scaffolds (inhomogeneous distribution and inefficient cell adhesion). The main objective of this technique is to simulate the structure of the ECM in vitro to allow cell growth and differentiation. This technique allows the preparation of semi-functional tissues like in the case of the group of Prof. Jorcano. This group developed a functional skin by 3D bioprinting using extrusion bioprinting to deposit different layers composed of fibrin and fibroblast or keratinocytes to form the dermis and the epidermis layers [167]. With this technique, the authors obtained a functional skin that can be used for implantation in burned tissues.

4.2.2. Directly Printing Tissue and Organs

The preparation of cell-loaded scaffolds does not assure the final functionality of the scaffold due to the dispersion of the cells and growth factors. For this reason, recent advances in 3D bioprinting are focused on the evolution of a direct-printing technology: tissue structures with physiological functions, containing seed cells, growth factors, and nutritional components [119]. One of the main goals is the pre-vascularization of the scaffolds, because the absence of a vasculature is one of the leading causes of failure for current 3D bioprinted scaffolds [168]. Recent advances have obtained functional vascularized tissues that showed better biointegration. Kim et al. developed a perfusable vascularized human skin formed by an epidermis, dermis, and hypodermis [169]. The system involves the preparation of a support of PCL and gelatin followed by the impression of the different skin layers: (1) The hypodermis layer was composed of fibrinogen and adipose-derived ECM with human adipocytes. (2) The vasculature was printed using a gelatin/glycerol/thrombin bioink with human umbilical vein endothelial cells. (3) The dermal layer was composed of fibrinogen, dermal-derived ECM, and human dermal fibroblast. (4) The final epidermis layer was composed of human keratinocytes. The result showed a fully functional skin with a microenvironment close to a real skin that can be used to test skin drugs or similar (Figure 8).

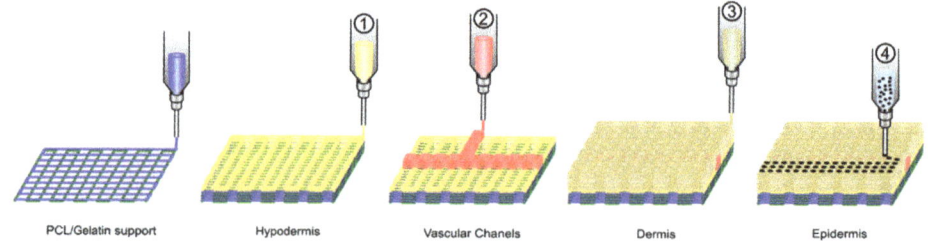

Figure 8. Schematic diagram exhibiting the 3D cell printing process for fabrication of a 3D full-thickness skin model.

4.3. Future Perspectives in 3D Printing

Considerable advances have been achieved in the use of polymers for 3D bioprinting. However, this field is still in the early stages of development. One of the principal key factors is the development of a bioink adequate to our system. The bioink needs to have a good printability and geometry accuracy, adequate mechanical properties, and good biocompatibility and needs to be biodegradable. Many bioinks have already been formulated and used, but researchers continue to develop new compositions, looking to obtain better properties and new biofunctionalities. These characteristics allow the 3D bioprinting of fully functional tissues and organs.

Recently, 4D bioprinting technology has emerged as a powerful platform to obtain stimulus responsive bioprinting. This new methodology is in its first stages. Only a proof of concept with smart polymers has been developed. In the next years, this technology will evolve into more sophisticated systems and will be used in the advanced biomedical field.

5. Summary and Future Direction

In this review, we have entered into the relationship between natural polymers and new biofabrication techniques. The use of 3D printing, microfluidics, and electrospinning techniques has been widely investigated for the biofabrication of naturally-derived polymer scaffolds with encapsulated cells. Important challenges must be addressed for the successful biofabrication of these cell-laden natural scaffolds. On the one hand, the poor mechanical strength of natural polymers makes the processing and material manipulation difficult. In this sense, different modifications and blends have been investigated to improve the mechanical properties of the processed scaffold. On the other hand, the application of these techniques must ensure that the viability and functionality of the encapsulated cells are not negatively affected during the processing. Controlling both material parameters (e.g., solvent, viscosity, or polymer concentration) and processing parameters (e.g., pressure, voltage, or feed rate) is essential to avoid stress damage of cells during the fabrication, and therefore to provide high cell viability, metabolic activity, and proliferation of the encapsulated cells. Therefore, it is one of the main challenges that need to be addressed. In the last years, different modifications, blends, and adaptations of the biofabrication process have been investigated, but we are still in the first stages of the development of these technologies. There are indeed still many technological issues and limitations that need to be solved. One of the most important concerns is the clinical implementation of the cell-laden scaffolds fabricated using these techniques. Some preliminary in vivo studies in animal models have been performed. However, more research studies regarding immunological response and vascularization (which are essential for effective tissue implantation) are still necessary to assess the real applicability of the materials for clinical applications. Despite all these limitations and challenges, many efforts are being made to develop more complex techniques to simulate the complexity of native tissues and overcome the processing limitations, so we cannot exclude the possibility that in a few years biofabrication techniques will evolve and allow obtaining fully functional organs and tissues.

Author Contributions: Conceptualization, A.M.-B., M.P.-B., L.G.-F.; investigation, A.M.-B., M.P.-B., L.G.-F.; resources, A.M.-B., M.P.-B., L.G.-F.; writing—original draft preparation, A.M.-B., M.P.-B., L.G.-F.; writing—review and editing, A.M.-B., M.P.-B., L.G.-F. All authors have read and agreed to the published version of the manuscript.

Funding: This research was funded by the Ministry of Science, Innovation, and Universities (Spain) (MAT2017-84277-R); Apadrina la Ciencia-Ford Foundations (Fellowship of A.M.-B)); and Leibniz ScienceCampus Living Therapeutic Materials (Saarbrücken, Germany).

Institutional Review Board Statement: Not Applicable.

Informed Consent Statement: Not Applicable.

Data Availability Statement: Not Applicable.

Acknowledgments: Luis García Fernández is a member of the SusPlast platform (Interdisciplinary Platform for Sustainable Plastics towards a Circular Economy) from the Spanish National Research Council (CSIC).

Conflicts of Interest: The authors declare no conflict of interest.

References

1. Tutar, R.; Motealleh, A.; Khademhosseini, A.; Kehr, N.S. Functional Nanomaterials on 2D Surfaces and in 3D Nanocomposite Hydrogels for Biomedical Applications. *Adv. Funct. Mater.* **2019**, *29*. [CrossRef]
2. Asadi, N.; Del Bakhshayesh, A.R.; Davaran, S.; Akbarzadeh, A. Common Biocompatible Polymeric Materials for Tissue Engineering and Regenerative Medicine. *Mater. Chem. Phys.* **2020**, *242*, 122528. [CrossRef]
3. Mora-Boza, A.; Puertas-Bartolomé, M.; Vázquez-Lasa, B.; San Román, J.; Pérez-Caballer, A.; Olmeda-Lozano, M. Contribution of Bioactive Hyaluronic Acid and Gelatin to Regenerative Medicine. Methodologies of Gels Preparation and Advanced Applications. *Eur. Polym. J.* **2017**, *95*, 11–26. [CrossRef]
4. Nolan, K.; Millet, Y.; Ricordi, C.; Stabler, C.L. Tissue Engineering and Biomaterials in Regenerative Medicine. *Cell Transpl.* **2008**, *17*, 241–243. [CrossRef] [PubMed]
5. Eiselt, P.; Yeh, J.; Latvala, R.K.; Shea, L.D.; Mooney, D.J. Porous Carriers for Biomedical Applications Based on Alginate Hydrogels. *Biomaterials* **2000**, *21*, 1921–1927. [CrossRef]
6. Yeo, M.G.; Kim, G.H. Fabrication of Cell-Laden Electrospun Hybrid Scaffolds of Alginate-Based Bioink and PCL Microstructures for Tissue Regeneration. *Chem. Eng. J.* **2015**, *275*, 27–35. [CrossRef]
7. Cook, M.T.; Tzortzis, G.; Charalampopoulos, D.; Khutoryanskiy, V.V. Production and Evaluation of Dry Alginate-Chitosan Microcapsules as an Enteric Delivery Vehicle for Probiotic Bacteria. *Biomacromolecules* **2011**, *12*, 2834–2840. [CrossRef] [PubMed]
8. Sideris, E.; Griffin, D.R.; Ding, Y.; Li, S.; Weaver, W.M.; Di Carlo, D.; Hsiai, T.; Segura, T. Particle Hydrogels Based on Hyaluronic Acid Building Blocks. *ACS Biomater. Sci. Eng.* **2016**, *2*, 2034–2041. [CrossRef]
9. Husain, S.; Al-Samadani, K.H.; Najeeb, S.; Zafar, M.S.; Khurshid, Z.; Zohaib, S.; Qasim, S.B. Chitosan Biomaterials for Current and Potential Dental Applications. *Materials* **2017**, *10*, 602. [CrossRef] [PubMed]
10. Qasim, S.B.; Zafar, M.S.; Najeeb, S.; Khurshid, Z.; Shah, A.H.; Husain, S.; Rehman, I.U. Electrospinning of Chitosan-Based Solutions for Tissue Engineering and Regenerative Medicine. *Int. J. Mol. Sci.* **2018**, *19*, 407. [CrossRef]
11. Chan, H.F.; Zhang, Y.; Leong, K.W. Efficient One-Step Production of Microencapsulated Hepatocyte Spheroids with Enhanced Functions. *Small* **2016**, *12*, 2720–2730. [CrossRef]
12. Zafar, M.S.; Al-Samadani, K.H. Potential Use of Natural Silk for Bio-Dental Applications. *J. Taibah Univ. Med. Sci.* **2014**, *9*, 171–177. [CrossRef]
13. Zafar, M.S.; Belton, D.J.; Hanby, B.; Kaplan, D.L.; Perry, C.C. Functional Material Features of *Bombyx mori* Silk Light versus Heavy Chain Proteins. *Biomacromolecules* **2015**, *16*, 606–614. [CrossRef]
14. Yang, C.Y.; Chiu, C.T.; Chang, Y.P.; Wang, Y.J. Fabrication of Porous Gelatin Microfibers Using an Aqueous Wet Spinning Process. *Artif. Cells Blood Substit. Biotechnol.* **2009**, *37*, 173–176. [CrossRef] [PubMed]
15. Cha, C.; Oh, J.; Kim, K.; Qiu, Y.; Joh, M.; Shin, S.R.; Wang, X.; Camci-Unal, G.; Wan, K.T.; Liao, R.; et al. Microfluidics-Assisted Fabrication of Gelatin-Silica Core-Shell Microgels for Injectable Tissue Constructs. *Biomacromolecules* **2014**, *15*, 283–290. [CrossRef]
16. Feng, Q.; Li, Q.; Wen, H.; Chen, J.; Liang, M.; Huang, H.; Lan, D.; Dong, H.; Cao, X. Injection and Self-Assembly of Bioinspired Stem Cell-Laden Gelatin/Hyaluronic Acid Hybrid Microgels Promote Cartilage Repair In Vivo. *Adv. Funct. Mater.* **2019**, *29*. [CrossRef]
17. Guo, Y.; Gilbert-Honick, J.; Somers, S.M.; Mao, H.Q.; Grayson, W.L. Modified Cell-Electrospinning for 3D Myogenesis of C2C12s in Aligned Fibrin Microfiber Bundles. *Biochem. Biophys. Res. Commun.* **2019**, *516*, 558–564. [CrossRef] [PubMed]
18. Jayani, T.; Sanjeev, B.; Marimuthu, S.; Uthandi, S. Bacterial Cellulose Nano Fiber (BCNF) as Carrier Support for the Immobilization of Probiotic, Lactobacillus Acidophilus 016. *Carbohydr. Polym.* **2020**, *250*, 116965. [CrossRef]
19. Aldana, A.A.; Valente, F.; Dilley, R.; Doyle, B. Development of 3D Bioprinted GelMA-Alginate Hydrogels with Tunable Mechanical Properties. *Bioprinting* **2021**, *21*, e00105. [CrossRef]
20. Lee, D.; Lee, K.; Cha, C. Microfluidics-Assisted Fabrication of Microtissues with Tunable Physical Properties for Developing an In Vitro Multiplex Tissue Model. *Adv. Biosyst.* **2018**, *2*. [CrossRef]
21. Seeto, W.J.; Tian, Y.; Pradhan, S.; Kerscher, P.; Lipke, E.A. Rapid Production of Cell-Laden Microspheres Using a Flexible Microfluidic Encapsulation Platform. *Small* **2019**, *15*, e1902058. [CrossRef]
22. Li, F.; Truong, V.X.; Thissen, H.; Frith, J.E.; Forsythe, J.S. Microfluidic Encapsulation of Human Mesenchymal Stem Cells for Articular Cartilage Tissue Regeneration. *ACS Appl. Mater. Interfaces* **2017**, *9*, 8589–8601. [CrossRef]
23. Lee, J.; Cuddihy, M.J.; Kotov, N.A. Three-Dimensional Cell Culture Matrices: State of the Art. *Tissue Eng. Part B Rev.* **2008**, *14*, 61–86. [CrossRef]
24. Fabbri, M.; García-Fernández, L.; Vázquez-Lasa, B.; Soccio, M.; Lotti, N.; Gamberini, R.; Rimini, B.; Munari, A.; San Román, J. Micro-Structured 3D-Electrospun Scaffolds of Biodegradable Block Copolymers for Soft Tissue Regeneration. *Eur. Polym. J.* **2017**, *94*, 33–42. [CrossRef]
25. Ribeiro, V.P.; Pina, S.; Costa, J.B.; Cengiz, I.F.; García-Fernández, L.; Fernández-Gutiérrez, M.D.M.; Paiva, O.C.; Oliveira, A.L.; San-Román, J.; Oliveira, J.M.; et al. Enzymatically Cross-Linked Silk Fibroin-Based Hierarchical Scaffolds for Osteochondral Regeneration. *ACS Appl. Mater. Interfaces* **2019**, *11*, 3781–3799. [CrossRef]

26. Puertas-Bartolomé, M.; Benito-Garzón, L.; Fung, S.; Kohn, J.; Vázquez-Lasa, B.; San Román, J. Bioadhesive Functional Hydrogels: Controlled Release of Catechol Species with Antioxidant and Antiinflammatory Behavior. *Mater. Sci. Eng. C* **2019**, *105*, 110040. [CrossRef] [PubMed]
27. Eltom, A.; Zhong, G.; Muhammad, A. Scaffold Techniques and Designs in Tissue Engineering Functions and Purposes: A Review. *Adv. Mater. Sci. Eng.* **2019**, *2019*, 3429527. [CrossRef]
28. Fu, J.; Li, X.B.; Wang, L.X.; Lv, X.H.; Lu, Z.; Wang, F.; Xia, Q.; Yu, L.; Li, C.M. One-Step Dip-Coating-Fabricated Core-Shell Silk Fibroin Rice Paper Fibrous Scaffolds for 3D Tumor Spheroid Formation. *ACS Appl. Bio Mater.* **2020**, *3*, 7462–7471. [CrossRef]
29. Tsukamoto, Y.; Akagi, T.; Akashi, M. Vascularized Cardiac Tissue Construction with Orientation by Layer-by-Layer Method and 3D Printer. *Sci. Rep.* **2020**, *10*. [CrossRef] [PubMed]
30. Siddiq, A.; Kennedy, A.R. Compression Moulding and Injection over Moulding of Porous PEEK Components. *J. Mech. Behav. Biomed. Mater.* **2020**, *111*. [CrossRef] [PubMed]
31. Van Bochove, B.; Grijpma, D.W. Mechanical Properties of Porous Photo-Crosslinked Poly(Trimethylene Carbonate) Network Films. *Eur. Polym. J.* **2021**, *143*. [CrossRef]
32. Peck, M.; Dusserre, N.; McAllister, T.N.; L'Heureux, N. Tissue Engineering by Self-Assembly. *Mater. Today* **2011**, *14*, 218–224. [CrossRef]
33. Chen, Q.; Chen, D.; Wu, J.; Lin, J.M. Flexible Control of Cellular Encapsulation, Permeability, and Release in a Droplet-Templated Bifunctional Copolymer Scaffold. *Biomicrofluidics* **2016**, *10*. [CrossRef] [PubMed]
34. Rossow, T.; Lienemann, P.S.; Mooney, D.J. Cell Microencapsulation by Droplet Microfluidic Templating. *Macromol. Chem. Phys.* **2017**, *218*. [CrossRef]
35. Zafar, M.; Najeeb, S.; Khurshid, Z.; Vazirzadeh, M.; Zohaib, S.; Najeeb, B.; Sefat, F. Potential of Electrospun Nanofibers for Biomedical and Dental Applications. *Materials* **2016**, *9*, 73. [CrossRef]
36. Zimmerling, A.; Yazdanpanah, Z.; Cooper, D.M.L.; Johnston, J.D.; Chen, X. 3D Printing PCL/NHA Bone Scaffolds: Exploring the Influence of Material Synthesis Techniques. *Biomater. Res.* **2021**, *25*, 1–12. [CrossRef]
37. Vyas, C.; Zhang, J.; Øvrebø, Ø.; Huang, B.; Roberts, I.; Setty, M.; Allardyce, B.; Haugen, H.; Rajkhowa, R.; Bartolo, P.; et al. 3D Printing of Silk Microparticle Reinforced Polycaprolactone Scaffolds for Tissue Engineering Applications. *Mater. Sci. Eng. C* **2021**, *118*, 111433. [CrossRef]
38. Jiang, W.; Li, M.; Chen, Z.; Leong, K.W. Cell-Laden Microfluidic Microgels for Tissue Regeneration. *Lab Chip* **2016**, *16*, 4482–4506. [CrossRef] [PubMed]
39. Agrawal, G.; Agrawal, R. Functional Microgels: Recent Advances in Their Biomedical Applications. *Small* **2018**, *14*, e1801724. [CrossRef]
40. Newsom, J.P.; Payne, K.A.; Krebs, M.D. Microgels: Modular, Tunable Constructs for Tissue Regeneration. *Acta Biomater.* **2019**, *88*, 32–41. [CrossRef]
41. Annabi, N.; Tamayol, A.; Uquillas, J.A.; Akbari, M.; Bertassoni, L.E.; Cha, C.; Camci-Unal, G.; Dokmeci, M.R.; Peppas, N.A.; Khademhosseini, A. 25th Anniversary Article: Rational Design and Applications of Hydrogels in Regenerative Medicine. *Adv. Mater.* **2014**, *26*, 85–124. [CrossRef]
42. Farjami, T.; Madadlou, A. Fabrication Methods of Biopolymeric Microgels and Microgel-Based Hydrogels. *Food Hydrocoll.* **2017**, *62*, 262–272. [CrossRef]
43. Tumarkin, E.; Kumacheva, E. Microfluidic Generation of Microgels from Synthetic and Natural Polymers. *Chem. Soc. Rev.* **2009**, *38*, 2161–2168. [CrossRef]
44. Huang, D.; Gibeley, S.B.; Xu, C.; Xiao, Y.; Celik, O.; Ginsberg, H.N.; Leong, K.W. Engineering Liver Microtissues for Disease Modeling and Regenerative Medicine. *Adv. Funct. Mater.* **2020**, *30*. [CrossRef] [PubMed]
45. Mora-Boza, A.; Mancipe Castro, L.M.; Schneider, R.S.; Han, W.M.; García, A.J.; Vázquez-Lasa, B.; San Román, J. Microfluidics Generation of Chitosan Microgels Containing Glycerylphytate Crosslinker for in Situ Human Mesenchymal Stem Cells Encapsulation. *Mater. Sci. Eng. C* **2021**, *120*, 111716. [CrossRef]
46. Zhang, H.; Tumarkin, E.; Sullan, R.M.A.; Walker, G.C.; Kumacheva, E. Exploring Microfluidic Routes to Microgels of Biological Polymers. *Macromol. Rapid Commun.* **2007**, *28*, 527–538. [CrossRef]
47. Velasco, D.; Tumarkin, E.; Kumacheva, E. Microfluidic Encapsulation of Cells in Polymer Microgels. *Small* **2012**, *8*, 1633–1642. [CrossRef]
48. Utech, S.; Prodanovic, R.; Mao, A.S.; Ostafe, R.; Mooney, D.J.; Weitz, D.A. Microfluidic Generation of Monodisperse, Structurally Homogeneous Alginate Microgels for Cell Encapsulation and 3D Cell Culture. *Adv. Healthc. Mater.* **2015**, *4*, 1628–1633. [CrossRef]
49. Mao, A.S.; Shin, J.W.; Utech, S.; Wang, H.; Uzun, O.; Li, W.; Cooper, M.; Hu, Y.; Zhang, L.; Weitz, D.A.; et al. Deterministic Encapsulation of Single Cells in Thin Tunable Microgels for Niche Modelling and Therapeutic Delivery. *Nat. Mater.* **2017**, *16*, 236–243. [CrossRef] [PubMed]
50. Cheng, Y.; Zhang, X.; Cao, Y.; Tian, C.; Li, Y.; Wang, M.; Zhao, Y.; Zhao, G. Centrifugal Microfluidics for Ultra-Rapid Fabrication of Versatile Hydrogel Microcarriers. *Appl. Mater. Today* **2018**, *13*, 116–125. [CrossRef]
51. Ma, T.; Gao, X.; Dong, H.; He, H.; Cao, X. High-Throughput Generation of Hyaluronic Acid Microgels via Microfluidics-Assisted Enzymatic Crosslinking and/or Diels-Alder Click Chemistry for Cell Encapsulation and Delivery. *Appl. Mater. Today* **2017**, *9*, 49–59. [CrossRef]

52. Jang, Y.; Cha, C.; Jung, J.; Oh, J. Interfacial Compression-Dependent Merging of Two Miscible Microdroplets in an Asymmetric Cross-Junction for In Situ Microgel Formation. *Macromol. Res.* **2018**, *26*, 1143–1149. [CrossRef]
53. Zhao, X.; Liu, S.; Yildirimer, L.; Zhao, H.; Ding, R.; Wang, H.; Cui, W.; Weitz, D. Injectable Stem Cell-Laden Photocrosslinkable Microspheres Fabricated Using Microfluidics for Rapid Generation of Osteogenic Tissue Constructs. *Adv. Funct. Mater.* **2016**, *26*, 2809–2819. [CrossRef]
54. Henke, S.; Leijten, J.; Kemna, E.; Neubauer, M.; Fery, A.; van den Berg, A.; van Apeldoorn, A.; Karperien, M. Enzymatic Crosslinking of Polymer Conjugates Is Superior over Ionic or UV Crosslinking for the On-Chip Production of Cell-Laden Microgels. *Macromol. Biosci.* **2016**, *16*, 1524–1532. [CrossRef] [PubMed]
55. Liu, H.T.; Wang, H.; Wei, W.B.; Liu, H.; Jiang, L.; Qin, J.H. A Microfluidic Strategy for Controllable Generation of Water-in-Water Droplets as Biocompatible Microcarriers. *Small* **2018**, *14*, e1801095. [CrossRef]
56. Siltanen, C.; Yaghoobi, M.; Haque, A.; You, J.; Lowen, J.; Soleimani, M.; Revzin, A. Microfluidic Fabrication of Bioactive Microgels for Rapid Formation and Enhanced Differentiation of Stem Cell Spheroids. *Acta Biomater.* **2016**, *34*, 125–132. [CrossRef] [PubMed]
57. Zhang, H.; Tumarkin, E.; Peerani, R.; Nie, Z.; Sullan, R.M.A.; Walker, G.C.; Kumacheva, E. Microfluidic Production of Biopolymer Microcapsules with Controlled Morphology. *J. Am. Chem. Soc.* **2006**, *128*, 12205–12210. [CrossRef]
58. Raz, N.; Li, J.K.; Fiddes, L.K.; Tumarkin, E.; Walker, G.C.; Kumacheva, E. Microgels with an Interpenetrating Network Structure as a Model System for Cell Studies. *Macromolecules* **2010**, *43*, 7277–7281. [CrossRef]
59. Tan, W.H.; Takeuchi, S. Monodisperse Alginate Hydrogel Microbeads for Cell Encapsulation. *Adv. Mater.* **2007**, *19*, 2696–2701. [CrossRef]
60. Kim, C.; Chung, S.; Kim, Y.E.; Lee, K.S.; Lee, S.H.; Oh, K.W.; Kang, J.Y. Generation of Core-Shell Microcapsules with Three-Dimensional Focusing Device for Efficient Formation of Cell Spheroid. *Lab Chip* **2011**, *11*, 246–252. [CrossRef]
61. Mao, A.S.; Özkale, B.; Shah, N.J.; Vining, K.H.; Descombes, T.; Zhang, L.; Tringides, C.M.; Wong, S.W.; Shin, J.W.; Scadden, D.T.; et al. Programmable Microencapsulation for Enhanced Mesenchymal Stem Cell Persistence and Immunomodulation. *Proc. Natl. Acad. Sci. USA* **2019**, *116*, 15392–15397. [CrossRef]
62. Singh, A. Biomaterials Innovation for next Generation Ex Vivo Immune Tissue Engineering. *Biomaterials* **2017**, *130*, 104–110. [CrossRef]
63. Hu, Y.; Wang, S.; Abbaspourrad, A.; Ardekani, A.M. Fabrication of Shape Controllable Janus Alginate/PNIPAAm Microgels via Microfluidics Technique and off-Chip Ionic Cross-Linking. *Langmuir* **2015**, *31*, 1885–1891. [CrossRef]
64. Karakasyan, C.; Mathos, J.; Lack, S.; Davy, J.; Marquis, M.; Renard, D. Microfluidics-Assisted Generation of Stimuli-Responsive Hydrogels Based on Alginates Incorporated with Thermo-Responsive and Amphiphilic Polymers as Novel Biomaterials. *Colloids Surf. B Biointerfaces* **2015**, *135*, 619–629. [CrossRef] [PubMed]
65. Mora-Boza, A.; López-Donaire, M.L.; Saldaña, L.; Vilaboa, N.; Vázquez-Lasa, B.; San Román, J. Glycerylphytate Compounds with Tunable Ion Affinity and Osteogenic Properties. *Sci. Rep.* **2019**, *9*, 11491. [CrossRef] [PubMed]
66. Li, W.; Zhang, L.; Ge, X.; Xu, B.; Zhang, W.; Qu, L.; Choi, C.H.; Xu, J.; Zhang, A.; Lee, H.; et al. Microfluidic Fabrication of Microparticles for Biomedical Applications. *Chem. Soc. Rev.* **2018**, *47*, 5646–5683. [CrossRef] [PubMed]
67. Yang, W.; Yu, H.; Li, G.; Wang, Y.; Liu, L. High-Throughput Fabrication and Modular Assembly of 3D Heterogeneous Microscale Tissues. *Small* **2017**, *13*. [CrossRef]
68. Hauck, N.; Seixas, N.; Centeno, S.P.; Schlüßler, R.; Cojoc, G.; Müller, P.; Guck, J.; Wöll, D.; Wessjohann, L.A.; Thiele, J. Droplet-Assisted Microfluidic Fabrication and Characterization of Multifunctional Polysaccharide Microgels Formed by Multicomponent Reactions. *Polymers* **2018**, *10*, 1055. [CrossRef] [PubMed]
69. Sugimura, R. Bioengineering Hematopoietic Stem Cell Niche toward Regenerative Medicine. *Adv. Drug Deliv. Rev.* **2016**, *99*, 212–220. [CrossRef]
70. Yanagawa, F.; Sugiura, S.; Kanamori, T. Hydrogel Microfabrication Technology toward Three Dimensional Tissue Engineering. *Regen. Ther.* **2016**, *3*, 45–57. [CrossRef]
71. Vedadghavami, A.; Minooei, F.; Mohammadi, M.H.; Khetani, S.; Rezaei Kolahchi, A.; Mashayekhan, S.; Sanati-Nezhad, A. Manufacturing of Hydrogel Biomaterials with Controlled Mechanical Properties for Tissue Engineering Applications. *Acta Biomater.* **2017**, *62*, 42–63. [CrossRef] [PubMed]
72. Doshi, J.; Reneker, D.H. Electrospinning Process and Applications of Electrospun Fibers. *J. Electrostat.* **1995**, *35*, 151–160. [CrossRef]
73. Agarwal, S.; Wendorff, J.H.; Greiner, A. Use of Electrospinning Technique for Biomedical Applications. *Polymer* **2008**, *49*, 5603–5621. [CrossRef]
74. Zanin, M.H.A.; Cerize, N.N.P.; de Oliveira, A.M. Production of Nanofibers by Electrospinning Technology: Overview and Application in Cosmetics. *Nanocosmet. Nanomed.* **2011**, 311–332. [CrossRef]
75. De Lima, G.G.; Lyons, S.; Devine, D.M.; Nugent, M.J.D. Electrospinning of Hydrogels for Biomedical Applications. *Hydrogels* **2018**, 219–258. [CrossRef]
76. Vasita, R.; Katti, D.S. Nanofibers and Their Applications in Tissue Engineering. *Int. J. Nanomed.* **2006**, *1*, 15–30. [CrossRef] [PubMed]
77. Pham, Q.P.; Sharma, U.; Mikos, A.G. Electrospinning of Polymeric Nanofibers for Tissue Engineering Applications: A Review. *Tissue Eng.* **2006**, *12*, 1197–1211. [CrossRef] [PubMed]
78. Dotti, F.; Varesano, A.; Montarsolo, A.; Aluigi, A.; Tonin, C.; Mazzuchetti, G. Electrospun Porous Mats for High Efficiency Filtration. *J. Ind. Text.* **2007**, *37*, 151–162. [CrossRef]

79. Tucker, N.; Hofman, K.; Stanger, J.; Staiger, M.; Hamid, N.A.; Torres, P.L. The History of the Science and Technology of Electrospinning from 1600 to 1995. *J. Eng. Fibers Fabr.* **2011**, *7*. [CrossRef]
80. Mercante, L.A.; Scagion, V.P.; Migliorini, F.L.; Mattoso, L.H.C.; Correa, D.S. Electrospinning-Based (Bio)Sensors for Food and Agricultural Applications: A Review. *TrAC Trends Anal. Chem.* **2017**, *91*, 91–103. [CrossRef]
81. Xue, J.; Wu, T.; Dai, Y.; Xia, Y. Electrospinning and Electrospun Nanofibers: Methods, Materials, and Applications. *Chem. Rev.* **2019**, *119*, 5298–5415. [CrossRef] [PubMed]
82. Azimi, B.; Maleki, H.; Zavagna, L.; de la Ossa, J.G.; Linari, S.; Lazzeri, A.; Danti, S. Bio-Based Electrospun Fibers for Wound Healing. *J. Funct. Biomater.* **2020**, *11*, 67. [CrossRef]
83. Tolba, E. Diversity of Electrospinning Approach for Vascular Implants: Multilayered Tubular Scaffolds. *Regen. Eng. Transl. Med.* **2020**, *6*, 383–397. [CrossRef]
84. Singh, M.K.; Kumar, P.; Behera, B.K. Scaffolds for Tissue Engineering. *Asian Text. J.* **2009**, *18*, 58–62. [CrossRef]
85. Chronakis, I.S. Novel Nanocomposites and Nanoceramics Based on Polymer Nanofibers Using Electrospinning Process—A Review. *J. Mater. Process. Technol.* **2005**, *167*, 283–293. [CrossRef]
86. Christopherson, G.T.; Song, H.; Mao, H.Q. The Influence of Fiber Diameter of Electrospun Substrates on Neural Stem Cell Differentiation and Proliferation. *Biomaterials* **2009**, *30*, 556–564. [CrossRef]
87. Wang, X.; Ding, B.; Li, B. Biomimetic Electrospun Nanofibrous Structures for Tissue Engineering. *Mater. Today* **2013**, *16*, 229–241. [CrossRef] [PubMed]
88. Braghirolli, D.I.; Steffens, D.; Pranke, P. Electrospinning for Regenerative Medicine: A Review of the Main Topics. *Drug Discov. Today* **2014**, *19*, 743–753. [CrossRef] [PubMed]
89. Ghanavi, J.; Farnia, P.; Velayati, A.A. Nano design of extracellular matrix for tissue engineering. In *Nanoarchitectonics in Biomedicine*; Elsevier: Amsterdam, The Netherlands, 2019; pp. 547–583. ISBN 9780128162002.
90. Townsend-Nicholson, A.; Jayasinghe, S.N. Cell Electrospinning: A Unique Biotechnique for Encapsulating Living Organisms for Generating Active Biological Microthreads/Scaffolds. *Biomacromolecules* **2006**, *7*, 3364–3369. [CrossRef] [PubMed]
91. Hong, J.; Yeo, M.; Yang, G.H.; Kim, G. Cell-Electrospinning and Its Application for Tissue Engineering. *Int. J. Mol. Sci.* **2019**, *20*, 6208. [CrossRef]
92. Jayasinghe, S.N.; Qureshi, A.N.; Eagles, P.A.M. Electrohydrodynamic Jet Processing: An Advanced Electric-Field-Driven Jetting Phenomenon for Processing Living Cells. *Small* **2006**, *2*, 216–219. [CrossRef]
93. Arumuganathar, S.; Irvine, S.; McEwan, J.R.; Jayasinghe, S.N. Pressure-Assisted Cell Spinning: A Direct Protocol for Spinning Biologically Viable Cell-Bearing Fibres and Scaffolds. *Biomed. Mater.* **2007**, *2*, 211–219. [CrossRef]
94. Jayasinghe, S.N. Cell Electrospinning: A Novel Tool for Functionalising Fibres, Scaffolds and Membranes with Living Cells and Other Advanced Materials for Regenerative Biology and Medicine. *Analyst* **2013**, *138*, 2215–2223. [CrossRef]
95. Sampson, S.L.; Saraiva, L.; Gustafsson, K.; Jayasinghe, S.N.; Robertson, B.D. Cell Electrospinning: An in Vitro and in Vivo Study. *Small* **2014**, *10*, 78–82. [CrossRef]
96. Chen, H.; Liu, Y.; Hu, Q. A Novel Bioactive Membrane by Cell Electrospinning. *Exp. Cell Res.* **2015**, *338*, 261–266. [CrossRef] [PubMed]
97. Onoe, H.; Takeuchi, S. Cell-Laden Microfibers for Bottom-up Tissue Engineering. *Drug Discov. Today* **2015**, *20*, 236–246. [CrossRef]
98. Bhardwaj, N.; Kundu, S.C. Electrospinning: A Fascinating Fiber Fabrication Technique. *Biotechnol. Adv.* **2010**, *28*, 325–347. [CrossRef] [PubMed]
99. Anu Bhushani, J.; Anandharamakrishnan, C. Electrospinning and Electrospraying Techniques: Potential Food Based Applications. *Trends Food Sci. Technol.* **2014**, *38*, 21–33. [CrossRef]
100. Bock, N.; Woodruff, M.A.; Hutmacher, D.W.; Dargaville, T.R. Electrospraying, a Reproducible Method for Production of Polymeric Microspheres for Biomedical Applications. *Polymers* **2011**, *3*, 131–149. [CrossRef]
101. Greig, D.; Jayasinghe, S.N. Genomic, Genetic and Physiological Effects of Bio-Electrospraying on Live Cells of the Model Yeast Saccharomyces Cerevisiae. *Biomed. Mater.* **2008**, *3*, 34125. [CrossRef]
102. Mongkoldhumrongkul, N.; Swain, S.C.; Jayasinghe, S.N.; Stürzenbaum, S. Bio-Electrospraying the Nematode Caenorhabditis Elegans: Studying Whole-Genome Transcriptional Responses and Key Life Cycle Parameters. *J. R. Soc. Interface* **2010**, *7*, 595–601. [CrossRef] [PubMed]
103. Poncelet, D.; de Vos, P.; Suter, N.; Jayasinghe, S.N. Bio-Electrospraying and Cell Electrospinning: Progress and Opportunities for Basic Biology and Clinical Sciences. *Adv. Healthc. Mater.* **2012**, *1*, 27–34. [CrossRef] [PubMed]
104. Vatankhah, E.; Prabhakaran, M.P.; Ramakrishna, S. Biomimetic Nanostructures by Electrospinning and Electrospraying. *Stem Cell Nanoeng.* **2015**, 123–141. [CrossRef]
105. Jeong, S.B.; Chong, E.S.; Heo, K.J.; Lee, G.W.; Kim, H.J.; Lee, B.U. Electrospray Patterning of Yeast Cells for Applications in Alcoholic Fermentation. *Sci. Rep.* **2019**, *9*, 1–7. [CrossRef] [PubMed]
106. Tycova, A.; Prikryl, J.; Kotzianova, A.; Datinska, V.; Velebny, V.; Foret, F. Electrospray: More than Just an Ionization Source. *Electrophoresis* **2021**, *42*, 103–121. [CrossRef] [PubMed]
107. Meireles, A.B.; Corrêa, D.K.; da Silveira, J.V.W.; Millás, A.L.G.; Bittencourt, E.; de Brito-Melo, G.E.A.; González-Torres, L.A. Trends in Polymeric Electrospun Fibers and Their Use as Oral Biomaterials. *Exp. Biol. Med.* **2018**, *243*, 665–676. [CrossRef] [PubMed]
108. Nam, J.; Huang, Y.; Agarwal, S.; Lannutti, J. Materials Selection and Residual Solvent Retention in Biodegradable Electrospun Fibers. *J. Appl. Polym. Sci.* **2008**, *107*, 1547–1554. [CrossRef]

109. Chee, B.S.; Nugent, M. Electrospun Natural Polysaccharide for Biomedical Application. *Nat. Polysaccharides Drug Deliv. Biomed. Appl.* **2019**, 589–615. [CrossRef]
110. Khorshidi, S.; Solouk, A.; Mirzadeh, H.; Mazinani, S.; Lagaron, J.M.; Sharifi, S.; Ramakrishna, S. A Review of Key Challenges of Electrospun Scaffolds for Tissue-Engineering Applications. *J. Tissue Eng. Regen. Med.* **2016**, *10*, 715–738. [CrossRef]
111. Liang, D.; Hsiao, B.S.; Chu, B. Functional Electrospun Nanofibrous Scaffolds for Biomedical Applications. *Adv. Drug Deliv. Rev.* **2007**, *59*, 1392–1412. [CrossRef] [PubMed]
112. McNamara, M.C.; Sharifi, F.; Wrede, A.H.; Kimlinger, D.F.; Thomas, D.G.; Vander Wiel, J.B.; Chen, Y.; Montazami, R.; Hashemi, N.N. Microfibers as Physiologically Relevant Platforms for Creation of 3D Cell Cultures. *Macromol. Biosci.* **2017**, *17*, 1700279. [CrossRef] [PubMed]
113. Xie, J.; Wang, C.H. Electrospray in the Dripping Mode for Cell Microencapsulation. *J. Colloid Interface Sci.* **2007**, *312*, 247–255. [CrossRef] [PubMed]
114. Yeo, M.; Kim, G.H. Anisotropically Aligned Cell-Laden Nanofibrous Bundle Fabricated via Cell Electrospinning to Regenerate Skeletal Muscle Tissue. *Small* **2018**, *14*, 1803491. [CrossRef] [PubMed]
115. Yeo, M.; Kim, G.H. Micro/Nano-Hierarchical Scaffold Fabricated Using a Cell Electrospinning/3D Printing Process for Co-Culturing Myoblasts and HUVECs to Induce Myoblast Alignment and Differentiation. *Acta Biomater.* **2020**, *107*, 102–114. [CrossRef]
116. Nosoudi, N.; Jacob, A.O.; Stultz, S.; Jordan, M.; Aldabel, S.; Hohne, C.; Mosser, J.; Archacki, B.; Turner, A.; Turner, P. Electrospinning Live Cells Using Gelatin and Pullulan. *Bioengineering* **2020**, *7*, 21. [CrossRef]
117. Ehler, E.; Jayasinghe, S.N. Cell Electrospinning Cardiac Patches for Tissue Engineering the Heart. *Analyst* **2014**, *139*, 4449–4452. [CrossRef] [PubMed]
118. Zhao, P.; Gu, H.; Mi, H.; Rao, C.; Fu, J.; Turng, L.S. Fabrication of Scaffolds in Tissue Engineering: A Review. *Front. Mech. Eng.* **2018**, *13*, 107–119. [CrossRef]
119. Yan, Q.; Dong, H.; Su, J.; Han, J.; Song, B.; Wei, Q.; Shi, Y. A Review of 3D Printing Technology for Medical Applications. *Engineering* **2018**, *4*, 729–742. [CrossRef]
120. Groll, J.; Burdick, J.A.; Cho, D.W.; Derby, B.; Gelinsky, M.; Heilshorn, S.C.; Jüngst, T.; Malda, J.; Mironov, V.A.; Nakayama, K.; et al. A Definition of Bioinks and Their Distinction from Biomaterial Inks. *Biofabrication* **2019**, *11*, 013001. [CrossRef]
121. Moldovan, N.I.; Hibino, N.; Nakayama, K. Tissue Engineering. *Tissue Eng.* **2017**, *23*, 237. [CrossRef]
122. Kuang, X.; Roach, D.J.; Wu, J.; Hamel, C.M.; Ding, Z.; Wang, T.; Dunn, M.L.; Qi, H.J. Advances in 4D Printing: Materials and Applications. *Adv. Funct. Mater.* **2019**, *29*, 1805290. [CrossRef]
123. Mandon, C.A.; Blum, L.J.; Marquette, C.A. Adding Biomolecular Recognition Capability to 3D Printed Objects: 4D Printing. *Procedia Technol.* **2017**, *27*, 1–2. [CrossRef]
124. Palmara, G.; Frascella, F.; Roppolo, I.; Chiappone, A.; Chiadò, A. Functional 3D Printing: Approaches and Bioapplications. *Biosens. Bioelectron.* **2021**, *175*, 112849. [CrossRef]
125. Nesaei, S.; Song, Y.; Wang, Y.; Ruan, X.; Du, D.; Gozen, A.; Lin, Y. Micro Additive Manufacturing of Glucose Biosensors: A Feasibility Study. *Anal. Chim. Acta* **2018**, *1043*, 142–149. [CrossRef]
126. Kirillova, A.; Maxson, R.; Stoychev, G.; Gomillion, C.T.; Ionov, L. 4D Biofabrication Using Shape—Morphing Hydrogels. *Adv. Mat.* **2017**, *29*, 1703443. [CrossRef] [PubMed]
127. Jungst, T.; Smolan, W.; Schacht, K.; Scheibel, T.; Groll, J. Strategies and Molecular Design Criteria for 3D Printable Hydrogels. *Chem. Rev.* **2016**, *116*, 1496–1539. [CrossRef]
128. Yue, Z.; Liu, X.; Coates, P.T.; Wallace, G.G. Advances in Printing Biomaterials and Living Cells: Implications for Islet Cell Transplantation. *Curr. Opin. Organ Transpl.* **2016**, *21*, 467–475. [CrossRef]
129. Seol, Y.J.; Kang, H.W.; Lee, S.J.; Atala, A.; Yoo, J.J. Bioprinting Technology and Its Applications. *Eur. J. Cardio Thorac. Surg.* **2014**, *46*, 342–348. [CrossRef] [PubMed]
130. Bhattacharyya, A.; Janarthanan, G.; Tran, H.N.; Ham, H.J.; Yoon, J.H.; Noh, I. Bioink Homogeneity Control during 3D Bioprinting of Multicomponent Micro/Nanocomposite Hydrogel for Even Tissue Regeneration Using Novel Twin Screw Extrusion System. *Chem. Eng. J.* **2021**, *415*. [CrossRef]
131. Muthukrishnan, L. Imminent Antimicrobial Bioink Deploying Cellulose, Alginate, EPS and Synthetic Polymers for 3D Bioprinting of Tissue Constructs. *Carbohydr. Polym.* **2021**, *260*, 117774. [CrossRef]
132. Butler, H.M.; Naseri, E.; MacDonald, D.S.; Andrew Tasker, R.; Ahmadi, A. Optimization of Starch- and Chitosan-Based Bio-Inks for 3D Bioprinting of Scaffolds for Neural Cell Growth. *Materialia* **2020**, *12*. [CrossRef]
133. Leucht, A.; Volz, A.C.; Rogal, J.; Borchers, K.; Kluger, P.J. Advanced Gelatin-Based Vascularization Bioinks for Extrusion-Based Bioprinting of Vascularized Bone Equivalents. *Sci. Rep.* **2020**, *10*. [CrossRef]
134. Rubio-Valle, J.F.; Perez-Puyana, V.; Jiménez-Rosado, M.; Guerrero, A.; Romero, A. Evaluation of Smart Gelatin Matrices for the Development of Scaffolds via 3D Bioprinting. *J. Mech. Behav. Biomed. Mater.* **2021**, *115*. [CrossRef] [PubMed]
135. Tan, E.Y.S.; Suntornnond, R.; Yeong, W.Y. High-Resolution Novel Indirect Bioprinting of Low-Viscosity Cell-Laden Hydrogels via Model-Support Bioink Interaction. *3D Print. Addit. Manuf.* **2020**, *8*, 69–78. [CrossRef]
136. Hauptstein, J.; Böck, T.; Bartolf-Kopp, M.; Forster, L.; Stahlhut, P.; Nadernezhad, A.; Blahetek, G.; Zernecke-Madsen, A.; Detsch, R.; Jüngst, T.; et al. Hyaluronic Acid-Based Bioink Composition Enabling 3D Bioprinting and Improving Quality of Deposited Cartilaginous Extracellular Matrix. *Adv. Healthc. Mater.* **2020**, *9*. [CrossRef] [PubMed]

137. Ma, L.; Li, Y.; Wu, Y.; Yu, M.; Aazmi, A.; Gao, L.; Xue, Q.; Luo, Y.; Zhou, H.; Zhang, B.; et al. 3D Bioprinted Hyaluronic Acid-Based Cell-Laden Scaffold for Brain Microenvironment Simulation. *BioDesign Manuf.* **2020**, *3*, 164–174. [CrossRef]
138. Wei, W.; Ma, Y.; Yao, X.; Zhou, W.; Wang, X.; Li, C.; Lin, J.; He, Q.; Leptihn, S.; Ouyang, H. Advanced Hydrogels for the Repair of Cartilage Defects and Regeneration. *Bioact. Mater.* **2021**, *6*, 998–1011. [CrossRef] [PubMed]
139. Moeinzadeh, S.; Park, Y.; Lin, S.; Yang, Y.P. In-Situ Stable Injectable Collagen-Based Hydrogels for Cell and Growth Factor Delivery. *Materialia* **2021**, *15*, 100954. [CrossRef]
140. Zhang, Y.; Ellison, S.T.; Duraivel, S.; Morley, C.D.; Taylor, C.R.; Angelini, T.E. 3D Printed Collagen Structures at Low Concentrations Supported by Jammed Microgels. *Bioprinting* **2021**, *21*. [CrossRef]
141. Redmond, J.; McCarthy, H.; Buchanan, P.; Levingstone, T.J.; Dunne, N.J. Advances in Biofabrication Techniques for Collagen-Based 3D in Vitro Culture Models for Breast Cancer Research. *Mater. Sci. Eng. C* **2021**, *122*, 111944. [CrossRef]
142. Pokusaev, B.; Vyazmin, A.; Zakharov, N.; Karlov, S.; Nekrasov, D.; Reznik, V.; Khramtsov, D. Thermokinetics and Rheology of Agarose Gel Applied to Bioprinting Technology. *Therm. Sci.* **2020**, *24*, 453. [CrossRef]
143. De Melo, B.A.G.; Jodat, Y.A.; Cruz, E.M.; Benincasa, J.C.; Shin, S.R.; Porcionatto, M.A. Strategies to Use Fibrinogen as Bioink for 3D Bioprinting Fibrin-Based Soft and Hard Tissues. *Acta Biomater.* **2020**, *117*, 60–76. [CrossRef]
144. Sarker, B.; Boccaccini, A.R. Alginate Utilization in Tissue Engineering and Cell Therapy. In *Alginates and their Biomedical Applications*; Springer: Singapore, 2018; pp. 121–155.
145. Soltan, N.; Ning, L.; Mohabatpour, F.; Papagerakis, P.; Chen, X. Printability and Cell Viability in Bioprinting Alginate Dialdehyde-Gelatin Scaffolds. *ACS Biomater. Sci. Eng.* **2019**, *5*, 2976–2987. [CrossRef] [PubMed]
146. Silva, R.; Singh, R.; Sarker, B.; Papageorgiou, D.G.; Juhasz, J.A.; Roether, J.A.; Cicha, I.; Kaschta, J.; Schubert, D.W.; Chrissafis, K.; et al. Hybrid Hydrogels Based on Keratin and Alginate for Tissue Engineering. *J. Mater. Chem. B* **2014**, *2*, 5441–5451. [CrossRef] [PubMed]
147. Peniche, H.; Reyes-Ortega, F.; Aguilar, M.R.; Rodríguez, G.; Abradelo, C.; García-Fernández, L.; Peniche, C.; Román, J.S. Thermosensitive Macroporous Cryogels Functionalized with Bioactive Chitosan/Bemiparin Nanoparticles. *Macromol. Biosci.* **2013**, *13*, 1556–1567. [CrossRef] [PubMed]
148. De La Mata, A.; Nieto-Miguel, T.; López-Paniagua, M.; Galindo, S.; Aguilar, M.R.; García-Fernández, L.; Gonzalo, S.; Vázquez, B.; Román, J.S.; Corrales, R.M.; et al. Chitosan-Gelatin Biopolymers as Carrier Substrata for Limbal Epithelial Stem Cells. *J. Mater. Sci. Mater. Med.* **2013**, *24*, 2819–2829. [CrossRef]
149. Fernández-Gutiérrez, M.; Pérez-Köhler, B.; Benito-Martínez, S.; García-Moreno, F.; Pascual, G.; García-Fernández, L.; Aguilar, M.R.; Vázquez-Lasa, B.; Bellón, J.M. Development of Biocomposite Polymeric Systems Loaded with Antibacterial Nanoparticles for the Coating of Polypropylene Biomaterials. *Polymers* **2020**, *12*, 1829. [CrossRef] [PubMed]
150. Nakal-Chidiac, A.; García, O.; García-Fernández, L.; Martín-Saavedra, F.M.; Sánchez-Casanova, S.; Escudero-Duch, C.; San Román, J.; Vilaboa, N.; Aguilar, M.R. Chitosan-Stabilized Silver Nanoclusters with Luminescent, Photothermal and Antibacterial Properties. *Carbohydr. Polym.* **2020**, *250*. [CrossRef] [PubMed]
151. Jin, J.; Song, M.; Hourston, D.J. Novel Chitosan-Based Films Cross-Linked by Genipin with Improved Physical Properties. *Biomacromolecules* **2004**, *5*, 162–168. [CrossRef]
152. Mora-Boza, A.; Włodarczyk-Biegun, M.K.; Del Campo, A.; Vázquez-Lasa, B.; Román, J.S. Glycerylphytate as an Ionic Crosslinker for 3D Printing of Multi-Layered Scaffolds with Improved Shape Fidelity and Biological Features. *Biomater. Sci.* **2020**, *8*, 506–516. [CrossRef]
153. He, Y.; Wang, F.; Wang, X.; Zhang, J.; Wang, D.; Huang, X. A Photocurable Hybrid Chitosan/Acrylamide Bioink for DLP Based 3D Bioprinting. *Mater. Des.* **2021**, *202*, 109588. [CrossRef]
154. Jian, Z.; Zhuang, T.; Qinyu, T.; Liqing, P.; Kun, L.; Xujiang, L.; Diaodiao, W.; Zhen, Y.; Shuangpeng, J.; Xiang, S.; et al. 3D Bioprinting of a Biomimetic Meniscal Scaffold for Application in Tissue Engineering. *Bioact. Mater.* **2021**, *6*, 1711–1726. [CrossRef]
155. Puertas-Bartolomé, M.; Włodarczyk-Biegun, M.K.; Del Campo, A.; Vázquez-Lasa, B.; Román, J.S. 3D Printing of a Reactive Hydrogel Bio-Ink Using a Static Mixing Tool. *Polymers* **2020**, *12*, 1986. [CrossRef]
156. Skardal, A.; Zhang, J.; McCoard, L.; Xu, X.; Oottamasathien, S.; Prestwich, G.D. Photocrosslinkable Hyaluronan-Gelatin Hydrogels for Two-Step Bioprinting. *Tissue Eng. Part A* **2010**, *16*, 2675–2685. [CrossRef] [PubMed]
157. Skardal, A.; Mack, D.; Kapetanovic, E.; Atala, A.; Jackson, J.D.; Yoo, J.; Soker, S. Bioprinted Amniotic Fluid-Derived Stem Cells Accelerate Healing of Large Skin Wounds. *Stem Cells Transl. Med.* **2012**, *1*, 792–802. [CrossRef]
158. Zou, Q.; Grottkau, B.E.; He, Z.; Shu, L.; Yang, L.; Ma, M.; Ye, C. Biofabrication of Valentine-Shaped Heart with a Composite Hydrogel and Sacrificial Material. *Mater. Sci. Eng. C* **2020**, *108*, 110205. [CrossRef] [PubMed]
159. Gopinathan, J.; Noh, I. Recent Trends in Bioinks for 3D Printing. *Biomater. Res.* **2018**, *22*, 11. [CrossRef]
160. Raphael, B.; Khalil, T.; Workman, V.L.; Smith, A.; Brown, C.P.; Streuli, C.; Saiani, A.; Domingos, M. 3D Cell Bioprinting of Self-Assembling Peptide-Based Hydrogels. *Mater. Lett.* **2017**, *190*, 103–106. [CrossRef]
161. Hassan, M.; Dave, K.; Chandrawati, R.; Dehghani, F.; Gomes, V.G. 3D Printing of Biopolymer Nanocomposites for Tissue Engineering: Nanomaterials, Processing and Structure-Function Relation. *Eur. Polym. J.* **2019**, *121*, 109340. [CrossRef]
162. Murata, D.; Kunitomi, Y.; Harada, K.; Tokunaga, S.; Takao, S.; Nakayama, K. Osteochondral Regeneration Using Scaffold-Free Constructs of Adipose Tissue-Derived Mesenchymal Stem Cells Made by a Bio Three-Dimensional Printer with a Needle-Array in Rabbits. *Regen. Ther.* **2020**, *15*, 77–89. [CrossRef]

163. Hernández-Arriaga, A.M.; del Cerro, C.; Urbina, L.; Eceiza, A.; Corcuera, M.A.; Retegi, A.; Auxiliadora Prieto, M. Genome Sequence and Characterization of the Bcs Clusters for the Production of Nanocellulose from the Low PH Resistant Strain Komagataeibacter Medellinensis ID13488. *Microb. Biotechnol.* **2019**, *12*, 620–632. [CrossRef] [PubMed]
164. Han, C.; Wang, X.; Ni, Z.; Ni, Y.; Huan, W.; Lv, Y.; Bai, S. Effects of Nanocellulose on Alginate/Gelatin Bio-Inks for Extrusion-Based 3D Printing. *BioResources* **2020**, *15*, 7357–7373. [CrossRef]
165. Gao, J.; Ding, X.; Yu, X.; Chen, X.; Zhang, X.; Cui, S.; Shi, J.; Chen, J.; Yu, L.; Chen, S.; et al. Cell-Free Bilayered Porous Scaffolds for Osteochondral Regeneration Fabricated by Continuous 3D-Printing Using Nascent Physical Hydrogel as Ink. *Adv. Healthc. Mater.* **2021**, *10*, 2001404. [CrossRef]
166. Ma, Y.; Zhang, C.; Wang, Y.; Zhang, L.; Zhang, J.; Shi, J.; Si, J.; Yuan, Y.; Liu, C. Direct Three-Dimensional Printing of a Highly Customized Freestanding Hyperelastic Bioscaffold for Complex Craniomaxillofacial Reconstruction. *Chem. Eng. J.* **2021**, *411*, 128541. [CrossRef]
167. Cubo, N.; Garcia, M.; Del Cañizo, J.F.; Velasco, D.; Jorcano, J.L. 3D Bioprinting of Functional Human Skin: Production and in Vivo Analysis. *Biofabrication* **2017**, *9*, 015006. [CrossRef] [PubMed]
168. Allsopp, B.J.; Hunter-Smith, D.J.; Rozen, W.M. Vascularized versus Nonvascularized Bone Grafts: What Is the Evidence? *Clin. Orthop. Relat. Res.* **2016**, *474*, 1319–1327. [CrossRef]
169. Kim, B.S.; Gao, G.; Kim, J.Y.; Cho, D.W. 3D Cell Printing of Perfusable Vascularized Human Skin Equivalent Composed of Epidermis, Dermis, and Hypodermis for Better Structural Recapitulation of Native Skin. *Adv. Healthc. Mater.* **2019**, *8*, 1801019. [CrossRef]

Review

Polymer Based Bioadhesive Biomaterials for Medical Application—A Perspective of Redefining Healthcare System Management

Nibedita Saha [1,*], Nabanita Saha [2,*], Tomas Sáha [1], Ebru Toksoy Öner [3], Urška Vrabič Brodnjak [4], Heinz Redl [5], Janek von Byern [5] and Petr Sáha [1,2]

[1] Footwear Research Centre, University Institute, Tomas Bata University in Zlin, University Institute & Tomas Bata University in Zlin, Nad Ovčírnou 3685, 76001 Zlín, Czech Republic; tsaha@utb.cz (T.S.); saha@utb.cz (P.S.)
[2] Faculty of Technology Polymer, Centre, Tomas Bata University in Zlin, University Institute, Centre of Polymer Systems & Tomas Bata University in Zlin, 76001 Zlín, Czech Republic
[3] Department of Bioengineering, IBSB. Marmara University, 34722 Istanbul, Turkey; ebru.toksoy@marmara.edu.tr
[4] Graphic Arts and Design, Department of Textiles, Faculty of Natural Sciences and Engineering, University of Ljubljana, 1000 Ljubljana, Slovenia; urska.vrabic@ntf.uni-lj.si
[5] Austrian Cluster for Tissue Regeneration, Ludwig Boltzmann Institute for Experimental and Clinical Traumatology, 1200 Vienna, Austria; office@trauma.lbg.ac.at (H.R.); janek.von.byern@univie.ac.at (J.v.B.)
* Correspondence: nibedita@utb.cz (N.S.); nabanita@utb.cz (N.S.); Tel.: +420-57603-8151 (N.S.); +420-57603-8156 (N.S.)

Received: 1 December 2020; Accepted: 13 December 2020; Published: 16 December 2020

Abstract: This article deliberates about the importance of polymer-based bioadhesive biomaterials' medical application in healthcare and in redefining healthcare management. Nowadays, the application of bioadhesion in the health sector is one of the great interests for various researchers, due to recent advances in their formulation development. Actually, this area of study is considered as an active multidisciplinary research approach, where engineers, scientists (including chemists, physicists, biologists, and medical experts), material producers and manufacturers combine their knowledge in order to provide better healthcare. Moreover, while discussing the implications of value-based healthcare, it is necessary to mention that health comprises three main domains, namely, physical, mental, and social health, which not only prioritize the quality healthcare, but also enable us to measure the outcomes of medical interventions. In addition, this conceptual article provides an understanding of the consequences of the natural or synthetic polymer-based bioadhesion of biomaterials, and its significance for redefining healthcare management as a novel approach. Furthermore, the research assumptions highlight that the quality healthcare concept has recently become a burning topic, wherein healthcare service providers, private research institutes, government authorities, public service boards, associations and academics have taken the initiative to restructure the healthcare system to create value for patients and increase their satisfaction, and lead ultimately to a healthier society.

Keywords: bioadhesion; biomaterials; biomedical application; healthcare system management; innovation; polymer based bioadhesive

JEL Classification: I1; I10; I11; I18; I21; I28; H51

1. Introduction

Currently, in the 21st century, healthcare management plays an important role in focusing and aligning the myriad continuous improvements that optimize the application of bioadhesion as related to innovative biomaterials' medical use. This article intends to reveal the importance of bioadhesive biomaterials' application in the healthcare system. Nowadays, the application of bioadhesion is one of greatest interests for various researchers who intend to develop new biomaterials, therapies and technological possibilities, such as biomedical application. Accordingly, progressive innovation in the bioadhesion of biomaterials has trended sharply upward, and is expected to double by 2020, especially with a focus on delivering quality healthcare. Although redefining health, the World Health Organization (WHO) defined 'health' as a state of complete physical, mental and social wellbeing that not only considers the illness, but prioritizes the concept of value-based healthcare [1]. On the other hand, from the functional perspective, bioadhesives can be considered as an identical material, which is biological in nature and holds together for extended periods of time by interfacial forces. Essentially, it is an area of active multidisciplinary research approach, wherein engineers, scientists (including chemists, physicists, biologists, and medical experts (supportive medical), materials producers, and manufacturers combine their knowledge [2]. Finally, from the practical point of view, this article proposes some research assumptions, which state that the bioadhesion of biomaterials for redefining healthcare management is not a new concept. Its implementation has been used for several years for medical applications, such as dentistry and orthopedics, and it is now entering new fields, for example, tissue sealing and directed drug delivery systems. In addition, the said issues and solutions affect and involve healthcare delivery organizations, health plans and employers, i.e., healthcare service providers, private research institutes, government authorities and public service boards, research institutes, associations and academics. The outcome will be, in the long-term, to restructure the healthcare system, which will not only create value for patients and increase satisfaction, but it will also improve the health effects through enabling new efficiencies and lowering costs.

1.1. Notion of Biomaterials

Regarding the notion of "biomaterials", it is necessary to mention that there are two significant topics that are inter-related with the concept of the word biomaterial. The first conceptual meaning of biomaterial deals with the term 'bio', which exemplifies, as a way of filling in the gaps where the question arises, whether we are discussing the process of taking out of life or putting into life. The second term, "material", has a broader sense, which indicates a substance. Now the question arises of how this material can enable us to keep our life more flexible. Research shows that from the healthcare benefit point of view, several scholars have made an effort to define the term "biomaterials" and its application as well as utility in our day-to-day life. In medical science, research has shown that it has ample potential to keep our life more flexible, in that it will easily enable us to respond to altered circumstances. Although, biomaterials' presentation in medical science did not get that recognition until the Consensus Conference on Definitions in Biomaterials Science, held in 1987. According to the European Society for Biomaterials, earlier, the term biomaterials and its medical application were not so profoundly known in the medical science, though its application was already existing [3], as the definition is a result of considered debate, which definitely has some reliability from a healthcare point of view. On the other hand, this conceptualization of biomaterials concludes that a biomaterial is "a non-viable material and its application in a medical device, is envisioned to interrelate with the biological systems" [4].

1.2. Overview of Bioadhesion

Bioadhesion may be defined as the binding of a natural or synthetic polymer or biological-origin adhesive to a biological substrate. When the substrate is a mucus layer, the term is known as mucoadhesion [5]. On the other hand, while referring to the application of bioadhesion in broad terms, it is necessary to mention that the terminology "bioadhesion" itself represents an extensively differentiated phenomena, as it covers the adhesive properties of both the synthetic components as well as the natural surfaces (such as cells). Furthermore, research shows that bioadhesion could also refer to the usage of bioadhesives in order to link the two surfaces together, especially in drug delivery, dental and surgical applications [6]. As such, the significance of bioadhesive biomaterial application has emerged and been recognized due to its consequences for the specific development of new biomaterials, therapies and technological products for redefining the healthcare sector.

2. Bioadhesion of Biomaterials

While discussing the significance of the bioadhesion of biomaterials, it is mandatory to highlight that in the contemporary world, healthcare is a fundamental issue in translational research, especially when it is innovative, as well as the fact that the bioadhesion of biomaterials application is being used in healthcare in order to fight against life-threatening diseases. In addition, over the past two decades, innovative biomaterials applications have been viewed as a significant issue in translational research in the field of regenerative medicine, where biomaterials have been extensively applied in numerous medical devices for the benefit of healthcare. In this regard, it is necessary to state that the study of biomaterials is essentially associated with the study of biocompatible materials, especially for biomedical applications, which encompasses not only the synthetic materials, such as metals, polymers, ceramics and composites, but also includes biological materials, for example proteins, cells and tissues. The below-mentioned Figure 1 shows examples of the bioadhesion of biomaterials.

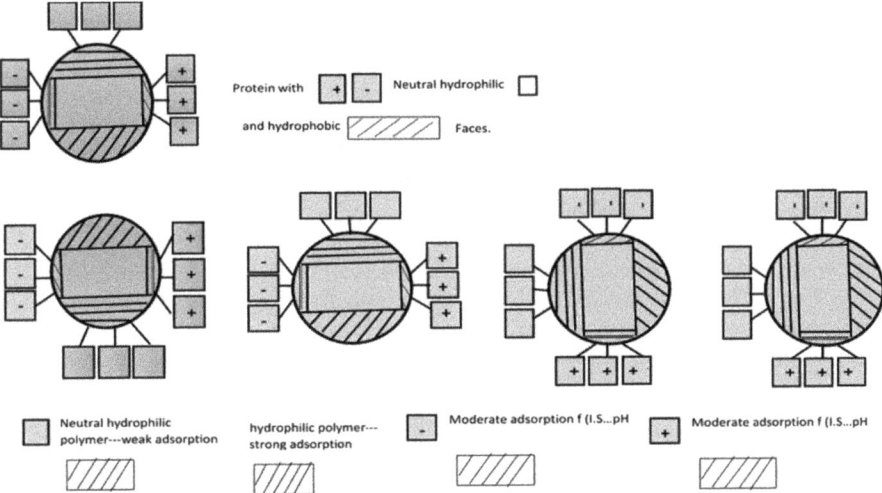

Figure 1. Bioadhesion of biomaterials (based on the idea from the *Bioadhesion of Biomaterials*, and Medical Devices, Springer Book [7]).

On the other hand, the term bioadhesion refers to the situation wherein natural and synthetic materials stick to each other, and especially to biological surfaces. Henceforth, the application of bioadhesive polymers in healthcare emerges, specifically with the use of medical devices for the effects on the biological exterior and crossing point. In this review article, the authors attempt to prove that, from the healthcare point of view, bioadhesion's presentation is advantageous. Considering the grafting of medical devices in the human body, it is necessary to remember that though this embedding procedure is a very useful and important aspect of healthcare, we cannot ignore the probability of high risks due to the interface for microorganisms. As implantable medical devices are the idyllic location for the growth of microbes, infections are triggered quickly by bacteria that mainly originate in the body itself. Consequently, some phases effect the bioadhesion of implantable medical devices, including surface topography, chemical interaction, mechanical interaction and physiological interactions.

Research shows that, considering these aspects, medical practitioners will likely try their best to control the medical devices through bioadhesion processes by enhancing the desirable interaction of bioadhesion and eliminating the adverse interactions. Therefore, to comprehend the debate on the bioadhesion of biomaterials in order to redefine healthcare management, it is necessary to mention some methods of the bioadhesion testing, which includes the evaluation of (a) surface roughness/surface morphology/surface topography, (b) chemical interactions, (c) physiological factors, (d) physical and mechanical effects, and (e) the contact angle and testing of biofilm formation [7,8]. In this conceptual article, focus has been placed on natural polymer-based bioadhesive biomaterials, i.e., polysaccharide/carbohydrate-based adhesives and protein-based adhesives. Carbohydrates in the form of polysaccharides are mostly available from plants (available in three different forms: cellulose, starch and natural gum), the exoskeleton of various marine animals, and/or are synthesized by some microorganisms. Cellulose is the principal structural material of the cell walls of plants. It is a homopolymer of β-D-hydroglucopyranose monomeric units that are linked via a linkage between the C-1 of the monomeric unit and the C-4 of the adjacent monomeric unit [8]. Due to the presence of the large number of hydroxyl groups, cellulose molecules readily form hydrogen bonds with other cellulose molecules so as to give highly crystalline structures, as the bonds are generally sensitive to water. These unique structural properties of cellulose are hindering its use as an adhesive itself.

As such, the future applications of these adhesives demand the modification of natural polymers so as to give components that can undergo further cross-linking to form water-insensitive bonds [8]. For example, cellulose converted to various cellulose derivatives in the form of ester and ether (e.g., cellulose acetate, carboxymethyl cellulose, hydroxyethyl cellulose, etc.) can be used in the formation of carbohydrate polymer as an adhesive. Instead, it is important to address the fact that cellulose adhesion performs at its best when connected through hydrogen bonds ranging from the macro level to the nano level. Regarding this matter, it is obligatory to mention that for knowledge about the application of these bioadhesive materials, in terms of composition, structural design and interactions with surfaces, it is crucial to expose the basic information about the biochemical and mechanical principles that are associated with the process of biological adhesion.

Similarly, protein-based bioadhesives are also recognized as one of the most significant and prolific categories of macromoecules in cells that facilitate the creation of bonding among microorganisms. In another way, it can be said that correspondingly, each protein molecule can be imagined as a polymer composed of amino acids, which are known as tiny macromolecules that contain an amine group, a carboxylic acid group and a variable side chain [9].

2.1. Polysaccharides-Based Adhesives

Even though cellulose, starch and gums are commercially available and used in and for adhesives, it is a challenge to establish novel adhesive polysaccharides which will be commercially available at low costs and are applicable in wet and dry states. Some interesting and praiseworthy polyssacharide-based biomaterials (bacterial cellulose, Levan, and chitosan) are stated below, which have excellent applications in the medical field.

Bacterial cellulose, as synthesized by *Acetobactor xylinum*, is a potential and promising natural polymer that has already been used quite successfully in several healthcare applications. It can be used in a wide variety of biomedical applications, from topical wound dressing to durable scaffolding that is useful in tissue engineering, and the regeneration of other tissues such as bone and cartilage. Although it was reported by Brown 1886, more attention to this biomaterial has been paid in the second half of the 20th century [10–12]. It is well organized in contrast to standard or plant cellulose, sometime referred as microbial cellulose. Bacterial cellulose and microbial cellulose have unique structural and mechanical properties compared to plant cellulose, but the molecular formulas ($C_6H_{10}O_5$) of both bacterial and plant cellulose are the same [11,12]. Intensive study on the production of bacterial cellulose was conducted by Herstrin and Schramn (H.S.) in 1954 [13].

They established that *Acetobacter xylinum* synthesized cellulose in the presence of glucose and oxygen. Moreover, the established H.S. medium is considered as a standard nutrient medium and *A. xylinum* as a model bacterium to produce bacterial cellulose [13]. However, it is an incompetent and expensive medium for bacterial cellulose production from the present point of view. The search for cost-effective alternatives is therefore a motivation. Agro-waste-based carbon sources (coconut water, pineapple juice, etc.) are reported as an alternative nutrient medium (as fruits contain abundant sugar in the form of glucose and fructose) for the production of bacterial cellulose [14] in an economic way. At the Tomas Bata University in Zlin, the second author optimized the production conditions of bacterial cellulose using "apple juice" as a nutrient medium and "*Gluconobacter xylinus* (CCM 3611T)" as the bacterial strain [15–17]. The bacterial cellulose once formed is deposited on the surface of a static liquid medium (as shown in Figure 2). It is reported that the most active layer of cellulose-producing bacteria is always in contact with the air. During the process of fermentation, the older layers of cellulose are pushed down by the newly formed cellulose fibrils.

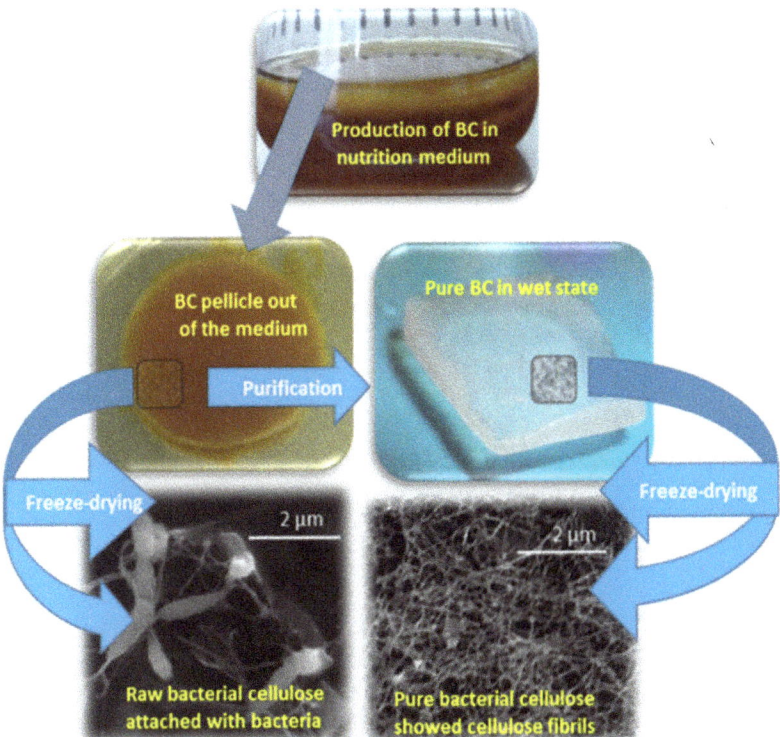

Figure 2. Scheme of production and purification of bacterial cellulose (BC), which exhibits the formation of cellulose fibrils by bacteria.

From a structural point of view, bacterial cellulose comprises a group of similar chains that are composed of D-glucopyranose units. Moreover, they are interlinked by intermolecular hydrogen bonds, which are identical in chemical composition to those of plant cellulose [17]. Such properties of BC and the lack of irregularities lead to both superior reinforcement and thermal expansion properties when used with matrix materials to form bacterial cellulose-based biocomposites [18]. From the degree of polymerization point of view, research shows that bacterial cellulose has a higher degree of purity and greater fibrousness, and the range of polymerization exists in the bacterial cellulose between 2000 and 6000. However, this relatively low stage of polymerization may limit the adhesion through interpenetrating networks or mechanical interlocking. On the other hand, in this circumstance it has been observed that in most of the cases, the adhesion in composite materials is limited to hydrogen bonding. Consequently, other applications of bioadhesion must be explored. The inter- and intra-molecular binding and/or adhesion is accomplished through hydrogen bonding and interactions with surfaces, and it is necessary to reveal the basic biochemical and mechanical principles involved in biological adhesion. According to a Vision and Technology Roadmap developed by Agenda 2020 [19], bacterial cellulose has a bright future as a renewable source of carbohydrate-based biopolymers. However, research still needs to be done for nanocellulose adhesion [19].

On the other hand, Levan is a fructan-type homopolysaccharide that is composed of fructose units joined by β-2,6 glycosidic linkages. It is widely present in nature and is produced by various microorganisms and plants from sucrose-based substrates (for a recent review, [20]), whereas microbial Levan is produced in the form of long-chained exopolysaccharides by the action of the Levan sucrase enzyme. Plant-derived Levan, instead, is shorter, and its biosynthesis takes place in the vacuoles and requires the action of several enzymes [21]. Levan stands out from other natural polysaccharides with its unusual properties such as high adhesive strength, very low intrinsic viscosity, several health benefits, and its ability to form gel alcohol and self-assembled structures. Recent efforts to associate these unique features with high-value medical applications have revived the interest in this underexplored polymer, and bring Levan into the focus of scientific and industrial interest. The applications of Levan in hair care products and whiteners, as well as its medical applications in healing wounds and burned tissue, anti-irritant, antioxidant and anti-inflammatory activities, weight loss and cholesterol control, are well documented [20].

Besides many mesophilic sources, the first extremophilic source of Levan was reported in 2009 [22]. Since then, Halomonas Levan (HL), produced by extremophilic Halomonas smyrnensis bacteria as well as its chemical derivatives, has been the subject of various high-value applications, ranging from laser-deposited bioactive surfaces to tissue engineering. These include its use in antioxidant [23] and anti-cancer [24,25] agents, as well as its suitability for the controlled delivery of peptide- and protein-based drugs [26,27] and as phosphonated HL in adhesive multilayer thin films obtained by the layer-by-layer (LbL) technique [28]. Moreover, HL was found to increase the biocompatibility and change the crystallinity in chitosan/levan/polyethyleneoxide ternary blend films [29]. It is also used as a crosslinker, and the obtained stimuli-responsive hydrogels were found to release 5-aminosalicylic acid (5-ASA) in a temperature-controlled manner [30]. Levan has also been reported as a suitable polymer for obtaining nanostructured bioactive surfaces by combinatorial matrix-assisted pulsed laser evaporation (C-MAPLE), and the obtained gradient surfaces were found to modulate the ERK signaling of osteoblasts [31,32]. Moreover, due to its high biocompatibility and heparin mimetic activity, a sulfated derivative of Halomonas Levan (SHL) has been reported to be a suitable functional biomaterial in designing engineered smart scaffolds with applications in cardiac tissue engineering [33,34]. Additionally, recently, SHL was found to not only improve the mechanical and adhesive properties of multilayered free-standing films, but also to allow myogenic differentiation, and it led to cytocompatible and myoconductive films [35]. All the above-mentioned studies make Levan polysaccharide a very promising bioadhesive for many medical applications.

Among polysaccharides, chitin and chitosan are among the most abundant natural compounds on earth, beside cellulose. Chitosan is obtained from crustaceans after the deacetylation of chitin or extraction from insects or fungi [36,37]. In view of the present scientific literature, chitosan is probably one of the most published polysaccharides. However, chitosan has not the same commercial success as cellulose. It is estimated that approximately 10 billion tons of chitin can be synthesized each year, where the main sources are crustaceans, insects, mollusks and fungi [38]. This biomaterial is still under investigation, and its adhesive properties present an industrial challenge as well as an important research area. In last decade, chitosan has gained significant attention as an adhesive biomaterial, due to its biodegradability, non-toxicity, biocompatibility and anti-microbial properties [39,40].

For the adequate adhesive properties of polymers, surface tension, ability of penetration and viscosity are the most important parameters. In a study, it was proven that the surface tension of chitosan decreases with increasing concentrations. The adhesive surface tension must be inferior at the material surface energy to obtain sufficient molecular interactions [41]. Kurtek et al. determined that 2% (w/v) of chitosan in a 1% (v/v) acetate solution exhibited 38.59 mN/m surface tension at the dispersive end and 1.10 mN/m in the polar part [42]. This has proven that acid-base Lewis interactions were dominating. Furthermore, chitosan with a low surface tension indicates that it is easily spread on many and different types of materials. On top of this, Bajaj et al. [43] obtained a viscosity of chitosan solution that increased with concentration, but decreased with temperature. Few researchers compared the viscosity of chitosan solutions with different molecular weights [43–45]. Moreover, the wide range of chitosan viscosity is an advantage in terms of its use as an adhesive. Since the adhesive viscosity depends on the application, it can be easily adapted as a chitosan solution.

Chitosan is the only cationic polysaccharide, due to the NH_{3+} group at an acidic pH [46]. The –OH, $-NH_{3+}$, $-NH_2$, $-CH_2OH$ and $-NHCOCH_3$ groups of chitosan are responsible for chemical modifications intended to improve cross-linking, and consequently improve adhesiveness.

For appropriate adhesive, high tensile strength (TS) is among the important parameters. Once the material is dried, it has to achieve good mechanical properties and also good resistance to water, moisture, temperature, etc. In our study, pure chitosan films and chitosan films in blends with rice starch were prepared. The determination of the physical–mechanical properties of films has been made [40]. Films were prepared with different concentrations of chitosan, rice starch, and as plasticizers when glycerol was added [38,40,47]. The addition of glycerol led to an increase in the elasticity of chitosan films and gave high resistance to mechanical constrains. At the same time, glycerol decreased the drying time of the films, since it acted as a hygroscopic agent. The results have shown that the tensile strength of chitosan films varied from 62.3 to 64.8 MPa. These differences can be explained by the influence of ultrasound as a pretreatment and the ratio of hydrogen bonds between hydroxyl and amino groups in chitosan films [39–41]. The temperature of decomposition was also determined in order to characterize adhesive thermal resistance [39,47]. The thermal degradation of chitosan films was at 253 °C, and this showed that it can be used at temperatures above room temperature and even more. The analysis of chitosan paper coating films has also been made in combination with rice starch and curdlan, in different amounts of components [39,47]. Based on the results, it was determined that chitosan improved the tensile properties, decreased water vapor permeability, and improved moisture content and surface appearance, which is for the paper coating very important. Apart from this, the cross-linking of other polysaccharides, such as rice starch and curdlan, with chitosan was also evaluated for bonding applications.

The literature shows that chitosan films are very good biomaterials when used as biomedical adhesives, such as for wound healing, tissue repair, etc. [38–43]. Some commercial applications of chitosan as an adhesive are already on the market, such as Axiostat® (Gujarat, India), HemConTM (Portland, OR, USA), Chitoflex® PRO (Portland, OR, USA), CeloxTM (Crewe, UK) and Surgilux (Delhi, India). Chitosan has become a popular biopolymer in the medical field. Due to its unique properties among polysaccharides, it has been shown as a competitive adhesive compared to some fossil sources. Nevertheless, progress in bioadhesives should be aimed at lowering the costs and the impact on the environment. This biotechnological challenge should be focused on the environmental assessment approach, especially for developing bio-based sustainable adhesives.

2.2. Protein-Based Adhesives

In order to discuss the practice of biological adhesive application for medical uses, it is necessary to emphasize one of the most important aspects that needs to be taken into consideration. That is, what are the most important requirements that the organisms must fulfill? All bio-based adhesives are superbly adapted, not only in view of chemical composition, biomechanical properties and gland morphology, but also in terms of being strongly optimized for the environment and for the requirements of the organism. Aquatic adhesives, for example, perform ideally under wet conditions, but mostly show no or weak bonding ability to dry surfaces. This guarantees that prospective applications under dry conditions (i.e., as skin sealant) are less favorable for such systems.

Currently, more than 100 marine and terrestrial organisms are known to produce bioadhesives [48], some of them for 500 million years. This high variety of adhesive systems with, e.g., permanent or temporary holdfast, the ability to bond on different surfaces and with curing times from milliseconds to minutes, surely offer a broad portfolio, suitable for every desirable medical application. In the following chapter, we aim to give a short overview of existing and prospective biological adhesive systems; further details could be found elsewhere [48–50].

The most well-known and best-established system is certainly fibrin. Fibrin and fibrinogen are components of the blood clotting system together with thrombin, calcium ions and factor XIII. Fibrin is the most biocompatible medical sealant available today on the market [51,52]. This is in view of its biocompatibility, biodegradability, lack of heavy metals or absence of volatile organic compounds in relation to other commercial medical sealants. However, its bonding strength (approx. 0.01 MPa) is about one magnitude lower than synthetic adhesives, such as gelatin–resorcinol–formalin adhesives (approx. 0.1 MPa), which dominate today's adhesive market [48].

One of the promising characteristics of the biological adhesives derived from marine species, such as Mytilus spec., is their curing time within seconds, strong bonding (35–75 MPa) [53] in different environments and on different surfaces (hard/soft, even Teflon® [54]) and their sustainability (biocompatible, biodegradable, non-toxic, etc.). No commercial product on the market to date is able to cover such a vast application range. In Mytilus, six different L-DOPA-rich proteins (mussel adhesive protein; MAP) maintain the holdfast. The catecholic amino acid, L-DOPA (L-3,4-dihydroxyphenylalanine), is currently the best-characterized key compound in marine adhesive proteins, produced not only by Mytilus but also used by Phragmatopoma and Sabella for a permanent holdfast [55,56]. The tissue adhesive Cell-TakTM (USA) was the first example (year 1986, TM-No. 73604754) of a marine-derived sealant, based on mussel adhesive proteins only. With the technical and scientific progress within the last few years, producing L-DOPA recombinantly, technological advances in particular in the biopolymer-DOPA engineering sector have been made, shown by the increasing number of publications [57,58] and technical possibilities [59].

Snail mucus is certainly one of the most exciting and promising, but also annoying, biomaterials in the animal kingdom. Gastropods produce a temporary viscoelastic mucus (see contribution in [56,60,61]) able to bind to any sharp or smooth surfaces, even extreme anti-adhesive non-slip materials and water-coated slippery hydrogels [60]. Moreover, snail mucus is proposed to have a promoting effect on skin cell migration, proliferation, survival and antiphotoaging [62–68]. Consequently, snail mucus is today sold in the cosmetics sector (see Patent US 5538740 A). Moreover, its viscoelastic properties make snail mucus promising for new biomimetic medical adhesives [69]. Still little is known about the composition of this biomaterial [56] and its bonding ability on different surfaces. Additionally, different to the l-DOPA in mussels, snail mucus proteins are still not produced recombinantly; instead, the mucus is still harvested from living animals.

While most frogs and salamanders use toxic or noxious secretions as defence, some species instead use adhesives [50,70]. Upon release through epidermal glands on the body and trunk [71], the secretion of those amphibians cures immediately, enabling an irreversible and strong bonding (tensile strength > 0.07 MPa, shear stress > 2.8 MPa on wood) to biological (human skin) and artificial (wood, glass, metal) surfaces [55,72].

Chemical analyses show that the glue in the frog Notaden spec. and the salamander Plethodon spec. is mainly protein based (55–78% dry weight), with a high amount of water (70–90%) and a low level of sugar (0.41–0.75% dry weight) [73]. Yet there is a clear difference between the two species. In Notaden, there is a wide range (13–500 kDa) of proteins, with a few prominent protein bands

(>8) and a dominant glycoprotein (Nb-1R) at 350–500 kDa [74,75]. The Plethodon glue, in contrast, contains a low range (15–120 kDa) of proteins, with a relatively high number of prominent bands (>18) and a pH from 5.0 to 8.0 [73]. Up to now, only a few biocompatibility studies for bioadhesives have been performed, probably due to their limited availability. In vitro studies on the adhesive secretions from the frog Notaden bennetti have shown that this adhesive not only shows a good cell compatibility [76], but also has a great potential for medical applications as a tissue glue [72,77,78]. Within the salamanders, some glue-producing species (i.e., *Ambystoma opacum*, *Plethodon shermani*) appear to be cell-compatible, having a probably proliferative effect on some primary cell lines [79]. The adhesive secretions of other species (*Ambystoma maculatum*, *Plethodon glutinosus*), however, have a cytotoxic effect on cell lines, making those glues less favorable candidates for potential medical applications [79].

Recently, researchers from the Ludwig Boltzmann Institute for Experimental and Clinical Traumatology have started to investigate the adhesive secretions of Chilopoda, or centipedes. The animals are known to use highly painful and lethal venoms [80], produced and secreted through glands in the forcipules (maxillipeds) to capture a wide variety of prey, including amphibians, reptiles and even mammals. As a defensive strategy, some species release on the ventral surface of each sternite [81] a fast-hardening glue droplet, which bonds strongly to glass and metal surfaces [50].

A comparison of the biochemical data of *Henia vesuviana* [82] with those of *Haplophilus subterraneus* reveals differences in the numbers and sizes of the protein bands. In the Henia glue, two major bands (12 and 130 kDa) were described [82], while in Haplophilus so far only three prominent bands, between 30 and 67 kDa, could be observed. In the adhesive defense secretion of other centipedes, cyanogenic components, such as hydrogen cyanide (HCN) and precursors (benzoyl nitrile, benzaldehyde, mandelonitrile and others) [83–86], are also present, to increase the repellent effect to predators. In the glue of Henia, such substances seem to be absent [82], and nothing is known so far of the glue of Haplophilus. A detailed and profound chemical and cytotoxic characterization of centipede glue is currently in progress, evaluating its potential as an alternative in convenient wound closure and for other tissue applications.

3. Bioadhesive Biomaterials' Biomedical Applications

Bioadhesives are generally used in wound healing and hemostasis, and their use is incipient in other biomedical applications such as tissue engineering and regeneration. The incoherence between the tissue and the biomaterial is connected using the tissue adhesives in tissue regeneration [87].

Furthermore, while discussing the practical applications of bioadhesive biomaterial research in medical aspects, it is necessary to mention that over the past decade, a growing amount of attention has been paid to bone tissue engineering for research and development in bioadhesive biomaterials' biomedical applications, and resource management, around the world to meet the societal challenges.

Accordingly, the progressive innovation in bioadhesive biomaterials has trended sharply upward, and is expected to double by 2020, especially with a focus on the application of bone tissue engineering. As such, to provide a quality healthcare service, microbially derived polysaccharides (MPs) are demanding, as they are sued for novel, multi-informant, operationally deployable, commercially exploitable and natural-origin raw materials for the production of commercially applicable products in the form of hydrogel and bio composites. These MPs are of bacterial origin (bacterial cellulose (*Acetobacter xylinum*); chitosan (*Aspergillus niger*) and Levan (*Microbacterium laevaniformans*)). Beside the applications of MP and MP-based bio-composites in the health and nano-biotechnology sectors (cell to-cell interactions, biofilm formation, and cell protection against environmental extremes), such polysaccharides are also used as thickeners, bioadhesives, stabilizers, probiotics, and gelling agents in the food and cosmetic industries, and as emulsifier, biosorbents and bioflocculants in the environmental sector.

Concerning the application of bacterial cellulose (BC), it is necessary to indicate that the application of BC has been observed in a broad spectrum, especially in different areas, such as the newspaper industry, electronics, and tissue engineering, due to its remarkable mechanical properties, conformability and porosity. This work has primarily focused on the issue of the biocompatibility of BC and BC nanocomposites and their biomedical aspects, such as surface modification for improving cell adhesion, and in vitro and in vivo studies that focus on the cellulose networks. In summation, the relevance of biocompatibility studies has also emphasized the development of BC-based biomaterials' medical applications in bone, skin and cardiovascular tissue engineering [88].

On the other hand, as regards the biological properties' influence on biomedical application, chitosan has many beneficial biomedical properties, such as biocompatibility, biodegradability, and no toxicity. Therefore, it has been observed that the biological activity of chitosan is closely related to its solubility. This also highlights the development and improvement of scaffolding, i.e., the support of biomaterials using a framework for regenerative medicine. Regarding biomaterials' medical applications, it is obligatory to remark that scaffolds are one of the crucial factors for tissue engineering, such as scaffolds containing natural polymers that have recently been developed more quickly and have gained more popularity. These include chitosan, a copolymer derived from the alkaline deacetylation of chitin. In order to provide a quality healthcare nowadays, the expectations for the use of these types of scaffolds are increasing as the knowledge regarding their chemical and biological properties expands, and new biomedical applications are being investigated [89,90].

In this review article, we emphasize the intrinsic properties offered by chitosan and its medical application in tissue engineering, which proffer it as a promising substitute for regenerative medicine as a bioactive polymer. Moreover, from the application point of view, Qasim et al. [91] showed that the electrospinning of chitosan and its composite formulations for creating fibers in combination with other natural polymers is actively working in tissue engineering. It shows that the favorable properties and biocompatibility of chitosan electrospun composite biomaterials can be used for a wide range of applications [92,93].

Simultaneously, Levan is also another important and useful biomaterial, known as an Exopolysaccharide (EPS), which is mainly covered by microorganisms. These types of microorganisms are natural, nontoxic, biocompatible and biodegradable polysaccharides, which are composed of fructose units joined by β-2,6 linkages. Apart from these characteristics, Levan is also an unconventional fructose polymer produced by extremophilic microorganisms that demonstrates hydroxyl groups and that has the capability to form strong adhesive bonds with various substrates. Therefore, considering the biomedical application of Levan, research shows that it has a strong bioadhesive property. As such, bioadhesives are important devices in both biomedical and tissue engineering applications. While medical adhesives and sealants require wound healing, the robust adhesion and protection against external injure in tissue engineering is performed to ensure the improvement of biomaterial/cell interactions. From the healthcare benefit point of view, a recent study has shown that the new findings concerning Levan's use in biomedical applications as surgical bandages and sealants and in tissue engineering mainly contribute to promoting and controlling the specific cellular responses related to their adhesion, metabolism and ideally stem cell differentiation mechanisms [94].

Apart from the above-mentioned discussion concerning some polymer-based bioadhesive biomaterials' medical applications, it is also necessary to highlight another important Exopolysaccharide (EPS), i.e., dextran, which is excreted from the cell having bacterial origin, and is also extensively used in different kinds of biomedical applications. It is mainly useful for the following healthcare issues: magnetic separation, magnetic resonance imaging, hyperthermia, magnetically guided drug delivery, tissue repair, and molecular diagnostics [95]. Consequently, from the healthcare point of view, this research shows that currently, several technological as well as medical challenges have been determined due to the advancement of nanotechnology and to the progress of materials sciences. The usage of nanotechnology in biomedical applications has significantly shown very promising and amazing outcomes at a global scale by developing new materials with controllable and reproducible properties [96].

The protein-based adhesives materials are basically from animal sources which trigger an inflammatory response compared with human derived materials. Nowadays, various protein-based bioadhesive products are under development for clinical trials (phase III and phase IV), for example, as hemostatic sealants in cardiac surgery as vascular graft attachments, valve attachments, etc., drug delivery systems (as for example in the gastrointestinal tract, nasal delivery and ocular drug delivery), wound-healing dressings and military applications [87].

4. Implementation of Bioadhesive Biomaterials in Healthcare

In this contemporary age, bioadhesive biomaterials are considered as an innovative property-oriented material that is able to build an intimate relationship with the living tissue. Currently, biomaterials are revolutionizing many aspects of preventive and therapeutic healthcare that play an important role, especially during the development of new medical devices, prostheses, tissue repair and replacement technologies, drug delivery systems and diagnostic techniques. As such, due to advanced biomaterials' promising opportunities, presently the application of biomaterials in health sectors is one of the main focuses of major research efforts around the world. Research shows that development in this field of research requires a multidisciplinary approach, whereby scientists interact with engineers, materials producers and manufacturers. On the other hand, it is necessary to mention that to face the recent challenges in healthcare management is often very demanding. Therefore, it has been observed that the required skills and resources are beyond the capabilities of a single organization, or even of a single country. Accordingly, collaborative research is thus becoming the key to achieving breakthrough results in order to bring leadership in the global marketplace [97]. "Bioadhesion of Biomaterials" covers the bioadhesion aspect of biomaterials as healthcare challenges via the research and development of effective and low-cost materials. However, their application as medical devices is limited given the degradation [7].

From the healthcare point of view, biomaterials can be demarcated as "materials that mainly clasp with some innovative properties that facilitate to emanate in immediate contact with the living tissue without eliciting any adverse immune rejection reactions." These types of biomaterials are envisioned for usage in healthcare, especially for the purpose of the diagnosis of disease and for the treatment or for the prevention of other diseases in the human body or other animals. Additionally, it is essential to express that this condition is normally not dependent upon being metabolized for the achievement of any of its principal intended purposes or not. Equally, these devices and/or any type of biomaterials are typically used for the physical replacement of some hard or soft tissue, which has suffered any accidental damage or destruction through some pathological processes [9].

In relation to biomaterials' applications in healthcare, it is known that biomaterials used for health purpose is not a new concept. The application of biomaterials in health issues started long ago. Although, the noticeable advancement of biomaterials application has been observed since the 1940s, but substantial development has been detected over the past 25 years, especially while applying therapeutic medical technologies and implant devices [9]. Furthermore, from the implementation of bioadhesive biomaterials' applications in healthcare, research shows that from ancient periods, tissue adhesives' and sealants' applications in healthcare have renovated a lot, especially in wound management and in traumatic and surgical injuries. For example, tissue adhesives' and sealants' applications in healthcare are well-known for treating disorders of hemostasis (the physiological process that stops bleeding at the site of an injury while maintaining the normal blood flow circulation within the body) [98]. Instead, various biologically driven glues and synthetic adhesives are clinically utilized either for the betterment of health as an adjunct to conventional hemostats and wound closure techniques, such as suturing, or for a replacement purpose. As a result, it can be said that this kind of bioadhesive biomaterial set-up in healthcare gradually improves the ability to effectively and quickly control bleeding. Consequently, it helps in reducing the risk of complications due to severe blood loss, which is an important implementation of medical adhesives, thus making it a highly suitable tool for wound management [99]. In order to provide more vibrant information about the polymer-based bioadhesive biomaterials' medical applications, the below-mentioned Table 1 demonstrates some examples of polymer-based bioadhesive biomaterials' medical applications.

Table 1. Types of polymer-based bioadhesive biomaterials' medical applications.

Polymer-Based Bioadhesive Biomaterials	Medical Applications
Bacterial Cellulose (BC)	Drug delivery, wound dressing, implantable devices (Scaffold) and BC-based biomaterials' medical applications in bone, skin and cardiovascular tissue engineering.
Chitosan	Tissue engineering and a promising substitute for regenerative medicine as a bioactive polymer.
Levan	Surgical bandages and sealants and in tissue engineering mainly contributing to promoting and controlling specific cellular responses related to their adhesion, and wound healing.

5. Redefining Healthcare Management in Relation to Bioadhesive Biomaterials' Medical Applications

To address the conceptualization of "redefining healthcare management", it is significant to discuss the idea of re-emerging "value-based healthcare" for healthy societal development. Currently, this value-based healthcare impression motivates researchers, mainly those who are interested in innovative bioadhesive biomaterial applications in healthcare due to the recent developments in their formulation. Here, engineers, scientists (i.e., chemists, physicists, biologists, and medical experts), material producers, and manufacturers combine their knowledge to reconsider all the aspects of healthcare management in order to provide and maintain the good health of a population. According to the report of the Economist Intelligence Unit [100], value-based healthcare can be considered as the formation and operation of a quality health system that explicitly prioritizes quality health products. In this regard, it is necessary to say that bioadhesive biomaterial applications in healthcare deliver quality health through integrated and technologically sophisticated heath care delivery systems. Modern healthcare also has four main principles, including the following: (i) evidence-based, patients-centered and inclusive care; (ii) community, continuous and coordinated; (iii) being ethically sound and (iv) having a regulated healthcare system [100–103]. This review article intends to describe in Figure 3 the contemporary understanding of the significance of bioadhesive biomaterials for biomedical applications in healthcare for redefining healthcare management as a novel approach.

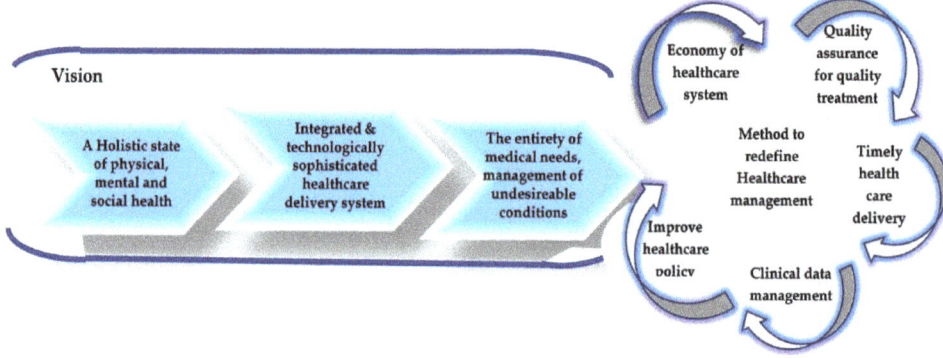

Figure 3. The conceptual approach of redefining healthcare management.

As such, the value-based healthcare concept, i.e., to redefine the healthcare system, particularly emphasizes the proper health objective in order to increase the value. Research shows that value is generated from health consequences, which are important for the following three reasons. The presented Figure 3 illustrates that for the conceptual approach to redefining healthcare, which demonstrated the way to enhance quality healthcare as well as to maintain a programmatic approach, it is necessary to have a holistic physical, mental and social health condition or environment, a need-integrated and technologically sophisticated healthcare delivery system to provide unique patient circumstances, and care for all-inclusive patients' medical needs, including critical and chronic disease prevention as well as the management of undesirable conditions [103].

However, to redefine healthcare, transformations must be done by both health providers and patients, as well through appropriate healthcare delivery and proper clinical data management by strengthening primary care, building integrated health systems, i.e., quality assurance for quality treatment, and implementing appropriate health payment schemes, i.e., the economy of the healthcare system that will promote the value and reduce moral hazards, enabling health information technology, and creating a policy appropriate for a healthy community [1].

The conceptual framework of redefining healthcare management in relation to bioadhesive biomaterials was developed based on the idea of the care management conceptual model [101]. In this research, the main highlighted point is intended to highlight the importance of innovative bioadhesive biomaterials' medical applications, so as to redefine all the aspects of health practice. This review article intended to raise the awareness of healthcare service providers, private research institutes, government authorities, public service boards, associations and academic initiatives to restructure the healthcare system in a way that will not only create value for patients and increase satisfaction, but it will also create a healthier society. Therefore, based on the idea of the care management conceptual model, this study develops a thematic diagram (Figure 4) to define the linkage of redefining healthcare management in relation to bioadhesives for medical applications. Figure 4 represents this connection between the healthcare service providers, patients and members, i.e., research institutions, associations and academics. This schematic diagram demonstrates the critical element of the patient in this connection, influencing medical issue factors. By including the patient element in the framework, this study considers the potential influence of patient characteristics, i.e., effective self-care and the relationships of patients with clinics/clinicians and community resources, i.e., high-quality clinical care.

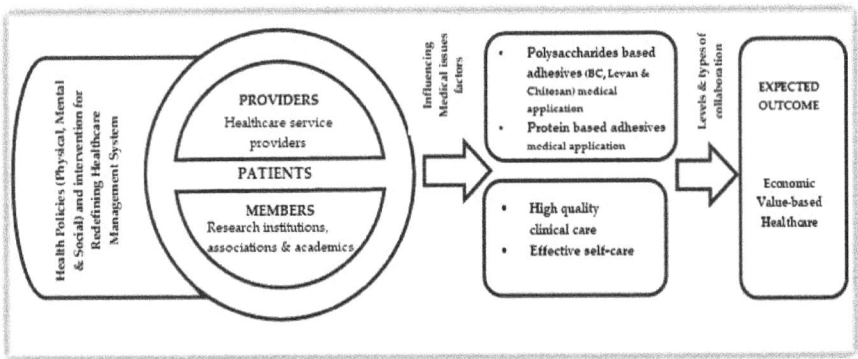

Figure 4. Thematic diagram of redefining healthcare management in relation to bioadhesive biomaterials (based on the idea of care management conceptual model) [101].

To define the relation to bioadhesive biomaterials' medical applications, it is necessary to state that biomaterials are widely used in many kinds of medical devices. The biomaterials used can be protein, metal, polymer, ceramic or composites. Similarly, bioadhesion will occur when the medical device contacts the biological surface. Figure 3 demonstrates that bacterial cellulose, Levan, and chitosan have excellent and praise-worthy applications in the medical field (already explained in an earlier part of this article).

Protein-based adhesives also play a vital role, especially when using biological adhesives for medical applications. The remarkable thing is that since primeval eras, tissue adhesives and sealant applications in healthcare have renovated a lot, particularly in wound management and in traumatic and surgical injuries. Thus, based on our previous discussion, it can be said that the processes of quality clinical care as well as patients' effective self-care have a close connection that redefines the existing healthcare in such a way that can avoid further risks and can receive the needed preventive services. A linkage, therefore, represents the combined influence of all seven basic factors (health policies, providers, patients, members, bioadhesive biomaterials' medical applications, quality clinical care and patients' self-care) and their levels of collaboration that enable one to achieve the expected outcome, i.e., economic value-based healthcare for the delivery of a preventive service.

6. Conclusions

Finally, it can be said that this review article delivers an understanding of the consequences of the bioadhesion of biomaterials and its implications for redefining healthcare management as a novel approach, even though some research has been performed in order to describe the polysaccharides-based adhesive application at a micro level or at a nano level, which has been done for the preparation of molecularly smooth films for healthcare resolution. As such, it is necessary to continue this research in this area in order to obtain a better understanding about the adhesive interactions beyond hydrogen bonding, including mechanical interlocking, interpenetrating networks, and covalent linkages, on a fundamental level to improve the interfacial properties of thermoplastics, thermosets and biopolymers. Relating to this issue of bioadhesive biomaterials' applications in the healthcare system, this study exposes the presentation of the progressive innovation in the bioadhesion of biomaterials. Meanwhile, today, innovative biomaterial applications tend sharply upward, and are expected to double by 2020, especially with a focus on delivering quality healthcare. While redefining health, it is necessary to mention that health consists of three main domains, namely, physical, mental, and social health, that are prioritized with a value-based healthcare concept.

The analyses revealed some important research assumptions that were predictive of both healthcare management and innovative biomaterials applications, which state that the bioadhesion of biomaterials for redefining healthcare management is not a new concept. Its implementation has been used for several years for medical applications, such as dentistry and orthopedics, and is now entering new fields, for example, tissue sealing and directed drug delivery systems. From the practical implication point of view, the results provide an important insight into the notion of involving healthcare delivery organizations, i.e., healthcare service providers, in medical science for resource management, which will help us to cope up with the socio-economic challenges of Horizon 2020. As an outcome, it is assumed that government authorities and public service boards, research institutes, associations and academics will aim to restructure healthcare systems, which will not only create value for patients and increase satisfaction, but will also improve health outcomes through enabling new efficiencies and lowering costs.

Author Contributions: Conceptualization: N.S. (Nibedita Saha), N.S. (Nabanita Saha); review of literature: N.S. (Nibedita Saha), N.S. (Nabanita Saha), T.S., E.T.Ö., U.V.B. and, J.v.B.; writing—original draft: N.S. (Nibedita Saha), N.S. (Nabanita Saha), E.T.Ö., U.V.B. and J.v.B.; writing—review and editing H.R., P.S. and N.S. (Nibedita Saha); conceptual/thematic diagram drawn: N.S. (Nibedita Saha). All authors have read and agreed to the published version of the manuscript.

Funding: The authors declare that there is no conflict of interest.

Acknowledgments: This work is supported by the Ministry of Education, Youth and Sports of the Czech Republic—DKRVO (RP/CPS/2020/005) and COST Action CA 15216 "European Network of Bioadhesion Expertise: Fundamental Knowledge to Inspire Advanced Bonding Technologies" ENBA (http://www.enba4.eu/). The first author is thankful to the Director of University Institute for providing management support system (MSS) and infrastructural facility to carry out this research. Additionally, the first author dedicated this article to her only beloved son "Kanishka Binayak Saha" and the first & second author dedicated this article to their beloved father "Chittaranjan Saha".

Conflicts of Interest: The authors declare that there is no conflict of interest.

References

1. Putera, I. Redefining Health: Implication for Value-Based Healthcare Reform. *Cureus* **2017**, *9*, 1067. [CrossRef]
2. Peled, H.B.; Pinhas, M.D. *Bioadhesion and Biomimetics: From Nature to Applications*; Pan Stanford: Boca Raton, FL, USA, 2015; 314p.
3. Williams, D.F. On the nature of biomaterials. *Biomaterials* **2009**, *30*, 5897–5909. [CrossRef] [PubMed]
4. Williams, D.F. *Definitions in Biomaterials*; Elsevier: Amsterdam, The Netherlands, 1987.
5. Brahmbhatt, D. Bioadhesive drug delivery systems: Overview and recent advances. *Int. J. Chem. Life Sci.* **2017**, *6*, 2016–2024. [CrossRef]
6. Palacio, M.L.B.; Bhushan, B. Bioadhesion: A review of concepts and applications. *Philos. Trans. R. Soc. A Math. Phys. Eng. Sci.* **2012**, *370*, 2321–2347. [CrossRef]
7. Sunarintyas, S. Bioadhesion of Biomaterials. In *Biomaterials and Medical Devices*; Mahyudin, F., Hermawan, H., Eds.; Springer: Cham, Switzerland, 2016; Volume 58, pp. 103–125.
8. Manuel, L.; Palacio, B.; Bhushan, B. Bioadhesion: A review of concepts and applications. *Phil. Trans. R. Soc. A* **2011**, *370*, 2321–2347. [CrossRef]
9. Zubay, G.L. *Biochemistry*, 4th ed.; W.C. Brown: Dubuque, IA, USA, 1998.
10. Brown, A.J. On an Acetic Ferment which form Cellulose. *J. Chem. Soc.* **1986**, *49*, 172–187. [CrossRef]
11. Mohite, B.V.; Patil, S.V. A novel biomaterial: Bacterial cellulose and its new era applications. *Biotechnol. Appl. Biochem.* **2014**, *61*, 101–110. [CrossRef]
12. Czaja, W.K.; Young, D.J.; Kawecki, M.; Brown, R.M. The Future Prospects of Microbial Cellulose in Biomedical Applications. *Biomacromolecules* **2007**, *8*, 1–12. [CrossRef]
13. Hestrin, S.; Schramm, M. Synthesis of cellulose by Acetobacter xylinum. 2. Preparation of freeze-dried cells capable of polymerizing glucose to cellulose. *Biochem. J.* **1954**, *58*, 345–352. [CrossRef]
14. Lestari, P.; Elfrida, N.; Suryani, A.; Suryadi, Y. Study on the Production of Bacterial Cellulose from Acetobacter Xylinum Using Agro—Waste. *Jordan J. Biol. Sci.* **2014**, *7*, 75–80. [CrossRef]
15. Saha, N.; Vyroubal, R.; Sáha, P. Apple Juice: An alternative feed-stock to enhance the production of Bacterial Nano Cellulose. In Proceedings of the 2nd International Symposium on Bacterial Nanocellulose, Gdańsk, Poland, 9–11 September 2015.
16. Zandraa, O.; Saha, N.; Shimoga, G.D.; Palem, R.R.; Saha, P. Bacterial Cellulose, An excellent biobased polymer produced from Apple, Book of Abstract Juice. In Proceedings of the 9th International Conference on Modification, Degradation and Stabilization of Polymers, Krakow, Poland, 4–8 September 2016.
17. Bandopadhyay, S.; Saha, N.; Zandraa, O.; Saha, P. Bacterial cellulose from apple juice—A polysaccharide based bioadditive for sustainable food packaging, Abstract Book, 35–36. In Proceedings of the 5th EPNOE International Polysaccharide Conference, Jena, Germany, 20–24 August 2017.
18. MohammadKazemi, F.; Azin, M.; Ashori, A. Production of bacterial cellulose using different carbon sources and culture media. *Carbohydr. Polym.* **2015**, *117*, 518–523. [CrossRef]
19. Gardner, J.D.; Oporto, S.G.; Mills, R.; Samir, A.S.A.M. Adhesion and surface Issues in Cellulose and Nanaocellulose. *J. Adhes. Sci. Technol.* **2008**, *22*, 545–567. [CrossRef]
20. Oner, E.T.; Hernández, L.; Combie, J. Review of Levan polysaccharide: From a century of past experiences to future prospects. *Biotechnol. Adv.* **2016**, *34*, 827–844. [CrossRef]
21. Versluys, M.; Kirtel, O.; Oner, E.T.; Ende, W.V.D. The fructan syndrome: Evolutionary aspects and common themes among plants and microbes. *Plant Cell Environ.* **2018**, *41*, 16–38. [CrossRef]
22. Poli, A.; Kazak, H.; Gürleyendağ, B.; Tommonaro, G.; Pieretti, G.; Oner, E.T.; Nicolaus, B. High level synthesis of levan by a novel Halomonas species growing on defined media. *Carbohydr. Polym.* **2009**, *78*, 651–657. [CrossRef]
23. Kazak, H.; Barbosa, A.M.; Baregzay, B.; da Cunha, M.A.A.; Oner, E.T.; Dekker, R.F.H.; Khaper, N. Biological activities of bacterial levan and three fungal β-glucans, botryosphaeran and lasiodiplodan under high glucose condition in the pancreatic β-cell line INS-1E. *Adapt. Biol. Med. New Dev.* **2014**, *7*, 105–115.

24. Queiroz, E.A.; Fortes, Z.B.; Da Cunha, M.A.; Sarilmiser, H.K.; Barbosa, A.M.; Oner, E.T.; Dekker, R.F.; Khaper, N. Levan promotes antiproliferative and pro-apoptotic effects in MCF-7 breast cancer cells mediated by oxidative stress. *Int. J. Biol. Macromol.* **2017**, *102*, 565–570. [CrossRef]
25. Sarilmiser, H.K.; Oner, E.T.; Sarılmışer, H.K. Investigation of anti-cancer activity of linear and aldehyde-activated levan from Halomonas smyrnensis AAD6T. *Biochem. Eng. J.* **2014**, *92*, 28–34. [CrossRef]
26. Sezer, A.D.; Sarılmışer, H.K.; Rayaman, E.; Çevikbaş, A.; Öner, E.T.; Akbuğa, J. Development and characterization of vancomycin-loaded levan-based microparticular system for drug delivery. *Pharm. Dev. Technol.* **2017**, *22*, 627–634. [CrossRef]
27. Sezer, A.D.; Kazak, H.; Oner, E.T.; Akbuğa, J. Levan-based nanocarrier system for peptide and protein drug delivery: Optimization and influence of experimental parameters on the nanoparticle characteristics. *Carbohydr. Polym.* **2011**, *84*, 358–363. [CrossRef]
28. Costa, R.R.; Neto, A.I.; Calgeris, I.; Correia, C.R.; De Pinho, A.C.M.; Fonseca, J.C.; Oner, E.T.; Mano, J.F. Adhesive nanostructured multilayer films using a bacterial exopolysaccharide for biomedical applications. *J. Mater. Chem. B* **2013**, *1*, 2367–2374. [CrossRef] [PubMed]
29. Bostan, M.S.; Mutlu, E.C.; Kazak, H.; Keskin, S.S.; Oner, E.T.; Eroglu, M.S. Comprehensive characterization of chitosan/PEO/levan ternary blend films. *Carbohydr. Polym.* **2014**, *102*, 993–1000. [CrossRef] [PubMed]
30. Osman, A.; Oner, E.T.; Eroglu, M.S. Novel levan and pNIPA temperature sensitive hydrogels for 5-ASA controlled release. *Carbohydr. Polym.* **2017**, *165*, 61–70. [CrossRef] [PubMed]
31. Axente, E.; Sima, F.; Sima, L.E.; Erginer, M.; Eroğlu, M.S.; Serban, N.; Ristoscu, C.; Petrescu, S.M.; Oner, E.T.; Mihailescu, I.N. Combinatorial MAPLE gradient thin film assemblies signalling to human osteoblasts. *Biofabrication* **2014**, *6*, 035010. [CrossRef] [PubMed]
32. Sima, F.; Axente, E.; Sima, L.E.; Tuyel, U.; Eroğlu, M.S.; Serban, N.; Ristoscu, C.; Petrescu, S.M.; Oner, E.T.; Mihailescu, I.N. Combinatorial matrix-assisted pulsed laser evaporation: Single-step synthesis of biopolymer compositional gradient thin film assemblies. *Appl. Phys. Lett.* **2012**, *101*, 233705. [CrossRef]
33. Avsar, G.; Agirbasli, D.; Agirbasli, M.A.; Gunduz, O.; Oner, E.T. Levan based fibrous scaffolds electrospun via co-axial and single-needle techniques for tissue engineering applications. *Carbohydr. Polym.* **2018**, *193*, 316–325. [CrossRef]
34. Erginer, M.; Akcay, A.; Coskunkan, B.; Morova, T.; Rende, D.; Bucak, S.; Baysal, N.; Ozisik, R.; Eroglu, M.S.; Agirbasli, M.; et al. Sulfated levan from Halomonas smyrnensis as a bioactive, heparin-mimetic glycan for cardiac tissue engineering applications. *Carbohydr. Polym.* **2016**, *149*, 289–296. [CrossRef]
35. Gomes, T.D.; Caridade, S.G.; Sousa, M.P.; Azevedo, S.; Kandur, M.Y.; Öner, E.T.; Alves, N.M.; Mano, J.F. Adhesive free-standing multilayer films containing sulfated levan for biomedical applications. *Acta Biomater.* **2018**, *69*, 183–195. [CrossRef]
36. Nemtsev, S.V.; Zueva, O.Y.; Khismatullin, M.R.; Albulov, A.I.; Varlamov, V.P. Isolation of Chitin and Chitosan from Honeybees. *Appl. Biochem. Microbiol.* **2004**, *40*, 39–43. [CrossRef]
37. Jayakumar, R.; Prabaharan, M.; Nair, S.V.; Tokura, S.; Tamura, H.; Selvamurugan, N. Novel carboxymethyl derivatives of chitin and chitosan materials and their biomedical applications. *Prog. Mater. Sci.* **2010**, *55*, 675–709. [CrossRef]
38. Yao, K.; Li, J.; Yao, F.; Yin, Y. (Eds.) *Chitosan-Based Hydrogels: Functions and Applications*; CRC Press: Boca Raton, FL, USA, 2011.
39. Vrabič Brodnjak, U. Influence of ultrasonic treatment on properties of bio-based coated paper. *Prog. Org. Coat.* **2017**, *103*, 93–100. [CrossRef]
40. Vrabič Brodnjak, U. Improvement of physical and optical properties of chitosan-rice starch films pre-treated with ultrasound. *Bulg. Chem. Commun.* **2017**, *49*, 859–867.
41. Mati-Baouche, N.; Elchinger, P.H.; De Baynast, H.; Pierre, G.; Delattre, C.; Michaud, P. Chitosan as an adhesive. *Eur. Polym. J.* **2014**, *60*, 198–212. [CrossRef]
42. Kurek, M.; Brachais, C.H.; Ščetar, M.; Voilley, A.; Galić, K.; Couvercelle, J.P.; Debeaufort, F. Carvacrol affects interfacial, structural and transfer properties of chitosan coatings applied onto polyethylene. *Carbohydr. Polym.* **2013**, *97*, 217–225. [CrossRef]
43. Bajaj, M.; Winter, J.; Gallert, C. Effect of deproteination and deacetylation conditions on viscosity of chitin and chitosan extracted from Crangon crangon shrimp waste. *Biochem. Eng. J.* **2011**, *56*, 51–62. [CrossRef]
44. Yanqiao, J.; Cheng, X.; Zheng, Z. Preparation and characterization of phenol–formaldehyde adhesives modified with enzymatic hydrolysis lignin. *Bioresour. Technol.* **2010**, *101*, 2046–2048.

45. Norström, E.; Fogelström, L.; Nordqvist, P.; Khabbaz, F.; Malmström, E. Gum dispersions as environmentally friendly wood adhesives. *Ind. Crop. Prod.* **2014**, *52*, 736. [CrossRef]
46. Patel, A.K. Chitosan: Emergence as potent candidate for green adhesive market. *Biochem. Eng. J.* **2015**, *102*, 74–81. [CrossRef]
47. Vrabič Brodnjak, U. Experimental investigation of novel curdlan/chitosan coatings on packaging paper. *Prog. Org. Coat.* **2017**, *112*, 86–92. [CrossRef]
48. Richter, K.; Grunwald, I.; von Byern, J. *Bioadhesives*; da Silva, L.F.M., Oechsner, A., Adams, R., Eds.; Springer: Berlin/Heidelberg, Germany, 2018; pp. 1–45.
49. Von Byern, J.; Grunwald, I. *Biological Adhesive Systems: From Nature to Technical and Medical Application*; Springer: New York, NY, USA, 2010.
50. Von Byern, J.; Müller, C.; Voigtländer, K.; Dorrer, V.; Marchetti-Deschmann, M.; Flammang, P.; Mayer, G. *Examples of Bioadhesives for Defence and Predation*; Gorb, S., Gorb, E., Eds.; Springer: Berlin/Heidelberg, Germany, 2017; pp. 141–191.
51. Ferguson, J.; Nürnberger, S.; Redl, H. *Fibrin: The Very First Biomimetic Glue—Still a Great Tool*; von Byern, J., Grunwald, I., Eds.; Springer: Berlin/Heidelberg, Germany, 2010; pp. 225–236.
52. Nürnberger, S.; Wolbank, S.; Peterbauer, A.; Morton, T.J.; Feichtinger, G.A.; Gugerell, A.; Meinl, A.; Labuda, K.; Bittner, M.; Pasteiner, W.; et al. *Properties and Potential Alternative Applications of Fibrin Glue*; von Byern, J., Grunwald, I., Eds.; Springer: Berlin/Heidelberg, Germany, 2010; pp. 237–259.
53. Bell, E.; Gosline, J. Mechanical design of mussel byssus: Material yield enhances attachment strength. *J. Exp. Biol.* **1996**, *199*, 1005–1017.
54. Harrington, M.J.; Waite, J.H. Holdfast heroics: Comparing the molecular and mechanical properties of Mytilus californianus byssal threads. *J. Exp. Biol.* **2007**, *210*, 4307–4318. [CrossRef]
55. Graham, L.D. *Biological Adhesives from Nature*; Bowlin, G.L., Wnek, G., Eds.; Taylor & Francis: Abingdon, UK, 2005; pp. 1–18.
56. Smith, A.M. *Biological Adhesives*; Springer International Publishing: Cham, Switzerland, 2016.
57. Krogsgaard, M.; Andersen, A.; Birkedal, H. Gels and threads: Mussel-inspired one-pot route to advanced responsive materials. *Chem. Commun.* **2014**, *50*, 13278–13281. [CrossRef]
58. Krogsgaard, M.; Behrens, M.A.; Pedersen, J.S.; Birkedal, H. Self-healing mussel-inspired multi-pH-responsive hydrogels. *Biomacromolecules* **2013**, *14*, 297–301. [CrossRef]
59. Shen, H.; Qian, Z.; Zhao, N.; Xu, J. *Preparation and Application of Biomimetic Materials Inspired by Mussel Adhesive Proteins*; Yang, G., Xiao, L., Lamboni, L., Eds.; Wiley & Sons: Hoboken, NJ, USA, 2018; pp. 103–118.
60. Smith, A.M. The Structure and Function of Adhesive Gels from Invertebrates. *Integr. Comp. Biol.* **2002**, *42*, 1164–1171. [CrossRef]
61. Smith, A.M. *Gastropod Secretory Glands and Adhesive Gels*; von Byern, J., Grunwald, I., Eds.; Springer: Berlin/Heidelberg, Germany, 2010; pp. 41–51.
62. Shirtcliffe, N.J.; McHale, G.; Newton, M.I. Wet Adhesion and Adhesive Locomotion of Snails on Anti-Adhesive Non-Wetting Surfaces. *PLoS ONE* **2012**, *7*, e36983. [CrossRef]
63. Bonnemain, B. Helix and Drugs: Snails for Western Health Care from Antiquity to the Present. *Evid.-Based Complement. Altern. Med.* **2005**, *2*, 25–28. [CrossRef]
64. Brieva, A.; Philips, N.; Tejedor, R.; Guerrero, A.; Pivel, J.; Alonso-Lebrero, J.; Gonzalez, S. Molecular Basis for the Regenerative Properties of a Secretion of the Mollusk Cryptomphalus aspersa. *Ski. Pharmacol. Physiol.* **2008**, *21*, 15–22. [CrossRef]
65. La Cruz, M.C.I.-D.; Sanz-Rodríguez, F.; Zamarrón, A.; Reyes, E.; Carrasco, E.; González, S.; Juarranz, A. A secretion of the mollusc Cryptomphalus aspersa promotes proliferation, migration and survival of keratinocytes and dermal fibroblasts in vitro. *Int. J. Cosmet. Sci.* **2012**, *34*, 183–189. [CrossRef]
66. Fabi, S.G.; Cohen, J.L.; Peterson, J.D.; Kiripolsky, M.G.; Goldman, M.P. The effects of filtrate of the secretion of the Cryptomphalus aspersa on photoaged skin. *J. Drugs Dermatol. JDD* **2013**, *12*, 453–457.
67. Meyer-Rochow, V.B.; Yamahama, Y.A. Comparison between the larval eyes of the dimly luminescent Keroplatus nipponicus and the brightly luminescent Arachnocampa luminosa (Diptera; Keroplatidae). *Luminescence* **2017**, *32*, 1072–1076. [CrossRef]
68. Tsoutsos, D.; Kakagia, D.; Tamparopoulos, K. The efficacy of Helix aspersa Müller extract in the healing of partial thickness burns: A novel treatment for open burn management protocols. *J. Dermatol. Treat.* **2009**, *20*, 219–222. [CrossRef]

69. Li, J.; Celiz, A.D.; Yang, J.; Yang, Q.; Wamala, I.; Whyte, W.; Seo, B.R.; Vasilyev, N.V.; Vlassak, J.J.; Suo, Z.; et al. Tough adhesives for diverse wet surfaces. *Science* **2017**, *357*, 378–381. [CrossRef]
70. Tyler, M.J. *Adhesive Dermal Secretions of the Amphibia, with Particular Reference to the Australian Limnodynastid Genus Notaden*; von Byern, J., Grunwald, I., Eds.; Springer: Berlin/Heidelberg, Germany, 2010; pp. 181–186.
71. Von Byern, J.; Dicke, U.; Heiss, E.; Grunwald, I.; Gorb, S.; Staedler, Y.; Cyran, N. Morphological characterization of the glue-producing system in the salamander Plethodon shermani (Caudata, Plethodontidae). *Zoology* **2015**, *118*, 334–347. [CrossRef]
72. Szomor, Z.L.; Murrell, G.A.C.; Appleyard, R.C.; Tyler, M.J. Meniscal repair with a new biological glue: An ex vivo study. *Tech. Knee Surg.* **2009**, *7*, 261–265. [CrossRef]
73. Von Byern, J.; Grunwald, I.; Kosok, M.; Saporito, R.A.; Dicke, U.; Wetjen, O.; Thiel, K.; Borcherding, K.; Kowalik, T.; Marchetti-Deschmann, M.; et al. Chemical characterization of the adhesive secretions of the salamander Plethodon shermani (Caudata, Plethodontidae). *Sci. Rep.* **2017**, *7*, 1–13. [CrossRef]
74. Graham, L.D.; Glattauer, V.; Huson, M.G.; Maxwell, J.M.; Knott, R.B.; White, J.W.; Vaughan, P.R.; Peng, Y.; Tyler, M.J.; Werkmeister, J.A.; et al. Characterization of a protein-based adhesive elastomer secreted by the Australian frog Notaden bennetti. *Biomacromolecules* **2005**, *6*, 3300–3312. [CrossRef]
75. Graham, L.D.; Glattauer, V.; Li, D.; Tyler, M.J.; Ramshaw, J.A. The adhesive skin exudate of Notaden bennetti frogs (Anura: Limnodynastinae) has similarities to the prey capture glue of Euperipatoides sp. velvet worms (Onychophora: Peripatopsidae). *Comp. Biochem. Physiol. Ser. B Biochem. Mol. Biol.* **2013**, *165*, 250–259. [CrossRef]
76. Graham, L.D.; Danon, S.J.; Johnson, G.; Braybrook, C.; Hart, N.K.; Varley, R.J.; Evans, M.; McFarland, G.A.; Tyler, M.J.; Werkmeister, J.A.; et al. Biocompatibility and modification of the protein-based adhesive secreted by the Australian frogNotaden bennetti. *J. Biomed. Mater. Res. Part A* **2009**, *93*, 429–441. [CrossRef]
77. Millar, N.L.; Bradley, T.A.; Walsh, N.A.; Appleyard, R.C.; Tyler, M.J.; Murrell, G.A. Frog glue enhances rotator cuff repair in a laboratory cadaveric model. *J. Shoulder Elb. Surg.* **2009**, *18*, 639–645. [CrossRef]
78. Tyler, M.J.; Ramshaw, J.A. An Adhesive Derived from Amphibian Skin Secretions. Australia Patent No. WO2002/022756, 18 September 2000.
79. Von Byern, J.; Mebs, D.; Heiss, E.; Dicke, U.; Wetjen, O.; Bakkegard, K.; Grunwald, I.; Wolbank, S.; Mühleder, S.; Gugerell, A.; et al. Salamanders on the bench—A biocompatibility study of salamander skin secretions in cell cultures. *Toxicon* **2017**, *135*, 24–32. [CrossRef]
80. Undheim, E.A.B.; Fry, B.G.; King, G.F. Centipede Venom: Recent Discoveries and Current State of Knowledge. *Toxins* **2015**, *7*, 679–704. [CrossRef]
81. Hopkin, S.P.; Anger, H.S. On the structure and function of the glue-secreting glands of Henia vesuviana (Newport, 1845) (Chilopoda: Geophilomorpha). *Ber. Nat.-Med. Ver. Innsbr. Suppl.* **1992**, *10*, 71–79.
82. Hopkin, S.P. Defensive secretion of proteinaceous glues by Henia (Chaetechelyne) vesuviana (Chilopoda Geophilomorpha). In Proceedings of the 7th International Congress of Myriapodology, Brill, The Netherlands, 1 March 1990; pp. 175–181.
83. Jones, T.H.; Conner, W.E.; Meinwald, J.; Eisner, H.E.; Eisner, T. Benzoyl cyanide and mandelonitrile in the cynogenetic secretion of a centipede. *J. Chem. Ecol.* **1976**, *2*, 421–429. [CrossRef]
84. Maschwitz, U.; Lauschke, U.; Würmli, M. Hydrogen cyanide-producing glands in a scolopender, Asanada n.sp. (Chilopoda, Scolopendridae). *J. Chem. Ecol.* **1979**, *5*, 901–907. [CrossRef]
85. Schildknecht, H.; Maschwitz, U.; Krauss, D. Blausäure im Wehrsekret des Erdläufers Pachymerium ferrugineum. *Die Nat.* **1968**, *55*, 230. [CrossRef]
86. Vujisić, L.; Vučković, I.M.; Makarov, S.E.; Ilić, B.; Antić, D.Ž.; Jadranin, M.B.; Todorović, N.M.; Mrkic, I.; Vajs, V.E.; Lučić, L.R.; et al. Chemistry of the sternal gland secretion of the Mediterranean centipede Himantarium gabrielis (Linnaeus, 1767) (Chilopoda: Geophilomorpha: Himantariidae). *Naturwissenschaften* **2013**, *100*, 861–870. [CrossRef]
87. Rathi, S.; Saka, R.; Domb, A.J.; Khan, W. Protein-based bioadhesives and bioglues. *Polym. Adv. Technol.* **2018**, 1–18. [CrossRef]
88. Torres, F.G.; Commeaux, S.; Troncoso, O.P. Biocompatibility of Bacterial Cellulose Based Biomaterials. *J. Funct. Biomater.* **2012**, *3*, 864–878. [CrossRef]
89. Rodríguez-Vázquez, M.; Vega-Ruiz, B.; Ramos-Zúñiga, R.; Saldaña-Koppel, D.A.; Quiñones-Olvera, L.F. Chitosan and Its Potential Use as a Scaffold for Tissue Engineering in Regenerative Medicine. *BioMed Res. Int.* **2015**, *2015*, 821279. [CrossRef]

90. Shi, C.; Zhu, Y.; Ran, X.; Wang, M.; Su, Y.; Cheng, T. Therapeutic Potential of Chitosan and Its Derivatives in Regenerative Medicine. *J. Surg. Res.* **2006**, *133*, 185–192. [CrossRef]
91. Qasim, S.S.B.; Zafar, M.S.; Najeeb, S.; Khurshid, Z.; Shah, A.H.; Husain, S.; Rehman, I.U. Electrospinning of Chitosan-Based Solutions for Tissue Engineering and Regenerative Medicine. *Int. J. Mol. Sci.* **2018**, *19*, 407. [CrossRef] [PubMed]
92. Stratakis, E. Novel Biomaterials for Tissue Engineering 2018. *Int. J. Mol. Sci.* **2018**, *19*, 3960. [CrossRef] [PubMed]
93. Wu, Q.; Zheng, C.; Ning, Z.X.; Yang, B. Modification of Low Molecular Weight Polysaccharides from Tremella Fuciformis and Their Antioxidant Activity In Vitro. *Int. J. Mol. Sci.* **2007**, *8*, 670–679. [CrossRef]
94. Larsson, T.F.; Martín Martínez, J.M.; Vallés, J.L. Biomaterials for Healthcare. In *A Decade of EU-Funded Research*; Office for Official Publications of the European Communities: Luxembourg, 2007.
95. Prodan, A.M.; Andronescu, E.; Truşcă, R.; Beuran, M.; Iconaru, S.L.; Barna, E.Ş.; Chifiriuc, M.C.; Marutescu, L. Anti-biofilm Activity of Dextran Coated Iron Oxide Nanoparticles. *Univ. Politeh. Buchar. Sci. Bull. Ser. B Chem. Mater. Sci.* **2014**, *76*, 81–90.
96. Iconaru, S.L.; Turculet, C.S.; Coustumer, P.L.; Bleotu, C.; Chifiriuc, M.; Lazar, V.; Surugiu, A.; Badea, M.; Iordache, F.; Soare, M.; et al. Biological Studies on Dextrin Coated Iron Oxide Nanoparticles. *Rom. Rep. Phys.* **2016**, *68*, 1536–1544.
97. Gale, A.J. Current Understanding of Hemostasis. *Toxicol. Pathol.* **2011**, *39*, 273–280. [CrossRef]
98. Mehdizadeh, M.; Yang, J. Design Strategies and Applications of Tissue Bioadhesives. *Macromol. Biosci.* **2013**, *13*, 271–288. [CrossRef]
99. *Value-Based Healthcare: A Global Assessment*; The Economist Intelligence Unit: London, UK, 2016.
100. Petrova, M.; Dale, J.; Fulford, B.K.W.M. Values-based practice in primary care: Easing the tensions between individual values, ethical principles and best evidence. *Br. J. Gen. Pract.* **2006**, *56*, 703–709.
101. Badash, I.; Kleinman, N.P.; Barr, S.; Jang, J.; Rahman, S.; Wu, B.W. Redefining Health: The Evolution of Health Ideas from Antiquity to the Era of Value-Based Care. *Cureus* **2017**, *9*, e1018. [CrossRef]
102. Ackermann, R. Evaluating state wide disease management programs. Regenstrief Institute for Healthcare. In Proceedings of the AHRQ Medicaid Care Management Learning Network, Rockville, NY, USA, 2 October 2006.
103. *Designing and Implementing Medicaid Disease and Care Management Programs*; Agency for Healthcare Research and Quality: Rockville, MD, USA, 2014. Available online: https://www.ahrq.gov/patient-safety/settings/long-term-care/resource/hcbs/medicaidmgmt/index.html (accessed on 25 November 2020).

Publisher's Note: MDPI stays neutral with regard to jurisdictional claims in published maps and institutional affiliations.

© 2020 by the authors. Licensee MDPI, Basel, Switzerland. This article is an open access article distributed under the terms and conditions of the Creative Commons Attribution (CC BY) license (http://creativecommons.org/licenses/by/4.0/).

Review

Recent Advances on Stimuli-Responsive Hydrogels Based on Tissue-Derived ECMs and Their Components: Towards Improving Functionality for Tissue Engineering and Controlled Drug Delivery

Julian A. Serna †, Laura Rueda-Gensini †, Daniela N. Céspedes-Valenzuela, Javier Cifuentes, Juan C. Cruz * and Carolina Muñoz-Camargo *

Department of Biomedical Engineering, Universidad de los Andes, Bogotá 111711, Colombia; ja.serna10@uniandes.edu.co (J.A.S.); l.ruedag@uniandes.edu.co (L.R.-G.); dn.cespedes@uniandes.edu.co (D.N.C.-V.); jf.cifuentes10@uniandes.edu.co (J.C.)
* Correspondence: jc.cruz@uniandes.edu.co (J.C.C.); c.munoz2016@uniandes.edu.co (C.M.-C.)
† Authors contributed equally to this work.

Abstract: Due to their highly hydrophilic nature and compositional versatility, hydrogels have assumed a protagonic role in the development of physiologically relevant tissues for several biomedical applications, such as in vivo tissue replacement or regeneration and in vitro disease modeling. By forming interconnected polymeric networks, hydrogels can be loaded with therapeutic agents, small molecules, or cells to deliver them locally to specific tissues or act as scaffolds for hosting cellular development. Hydrogels derived from decellularized extracellular matrices (dECMs), in particular, have gained significant attention in the fields of tissue engineering and regenerative medicine due to their inherently high biomimetic capabilities and endowment of a wide variety of bioactive cues capable of directing cellular behavior. However, these hydrogels often exhibit poor mechanical stability, and their biological properties alone are not enough to direct the development of tissue constructs with functional phenotypes. This review highlights the different ways in which external stimuli (e.g., light, thermal, mechanical, electric, magnetic, and acoustic) have been employed to improve the performance of dECM-based hydrogels for tissue engineering and regenerative medicine applications. Specifically, we outline how these stimuli have been implemented to improve their mechanical stability, tune their microarchitectural characteristics, facilitate tissue morphogenesis and enable precise control of drug release profiles. The strategic coupling of the bioactive features of dECM-based hydrogels with these stimulation schemes grants considerable advances in the development of functional hydrogels for a wide variety of applications within these fields.

Keywords: extracellular matrix; hydrogels; external stimuli; tissue maturation; drug delivery

1. Introduction

Tissue engineering (TE) and regenerative medicine (RM) are closely related research fields where biology, medicine, and engineering converge towards the development of solutions for in vitro or clinical applications. The most recurrent include repairing or replacing tissues whose function is impaired, and developing relevant tissue models for testing drugs or studying pertinent mechanisms of prevalent and rare diseases [1]. Over the past 20 years, hydrogels have been key to the great advances made in these fields to manufacture functional tissue-like structures and the implementation of minimally invasive therapeutics for regenerative and drug delivery purposes. In particular, their highly hydrophilic nature and compositional versatility makes them ideal building blocks for designing cell-friendly and multifunctional microenvironments suitable for engineering tissues [2].

Recent efforts have been focused on designing hydrogels whose physicochemical and mechanical characteristics are ideal for their use as enablers of currently emerging biofabrication technologies [3]. This field focuses on the automated generation of structurally organized and biologically functional products for TE and RM applications. The main goal of this approach is to develop in vitro tissue models or transplantable constructs for in vivo repairing or replacement of tissues [4]. Among the ample variety of 3D bioprinting techniques, the extrusion-based ones have proven the most popular and with the highest potential to fabricate anatomically relevant, multi-material, tissue-like constructs [5]. Hydrogels suitable for extrusion-based 3D bioprinting should exhibit physical properties such that they can be extruded through a nozzle or needle in a controlled manner without losing their integrity (i.e., printability) [6–8]. This can be accomplished by properly tuning their mechanical properties, specifically their rheological behavior and viscosity profiles. In this regard, mechanically robust hydrogels must be viscous enough to be able to form a filament as they are extruded and deposited on a surface, and chemically suited to build structurally sTable 3D constructs upon crosslinking [9–11]. Another relevant application of hydrogels is in drug delivery, where they have been widely implemented as depots for the localized and controlled release of therapeutic molecules, growth factors, and cells [12–14]. In particular, the encapsulation of these agents within hydrogels provides improved bioavailability, high and controlled release rates, and protection from early enzymatic degradation. This makes hydrogels powerful carriers for developing highly biocompatible and tunable delivery systems for the treatment of various diseases [15,16].

Depending on their origin, the materials commonly employed in the formulation of hydrogels can be classified into either synthetic or natural. Synthetic materials are known to have exceptional mechanical and physicochemical properties, but lack the bioactivity exhibited by natural materials [2]. Recently, hydrogels derived from decellularized extracellular matrices (dECMs) have been explored as the most compelling candidates for mimicking the microenvironments of native tissues [17–19]. This has been attributed to the fact that, unlike other natural or synthetic materials, extracellular matrices (ECMs) exhibit a cocktail of bioactive molecules that interact synergistically to create the adequate environment for cellular development. ECMs hold a complex of hierarchically assembled fibrous proteins that comprise the structural backbone of tissues and grant adequate mechanical stability for supporting cellular adhesion, chemotaxis, and migration in vivo [20]. The most abundant of these fibrous proteins are various types of collagen, which are the main structural components of ECMs due to their ability to self-assemble into hierarchically organized fibers or networks [21]. These collagen structures are responsible for providing stiffness and tensile strength, as well as for most of the mechanotransduction cues produced during tissue development [22]. Moreover, different types of collagen can also interact with cells through membrane receptors, thus actively participating in cell growth, differentiation, and migration [23]. Other fibrous proteins, such as elastin and fibronectin, associate with collagen structures to counter their characteristic stiffness and are responsible for tissue elasticity and mechano-regulation. In turn, the relative ratio of these proteins within ECMs yields different mechanical characteristics, which are closely related to native tissue functionality [20]. Another large family of ECM molecules are proteoglycans, which comprise glycosaminoglycan (GAG) chains covalently linked to a protein core. An exception to this is hyaluronic acid (HA), a high-molecular weight GAG that, instead, interacts noncovalently with ECM proteins [24]. GAGs are extremely hydrophilic and adopt extended conformations essential for hydrogel formation and for withstanding high compressive forces [25]. Accordingly, this hydrated network of GAGs and proteoglycans comprises the majority of the extracellular space and closely interacts with fibrous proteins to create a versatile network for cellular scaffolding. As several of the growth factors secreted by cells can bind to their surrounding matrix, an easily accessible repository for such factors is enabled by this hydrated network. This ultimately helps direct essential morphological organization and physiological function [20]. Accordingly, the heterogeneous composition of ECMs, rich in essential molecules implicated in native tissue development, makes them ideal materials for the development of hydrogels intended for TE and RM applications. dECM-based hydrogel formulations are, therefore, not only biocompatible, but also provide superior bioactive cues for guiding cellular behavior when compared to other natural materials [26].

To isolate ECMs and formulate hydrogels with them, tissue samples are exposed to decellularization procedures and subsequently solubilized to form pre-gels, which conserve the majority of the original ECM components and can later be patterned into constructs with desired geometries [27]. However, decellularization and solubilization procedures involve mechanical and chemical processing, as well as the use of proteolytic enzymes, which largely disrupt the hierarchical organization of ECM components and alter their native architecture [27]. Although the components of dECM hydrogels have been shown to partially re-assemble under physiological conditions (i.e., 37 °C, pH 7.0), the resulting conformations are not equivalent to those normally found in the native tissue [28]. Consequently, hydrogels formulated solely from dECMs often exhibit poor architectural characteristics and mechanical stability [17,28,29], which translates into suboptimal tissue maturation processes. To overcome these impediments, a variety of external stimuli have been explored to improve the performance of dECM hydrogels. Several physical stimuli (e.g., temperature, light, strain, stress, electric, magnetic, and acoustic) have been implemented to improve the mechanical stability of dECM-based hydrogels through alternative crosslinking mechanisms or by tuning their microarchitecture according to specific tissue hallmarks. The microarchitecture is, in fact, one of the most important performance parameters when these hydrogels are intended for the engineering of physiologically relevant tissues [30,31]. Moreover, considering that the functionality of many native tissues is dependent on built-in mechanisms that promote dynamic behaviors in response to external stimuli (e.g., muscle contraction, neuronal electrical transduction, and articular loading), construct stimulation during maturation is thought to help guide the development of tissues with functional phenotypes [32]. On this matter, several stimuli-responsive materials and biomolecules have been combined with dECM-based hydrogels to guide their response to external stimuli. This approach has also enabled numerous applications in the localized delivery of therapeutics, considering that the release of loaded pharmacological agents can be precisely controlled through finely tuned responses [33].

In this review, we outline recent advances in the development of dECM-based hydrogels, understood as those derived from whole ECMs or based on their major components (e.g., collagen, HA, and fibronectin) for TE and RM applications. We particularly focus on elucidating how different types of external stimuli have been employed for improving their performance, both for cellular scaffolding and for the controlled delivery of therapeutics (summarized in Table 1). Our aim is to highlight the protagonism of biomimetic materials with tunable stimuli-responsive properties on recent advances in biofabrication and drug delivery, as well as to address their potential for future developments in these fields.

Table 1. Most relevant works implementing external stimuli for improving cellular scaffolding and localized drug delivery from dECM-derived hydrogels.

Stimulus	Stimulation Parameters	Stimulus Enhancer	Hydrogel	Embedded Cells/Cargo	Effect	Ref.
Light	UV light (365 nm) at 18 W/cm^2 intensity for 'several' seconds.	Irgacure 2959	Porcine liver dECM, thiolated-HA, thiolated-gelatin, PEG-acrylate, and PEG-alkyne	Primary human hepatocytes, primary human stellate cells, and primary human Kupffer cells spheroids.	Bioink stiffness increases by more than one order of magnitude upon light stimulus.	[34]
	White light exposure for 5 min post-bioprinting.	Eosin Y	Cardiac dECM-GelMA	Neonatal human cardiac progenitor cells	Improved mechanical stiffness upon photocrosslinking while maintaining cell viability above 75%.	[35]
	Blue-light exposure (405 nm) with 1.5 W/cm^2 intensity for 40 s.	Riboflavin phosphate (RFP)	Thiol- and methacryloyl-modified HA	Primary rabbit corneal fibroblasts	Improved hydrogel crosslinking and mechanical stiffness in an RFP-dependent manner. Crosslinking degree dictated bovine serum albumin release profiles and hydrogel degradation.	[36]
	NIR laser excitation (808 nm) at 5.6 and 8.3 mW/cm^2 power densities for 3 min every other day	Carbon dot nanoparticles	Type I collagen	Bone marrow-derived stem cells (BMSCs)	Increased proliferation and chondrogenic differentiation of BMSCs as a result of non-lethal doses of ROS produced with photodynamic therapy.	[37]
	UV exposure (365 nm) for 120 s.	Irgacure 2959	Methacryloyl-modified kidney dECM, HA, gelatin, and glycerol	Human primary kidney cells	Almost 2-fold increase in storage modulus after photocrosslinking when compared to unmodified kidney dECM-based hydrogels.	[38]
	Blue light exposure (405 nm) at 20 W/cm^2 for 15, 30, and 45 s.	LAP	Methacryloyl-modified bone dECM	Human dental pulp stem cells, hMSCs, and HUVECs	Storage modulus increase in a dose-dependent manner.	[39]
	Blue light exposure (405 nm) at 62 W/cm^2 for 1 min	Riboflavin	Methacryloyl-modified SIS dECM	Adipose-derived MSCs	Two-fold increase in storage modulus upon photocrosslinking.	[40]

Table 1. *Cont.*

Stimulus	Stimulation Parameters	Stimulus Enhancer	Hydrogel	Embedded Cells/Cargo	Effect	Ref.
	Blue light exposure (405 nm) at 30 mW/cm^2.	Ruthenium/sodium persulfate	Corneal and heart dECM	Human bone marrow- and turbunate-derived MSCs, hiPSC-derived cardiomyocytes	A 2.55- and 3.79-fold increase in compressive and storage modulus, respectively, upon photocrosslinking. High shape fidelity allowed bioprinting of multi-layered and complex anatomical structures.	[41]
	Blue light exposure (laser at 488 and LED at 460 nm).	tris (2,2′-bipyridyl) dichlororuthenium (II) hexahydrate/sodium persulfate.	Fibrin	Normal human dermal fibroblasts	The stiffness of fibrin hydrogels was patterned at the micron scale by spatio-selective irradiation with a blue light laser.	[42]
	NIR laser irradiation (635 nm) at 169.85 mW/cm^2 for 10 min.	meso-Tetra (N-methyl-4-pyridyl) porphine tetrachloride (photosensitive drug) and gold nanoparticles	Type I collagen	meso-Tetra (N-methyl-4-pyridyl) porphine tetrachloride	After a single injection, but multiple treatments with NIR, the combined photothermal and photodynamic therapies achieved complete tumor eradication on a breast tumor xenograft mice model.	[14]
	NIR laser irradiation (633 nm) at 50 mW/cm^2 for 10 min.	Protophorphyrin IX (PpIX)	Adipic dihydrazide-modified hyaluronic acid conjugated with PpIX, and dialdehyde-functionalized thioketal containing a ROS-cleavable thioketal linker	Doxorubicin (DOX)	Combined photodynamic therapy and chemotherapy that nearly suppressed tumor growth on a tumor xenograft mice model.	[43]
	NIR laser irradiation (760 nm) at 0.47 W/cm^2 for 3 min.	Indocyanine green	Type I collagen, poly (gamma-glutamic acid)	DOX or granulocyte macrophage colony-stimulating factor	Upon local temperature increase induced by NIR irradiation, the hydrogel network was disrupted, and payload release was increased up to 3-fold in vitro.	[44]

Table 1. Cont.

Stimulus	Stimulation Parameters	Stimulus Enhancer	Hydrogel	Embedded Cells/Cargo	Effect	Ref.
	NIR laser irradiation at 660 nm (0.5 W/cm^2) for 20 min and, three days later, at 915 nm (0.5 W/cm^2) for 10 min.	Perylene diimide zwitterionic polymer	Benzoxaborole-modified HA, fructose-based glycopolymer	DOX and photothermal polymeric nanoparticles	In a 4T1 tumor bearing mice model, tumors were almost eradicated with a combined chemo- and photothermal therapy. First, NIR irradiation at 660 nm led to enzymatic degradation of the hydrogel, thus releasing the DOX and the photothermal nanoparticles. Then, after 3 days, a second NIR irradiation, but at 915 nm, led to localized hyperthermia for killing tumor cells.	[45]
Electric	Biphasic waveforms with 50 ms pulses of 3–5 V/cm at 0.5, 1, 2, and 3 Hz.	rGO	GelMA	Primary neonatal rat cardiomyocytes	Increased contractility of rGO-GelMA constructs with respect to GelMA.	[46]
	(i) Dielectrophoresis before crosslinking: sinusoidal electric field of 1 MHz and 20 V for 10 s. (ii) Biphasic waveforms with 10 ms pulses of 3 V at 1 Hz applied continuously from day 2 to day 4.	MW-CNTs	GelMA	129/SVE-derived mouse stem cells (embryoid bodies)	Enhanced electrical conductivity, mechanical stiffness, and cardiac differentiation in aligned CNT-GelMA constructs when compared to unaligned CNT-GelMA and pristine GelMA.	[47]
	Rectangular electrical pulses of 2 ms at 1, 2, and 3 Hz applied continuously from days 3 to day 7.	Dopamine-rGO	GelMA	Cardiomyocytes	Improved orientational order of sarcomeres, propagation of intercellular pacing signals, and calcium handling.	[48]
	Square-wave pulses of 5 V at 1.5 Hz.	rGO	Myocardial dECM	hiPSC-derived cardiomyocytes	Improved electrophysiological function (calcium handling, action potential duration, and conduction velocity) and physiologically relevant drug responses in rGO-GelMA.	[49]

Table 1. Cont.

Stimulus	Stimulation Parameters	Stimulus Enhancer	Hydrogel	Embedded Cells/Cargo	Effect	Ref.
	Constant electric field of 100 mV/cm applied 20 min daily for 1 week.	Polydopamine-modified black phosphorus (PDA-BP) nanosheets	GelMA	Rat bone marrow derived MSCs	Enhanced neural differentiation of MSCs in stimulated PDA-BP@GelMA hydrogels when compared to unstimulated PDA-BP@GelMA and GelMA.	[50]
	External electric field of 5 V AC pulses at 1 Hz applied continuously for 21 days.	GNWs	Type I collagen	C2C12 myoblasts	Aligned GNWs in electric field direction favored myoblast alignment and enhanced myotube formation.	[51]
	DC stimulation at 50 mV/mm applied continuously for 8 hrs.	SW-CNTs	Matrigel-type I collagen	Neonatal rat dorsal root ganglia cells	Enhanced neurite outgrowth on stimulated SWCNT-loaded hydrogels when compared to unstimulated SWCNT-loaded hydrogels and SWCNT-free hydrogels.	[52]
	Electric current below 1 mA generated with an electric potential of 1 V. Applied for 15, 30, 45, and 60 min.	Ag nanowires	GelMA-Collagen	Fluorescein isothiocyanate (FITC)-dextran	Ion currents during electrical stimulation created osmotic gradients that caused periodic hydrogel contractions and facilitated drug release.	[53]
Magnetic	2 mT cylindrical magnet placed beneath the constructs.	Streptavidin-coated IONs	Type I collagen-agarose	Human primary knee articular chondrocytes	Collagen fiber alignment in magnetic field direction, as a result of ION motion inside hydrogel, improved type II collagen secretion by chondrocytes.	[54]
	100 mT magnetic field.	Rod-shaped acrylate-modified poly(ethylene oxide-stat-propylene oxide) microgels embedded with IONs	Fibrin	Chicken-derived primary dorsal root ganglions	Unidirectional alignment of rod-shaped microgels in field direction-oriented neurite outgrowth inside hydrogel.	[55]

Table 1. *Cont.*

Stimulus	Stimulation Parameters	Stimulus Enhancer	Hydrogel	Embedded Cells/Cargo	Effect	Ref.
	External alternating magnetic field (1478 Hz, 10 A, 10 min).	IONs	Dopamine-conjugated hyaluronan and IONs	DOX	IONs serve as structural crosslinkers and facilitate hyperthermia treatment and on-demand release of DOX under alternating magnetic fields.	[56]
	Permanent magnet placed against the hydrogel immersed in water.	IONs	Collagen	Fluorescein sodium salt	Cargo release was triggered and accurately controlled upon hydrogel deformation induced by external magnetic field.	[57]
Acoustic	2.5 MHz ultrasound at either 3.3 or 8.8 MPa peak rarefactional pressures with 4.3 or 17.2 MPa peak compressional pressures, respectively.	Perfluorocarbon (PFC) double emulsion droplets	Fibrin	bFGF	Ultrasound stimulus induced vaporization of PFC in the double droplet emulsions, thereby causing the release of encapsulated bFGF. The timely release of this factor promoted angiogenic sprouting of HUVECs within the outer layer of the fibrin hydrogel.	[58]
	Acoustic standing wave fields of 3.25 MHz at day 0 and of 8.6 MHz at day 4	Two PFC double emulsion droplets based on perfluoropentane and perfluorohexane	Fibrin	bFGF and PDGF-BB	Sequential payload release induced by vaporization of PFC double droplet emulsions at different pressure thresholds modulated with ultrasound standing wave fields. bFGF was released at day 0 and PDGF-BB at day 4.	[59]

2. Enhancing the Mechanical Properties of dECM-Based Hydrogels

As the main component of the majority of ECMs is type I collagen, the most straightforward approach for inducing gelation of dECM-derived hydrogels is through thermal stimuli [60]. Type I collagen molecules are triple-helical structures that, under physiological conditions of temperature and pH, are able to assemble into hierarchically organized structures that interact through hydrogen bonding forces (Figure 1A) [61,62]. In native tissues, collagen molecules assemble into fibrillar structures, which, in turn, organize into fibers and create the characteristic fibrous backbone of ECMs. However, collagen crosslinking dynamics are not fully preserved on ECM-derived hydrogels, mainly as the decellularization and enzymatic solubilization processes to produce them largely disrupt their organizational hierarchy and therefore compromise their ability to reassemble [21,63]. Moreover, thermal crosslinking is considerably limited by the heat transport rate within the surrounding environment, which results in slow gelation processes that, in turn, lead to insufficient structural stability over time, as evidenced by reversible sol–gel transitions [64]. For this reason, when extruded through 3D bioprinting devices, thermally crosslinked dECM hydrogels normally exhibit poor shape fidelity [65]. This is concerning from the tissue engineering viewpoint, as the degradation kinetics of dECM-based hydrogels are usually faster than the required matrix remodeling rates for the timely maturation of biofabricated constructs and for in vivo regeneration [62,66]. Accordingly, there is a critical need to improve the biomechanical performance of dECM-based hydrogels such that they can be successfully implemented in tissue engineering applications with the ultimate goal of clinical translation. In order to address this limitation, several methods have been employed, which we discuss in detail below [61,65,67].

2.1. Improving Thermal Gelation Dynamics

With the purpose of reducing the gelation time of a porcine skin dECM bioink upon deposition, a 3D cell printing system equipped with two heating modules located above and below the construct was devised by Anne and colleagues [68]. They found that this setup, combined with an increase in the dECM concentration of the hydrogels, improved the structural stability of dECM constructs over time. This was due to the faster gelation kinetics triggered by the setup and the denser hydrogel networks, which facilitated the self-assembly of ECM proteins into tissue-like patterns. However, although a 2.5% (w/v) concentration presented the best mechanical performance, increasing hydrogel viscosity inadvertently reduced cell viability, which hampered the possibility of fabricating more stable constructs [68]. Alternatively, several researchers have incorporated polymeric frameworks during dECM-based hydrogel deposition to add structural support without the need of high dECM concentrations [18,61,69]. For instance, Pati and colleagues bioprinted adipose tissue and cartilage-derived dECM bioinks on a polycaprolactone (PCL) framework that provided sufficient structural support to allow adequate thermal gelation and tissue maturation without significantly reducing the viability of embedded human adipose tissue-derived stem cells (hASCs) and human inferior turbinate-derived stromal cells (hTMSCs) [70]. Their bioprinting process consisted of the sequential layer-by-layer deposition of the PCL frame and of the dECM bioink in every alternating gap between PCL filaments. By acting as an external structure that supported the constructs, the PCL helped to relieve the low mechanical properties of the dECM bioink before thermal crosslinking and during construct maturation. Likewise, Lee and colleagues bioprinted a liver dECM hydrogel embedded with human bone marrow derived mesenchymal stem cells (hBMM-SCs) within a PCL framework in a similar fashion [71]. In both studies, the use of a frame granted a structurally stable and biocompatible microenvironment for the proliferation and differentiation of embedded human stem cells into tissue-specific lineages.

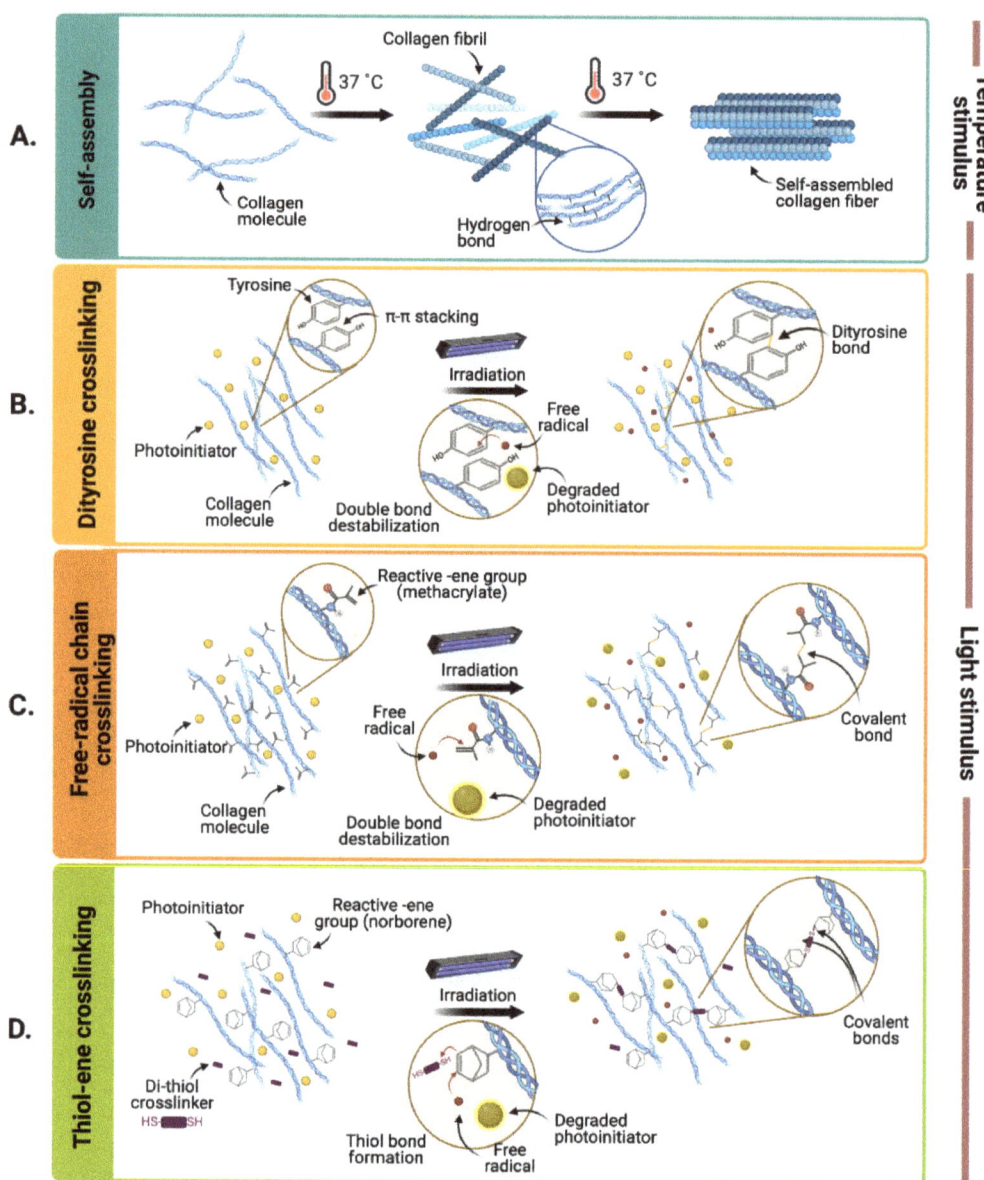

Figure 1. Hydrogel crosslinking upon thermal and light stimuli. (**A**) Self-assembly of collagen molecules upon thermal stimulus. (**B**) Photo-induced crosslinking between tyrosine residues, (**C**) methacryloyl, or (**D**) norbornene moieties (Created with BioRender.com).

2.2. Photocrosslinking

Stimulation with light has been an alternative strategy for strengthening the mechanical properties of hydrogels, as it can initiate photocrosslinking reactions in a wide variety of biomaterials [72–74]. These reactions typically occur in the presence of a photoinitiator molecule (e.g., riboflavin (RF), riboflavin phosphate (RFP), lithium phenyl-2,4,6-trimethylbenzoylphosphinate (LAP), 2-hydroxy-4′-(2-hydroxyethoxy)-2-methylpropiophenone (Irgacure 2559), ruthenium/sodium persulfate (Ru/SPS)), which forms reactive radical species upon light-mediated degradation [67,74]. Once generated, these species can react with specific functional groups on the main backbone chain of ECM proteins to form covalent bonds or reactive radical intermediates [74]. Ultraviolet (UV) light (100–400 nm), blue light (400–490 nm), and white light (400–700 nm) are the most common light stimuli for inducing photocrosslinking reactions. However, as UV light is known to induce DNA mutations and cell death, it is the least preferred wavelength range for photocrosslinking reaction schemes [35,75].

Notably, dECM hydrogels have shown photosensitive properties in the presence of photoinitiators due to the destabilization of nearby collagen or proteoglycan residues. Although the precise mechanisms of these reactions are elusive, it has been reported that tyrosine and histidine residues may be involved in this light-induced crosslinking approach [76–78]. This is as tyrosine residues form π–π complexes that facilitate dityrosine bonds in the presence of singlet oxygen (Figure 1B) [76], and the imidazole group of histidine residues is easily oxidized to transient intermediates that can lead to new crosslinked products [77].

Numerous studies exploring the photocrosslinking dynamics of collagen hydrogels in the presence of RF [79–82], have demonstrated a consistent increase in elastic and compressive moduli upon blue-light [81] or UV [79] exposure. Similarly, our research group developed a photoresponsive hydrogel by mixing small intestinal submucosa (SIS) dECM with RF that photocrosslinks in response to blue light. We reported that an increase in RF (up to 0.5% (w/v)) led to a proportional increase in the storage modulus from 2 kPa to 4 kPa [75]. Jang and colleagues developed a heart dECM hydrogel, also with RF, which demonstrated increased mechanical properties upon UVA light exposure. Additionally, the material exhibited an entropy-driven self-assembly of collagen with an increase in temperature to 37 °C [66]. With this dual crosslinking scheme, the compressive modulus of these hydrogels increased from 0.18 kPa to 15.74 kPa, and the dynamic complex modulus from 0.33 kPa to 10.58 kPa.

Recently, Kim and colleagues showed that using Ru/SPS as a photoinitiator system potentiated dityrosine crosslinking in corneal and heart dECMs upon blue-light exposure [41]. Due to its high absorptivity within this light range (400–450 nm), high molar extinction coefficient, and high chemical stability in its excited state, Ru/SPS allowed efficient curing reactions at relatively low concentrations (0.5–2 mM) and exposure time (5 s), which demonstrates better crosslinking dynamics than RF. In particular, they show that Ru/SPS-mediated dityrosine crosslinking increased the compressive modulus of corneal and heart dECMs up to 60 and 70 kPa, respectively. However, their complex modulus remained below 1 and 0.5 kPa. Moreover, with extrusion-based and digital-light processing bioprinting technologies, they built multi-layered and complex geometries with both bioinks and demonstrated that their method allows the additive manufacturing of constructs with high shape fidelity.

However, beyond the utilized photoinitiator, the efficiency of these photocrosslinking reactions is highly limited by the low concentration of reactive tyrosine or histidine residues on dECM protein backbones [78]. Therefore, recent efforts have focused on increasing the number of photosensitive groups that can participate in these crosslinking reactions.

Biochemically Modified dECM-Based Hydrogels with Augmented Photosensitivity

Several research groups have potentiated photocrosslinking in dECM-based hydrogels by the addition of biochemically modified natural polymers with enhanced photosensitivity. These biochemical modifications introduce pendant groups with reactive double bonds (e.g., methacrylate, acrylate, norbornene, vinyl ethers, and N-vinyl amide) [74] that can be easily destabilized in the presence of free radicals, thus facilitating the formation of covalent bonds and increasing the efficiency of photocrosslinking (Figure 1C,D). Methacryloyl-modified gelatin (GelMA) is perhaps the most commonly implemented biochemically modified material due to its high biocompatibility, ease of processing and rapid crosslinking upon light exposure [18,83]. Bejleri and colleagues, for example, developed a photoresponsive GelMA (0.1 mM)-based bioink mixed with cardiac dECM and eosin Y as a photoinitiator (5% (w/v)) for bioprinting cardiac patches [35]. This crosslinking scheme yielded a homogeneous distribution of dense fibers within the bioprinted constructs without reducing cell viability. The incorporation of GelMA was reported to significantly increase the storage modulus of the bioprinted constructs after the photocrosslinking process. Furthermore, Skardal and colleagues developed a porcine liver dECM-based bioink mixed with thiolated HA and thiolated gelatin, two different PEG-based crosslinkers (PEG-acrylate and PEG-alkyne) and a photoinitiator (Irgacure 2959) [34]. This bioink spontaneously crosslinked at neutral pH due to the formation of thiol-acrylate bonds between the PEG-acrylate and the thiolated polymers. However, further crosslinking was achieved upon pendant thiol group exposure to UV light, due to the photo-induced near-instantaneous thiol-alkyne polymerization reaction with the PEG-alkyne crosslinkers. This study also reported that the stiffness increased by nearly 200-fold (i.e., from 0.1 to 19.8 kPa) by varying the concentration, molecular weight, and geometry of the alkyne-modified PEG crosslinker.

Direct biochemical modifications to dECM proteins have also become recurrent approaches as they eliminate the need of incorporating additional materials. For instance, thiol- and methacryloyl-modified HA hydrogels were investigated by Lee and colleagues for developing prolonged-action delivery vehicles to the ocular surface [36]. The modification of HA with these pendant groups, in combination with the addition of 0.01% (w/v) RF as photoinitiator, allowed forming dense photocrosslinked networks upon blue light exposure, which can be attributed to thiol–ene reactions between methacryloyl and thiol groups. They demonstrated that the photocrosslinking process prolonged the stability of the hydrogel, and dictated the release of loaded bovine serum albumin (BSA). Wu and colleagues also developed a photocrosslinkable construct to fabricate personalized pharmaceutical tablets with controlled dosages of active pharmaceutical ingredients, both hydrophilic and hydrophobic (e.g., lisinopril and spironolactone) [84]. The manufactured constructs consisted of two layers. First, a lisinopril-loaded hydrophilic hydrogel based on 3% (w/w) norbornene-functionalized HA mixed with PEG dithiol (PEGDT) supplemented with 10% (v/v) eosin Y as photoinitiator and 10% (v/v) PEG. Second, a spironolactone-loaded hydrophobic hydrogel based on 30% (w/w) PEG diacrylate (PEGDA), 50% (w/w) PEG, and 20% (v/v) ethanol supplemented with both 1 mM eosin Y as photoinitiator and 0.05 M mPEG-amine as co-initiator. The layers were dually photocrosslinked upon visible light exposure due to thiol–ene reactions between thiol groups of PEGDT and norbornene groups of HA, and the covalent bonds formed between destabilized acrylate groups of PEGDA, respectively. They demonstrated that, irrespective of drug loading concentration, but with the appropriate light dosage, these tablets achieved optimal mechanical properties and dual sustained release profiles over time [84].

Direct biochemical modifications to dECM hydrogels have also been conducted. For instance, Ali and colleagues successfully bioprinted a bioink based on methacryloyl-modified kidney dECM (KdECMMA) and mixed with HA, glycerol, and gelatin [38]. The storage modulus of irradiated KdECMMA constructs increased 1.7-fold with respect to constructs fabricated with unmodified kidney dECM (KdECM). Moreover, the stiffness of KdECMMA hydrogels increased proportionally to dECM concentration, but remained unaltered at a low level for KdECM hydrogels. Consequently, KdECMMA hydrogels presented a lower degradation rate and helped maintain the stability of entrapped bioactive molecules by protecting them from early enzymatic degradation. Similarly, our research group developed methacryloyl-modified SIS dECM hydrogels (SISMA) mixed with 0.05% (w/v) RF, which significantly potentiated their mechanical properties upon blue-light irradiation without the need for excipient materials. As shown in Figure 2A, the storage modulus of SISMA hydrogels is around 3 and 3.5 times greater than that of unmodified hydrogels before and after blue light exposure for 5 min. Furthermore, the photocrosslinking reaction in unmodified hydrogels saturates rapidly, yielding a similar storage modulus after 2 min and 5 min of exposure, while that of SISMA hydrogels increases according to exposure time. These results support the notion that this biochemical functionalization significantly improves the overall mechanical stability of hydrogels and their crosslinking dynamics, resulting in superior printability and enhanced shape fidelity of constructs (Figure 2B,C).

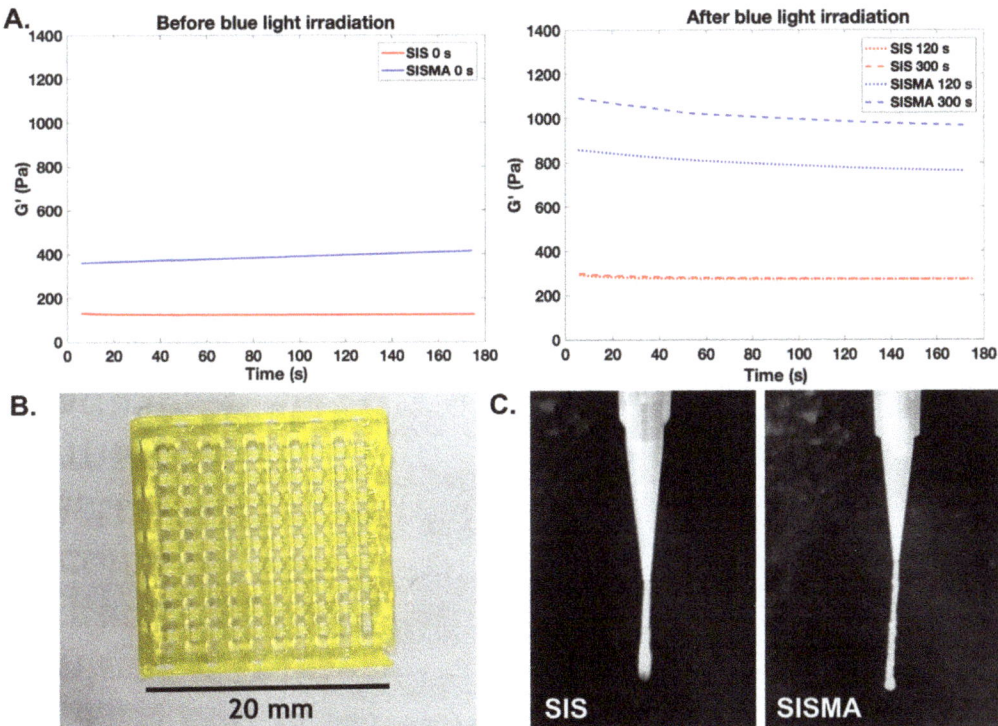

Figure 2. (A) Methacryloyl-modified small intestine submucosa dECM (SISMA) hydrogel bioprinted into a square lattice with high shape fidelity. (B) Filament formation of unmodified SIS and SISMA hydrogels made at the same ECM concentration (25 mg/mL) upon extrusion as a key feature for adequate printability. (C) Storage modulus of unmodified-SIS and SISMA hydrogels before and after photocrosslinking with varying exposure time. SISMA hydrogels exhibit superior mechanical stability, before and after blue light irradiation, and superior crosslinking dynamics.

Similarly, Parthiban and colleagues developed a methacryloyl-modified bone dECM hydrogel (BoneMA) mixed with 0.15% (*w*/*v*) LAP which, upon photocrosslinking, supported the formation of interconnected vascular networks from embedded human umbilical vein endothelial cells (HUVECs) [39]. The elastic modulus of photocrosslinked BoneMA hydrogels also increased according to exposure time, which demonstrates that this crosslinking method allows highly tunable mechanical properties [85]. BoneMA constructs showed a higher total vessel length and visibly faster network formation than GelMA constructs (which have shown vasculogenic potential), possibly due to the retention of pro-angiogenic growth factors from the tissue of origin, which promoted faster vascularization in vitro [39].

3. Tunning Microarchitectural Characteristics of dECM-Based Hydrogels

Although the composition of dECM-based hydrogels is a major contributor to their superior bioactivity, biomimicry is not only important composition-wise, but also architecture-wise [86]. The structural organization of ECM components in vivo imparts highly specialized biomechanical characteristics that are specific to tissue location and function. For instance, a linear orientation of ECM fibrillar components predominates within tissues exposed to high tensile stresses (e.g., tendons and some ligaments), whereas circumferential organization predominates within tissues that experience multiaxial tension, compression, and shear (e.g., annulus fibrosus and meniscus) [87]. The spatial distribution of cells within tissues is a major contributor to the microstructure as well, considering that organized cellular interactions are often necessary for specific tissue functions. This is typically observed within muscular tissues, as muscle cells must hierarchically organize into myotubular structures and then muscle fibers to ensure cooperative interactions during contraction [88]. The precise structural organization of hydrogel microenvironments upon deposition is therefore an essential feature for developing constructs with tissue-specific functionalities. Controlling the initial structural properties and distribution of embedded cells can dictate cellular fate (e.g., from pluripotency to specific differentiation profiles) and posterior matrix remodeling in uniquely prepared structural patterns.

Although the controlled patterning of hydrogels is highly effective in recapitulating the macro-geometry of tissues, it often fails in replicating micro-architectural characteristics. A wide variety of tissue architectures have been induced in dECM-based hydrogels through several types of stimuli including mechanical, magnetic, electric, and acoustic. These approaches are discussed below.

3.1. Mechanical Stimuli

As many tissue architectures require the anisotropic alignment of fibrous ECM proteins (mainly collagen) [86], the most straightforward approach has been to tune the shear stress during extrusion to induce their alignment in the direction of flow. This mechanical stimulus was exploited by Kim and colleagues for the unidirectional alignment of the collagen fibrils of a corneal stroma-derived dECM (Co-dECM) bioink to reproduce the architecture of native corneal stroma after tissue maturation [89]. They found that by increasing the nozzle gauge up to 25 G it was possible to induce collagen fibril alignment and create anisotropic structures that guide keratocyte orientation into linear patterns. This initial cellular distribution led to the secretion of collagen fibrils during keratocyte maturation in perpendicular directions, which, in turn, yielded the characteristic crisscrossed pattern of native corneal stroma that is fundamental for corneal transparency (see Figure 3A) [90]. The shear-induced alignment of collagen fibrils was also performed by Schwab and colleagues to reproduce the architecture of knee articular cartilage [91], which comprises three distinct layers: a superficial layer with fibrils tangentially oriented to the joint surface, a middle layer with random orientation, and an internal layer with columnar alignment (see Figure 3C) [86]. With a computer-aided design (CAD), they printed a tyramine-HA (THA)-collagen I bioink with embedded human mesenchymal stem cell (hMSC) spheroids. The superficial layer was printed in horizontal filaments while the internal layers were formed by circular-vertical structures. As expected, the initial shear-induced alignment of collagen fibrils guided the unidirectional migration of hMSCs from spheroid micropellets and significantly increased cytoskeleton alignment along the direction of fibrils, which was not observed in isotropic THA-collagen I hydrogels. Recently, another research group reported an alternative method to increase the shear-induced cellular alignment in bioprinted constructs by pre-incubating methacryloyl-modified collagen (ColMA) bioinks with myoblasts (C2C12) [92]. Their rationale was that the pre-incubation period promoted cell–matrix interactions before printing, which induced cell transition from spherical to elongated morphology and favored their alignment during extrusion. Moreover, according to their findings, pre-incubated constructs had significantly higher cytoskeleton alignment and superior myotube formation after 21 days, when compared with constructs that were not pre-incubated.

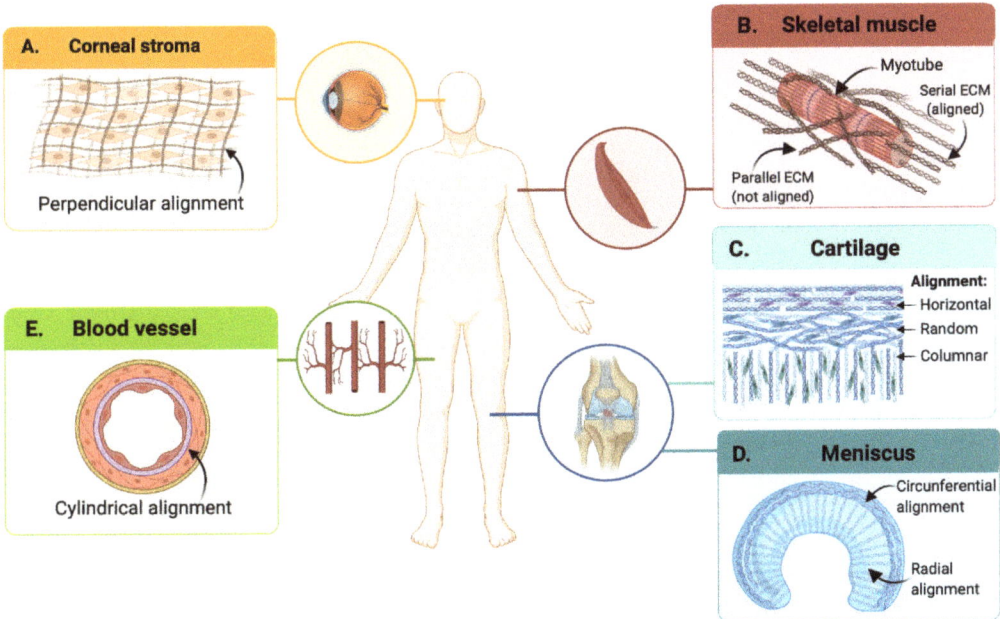

Figure 3. Examples of hierarchical cellular organization and ECM architecture in different native tissues. (**A**) Corneal stroma, (**B**) skeletal tissue, (**C**) cartilage, (**D**) meniscus, and (**E**) blood vessels (Created with BioRender.com).

In addition to shear-stress, other mechanical stimuli, such as strain, have been explored for the microarchitectural remodeling of bioprinted constructs [93,94]. For instance, Puetzer and colleagues induced circumferential alignment of collagen fibers within a meniscus-replicating construct [95]. They anchored the meniscus model at the horns to mimic the native tibial attachment sites in vivo, and to restrict the compaction of the collagen hydrogel, a phenomenon commonly observed during the maturation of dECM-based constructs. This geometrical constraint induced residual hoop stress circumferentially and encouraged the alignment of fibrils. Clamped menisci developed native-like sized and circumferential fibers in the outermost surface, as well as radially aligned fibers in the innermost surface, thereby resembling native tissue after 8 weeks of culture (see Figure 3D). This approach was also explored by Choi and colleagues for the development of anisotropically aligned skeletal muscle constructs. Accordingly, they bioprinted a skeletal muscle-derived dECM bioink with embedded myoblasts (C2C12) in a PCL anchoring set up that geometrically constrained opposite ends of the construct. This generated a longitudinal constant strain by opposing hydrogel compaction during the maturation process [94]. With this mechanical stimulus, they reported that 76% of cells were unidirectionally aligned after 7 days and with an elastic modulus comparable to that of native skeletal muscle within 14 days of incubation. Most importantly, striated band patterns were observed within myotubular structures, which indicated that the contractile apparatus of native muscle was successfully formed, implying structural and functional maturity.

3.2. Electrical Stimuli

Other approaches have coupled mechanical with electrical stimulation for the engineering of hierarchically organized and functional muscle tissue, considering that beyond mechanical response it also displays inherent electrosensitive properties. In addition to anisotropic cellular alignment, mature muscular phenotypes have specific ECM patterns around myotubular structures to allow adequate muscle contraction. Collagen IV fibers, which attach directly to sarcolemmas and connect them to other types of collagen [96], exhibit a lower degree of alignment/stiffness in parallel locations to myotubes to minimally interfere with myotube contraction. In contrast, fibers located in series with myotubes have a higher degree of alignment to bear and transmit the contraction forces to the imposed load (see Figure 3B) [97]. Accordingly, Kim and colleagues demonstrated that by applying an out-of-phase mechanical and electrical co-stimulation scheme (for short exposure periods of 20 min) to myoblast-laden Matrigel/fibrinogen constructs previously matured to skeletal muscle tissue, they remodeled into the characteristic muscular structural patterns [98]. The rationale behind this approach is that mechanical stretching generates tensile stress along the load direction, while muscle contraction triggered by an electric potential induces shear and compressive stress on collagen IV fibers parallel to myotubes and tensile stress at intrafascicularly terminating ends [98]. They showed that an alternating application of these two stimuli yielded a superior contraction force in the matured constructs compared to applying either of the two stimuli separately or simultaneously. The synergistic stimulation scheme effectively maximized fiber alignment in serial ECM locations and minimized it in parallel ECM locations.

Electrical stimulation alone, for longer time periods, has also been employed for controlling cell alignment within ECM-based hydrogels [99]. Electrosensitive cell types (e.g., cardiomyocytes, myoblasts, and neural stem cells) align parallel to the electric field vector to maximize the field gradient across them, whereas other types of cells (e.g., adipose-derived stromal cells, or endothelial progenitor cells) align perpendicularly to minimize it [100]. However, the effectiveness of electrically induced cellular alignment has proven to be low and to require long stimulation periods that, ultimately, may reduce cell viability [101]. The incorporation of rod-shaped electrosensitive nanostructures (e.g., carbon nanotubes (CNTs) and gold nanowires (GNWs)) into ECM-based hydrogels has emerged as a plausible alternative to guide their electrical response as they can easily align with the direction of the applied electric field and therefore facilitate the alignment of surrounding microstructures [51,102]. Kim and colleagues, for example, embedded GNWs in a myoblast-laden collagen bioink and demonstrated that the electrically induced alignment of these nanostructures significantly narrowed the fiber orientation distribution. This, in turn, favored myogenic differentiation compared with electro-stimulated myoblast-laden collagen constructs in the absence of GNWs and non-electrostimulated constructs [51].

3.3. Magnetic Stimuli

The incorporation of magnetic nanostructures, specifically iron oxide nanoparticles (IONs), has also shown great promise for the alignment of ECM fibers due to their high sensitivity to externally applied magnetic fields. A recent work by Betsch and colleagues showed that streptavidin-coated IONs incorporated within collagen–agarose hydrogels successfully guided the anisotropic alignment of collagen fibers for the recreation of the cartilage microarchitecture [54]. Upon exposure to an external magnetic field during the temperature gelation process, collagen fibers were forced to align unidirectionally due to the traveling motion of these nanoparticles across the hydrogel. The authors showed that collagen fiber alignment in all formulated hydrogels more than doubled their compressive tangent modulus and yielded mechanical features that were similar to those of the native cartilage. Moreover, the fabrication of a two-layered construct, consisting of an anisotropically aligned superficial layer and a randomly oriented internal layer led to remarkable collagen I and II secretion by embedded primary knee articular cartilage cells than in any of the bioprinted samples of either of the two individual layers. This suggests that the combined microarchitectural characteristics of the two-layered construct, which is similar to native cartilage, significantly improves chondrocyte maturation. Similarly, Rose and colleagues developed a magnetically responsive hybrid hydrogel consisting of (i) rod-shaped acrylate-modified poly(ethylene oxide-*stat*-propylene oxide) microgels loaded with monodispersed IONs in (ii) a surrounding fibrin (i.e., polymerized fibrinogen) matrix with embedded chicken-derived primary dorsal root ganglions (DRGs) [55]. Upon exposure to a low-intensity magnetic field (in the mT order), the rod-shaped microgels aligned well along the field direction. Their orientation was subsequently fixed with an in situ fibrin crosslinking treatment. Although embedded microgels fail to provide cell adhesion focal points and fibrin fibers are not directly aligned, microgel alignment was sufficient to guide neurite growth parallel to their direction. This novel approach demonstrated that minimal structural guidance can trigger nerves to grow unidirectionally.

3.4. Acoustic Stimuli

The development of complex microarchitectures with non-linear patterns within ECM-based hydrogels is an ongoing challenge due to the requirements in terms of means for high spatial control. In this regard, acoustophoretic systems have shown the most promising results as demonstrated by the guided migration of cells into specific patterns that follow the pressure fields formed by standing surface and standing bulk acoustic waves [103,104]. In addition to favoring the development of hydrogel constructs with high cellular alignment, acoustophoresis has been used to organize embedded cells within hydrogels into the cylindrical structures needed to guide the fabrication of vascular networks. As reported by Kang and colleagues, HUVECs and human adipose-derived stem cells (hADSCs) embedded within a catechol-functionalized HA hydrogel were successfully co-aligned into collateral cylindrical patterns (see Figure 3E) [105]. This was enabled by employing surface acoustic waves of 280 μm wavelength, which concentrated the cylindrical patterns at intervals similar to the intercapillary distance of human skeletal muscle. This initial cell patterning led to both interconnected capillary networks due to the enhanced branching between parallel structures, and the organized formation of functional endothelial barriers. Moreover, these mature constructs were able to successfully integrate with host vasculature after subcutaneous transplantation into dorsal regions of mice, forming perfusable and interconnected networks with native microvessels. This was in contrast with constructs that were not patterned.

Similarly, Petta and colleagues proposed an acoustic-based technology, termed sound induced morphogenesis (SIM), in which they combine acoustic patterning with physiological self-assembly to generate multi-scale and perfusable vascular networks [106]. With vertical mechanical vibrations (80 Hz and 0.5 g acceleration amplitude), they generated acoustic surface standing waves that patterned HUVEC and hMSC spheroids in concentric rings within a fibrin hydrogel. They showed that the small distance between adjacent rings favored HUVEC sprouting from the patterned spheroids in capillary-like structures that fused with sprouts from adjacent rings, thus creating microvessel-like morphologies. Moreover, when embedding an endothelialized macrochannel through the center of the hydrogel, simulating a macrovessel, the capillary-like sprouts from patterned spheroids were able to fuse with this structure as well, and create perfusable multiscale vascular networks after only 5 days.

Despite these efforts on recreating vascular networks, these works employ simple and symmetric geometries that may not recapitulate the arbitrary and non-symmetrical arrangements of vasculature in most biological tissues. To circumvent this, acoustic holography was recently proposed by Ma and colleagues for constructing sophisticated 3D cell patterns according to complex pressure fields [107]. In this technique, complex patterning is enabled by propagating a 5 MHz acoustic plane through a 3D-printed topography (termed the 'hologram'). This is precisely designed to induce a specific phase field, which is then projected towards the hydrogel to induce cell arrangement according to the generated pressure field. Accordingly, by changing the topography of the hologram, it is possible to obtain desired patterns on demand. As a proof-of-concept, they demonstrated that human colon cancer cells (HCT-116) embedded in collagen hydrogels could be patterned with this technique into complex structures (e.g., asymmetric geometric patterns, human profiles), prior to temperature gelation and without significant impact on cell viability. Although acoustic holography is yet to be employed for the development of physiologically relevant structures, their work demonstrated the potential of this technology in tissue engineering and mechanobiology.

4. Improving Morphogenesis and Functionality of dECM-Based 3D Cultures with External Stimuli

Although the mechanical and micro-architectural characteristics of dECM-based hydrogels significantly influence cell behavior [108,109], in most cases, they are not enough for ensuring full functionality in vitro. This is as they fail to replicate the myriad of biochemical signals originating from the native extracellular environment, which are essential for directing cellular fate [110]. During maturation, cells must be able to communicate with each other and their niche through biochemical and mechanotransductional cues, as these interactions induce morphogenetic signals that ultimately lead to tissue functionality [111]. Therefore, ensuring an adequate maturation process is key for directing morphogenesis and the hierarchical organization of tissues at the macro and micro scales. Several strategies that employ external stimuli for directing specific maturation profiles in cell laden dECM-derived hydrogels are discussed below.

4.1. Directing Stem Cell Differentiation

One of the most valuable applications of external stimuli in TE and RM is directing stem cell differentiation into specialized and adult-like phenotypes [112]. Mechanical and electrical stimulation, for example, have been widely explored for directing the differentiation of human mesenchymal stem cells (hMSCs) towards osteochondral lineages [113]. This has been attributed to the modulating effect of such stimuli over calcium (Ca^{2+}) influx towards intracellular compartments. In this regard, specific intracellular Ca^{2+} profiles have shown to direct chondrogenesis and osteogenesis in vivo, as calcium channels participate in the transduction of mechanical signals that direct these differentiation profiles during embryonic development [114,115]. Notably, brief and pulsed Ca^{2+} influx periods during early differentiation stages have shown to promote MSC chondrogenesis, whereas prolonged Ca^{2+} exposure periods promote osteogenesis [116,117]. Accordingly, these physiological hallmarks have been harnessed to promote the osteochondral differentiation of MSC-laden dECM-derived hydrogels by modulating the activity of these channels through mechanical stimuli [118]. Alternatively, it is possible to control the activity of voltage-gated calcium channels with electrical stimuli (Figure 4A,B) [116,119,120].

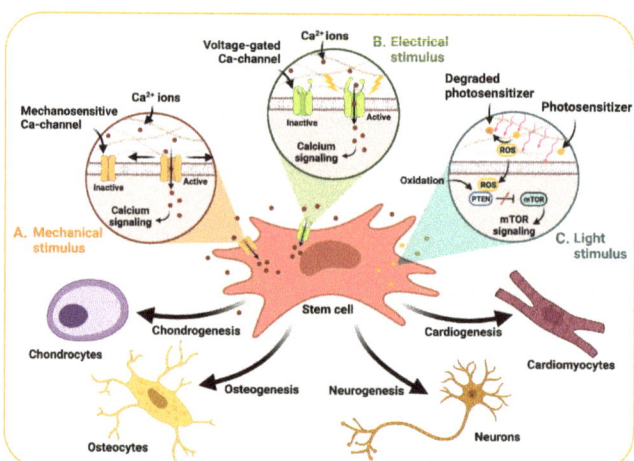

Figure 4. Schematic of the modulating effect of (**A**) mechanical, (**B**) electrical, and (**C**) light stimuli on the activation of specific signaling pathways that direct differentiation profiles of stem cells. Specific stimulation regimes that mimic mechanotransductional or electrical cues in embryonic development can activate calcium signaling. Similarly, ROS production upon light stimulation can promote the oxidation of PTEN, a strong inhibitor of mTOR, and thus upregulate mTOR signaling [121]. (Created with BioRender.com).

For instance, Aisenbrey and colleagues studied the chondrogenic effect of mechanical stimuli on cartilage mimetic hydrogels with embedded human induced pluripotent stem cell (hiPSC)-derived mesenchymal progenitor cells (MP-iPSCs) [122]. Their mimetic hydrogels consisted of 8-arm PEG functionalized with norbornene, PEG dithiol crosslinkers, and thiolated chondroitin sulfate, which is the main proteoglycan of cartilaginous tissues. The obtained cylinder-like constructs were subjected to mechanical compressive strains daily, during 1 h for 3 weeks, and the combined effect of this dynamic culturing with the presence or absence of differentiating growth factors was explored (e.g., TGFβ2 and BMP2). Their results showed that the mimetic hydrogel alone induced morphogenetic signatures, as evidenced by chondrogenic gene expression and immunohistochemical analyses. However, the combination of dynamic culturing and growth factor supplementation synergistically induced chondrogenesis while limiting tissue hypertrophy, which has been a main drawback observed on static culturing in the absence of growth factors. Similarly, Choi and colleagues showed that simultaneous growth factor supplementation and mechanical stimulation with low-intensity ultrasound (10 min/day, 1 MHz) also promoted chondrogenic differentiation in MSC-embedded fibrin-HA hydrogels. After 28 days, an increased sulfated GAG and collagen synthesis was observed in the stimulated constructs, which is a characteristic behavior of mature chondrocytes [118].

Regarding electrical stimuli, Vaca-Gonzalez and colleagues recently showed that intermittent 60 kHz electrical stimulation of hMSCs-embedded HA-gelatin hydrogels (30 min, 4 times per day for 21 days) not only enhanced GAG and type II collagen synthesis, but significantly increased the expression of chondrogenic markers (e.g., SOX-9 and aggrecan) without the need for differentiating growth factors [123]. Moreover, longer electrical stimulation periods of 4 h performed 3 times/day (7.6 mV/cm, 10 Hz) on type I collagen- and HA-coated PCL scaffolds, seeded with MSCs, significantly promoted osteogenesis [124]. This was demonstrated, after 28 days of stimulation, by the increased alkaline phosphatase (ALP) activity and increased expression of osteogenic markers (e.g., osteopontin, osteocalcin, and ALP).

The modulation of Ca^{2+} signaling has also been harnessed for the differentiation of other types of stem cells towards electrically active lineages, such as neural, cardiac, or skeletal muscle cells [125,126]. In this case, the increased Ca^{2+} handling induced by external stimuli has shown to promote early electrical/contractile activity by upregulating expression profiles of genes related to voltage-gated ion channels, which ultimately directs their maturation towards electrically active phenotypes [127]. However, as these differentiation routes require higher intensity and sustained electrical currents, most approaches have employed nanocomposite hydrogels embedded with electroconductive nanostructured materials to achieve superior electrical percolation networks within the hydrogel. This, in turn, ensures an early electrical connectivity within embedded cells, and facilitates their differentiation towards electrically active lineages. In this regard, carbon-based nanostructures such as graphene, reduced graphene oxide (rGO) [46,128], and carbon nanotubes (CNTs) [47,129], have been widely explored for their exceptional electroconductivity [130]. Nonetheless, a major setback in their implementation has been their limited cytocompatibility and poor dispersibility in hydrophilic media, which result from their inert sp^2 hybridized carbon backbone [131].

To circumvent this, several studies have functionalized their surfaces with hydrophilic agents [48,132,133], controlled the reduction degree of graphene oxide (GO) by using weaker reducing agents [49], or included hydrophobic moieties within the hydrogels to promote hydrophobic or π–π stacking interactions with these nanostructured materials [129]. For instance, Shin and colleagues devised a catechol-functionalized HA hydrogel where CNTs were stably dispersed along with polypyrrole nanocomposites through hydrophobic interactions with the present catechol moieties [129]. They showed that these electroconductive motifs significantly promoted the differentiation of hiPSC-derived neural progenitor cells and human fetal neural stem cells with improved electrophysiological functionality, as demonstrated by the upregulation of calcium channel expression and improved depolarization dynamics. Alternatively, our group proposed a biocompatible, in situ reduction scheme for GO embedded within SISMA bioinks, which harnessed the enhanced dispersibility and bioactive properties of GO at initial maturation stages, and subsequently increased its electroconductivity by two orders of magnitude upon partial reduction aided by ascorbic acid exposure [40]. With this maturation scheme, embedded human adipose-derived MSCs maintained high viability levels, as well as enhanced cell adhesion and proliferation, which demonstrated the enormous potential of this scaffold and maturation scheme for facilitating biocompatible electrical stimulation regimes that can potentially guide their differentiation towards neurogenic or myogenic phenotypes. Carboxyl-functionalization of CNTs was also proposed by Ahadian and colleagues to facilitate their stable dispersion in GelMA hydrogels [47]. Their incorporation significantly enhanced the differentiation of mouse embryonic bodies into cardiogenic phenotypes, as demonstrated by the upregulation of cardiogenic markers (e.g., Tnnt2, Nkx2-5, and Actc1) and increased contractile function upon electrical pulse stimulation of 3V at 1 Hz for 2 days. Although GelMA hydrogels are not strictly ECM-derived, this dispersion method can be easily extrapolated to ECM-derived hydrogels considering that gelatin is a hydrolyzed form of collagen and, in turn, retains some of its physical and chemical properties [134].

The electrically guided differentiation of stem cells, through the incorporation of other electroconductive nanostructured materials, has also been studied with GelMA hydrogels. Black phosphorus (BP) nanosheets, for example, were explored by Xu and colleagues to enhance the neural differentiation of seeded MSCs [50]. They showed that the electrical stimulation (100 mV/cm) of BP-embedded GelMA hydrogels for a 7-day period significantly increased the expression of neuronal markers (e.g., nestin, Tuj1) when compared to electrostimulated and non-electrostimulated GelMA hydrogels. Heo and colleagues similarly showed that the potentiating effect of gold nanoparticles (GNPs) on hydrogel electroconductivity significantly improved osteogenic differentiation of MSCs. They demonstrated that GNP addition to MSC-embedded GelMA hydrogels induced a marked increase in osteogenic markers (e.g., BSP, OCN, COL1, and Runx2) and ALP activity [135]. The observed changes occurred in a dose-dependent manner after 14 days of incubation. These approaches exemplify other routes that could be considered for engineering electroconductive ECM-derived hydrogels that direct stem cell differentiation.

In addition to electrically or mechanically induced calcium signaling, intracellular ROS modulation has also shown to contribute to stem cell differentiation processes (Figure 4C) [121,136,137]. In this regard, light stimuli have been employed to induce ROS production by degrading embedded photosensitizers within the hydrogels. Lu and colleagues, for instance, demonstrated that the implementation of photodynamic therapy promoted the chondrogenic differentiation of bone marrow stem cells (BMSCs) embedded in type I collagen hydrogels conjugated with carbon dots through genipin crosslinkers [37]. The hydrogels were injected into a joint cartilage defect created on mice, followed by stimulation with a near-infrared (NIR) laser for 3 min every other day for 8 weeks. The stiffness of the crosslinked hydrogel and the non-lethal doses of intracellular ROS after stimulation synergistically upregulated cartilage-specific genes (e.g., SOX9, ACAN, and COL2A1) and enhanced GAG secretion. Moreover, the authors proved that increased ROS levels significantly promoted the activation of mTOR signaling, which is an important pathway in chondrogenic differentiation [138].

4.2. Maturation of Electrosensitive Tissues

Beyond its effect on directing differentiation profiles, electrical stimulation has also proved useful for achieving electrically mature phenotypes in differentiated myocardial and neural cells. The post-mitotic nature of terminally differentiated cardiomyocytes and neurons has hampered their use within 3D culture systems that need to be dynamically remodeled for proper functionality. Therefore, the use of phenotypically immature precursors has been the most straightforward and cost-effective solution, as it eliminates the need of acquiring pluripotent stem cells and implementing precisely controlled differentiation regimes every time [139]. Accordingly, several electroconductive dECM-based hydrogels have been envisioned for improving the maturation of these tissues by promoting electrically active phenotypes in biomimetic microenvironments [49]. Tsui and colleagues, for instance, developed an rGO-embedded myocardial dECM hydrogel that showed high hiPSC-derived cardiomyocyte viability after 35 days and increased contractile and electrophysiological function by tuning rGO reduction degree and concentration within the bioink [49]. Alternatively, Roshanbinfar and colleagues developed an electroconductive hydrogel based on pericardial tissue-derived dECM and carbodihydrazide-functionalized multi-walled CNTs (MWCNTs) with embedded hiPSC-derived cardiomyocytes for engineering cardiac tissue [140]. The embedded cells showed enhanced unidirectional orientation, greater contraction amplitude and speed, and improved calcium handling, when compared to the non-electroconductive counterparts. As a result, the presence of MWCNTs improved the beating properties of the constructs, as evidenced by the synchronous contraction of the engineered tissues without signs of arrhythmia.

Similarly, Koppes and colleagues developed a CNT-based electroconductive nanocomposite hydrogel but, instead of relying solely on the electric activity of the embedded cells, they explored the effect of a low-voltage direct current (DC) stimulation on the maturation of nerve tissue constructs [52]. They investigated this by manually casting tissue constructs from a Matrigel and type I collagen hydrogel embedded with carboxyl-functionalized single walled CNTs (SWCNTs) and dorsal root ganglia cells isolated from neonatal rats. The constructs were electrically stimulated for 8 h, incubated for 48 h, and subsequently fixed for studying neurite outgrowth. They found that the sole presence of the SWCNTs led to a 1.6-fold increase in the neural length and a 3.3-fold increase in the total outgrowth of cells, when compared to nanomaterial-free hydrogels. Moreover, relative to nanomaterial-free and non-stimulated tissue constructs, the electrical stimulation resulted in an exceptional 2.1-fold and 7-fold increase in neural length and total neurite outgrowth, respectively.

Interestingly, other studies have explored the effect of electrical stimuli on non-electroconductive hydrogels, and have shown that its coupling with mechanical stimulation compensates for the possible loss in electrical conductivity by reduced percolation. Ronaldson-Bouchard and colleagues developed a protocol for engineering mature human cardiac muscle based on the electromechanical stimulation of hiPSC-derived cardiomyocytes embedded on a fibrin hydrogel obtained after crosslinking of fibrinogen with thrombin [141]. With a custom-made bioreactor, newly formed cardiac tissue constructs were stimulated with a mechanical preload while applying electrical signals with the aid of two carbon rods placed in parallel to the tissue. Maturation was enhanced by slowly increasing the electrical stimulation frequency over the course of 2 weeks, which resulted in what the authors termed "intensity training". This work, and the others mentioned previously, illustrate the importance of external stimuli during maturation stages for achieving superior morphogenesis in electrically active tissues.

4.3. Maturation of Load-Bearing Tissues

Mechanical stimuli have also been widely exploited for inciting mechanotransductional cues that guide functional phenotypes in load-bearing tissues. Typically, dynamic culturing of tissue constructs is employed to induce different types of stress on the materials, which can ultimately direct cell fate. Tensile and compressive forces, as well as fluid-induced shear stress, are among the most common stimuli for inducing such responses [142]. Goldfracht and colleagues explored the effect of dynamic culture conditions on cardiac tissue constructs fabricated with a composite hydrogel made of porcine heart dECM and chitosan [143]. Specifically, ring-shaped constructs were casted on molds, incubated for 24–48 h, and subsequently transferred to a passive stretcher device that exerted a constant tensile stress on them. Their results demonstrated enhanced morphogenesis as evidenced by continuous spontaneous contractions for up to six months. After 30 days under these dynamic conditions, embedded hiPSC-derived cardiomyocytes were arranged anisotropically, along the axis of stretch, and exhibited an organized sarcomeric pattern with distinguishable aligned Z bands. Additionally, the cardiomyocytes showed an upregulation in the expression of genes related to the contractile apparatus. Furthermore, the authors used programmed electrical stimulation to induce and map the development of arrhythmias, which demonstrated that the developed tissue models can be employed for in vitro disease modeling.

Similarly, the biosynthetic activity of primary human articular chondrocytes, embedded in methacryloyl-modified HA and GelMA (GelMA–HAMA) hydrogels, was shown to be significantly improved with the application of uniaxial and biaxial loads within precise mechanical stimulation regimes. As shown by Meinert and colleagues, uniaxial compressive and shear loads, applied independently and biaxially for 1 h after 14 days of 3D culturing, induced the immediate upregulation of hyaline cartilage-specific genes (e.g., ACAN, COL2A1, and PRG4) in an amplitude dependent manner [144]. Moreover, they show that intermittent biaxial loading applied at regular intervals (1 h daily at 1 Hz for 14 more days) promoted the secretion of type II collagen and the generation of a hyaline cartilage-like matrix in comparison to unstimulated constructs, without compromising cell viability. They suggest that these mechanical stimuli generate the necessary mechanotransductional cues for ensuring cartilage development and function. Likewise, Vasquez and colleagues showed that the mechanical loading of osteocytes controls osteoblast bone formation dynamics in a type I collagen 3D co-culture model [145]. In particular, they demonstrate that by applying a loading regime that consisted of cyclic compressions (2.5 N, 10 Hz, 5 min) to models pre-cultured for 7 days, prostaglandin E_2 (PE_2) release increased by about 4-fold, 30 min post-load; they also noted increased type I pro-collagen secretion for up to 5 days post-load. The secretion of these two agents is a clear indicator of bone formation, as PE_2 is a crucial regulator of osteoblast proliferation and differentiation [146]. Additionally, type I pro-collagen synthesis has been correlated with bone collagen synthesis and bone formation rate [147].

5. Stimuli-Responsive dECM-Based Hydrogels for the Delivery of Therapeutics

Hydrogels have been widely implemented for the localized and controlled delivery of therapeutics due to their versatility and high biocompatibility. Several design parameters such as pore size, backbone charge, hydrophilicity, and crosslinking density, can be strategically tuned to modulate the solubility/dispersibility of hydrophobic and hydrophilic molecules within hydrogels [13]. Moreover, their release profiles can be precisely controlled with specific changes in hydrogel structure induced by swelling, dissolution, or degradation, as a consequence of exposure to external stimuli (e.g., light, ultrasound, and electric and magnetic fields) [148]. The strategic design of drug-loaded hydrogels has therefore been directed towards the inclusion of stimuli-responsive molecules (e.g., polymers, proteins and peptides) or nanostructured materials (e.g., iron oxide, gold nanoparticles, carbon nanotubes, graphene oxide, and reduced graphene oxide) that respond to individual or combined stimuli [15]. Recent approaches that have employed external stimuli for controlling drug release profiles in rationally designed dECM-derived hydrogels are discussed below and summarized in Figure 5.

5.1. Light-Triggered Drug Release

As evidenced in previous sections, the incorporation of light-sensitive nanostructured materials, molecules, or polymers into ECM-based hydrogels has been extensively investigated in recent years. However, the use of these hydrogels as vehicles for controlled drug delivery upon exposure to light is still largely unexplored.

As NIR laser irradiation is undoubtedly the most extensively studied light stimulus for controlled drug delivery on ECM-based hydrogels, a broad variety of release mechanisms triggered by this stimulus have been considered in several pharmacological applications. NIR-triggered photodynamic therapy (PDT), a process in which cell death is induced by an increase in ROS generation with the activation of a photosensitizer, is one of the emerging anticancer therapies with the highest potential for clinical translation [149]. This is as, through this approach, it is possible to simultaneously harness the controlled activation of anticancer drugs and their release rates from the hydrogel matrix. For example, Xu and colleagues relied on the elevated levels of ROS during PDT to induce hydrogel degradation and the subsequent release of the chemotherapeutic drug doxorubicin (DOX) [43]. In this work, the researchers developed an injectable and photodegradable HA-based hydrogel loaded with DOX for achieving NIR light-tunable and on-demand drug release for combined PDT and chemotherapy. The developed hydrogel consisted of adipic dihydrazide-modified HA (HA-ADH) conjugated with the photosensitizer protoporphyrin IX (PpIX) (HA-ADH-PpIX). Moreover, a dialdehyde-functionalized thioketal (TK−CHO) crosslinker was incorporated into the hydrogel for introducing ROS-cleavable acylhydrazone bonds between the hydrazide groups of HA-ADH-PpIX and the benzaldehyde groups of TK-CHO. Upon irradiation with NIR light (633 nm), PpIX was activated, and the resulting high local production of ROS facilitated hydrogel degradation by cleaving the acylhydrazone bonds within the hydrogel network. The controlled release of DOX, combined with the induced cell death by the high ROS levels, demonstrated an outstanding anticancer activity in both an in vitro breast cancer model (MCF-7 cells) and an in vivo 4T1 tumor-bearing mouse model [43].

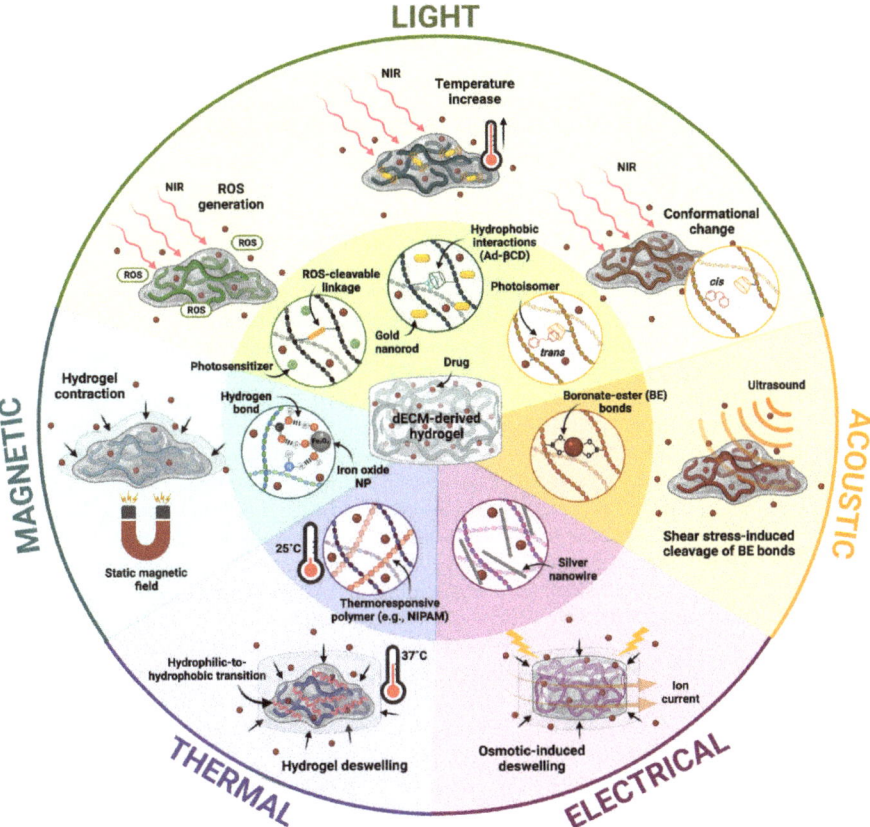

Figure 5. General schematic of drug release mechanisms from dECM-derived hydrogels induced by external stimuli, namely light, temperature, or magnetic, acoustic, and electric fields. Hydrogel destabilization, degradation, or deswelling as a consequence of these stimuli mediates the release profiles of encapsulated molecules or drugs (Created with BioRender.com).

Alternatively, Xing and coworkers potentiated the anticancer effects of PDT with the localized release of NIR light-activated anticancer drugs and a local temperature increase [14]. They developed an injectable nanocomposite hydrogel by incorporating gold nanoparticles (AuNPs) within a collagen hydrogel loaded with the photosensitive drug meso-Tetra (N-methyl-4-pyridyl) porphine tetrachloride (TMPyP), which becomes cytotoxic and produces ROS upon NIR laser irradiation (635 nm). Moreover, the presence of AuNPs in the hydrogel permitted a simultaneous photothermal therapy (PTT), as these nanoparticles exhibit a high photon-to-heat conversion efficiency upon irradiation with NIR light. This led to a remarkable local temperature increase which, in combination with the cytotoxic effect of activated TMPyP, resulted in an exceptional antitumor efficacy in a breast (MCF-7) tumor-xenograft mouse model. In addition, the researchers demonstrated that the hydrogel significantly improved the bioavailability of the drug by protecting it from rapid body clearance, as well as minimized the collateral effects by allowing a localized retention of the treatment within the tumor region [14].

PTT alone has also been employed for controlling cargo release via degradation of hydrogel networks in response to local temperature increase. For instance, Highley and coworkers developed an NIR-responsive nanocomposite hydrogel by mixing gold nanorods with a β-cyclodextrin (βCD)- and adamantane (Ad)-modified HA hydrogel [150]. They exploited the light-absorbing properties of gold to induce local heating upon irradiation and, in turn, destabilize the hydrophobic interactions between βCD and Ad. This resulted in hydrogel degradation and subsequent payload release. Their results showed that NIR irradiation is well-suited to induce plasmonic heating and increase the release of encapsulated molecules between 0.37–500 kDa by more than 2-fold [150]. Alternatively, Sun and colleagues developed an NIR-responsive hydrogel based on benzoxaborole-modified HA (BOB-HA) and a fructose-based glycopolymer, supplemented with perylene diimide zwitterionic polymer (PDS) as photosensitizer, photothermal polymeric nanoparticles, ascorbic acid, and DOX [45]. By irradiating with NIR at 660 nm, PDS and ascorbic acid interacted for the conversion of oxygen into hydrogen peroxide, which was able to cleave dynamic covalent bonds based on benzoxaborole-carbohydrate interactions within the hydrogel network. This controlled degradation led to the site-specific release of the embedded photothermal nanoparticles and the DOX for chemotherapy. Moreover, upon irradiation with 915 nm NIR 3 days after the first irradiation, the photothermal nanoparticles were able to locally increase temperature for PTT. As a result, this combined therapy nearly eradicated tumors in a 4T1 mouse model, while those untreated kept growing.

Lastly, light stimuli have also been shown useful in drug delivery applications by inducing conformational changes in hydrogels. This is exemplified in ECM-based hydrogels with the work reported by Wu and colleagues [151]. Briefly, the authors developed NIR-responsive core-shell hybrid nanogels for fluorescence imaging and combined chemo-photothermal cancer therapy. The hybrid nanogels were assembled by coating Ag-Au bimetallic nanoparticles with a thermo-responsive nonlinear PEG-based hydrogel shell decorated with superficial semi-interpenetrating HA chains. The temperature increase in Au cores upon NIR irradiation, as well as the local temperature increase during incubation, induced a hydrophilic to hydrophobic transition of the PEG chains in the shell, which destabilized its interactions with the loaded anticancer drug (temozolomide) and allowed its release. These hybrid nanogels exhibited high anticancer activity against mouse melanoma B16F10 cells as it allowed a multimodal therapy that combined localized chemotherapy with NIR-triggered photothermal treatment [151]. Likewise, Rosales and colleagues designed a drug delivery system based on a UV light-responsive azobenzene- and βCD-modified HA hydrogel, crosslinked via azobenzene-βCD hydrophobic interactions. Azobenzene is a photoisomer that transitions from its trans to cis isomeric configuration upon exposure to light in the 350–550 nm wavelength range [152]. This transition induces a conformational change within the hydrogel that reduces azobenzene's affinity for the hydrophobic cavity of βCD and, in turn, the reduced crosslinking density within the hydrogel allows the release of the encapsulated molecules. However, when the light stimulus is removed, the azobenzene moieties return to its *trans* state and the hydrogel recovers its original conformation. This reversible behavior grants a high tunability of its mechanical properties and makes it suitable for different applications, such as drug delivery or mechanobiology studies, in which temporal regulation of material properties is key for robust and more compelling studies [153].

5.2. Magnetic-Triggered Release

Magnetic forces, induced by static or dynamic magnetic fields, have been used to trigger the controlled release of molecules from hydrogel depots. This release can be mediated by reversible or non-reversible structural changes that disrupt the hydrogel network itself or the bonds that entrap the molecules within the hydrogel. Bettini and colleagues, for example, developed an ION-embedded collagen hydrogel scaffold with drug-retaining and paramagnetic properties [57]. IONs were covalently conjugated to the collagen network by implementing a dehydrothermal (DHT) treatment at 120 °C for 48 h. As controls, the authors synthesized collagen hydrogels without IONs and crosslinked either by a chemical treatment with formaldehyde and DHT or solely by DHT treatment. Surprisingly, ION-conjugated hydrogels exhibited a higher crosslinking degree, as demonstrated by smaller pore size and a lower swelling degree. Moreover, only the ION-conjugated and formaldehyde-crosslinked hydrogels were capable of retaining loaded cargo, i.e., fluorescein molecules. After stimulating the loaded ION-conjugated hydrogels with a permanent magnet, deformation of the scaffold was induced, thus facilitating the flow of water and dissolved fluorescein out of the gel by means of pore collapse. By applying consecutive magnetic stimulation cycles, they demonstrated the controlled release of fluorescein.

Similarly, Dai and colleagues proposed dopamine-conjugated hyaluronan (HA-DOPA) hydrogel embedded with IONs as a drug delivery platform for anticancer therapies [56]. In this hydrogel, crosslinking was facilitated by the interaction between catechol groups of dopamine and iron(III) of IONs, and DOX was then loaded and retained by means of electrostatic interactions with the positively charged HA-DOPA. Moreover, besides serving as crosslinking agents, IONs also played an important role in the on-demand release of DOX. The authors demonstrated that the release rate of the entrapped drug could be accelerated with an external alternating magnetic field (AMF), as it generated local hyperthermia and subsequent hydrogel destabilization. The authors also showed that the release rate could be slowed down when removing the applied magnetic stimulus. In vivo experiments were performed on a A375-xenografted tumor mice model, where the combined chemotherapy (DOX) and hyperthermia (IONs) therapy resulted in significantly smaller tumor volume after 18 days when compared to the control group treated only with PBS. Moreover, single modality treatment (either chemotherapy or hyperthermia) failed to show an equivalent efficacy of the combined therapy at the same dose level of the chemotherapeutic agent.

5.3. Ultrasound-Triggered Release

Mechanical stress generated by ultrasound has also been employed to disrupt the bonds between drugs and hydrogels for enabling drug release. Sun and coworkers designed a dual-crosslinked hydrogel based on methacrylic HA and a four-armed PEG acrylate (4arm-PEG-Aclt) for the ultrasound-induced delivery of tannic acid as a drug model [154]. The crosslinked hydrogel network was constructed through free radical polymerization of 4arm-PEG-Aclt with HA previously modified with 4-(aminomethyl) phenylboronic acid and methacrylic anhydride (PhB-mHA). Then, a dynamic crosslinking was achieved by forming boronate ester bonds between phenylboronic acid (conjugated along the polymeric network) and tannic acid. This dual-crosslinking endowed the hydrogel with remarkable mechanical properties, thereby making it resistant to deformation by external mechanical forces (i.e., compression). Moreover, boronate ester bonds dynamically respond to ultrasound, thereby allowing the non-invasive and on-demand release of tannic acid. This can be attributed to the dynamic shear force generated by this stimulus, which is enough to disrupt such bonds [154].

Alternatively, ultrasound-induced drug release has been achieved by destabilizing encapsulating agents with acoustic droplet vaporization (ADV) mechanisms, which mediate their transition from liquid to gas phase in response to pressure induced by acoustic stimuli. In this regard, Dong and colleagues developed an acoustically responsive hydrogel consisting of sonosensitive micron-sized emulsions embedded within a fibrin matrix which, under 2.5 MHz sound waves, released encapsulated basic fibroblast growth factor (bFGF) [58]. The emulsion entails a water-in-perfluorocarbon (PFC)-in-water ($W_1/PFC/W_2$) double emulsion, with bFGF in the W_1 phase, and its release occurs when PFC within each droplet phase transitions from liquid to gas, thereby disrupting the droplet morphology. They showed that this hydrogel-emulsion system allowed a non-invasive and on-demand release of bFGF, as hydrophobic PFC acted as a diffusion barrier for bFGF until its removal by ADV. To further demonstrate the applicability of this hydrogel-emulsion system, they added an external layer of a HUVEC-laden fibrin hydrogel to test whether angiogenesis is promoted by bFGF release. In this regard, HUVEC angiogenic sprouting was significantly enhanced after ADV-mediated bFGF release, specifically when increasing acoustic pressure and emulsion volume fraction. A recent work by the same group demonstrated that release profiles of two or more payloads could be specifically modulated by loading them within emulsion droplets containing PFCs with different ADV thresholds (i.e., with varying carbon chain lengths in the PFC) [59]. Accordingly, by tuning the frequency of ultrasound standing wave fields (SWFs), they could precisely control the release of two different factors relevant in angiogenesis, namely bFGF and platelet-derived growth factor BB (PDGF-BB), at specific time points during maturation.

5.4. Electric-Triggered Release

Although electrical stimuli have been largely unexplored for controlling drug release in dECM-derived hydrogels, a recent work by Ha and colleagues harnessed the ionic currents induced by this type of stimulus for modulating release profiles of model drugs within collagen-GelMA hydrogels embedded with silver nanowires (AgNWs) [53]. They showed that when applying a current with electric potential of 1V, a consistent weight loss was observed as a function of exposure time in the conductive hydrogel, while collagen and GelMA hydrogels remained unchanged. They attributed this weight loss to the enhanced electrical conductivity of AgNW, which facilitated the generation of an outward osmotic gradient as a consequence of the flow of ions crossing the hydrogel. Moreover, this weight loss was highly correlated to the release of encapsulated fluorescein isothiocyanate (FITC)-dextran particles, which suggests that electrically induced structural changes and promote outward flows of fluids that facilitate the release of loaded molecules [53].

5.5. Temperature-Triggered Release

Temperature-responsive materials have also been incorporated within dECM-hydrogels for directing the release of loaded cargoes as most delivery applications must be performed at physiological temperature (37 °C). As an example, Ravichandran and colleagues described the synthesis of a temperature-responsive injectable hydrogel for the controlled release of drugs and proteins, based on ColMA building blocks and the thermo-responsive polymer N-isopropyl acrylamide (NIPAm) [155]. In this case, drug release is attributed to temperature-induced hydrogel deswelling, as well as changes in the drug/molecule affinity to the matrix, as NIPAm is hydrophilic at 25 °C but becomes hydrophobic at 37 °C. In consequence, as temperature approaches physiological conditions, the hydrogel shrinks due to conformational changes from elongated to coiled morphologies, which induces an outward flow of water and alters drug-hydrogel affinity. Moreover, hydrogel collapse is prevented due to the pH-responsiveness of ColMA. As carboxyl groups of collagens are converted into carboxylate anions ($-COO^-$) at physiological pH (~7.4), the electrostatic repulsion between them grants sufficient hydrogel porosity for facilitating drug release.

The coupled effect of NIPAm solubility change and ColMA swelling favored significantly the release of the model drugs bovine serum albumin (BSA) and vitamin E, yielding over 80% release after 5 days [155].

6. Perspectives on Clinical Translation

Despite the considerable advances in the engineering of functional ECM-based hydrogels, their translation into clinical scenarios is still in its infancy. In particular, few preclinical models have been developed for evaluating their performance beyond in vitro setups. ECM-based hydrogel constructs destined for replacing injured tissue or organ defects have been limited to in vitro characterizations and monitoring of their maturation dynamics, mainly as true biomimicry has not been fully accomplished yet. Accordingly, only a handful of these hydrogels have been implanted in animal models, and these usually comprise simple cellular patterns that can be easily integrated with host tissues [105]. This is the case of the HA-based constructs developed by Kang and colleagues, in which acoustically patterned cells unidirectionally aligned in cylindrical tubes were able to integrate with mouse vasculature and form perfusable and interconnected networks [105]. Similarly, of the thirteen works mentioned in Section 5 where ECM-based hydrogels have been implemented for the localized delivery of therapeutic agents, only four have evaluated drug release dynamics and effects on animal models: [14,43,45,56]. This, in turn, could have been the result of the low outreach and reception that ECM-based hydrogels have had on drug delivery fields, where highly accessible and easier to use synthetic or natural biodegradable materials are commonly employed [156].

Moreover, according to the ClinicalTrials.gov database by the U.S. National Library of Medicine, there has been only one clinical trial where tissue-derived ECM hydrogels have been studied. This trial was conducted in the U.S., and its main aim was to evaluate the safety and feasibility of a porcine myocardial tissue-derived ECM hydrogel delivered trans-endocardially to human subjects following myocardial infarction [157]. Although the study was not intended to evaluate efficacy, physicians reported myocardial function improvement for all intervened patients ($n = 15$). This clinical trial demonstrated the significant advances that the tissue engineering industry and academic research groups have accomplished towards overcoming translational challenges faced by tissue-derived ECM hydrogels, particularly in terms of shelf-life and scalability [19]. It also further demonstrated the biomimetic potential of ECM-based hydrogels that could be further exploited for facilitating the translation of TE and RM into clinical scenarios.

The translation of ECM-based hydrogels into clinical settings is especially appealing, as they hold much promise for the emerging field of precision medicine. For instance, Noor and colleagues demonstrated the feasibility of personalized regenerative medicine by developing cardiac patches using patient-derived ECM hydrogels and cells [158]. From a biopsy of omental tissue, the authors isolated and reprogramed the patient's own cells into stem cells, while the ECM was processed into a personalized hydrogel. The reprogrammed stem cells were further differentiated into cardiomyocytes and endothelial cells, embedded into either the ECM hydrogel or a sacrificial material, and bioprinted into thick and perfusable cardiac patches. Although they only addressed the formulation and bioprinting of these constructs, the diverse strategies described in this review for improving tissue organization and morphogenesis could be exploited for improving construct maturation, taking the field one step closer towards the biofabrication of functional personalized tissues suitable for transplantation. Alternatively, the isolation and biointegration of diseased human cells into ECM-derived hydrogels could potentiate the development of in vitro models for studying the involved physiological and pathophysiological hallmarks. Goldfracht and colleagues, for example, showed that in vitro heart tissue models with patient-derived hiPSC-cardiomyocytes embedded in ECM-based hydrogels (discussed in detail in Section 4.3) could recapitulate the abnormal phenotypes of arrhythmogenic disorders by using diseased cells from patients [143]. They demonstrated that this disease model could be used even further to create a personalized model to evaluate the efficacy of therapeutics against patient-specific cardiac pathologies. The implementation of biofabrication technologies appears critical for developing robust in vitro models of disease that could be eventually translated into the clinical practice for the high-throughput screening of therapeutics. We believe that this is a route that enables a foreseeable future of favorable outcomes for diseased patients.

7. Concluding Remarks

ECMs have been devised as the next-generation materials for alleviating the limitations regarding functionality and cell development faced with advanced TE technologies, such as 3D bioprinting and RM therapies. The utilization of these materials for the formulation of hydrogels is considered one of the most promising approaches for achieving superior tissue functionality due to their inherent biochemical composition that can guide cell thrive. However, their heterogeneous and biomimetic biochemical composition alone has been demonstrated insufficient for guaranteeing morphogenesis of engineered tissues in vitro or in vivo, nor to be appropriate for their use in advanced biomanufacturing technologies that require particular mechanical properties.

As shown in this review, external stimuli have been pivotal for facilitating the recapitulation of biologically relevant micro- and macro-environments in bioengineered tissues. For instance, light stimuli between the UV and visible ranges have become the gold standard for facilitating crosslinking schemes that ultimately improve the mechanical properties of ECM-based hydrogels. Directed electric, magnetic and pressure fields have been crucial for sculpting microarchitectural characteristics that drive cellular organization in tissue-specific patterns. Their effects on electrochemical gradients have been fundamental, as well, for directing stem cell differentiation profiles and dictating morphogenetic cues that mediate tissue maturation. Beyond their contribution in the biofabrication of biomimetic constructs, external stimuli have proven to be critical for controlling the delivery of therapeutics from hydrogel depots. This comprises an advanced strategy for accelerating tissue regeneration, both in vitro and in vivo, with specific release profiles of embedded biomolecules or drugs.

However, this review demonstrates that among the limited collection of papers that developed and evaluated stimuli-responsive hydrogels based on ECM components, only a handful really implemented tissue-derived ECMs as their main hydrogel component. Collagen and HA hydrogels are undoubtedly the most commonly used for implementing the referred schemes, probably due to their accessibility and ease of use. Nevertheless, we believe that the synergistic action of the native ECM components in tissue-derived ECM hydrogels, coupled with the mentioned stimulation regimes, could significantly potentiate these results, leading us one step closer towards functional and biomimetic constructs for TE and RM.

Author Contributions: L.R.-G. and J.A.S. wrote the manuscript with help from D.N.C.-V. and J.C., C.M.-C. and J.C.C. revised the manuscript. All authors have read and agreed to the published version of the manuscript.

Funding: This research was funded by the Department of Biomedical Engineering, Vice-presidency of Research and creation through Fondo de Apoyo a Profesores Asistentes (FAPA) grant to Carolina Muñoz-Camargo at the Universidad de los Andes and by the Colciencias Grant Contract #689-2018" and the APC was funded by the Department of Biomedical Engineering and Vice-presidency of Research and creation through Fondo de Apoyo a Profesores Asistentes (FAPA) grant to Carolina Muñoz-Camargo at the Universidad de los Andes.

Institutional Review Board Statement: Not applicable.

Informed Consent Statement: Not applicable.

Conflicts of Interest: The authors declare no conflict of interest.

References

1. Li, Z.; Xie, M.-B.; Li, Y.; Ma, Y.; Li, J.-S.; Dai, F.-Y. Recent progress in tissue engineering and regenerative medicine. *J. Biomater. Tissue Eng.* **2016**, *6*, 755–766. [CrossRef]
2. Lee, J.-H.; Kim, H.-W. Emerging properties of hydrogels in tissue engineering. *J. Tissue Eng.* **2018**, *9*, 2041731418768285. [CrossRef]
3. Zhang, Y.S.; Khademhosseini, A. Advances in engineering hydrogels. *Science* **2017**, *356*, 6337. [CrossRef] [PubMed]
4. Moroni, L.; Boland, T.; Burdick, J.A.; De Maria, C.; Derby, B.; Forgacs, G.; Grol, J.; Li, Q.; Malda, J.; Mironov, V.A.; et al. Biofabrication: A guide to technology and terminology. *Trends Biotechnol.* **2018**, *36*, 384–402. [CrossRef]
5. Ozbolat, I.T.; Hospodiuk, M. Current advances and future perspectives in extrusion-based bioprinting. *Biomaterials* **2016**, *76*, 321–343. [CrossRef] [PubMed]
6. Kyle, S.; Jessop, Z.M.; Al-Sabah, A.; Whitaker, I.S. Printability of candidate biomaterials for extrusion based 3d printing: State-of-the-Art. *Adv. Healthc. Mater.* **2017**, *6*, 1–16. [CrossRef] [PubMed]
7. He, Y.; Yang, F.; Zhao, H.; Gao, Q.; Xia, B.; Fu, J. Research on the printability of hydrogels in 3D bioprinting. *Sci. Rep.* **2016**, *6*, 29977. [CrossRef]
8. Schwab, A.; Levato, R.; D'Este, M.; Piluso, S.; Eglin, D.; Malda, J. Printability and shape fidelity of bioinks in 3d bioprinting. *Chem. Rev.* **2020**, *120*, 11028–11055. [CrossRef]
9. Theus, A.S.; Ning, L.; Hwang, B.; Gil, C.; Chen, S.; Wombwell, A.; Mehta, R.; Srepooshan, V. Bioprintability: Physiomechanical and biological requirements of materials for 3d bioprinting processes. *Polymers* **2020**, *12*, 2262. [CrossRef]
10. Ouyang, L.; Yao, R.; Zhao, Y.; Sun, W. Effect of bioink properties on printability and cell viability for 3D bioplotting of embryonic stem cells. *Biofabrication* **2016**, *8*, 035020. [CrossRef]
11. Hölzl, K.; Lin, S.; Tytgat, L.; van Vlierberghe, S.; Gu, L.; Ovsianikov, A. Bioink properties before, during and after 3D bioprinting. *Biofabrication* **2016**, *8*, 032002. [CrossRef] [PubMed]
12. Li, J.; Mooney, D.J. Designing hydrogels for controlled drug delivery. *Nat. Rev. Mater.* **2016**, *1*, 16071. [CrossRef] [PubMed]
13. Dimatteo, R.; Darling, N.J.; Segura, T. In situ forming injectable hydrogels for drug delivery and wound repair. *Adv. Drug Deliv. Rev.* **2018**, *127*, 167–184. [CrossRef]
14. Xing, R.; Liu, K.; Jiao, T.; Zhang, N.; Ma, K.; Zhang, R.; Zou, Q.; Ma, G.; Yan, X. An injectable self-assembling collagen-gold hybrid hydrogel for combinatorial antitumor photothermal/photodynamic therapy. *Adv. Mater.* **2016**, *28*, 3669–3676. [CrossRef] [PubMed]
15. Merino, S.; Martín, C.; Kostarelos, K.; Prato, M.; Vázquez, E. Nanocomposite Hydrogels: 3D Polymer–Nanoparticle Synergies for On-Demand Drug Delivery. *ACS Nano* **2015**, *9*, 4686–4697. [CrossRef]
16. Hamidi, M.; Azadi, A.; Rafiei, P. Hydrogel nanoparticles in drug delivery. *Adv. Drug Deliv. Rev.* **2008**, *60*, 1638–1649. [CrossRef]
17. Saldin, L.T.; Cramer, M.C.; Velankar, S.S.; White, L.J.; Badylak, S.F. Extracellular matrix hydrogels from decellularized tissues: Structure and function. *Acta Biomater.* **2017**, *49*, 1–15. [CrossRef]
18. Yu, C.; Ma, X.; Zhu, W.; Wang, P.; Miller, K.L.; Stupin, J.; Koroleva-Maharajh, A.; Hairabedian, A.; Chen, S. Scanningless and continuous 3D bioprinting of human tissues with decellularized extracellular matrix. *Biomaterials* **2018**, *194*, 1–13. [CrossRef]
19. Spang, M.T.; Christman, K.L. Extracellular matriz hydrogel therapies: In vivo applications and development. *Acta Biomater.* **2018**, *1*, 1–14. [CrossRef]
20. Frantz, C.; Stewart, K.M.; Weaver, V.M. The extracellular matrix at a glance. *J. Cell Sci.* **2010**, *123*, 4195–4200. [CrossRef]
21. Wess, T.J. Collagen fibrilar structures and hierarchies. In *Collagen*; Fratzl, P., Ed.; Springer: Boston, MA, USA, 2008; pp. 49–80.
22. Ricard-Blum, S. The Collagen Family. *Cold Spring Harb. Perspect. Biol.* **2011**, *3*, 1–19. [CrossRef] [PubMed]
23. Hynes, R.O. The extracellular matrix: Not just pretty fibrils. *Science* **2009**, *326*, 1216–1219. [CrossRef]
24. Nusgens, B.-V. Acide hyaluronique et matrice extracelulaire: Une molécule primitive? *Ann. Dermatol. Venereol.* **2010**, *137*, S3–S8. [CrossRef]
25. Schaefer, L.; Schaefer, R.M. Proteoglycans: From structural compounds to signaling molecules. *Cell Tissue Res.* **2010**, *339*, 237–246. [CrossRef] [PubMed]
26. Giobbe, G.G.; Crowley, C.; Luni, C.; Campinoti, S.; Khedr, M.; Kretzschmar, K.; de Santis, M.M.; Zambaiti, E.; Michielin, F.; Meran, L.; et al. Extracellular matrix hydrogel derived from decellularized tissues enables endodermal organoid culture. *Nat. Commun.* **2019**, *10*, 5658. [CrossRef]
27. Yamaoka, T. Preparation Methods for Tissue/Organ-derived dECMs–Effects on Cell Removal and ECM Changes. In *Decellularized Extracellular Matrix: Characterization, Fabrication and Applications*; Hoshiba, T., Yamaoka, T., Eds.; Royal Society of Chemistry: London, UK, 2019; pp. 15–28.
28. Fernández-Pérez, J.; Ahearne, M. The impact of decellularization methods on extracellular matrix derived hydrogels. *Sci. Rep.* **2019**, *9*, 14933. [CrossRef]
29. Dzobo, K.; Motaung, K.S.C.M.; Adesida, A. Recent Trends in Decellularized Extracellular Matrix Bioinks for 3D Printing: An Updated Review. *Int. J. Mol. Sci.* **2019**, *20*, 4628. [CrossRef]
30. Annabi, N.; Nichol, J.W.; Zhong, X.; Ji, C.; Koshy, S.; Khademhosseini, A.; Dehghani, F. Controlling the porosity and microarchitecture of hydrogels for tissue engineering. *Tissue Eng. Part B Rev.* **2010**, *16*, 371–383. [CrossRef]
31. Aubin, H.; Nichol, J.W.; Hutson, C.B.; Bae, H.; Sieminski, A.L.; Cropek, D.M.; Akhyari, P.; Khademhosseini, A. Directed 3D cell alignment and elongation in microengineered hydrogels. *Biomaterials* **2010**, *31*, 6941–6951. [CrossRef] [PubMed]

32. Li, Y.-C.; Zhang, Y.S.; Akpek, A.; Shin, S.R.; Khademhosseini, A. 4D bioprinting: The next-generation technology for biofabrication enabled by stimuli-responsive materials. *Biofabrication* **2016**, *9*, 012001. [CrossRef] [PubMed]
33. Kasiński, A.; Zielińska-Pisklak, M.; Oledzka, E.; Sobczak, M. Smart hydrogels–synthetic stimuli-responsive antitumor drug release systems. *Int. J. Nanomed.* **2020**, *15*, 4541–4572. [CrossRef]
34. Skardal, A.; Devarasetty, M.; Kang, H.-W.; Mead, I.; Bishop, C.; Shupe, T.; Lee, S.L.; Jackson, J.; Yoo, J.; Soker, S. A hydrogel bioink toolkit for mimicking native tissue biochemical and mechanical properties in bioprinted tissue constructs. *Acta Biomater.* **2015**, *25*, 24–34. [CrossRef]
35. Bejleri, D.; Streeter, B.W.; Nachlas, A.L.; Broen, M.E.; Gaetani, R.; Christman, K.L.; Davis, M.E. A bioprinted cardiac patch composed of cardiac-specific extracellular matrix and progenitor cells for heart repair. *Adv. Healthc. Mater.* **2018**, *7*, 1800672. [CrossRef] [PubMed]
36. Lee, H.J.; Fernandes-Cunha, G.M.; Myung, D. In situ-forming hyaluronic acid hydrogel through visible light-induced thiol-ene reaction. *React. Funct. Polym.* **2018**, *131*, 29–35. [CrossRef] [PubMed]
37. Lu, Z.; Liu, S.; Le, Y.; Qin, Z.; He, M.; Xu, F.; Zhu, Y.; Zhao, J.; Mao, C.; Zheng, L. An injectable collagen-genipin-carbon dot hydrogel combined with photodynamic therapy to enhance chondrogenesis. *Biomaterials* **2019**, *218*, 119190. [CrossRef] [PubMed]
38. Ali, M.; Pr, A.K.; Yoo, J.J.; Zahran, F.; Atala, A.; Lee, S.J. A photo-crosslinkable kidney ecm-derived bioink accelerates renal tissue formation. *Adv. Healthc. Mater.* **2019**, *8*, 1800992. [CrossRef] [PubMed]
39. Parthiban, S.P.; Athirasala, A.; Tahayeri, A.; Abdelmoniem, R.; George, A.; Bertassoni, L.E. BoneMA—synthesis and characterization of a methacrylated bone-derived hydrogel for bioprinting of in vitro vascularized tissue constructs. *Biofabrication* **2021**, *13*, 035031. [CrossRef]
40. Rueda-Gensini, L.; Serna, J.A.; Cifuentes, J.; Cruz, J.C.; Muñoz-Camargo, C. Graphene oxide-embedded extracellular matrix-derived hydrogel as a multiresponsive platform for 3d bioprinting applications. *Int. J. Bioprint.* **2021**, *7*. [CrossRef]
41. Kim, H.; Kang, B.; Cui, X.; Lee, S.-H.; Lee, K.; Cho, D.-W.; Hwang, W.; Woodfield, T.B.F.; Lim, K.S.; Jang, J. Light-activated decellularized extracellular matrix-based bioinks for volumetric tissue analogs at the centimeter scale. *Adv. Funct. Mater.* **2021**, *31*, 2011252. [CrossRef]
42. Keating, M.; Lim, M.; Hu, Q.; Botvinick, E. Selective stiffening of fibrin hydrogels with micron resolution via photocrosslinking. *Acta Biomater.* **2019**, *87*, 88–96. [CrossRef]
43. Xu, X.; Zeng, Z.; Huang, Z.; Sun, Y.; Huang, Y.; Chen, J.; Ye, J.; Yang, H.; Yang, C.; Zhao, C. Near-infrared light-triggered degradable hyaluronic acid hydrogel for on-demand drug release and combined chemo-photodynamic therapy. *Carbohydr. Polym.* **2020**, *229*, 115394. [CrossRef]
44. Cho, S.-H.; Kim, A.; Shin, W.; Heo, M.B.; Noh, H.J.; Hong, K.S.; Cho, J.-H.; Lim, Y.T. Photothermal-modulated drug delivery and magnetic relaxation based on collagen/poly(γ-glutamic acid) hydrogel. *Int. J. Nanomed.* **2017**, *12*, 2607–2620. [CrossRef]
45. Sun, P.; Huang, T.; Wang, X.; Wang, G.; Liu, Z.; Chen, G.; Fan, Q. Dynamic-Covalent Hydrogel with NIR-Triggered Drug Delivery for Localized Chemo-Photothermal Combination Therapy. *Biomacromolecules* **2020**, *21*, 556–565. [CrossRef] [PubMed]
46. Shin, S.R.; Zihlmann, C.; Akbari, M.; Assawes, P.; Cheung, L.; Zhang, K.; Manoharan, V.; Zhang, Y.S.; Yüksekkaya, M.; Wan, K.-T. Reduced Graphene Oxide-GelMA Hybrid Hydrogels as Scaffolds for Cardiac Tissue Engineering. *Small* **2016**, *12*, 3677–3689. [CrossRef]
47. Ahadian, S.; Yamada, S.; Ramón-Azcón, J.; Estili, M.; Liang, X.; Nakajima, K.; Shiku, H.; Khademhosseini, A.; Matsue, T. Hybrid hydrogel-aligned carbon nanotube scaffolds to enhance cardiac differentiation of embryoid bodies. *Acta Biomater.* **2016**, *31*, 134–143. [CrossRef] [PubMed]
48. Li, X.-P.; Qu, K.-Y.; Zhou, B.; Zhang, F.; Wang, Y.-Y.; Abodunrin, O.D.; Zhu, Z.; Huang, N.-P. Electrical stimulation of neonatal rat cardiomyocytes using conductive polydopamine-reduced graphene oxide-hybrid hydrogels for constructing cardiac microtissues. *Colloids Surf. B Biointerfaces* **2021**, *205*, 111844. [CrossRef]
49. Tsui, J.H.; Leonard, A.; Camp, N.D.; Long, J.T.; Nawas, Z.Y.; Chavanachat, R.; Smith, A.S.T.; Choi, J.S.; Dong, Z.; Ahn, E.H.; et al. Tunable electroconductive decellularized extracellular matrix hydrogels for engineering human cardiac microphysiological systems. *Biomaterials* **2021**, *272*, 120764. [CrossRef] [PubMed]
50. Xu, C.; Xu, Y.; Yang, M.; Chang, Y.; Nie, A.; Liu, Z.; Wang, J.; Luo, Z. Black-phosphorus-incorporated hydrogel as a conductive and biodegradable platform for enhancement of the neural differentiation of mesenchymal stem cells. *Adv. Funct. Mater.* **2020**, *30*, 2000177. [CrossRef]
51. Kim, W.; Jang, C.H.; Kim, G.H. A myoblast-laden collagen bioink with fully aligned au nanowires for muscle-tissue regeneration. *Nano Lett.* **2019**, *19*, 8612–8620. [CrossRef] [PubMed]
52. Koppes, A.N.; Keating, K.W.; McGregor, A.L.; Koppes, R.A.; Kearns, K.R.; Ziemba, A.M.; McKay, C.A.; Zuidema, J.M.; Rivet, C.J.; Gilbert, R.J.; et al. Robust neurite extension following exogenous electrical stimulation within single walled carbon nanotube-composite hydrogels. *Acta Biomater.* **2016**, *39*, 34–43. [CrossRef] [PubMed]
53. Ha, J.H.; Lim, J.H.; Kim, J.W.; Cho, H.-Y.; Jo, S.G.; Lee, S.H.; Eom, J.Y.; Lee, J.M.; Chung, B.G. Conductive gelma–collagen–agnw blended hydrogel for smart actuator. *Polymers* **2021**, *13*, 1217. [CrossRef] [PubMed]
54. Betsch, M.; Cristian, C.; Lin, Y.-Y.; Blaeser, A.; Schöneberg, J.; Vogt, M.; Buhl, E.M.; Fischer, H.; Duarte Campos, D.F. Incorporating 4d into bioprinting: Real-time magnetically directed collagen fiber alignment for generating complex multilayered tissues. *Adv. Healthc. Mater.* **2018**, *7*, 1800894. [CrossRef] [PubMed]

55. Rose, J.C.; Cámara-Torres, M.; Rahimi, K.; Köhler, J.; Möller, M.; de Laporte, L. Nerve Cells Decide to Orient inside an injectable hydrogel with minimal structural guidance. *Nano Lett.* **2017**, *17*, 3782–3791. [CrossRef]
56. Dai, G.; Sun, L.; Xu, J.; Zhao, G.; Tan, Z.; Wang, C.; Sun, X.; Xu, K.; Zhong, W. Catechol–metal coordination-mediated nanocomposite hydrogels for on-demand drug delivery and efficacious combination therapy. *Acta Biomater.* **2021**, *129*, 84–95. [CrossRef]
57. Bettini, S.; Bonfrate, V.; Syrgiannis, Z.; Sannino, A.; Salvatore, L.; Madaghiele, M.; Valli, L.; Giancane, G. Biocompatible collagen paramagnetic scaffold for controlled drug release. *Biomacromolecules* **2015**, *16*, 2599–2608. [CrossRef]
58. Dong, X.; Lu, X.; Kingston, K.; Brewer, E.; Juliar, B.A.; Kripfgans, O.D.; Fowlkes, J.B.; Franceschi, R.T.; Putnam, A.J.; Liu, Z.; et al. Controlled delivery of basic fibroblast growth factor (bFGF) using acoustic droplet vaporization stimulates endothelial network formation. *Acta Biomater.* **2019**, *97*, 409–419. [CrossRef]
59. Aliabouzar, M.; Jivani, A.; Lu, X.; Kripfgans, O.D.; Fowlkes, J.B.; Fabiilli, M.L. Standing wave-assisted acoustic droplet vaporization for single and dual payload release in acoustically responsive scaffolds. *Ultrason. Sonochem.* **2020**, *66*, 105109. [CrossRef]
60. Chameettachal, S.; Sasikumar, S.; Sethi, S.; Sriya, Y.; Pati, F. Tissue/organ-derived bioink formulation for 3D bioprinting. *J. 3D Print. Med.* **2019**, *3*, 39–54. [CrossRef]
61. Unagolla, J.M.; Jayasuriya, A.C. Hydrogel-based 3D bioprinting: A comprehensive review on cell-laden hydrogels, bioink formulations, and future perspectives. *Appl. Mater. Today* **2020**, *18*, 100479. [CrossRef] [PubMed]
62. Nam, S.Y.; Park, S.-H. ECM Based bioink for tissue mimetic 3d bioprinting. In *Biomimetic Medical Materials*; Springer: Berlin/Heidelberg, Germany, 2018; pp. 335–353.
63. Kabirian, F.; Mozafari, M. Decellularized ECM-derived bioinks: Prospects for the future. *Methods* **2020**, *171*, 108–118. [CrossRef] [PubMed]
64. Yang, C. Enhanced physicochemical properties of collagen by using EDC/NHS-crosslinking. *Bull. Mater. Sci.* **2012**, *35*, 913–918. [CrossRef]
65. Chimene, D.; Lennox, K.K.; Kaunas, R.R.; Gaharwar, A.K. Advanced Bioinks for 3D Printing: A Materials Science Perspective. *Ann. Biomed. Eng.* **2016**, *44*, 2090–2102. [CrossRef]
66. Jang, J.; Kim, T.G.; Kim, B.S.; Kim, S.W.; Kwon, S.M.; Cho, D.W. Tailoring mechanical properties of decellularized extracellular matrix bioink by vitamin B2-induced photo-crosslinking. *Acta Biomater.* **2016**, *33*, 88–95. [CrossRef] [PubMed]
67. Hospodiuk, M.; Dey, M.; Sosnoski, D.; Ozbolat, I.T. The bioink: A comprehensive review on bioprintable materials. *Biotechnol. Adv.* **2017**, *35*, 217–239. [CrossRef]
68. Ahn, G.; Min, K.-H.; Kim, C.; Lee, J.-S.; Kang, D.; Won, J.-Y.; Cho, D.-W.; Kim, J.-Y.; Jin, S.; Yun, W.-S.; et al. Precise stacking of decellularized extracellular matrix based 3D cell-laden constructs by a 3D cell printing system equipped with heating modules. *Sci. Rep.* **2017**, *7*, 8624. [CrossRef] [PubMed]
69. Kuss, M.A.; Harms, R.; Wu, S.; Wang, Y.; Untrauer, J.B.; Carlson, M.A.; Duan, B. Short-term hypoxic preconditioning promotes prevascularization in 3D bioprinted bone constructs with stromal vascular fraction derived cells. *RSC Adv.* **2017**, *7*, 29312–29320. [CrossRef]
70. Pati, F.; Jang, J.; Ha, D.-H.; Won Kim, S.; Rhie, J.-W.; Shim, J.-H.; Kim, D.-H.; Cho, D.-W. Printing three-dimensional tissue analogues with decellularized extracellular matrix bioink. *Nat. Commun.* **2014**, *5*, 1–11. [CrossRef]
71. Lee, H.; Han, W.; Kim, H.; Ha, D.-H.; Jang, J.; Kim, B.S.; Cho, D.-W. Development of Liver Decellularized Extracellular Matrix Bioink for Three-Dimensional Cell Printing-Based Liver Tissue Engineering. *Biomacromolecules* **2017**, *18*, 1229–1237. [CrossRef]
72. Pereira, R.F.; Bártolo, P.J. 3D bioprinting of photocrosslinkable hydrogel constructs. *J. Appl. Polym. Sci.* **2015**, *132*. [CrossRef]
73. Lee, M.; Rizzo, R.; Surman, F.; Zenobi-Wong, M. Guiding Lights: Tissue Bioprinting Using Photoactivated Materials. *Chem. Rev.* **2020**, *120*, 10950–11027. [CrossRef]
74. Lim, K.S.; Galarraga, J.H.; Cui, X.; Lindberg, G.C.J.; Burdick, J.A.; Woodfield, T.B.F. Fundamentals and Applications of Photo-Cross-Linking in Bioprinting. *Chem. Rev.* **2020**, *120*, 10662–10694. [CrossRef]
75. Serna, J.A.; Florez, S.L.; Talero, V.A.; Briceño, J.C.; Muñoz-Camargo, C.; Cruz, J.C. Formulation and characterization of a SIS-Based photocrosslinkable bioink. *Polymers* **2019**, *11*, 569. [CrossRef]
76. Kato, Y.; Uchida, K.; Kawakishi, S. Aggregation of collagen exposed to UVA in the presence of riboflavin: A plausible role of tyrosine modification. *Photochem. Photobiol.* **1994**, *59*, 343–349. [CrossRef]
77. Au, V.; Madison, S.A. Effects of singlet oxygen on the extracellular matrix protein collagen: Oxidation of the collagen crosslink histidinohydroxylysinonorleucine and histidine. *Arch. Biochem. Biophys.* **2000**, *384*, 133–142. [CrossRef] [PubMed]
78. Zhang, Y.; Mao, X.; Schwend, T.; Littlechild, S.; Conrad, G.W. Resistance of Corneal RFUVA–Cross-Linked Collagens and Small Leucine-Rich Proteoglycans to Degradation by Matrix Metalloproteinases. *Investig. Opthalmol. Vis. Sci.* **2013**, *54*, 1014. [CrossRef] [PubMed]
79. Heo, J.; Koh, R.H.; Shim, W.; Kim, H.D.; Yim, H.-G.; Hwang, N.S. Riboflavin-induced photo-crosslinking of collagen hydrogel and its application in meniscus tissue engineering. *Drug Deliv. Transl. Res.* **2016**, *6*, 148–158. [CrossRef] [PubMed]
80. Tirella, A.; Liberto, T.; Ahluwalia, A. Riboflavin and collagen: New crosslinking methods to tailor the stiffness of hydrogels. *Mater. Lett.* **2012**, *74*, 58–61. [CrossRef]

81. Diamantides, N.; Wang, L.; Pruiksma, T.; Siemiatkoski, J.; Dugopolski, C.; Shortkroff, S.; Kennedy, S.; Bonassar, L.J. Correlating rheological properties and printability of collagen bioinks: The effects of riboflavin photocrosslinking and pH. *Biofabrication* 2017, *9*, 34102. [CrossRef]
82. Rich, H.; Odlyha, M.; Cheema, U.; Mudera, V.; Bozec, L. Effects of photochemical riboflavin-mediated crosslinks on the physical properties of collagen constructs and fibrils. *J. Mater. Sci. Mater. Med.* 2014, *25*, 11–21. [CrossRef]
83. Mao, Q.; Wang, Y.; Li, Y.; Juengpanich, S.; Li, W.; Chen, M.; Yin, J.; Fu, J.; Cai, X. Fabrication of liver microtissue with liver decellularized extracellular matrix (dECM) bioink by digital light processing (DLP) bioprinting. *Mater. Sci. Eng. C* 2020, *109*, 110625. [CrossRef]
84. Acosta-Vélez, G.; Linsley, C.; Zhu, T.; Wu, W.; Wu, B. Photocurable Bioinks for the 3D Pharming of Combination Therapies. *Polymers* 2018, *10*, 1372. [CrossRef]
85. Yue, K.; de Santiago, G.T.; Alvarez, M.M.; Tamayol, A.; Annabi, N.; Khademhosseini, A. Synthesis, properties, and biomedical applications of gelatin methacryloyl (GelMA) hydrogels. *Biomaterials* 2015, *73*, 254–271. [CrossRef]
86. Datta, P.; Vyas, V.; Dhara, S.; Chowdhury, A.R.; Barui, A. Anisotropy properties of tissues: A basis for fabrication of biomimetic anisotropic scaffolds for tissue engineering. *J. Bionic Eng.* 2019, *16*, 842–868. [CrossRef]
87. Mauck, R.L.; Baker, B.M.; Nerurkar, N.L.; Burdick, J.A.; Li, W.-J.; Tuan, R.S.; Elliott, D.M. Engineering on the Straight and Narrow: The Mechanics of Nanofibrous Assemblies for Fiber-Reinforced Tissue Regeneration. *Tissue Eng. Part B Rev.* 2009, *15*, 171–193. [CrossRef] [PubMed]
88. Roy, R.R.; Edgerton, V.R. Skeletal Muscle Architecture. In *Encyclopedia of Neuroscience*; Springer: Berlin/Heidelberg, Germany, 2009; pp. 3702–3707.
89. Kim, H.; Jang, J.; Park, J.; Lee, K.-P.; Lee, S.; Lee, D.-M.; Kim, K.H.; Kim, H.K.; Cho, D.-W. Shear-induced alignment of collagen fibrils using 3D cell printing for corneal stroma tissue engineering. *Biofabrication* 2019, *11*, 035017. [CrossRef] [PubMed]
90. Torricelli, A.A.M.; Wilson, S.E. Cellular and extracellular matrix modulation of corneal stromal opacity. *Exp. Eye Res.* 2014, *129*, 151–160. [CrossRef]
91. Schwab, A.; Hélary, C.; Richards, R.G.; Alini, M.; Eglin, D.; D'Este, M. Tissue mimetic hyaluronan bioink containing collagen fibers with controlled orientation modulating cell migration and alignment. *Mater. Today Bio* 2020, *7*, 100058. [CrossRef]
92. Kim, W.; Kim, G. 3D bioprinting of functional cell-laden bioinks and its application for cell-alignment and maturation. *Appl. Mater. Today* 2020, *19*, 100588. [CrossRef]
93. Das, S.; Kim, S.-W.; Choi, Y.-J.; Lee, S.; Lee, S.-H.; Kong, J.-S.; Park, H.-J.; Cho, D.-W.; Jang, J. Decellularized extracellular matrix bioinks and the external stimuli to enhance cardiac tissue development in vitro. *Acta Biomater.* 2019, *95*, 188–200. [CrossRef] [PubMed]
94. Choi, Y.-J.; Kim, T.G.; Jeong, J.; Yi, H.-G.; Park, J.W.; Hwang, W.; Cho, D.-W. 3D Cell Printing of Functional Skeletal Muscle Constructs Using Skeletal Muscle-Derived Bioink. *Adv. Healthc. Mater.* 2016, *5*, 2636–2645. [CrossRef]
95. Puetzer, J.L.; Koo, E.; Bonassar, L.J. Induction of fiber alignment and mechanical anisotropy in tissue engineered menisci with mechanical anchoring. *J. Biomech.* 2015, *48*, 1436–1443. [CrossRef]
96. Murphy, S.; Ohlendieck, K. The extracellular matrix complexome from skeletal muscle. In *Composition and Function of the Extracellular Matrix in the Human Body*; Travacio, F., Ed.; InTech: London, UK, 2016.
97. Gillies, A.R.; Lieber, R.L. Structure and function of the skeletal muscle extracellular matrix. *Muscle Nerve* 2011, *44*, 318–331. [CrossRef]
98. Kim, H.; Kim, M.-C.; Asada, H.H. Extracellular matrix remodelling induced by alternating electrical and mechanical stimulations increases the contraction of engineered skeletal muscle tissues. *Sci. Rep.* 2019, *9*, 2732. [CrossRef]
99. Yang, G.; Long, H.; Ren, X.; Ma, K.; Xiao, Z.; Wang, Y.; Guo, Y. Regulation of adipose-tissue-derived stromal cell orientation and motility in 2D- and 3D-cultures by direct-current electrical field. *Dev. Growth Differ.* 2017, *59*, 70–82. [CrossRef]
100. Chen, C.; Bai, X.; Ding, Y.; Lee, I.-S. Electrical stimulation as a novel tool for regulating cell behavior in tissue engineering. *Biomater. Res.* 2019, *23*, 25. [CrossRef] [PubMed]
101. Tanaka, T.; Hattori-Aramaki, N.; Sunohara, A.; Okabe, K.; Sakamoto, Y.; Ochiai, H.; Hayashi, R.; Kishi, K. Alignment of Skeletal Muscle Cells Cultured in Collagen Gel by Mechanical and Electrical Stimulation. *Int. J. Tissue Eng.* 2014, *2014*, 1–5. [CrossRef]
102. Sun, H.; Zhou, J.; Huang, Z.; Qu, L.; Lin, N.; Liang, C.; Dai, R.; Tang, L.; Tian, F. Carbon nanotube-incorporated collagen hydrogels improve cell alignment and the performance of cardiac constructs. *Int. J. Nanomed.* 2017, *12*, 3109–3120. [CrossRef] [PubMed]
103. Lenshof, A.; Laurell, A. Acoustophoresis. In *Encyclopedia of Nanotechnology*; Springer: Dordrecht, The Netherlands, 2015; pp. 1–6.
104. Cohen, S.; Sazan, H.; Kenigsberg, A.; Schori, H.; Piperno, S.; Shpaisman, H.; Shefi, O. Large-scale acoustic-driven neuronal patterning and directed outgrowth. *Sci. Rep.* 2020, *10*, 4932. [CrossRef] [PubMed]
105. Kang, B.; Shin, J.; Park, H.-J.; Rhyou, C.; Kang, D.; Lee, S.-J.; Yoon, Y.; Cho, S.-W.; Lee, H. High-resolution acoustophoretic 3D cell patterning to construct functional collateral cylindroids for ischemia therapy. *Nat. Commun.* 2018, *9*, 5402. [CrossRef]
106. Petta, D.; Basoli, V.; Pellicciotta, D.; Tognato, R.; Barcik, J.; Arrigoni, C.; Bella, E.D.; Armiento, A.R.; Candrian, C.; Richards, R.G.; et al. Sound-induced morphogenesis of multicellular systems for rapid orchestration of vascular networks. *Biofabrication* 2020, *13*, 015004. [CrossRef]
107. Ma, Z.; Holle, A.W.; Melde, K.; Qiu, T.; Poeppel, K.; Kadiri, V.M.; Fischer, P. Acoustic Holographic Cell Patterning in a Biocompatible Hydrogel. *Adv. Mater.* 2020, *32*, 1904181. [CrossRef] [PubMed]

108. Malda, J.; Visser, J.; Melchels, F.P.; Jüngst, T.; Hennink, W.E.; Dhert, W.J.A.; Groll, J.; Hutmacher, D.W. 25th anniversary article: Engineering hydrogels for biofabrication. *Adv. Mater.* **2013**, *25*, 5011–5028. [CrossRef]
109. Williams, D.; Thayer, P.; Martinez, H.; Gatenholm, E.; Khademhosseini, A. A perspective on the physical, mechanical and biological specifications of bioinks and the development of functional tissues in 3D bioprinting. *Bioprinting* **2018**, *9*, 19–36. [CrossRef]
110. Lutolf, M.P.; Hubbell, J.A. Synthetic biomaterials as instructive extracellular microenvironments for morphogenesis in tissue engineering. *Nat. Biotechnol.* **2005**, *23*, 47–55. [CrossRef]
111. Groll, J.; Boland, T.; Blunk, T.; Burdick, J.A.; Cho, D.-W.; Dalton, P.D.; Derby, B.; Forgacs, G.; Li, Q.; Mironov, V.A.; et al. Biofabrication: Reappraising the definition of an evolving field. *Biofabrication* **2016**, *8*, 013001. [CrossRef]
112. Farhat, W.; Hasan, A.; Lucia, L.; Becquart, F.; Ayoub, A.; Kobeissy, F. Hydrogels for Advanced Stem Cell Therapies: A Biomimetic Materials Approach for Enhancing Natural Tissue Function. *IEEE Rev. Biomed. Eng.* **2019**, *12*, 333–351. [CrossRef]
113. Hiemer, B.; Krogull, M.; Bender, T.; Ziebart, J.; Krueger, S.; Bader, R.; Jonitz-Heincke, A. Effect of electric stimulation on human chondrocytes and mesenchymal stem cells under normoxia and hypoxia. *Mol. Med. Rep.* **2018**, *18*, 2133–2141. [CrossRef]
114. Matta, C. Calcium signalling in chondrogenesis implications for cartilage repair. *Front. Biosci.* **2013**, *S5*, S374. [CrossRef] [PubMed]
115. Eijkelkamp, N.; Quick, K.; Wood, J.N. Transient Receptor Potential Channels and Mechanosensation. *Annu. Rev. Neurosci.* **2013**, *36*, 519–546. [CrossRef] [PubMed]
116. Parate, D.; Franco-Obregón, A.; Fröhlich, J.; Beyer, C.; Abbas, A.A.; Kamarul, T.; Hui, J.H.P.; Yang, Z. Enhancement of mesenchymal stem cell chondrogenesis with short-term low intensity pulsed electromagnetic fields. *Sci. Rep.* **2017**, *7*, 9421. [CrossRef] [PubMed]
117. Viti, F.; Landini, M.; Mezzelani, A.; Petecchia, L.; Milanesi, L.; Scaglione, S. Osteogenic Differentiation of MSC through Calcium Signaling Activation: Transcriptomics and Functional Analysis. *PLoS ONE* **2016**, *11*, e0148173. [CrossRef]
118. Choi, J.W.; Choi, B.H.; Park, S.-H.; Pai, K.S.; Li, T.Z.; Min, B.-H.; Park, S.R. Mechanical Stimulation by Ultrasound Enhances Chondrogenic Differentiation of Mesenchymal Stem Cells in a Fibrin-Hyaluronic Acid Hydrogel. *Artif. Organs* **2013**, *37*, 648–655. [CrossRef] [PubMed]
119. Zhu, S.; Jing, W.; Hu, X.; Huang, Z.; Cai, Q.; Ao, Y.; Yang, X. Time-dependent effect of electrical stimulation on osteogenic differentiation of bone mesenchymal stromal cells cultured on conductive nanofibers. *J. Biomed. Mater. Res. Part A* **2017**, *105*, 3369–3383. [CrossRef]
120. Leppik, L.; Zhihua, H.; Mobini, S.; Thottakkattumana Parameswaran, V.; Eischen-Loges, M.; Slavici, A.; Helbing, J.; Pindur, L.; Oliveira, K.M.C.; Bhavsar, M.B.; et al. Combining electrical stimulation and tissue engineering to treat large bone defects in a rat model. *Sci. Rep.* **2018**, *8*, 6307. [CrossRef] [PubMed]
121. Kim, J.-H.; Choi, T.G.; Park, S.; Yun, H.R.; Nguyen, N.N.Y.; Jo, Y.H.; Jang, M.; Kim, J.; Kim, J.; Kang, I.; et al. Mitochondrial ROS-derived PTEN oxidation activates PI3K pathway for mTOR-induced myogenic autophagy. *Cell Death Differ.* **2018**, *25*, 1921–1937. [CrossRef] [PubMed]
122. Aisenbrey, E.A.; Bilousova, G.; Payne, K.; Bryant, S.J. Dynamic mechanical loading and growth factors influence chondrogenesis of induced pluripotent mesenchymal progenitor cells in a cartilage-mimetic hydrogel. *Biomater. Sci.* **2019**, *7*, 5388–5403. [CrossRef]
123. Vaca-González, J.J.; Clara-Trujillo, S.; Guillot-Ferriols, M.; Ródenas-Rochina, J.; Sanchis, M.J.; Ribelles, J.L.G.; Garzón-Alvarado, D.A.; Ferrer, G.G. Effect of electrical stimulation on chondrogenic differentiation of mesenchymal stem cells cultured in hyaluronic acid–Gelatin injectable hydrogels. *Bioelectrochemistry* **2020**, *134*, 107536. [CrossRef] [PubMed]
124. Hess, R.; Jaeschke, A.; Neubert, H.; Hintze, V.; Moeller, S.; Schnabelrauch, M.; Wiesmann, H.-P.; Hart, D.A.; Scharnweber, D. Synergistic effect of defined artificial extracellular matrices and pulsed electric fields on osteogenic differentiation of human MSCs. *Biomaterials* **2012**, *33*, 8975–8985. [CrossRef]
125. Maxwell, J.T.; Wagner, M.B.; Davis, M.E. Electrically induced calcium handling in cardiac progenitor cells. *Stem Cells Int.* **2016**, *2016*, 1–11. [CrossRef]
126. Ma, R.; Liang, J.; Huang, W.; Guo, L.; Cai, W.; Wang, L.; Paul, C.; Yang, H.-T.; Kim, H.W.; Wang, Y. Electrical Stimulation Enhances Cardiac Differentiation of Human Induced Pluripotent Stem Cells for Myocardial Infarction Therapy. *Antioxid. Redox Signal.* **2018**, *28*, 371–384. [CrossRef]
127. Spitzer, N.C. Electrical activity in early neuronal development. *Nature* **2006**, *444*, 707–712. [CrossRef] [PubMed]
128. Guo, W.; Wang, S.; Yu, X.; Qiu, J.; Li, J.; Tang, W.; Li, Z.; Mou, X.; Liu, H.; Wang, Z. Construction of a 3D rGO–collagen hybrid scaffold for enhancement of the neural differentiation of mesenchymal stem cells. *Nanoscale* **2016**, *8*, 1897–1904. [CrossRef]
129. Shin, J.; Choi, E.J.; Cho, J.H.; Cho, A.-N.; Jin, Y.; Yang, K.; Song, C.; Cho, S.-W. Three-Dimensional Electroconductive Hyaluronic Acid Hydrogels Incorporated with Carbon Nanotubes and Polypyrrole by Catechol-Mediated Dispersion Enhance Neurogenesis of Human Neural Stem Cells. *Biomacromolecules* **2017**, *18*, 3060–3072. [CrossRef] [PubMed]
130. Mostafavi, E.; Medina-Cruz, D.; Kalantari, K.; Taymoori, A.; Soltantabar, P.; Webster, T.J. Electroconductive Nanobiomaterials for Tissue Engineering and Regenerative Medicine. *Bioelectricity* **2020**, *2*, 120–149. [CrossRef]
131. Yi, J.; Choe, G.; Park, J.; Lee, J.Y. Graphene oxide-incorporated hydrogels for biomedical applications. *Polym. J.* **2020**, *52*, 823–837. [CrossRef]
132. Chen, X.; Ranjan, V.D.; Liu, S.; Liang, Y.N.; Lim, J.S.K.; Chen, H.; Hu, X.; Zhang, Y. In Situ Formation of 3D Conductive and Cell-Laden Graphene Hydrogel for Electrically Regulating Cellular Behavior. *Macromol. Biosci.* **2021**, *21*, 2000374. [CrossRef] [PubMed]

133. Liu, X.; Miller, A.L.; Park, S.; Waletzki, B.E.; Zhou, Z.; Terzic, A.; Lu, L. Functionalized Carbon Nanotube and Graphene Oxide Embedded Electrically Conductive Hydrogel Synergistically Stimulates Nerve Cell Differentiation. *ACS Appl. Mater. Interfaces* **2017**, *9*, 14677–14690. [CrossRef]
134. Bello, A.B.; Kim, D.; Kim, D.; Park, H.; Lee, S.-H. Engineering and Functionalization of Gelatin Biomaterials: From Cell Culture to Medical Applications. *Tissue Eng. Part B Rev.* **2020**, *26*, 164–180. [CrossRef]
135. Heo, D.N.; Ko, W.-K.; Bae, M.S.; Lee, J.B.; Lee, D.-W.; Byun, W.; Lee, C.H.; Kim, E.-C.; Jung, B.-Y.; Kwon, I.K. Enhanced bone regeneration with a gold nanoparticle–hydrogel complex. *J. Mater. Chem. B* **2014**, *2*, 1584–1593. [CrossRef]
136. Dayem, A.A.; Kim, B.; Gurunathan, S.; Choi, H.Y.; Yang, G.; Saha, S.K.; Han, D.; Han, J.; Kim, K.; Kim, J.-H.; et al. Biologically synthesized silver nanoparticles induce neuronal differentiation of SH-SY5Y cells via modulation of reactive oxygen species, phosphatases, and kinase signaling pathways. *Biotechnol. J.* **2014**, *9*, 934–943. [CrossRef]
137. Kim, K.S.; Choi, H.W.; Yoon, H.E.; Kim, I.Y. Reactive Oxygen Species Generated by NADPH Oxidase 2 and 4 Are Required for Chondrogenic Differentiation. *J. Biol. Chem.* **2010**, *285*, 40294–40302. [CrossRef] [PubMed]
138. Hino, K.; Horigome, K.; Nishio, M.; Komura, S.; Nagata, S.; Zhao, C.; Jin, Y.; Kawakami, K.; Yamada, Y.; Ohta, A.; et al. Activin-A enhances mTOR signaling to promote aberrant chondrogenesis in fibrodysplasia ossificans progressiva. *J. Clin. Investig.* **2017**, *127*, 3339–3352. [CrossRef] [PubMed]
139. Yang, X.; Pabon, L.; Murry, C.E. Engineering Adolescence. *Circ. Res.* **2014**, *114*, 511–523. [CrossRef] [PubMed]
140. Roshanbinfar, K.; Mohammadi, Z.; Sheikh-Mahdi Mesgar, A.; Dehghan, M.M.; Oommen, O.P.; Hilborn, J.; Engel, F.B. Carbon nanotube doped pericardial matrix derived electroconductive biohybrid hydrogel for cardiac tissue engineering. *Biomater. Sci.* **2019**, *7*, 3906–3917. [CrossRef]
141. Ronaldson-Bouchard, K.; Yeager, K.; Teles, D.; Chen, T.; Ma, S.; Song, L.; Morikawa, K.; Wobma, H.M.; Vasciaveo, A.; Ruiz, E.C. Engineering of human cardiac muscle electromechanically matured to an adult-like phenotype. *Nat. Protoc.* **2019**, *14*, 2781–2817. [CrossRef]
142. Wittkowske, C.; Reilly, G.C.; Lacroix, D.; Perrault, C.M. In Vitro Bone Cell Models: Impact of Fluid Shear Stress on Bone Formation. *Front. Bioeng. Biotechnol.* **2016**, *4*, 87. [CrossRef]
143. Goldfracht, I.; Efraim, Y.; Shinnawi, R.; Kovalev, E.; Huber, I.; Gepstein, A.; Arbel, G.; Shaheen, N.; Tiburcy, M.; Zimmermann, W.H.; et al. Engineered heart tissue models from hiPSC-derived cardiomyocytes and cardiac ECM for disease modeling and drug testing applications. *Acta Biomater.* **2019**, *92*, 145–159. [CrossRef]
144. Meinert, C.; Schrobback, K.; Hutmacher, D.W.; Klein, T.J. A novel bioreactor system for biaxial mechanical loading enhances the properties of tissue-engineered human cartilage. *Sci. Rep.* **2017**, *7*, 16997. [CrossRef]
145. Vazquez, M.; Evans, B.A.J.; Riccardi, D.; Evans, S.L.; Ralphs, J.R.; Dillingham, C.M.; Mason, D.J. A New Method to Investigate How Mechanical Loading of Osteocytes Controls Osteoblasts. *Front. Endocrinol.* **2014**, *5*, 208. [CrossRef]
146. Li, L.; Yang, Z.; Zhang, H.; Chen, W.; Chen, M.; Zhu, Z. Low-intensity pulsed ultrasound regulates proliferation and differentiation of osteoblasts through osteocytes. *Biochem. Biophys. Res. Commun.* **2012**, *418*, 296–300. [CrossRef]
147. Kuo, T.-R.; Chen, C.-H. Bone biomarker for the clinical assessment of osteoporosis: Recent developments and future perspectives. *Biomark. Res.* **2017**, *5*, 18. [CrossRef]
148. Huang, W.-C.; Shen, M.-Y.; Chen, H.-H.; Lin, S.-C.; Chiang, W.-H.; Wu, P.-H.; Chang, C.-W.; Chiang, C.-S.; Chiu, H.-C. Monocytic delivery of therapeutic oxygen bubbles for dual-modality treatment of tumor hypoxia. *J. Control. Release* **2015**, *220*, 738–750. [CrossRef]
149. Niculescu, A.-G.; Grumezescu, A.M. Photodynamic Therapy—An Up-to-Date Review. *Appl. Sci.* **2021**, *11*, 3626. [CrossRef]
150. Highley, C.B.; Kim, M.; Lee, D.; Burdick, J.A. Near-infrared light triggered release of molecules from supramolecular hydrogel-nanorod composites. *Nanomedicine* **2016**, *11*, 1579–1590. [CrossRef] [PubMed]
151. Wu, W.; Shen, J.; Banerjee, P.; Zhou, S. Core-shell hybrid nanogels for integration of optical temperature-sensing, targeted tumor cell imaging, and combined chemo-photothermal treatment. *Biomaterials* **2010**, *31*, 7555–7566. [CrossRef] [PubMed]
152. Beharry, A.A.; Woolley, G.A. Azobenzene photoswitches for biomolecules. *Chem. Soc. Rev.* **2011**, *40*, 4422. [CrossRef] [PubMed]
153. Rosales, A.M.; Rodell, C.B.; Chen, M.H.; Morrow, M.G.; Anseth, K.S.; Burdick, J.A. Reversible Control of Network Properties in Azobenzene-Containing Hyaluronic Acid-Based Hydrogels. *Bioconjug. Chem.* **2018**, *29*, 905–913. [CrossRef] [PubMed]
154. Sun, W.; Jiang, H.; Wu, X.; Xu, Z.; Yao, C.; Wang, J.; Qin, M.; Jiang, Q.; Wang, W.; Shi, D.; et al. Strong dual-crosslinked hydrogels for ultrasound-triggered drug delivery. *Nano Res.* **2019**, *12*, 115–119. [CrossRef]
155. Ravichandran, R.; Astrand, C.; Patra, H.K.; Turner, A.P.F.; Chotteau, V.; Phopase, J. Intelligent ECM mimetic injectable scaffolds based on functional collagen building blocks for tissue engineering and biomedical applications. *RSC Adv.* **2017**, *7*, 21068–21078. [CrossRef]
156. Shoukat, H.; Buksh, K.; Noreen, S.; Pervaiz, F.; Maqbool, I. Hydrogels as potential drug-delivery systems: Network design and applications. *Ther. Deliv.* **2021**, *12*, 375–396. [CrossRef]
157. Traverse, J.H.; Henry, T.D.; Dib, N.; Patel, A.N.; Pepine, C.; Schaer, G.L.; DeQuach, J.A.; Kinsey, A.M.; Chamberlin, P.; Christman, K.L. First-in-Man Study of a Cardiac Extracellular Matrix Hydrogel in Early and Late Myocardial Infarction Patients. *JACC Basic Transl. Sci.* **2019**, *4*, 659–669. [CrossRef]
158. Noor, N.; Shapira, A.; Edri, R.; Gal, I.; Wertheim, L.; Dvir, T. 3D Printing of Personalized Thick and Perfusable Cardiac Patches and Hearts. *Adv. Sci.* **2019**, *6*, 1900344. [CrossRef] [PubMed]

Article

Design and Fabrication of Nanofibrous Dura Mater with Antifibrosis and Neuroprotection Effects on SH-SY5Y Cells

Zhiyuan Zhao [1,2,3,†], Tong Wu [2,3,†], Yu Cui [2,3], Rui Zhao [1,2,3], Qi Wan [2,3,*] and Rui Xu [1,3,*]

1. Department of Interventional Radiology, The Affiliated Hospital of Qingdao University, Jiangsu Road 16, Qingdao 266000, China; qduzzy@126.com (Z.Z.); zhrui97@163.com (R.Z.)
2. Institute of Neuroregeneration and Neurorehabilitation, Qingdao University, Qingdao 266071, China; twu@qdu.edu.cn (T.W.); cuiyu1216@126.com (Y.C.)
3. Qingdao Medical College, Qingdao University, Qingdao 266071, China
* Correspondence: qiwan1@hotmail.com (Q.W.); xray3236@126.com (R.X.)
† These authors contributed equally to this work.

Abstract: The development and treatment of some diseases, such as large-area cerebral infarction, cerebral hemorrhage, brain tumor, and craniocerebral trauma, which may involve the injury of the dura mater, elicit the need to repair this membrane by dural grafts. However, common dural grafts tend to result in dural adhesions and scar tissue and have no further neuroprotective effects. In order to reduce or avoid the complications of dural repair, we used PLGA, tetramethylpyrazine, and chitosan as raw materials to prepare a nanofibrous dura mater (NDM) with excellent biocompatibility and adequate mechanical characteristics, which can play a neuroprotective role and have an antifibrotic effect. We fabricated PLGA NDM by electrospinning, and then chitosan was grafted on the nanofibrous dura mater by the EDC-NHS cross-linking method to obtain PLGA/CS NDM. Then, we also prepared PLGA/TMP/CS NDM by coaxial electrospinning. Our study shows that the PLGA/TMP/CS NDM can inhibit the excessive proliferation of fibroblasts, as well as provide a sustained protective effect on the SH-SY5Y cells treated with oxygen–glucose deprivation/reperfusion (OGD/R). In conclusion, our study may provide a new alternative to dural grafts in undesirable cases of dural injuries.

Keywords: nanofibrous dura mater; antifibrosis; neuroprotection; PLGA; tetramethylpyrazine

1. Introduction

The dura mater surrounds the brain and retains the cerebrospinal fluid [1]. The dura mater may be damaged in the development and treatment of neurosurgical diseases, such as large-area cerebral infarction, cerebral hemorrhage, brain tumor, and craniocerebral trauma, which need to be repaired in time [2]. For dura mater that is difficult to complete in a one-stage repair process, it can be repaired with dural grafts through duroplasty [3]. Autologous dura grafts obtained from the periosteum and fascia have no immune rejection, but their clinical application is greatly limited due to insufficient quantities, difficult sampling, postoperative pain, and other shortcomings [4,5]. Dural grafts have several important features: supporting tissue regeneration, avoiding an immune inflammatory response, good watertight confinement, anti-adhesion, inhibiting scar tissue formation, biodegradability, and releasing therapeutic drugs to promote recovery [6,7].

Many synthetic polymers, such as polylactic acid (PLA), polyglycolic acid (PGA), and their copolymer, polylactic-co-glycolic acid (PLGA), have been widely used in biomedical engineering due to their good biocompatibility, non-toxicity, film-forming properties, and biodegradability [8,9]. In addition, the use of PLGA as a carrier for sustained drug release is also a hot topic of current research due to the controlled degradation rate of PLGA [10–12]. However, PLGA also has defects as a graft. The degradation products of PLGA are lactic acid and hydroxyacetic acid, which are by-products of human metabolism [13]. The accumulation of acidic degradation products can decrease the local pH value, trigger inflammatory reactions, and affect the rate of polymer degradation [14].

As a natural polysaccharide with good biocompatibility, degradability, antibacterial properties, and mechanical strength, chitosan (CS) is a kind of good implant material [15]. Previous studies have shown that CS can inhibit the proliferation of fibroblasts and suppress type I and III procollagen production, and it has anti-fibrosis as well as anti-adhesive effects [16–19]. In addition, chitooligosaccharides (COS), the degradation products of CS, have been proven to promote nerve regeneration [20,21]. CS is an alkaline polysaccharide among natural polysaccharides that can chemically bind to other substances through some primary amino groups carried by the CS main chain. It has been reported that CS can buffer the acidity produced from the degradation of PLLA [22,23].

2,3,5,6-tetramethylpyrazine (TMP) is an active compound extracted from the herb Chuanxiong Ligusticum, which exhibits neuroprotective effects [24,25]. TMP can protect neurons by scavenging oxygen free radicals, protecting mitochondrial function, inhibiting calcium inward flow and glutamate release, as well as attenuating ischemia-induced neuronal death by regulating the expression of bcl-2 and bax proteins [26–30]. In addition, TMP affects neurogenesis, for example, it can promote the differentiation of neural stem cells into neurons [31,32]. At present, TMP has been widely used in the clinical treatment and basic research of cardiovascular and cerebrovascular diseases. However, due to its poor water solubility, short half-life, and low concentration of distribution at the injury site, it requires high doses and multiple administrations to maintain therapeutic concentrations, which is a drawback for clinical treatment [33–35].

To overcome the above drawbacks and limitations, in this study, we designed and fabricated PLGA/TMP nanofibrous dura mater (NDM) with a coaxial electrospinning technique, and then generated PLGA/TMP/CS NDM by chemically binding a CS coating to the surface of the NDM. Our study found that the PLGA/TMP/CS NDM could inhibit the excessive proliferation of fibroblasts in vitro, thus exerting anti-adhesive effects and inhibiting the formation of scar tissue. Through the degradation of the PLGA and CS, TMP was released into the cell matrix, which could promote the survival of OGD/R-treated SH-SY5Y cells, as well as facilitate the regeneration of SH-SY5Y cells, and finally, exert a sustained neuroprotective effect.

2. Materials and Methods
2.1. Materials

PLGA (50:50, Mw: 60,000–80,000) was purchased from Match Biomaterials (Shenzhen, China), TMP and MES Buffered Solution from Aladdin (Shanghai, China), ethylcarbodiimide hydrochloride (EDC), N-hydroxysuccinimide (NHS), acetic acid, hexafluoroisopropanol (HFIP), and N,N-dimethylformamide (DMF) from Macklin (Shanghai, China), and CS from Sigma-Aldrich (America). Fibroblasts were a gift from Ziyi Zhou from the Medical Cosmetology Center at the Affiliated Hospital of Qingdao University. Human neuroblastoma SH-SY5Y cells were provided by the Institute of Neuroregeneration and Neurorehabilitation, Qingdao University. Dulbecco's modified eagle medium (DMEM) and antibiotic-antimycotic were purchased from Solarbio (Beijing, China). Fetal bovine serum (FBS) was purchased from Pan (Adenbach, Germany). Phalloidin-iFluor 488 and DAPI staining solution were acquired from Abcam (Shanghai, China). The following antibodies from ABclonal (Wuhan, China) were used: Ki67 Rabbit pAb and GAP43 Rabbit pAb. Goat Anti-rabbit IgG H&L/Alexa Fluor 555 as a secondary antibody was from Bioss (Beijing, China). A Cell Counting Kit-8 (CCK8) was supplied from Targetmol (Boston, MA, USA).

2.2. Electrospinning

PLGA/TMP/CS NDM was fabricated by the electrospinning technique. Briefly, PLGA was dissolved in HFIP via magnetic stirring at room temperature for 4 h to form an electrospinning solution with a concentration of 20 wt.% [36]. The solution flowed out from the syringe at a rate of 2 mL/h. PLGA NDM was fabricated by electrospinning for

1 h with +12 kV, and the acceptance distance was 15 cm. The PLGA NDM was immersed in EDC/NHS solution (0.96 g of EDC and 0.14 g of NHS in 50 mL of MES buffer) at 4 °C for 12 h. The CS solution (3 wt.%) at an equal volume to the EDC/NHS solution was added to ensure an excess of CS, and the mixed PLGA-CS system was kept for 24 h at room temperature until the cross-linking between the PLGA and CS components (via the coupling reaction between the NHS-ester groups belonging to PLGA and some primary amine groups of CS) was accomplished [12]. Then, the cross-linked product was rinsed repeatedly with deionized water and dried to obtain PLGA/CS NDM. For coaxial electrospinning, the shell solution consisted of 20% PLGA in HFIP, and the core of the fibers was 10 mg/mL TMP dissolved in ethanol. The prepared solutions were delivered to the outer and inner coaxial needle at 2.0 and 0.1 mL/h feeding ratios, respectively, with a programmable syringe pump. The applied voltage was 12 kV and the acceptance distance is 15 cm. The fibers were collected in a layer-by-layer manner during a 4 h period to obtain PLGA/TMP/CS NDM [36,37].

2.3. NDM Characterization

The morphology of the nanofiber was observed using scanning electron microscopy (SEM, VEGA 3 SBH, TESCAN, Shanghai, China) and the nanofiber diameter was measured by applying Nano Measurer software. The chemical structure and composition of the fibers were characterized by Fourier infrared spectroscopy (Nicolet Is50, Thermo Electron Corporation, Waltham, MA, USA). All NDMs were prepared for the same size (5 cm × 1 cm, about 0.02 mm in thickness) to characterize their tensile mechanical property by employing an Electro-mechanical Universal Testing Instrument (CMT6103, Mechanical Testing & Simulation, Eden Prairie, MN, USA). Thermogravimetric analysis (TGA) of the NDMs was performed using a Thermal Gravimetric Analyzer (TASDT650, TA INSTRUMENTS, New Castle, DE, USA).

2.4. Encapsulation Efficiency (EE)

The absorbance value of the TMP was detected using a full-function microplate detector (Synergy Neo2, Bio Tek, Vermont, USA) to determine the maximum UV absorption peak at a wavelength of 280 nm (Supplementary Figure S1). The PLGA/TMP/CS NDM samples were immersed in DCM until the TMP was completely dissolved, and then its actual content was determined spectrophotometrically using a corresponding calibration curve. Comparatively, the theoretical content of the TMP was considered to be the overall TMP amount consumed during the electrospinning. The calculation of the EE (in %) of the TMP was performed as follows:

$$EE = (actual\ content/theoretical\ content) \times 100 \quad (1)$$

2.5. In Vitro TMP Release Profiles

The PLGA/TMP/CS NDM was immersed in 5 mL of PBS solution (pH = 7.4) as a release medium and placed in a thermostatic shaker for gentle shaking. For current measurements, 1 mL of the PBS solution was taken out at different times to spectrophotometrically determine the TMP release (using the corresponding calibration curve), and then 1 mL of fresh PBS solution was added to the release medium contained in the shaker. The cumulative TMP release (C, in %) was calculated according to the following equation:

$$C = (m_1 + m_2 + \ldots + m_n)/m_0 \times 100\% \quad (2)$$

where m_1, m_2, and m_n are the weights determined at the times t_1, t_2, and t_n, respectively, and m_0 is the total weight of TMP to be released.

2.6. Extraction of NDM Immersion Solution

Multiple different NDMs with the same mass were sterilized by ethanol fumigation for 3 h and UV irradiation for half an hour. The NDMs were immersed in 8 mL of fresh

DMEM complete medium (DMEM with 10% FBS) under aseptic conditions at 37 °C, while a separate fresh DMEM complete medium was used as a control. The NDMs were removed after 1, 4, 7, and 14 days of immersion from the medium, and the remaining medium was the NDM immersion solution.

2.7. Cell Culture

Fibroblasts and SH-SY5Y cells were cultured using DMEM complete medium containing 10% FBS at 37 °C and 5% CO_2. Depending on the needs of the experiment, fibroblasts were seeded evenly on glass slides and different NDMs of a certain number, while SH-SY5Y cells were grown directly onto well plates and glass slides of a certain number. The culture medium was replaced every two days during the culture. Prior to cell seeding, all the glass slides and NDMs were sterilized by ethanol fumigation for 3 h and UV irradiation for 30 min.

2.8. OGD Challenge

The complete medium was replaced with deoxygenated glucose-free extracellular solution (in mM: 116 NaCl, 5.4 KCl, 0.8 $MgSO_4$, 1.0 NaH_2PO_4, 1.8 $CaCl_2$, and 26 $NaHCO_3$) [38]. The cells were cultured in a dedicated chamber (Plas-Labs, Lansing, MI, USA) with 95% N_2/5% CO_2 at 37 °C for 3 h. Then, the cells were reperfused using fresh complete medium containing different immersion solutions according to the experimental requirements and transferred to normal conditions for culture.

2.9. Cell Viability

A cell viability assay for fibroblasts and SH-SY5Y cells was performed following an algorithm reported elsewhere [39]. The fibroblasts were seeded evenly at a density of 5000 cells/well on glass slides and different NDM amounts in 24-well plates for culture. CCK8 solution was added and incubated with the cells for 3 h at 1, 3, and 5 days after culture, respectively. The absorbance value at 450 nm was measured using a 96-well plate reader. The SH-SY5Y cells were seeded in 48-well plates at a density of 5000 cells/well, and after 24 h of culture, the cells were treated with OGD/R, after which CCK8 solution was added and incubated for 2 h. The absorbance value at 450 nm was measured using a 96-well plate reader. The absorbance values at 450 nm corresponded to the amounts of formazan dye that resulted under the action of cellular dehydrogenases exerted on the tetrazolium salt present in the initial CCK8 solution which, in turn, was proportional to the number of living cells.

2.10. Lactate Dehydrogenase (LDH) Release Assay

According to the manufacturer's instructions (Beyotime, Shanghai, China), the supernatants from the OGD/R-treated cell culture medium were harvested, and the absorbance value at 490 nm was measured using a 96-well plate reader with the absorbance value at 600 nm as reference. The LDH release was calculated according to the manufacturer's formula. The absorbance values at 490 nm were proportionally correlated with the LDH release.

2.11. Cell Morphology

The fibroblasts were seeded at a density of 5000 cells/well on glass slides, the PLGA NDM and PLGA/CS NDM in 24-well plates for culture. After 5 days of culture, the fibroblasts were stained with Phalloidin-iFluor 488 and DAPI to observe the morphology using fluorescence microscopy. For the SH-SY5Y cells, they were stained and observed in the same way after OGD/R treatment.

2.12. Immunofluorescence

Cells on the NDM and glass slides were fixed in 4% paraformaldehyde, then permeabilized with 0.5% Triton X-100 for 15 min and blocked in 5% FBS for 2 h. The primary antibodies against Ki-67 (1:200) and Gap43 (1:200) diluted in the blocked buffer were added to incubate with cells at 4 °C overnight. Then the cells were labeled by a secondary antibody and stained with DAPI. The samples were observed using fluorescence microscopy and analyzed using Image J software.

2.13. Statistical Analysis

The statistical analysis was performed using GraphPad Prism software. The results are presented in the form of the mean ± standard deviation. Statistical comparisons between groups were performed using one-way ANOVA. Values of * $p < 0.05$, ** $p < 0.01$, and *** $p < 0.001$ are considered statistically significant.

3. Results and Discussion

3.1. NDM Characterization

We fabricated PLGA NDM and PLGA/TMP NDM using an electrospinning device according to the previously described conditions and coated them with CS by EDC/NHS cross-linking to prepare PLGA/TMP/CS NDM. We observed the morphology of the NDM by SEM. A smooth surface structure was shown on the fibers of the PLGA NDM, which were arranged in a disordered manner and interwoven into a network. The mean diameter of the fibers was 646 ± 103 nm, which was relatively uniform (Figure 1A). The PLGA/TMP NDM was morphologically similar to the PLGA NDM, with 692 ± 97 nm in mean fiber diameter (Figure 1B). For the PLGA/CS NDM, shiny granular agglomerates could be seen on the fiber surfaces (Figure 1C). To determine whether these were made of CS, structural analysis on the molecular scale of the PLGA/CS NDM was performed via infrared spectroscopy (FTIR) (Figure 1D). In the PLGA spectrum, the peak at 1087 cm^{-1} was assigned to C-O stretching, that located at 1748 cm^{-1} to C=O stretching vibrations, and the peaks placed between approximately 2950 and 3000 cm^{-1} to C-H stretching vibrations mainly involving methyl (CH3) and methylene (CH2) groups [40]. Instead, the infrared spectrum of the CS displayed the two vibrations of amide I and amide II at 1646 and 1587 cm^{-1}, respectively. These weak peaks ascribed to the vibrational mode of amide I and II are due to a high degree of N-deacetylation associated with the CS used. In addition, the band at 2870 cm^{-1} was attributed to C-H stretching involving the carbon atoms of the sugar rings. The broad band centered at ca. 3500 cm^{-1} corresponded to the N-H stretch vibrations overlapped by the O-H stretches of the OH groups [41,42]. On the other hand, the simultaneous presence of the bands at 3500, 1635, and 1536 cm^{-1} (for CS) and at 1748 cm^{-1} (for PLGA) on the PLGA/CS NDM spectrum confirms the coexistence of CS and PLGA in the same mixed system (PLGA/CS NDM). Moreover, the strengthening of the bands at 1635 and 1536 cm^{-1} on the IR spectrum of the PLGA/CS NDM is consistent with the increased number of amide cross-linkages newly formed during the grafting of CS onto PLGA via the EDC/NHS coupling reaction [12].

The tensile results of all NDMs are shown in Figure 2A as the stress–strain curves of the PLGA NDM and PLGA/CS NDM. The tensile strength of the PLGA NDM was 6.27 ± 0.96 MPa, while that of the PLGA/CS NDM is 8.71 ± 1.03 MPa. Correspondingly, the values of the maximum strain (at break) are ca. 216% for the PLGA NDM and 161% for the PLGA/CS NDM. It is obvious that the CS improved the mechanical stress of the PLGA NDM but had a negative effect on the flexibility. According to experimental data reported elsewhere [43], the tensile strength of the human dura mater is about 7 MPa, and the maximum strain is 11%. The result suggests that the PLGA/CS NDM is more suitable for a dural graft than the PLGA NDM.

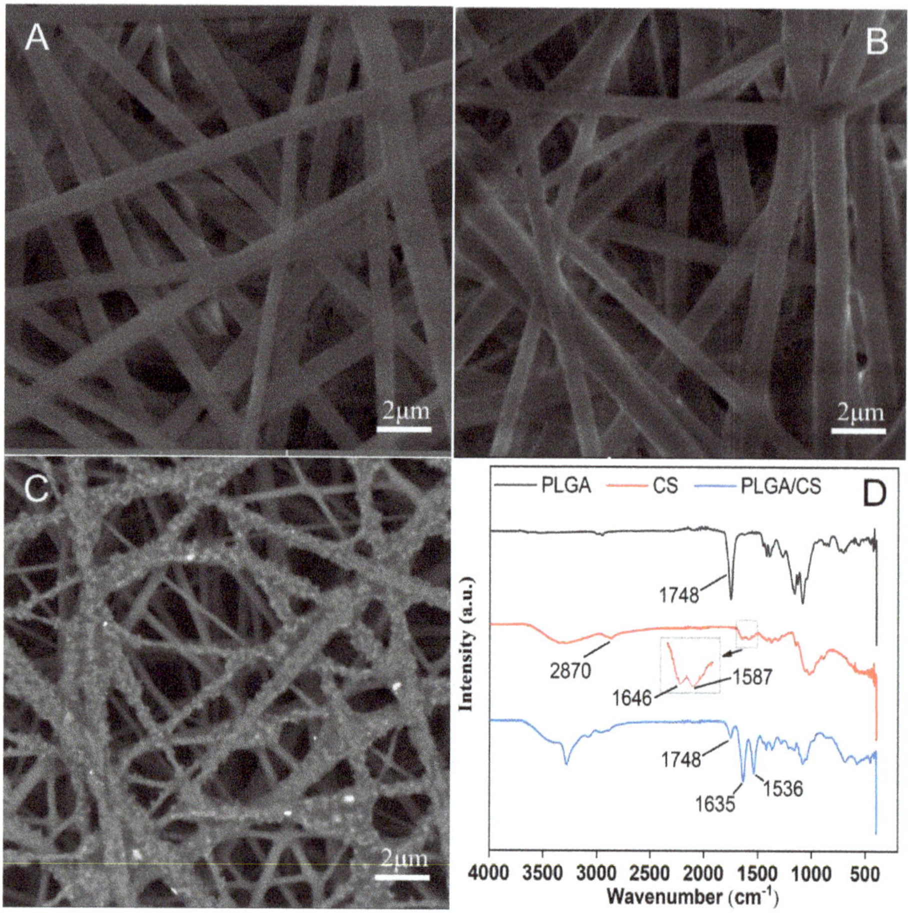

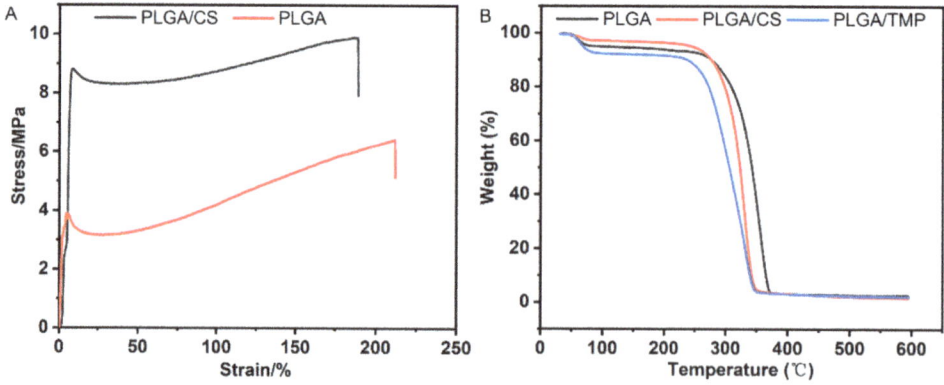

Figure 1. (**A**–**C**) SEM images showing the PLGA NDM (**A**), the PLGA/TMP NDM (**B**), and the PLGA/CS NDM (**C**). (**D**) FTIR spectra of the PLGA NDM, CS, and PLGA/CS NDM.

Figure 2. (**A**) Stress–strain curves of PLGA NDM and PLGA/CS NDM. (**B**) TGA profiles of PLGA NDM, PLGA/CS NDM, and PLGA/TMP NDM.

The thermal stability of the NDMs was characterized using thermogravimetry (Figure 2B). The NDMs showed two stages of weight loss. From 30 to 200 °C, the PLGA NDM showed a gradual weight loss of 7.9 wt.%, while the PLGA/CS showed a 6.52% loss, and the PLGA/TMP NDM showed a 9.32% loss due to the desorption of solvent, adsorbed, and bound water. Then, the PLGA/TMP NDM began to decompose at 230 °C, while the PLGA NDM and PLGA/CS NDM did so at about 260 °C. The decomposition of the PLGA/CS NDM and PLGA/TMP NDM was completed at around 340 °C, while that of the PLGA NDM was done at about 360 °C. The ash residue of the NDM was around 2%. This shows that the CS and TMP had a weak influence on the thermal stability of the PLGA. However, the physiological temperature of the human body cannot interfere at all with the thermal stability of the systems investigated by thermogravimetry.

3.2. Encapsulation Efficiency and In Vitro TMP Release Profiles

According to our calculations, the EE (%) of TMP is (57.95 ± 2.46) %, and the working concentration of TMP can be reached after release (the standard curve used to obtain the value of the EE is plotted in the Supplementary Materials, Figure S2A). Figure 3 shows the in vitro release profile of the TMP (the associated calibration curve is displayed in the Supplementary Materials, Figure S2B). During the first 8 h, the TMP exhibited a burst release of more than 50%, reaching the working concentration that could play a neuroprotective role in the early stage. After this steep increase, the release of the TMP occurred. It was released moderately and sustainedly until the 14th day, when the process leveled off and the cumulative release ratio reached about 80%.

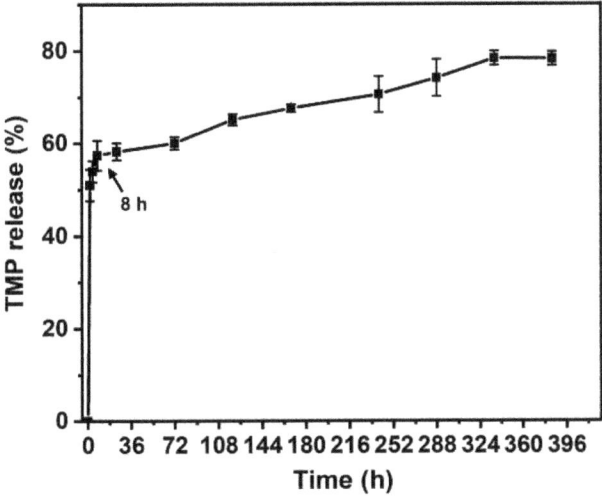

Figure 3. TMP release profiles in vitro from PLGA/TMP/CS NDM (measurements were performed in triplicate). During the first 8 h, the TMP exhibited a burst release of more than 50%.

3.3. PLGA/CS NDMs Inhibit the Excessive Proliferation of Fibroblasts

During wound healing, fibroblasts are activated. Activated fibroblasts have higher cytoskeleton tension, indicated by obvious stress fibers and contractile phenotype, which enhance the secretion of ECM and promote wound healing and tissue regeneration. A balanced secretion of ECM is essential for wound healing, as the accumulation of excessive ECM can lead to the development of tissue adhesions and cause tissue fibrosis and scar tissue formation [44]. It has been reported that PLGA can reduce the formation of epidural fibrosis [45]. To investigate the effect of the PLGA NDM and PLGA/CS NDM on fibroblasts, we seeded fibroblasts on NDMs for culture and performed cell viability assays on the fibroblasts after 1, 3, and 5 days of cell culture. As shown in Figure 4A, cell viability was increased from day 1 to day 5, regardless of the NDM used for culture. However, cell viability was decreased in the PLGA NDM and PLGA/CS NDM compared to the control group (TCP), in which the cells were cultured under normal conditions, and the PLGA/CS NDM induced lower cell viability than that observed in the PLGA NDM, although there were no statistical differences between the two groups. This is consistent with the previous reports that CS can progressively inhibit the proliferation of fibroblasts in a CS dose-dependent manner [17].

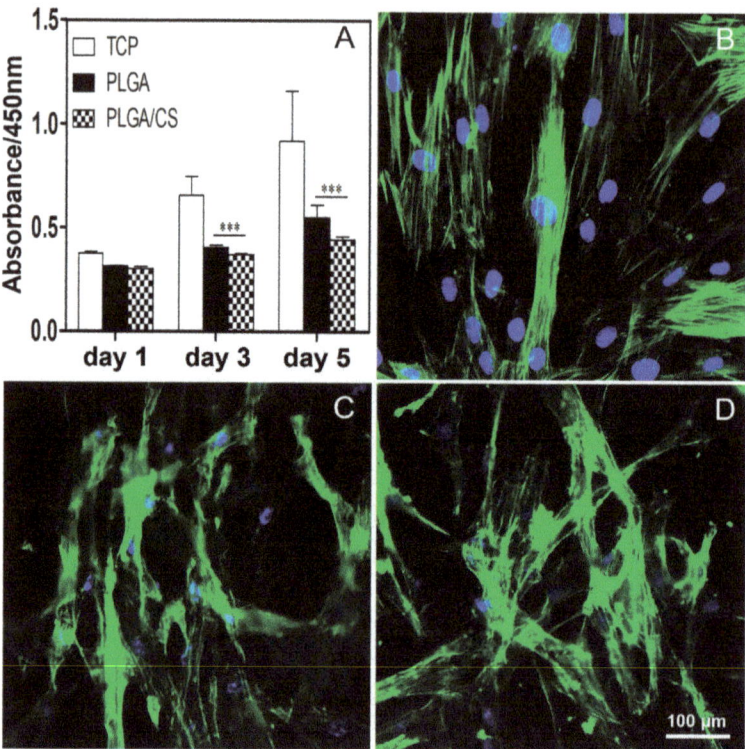

Figure 4. (A) Cell viability of fibroblasts seeded on glass slides as the control group (TCP), on PLGA NDM and PLGA/CS NDM after culture of 1, 3, and 5 days. *** $p < 0.001$ as compared with TCP. (B–D) Fluorescence micrographs showing the morphology of the fibroblasts after 5 days of culture. (B) TCP cells. (C) Cells seeded on PLGA NDM. (D) Cells seeded on PLGA/CS NDM. The cell nuclei were stained with DAPI (blue), and the actin cytoskeleton was stained with Phalloidin-iFluor 488 (FITC, green). Data of 3 replicates are plotted in Figure 4A.

From a morphological point of view, the fibroblasts of the TCP stained with FITC-Phalloidin were compared to the fibroblasts similarly stained and cultured in the presence of NDMs. Fibroblasts on the TCP group showed a normal long spindle shape, while fibroblasts on the PLGA NDM and PLGA/CS NDM groups became more elongated, grew more branched, and overall became irregular in morphology (Figure 4B–D), which means that both the PLGA NDM and PLGA/CS NDM could have affected the growth and morphology of the fibroblasts. These results are in agreement with those of the CCK8.

To further determine whether NDMs reduced fibroblast cell viability by decreasing cell proliferation, the fibroblasts were labeled as Ki-67, and the cell proliferation capacity was analyzed. The percentage of Ki-67-positive cells was significantly decreased in the PLGA NDM and PLGA/CS NDM groups compared to the TCP group, and this phenomenon was more pronounced in the PLGA/CS NDM group (Figure 5 and Supplementary Materials,

Figure S3), which was consistent with the previous CCK8 results. This suggests that the PLGA/CS NDM repressed cell viability by inhibiting the excessive proliferation of fibroblasts, which is in line with a previous report [46].

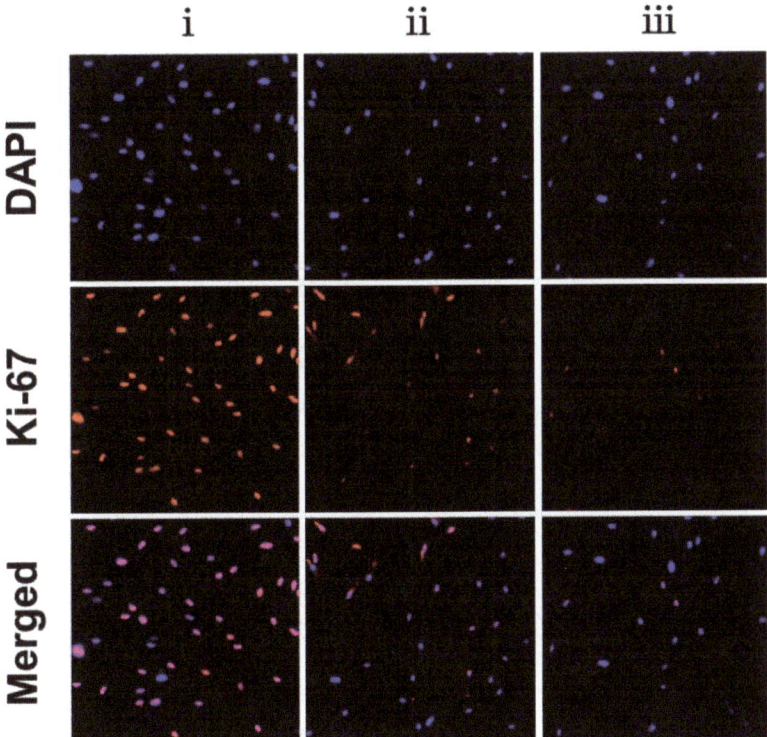

Figure 5. Fluorescence micrographs displaying Ki-67-positive fibroblasts seeded on TCP (i), PLGA NDM (ii), and PLGA/CS NDM (iii) after 5 days of culture. The cell nuclei were stained with DAPI (blue) and the cell nuclei with Ki-67-positive were labeled with red, meaning the cells were proliferating.

The cell viability of the fibroblasts showed an increasing trend in the PLGA/CS NDM group with the passage of culture time, meaning that the PLGA/CS NDM was capable of supporting tissue regeneration in our culture system. However, the PLGA/CS NDM was able to inhibit the excessive proliferation of fibroblasts and, thus, maintain a balanced secretion of ECM. During wound healing, an excessive proliferation of fibroblasts and excessive secretion of collagen type I can cause wound adhesion and ultimately lead to the formation of scar tissue. It has been reported that CS can inhibit the secretion of collagen type I with fibroblasts in scar tissue but has no effect on normal fibroblasts, which also prevents the formation of scar tissue [44,47]. In a few words, these results suggest that PLGA/CS NDM has the potential to both support wound healing and prevent dural adhesion and scar tissue formation.

3.4. PLGA/CS NDMs Promote the Survival of OGD-Treated SH-SY5Y Cells

Previous studies have shown that CS degradation products (COS) can promote nerve repair by improving the local microenvironment (in particular, by stimulating Schwann cell proliferation), regulating macrophage migration, and alleviating the cell apoptosis of cortical neurons treated with glucose deprivation [21,48]. In order to investigate whether PLGA/CS NDMs have neuroprotective effects on ischemia–reperfusion brain injury, SH-

SY5Y cells underwent an OGD treatment, and the immersion solution of the PLGA NDM and PLGA/CS NDM were mixed 1:1 with fresh complete medium for reperfusion to avoid the direct influence of the topological structure of the NDMs on SH-SY5Y cells. Cell viability was detected with CCK8 after 24 h of reperfusion. In Figure 6A, there was no significant difference in the cell viability between the PLGA NDM group and the OGD/R group, while the cell viability of the PLGA/CS NDM group was significantly improved. This result suggested that PLGA/CS NDM can promote the survival of SH-SY5Y cells treated with OGD/R. In view of these results, we believe that PLGA/CS NDMs are more suitable for neuroprotection as a sustained release carrier.

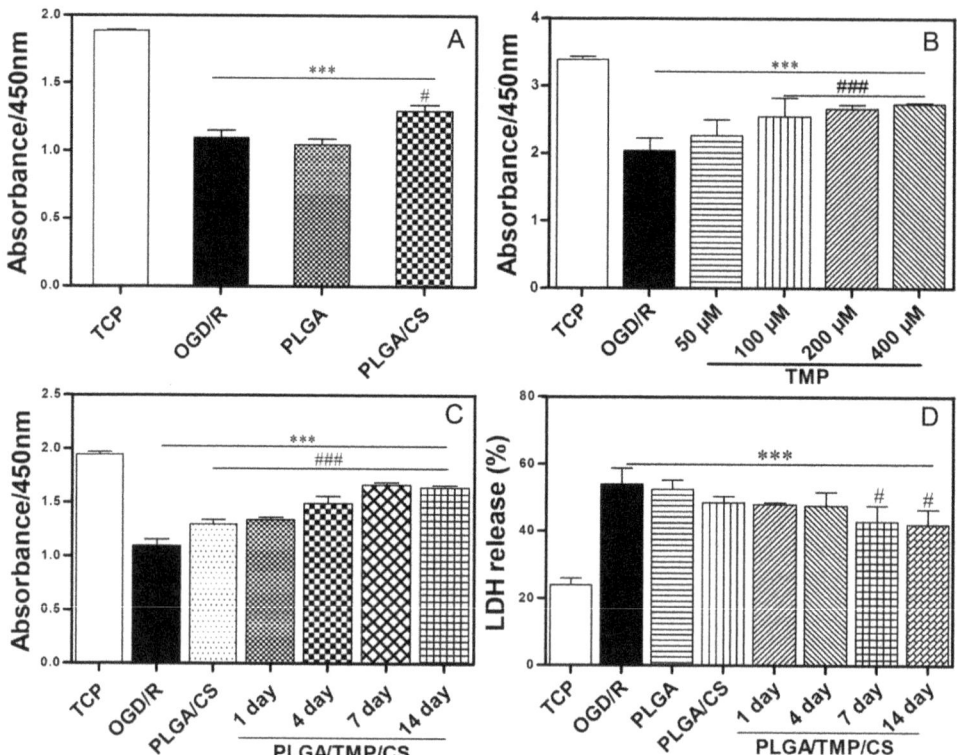

Figure 6. (**A**) Cell viability of OGD-treated SH-SY5Y cells—treated for 3 h and reperfused with immersion solution of PLGA NDM and PLGA/CS NDM mixed 1:1 with fresh complete medium. (**B**) Cell viability of OGD-treated SH-SY5Y cells—treated for 3 h and reperfused with complete medium containing different concentrations of TMP. *** $p < 0.001$ as compared with the TCP. # $p < 0.05$ and ### $p < 0.001$ as compared with the cells reperfused only with complete medium. (**C**) Cell viability of OGD-treated SH-SY5Y cells—treated for 3 h and reperfused with immersion solution of PLGA/CS NDM and PLGA/TMP/CS NDM. The durations of the PLGA/TMP/CS NDM immersion were 1 day, 3 days, and 5 days, respectively, to obtain the corresponding immersion solutions. (**D**) LDH release assay of OGD-treated SH-SY5Y cells—treated for 3 h and reperfused with immersion solution of PLGA NDM, PLGA/CS NDM, and PLGA/TMP/CS NDM.

3.5. The Working Concentration of TMP with Neuroprotective Effect

It has been reported that TMP shows good neuroprotective effects on OGD/R-treated SH-SY5Y cells and neurons in in vitro experiments. However, the working concentration of TMP is not identical in different reported investigations, which may be related to the state of the cells, the batch of the drug, the laboratory environment, or the treatment procedure [49–51]. To determine the working concentration of TMP upon OGD/R challenge, we used OGD-treated SH-SY5Y cells that were then reperfused with complete medium with different concentrations of TMP. After 24 h of reperfusion, the cell viability was detected with CCK8. The results show that the OGD/R treatment significantly reduced the cell viability of SH-SY5Y cells compared to the control group. TMP (50 μM) did not have an obvious protective effect against OGD/R-induced injury, while 100 μM, 200 μM, and 400 μM of TMP significantly improved the cell viability of OGD/R-treated SH-SY5Y cells, the last two concentrations having almost the same effect (Figure 6B). Therefore, we concluded that TMP has a neuroprotective effect on our culture system when the concentration in the culture medium exceeds 100 μM.

3.6. PLGA/TMP/CS NDMs Promote the Survival of OGD-Treated SH-SY5Y Cells

TMP has been shown to have significant neuroprotective effects on ischemia–reperfusion-induced brain injury and has been applied in clinical and basic research. In addition, TMP can diminish the proliferation of fibroblasts [52]. However, the clinical application of TMP has been limited due to its poor water solubility, short half-life, and low concentration of distribution at the injury site. To address these drawbacks, we fabricated a PLGA/TMP NDM with the coaxial electrospinning technique, in which the PLGA serves as a slow-release carrier for the TMP. The TMP was released into the medium by the degradation of the PLGA. To enhance the neuroprotective effect of the NDM, we fabricated a PLGA/TMP/CS NDM by grafting CS onto the PLGA/TMP NDM. However, we did not observe that the PLGA/TMP/CS NDM further reduced the proliferation of fibroblasts compared to the PLGA/CS NDM (Supplementary Materials, Figure S4), possibly because the TMP concentration released from the NDM did not reach the working concentration of 400 μM to diminish the proliferation of fibroblasts, according to the report. To investigate the neuroprotective effects of the PLGA/TMP/CS NDM on ischemia–reperfusion-induced brain injury, SH-SY5Y cells were subjected to OGD treatment. We extracted the immersion solution of the PLGA/TMP/CS NDM after 1, 4, 7, and 14 days, and mixed the immersion solution 1:1 with fresh complete medium for reperfusion after the OGD treatment. After 24 h of reperfusion, the cell viability was tested with CCK8, and the LDH release was assayed. The CCK8 results show that the PLGA/CS NDM promoted the survival of SH-SY5Y as before. The cell viability was further increased with the 1-, 4-, 7-, and 14-day immersion solution of the PLGA/TMP/CS NDM groups compared to the PLGA/CS NDM group. From day 1 to day 14, the cell viability continued to increase (Figure 6C), which may have been due to the COS produced with the CS degradation and the TMP released from the NDM. However, the LDH release assay revealed that the effects of the NDM on LDH release were not obvious. Only 7- and 14-day immersion solution of the PLGA/TMP/CS NDM could significantly reduce the release of the LDH compared to the OGD/R group. (Figure 6D). In conclusion, these results indicate that PLGA/TMP/CS NDM may have long-term neuroprotective effects for ischemia–reperfusion-induced brain injury.

3.7. PLGA/TMP/CS NDM Promote Nerve Repair

TMP not only has neuroprotective effects, but it also enhances neurogenesis. COS can facilitate nerve regeneration. To investigate whether PLGA/TMP/CS NDM can promote the regeneration of neural tissue subjected to ischemia–reperfusion injury, we used OGD-treated SH-SY5Y cells and reperfused them with the previous immersion solution for different durations, as described in Figure 6C. The cells were labeled as GAP43. GAP43 is an axonal membrane protein involved in neural outgrowth, synapse development formation, and neural cell regeneration. The expression of GAP43 means that OGD-treated SH-SY5Y cells are undergoing neural regeneration. The labeling results show that the OGD/R treatment slightly increased the expression of the GAP43 protein, but there was no statistical difference. Both the PLGA/CS NDM and PLGA/TMP/CS NDM further increased the expression of the GAP43 protein, while the PLGA NDM did not have such an effect (Figure 7 and Supplementary Materials, Figure S5). The results suggest that COS and TMP may promote the expression of GAP43 during NDM degradation. Regarding the effect of OGD/R treatment on GAP43 protein expression, the results of different studies were inconsistent. This may be related to factors such as the cell status or treatment process. Some researchers consider that OGD injury can activate the endogenous mechanisms of neuroprotection and neuroplasticity, which may promote the expression of GAP43 [53]. We believe that when cells grow well or when the OGD treatment time is short, GAP43 protein expression may be up-regulated. On the contrary, when the cell growth status is poor or the OGD treatment is severe, the expression of the GAP43 protein may be down-regulated.

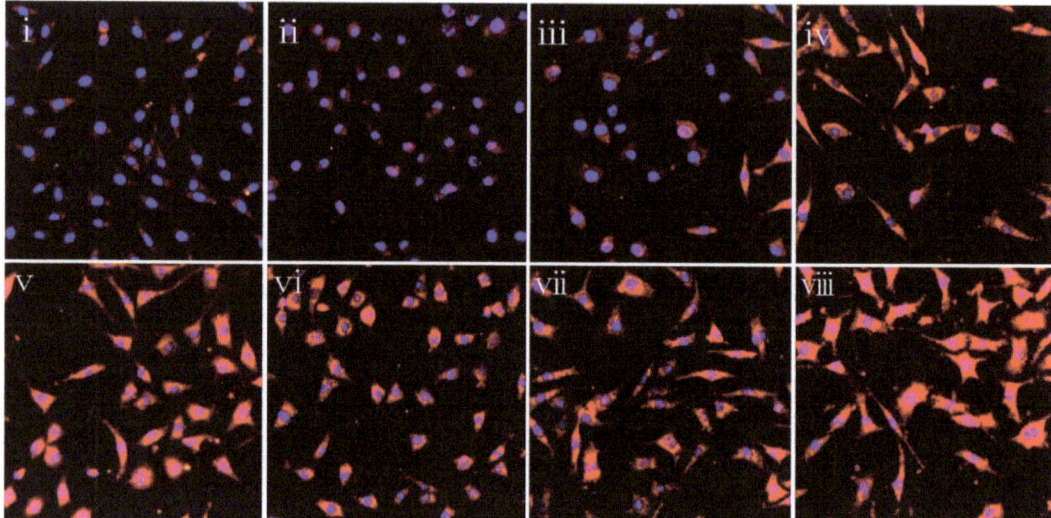

Figure 7. Fluorescence micrographs showing the expression of GAP43 of SH-SY5Y cells. (**i**) TCP cells. (**ii**) The cells treated with OGD for 3 h and reperfused only with complete medium. (**iii,iv**) The OGD-treated cells reperfused with immersion solutions of PLGA NDM (**iii**) and PLGA/CS NDM (**iv**). (**v–viii**) The OGD-treated cells reperfused with immersion solution of PLGA/TMP/CS NDM, with durations of NDM immersion of 1 day (**v**), 4 days (**vi**), 7 days (**vii**), and 14 days (**viii**).

To further clarify the influence of the PLGA/TMP/CS NDM on nerve repair, we stained the actin cytoskeleton and observed the cell morphology. We found that approximately 35% of the SH-SY5Y cells showed pyknosis and lost neurites most likely caused by OGD/R injury. The PLGA NDM and PLGA/CS NDM did not improve the cell morphology. The immersion solution of the PLGA/TMP/CS NDM reduced the damage to the neurites, and the proportion of cells without neurites gradually decreased from day 1 to 14 of the immersion solution (Figure 8A,B). These results indicate that PLGA/TMP/CS NDM can protect the neurites of OGD/R-treated SH-SY5Y cells during degradation. As for the reason the PLGA/CS NDM had no effect, we suppose that the level of COS could not reach the working concentration due to the short immersion time of the NDM, and the time of the cell culture was not enough. We next measured the length of the remaining neurites. As shown in Figure 8C, the lengths of the remaining neurites of the OGD-treated cells reperfused with the PLGA NDM immersion solution were significantly reduced, while they were significantly recovered by those reperfused with the PLGA/TMP/CS NDM immersion solution. We also noted that the PLGA/CS NDM seemed to have had a good effect on the recovery of the lengths of the neurites. Therefore, PLGA/TMP/CS NDM may promote nerve repair with COS and TMP.

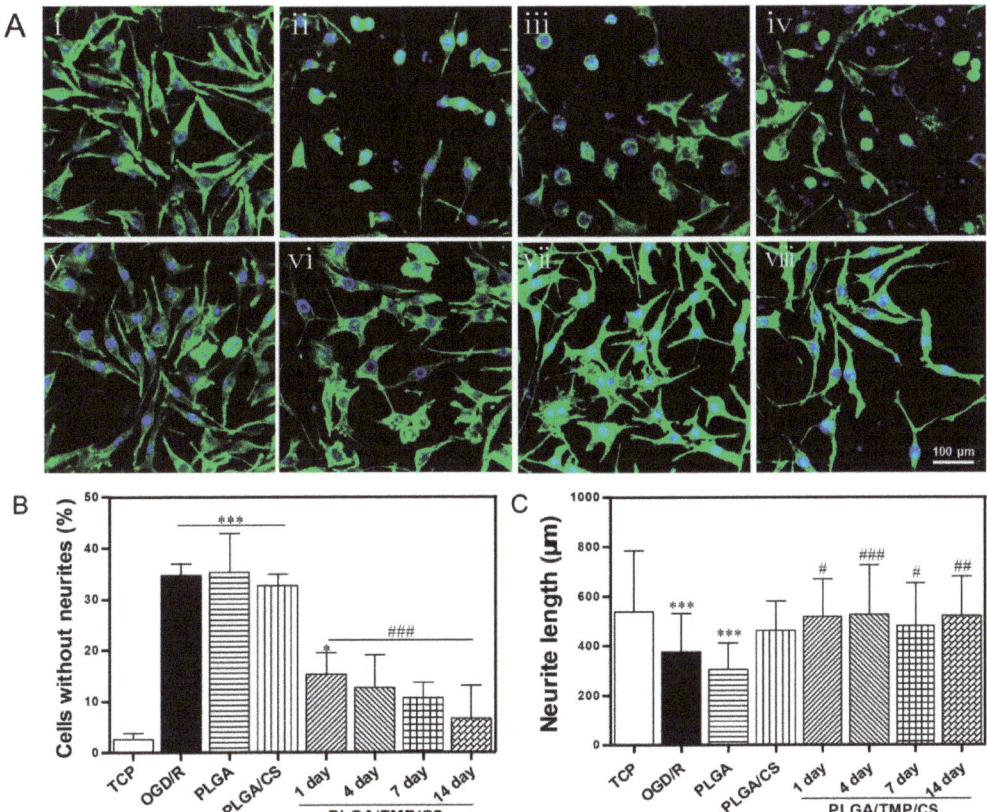

Figure 8. (**A**) Fluorescence micrographs showing the morphology of SH-SY5YS cells treated with OGD/R for 2 days. (**i**) TCP cells. (**ii**) The cells treated with OGD for 3 h and reperfused only with complete medium. (**iii,iv**) The cells treated with OGD and reperfused with immersion solution of PLGA NDM (**iii**) and PLGA/CS NDM (**iv**). (**v–viii**) The cells treated with OGD and reperfused with immersion solution of PLGA/TMP/CS NDM, the durations of which were 1 day (**v**), 4 days (**vi**), 7 days (**vii**), and 14 days (**viii**). (**B**) The percentage of cells after 2 days of OGD/R treatment. (**C**) Lengths of neurites of the cells after 2 days of OGD/R treatment. * $p < 0.05$ and *** $p < 0.001$ as compared to TCP. # $p < 0.05$, ## $p < 0.01$ and ### $p < 0.001$ as compared to OGD/R.

Interestingly, we found that the results of the nerve repair did not exactly match the CCK8 results. According to our statistical results, the morphology of the OGD/R-treated SH-SY5Y cells seemed to be almost completely restored with the PLGA/TMP/CS NDM, but the cell viability was not regained to normal levels, which might have been a consequence of an insufficient analysis of the cell morphology. Furthermore, no in vivo experiments were carried out, so the physiological significance of PLGA/TMP/CS NDM needs to be further studied. In conclusion, our results suggest that PLGA/TMP/CS NDM systems inhibit the excessive proliferation of fibroblasts in vitro and promote the survival and neural repair of OGD/R-treated SH-SY5Y cells in our culture system. Thus, PLGA/TMP/CS NDM may be an alternative for dural grafts.

4. Conclusions

In this paper, we report some PLGA/TMP NDM structures with antifibrotic and neuroprotective effects fabricated with the coaxial electrospinning technique, where the PLGA component was chosen as a slower release carrier. To neutralize the acidic environment generated by PLGA degradation and to enhance the beneficial effect of the NDM (antifibrotic, neuroprotective), we prepared PLGA/TMP/CS NDM by grafting CS on the NDM surface via the EDC/NHS cross-linking method. We found that all NDMs inhibited the excessive proliferation of fibroblasts, but the PLGA/TMP/CS NDM systems were more effective. In terms of neuroprotection, the PLGA/TMP/CS NDM structures were able to promote the survival of OGD/R-treated SH-SY5Y cells as well as nerve repair. In conclusion, the study suggests that PLGA/TMP/CS NDMs may have long-lasting neuroprotective effects and prevent tissue adhesion, fibrosis, and scar tissue formation as dural grafts in undesirable cases of dural injuries.

Supplementary Materials: The following supporting information can be downloaded at: https://www.mdpi.com/article/10.3390/polym14091882/s1, Figure S1: The ultraviolet full-wavelength scanning spectrum of TMP; Figure S2: (A) TMP standard curve made with DCM as solvent for calculation of EE. (B) TMP standard curve made with PBS as the solvent for calculation of TMP release amount; Figure S3: Percentage of Ki-67 positive cells as in Figure 4; Figure S4: Cell viability of fibroblasts seeded on seeded on glass slides as the control group (TCP), on PLGA NDM, PLGA/CS NDM, PLGA/TMP NDM and PLGA/TMP/CS NDM after culture of 1, 3 and 5 days; Figure S5: Expression of GAP43 of OGD/R-treated-SH-SY5Y cells as in Figure 6.

Author Contributions: Z.Z. performed most experiments, analyzed the data, and prepared the manuscript; T.W. conceptualized the idea, directed the study, provided resources, and acquired funding; Y.C. analyzed the data and prepared the manuscript; R.Z. helped with the electrospinning and cell culture; Q.W. acquired funding, provided resources, and prepared the manuscript; R.X. acquired funding, administrated the project, and prepared the manuscript. All authors have read and agreed to the published version of the manuscript.

Funding: This study was supported by the Natural Science Foundation of Shandong Province, China (ZR202102190696) and the Clinical Medicine + X Scientific Research Project of the Affiliated Hospital of Qingdao University to R.X., the National Natural Science Foundation of China (82001970, 32171322), the Natural Science Foundation of Shandong Province, China (ZR2021YQ17), the Young Elite Scientists Sponsorship Program by CAST (No. YESS20200097) to T.W., and the National Natural Science Foundation of China (31900634) to Y.C. This work was also supported by the National Key R&D Program of China (2019YFC0120000; 2018YFC1312300), the National Natural Science Foundation of China (NSFC: 82071385), and the Key Research and Development Project of Shandong (2019JZZY021010) to Q.W.

Institutional Review Board Statement: This study did not require ethical approval.

Informed Consent Statement: This study did not involve humans.

Data Availability Statement: The original contributions presented in the study are included in the article/Supplementary Materials; further inquiries can be directed to the corresponding author.

Conflicts of Interest: The authors declare no competing financial interests.

References

1. Suwanprateeb, J.; Luangwattanawilai, T.; Theeranattapong, T.; Suvannapruk, W.; Chumnanvej, S.; Hemstapat, W. Bilayer oxidized regenerated cellulose/poly ε-caprolactone knitted fabric-reinforced composite for use as an artificial dural substitute. *J. Mater. Sci. Mater. Electron.* **2016**, *27*, 122. [CrossRef] [PubMed]
2. Xu, C.; Ma, X.; Chen, S.; Tao, M.; Yuan, L.; Jing, Y. Bacterial Cellulose Membranes Used as Artificial Substitutes for Dural Defection in Rabbits. *Int. J. Mol. Sci.* **2014**, *15*, 10855–10867. [CrossRef] [PubMed]
3. Ramot, Y.; Harnof, S.; Klein, I.; Amouyal, N.; Steiner, M.; Manassa, N.N.; Bahar, A.; Rousselle, S.; Nyska, A. Local Tolerance and Biodegradability of a Novel Artificial Dura Mater Graft Following Implantation Onto a Dural Defect in Rabbits. *Toxicol. Pathol.* **2020**, *48*, 738–746. [CrossRef] [PubMed]
4. Yamada, K.; Miyamoto, S.; Nagata, I.; Kikuchi, H.; Ikada, Y.; Iwata, H.; Yamamoto, K. Development of a dural substitute from synthetic bioabsorbable polymers. *J. Neurosurg.* **1997**, *86*, 1012–1017. [CrossRef] [PubMed]
5. Parlato, C.; Di Nuzzo, G.; Luongo, M.; Parlato, R.S.; Accardo, M.; Cuccurullo, L.; Moraci, A. Use of a collagen biomatrix (TissuDura®) for dura repair: A long-term neuroradiological and neuropathological evaluation. *Acta Neurochir.* **2010**, *153*, 142–147. [CrossRef]

6. Wang, Y.-F.; Guo, H.-F.; Ying, D.-J. Multilayer scaffold of electrospun PLA-PCL-collagen nanofibers as a dural substitute. *J. Biomed. Mater. Res. Part B Appl. Biomater.* **2013**, *101*, 1359–1366. [CrossRef]
7. Pogorielov, M.; Kravtsova, A.; Reilly, G.C.; Deineka, V.; Tetteh, G.; Kalinkevich, O.; Pogorielova, O.; Moskalenko, R.; Tkach, G. Experimental evaluation of new chitin–chitosan graft for duraplasty. *J. Mater. Sci. Mater. Med.* **2017**, *28*, 34. [CrossRef]
8. Maksimenko, O.; Malinovskaya, J.; Shipulo, E.; Osipova, N.; Razzhivina, V.; Arantseva, D.; Yarovaya, O.; Mostovaya, U.; Khalansky, A.; Fedoseeva, V.; et al. Doxorubicin-loaded PLGA nanoparticles for the chemotherapy of glioblastoma: Towards the pharmaceutical development. *Int. J. Pharm.* **2019**, *572*, 118733. [CrossRef]
9. Mooney, D. Stabilized polyglycolic acid fibre-based tubes for tissue engineering. *Biomaterials* **1996**, *17*, 115–124. [CrossRef]
10. Gelperina, S.; Maksimenko, O.; Khalansky, A.; Vanchugova, L.; Shipulo, E.; Abbasova, K.; Berdiev, R.; Wohlfart, S.; Chepurnova, N.; Kreuter, J. Drug delivery to the brain using surfactant-coated poly(lactide-co-glycolide) nanoparticles: Influence of the formulation parameters. *Eur. J. Pharm. Biopharm.* **2010**, *74*, 157–163. [CrossRef]
11. Wohlfart, S.; Khalansky, A.S.; Gelperina, S.; Maksimenko, O.; Bernreuther, C.; Glatzel, M.; Kreuter, J. Efficient Chemotherapy of Rat Glioblastoma Using Doxorubicin-Loaded PLGA Nanoparticles with Different Stabilizers. *PLoS ONE* **2011**, *6*, e19121. [CrossRef]
12. Wang, Y.; Li, P.; Kong, L. Chitosan-Modified PLGA Nanoparticles with Versatile Surface for Improved Drug Delivery. *AAPS PharmSciTech* **2013**, *14*, 585–592. [CrossRef]
13. Naahidi, S.; Jafari, M.; Logan, M.; Wang, Y.; Yuan, Y.; Bae, H.; Dixon, B.; Chen, P. Biocompatibility of hydrogel-based scaffolds for tissue engineering applications. *Biotechnol. Adv.* **2017**, *35*, 530–544. [CrossRef]
14. Liu, H.; Slamovich, E.B.; Webster, T.J. Less harmful acidic degradation of poly(lactic-co-glycolic acid) bone tissue engineering scaffolds through titania nanoparticle addition. *Int. J. Nanomed.* **2006**, *1*, 541–545. [CrossRef]
15. Nawrotek, K.; Tylman, M.; Rudnicka, K.; Gatkowska, J.; Wieczorek, M. Epineurium-mimicking chitosan conduits for peripheral nervous tissue engineering. *Carbohydr. Polym.* **2016**, *152*, 119–128. [CrossRef]
16. Shin, S.-Y.; Park, H.-N.; Kim, K.-H.; Lee, M.-H.; Choi, Y.S.; Park, Y.-J.; Lee, Y.-M.; Ku, Y.; Rhyu, I.-C.; Han, S.-B.; et al. Biological Evaluation of Chitosan Nanofiber Membrane for Guided Bone Regeneration. *J. Periodontol.* **2005**, *76*, 1778–1784. [CrossRef]
17. Zhang, J.; Dai, H.; Lv, C.; Xing, X. The systematic effects of chitosan on fibroblasts derived from hypertrophic scars and keloids. *Indian J. Dermatol. Venereol. Leprol.* **2012**, *78*, 520. [CrossRef]
18. Zhou, J. Reduction in postoperative adhesion formation and re-formation after an abdominal operation with the use of N, O-carboxymethyl chitosan. *Surgery* **2004**, *135*, 307–312. [CrossRef]
19. Kim, Y.-S.; Li, Q.; Youn, H.-Y.; Kim, D.Y. Oral Administration of Chitosan Attenuates Bleomycin-induced Pulmonary Fibrosis in Rats. *In Vivo* **2019**, *33*, 1455–1461. [CrossRef]
20. Xu, Y.; Huang, Z.; Pu, X.; Yin, G.; Zhang, J. Fabrication of Chitosan/Polypyrrole-coated poly(L-lactic acid)/Polycaprolactone aligned fibre films for enhancement of neural cell compatibility and neurite growth. *Cell Prolif.* **2019**, *52*, e12588. [CrossRef]
21. Zhao, Y.; Wang, Y.; Gong, J.; Yang, L.; Niu, C.; Ni, X.; Wang, Y.; Peng, S.; Gu, X.; Sun, C.; et al. Chitosan degradation products facilitate peripheral nerve regeneration by improving macrophage-constructed microenvironments. *Biomaterials* **2017**, *134*, 64–77. [CrossRef] [PubMed]
22. Lou, T.; Wang, X.; Yan, X.; Miao, Y.; Long, Y.-Z.; Yin, H.-L.; Sun, B.; Song, G. Fabrication and biocompatibility of poly(l-lactic acid) and chitosan composite scaffolds with hierarchical microstructures. *Mater. Sci. Eng. C* **2016**, *64*, 341–345. [CrossRef] [PubMed]
23. Ehterami, A.; Masoomikarimi, M.; Bastami, F.; Jafarisani, M.; Alizadeh, M.; Mehrabi, M.; Salehi, M. Fabrication and Characterization of Nanofibrous Poly (L-Lactic Acid)/Chitosan-Based Scaffold by Liquid–Liquid Phase Separation Technique for Nerve Tissue Engineering. *Mol. Biotechnol.* **2021**, *63*, 818–827. [CrossRef] [PubMed]
24. Pang, P.K.T.; Shan, J.J.; Chiu, K.W. Tetramethylpyrazine, a Calcium Antagonist. *Planta Med.* **1996**, *62*, 431–435. [CrossRef]
25. Yan, Y.; Zhao, J.; Cao, C.; Jia, Z.; Zhou, N.; Han, S.; Wang, Y.; Xu, Y.; Cui, H. Tetramethylpyrazine promotes SH-SY5Y cell differentiation into neurons through epigenetic regulation of Topoisomerase IIβ. *Neuroscience* **2014**, *278*, 179–193. [CrossRef]
26. Yang, Z.; Zhang, Q.; Ge, J.; Tan, Z. Protective effects of tetramethylpyrazine on rat retinal cell cultures. *Neurochem. Int.* **2008**, *52*, 1176–1187. [CrossRef]
27. Chen, Z.; Pan, X.; Georgakilas, A.G.; Chen, P.; Hu, H.; Yang, Y.; Tian, S.; Xia, L.; Zhang, J.; Cai, X.; et al. Tetramethylpyrazine (TMP) protects cerebral neurocytes and inhibits glioma by down regulating chemokine receptor CXCR4 expression. *Cancer Lett.* **2013**, *336*, 281–289. [CrossRef]
28. Fu, Y.-S.; Lin, Y.-Y.; Chou, S.-C.; Tsai, T.-H.; Kao, L.-S.; Hsu, S.-Y.; Cheng, F.-C.; Shih, Y.-H.; Cheng, H.; Wang, J.-Y. Tetramethylpyrazine inhibits activities of glioma cells and glutamate neuro-excitotoxicity: Potential therapeutic application for treatment of gliomas. *Neuro-Oncol.* **2008**, *10*, 139–152. [CrossRef]
29. Li, S.-Y.; Jia, Y.-H.; Sun, W.-G.; Tang, Y.; An, G.-S.; Ni, J.-H.; Jia, H.-T. Stabilization of mitochondrial function by tetramethylpyrazine protects against kainate-induced oxidative lesions in the rat hippocampus. *Free Radic. Biol. Med.* **2010**, *48*, 597–608. [CrossRef]
30. Liang, Y.; Yang, Q.-H.; Yu, X.-D.; Jiang, D.-M. Additive effect of tetramethylpyrazine and deferoxamine in the treatment of spinal cord injury caused by aortic cross-clamping in rats. *Spinal Cord* **2010**, *49*, 302–306. [CrossRef]
31. Xiao, X.; Liu, Y.; Qi, C.; Qiu, F.; Chen, X.; Zhang, J.; Yang, P. Neuroprotection and enhanced neurogenesis by tetramethylpyrazine in adult rat brain after focal ischemia. *Neurol. Res.* **2010**, *32*, 547–555. [CrossRef]

32. Tian, Y.; Liu, Y.; Chen, X.; Zhang, H.; Shi, Q.; Zhang, J.; Yang, P. Tetramethylpyrazine promotes proliferation and differentiation of neural stem cells from rat brain in hypoxic condition via mitogen-activated protein kinases pathway in vitro. *Neurosci. Lett.* **2010**, *474*, 26–31. [CrossRef]
33. Feng, J.; Li, F.; Zhao, Y.; Feng, Y.; Abe, Y. Brain pharmacokinetics of tetramethylpyrazine after intranasal and intravenous administration in awake rats. *Int. J. Pharm.* **2009**, *375*, 55–60. [CrossRef]
34. Xia, H.; Cheng, Z.; Cheng, Y.; Xu, Y. Investigating the passage of tetramethylpyrazine-loaded liposomes across blood-brain barrier models in vitro and ex vivo. *Mater. Sci. Eng. C* **2016**, *69*, 1010–1017. [CrossRef]
35. Xia, H.; Jin, H.; Cheng, Y.; Cheng, Z.; Xu, Y. The Controlled Release and Anti-Inflammatory Activity of a Tetramethylpyrazine-Loaded Thermosensitive Poloxamer Hydrogel. *Pharm. Res.* **2019**, *36*, 52. [CrossRef]
36. Wu, T.; Huang, C.; Li, D.; Yin, A.; Liu, W.; Wang, J.; Chen, J.; Ei-Hamshary, H.; Al-Deyab, S.S.; Mo, X. A multi-layered vascular scaffold with symmetrical structure by bi-directional gradient electrospinning. *Colloids Surf. B Biointerfaces* **2015**, *133*, 179–188. [CrossRef]
37. Reis, K.; Sperling, L.; Teixeira, C.; Sommer, L.; Colombo, M.; Koester, L.; Pranke, P. VPA/PLGA microfibers produced by coaxial electrospinning for the treatment of central nervous system injury. *Braz. J. Med Biol. Res.* **2020**, *53*, e8993. [CrossRef]
38. Cui, Y.; Zhang, Y.; Zhao, X.; Shao, L.; Liu, G.; Sun, C.; Xu, R.; Zhang, Z. ACSL4 exacerbates ischemic stroke by promoting ferroptosis-induced brain injury and neuroinflammation. *Brain Behav. Immun.* **2021**, *93*, 312–321. [CrossRef]
39. Liu, Y.; Xie, L.; Gao, M.; Zhang, R.; Gao, J.; Sun, J.; Chai, Q.; Wu, T.; Liang, K.; Chen, P.; et al. Super-Assembled Periodic Mesoporous Organosilica Frameworks for Real-Time Hypoxia-Triggered Drug Release and Monitoring. *ACS Appl. Mater. Interfaces* **2021**, *13*, 50246–50257. [CrossRef]
40. Yang, S.; Han, X.; Jia, Y.; Zhang, H.; Tang, T. Hydroxypropyltrimethyl Ammonium Chloride Chitosan Functionalized-PLGA Electrospun Fibrous Membranes as Antibacterial Wound Dressing: In Vitro and In Vivo Evaluation. *Polymers* **2017**, *9*, 697. [CrossRef]
41. Yan, S.; Rao, S.; Zhu, J.; Wang, Z.; Zhang, Y.; Duan, Y.; Chen, X.; Yin, J. Nanoporous multilayer poly(l-glutamic acid)/chitosan microcapsules for drug delivery. *Int. J. Pharm.* **2012**, *427*, 443–451. [CrossRef] [PubMed]
42. Shi, Y.; Xue, J.; Jia, L.; Du, Q.; Niu, J.; Zhang, D. Surface-modified PLGA nanoparticles with chitosan for oral delivery of tolbutamide. *Colloids Surfaces B Biointerfaces* **2018**, *161*, 67–72. [CrossRef] [PubMed]
43. Zwirner, J.; Scholze, M.; Waddell, J.N.; Ondruschka, B.; Hammer, N. Mechanical Properties of Human Dura Mater in Tension—An Analysis at an Age Range of 2 to 94 Years. *Sci. Rep.* **2019**, *9*, 16655. [CrossRef] [PubMed]
44. Xu, Y.; Shi, G.; Tang, J.; Cheng, R.; Shen, X.; Gu, Y.; Wu, L.; Xi, K.; Zhao, Y.; Cui, W.; et al. ECM-inspired micro/nanofibers for modulating cell function and tissue generation. *Sci. Adv.* **2020**, *6*, eabc2036. [CrossRef]
45. Li, X.; Chen, L.; Lin, H.; Cao, L.; Cheng, J.; Dong, J.; Yu, L.; Ding, J. Efficacy of Poly(d,l-Lactic Acid-co-Glycolic acid)-Poly(Ethylene Glycol)-Poly(d,l-Lactic Acid-co-Glycolic Acid) Thermogel As a Barrier to Prevent Spinal Epidural Fibrosis in a Postlaminectomy Rat Model. *Clin. Spine Surg. A Spine Publ.* **2017**, *30*, E283–E290. [CrossRef]
46. Li, C.; Wang, H.; Liu, H.; Yin, J.; Cui, L.; Chen, Z. The prevention effect of poly (l-glutamic acid)/chitosan on spinal epidural fibrosis and peridural adhesion in the post-laminectomy rabbit model. *Eur. Spine J.* **2014**, *23*, 2423–2431. [CrossRef]
47. Chen, X.-G.; Wang, Z.; Liu, W.-S.; Park, H.-J. The effect of carboxymethyl-chitosan on proliferation and collagen secretion of normal and keloid skin fibroblasts. *Biomaterials* **2002**, *23*, 4609–4614. [CrossRef]
48. Wang, Y.; Zhao, Y.; Sun, C.; Hu, W.; Zhao, J.; Li, G.; Zhang, L.; Liu, M.; Liu, Y.; Ding, F.; et al. Chitosan Degradation Products Promote Nerve Regeneration by Stimulating Schwann Cell Proliferation via miR-27a/FOXO1 Axis. *Mol. Neurobiol.* **2014**, *53*, 28–39. [CrossRef]
49. Shao, Z.; Wang, L.; Liu, S.; Wang, X. Tetramethylpyrazine Protects Neurons from Oxygen-Glucose Deprivation-Induced Death. *Med Sci. Monit.* **2017**, *23*, 5277–5282. [CrossRef]
50. Zhao, T.; Fu, Y.; Sun, H.; Liu, X. Ligustrazine suppresses neuron apoptosis via the Bax/Bcl-2 and caspase-3 pathway in PC12 cells and in rats with vascular dementia. *IUBMB Life* **2017**, *70*, 60–70. [CrossRef]
51. Zhang, G.; Zhang, T.; Wu, L.; Zhou, X.; Gu, J.; Li, C.; Liu, W.; Long, C.; Yang, X.; Shan, L.; et al. Neuroprotective Effect and Mechanism of Action of Tetramethylpyrazine Nitrone for Ischemic Stroke Therapy. *Neuromol. Med.* **2018**, *20*, 97–111. [CrossRef]
52. Cai, X.; Yang, Y.; Chen, P.; Ye, Y.; Liu, X.; Wu, K.; Yu, M. Tetramethylpyrazine Attenuates Transdifferentiation of TGF-β2–Treated Human Tenon's Fibroblasts. *Investig. Opthalmology Vis. Sci.* **2016**, *57*, 4740–4748. [CrossRef]
53. Olivares-Hernández, J.D.; Balderas-Márquez, J.E.; Carranza, M.; Luna, M.; Martínez-Moreno, C.G.; Arámburo, C. Growth Hormone (GH) Enhances Endogenous Mechanisms of Neuroprotection and Neuroplasticity after Oxygen and Glucose Deprivation Injury (OGD) and Reoxygenation (OGD/R) in Chicken Hippocampal Cell Cultures. *Neural Plast.* **2021**, *2021*, 9990166. [CrossRef]

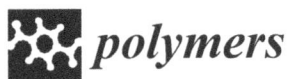

Article

Medical-Grade Silicone Rubber–Hydrogel-Composites for Modiolar Hugging Cochlear Implants

Suheda Yilmaz-Bayraktar [1,2,*], Katharina Foremny [1,2], Michaela Kreienmeyer [1,2], Athanasia Warnecke [1,2] and Theodor Doll [1,2,3]

1   Department of Otolaryngology, Hannover Medical School, Carl-Neuberg-Straße 1, 30625 Hannover, Germany; foremny.katharina@mh-hannover.de (K.F.); kreienmeyer.michaela@mh-hannover.de (M.K.); warnecke.athanasia@mh-hannover.de (A.W.); doll.theodor@mh-hannover.de (T.D.)
2   Cluster of Excellence Hearing4All, Carl-Neuberg-Straße 1, 30625 Hannover, Germany
3   Fraunhofer Institute for Toxicology and Experimental Medicine (ITEM), Nikolai-Fuchs-Straße 1, 30625 Hannover, Germany
*   Correspondence: yilmaz-bayraktar.suheda@mh-hannover.de

Abstract: The gold standard for the partial restoration of sensorineural hearing loss is cochlear implant surgery, which restores patients' speech comprehension. The remaining limitations, e.g., music perception, are partly due to a gap between cochlear implant electrodes and the auditory nerve cells in the modiolus of the inner ear. Reducing this gap will most likely lead to improved cochlear implant performance. To achieve this, a bending or curling mechanism in the electrode array is discussed. We propose a silicone rubber–hydrogel actuator where the hydrogel forms a percolating network in the dorsal silicone rubber compartment of the electrode array to exert bending forces at low volume swelling ratios. A material study of suitable polymers (medical-grade PDMS and hydrogels), including parametrized bending curvature measurements, is presented. The curvature radii measured meet the anatomical needs for positioning electrodes very closely to the modiolus. Besides stage-one biocompatibility according to ISO 10993-5, we also developed and validated a simplified mathematical model for designing hydrogel-actuated CI with modiolar hugging functionality.

Keywords: sensorineural hearing loss; cochlear implants; self-bending electrode arrays; silicone rubber–hydrogel composites; actuators; swelling behavior; curvature; biocompatibility

1. Introduction

Cochlear Implants (CIs) are currently the best solution in compensating for sensorineural hearing loss by direct electrical stimulation of the auditory nerve cells in the inner ear [1]. CIs consist of electrode shafts made from silicone rubber with platinum electrode contacts that are connected via platinum wires with the receiver–stimulator of the implant [2]. Even though this implant is the gold standard for sensorineural hearing loss, a series of limitations, e.g., listening to music or communicating in noisy environments, are yet to be overcome [3,4]. One reason for these limitations is the low effective number of stimulating channels [5], which is due to the wide distance between electrodes and the nerve ganglion cells to be stimulated [6]. Since these nerve cells are situated in the modiolus, a reduction in the distance between implant and cells can be achieved using a CI electrode bent towards the modiolus. Available products are pre-bent shafts that are released from straight stylets during insertion, known as the Contour Advance® (CI612) electrode shaft from the Cochlear Nucleus Profile™ (Cochlea Ltd., Sydney Australia), the Hi-Focus electrode (Advanced Bionics, Valencia USA), and the peri-modiolar push-wire-based electrode Combi 40PM (MED-El, Innsbruck Austria) [5,7]. Another research approach involves using smart alloys which, in contact with tissue at body temperature, change from a stretched shape into a previously impressed curved shape [8]. However, implanting these smart alloys as straight electrodes was obstructed by the reaction of the immediate materials to, e.g., body heat (surgeon) or other heat sources in the operating room, such as lamps.

To overcome the distance between the electrodes and the modiolus, the integration of an actuator into CI arrays using the swelling properties of hydrogels might present a promising alternative.

1.1. Fundamentals of Material and Components

Polydimethylsiloxanes (PDMS) are silicone elastomers used in medical applications since their material properties, e.g., flexibility, biocompatibility, stability, hemocompatibility, and sterilization resistance, are advantageous for longtime implantation. For neuronal implants, liquid silicone rubbers (LSR) or room-temperature polymerized silicone rubbers (RTV) are typically used [9,10] due to favorable characteristics for implants, such as a high degree of physical or chemical purity and thus a very low negative tissue response [11]. PDMS has proven to be a suitable material for integrations in manufacturing CI-electrode arrays.

Hydrogels are three-dimensional hydrophilic crosslinked polymer networks able to swell and shrink with specific characteristics of deformation. They are increasingly used in biomedical applications [12] as mechano-active implants, for example, for tissue repair and regeneration, such as arthrosis, by using mechanical stimuli in order to control cell or drug delivery, accelerating tissue remodeling or healing processes [13]. Examples of synthetic hydrogels are crosslinked poly(methacrylic acid), poly(vinyl alcohol), and polyacrylamide (PAM) [14]. The first silicone rubber–hydrogel composite in the context of CIs under investigation was polyacrylic acid (PAA), used as filler material in a silicone rubber called Silastic. In vivo and in vitro biocompatibility tests showed positive cellular and tissue responses, and an intracochlear curled array could be achieved [15]. Abbasi et al. continued developing an intra-cochlear electrode by using PDMS-PAA as interpenetrating polymer networks proven to have hydrophilic characteristics. Positive cytotoxicity tests were shown, but significantly different cell-adhesion behavior was observed. The best adhesion was achieved for composite samples consisting of 20% (w/v) PAA content, which was suggested to result from an increased wettability of the PDMS-PAA network [16]. To increase the possibility of usage in vivo, our investigation into this field was completed using medical-grade silicone rubber.

In this study, we compare the swelling behavior of medical-grade silicone rubbers Nusil Med 4850 (Avantor Sciences, Radnor, PA, USA) [17] and Silpuran 2430 (Wacker Chemie AG, München, Germany) [18] as well as the hydrogel polyacrylamide/acrylate (PAMAA) as a swelling filler as an approach towards a curved CI electrode array. PAMAA is a hydrophilic copolymer of acrylamide and acrylic acid but crosslinked enough to be insoluble in water. It can absorb up to 400% of its own weight in water and swells proportionately [19]. Based on our previous work presented by Stieghorst et al., the general feasibility of the actuating effect with hydrogel–silicone rubber composites was shown by using the PDMS Sylgard 184 and the hydrogel PAM [20–22]. Due to further experimental results with PAM that negatively influenced the stability of manufactured electrode arrays, we extended the concept by testing different material combinations with the hydrogel PAMAA in different sample designs. In this context, the usability of two medical-grade PDMS (restricted: Silpuran 2430; unrestricted: Nusil Med 4850) in combination with PAMAA is shown in the present study. The general principle of the curling and hugging actuation results from the use of a bimorph actuator consisting of a pure silicone rubber shaft containing electrodes and wires and a silicone rubber–hydrogel composite attached on top of the electrode shaft. The dehydrated hydrogel particles that are compounded into the silicone rubber may swell under water uptake, which leads to the desired curling, as reported by Walling et al. [23]. In our approach, the Ringer solution was used as a substitute for human perilymph [24,25]. The experiments were designed to evaluate the biocompatibility and swelling behavior as well as the functionality of the actuator in CI-like samples.

1.2. Fundamental Physical Processes

This section derives a general description of bending actuation based on swelling tests and visual inspection of the bent test specimen. For the free swelling tests, hydrogel–PDMS composite samples were used, as described in Section 2. For this bending study, bimorph actuators, as given in design B of the same figure, were considered. From literature and data sheets, the following parameters were available: the Young's modulus of the pure silicone rubber (E_{PDMS}) and its density (ρ_{PDMS}), the density of the dry hydrogel ($\rho^{H,dry}$), and the weight fraction of the dry hydrogel contained in the silicone rubber matrix (f_H). As the considerations presented in the following sections only address the final steady states, we do not take the particle size into account, which has an influence on the dynamic behavior.

The measurable quantities obtained by visual inspection are the thickness of the PDMS layer in design B (h_{PDMS}) and the corresponding thickness of the composite layer in its dry, initial state ($h_{comp,init}$). After swelling, the quantity $h_{comp,init}$ reaches the $h_{comp,fin}$ value according to the newly swollen state. The bending radius R of such a swollen test specimen may also be assessed by visual inspection and fitting. The swelling process can be described by analogy with osmosis, where the hydrogel acts as both a semi-permeable membrane and receptacle. The osmotic pressure in the hydrogel is limited by the mechanical stress corresponding to the maximum elastic deformation of the PDMS network during hydrogel swelling, which must be balanced with the external osmotic pressure of the perilymph/Ringer solution [26]. The amount of swelling can be calculated from the measurements of the Ringer solution uptake of the free swelling composite samples (design A), with m_i indicating the initial weight of the sample in a dry state and m_f indicating the final weight of the same sample in a swollen state.

The physical bending model is based on the following considerations. It is assumed that in the free-swollen composite material in specimens of design A, the effective net force is zero. Actually, the internal osmotic pressure of the hydrogel phase in a swollen composite sample is balanced by the opposite stress generated by the elongation of the interweaving PDMS matrix. In contrast, the PDMS–hydrogel composite material in design B will have regions of mechanical limitation of swelling. In particular, in the regions near the interface of the PDMS/composite layer, an effective pressure will predominate, which arises from the internal osmotic pressure exceeding the reduced elongation of the PDMS but is balanced by the remaining elongation induced into the pure, adjacent PDMS layer. The distribution of this internal pressure from the PDMS/composite interface towards the freestanding end of the composite is initially unknown and must be clarified by additional investigations.

In order to make stress and pressure accessible, the Ringer solution uptake of the composite is transformed into the corresponding volumes. Starting from the known quantities m_i, m_f, and f_H, the volumes V_i and V_f can be calculated via the densities of the PDMS (ρ_{PDMS}) and the hydrogel phase in its dry ($\rho_{H,dry}$) and swollen state ($\rho_{H,sw}$)

$$V_i = m_i \left(\frac{1 - f_H}{\rho_{PDMS}} + \frac{f_H}{\rho_{H,dry}} \right) \quad (1)$$

$$V_f = \frac{m_i(1 - f_H)}{\rho_{PDMS}} + \frac{m_f - m_i(1 - f_H)}{\rho_{H,sw}} \quad (2)$$

which yields the relative volume expansion v (expressed as $(V_f - V_i)/V_i$):

$$v = \frac{\frac{1 - f_H}{\rho_{PDMS}} + \frac{m_f/m_i - 1 + f_H}{\rho_{H,sw}}}{\frac{1 - f_H}{\rho_{PDMS}} + \frac{f_H}{\rho_{H,dry}}} \quad (3)$$

Due to the bending of the test specimen, we assume a radially increasing value of the relative volume expansion $v(r)$.

In a free swelling PDMS–hydrogel composite, the swelling is always isotropic, which means that the linear expansion factor $\Delta l / l_0 = \sqrt[3]{v}$ is the same irrespective of the spatial direction; we have to check this for the complex stress and pressure balance in design B. However, it might be reasonable to expect that in the vicinity of the interface from the composite to the surrounding fluid, almost free swelling is reached.

An approach with finite differences is sketched in Figure 1. On the upper left side, a cross-section of design B is given. The bending here takes place out of the drawing plane along the x-axis. The small inset shows that bending along the section plane and a small segment of this circular bending was magnified and related to the bending radius R. This illustrates that the bending of the PDMS composite is different from a classic bimorph, as both layers undergo elongations Δx_i, as can be seen in Figure 1. As indicated in the graph to the right, the stress σ in pure PDMS must increase from the fluid–PDMS interface (position R) towards the internal PDMS composite interface (position $R + h_{PDMS}$). In the swollen composite material, on the other hand, it seems to obvious that there is an osmotic pressure p that decreases towards the composite–fluid interface (position $R + h_{PDMS} + h_{comp,fin}$). Thus, the elongations along the X-axis of the composite in its swollen state are Δx_R, $\Delta x_{R+hPDMS}$, and $\Delta x_{R+hPDMS+hcomp,fin}$, as graphically illustrated in Figure 1c. From the similar triangles characterized by the hypotenuses denoted by R, $R + h_{PDMS}$, and $R + h_{PDMS} + h_{comp,fin}$, the following equalities result:

$$\frac{x_0 + \Delta x_R}{R} = \frac{x_0 + \Delta x_{R+hPDMS}}{R + h_{PDMS}} = \frac{x_0 + \Delta x_{R+hPDMS+hcomp,fin}}{R + h_{PDMS} + h_{comp,fin}} \quad (4)$$

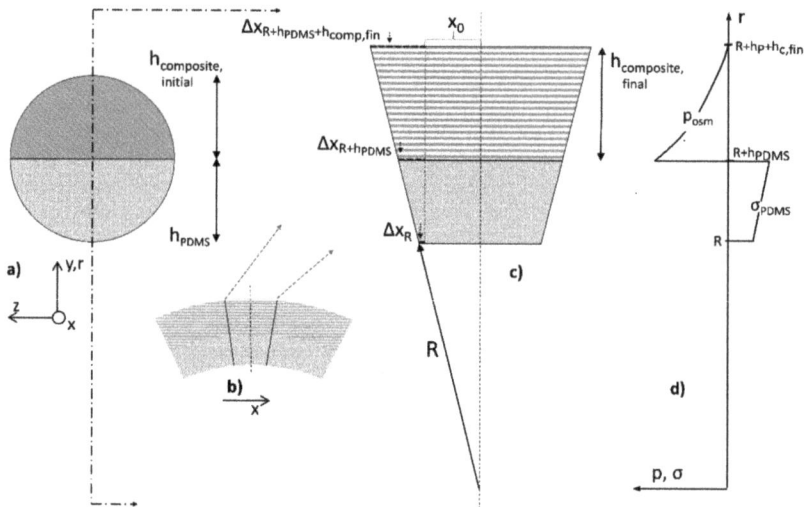

Figure 1. Swelling of cylindrical specimens "Design B": (**a**) cross-section with pure silicone rubber (gray) and dry, initial silicone–hydrogel composite (hatched) compartments, (**b**) cross-section showing bending in (x-) direction due to swelling of the silicone–hydrogel composite, (**c**) calculation of elongations in x-direction for given bending radius R and final height of the composite layer and (**d**) schematic distribution of stress and pressure in PDMS and composite compartments.

Equation (4) can be rewritten as:

$$R = (1 + \varepsilon(R))\frac{R + h_{PDMS}}{1 + \varepsilon(R + h_{PDMS})} = (1 + \varepsilon(R))\frac{R + h_{PDMS} + h_{comp,fin}}{1 + \varepsilon\left(R + h_{PDMS} + h_{comp,fin}\right)} \quad (5)$$

where ε signifies the extensional strain or elongation at different interfaces of the composite expressed as follows: $\varepsilon(R) = \Delta x_R/\Delta x_0$, $\varepsilon(R + h_{PDMS}) = \Delta x_R + h_{PDMS}/\Delta x_0$ and $\varepsilon(R + h_{PDMS} + h_{comp,fin}) = \Delta x_R + h_{PDMS} + h_{comp,fin}/\Delta x_0$. At the same time, it is assumed that an isotropic swelling at the composite–fluid edge can provide the necessary boundary condition for the determination of all further parameters. This holds if shear moduli and transverse contractions can be neglected, which leaves the sectional planes in the radial direction free of forces. Furthermore, if the modulus of elasticity of the hydrogel in its swollen state makes a negligible contribution to the Young's modulus of the composite body compared to the PDMS, one obtains $E_{comp,fin} \cong (1-f_H)(\rho_{PDMS}/\rho_{H,dry}) E_{PDMS}$ for the freely swollen composite layer. Under the variable volume expansion $v(r)$, this leads to:

$$E_{comp,sw}(r) = E_{PDMS} \frac{1-f_H}{v(r)} \times \frac{\rho_{PDMS}}{\rho_{H,dry}} \tag{6}$$

Basically, both the stress in the PDMS and the pressure in the composite act like torques on a one-armed lever with a pivot point at the center of the bending circle. These both sum to zero, as otherwise, the pivoting point would become displaced. While the torques for the PDMS layer can be easily specified, the pressure curve in the composite actually requires the functional for equilibrium pressure via the volume-limited swelling of the hydrogel. In this generalized form, one obtains the final relationship:

$$0 = \left(R + \frac{h_{PDMS}}{2}\right) \frac{\varepsilon(R) + \varepsilon(R + h_{PDMS})}{2} E_{PDMS} - \int_{h_{PDMS}}^{h_{comp,fin}} p(r, \varepsilon(r)) dr \tag{7}$$

2. Materials and Methods

2.1. Sample Preparation

Two silicone rubbers were used to evaluate their suitability for the envisioned application. Silpuran 2430 (Wacker Chemie AG, Burghausen, Germany) is a two-component (RTV2) silicone rubber approved for human use in a restricted interval not exceeding 28 days. The two-component medical LSR NuSil® MED-4850 (Avantor Sciences, Radnor, PA, United States) is approved for use in human implantation for a period of greater than 28 days. The components of both PDMS are mixed in a ratio of 1:1. Relevant PDMS properties are shown in Table 1. Nusil Med 4850 has a tensile strength of 10.17 MPa and a maximum elongation of 675%, while Silpuran 2430 has a tensile strength of 6 MPa and a 540% elongation. Therefore, mechanical stability after the expansion is expected for both PDMSs [17,18].

Table 1. Material properties of Silpuran 2430 and Nusil Med 4850 [17,18].

	Silpuran 2430	Nusil Med 4850
Curing	10 min at 135 °C	5 min at 150 °C
Biocompatibility	Restricted (<28 days)	Unrestricted
Polymerization	Room temperature polymerized (RTV 2)	Liquid Silicone Rubber (LSR)
Hardness Shore	20	50
Tensile strength	6 MPa	10.17 MPa
Elongation at rupture	540%	675%

1 ISO 888; 2 According to ISO 10993-5 no cytotoxicity.

Samples consist of either Silpuran 2430 or Nusil Med 4850 mixed with the hydrogel powder (polyacrylamide/-acrylate (PAMAA) (AC33-GL-000110, Goodfellow, Hamburg, Germany)). Three sample designs were used for the investigation (see Figure 2). Sample design A was used to investigate basic material properties. Sample designs B and C were exclusively produced with the unrestricted medical LSR Nusil Med 4850 and were used for proof-of-principle observations. For samples B and C, the compound was added onto a silicone rubber layer to investigate the interactive material behavior in the course of the swelling process and to evaluate the compounds' readiness to use as an actuator on CIs. The sample design A and B size (see Figure 2) results in a surface-to-volume ratio of 1.7 mm, which is close to that of CI electrode array, as can be seen in Figure 2. Sample design C had measures in accordance with cochlea measurements. The molds for sample designs A and B were 3D-printed (Formlabs, Berlin, Germany) with artificial resin (Clear resin, Formlabs, Berlin, Germany). The mold for sample type C was produced manually with acrylic glass.

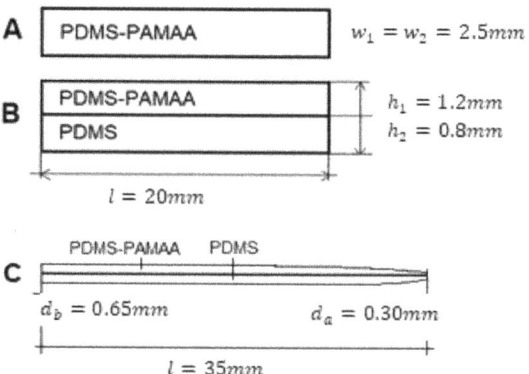

Figure 2. Schematic of the three sample designs, where w, h, and l are the width, height, and length of the outlined samples, respectively. (**A**): Rectangular design consisting of one-layer composite compound material; (**B**): rectangular piggy-back design: the bottom layer is the silicone rubber, and the upper layer is the compound material; (**C**): CI-shape design with an additional layer of composite material.

Since the grain diameter is 2.5 mm, PAMAA granules were ground and sieved into different grain-size fractions to evaluate the influence of grain size on the actuator's performance (see Appendix A.1). Three grain-size fractions were produced using a ball mill (Pulverisette 23, Fritsch, Idar-Oberstein, Germany). The first fraction contained all grains with a diameter ($\varnothing$) $\leq$ 20 µm, the second fraction consisted of 20 µm $< \varnothing \leq$ 50 µm (diameter) grains, and the third fraction contained 50 µm $< \varnothing \leq$ 100 µm (diameter) grains. The particle size distribution of these fractions was determined statistically using a helium-neon laser for optical spectrometry (HELOS) sensor (Sympatec GmbH, Clausthal-Zellerfeld, Germany).

In preliminary tests, several PAMAA contents from 10 wt% up to 40 wt% were investigated. Samples with initial hydrogel percentage of >30 wt% exhibited very high volume expansions, unfit for the dimension in the inner ear. Therefore, this study focused on PAMAA hydrogel fractions between 20 and 30 wt%, as given in Table 2. Silicone rubber and hydrogel particles were mixed together by means of a speed mixer (Hauschild Engineering, Hamm, Germany) at 3500 rpm. As displayed in Table 2, sample designs B and C were not prepared with the smallest grain-size fractions. This was the result of the low amount of the respective powder after long grinding and sieving processes. It was decided to first evaluate the other fractions. Sample design C was then only prepared with the medium grain-size fraction due to the results from sample design A.

2.2. Swelling Tests

To investigate the hydrogel swelling, samples were stored in Ringer solution for up to 28 days in a dry cabinet at 37 °C. Test containers (microboxes made of PP) were filled with 10 mL Ringer solution (Berlin-Chemie AG, Berlin, Germany), then dry samples were added after measuring the initial weight. The samples' weight change was measured eight times in one-hour steps after initial weighing. Then, measurements were completed in 24 h steps for seven days, followed by one measurement per week. Before weighing, samples were dabbed lightly on a tissue wipe to remove any adhering liquid drops. The swelling ratio was plotted vs. the square root of time. For sample designs B and C, the values of radii of each sample and their corresponding median were determined using GraphPad Prism.

Table 2. List of manufactured samples with designs according to Figure 2. The investigated silicone rubbers are given for each design, as well as the PAMAA particle size and the respective weight percentage used in the compounds.

PAMAA Particle Size	Layer Height	Material Combinations	
Samples of design A <20 µm 20–50 µm 50–100 µm		Silpuran 2430 + PAMAA | 20 wt%, 25 wt%, 30 wt% $n = 6$ samples each	Nusil Med 4850 + PAMAA
Samples of design B 20–50 µm 50–100 µm		Nusil Med 4850 + PAMAA 20 wt% | $n = 6$ samples 20 wt%, 25 wt%, 30 wt% | $n = 6$ samples each	
Samples of design C 20–50 µm	Basal 0.65 mm Apical 0.30 mm	Nusil Med 4850 + PAMAA 20 wt% | $n = 3$ samples	

2.3. Biocompatibility Tests

For first biocompatibility information on the silicone rubber–hydrogel, the water-soluble tetrazolium dye assay (WST-1 assay, Cell proliferation Reagent WST-1, Roche GmbH, Basel, Switzerland) was used in accordance with EN ISO 10993-1-19. Further information can be found in Appendix A.2. If the cell viability was confirmed, consecutive tests need to be performed in later steps of the conformity-evaluation procedure.

3. Results and Discussion

3.1. Hydrogel Processing and Particle Size Distribution

Hydrogel with an initial particle size of 2.5 mm was ground and sieved to generate three particle-size fractions. Several iterations of the grinding process were necessary to obtain sufficient material for $n = 6$ samples for each of the three weight percentages under evaluation (20 wt%, 25 wt%, 30 wt%). Only small amounts of particle size fraction <20 µm were produced due to rapid reagglomeration of the hydrogel particles.

Optical spectroscopy (laser scattering/diffraction) was performed to determine the particle size distribution in the investigated grain-size fractions. Both cumulative and density distributions were calculated. For each sample, three measurements were taken that showed varying median values of particle sizes depending on the sieve mesh size.

Figure 3 shows a distribution diagram with both cumulative and density distributions obtained through laser diffraction for hydrogel powder of the grain-size fraction <20 µm.

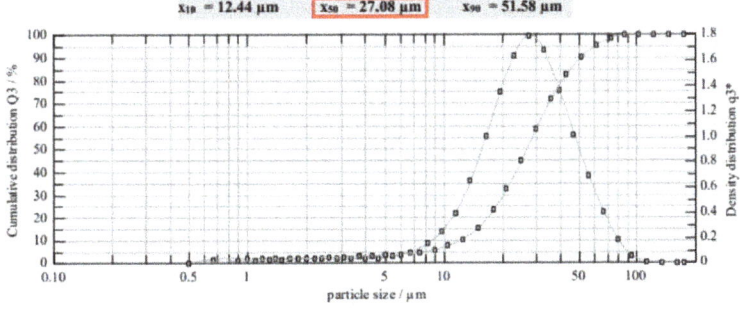

Figure 3. Exemplary distribution diagram of mean particle size distribution measured via spectroscopy (laser scattering) showing cumulative and density distributions for one sample of the grain-size fraction <20 µm. Measurement was repeated three times (see Table 3).

Table 3. Laser diffraction results of hydrogel particle size distribution of the grain size mass fractions <20 µm, 20–50 µm, and 50–100 µm. Three averaged median values (n = 3) for each size distribution are shown.

	50–100 (µm)	20–50 (µm)	0–20 (µm)
Median (x_{50})	77.39	48.00	28.34
Standard deviation	6.38	1.01	1.26

After sieving with a mesh size of 20 µm, the particle fraction was expected to consist of particles <20 µm, while spectroscopy results showed a median for one sample of 27.08 µm (marked red). As can be seen in Figure 3, a large number of particles with higher grain size was detected, while a lower quantity of particles with grain size smaller than <20 µm was recorded. Particle sizes as median and standard deviation values averaged over three measurements are listed in Table 3. With the exception of the smallest fraction, the PAMAA has median values that correspond to the grain-size mass fractions expected after sieving. In the following, the ranges of the grain size distribution are replaced by the measured grain-size fraction median values of 77 µm (replacing 50–100 µm), 48 µm (replacing 20–50 µm), and 28 µm (replacing 0–20 µm).

It is known that the process of mixing silicone rubber with crosslinked fillers in elastomers requires the break-down of the pellets for efficient dispersion of pellet fragments as small micro-scale particles [27]. The smallest possible grain-size fraction proposed in our approach was 20 µm, with a measured median value of 28 µm. Assuming the 20 µm sieve is flawless, this result confirms a particle aggregation after the sieving procedure. During hydrogel manufacturing, reagglomeration is a known problem caused by, e.g., residual water [28]. Reagglomeration is likely to occur due to water uptake of the hydrogel particles in our processing and storage environment. Yet, the agglomeration mainly seems to occur for particle sizes <20 µm. Yanagioka et al. reported that the agglomeration of inorganic particles in polymeric composite materials is fundamentally important but not yet completely understood [29]. Their study also reported a relationship between particle distribution within a polymer matrix and some composite mechanical properties, such as agglomeration and swelling [29].

3.2. Swelling Tests

3.2.1. Design A—Hydrogel–Silicone Rubber-Compound Properties

The sample weight measurements indicate a rapid mass increase during the first 24 h after immersing the sample in Ringer solution. An increase in hydrogel concentration induces a significant increase in the degree of swelling, which is, however, influenced by the PDMS matrix. Figure 4 shows a sample of design A, which exhibits a typically uneven surface structure after the swelling process. This structure is due to inhomogeneous particle distribution in the initial compound. Based on the fact that swelling behavior is dependent on the hydrogel fraction, the crosslinking density, and the particle grain size [30], three parameters were evaluated in our study, including the influence of the silicone rubber matrix, the initial hydrogel content, as well as the hydrogel particle size, and will be discussed in the following. In general, the samples showed a rapid weight increase during the first 24 h after being placed in the Ringer solution. PAMAA-Silpuran 2430 samples all show a rapid water uptake in the first 24 h. Regarding the swelling characteristics, samples with the highest particle content (30 wt%) dispersed into Silpuran 2430 show the highest volume swelling ratio (max. 370%), while the compound with the lowest amount of dry hydrogel (20 wt%) led to the lowest volume swelling ratio (max. 156%) (for comparison, see Figure 5). For the initial hydrogel percentage of 20 wt% and 25 wt%, the maximum volume swelling increases with an increase in grain size. Thus, for an initial hydrogel amount of 25 wt%, the highest results are 164% (28 µm mean grain size), 214% (48 µm mean grain size), and 256% (77 µm mean grain size). However, this fact could not be observed for the 30 wt% hydrogel. For this initial hydrogel amount, the highest volume swelling ratio was obtained for a grain-size fraction of 48 μm. The grain-size fraction influences the duration of water uptake until the saturation point is reached.

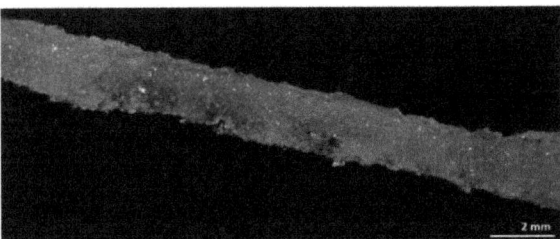

Figure 4. Optical microscopy image of sample design A representing Nusil Med 4850-PAMAA in swollen state (672 h). Hydrogel particle size fraction was 77 μm, and hydrogel content was 30 wt% prior to swelling.

Nusil Med 4850-PAMAA samples showed a steep increase in swelling ratio over the first hours of tests until they reached a saturation point. After this point, the swelling evolved much slower and mainly depended on the hydrogel particle size. Overall, for the smallest and the largest grain-size fractions, the swelling ratio was leveled down around the saturation point. In contrast to the compounds made of Silpuran 2430, the smallest grain-size fraction contained in the Nusil Med 4850-PAMAA composite led to the highest swelling ratios (187%/30 wt% PAMAA), and the highest grain-size fractions (77 μm) led to the lowest maximum swelling ratios (113%/25 wt%). In the case of hydrogel particles with a mean grain size of 48 μm, the swelling ratio followed a descending tendency to an equilibrium value after reaching the saturation point. For the smallest and medium grain-size fractions of hydrogel, an increased amount of PAMAA led to an increased swelling ratio. This was not true for the samples containing the highest grain-size fractions, where the highest initial weight percentage of PAMAA exhibited the lowest swelling ratio.

For test results with Silpuran 2430-PAMAA, the lowest grain-size fraction of 28 μm and the highest fraction of 77 μm displayed lower swelling ratios compared to the middle grain-size fractions with 48 μm particles. However, this trend cannot be observed for tests with Nusil Med 4850-PAMAA. Here, the lowest grain-size fraction indicated the highest swelling ratios, while the middle and greatest grain-size fractions had similarly low swelling ratios. As can be observed for several material combinations (see Figures 5 and 6), the swelling ratio decreased after reaching equilibrium. This is assumed to be due to particle loss. In general, the Nusil Med 4850-PAMAA composite samples achieved a lower equilibrium swelling compared to samples composed of Silpuran 2430 (see Figures 5 and 6) for each grain-size fraction. The time until a saturation point was reached differed significantly for both silicone rubbers, irrespective of the amount of hydrogel or grain-size fraction. A possible explanation goes in parallel with the leaching as seen above: Nusil has a weaker interaction with the hydrogel, which speeds up the water uptake. Silpuran holds the particles tighter, which slows down the swelling. This indicates that the experiments can be modeled with the described equations in Section 1.2. Another difference in the swelling behavior between the silicone rubbers is that PAMAA with Silpuran 2430 exhibits higher swelling ratios than PAMAA with Nusil Med 4850. In the context of sodium polyacrylate, the dependence of the shear modulus on swelling behavior results in an ascending tendency. Additionally, the osmotic swelling pressure and the elastic modulus strongly depend on the ionic composition of the surrounding fluid [26]. Among many things, the Young's modulus and Shore hardness seem to influence the swelling ratio. Our performed swelling measurements show the following: the Nusil Med 4850 (hardness Shore 50)-based composites displayed lower swelling ratios compared to those obtained for Silpuran 2430-based hydrogels, for which the rubber component was characterized by a lower Shore hardness (20) and, consequently, by an augmented elasticity. This leads to the conclusion that increased Shore hardness leads to decreased elasticity and, therefore, to reduced hydrogel swelling in the network.

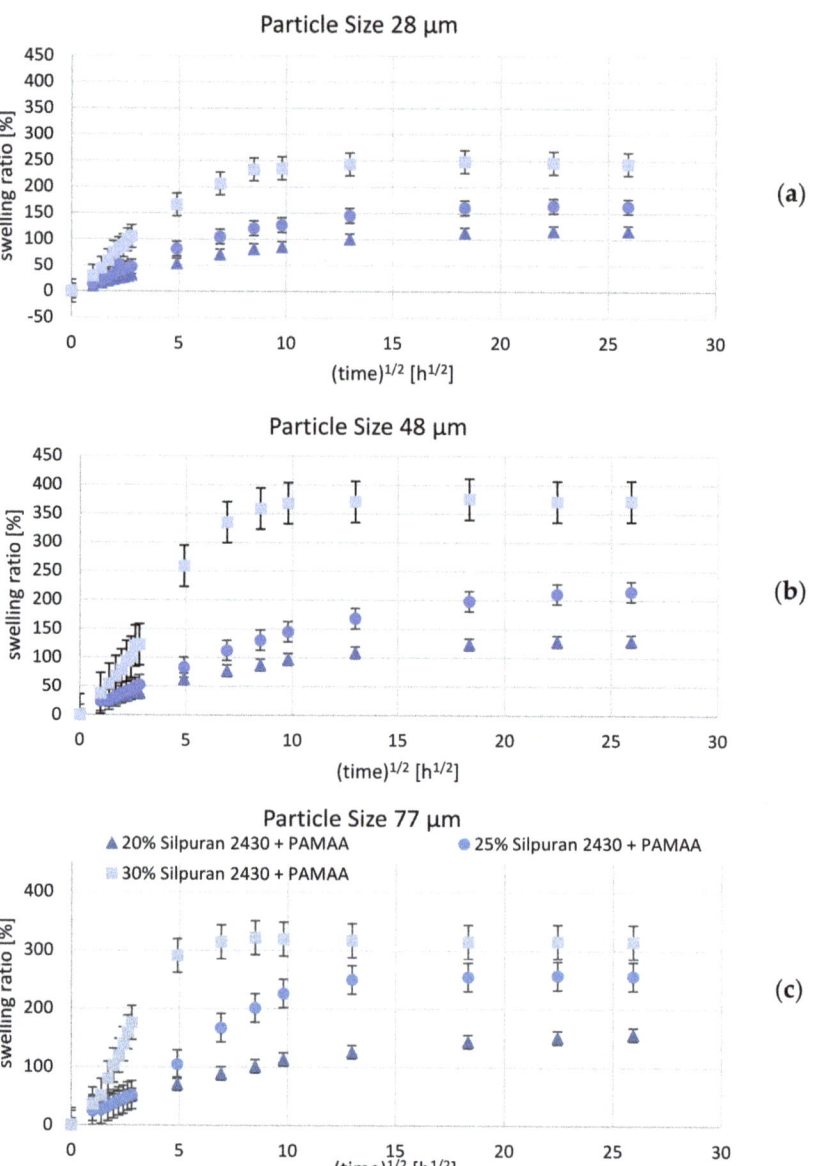

Figure 5. Mean swelling ratio over square root of time (total swelling time: 672 h) for Silpuran 2430-PAMAA composite samples with initial hydrogel content of 20 wt%, 25 wt%, and 30 wt%. Error bars indicate standard deviation for test samples with hydrogel particle size (median) of (**a**) 28 μm, (**b**) 48 μm, and (**c**) 77 μm. For each material combination, $n = 6$ samples were investigated.

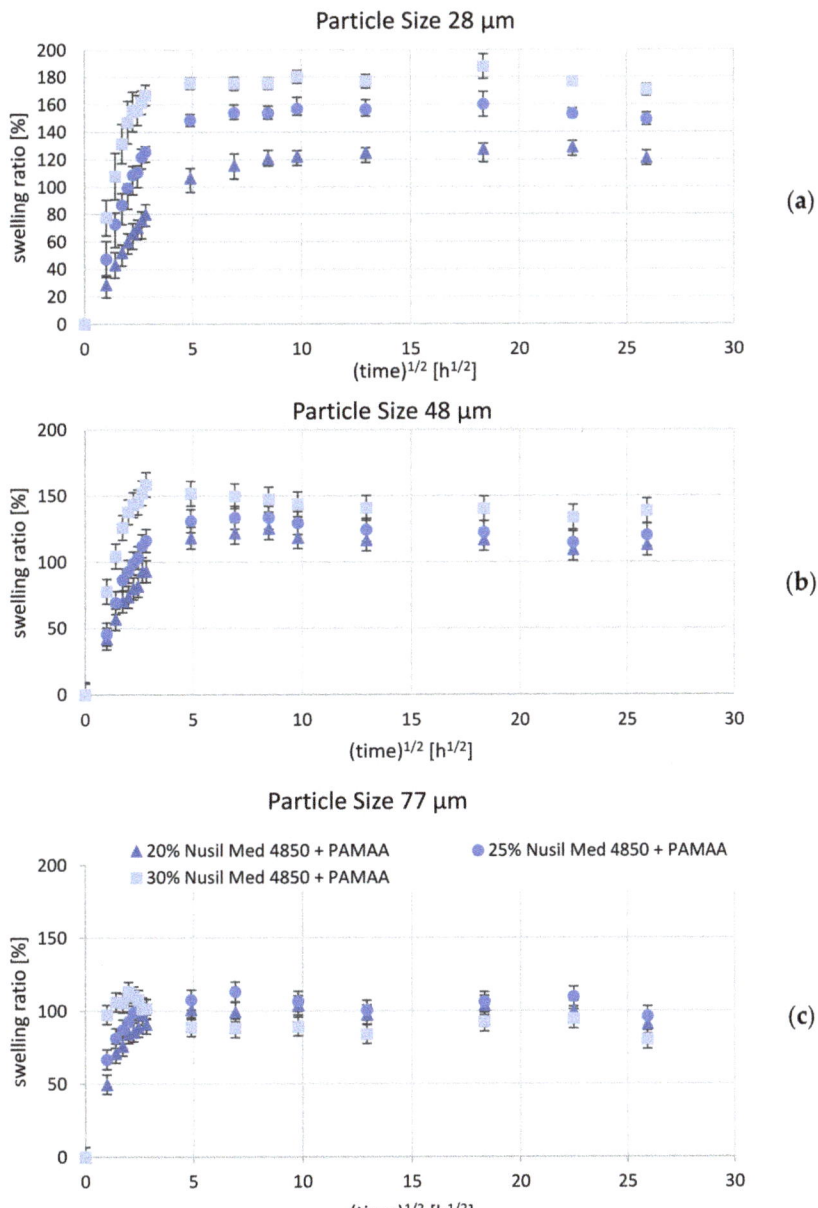

Figure 6. Mean swelling ratio over square root of time (total swelling hours: 672) for Nusil Med 4850-PAMAA composite samples with initial hydrogel content of 20 wt%, 25 wt%, and 30 wt%. Error bars indicate standard deviation for test samples with hydrogel particle size (median) of (**a**) 28 µm, (**b**) 48 µm, and (**c**) 77 µm. For each material combination, n = 6 samples were investigated.

3.2.2. Design B—Bimorph Hydrogel–Silicone Rubber-Composite on Silicone Rubber Base

The casting mold for manufacturing samples with design B proved to be difficult since the composite layer had to be applied onto a layer of pre-vulcanized silicone rubber. Nevertheless, the resulting interface proved to be mechanically stable enough to withstand delamination after water uptake of the hydrogel phase (see Figure 7). Figure 8 shows the mean swelling ratio as a function of the square root of time for Nusil Med 4850-PAMAA samples with a grain-size fraction of 77 µm.

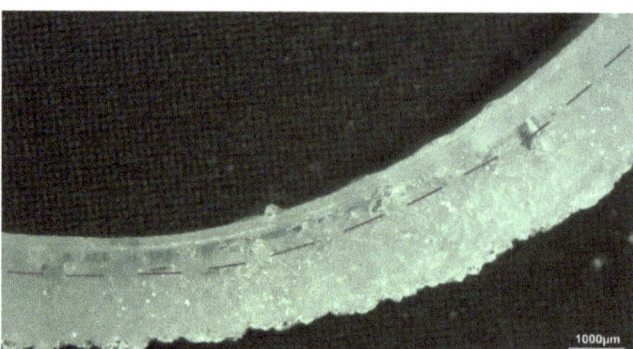

Figure 7. Optical microscopy image of swollen sample design B using Nusil Med 4850-PAMAA: swelling medium—Ringer solution; swelling time—168 h; hydrogel particle size fraction—77 µm; initial hydrogel content—20 wt%.

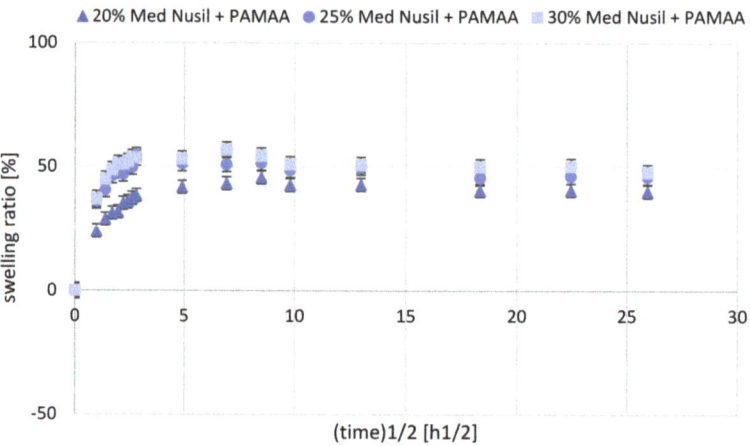

Figure 8. Mean swelling ratio over square root of time (total swelling time: 672 h) for Nusil Med 4850-PAMAA composite design B samples with initial hydrogel concentration of 20 wt%, 25 wt%, and 30 wt%. Error bars indicate standard deviation for test samples with hydrogel particle size (median) of 77 µm. For each material combination, n = 6 samples were investigated.

In comparison to samples of design A, the samples of design B displayed a distinctly lower swelling ratio with a Ringer solution absorption, attaining a maximum at a relatively early stage of swelling and then slightly deswelling until equilibrium was reached. The swelling ratio of sample design B with an initial hydrogel content of 30 wt% did not reach 60% during the first 24 h. In the course of the next measurements until 336 h, the maximum value of the swelling ratio declined by up to 45%. Samples with 25 wt% PAMAA reached a maximum swelling ratio of 55% after 8 h, and Ringer solution saturation was detected after 336 h at 45%. Samples with 20 wt% hydrogels have a maximum swelling ratio of 45% after 72 h. After 336 h, saturation was measured at 41%. Compared to the results of design

A with the same material composition and grain size, a reduction in half of the swelling ratio from 100% to 56% was found. In general, the lower swelling ratios were caused by the silicone rubber's retarding force given by an additional PDMS layer, which is the solid part within the process for inducing the matrix bending apart from swelling. This is also confirmed and was expected according to the theoretical description of the swelling process in Section 1.2.

To evaluate the swelling calculations mentioned in Section 1.2, cross-sections of samples of design B were evaluated and measured, as seen in Figure 9.

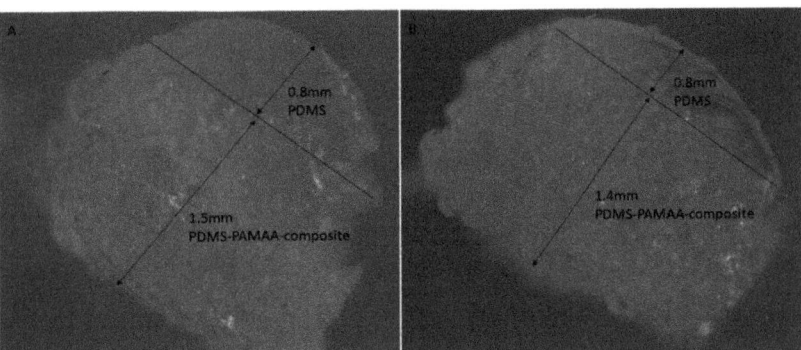

Figure 9. Cross-sections of design B samples with (**A**) initial hydrogel content of 20 wt% and a hydrogel particle size of 48 µm, and (**B**) initial hydrogel content of 30 wt% and a hydrogel particle size of 77 µm.

Based on the considerations in Section 1.2, a computational model was set up in which all known parameters were first included. These were the density of the silicone elastomers ρ_{PDMS} = 0.965 g/mL and their Young's modulus of E_{PDMS} = 6 MPa, as well as the density of PAMAA in the dry state $\rho_{H,dry}$ = 0.75 g/mL. The density of PAMAA in the swollen state reaches that of water $\rho_{H,sw}$ = 1.0 g/mL. For further comparison, data on free-swelling Nusil specimen A were evaluated and converted to volume increases. The mass increases m_f/m_i, which is in the range of 2.0–2.8, were thus transferred to enlarged volumes towards 149%–198%. A rough comparison of the linearly achieved elongations in design B (see Figure 9) shows rather larger values compared to those resulting from the measurements with specimen A. From this, it was concluded that free swelling of specimen B at the free end of the composite layer ($h_{comp,fin}$) was achieved with high probability, and therefore the main working hypothesis of the approach in Section 1.2 holds. Thus, for the specimen shown in Figure 9a (f_H = 20%, 48 µm size fraction) with h_{PDMS} = 0.8 mm and $h_{comp,i}$ = 1.2 mm or $h_{comp,fin}$ = 1.5 mm, a radius of curvature of 7.6 mm was achieved. The elongation at the free end of the composite ε ($R + h_{PDMS} + h_{comp,fin}$) was subsequently found to be 0.20 and was very close to the value of 0.186 obtained from the associated data of the A designs. The corresponding Young's modulus of the swollen silicone–hydrogel composite averaged $E_{comp,sw}$ = 3.7 MPa. The mean pressure in the swollen hydrogel composite was thus 0.68 MPa. The mean stress in a layer of pure PDMS was 0.97 MPa. The stress in the PDMS composite interface reached a value of 1.67 MPa, assuming a linear pressure profile in the composite.

This very consistent picture achieved with the 48 µm particles in design B is no longer reached for the 77 µm fraction. For $h_{comp,fin}$, only 1.4 mm is measured, which means that the elongation of 0.14 was below the maximum value of 0.15 from the measurements with design A. Additionally, the bending radii reaching 14 mm indicated the partial detachment of the hydrogel. Nevertheless, the calculated stress along the inner interface still reached values of 1.2 MPa.

3.2.3. Design C—CI Shape with Silicone Rubber–Hydrogel Top Layer

Figure 10 shows a sample consisting of Nusil Med 4850-PAMAA in a dry state, while Figure 11 displays two examples of Nusil Med 4850-PAMAA after 168 h and 336 h of swelling in Ringer solution, proving excellent interlayer adherence. The samples proved to be mechanically stable even after a total swelling time of 672 h. The expected curving was achieved. The achieved curvature radius for one sample of design C (Figure 11b) was 3.9 mm for the basal and middle section and 3.1 mm for the apical section. The determined radii for all samples are listed in Table 4.

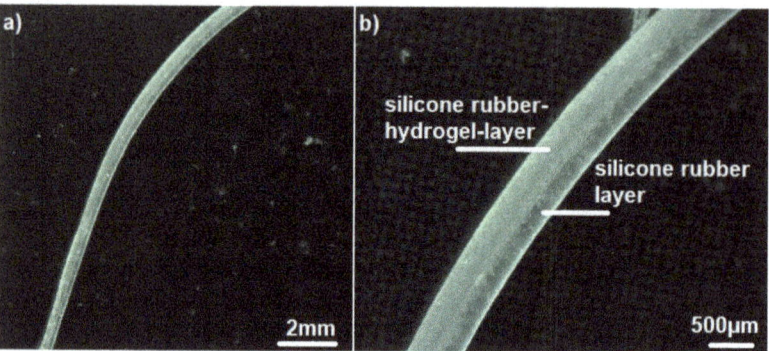

Figure 10. (a) Optical microscopy images of sample design C of Nusil Med 4850-PAMAA (25 wt% hydrogel particle size fraction of 48 μm) in dry state deposited on Nusil Med 4850 imitating a CI. (b) is an enlargement of image (a).

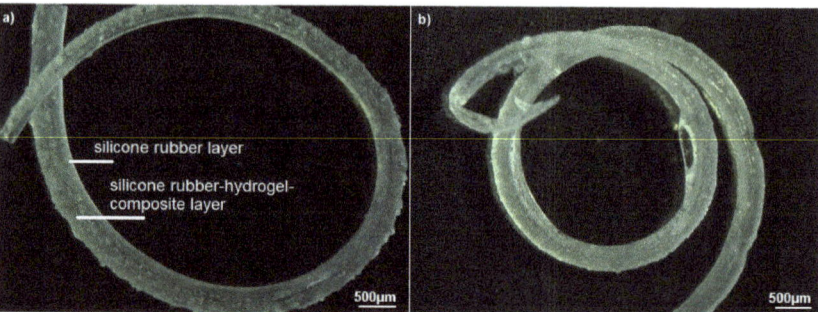

Figure 11. Optical microscopy images of two samples of design C manufactured with Nusil Med 4850-PAMAA (20 wt% hydrogel particles; grain-size fraction 48 μm) deposited on Nusil Med 4850 as a substrate. Samples are shown after (a) 168 h and (b) 336 h in Ringer solution.

The volume swelling ratios of design C being lower than those of sample design A was likely due to the same effect as for sample design B. However, the swelling ratios of the CI shapes (about 40%) and the samples of design B (between 40 and 55%) achieved quite similar swelling values. As listed in Table 4, the determined radii of samples of design B (10.9–16.6 mm) are considerably higher than those of the samples of design C (between ~2.5 and 3.9 mm). This is assumed to be due to the different sample geometries. Samples of design B consist of a PDMS layer with a significantly higher thickness and therefore stiffness, which inhibits the curvature. In contrast, the CI-shape design facilitates an increased curvature due to its thinner PDMS layer and thin diameter towards the apex of the array.

Table 4. Curvature radii determined for samples of designs B and C.

Hydrogel Content (wt%)	Grain-Size Fraction (µm)	Individual Radii for the Samples Prepared (S1–S6) (mm)			Mean Radius (mm)
		Design B			
20	48	S1—17.4 S4—8.7	S2—7.7 S5—9.8	S3—7.5 S6—14.4	10.9
20	77	S1—15.9 S4—5.6	S2—12.0 S5—12.3	S3—12.6 S6—14.8	12.2
25	77	S1—14.4 S4—16.5	S2—21.2 S5—16.2	S3—16.2 S6—15.2	16.6
30	77	S1—13.7 S4—14.0	S2—17.5 S5—11.4	S3—13.3 S6—18.8	14.8
		Design C			
20	48	S1		basal section—3.9 middle section—3.9 apical section—3.1	3.6
20	48	S2		basal section—2.5 middle section—1.25 apical section—1.87	1.9

In design C, only two test samples were fabricated with 20% hydrogel content in the 48 µm particle size fraction. In the Ringer solution, both samples showed 360° full-circle bending after 168 h and even achieved two complete turns after 336 h. One of the samples reached a mean bending radius of 3.6 mm and showed a loss of swollen layer thickness of about 100 µm in the visual inspection. The second sample, on the other hand, achieved a mean bend radius of 1.9 mm and the best value of 1.25 mm in the middle range. At 300 µm PDMS thickness and initial composite thickness of 350 µm in the basal region, the linear swell increase was 17%, and the corresponding isotropic volume increase was 60%, respectively. The elongation at the free-swelling end was 14.6%, well below the theoretically expected value of 18.6%. The stress values were 0.63 MPa on average, 1.1 MPa at the inner interface, and the mean pressure prevailing in the composite was 0.49 MPa. Overall, these values were very similar to the test specimen described above according to Design B with a 77 µm particle size fraction. That is, a loss of hydrogel particles may still have occurred here as well. Overall, the achievement of two full circles exceeds the minimum requirements of 1.5 revolutions or 540° as established for CI. The local achievement of a 1.25 mm bending radius would fit even the narrowest apical turn in the human cochlea, whose radius is about 1.3 mm. Even though the number of specimens is statistically small and no further parameter variations are available, it is clear that as the specimen dimensions decrease, the bending radii also decrease while the internal stress and pressure ratios remain the same. The possibility that leaching also occurs in the 48 µm particle size fraction with thinner coatings requires further investigation.

In addition, the observed volume increase in such actuating implant shafts results in diameters lower than the dimensions of the human scala tympani [31]. This indicates that the usage of a swelling actuator does not bear the danger of damaging the basilar membrane by shear force through swelling. Additionally, the implant will not completely fill the space in the scala tympani and therefore will not induce the risk of displacing the remaining perilymph by filling space, which could affect homeostasis.

3.3. Biocompatibility Tests

The tested composite samples displayed an open-pore and flexible surface. In the first in vitro investigations, cells were reluctant to grow on the surface, and microscopic evaluation was complicated due to cells growing in different dimensions. Furthermore, the absorption of the cell culture media by the samples quickly left no conditions for cell proliferation. Since the aim of the tests was to evaluate the materials' biocompatibility and not the cell growth on the material surface, a WST test was completed using cell culture media conditioned with the composites.

Figure 12 shows light microscopy images of the cell morphology and proliferation in a well with conditioned media (Figure 12a) and the negative control (Figure 12b). Due to good cell proliferation, a cell layer of 100% confluence was developed.

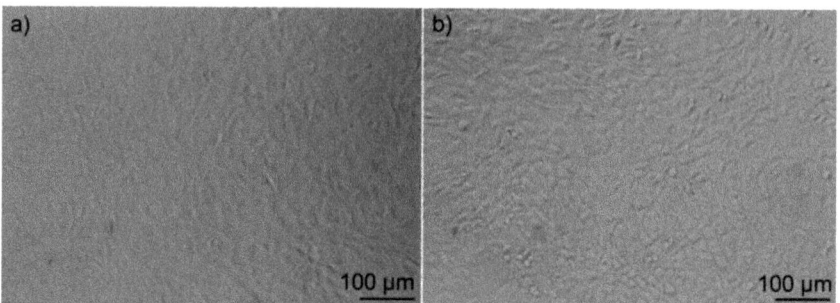

Figure 12. Light microscopy images of a layer of NIH-3Te fibroblast cells grown in well plates for evaluation of cell morphology and proliferation: (**a**) cells in media conditioned with Nusil Med 4850-PAMAA composite samples and (**b**) the negative control. Cells in both wells showed normal morphology and proliferated as expected.

The cell viability results of each test repetition with $n = 6$ samples were evaluated. Results of the third repetition were considerably lower, with $51 \pm 5.1\%$ for Silpuran 2430 + PAMAA and $58.2 \pm 7.6\%$ for Nusil Med 4850 + PAMAA when compared to the general results obtained. This could have been due to unintentional experimental faults. Except for the situation mentioned above, Silpuran 2430-PAMAA samples showed cell viability higher than 70%. In Figure 13, a summary of WST-1 tests with each material combination and the negative control is plotted. While cells cultured in Nusil 4850-PAMAA-conditioned medium exhibited a mean cell viability of about 80%, the Silpuran composite showed a slightly higher value of 83%.

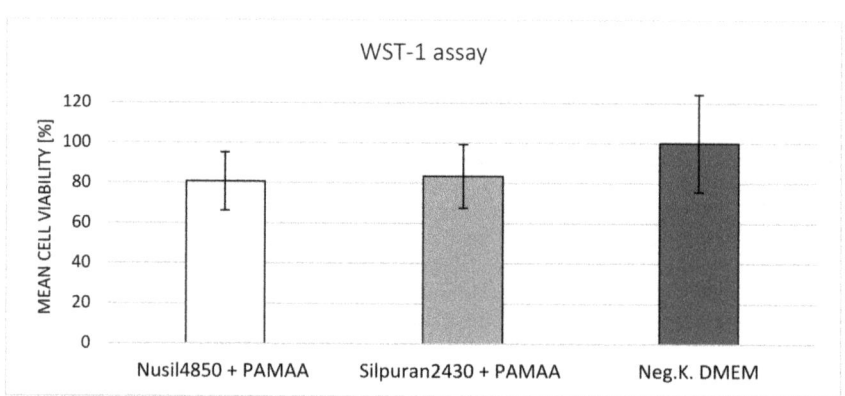

Figure 13. WST-1 biocompatibility results of Silpuran 2430-PAMAA and Nusil 4850-PAMAA samples repeated four times with $n = 6$ samples, mean ± standard deviation, and negative control. Mean cell viability for all repetitions corresponding to each material combination was calculated. All composites, irrespective of particle size and percentage of the hydrogel phase, yielded the same result of over 70% cell viability.

These high cell viabilities of more than 70% are generally declared to be non-toxic sample reactions and are positive for a first biocompatibility evaluation. However, further testing of possible interaction reactions of silicone rubber with PAMAA has to be performed according to ISO 10993-5. In addition, long-term biocompatibility tests also have to be performed in future experiments.

To our knowledge, reports in the literature on the biocompatibility of pure PAMAA do not exist. However, nanocomposite hydrogels based on PAMAA have been investigated as biomaterials for tissue engineering. When implanted subcutaneously in mice, these nanocomposite hydrogels showed good biocompatibility with the absence of an immune response [32].

4. Conclusions and Outlook

A curved unwired cochlear implant shaft based on a PAMAA hydrogel–PDMS rubber composite as a swelling actuator with bending radii close to the typical apical geometry of human cochlea was successfully manufactured. Even though the crosslinked PAMAA hydrogel is solely characterized by a maximum water uptake of 400%, its mixture with PDMS (with an initial hydrogel content ranging from 20 to 30 wt%) led to swelling ratios of up to 375% (Silpuran 2430) and up to 190% (Nusil Med 4850). However, in a swollen state, the hydrogel–PDMS composite was still able to exert the necessary bending forces on a PDMS rubber substrate to which the composite was deposited and adhered. The internal interfaces of swelling and stretched materials withstand these forces well and steadily. In addition, the only slight volume increase suggests no danger of damaging the basilar membrane by too much swelling nor misbalancing the homeostasis by reducing the free space in the scala tympani. In the final design, CI arrays with the possible variants of internal connecting wires must still be tested. Here, there is a chance that the fine wires could minimize the elongation of the pure silicone rubber, which, according to our estimations, could further improve the curvature properties of the electrode shafts. Additional potential for optimization lies in the use of higher particle concentrations with smaller particle size fractions if these become able to be processed. However, further challenges of biocompatibility have to be considered. Modified hydrogels with similar swelling capacities have to be developed as well as methods to process hydrogel fractions with a very low particle size (in our case, <28 μm) to avoid agglomeration as an undesirable phenomenon.

Author Contributions: S.Y.-B., K.F. and T.D. conceptualized the study. S.Y.-B., K.F. and M.K. performed and supervised sample preparation and measurements, M.K. performed media preparation and cell culture tests. S.Y.-B., K.F. and T.D. performed the data analysis. T.D. developed the mathematical model. S.Y.-B. and K.F. prepared the figures and wrote the manuscript draft. S.Y.-B., K.F. and T.D. reviewed and edited the manuscript. A.W. and T.D. supervised the project; S.Y.-B., K.F. and T.D. administrated the project. T.D. acquired the funding. All authors have read and agreed to the published version of the manuscript.

Funding: The study presented in this paper is funded by "the Cluster of Excellence Hearing4All (grant 580 number EXC 1077/1. and EXC2077) and by the AiF Projekt GmbH represented by the 581 Federal ministry for Economic Affairs and Energy (grant number ZF4412702SL7)".

Institutional Review Board Statement: Not applicable.

Informed Consent Statement: Not applicable.

Data Availability Statement: Not applicable.

Acknowledgments: The authors are grateful for the support from Filip Jakimovski and Anna Lena Weichaus during the swelling tests.

Conflicts of Interest: The authors declare no conflict of interest.

Appendix A

Appendix A.1 Sample Preparation

The grinding process was started with two grinding balls with a 15 mm diameter for 30 min with a frequency of 30 s^{-1}. Subsequently, the material was ground with 30 grinding balls with a 5 mm diameter at a frequency of 30 s^{-1} for 50 min. After grinding, the powder was sieved using a vibratory screening machine (Analysette 3 Pro Fritsch, Idar-Oberstein, Germany) with an amplitude of 1 mm and a sieving duration of 60 min. Sieve mesh sizes of 100 µm, 50 µm, and 20 µm were used. Due to the rapid agglomeration of the hydrogel particles, the sieves became clogged often and had to be cleaned with purified water in an ultrasonic bath. All powder with the desired grain-size fraction was stored. The rest was ground again following the described procedure.

Appendix A.2 Biocompatibility Tests

By determining the metabolic activity of native murine NIH-3T3 fibroblasts, the samples were evaluated in line with ISO 10993-5. Due to the high water absorption of the hydrogel, tests were performed with conditioned media instead of direct sample contact. For each material (Silpuran 2430-PAMAA and Nusil Med4850-PAMAA), six samples consisting of 0.8 mm^2 silicone rubber–hydrogel composite were tested. The samples were autoclaved at 121 °C for 20 min in a 100 mL wide-necked Schott bottle. Under sterile conditions, 20 mL of cell culture medium consisting of Dulbecco's modified Eagle's medium (DMEM, Biochrom, Berlin, Germany) with 10% fetal calf serum (FCS, Biochrom, Berlin, Germany) and 0.5% L-Glutamin (Biochrom, Berlin, Germany) was added to the bottle and incubated with the sample under cell culture conditions (37 °C/5% CO_2) for seven days in a slightly open bottle.

The conditioned cell culture medium was then aliquoted and frozen at −20 °C. For the WST-1-Assay, 10,000 fibroblasts per well were cultivated for 48 h at 37 °C and 5% CO_2 in the conditioned cell culture media (100 µL per well), the negative control (unconditioned cell culture medium, proliferation under normal conditions), and the positive control (DMSO, induces cell death). Then, 10 µL of WST-1 solution was added to each well and incubated again for 30 min (37 °C, 5% CO_2). After the incubation period, the cleavage of the tetrazolium salt WST-1 to formazan was measured with a multi-well spectrophotometer (ELISA reader) and determined quantitatively. The measured absorbance correlated directly with the number of viable cells, calculated by the mean optical density of a certain sample and the negative control wells, which were set at 100% cell viability.

References

1. Wilson, B.S.; Dorman, M.F. Cochlear Implants: A Remarkable Past and a Brilliant Future. *Hear. Res.* **2008**, *242*, 3–21. [CrossRef] [PubMed]
2. Lenarz, T. Funktionsersatz des Innenohres. In *Medizintechnik*; Springer: Berlin/Heidelberg, Germany, 2009; pp. 1933–1949.
3. McDermott, H.J. Music Perception with Cochlear Implants: A Review. *Trend Amplif.* **2004**, *8*, 49–82. [CrossRef] [PubMed]
4. Friesen, L.M.; Shannon, R.V.; Baskent, D.; Wang, X. Speech Recognition in Noise as a Function of the Number of Spectral Channels: Comparison of Acoustic Hearing and Cochlear Implants. *J. Acoust. Soc. Am.* **2001**, *110*, 1150–1163. [CrossRef] [PubMed]
5. Gstoettner, W.K.; Adunka, O.; Franz, P.; Hamzavi, J.; Plenk, H., Jr.; Susani, M.; Baumgartner, W.; Kiefer, J. Perimodiolar Electrodes in Cochlear Implant Surgery. *Acta Otolaryngol.* **2001**, *121*, 216–219. [CrossRef] [PubMed]
6. Shepherd, R.K.; Hatsushika, S.; Clark, G.M. Electrical Stimulation of the Auditory Nerve: The Effect of Electrode Position on Neural Excitation. *Hear. Res.* **1993**, *66*, 108–120. [CrossRef]
7. Eshraghi, A.A.; Yang, W.N.; Balkany, T.J. Comparative Study of Cochlear Damage with Three Perimodiolar Electrode Designs. *Am. Laryngol. Rhinol. Otol. Soc. Inc.* **2003**, *113*, 415–419. [CrossRef]
8. Majdani, O.; Lenarz, T.; Pawsey, N.; Risi, F.; Sedlmayr, G.; Rau, T. First Results with a Prototype of a New Cochlear Implant Electrode Featuring Shape Memory Effect. *Biomed. Eng. Biomed. Tech.* **2013**, *58*. [CrossRef]
9. Röthemeyer, F.; Sommer, F. *Kautschuktechnologie: Werkstoffe—Verarbeitung—Produkte*; Carl Hanser Verlag GmbH & Co KG: Munich, Germany, 2013.
10. Bargel, G.S.H. *Werkstoffkunde*; Springer: Berlin/Heidelberg, Germany, 2008.
11. Eyerer, P. Einführung in Polymer Engineering. In *Die Kunststoffe und Ihre Eigenschaften*; Springer: Berlin/Heidelberg, Germany, 2005; pp. 1–449.

12. Hoffman, A.S. Hydrogels for Biomedical Applications. *Adv. Drug Deliv. Rev.* **2012**, *64*, 18–23. [CrossRef]
13. Lin, X.; Bai, Y.; Zhou, H.; Yang, L. Mechano-Active Biomaterials for Tissue Repair and Regeneration. *J. Mater. Sci. Technol.* **2020**, *59*, 227–233. [CrossRef]
14. Kazanskii, K.S.; Dubrovskii, S.A. Chemistry and physics of "agricultural" hydrogels. In *Polyelectrolytes Hydrogels Chromatographic Materials*; Spinger: Berlin/Heidelberg, Germany, 1992; pp. 97–133.
15. Seldon, H.L.; Dahm, M.C.; Clark, G.M.; Crowe, S. Silastic with Polyacrylic Acid Filler: Swelling Properties, Biocompatibility and Potential Use in Cochlear Implants. *Biomaterials* **1994**, *15*, 1161–1169. [CrossRef]
16. Abbasi, F.; Mirzadeh, H.; Simjoo, M. Hydrophilic Interpenetrating Polymer Networks of Poly(Dimethyl Siloxane) (PDMS) as Biomaterial for Cochlear Implants. *J. Biomater. Sci. Polym. Ed.* **2006**, *17*, 341–355. [CrossRef] [PubMed]
17. Avantor Sciences Datasheet-Biomaterials Implant Line Liquid Silicone Rubber-Nusil Med 4850. Available online: https://www.avantorsciences.com/assetsvc/asset/en_US/id/29019577/contents/en_us_tds_nusimed-4850.pdf (accessed on 21 April 2022).
18. Wacker Chemie AG Datasheet-SILPURAN® 2430 A/B Room Temperature Curing Silicone Rubber (RTV-2). 2021.
19. The Goodfellow Group Polyacrylamide/Acrylate Material Information AC33-GL-000110. Available online: https://www.goodfellow.com/de/de/displayitemdetails/p/ac33-gl-000110/polyacrylamide-acrylate-granule (accessed on 21 April 2022).
20. Stieghorst, J.; Tegtmeier, K.; Aliuos, P.; Zernetsch, H.; Glasmacher, B.; Doll, T. Self-Bending Hydrogel Actuation for Electrode Shafts in Cochlear Implants. *Phys. Status Solidi* **2014**, *211*, 1455–1461. [CrossRef]
21. Stieghorst, J.; Doll, T. Dispersed Hydrogel Actuator for Modiolar Hugging Cochlear Implant Electrode Arrays. *IEEE Trans. Biomed. Eng.* **2016**, *63*, 2294–2300. [CrossRef] [PubMed]
22. Stieghorst, J.; Tran, B.N.; Hadeler, S.; Beckmann, D.; Doll, T. Hydrogel-Based Actuation for Modiolar Hugging Cochlear Implant Electrode Arrays. *Proc. Eng.* **2016**, *168*, 1529–1532. [CrossRef]
23. Walling, N.; Philamore, H.; Matsuno, F. A Soft Bimorph Actuator Using Silicone-Encapsulated Hydrogels. In *Robosoft 2019*; Late Breaking Result: Seoul, Korea, 2019. [CrossRef]
24. Anniko, M.; Wróblewski, R. Ionic Environment of Cochlear Hair Cells. *Hear. Res.* **1986**, *22*, 279–293. [CrossRef]
25. Kara, A.; Salt, A.N.; Thalmann, R. Perilymph Composition in Scala Tympani of the Cochlea: Influence of Cerebrospinal Fluid. *Hear. Res.* **1989**, *42*, 265–271. [CrossRef]
26. Ferenc, H.; Tasaki, I.; Basser, P.J. Osmotic Swelling of Polyacrylate Hydrogels in Physiological Salt Solutions. *Biomacromolecules* **2000**, *1*, 84–90.
27. Bokobza, L. Elastomeric Composites. I. Silicone Composites. *J. Appl. Polym. Sci.* **2004**, *93*, 2095–2104. [CrossRef]
28. Hemingway, M.G.; Gupta, R.B.; Elton, D.J. Hydrogel Nanopowder Production by Inverse-Miniemulsion Polymerization and Supercritical Drying. *Ind. Eng. Chem. Res.* **2010**, *49*, 10094–10099. [CrossRef]
29. Yanagioka, M.; Frank, C.W. Effect of Particle Distribution on Morphological and Mechanical Properties of Filled Hydrogel Composites. *Macromolecules* **2008**, *41*, 5441–5450. [CrossRef]
30. Andreopoulos, A.; Diakoulaki, D. The Diffuse Properties of Crosslinked Polyethylene. *Mater. Sci. Monogr.* **1984**, *31*, 569–574.
31. Biedron, S.; Prescher, A.; Ilgner, J.; Westhofen, M. The Internal Dimensions of the Cochlear Scalae with Special Reference to Cochlear Electrode Insertion Trauma. *Otol. Neurotol.* **2010**, *31*, 731–737. [CrossRef] [PubMed]
32. Jing, Z.; Xian, X.; Huang, Q.; Chen, Q.; Hong, P.; Li, Y.; Shi, A. Biocompatible Double Network Poly(Acrylamide-Co-Acrylic Acid)–Al^{3+}/Poly(Vinyl Alcohol)/Graphene Oxide Nanocomposite Hydrogels with Excellent Mechanical Properties, Self-Recovery and Self-Healing Ability. *New J. Chem.* **2020**, *44*, 10390–10403. [CrossRef]

Article

Tribological and Antibacterial Properties of Polyetheretherketone Composites with Black Phosphorus Nanosheets

Xuhui Sun [1], Chengcheng Yu [2], Lin Zhang [1,*], Jingcao Cao [1], Emrullah Hakan Kaleli [3] and Guoxin Xie [1,*]

[1] State Key Laboratory of Tribology, Department of Mechanical Engineering, Tsinghua University, Beijing 100084, China; sun-xh20@mails.tsinghua.edu.cn (X.S.); cjc2021@mail.tsinghua.edu.cn (J.C.)
[2] Jihua Laboratory, Foshan 528200, China; yccycc9999@163.com
[3] Faculty of Mechanical Engineering, Automotive Division, Yildiz Technical University, Besiktas, Yildiz, 34349 Istanbul, Turkey; kaleli@yildiz.edu.tr
* Correspondence: zhanglin2020@mail.tsinghua.edu.cn (L.Z.); xgx2014@mail.tsinghua.edu.cn (G.X.); Tel.: +86-010-62771438 (G.X.)

Abstract: Over the past few decades, polyetheretherketone (PEEK) artificial bone joint materials faced problems of poor wear resistance and easy infection, which are not suitable for the growing demand of bone joints. The tribological behavior and wear mechanism of polyetheretherketone (PEEK)/polytetrafluoroethylene (PTFE) with black phosphorus (BP) nanosheets have been investigated under dry sliding friction. Compared with pure PEEK, the COF of PEEK/10 wt% PTFE/0.5 wt% BP was reduced by about 73% (from 0.369 to 0.097) and the wear rate decreased by approximately 95% (from 1.0×10^{-4} mm^3/(N m) to 5.1×10^{-6} mm^3/(N m)) owing to the lubrication of the BP transfer film. Moreover, BP can endow the PEEK composites with excellent biological wettability and antibacterial properties. The antibacterial rate of PEEK/PTFE/BP was assessed to be over 99.9%, which might help to solve the problem of PEEK implant inflammation. After comprehensive evaluation in this research, 0.5 wt% BP nanosheet-filled PEEK/PTFE material displayed the optimum lubrication and antibacterial properties, and thus could be considered as a potential candidate for its application in biomedical materials.

Keywords: black phosphorus; polyetheretherketone; lubrication properties; antibacterial properties

1. Introduction

For decades, the demand for artificial joints has been growing dramatically [1–3], while the research and development of new artificial bone joints has attracted extensive attention [4–6]. However, secondary injury after implantation, which is mainly caused by postoperative infection and wear of materials [7], is a key issue.

Polyetheretherketone (PEEK), a thermoplastic material with outstanding comprehensive properties, has been widely used in the manufacturing of aerospace items, electronic information products, automobile manufacturing, pharmaceutical and medical devices, etc., owing to its excellent biocompatibility, self-lubricating properties, chemical resistance, and good formability [8–10]. Since PEEK was first introduced into orthopedic joints in 1987 [11], it has been extensively used in the manufacture of artificial bone joints [12,13] and has been recognized by many medical device manufacturers and orthopedic surgeons owing to its biomechanical properties similar to those of human bones [14–16]. Even though its bioinertness remains a limitation for bone graft applications [12,17], PEEK implant modification by adding zirconia [18–20], cobalt alloy [21], titanium alloy [22], and so on improves the mechanical properties, biological activity, treatment of postoperative infection, and inflammation regulation of bone graft. Although PEEK possesses high raw material cost and high molding energy consumption [23], it exhibits the advantages of high strength, high temperature resistance, and chemical corrosion resistance [24], which are

difficult to be matched by other polymers. PEEK can also be applied to manufacturing high value-added products [25,26].

Polytetrafluoroethylene (PTFE), an excellent solid lubricating material [27–29], has good self-lubricating performance when combined with PEEK [13,30,31]. Lin et al. [30] reported that the tribological characteristics of PEEK were enhanced by using a PTFE composite as a sacrificial tribofilm-generating part in a dual pin-on-disk tribometer. Haidar et al. [31] reported that the physical nature of transfer film adhesion by PEEK/PTFE could increase its wear tolerance to changes in environmental moisture.

Black phosphorus (BP), a novel graphene-like two-dimensional material, has been successfully applied in the solid lubrication technology field [32,33] because of its unique fold structure and interlayer interaction by the van der Waals force [34–37]. Moreover, BP has broad prospects in biomedicine in virtue of the effect of bacteriostatic/bactericidal effects without cytotoxicity [38–41]. However, research results of the dual functions of the tribological and antibacterial properties mainly focused on the modification of alloy and ceramic materials [42–45], and only a few studies concentrated on the evaluation of the dual functions of PEEK and other polymers [7,46]. The main objective of this work is to develop a new type of PEEK composite material formulation with double functions of wear-resistance and antibacterial properties and dedicated to providing a new reference for solving the problem of secondary injury of artificial bone joints (see Figure 1). Therefore, the PEEK/PTFE/BP composite is evaluated from tribological and antibacterial aspects in this work.

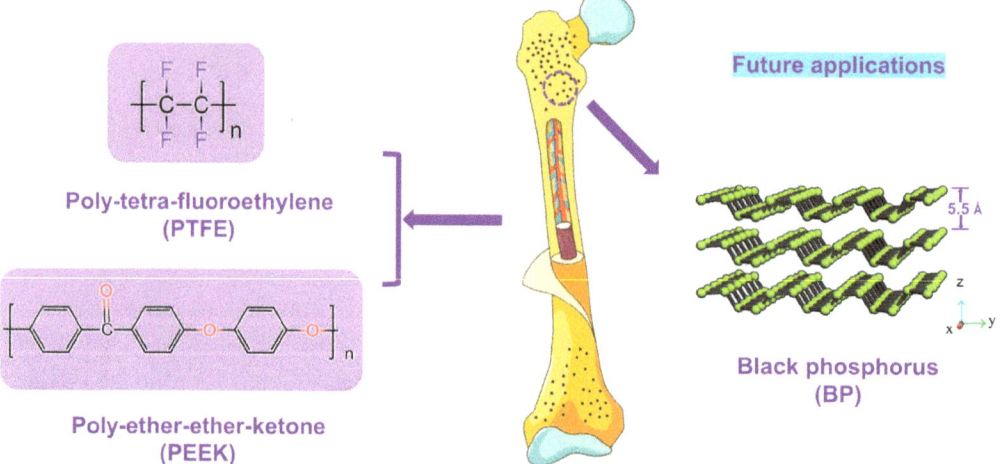

Figure 1. The schematic diagram and future applications of the PEEK/PTFE/BP artificial bone joint.

2. Materials and Methods

2.1. Materials

The PEEK purchased from Victrex (450P), Lancashire, UK was used as the matrix, and the average particle diameter was 15 μm. The PTFE with a mean particle diameter of 12 μm was supplied by DuPont (MP1300), Wilmington, DE, USA. Red phosphorus (RP) powder (Aladdin, Shanghai, China, AR > 98.5%) with an average grain size of 15 μm was used as the raw material to prepare the BP nanosheets. The materials used in the antibacterial experiment were LB broth medium (Item No: A507002, Sangon, Shanghai, China), agar powder (Item No: A505255-0250, Sangon, Shanghai, China), *Staphylococcus aureus* (control No: ATCC29213, CGMCC), and phosphate buffered saline (Hopebio, Qingdao, China).

2.2. Preparation of Black Phosphorus Nanosheets

BP nanosheets were prepared using the high-energy ball-milling technique with RP as the raw material by using a planetary ball mill (Pulverisette 7, FRITSCH, Germany) at a speed of 800 rpm for 36 h. The ball-milling process was carried out alternately in the order of 25 min milling and 5 min suspension. The weight ratio of ball-to-powder was 20:1. Stainless steel balls with diameters of 4.5 mm, 6.5 mm, and 10.0 mm were employed as the ball-milling medium, and the weight ratio of the ball grinding medium was 1:3:1. After cooling down to room temperature naturally, the ball mill tank was unscrewed in a nitrogen-filled glove box to collect BP and grind it for later use.

2.3. Preparation of PEEK/PTFE/BP Composite

Firstly, the powder mixtures of PEEK, PTFE, and BP were put into the ethanol (liquid-solid mass ratio: 1:1), and the mixed solution was stirred for 0.5 h at room temperature. After that, the mixed solution was poured into a ball mill tank and mechanically ground at a speed of 300 rpm for 2 h. Then, the mixed powder was placed into a vacuum drying oven at a temperature of 70 °C for 6 h. Afterwards, it was ground and filtered through an 800-mesh screen.

The mixed powder was pressed under a pressure of 100 MPa and sintered in a vacuum muffle furnace at 360 °C. The sintering procedure is shown in Figure 2. Finally, the prepared composite materials were pure PEEK (P1), PEEK/10 wt% PTFE (P2), PEEK/10 wt% PTFE/0.5 wt% BP (P3).

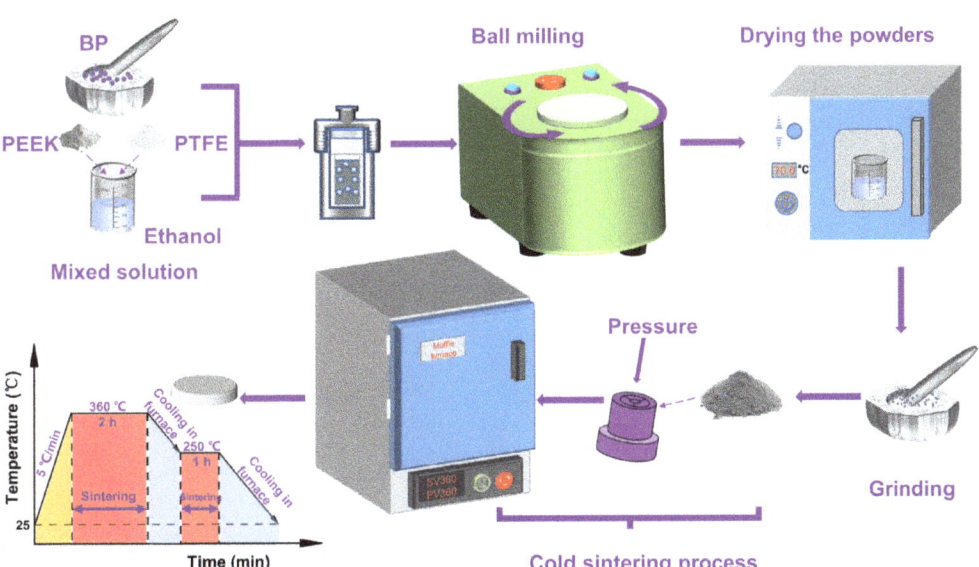

Figure 2. The schematic diagram of the preparation of PEEK/PTFE/BP composites through the cold sintering process.

2.4. Frictional Tests and Characterizations

2.4.1. Characterizations of Red Phosphorous (RP) and Black Phosphorous (BP) Nanosheets

The use of X-ray diffraction (D/Max 2550, Rigaku, Akishima, Japan) and Raman spectra (LabRAM HR Evolution, Horiba, Japan) with a laser of 532 nm and a power of 50 mW were adopted for the characterization of the RP and the prepared BP. The high resolution transmission electron microscope (HR-TEM, FEI Tecnai G2-F20, Hillsboro, OR, USA) was used to examine the morphology of the BP. Atomic force microscopy (AFM,

Bruker Dimension ICON, Santa Barbara, CA, USA) was used to determine the thickness of the BP.

2.4.2. Frictional Tests and Characterizations of PEEK Composite Materials

The tribotests of the composites were conducted with a universal mechanical tester (UMT-5, CETR, Campbell, CA, USA) with a ball-on-disk configuration. GCr15 bearing steel balls with a diameter of 4.68 mm and a surface roughness S_a of 50 nm were chosen as the tribopair materials, and they were rinsed with alcohol prior to each experiment. Tests were performed under normal loads of 3 N at a frequency of 5 Hz, and the reciprocating friction stroke was 5.0 mm.

The morphologies of worn tribopair surfaces were investigated by using the scanning electron microscopy (ZEISS, Jena, Germany, GeminiSEM 300, Signal A = SE2, Vac < 10^{-3} Pa, beam current = 72.9 µA) and energy dispersive X-ray spectroscopy (Oxford Xplore30, Oxford, UK). The wear volumes of the worn composite were measured using a 3D white light interference surface topography device (Nexview NX2, Zygo, Middlefield, CT, USA). The static contact angles of the composites were measured with 50 µL distilled water, normal saline (AS-ONE, Osaka, Japan) and calf serum (Pingrui Biotechnology, Beijing, China) by using a contact angle instrument (OCA-25, DataPhysics, Filderstadt, Germany).

2.5. Antibacterial Experiment by Film Sticking Method

The sample of PEEK/10 wt% PTFE, PEEK/10 wt% PTFE/0.5 wt% BP, and polypropylene covering film were sterilized in alcohol for 30 min and dried to reserve. After dropping 100 µL *S. aureus* suspension cultivated by LB medium with a concentration of 10^6 CFU/mL to the sample (size: 20 mm × 20 mm × 2 mm), the sample was covered with PP film using sterile forceps to ensure that the bacteria contacted the sample evenly, and then kept in a 37 °C incubator (SLI-1200, Sanyo, Osaka, Japan) for 24 h. The sample and the covering film were washed with 9.9 mL sterile PBS solution (diluted 100 times), and the collected solution was continuously diluted 10 times with the PBS solution. Then, the appropriate dilution ratio of 100 µL bacteria solution was applied to the surface of the medium and kept in a 37 °C incubator for 24 h. After culture, photos were taken in order to count the colony values of all the plates. The antibacterial experiments (the procedure as described in Figure 3) were repeated 3 times to reduce the experimental error. According to HG/T 3950-2007, the bacterial inhibition rate (%) was represented in Equation (1):

$$R\ (\%) = (C - E)/E \times 100\% \tag{1}$$

$$R: \text{Antibacterial rate, \%} \tag{2}$$

$$C: \text{Concentration of control group, CFU/mL} \tag{3}$$

$$E: \text{Concentration of experimental group, CFU/mL} \tag{4}$$

$$C(E): \text{CFU} \times \text{dilution ratio} \times 10/\text{mL, CFU/mL} \tag{5}$$

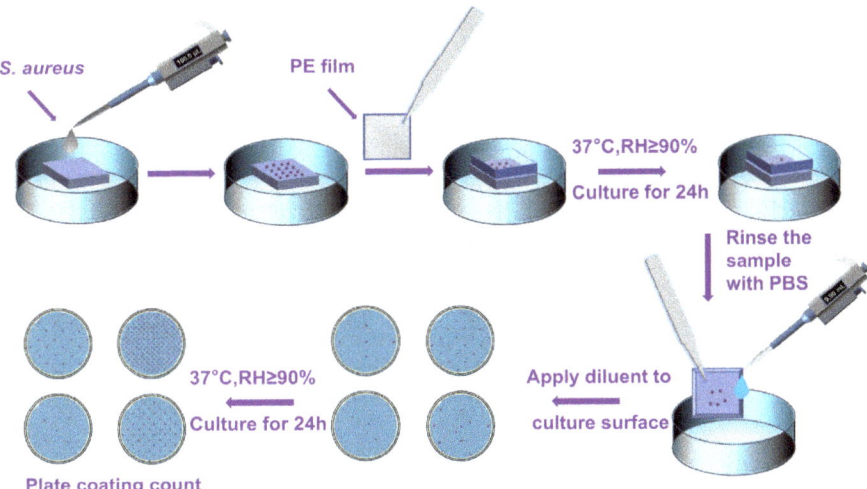

Figure 3. The schematic diagram of the antibacterial experiment process.

3. Results and Discussion

3.1. Characterizations of RP and BP

According to the XRD analysis of Figure 4a, the XRD peaks of BP prepared by high-energy ball milling were consistent with those of pure BP (JCPDS no.73-1358), with sharp diffraction peaks and good crystallinity. Raman spectra confirmed the information about the BP, as shown in Figure 4b. The characteristic peaks of BP appeared at 359.5 cm^{-1}, 434.0 cm^{-1}, and 462.3 cm^{-1}, corresponding to three atomic vibration modes A_g^1, B_{2g}, and A_g^2 of the phosphorus atom, respectively [2]. Software digital micrograph analysis of the HR-TEM image in Figure 4c showed that the lattice spacing of the prepared BP nanosheet was 0.53 nm, which was attributed to d-spacing of (020) crystal lattices as reported in the literature [32,47]. The lamella thickness of the BP nanosheet (see Figure 4d) was 5 nm, which proved the ultra-thin morphological characteristics of the prepared nanosheets.

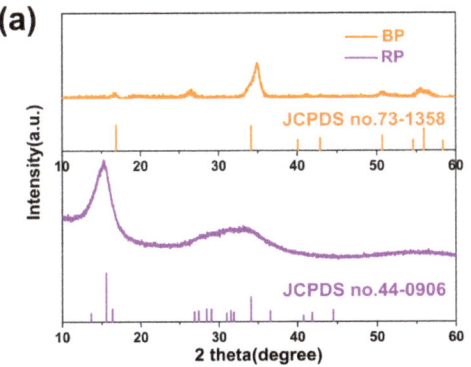
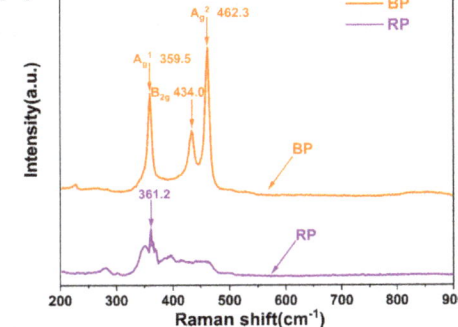

Figure 4. *Cont.*

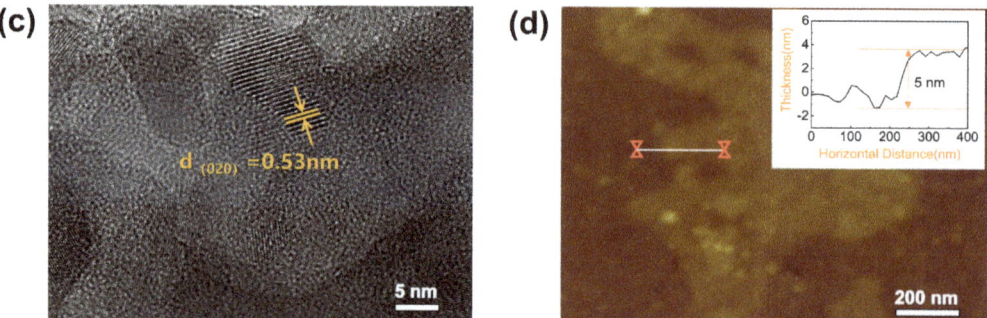

Figure 4. The Raman spectra (**a**) and X-ray diffraction (**b**) of RP and BP nanosheets; the HR-TEM image (**c**) and AFM image (**d**) of the BP nanosheets.

3.2. Tribological Properties and Analysis of Composite Materials

The coefficients of friction (COFs) of the samples under 3 N at the sliding speed of 50.0 mm/s were summarized and shown in Figure 5. The COF of pure PEEK was relatively low at first, and then it gradually increased to 0.369. After the incorporation of 10 wt% PTFE, the COF reduced from 0.369 to 0.178. After 0.5 wt% BP nanosheets were added into the PEEK/PTFE composite, the lubrication performance of the composite was greatly improved, and the COF reduced significantly, with a minimum COF of 0.097. Figure 5c presented the wear rates of three samples P1, P2, and P3. The wear marks of the polymers were summarized by using a three-dimensional interference surface topography device, and the wear rates were calculated using ZYGO MetroProX software. Compared with pure PEEK, the wear rate of the composite after the addition of 10 wt% PTFE filler obviously decreased from 1.0×10^{-4} mm^3/(N m) to 4.3×10^{-5} mm^3/(N m). Similarly, the incorporation of BP significantly reduced the wear rate of the composite to 5.1×10^{-6} mm^3/(N m). On the whole, compared with pure PEEK, the COF reduced by about 73%, and the wear rate decreased by approximately 95% for PEEK/10 wt% PTFE/0.5 wt% BP.

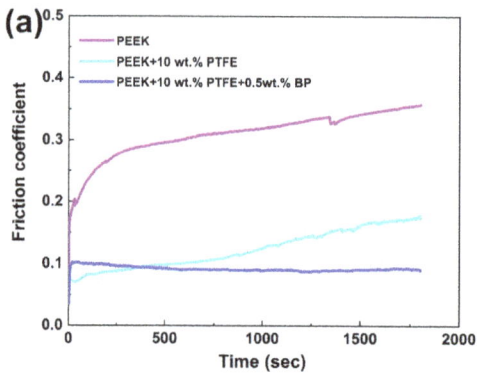
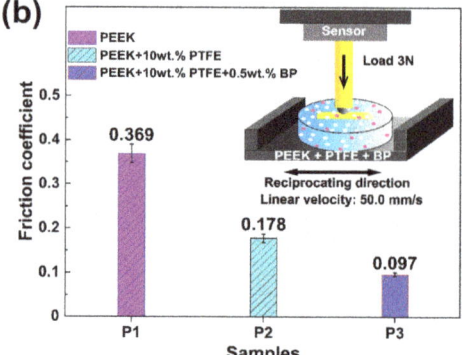

Figure 5. *Cont.*

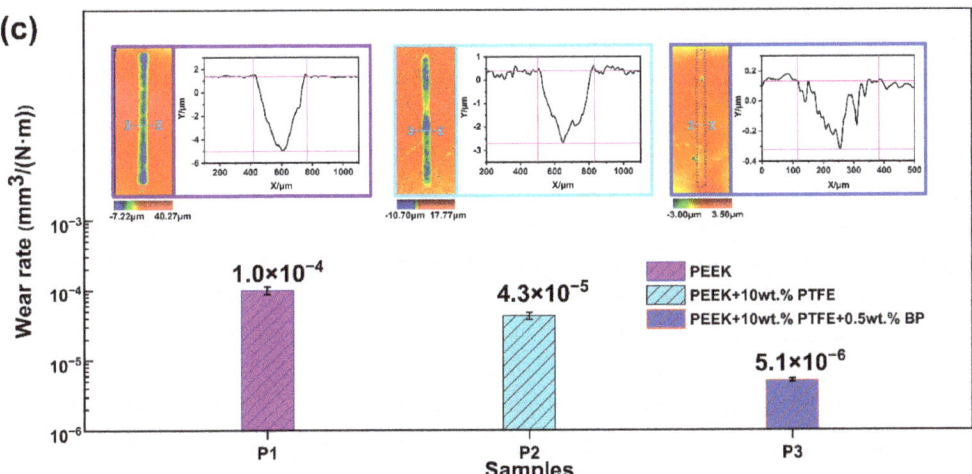

Figure 5. The friction coefficients and wear rates of the samples. (**a**) Variation curves in the COFs as a function of time for three samples; (**b**) The average COFs of three samples; (**c**) The wear rates of three samples.

The SEM-EDX micrographs of the bearing steel ball surfaces after the dry friction condition were presented in Figure 6. As can be seen from Figure 6a, the thick transfer film formed on the spherical surface of the PEEK had so poor wear resistance that it easily fell off and produced fragments during the reciprocating motion. The double formation of PEEK and PTFE mixed transfer film (see Figure 6b) significantly reduced the COF. The transfer film (see Figure 6c) formed on the spherical surface of the PEEK/PTFE/BP composite was relatively smooth, thin, and uniform. The transfer film contained higher phosphorus (P) and oxygen (O) elements (see Figure 6f), which proved that BP can form a stable transfer film at the sliding interface. Because the BP layers were combined by van der Waals force, the BP transfer film had good adhesion with the tribopairs, which can protect the tribopairs from wear.

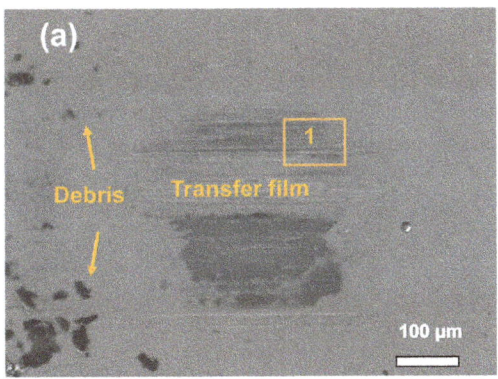
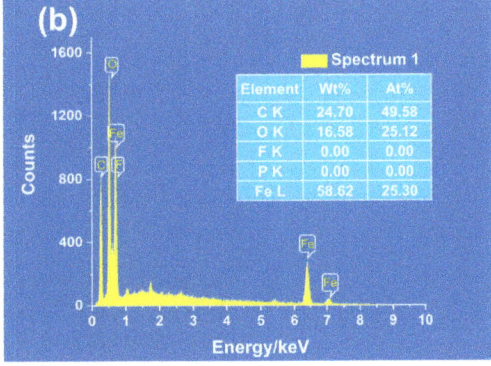

Figure 6. *Cont.*

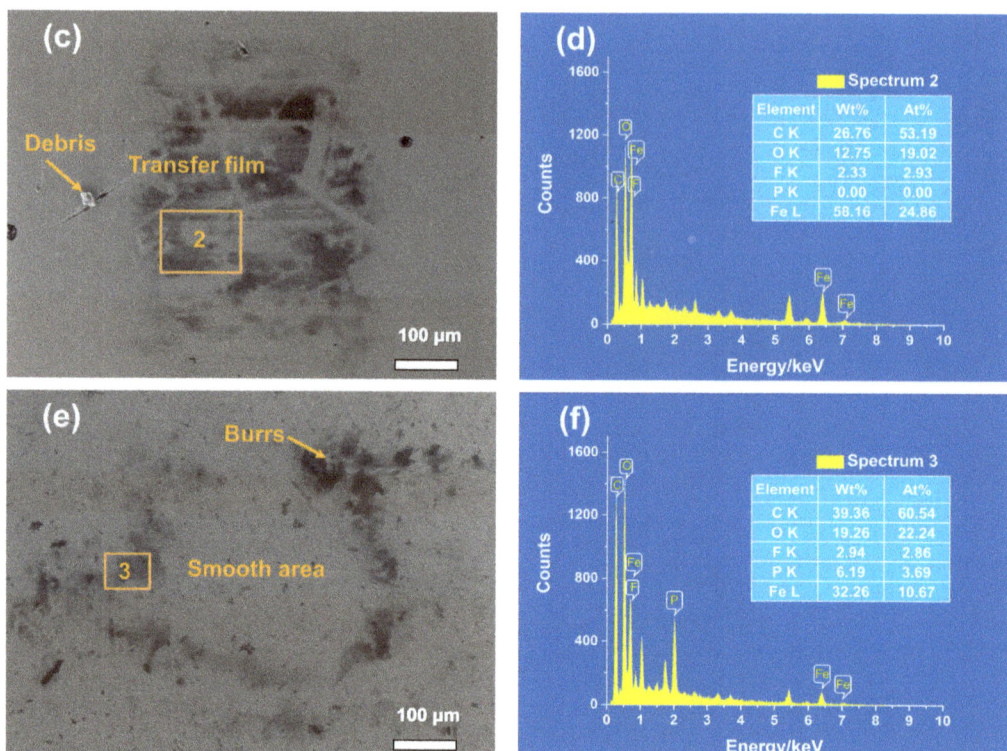

Figure 6. SEM-EDX micrographs of the bearing steel ball surfaces sliding against different composites: (**a**) pure PEEK; (**b**) EDX image of area 1; (**c**) PEEK/10 wt% PTFE; (**d**) EDX image of area 2; (**e**) PEEK/10 wt% PTFE/0.5 wt% BP; (**f**) EDX image of area 3. (EHT = 10.00 kV, Mag = 160×).

The SEM micrographs of the worn surfaces of different composites were shown in Figure 7. Rough furrows were parallelly distributed on the PEEK worn surface (see Figure 7a), showing severe adhesive wear. This was due to the dimensional cutting of the PEEK materials by the micro convex body on the surface of the steel ball, resulting in the plastic deformation and furrow effect of PEEK. The addition of PTFE (see Figure 7b) reduced the mechanical wear to a certain extent. However, the worn surface of the PEEK/PTFE/BP (see Figure 7c) was relatively smooth with very slight furrows. Therefore, BP reduced the COF and improved the wear resistance of the PEEK composites.

3.3. Biological Wettability and Antibacterial Property

The static contact angle values of different composites under distilled water, normal saline, and calf serum were shown in Figure 8. It can be clearly seen that there was no significant difference between the contact angle values of pure PEEK and PEEK/10 wt% PTFE for the three liquids mentioned. The PEEK/10 wt% PTFE/0.5 wt% BP showed the optimum wettability with contact angle values of 76.9°, 69.3°, and 53.1° under distilled water, normal saline, and calf serum, respectively. The addition of BP could diminish the surface tension of the liquid and improve the biological wettability of PEEK composites, which might reduce the biological responses for the implant materials.

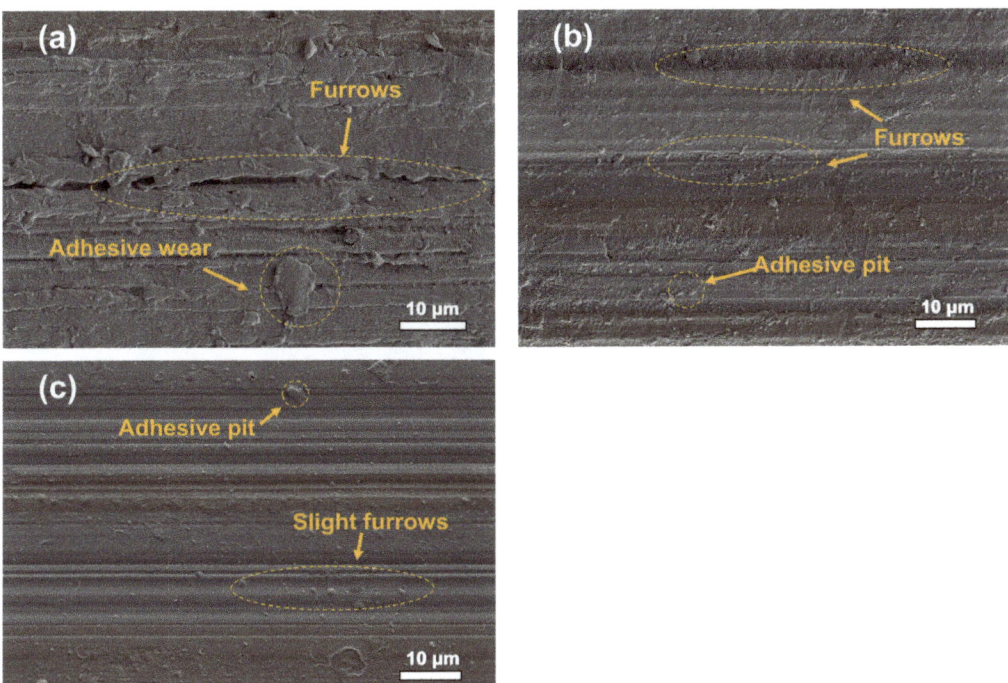

Figure 7. The SEM micrographs of the worn surfaces of different composites: (**a**) pure PEEK; (**b**) PEEK/10 wt% PTFE; (**c**) PEEK/10 wt% PTFE/0.5 wt% BP. (EHT = 3.00 kV, Mag = 1600×).

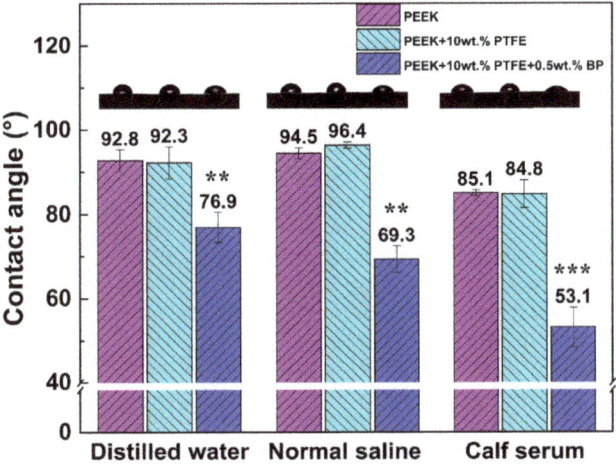

Figure 8. The contact angles of the composites with distilled water, normal saline, and calf serum. (**: $p < 0.01$, ***: $p < 0.001$, statistically significant difference; PEEK vs. PEEK/PTFE, PEEK/PTFE/BP).

The photographs of the agar plates of *S. aureus* after incubation with PEEK/10 wt% PTFE and PEEK/10 wt% PTFE/0.5 wt% BP were presented in Figure 9. The group of composites added to BP can effectively inhibit the reproduction of *S. aureus*. The antibacterial rates of PEEK/PTFE/BP calculated by three repeated experiments were 99.9% (see Table 1), which were evaluated as class I strong antibacterial material based on HG/T 3950-2007.

Many studies on BP nanosheets as antibacterial materials have been conducted [48–51], indicating that the surface of BP nanosheets could produce reactive oxygen species, destroy bacterial cell membranes, and inhibit bacterial reproduction.

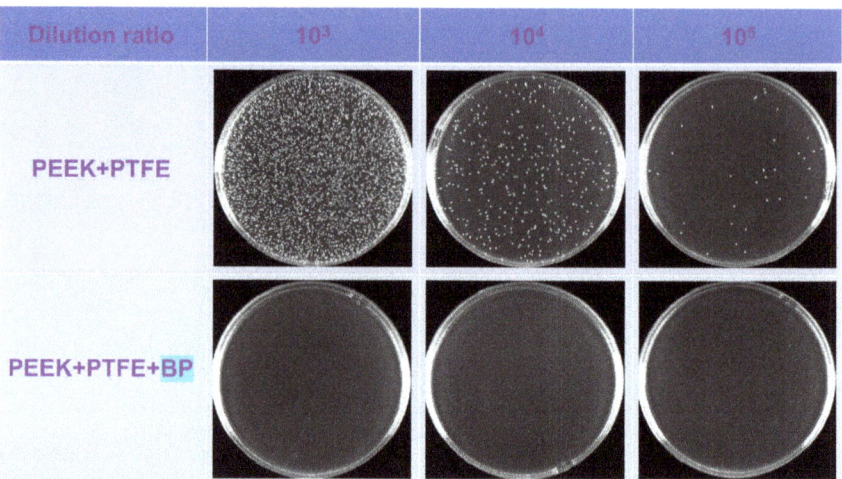

Figure 9. Photographs of agar plates of *S. aureus* after incubation with PEEK/10 wt% PTFE and PEEK/10 wt% PTFE/0.5 wt% BP.

Table 1. A summary of the antibacterial rates of three repeated antibacterial experiments.

Serial Number	Concentration of the Colon (CFU/mL)		Antibacterial Rate (%)
	Control Group	Experimental Group	
1	3.4×10^8	$<10^4$	99.9
2	5.8×10^9	$<10^4$	99.9
3	1.5×10^9	$<10^4$	99.9

4. Conclusions

In summary, this research investigated the tribological behavior and wear mechanism of PEEK/PTFE with the addition of BP, and conducted biological wettability and the antibacterial experiments. Compared with pure PEEK, the COF of PEEK/10 wt% PTFE/0.5 wt% BP was reduced by about 73% (from 0.369 to 0.097) and the wear rate decreased by approximately 95% (from 1.0×10^{-4} mm^3/(N m) to 5.1×10^{-6} mm^3/(N m)) owing to the lubrication of the BP transfer film, making PEEK composite materials more wear-resisting for use in artificial joint implants.

In addition, BP endowed the PEEK composites with excellent biological wettability and antibacterial properties. It was measured that PEEK/PTFE/BP was considered as class I antibacterial material owing to its antibacterial rate above 99.9%, which was helpful to solve the problem of adverse infection reaction caused by PEEK materials implanted in the body.

PEEK/PTFE/BP composites can realize blending and granulation, and are suitable for 3D printing and injection molding. Thus, it is expected that the research results will provide a potential opportunity for an extensive range of applications for PEEK artificial joint materials. The formulation is prior to commercial PEEK production, which still requires improved mechanical properties and extensive clinical biological tests.

Author Contributions: Experiments, X.S. and J.C.; methodology, C.Y.; data analysis, X.S., C.Y. and L.Z.; writing—original draft preparation, X.S.; writing—review and editing, G.X. and E.H.K.; project administration, G.X. and L.Z.; funding acquisition, G.X. and L.Z. All authors have read and agreed to the published version of the manuscript.

Funding: This research was funded by the Beijing Natural Science Foundation of China (grant no. JQ21008) and the Tsinghua–Foshan Innovation Special Fund (TFISF) (grant no. 2020THFS0127).

Institutional Review Board Statement: Not applicable.

Informed Consent Statement: Not applicable.

Data Availability Statement: Not applicable.

Conflicts of Interest: The authors declare no conflict of interest.

References

1. Yang, Y.; He, C.; Dianyu, E.; Yang, W.; Qi, F.; Xie, D.; Shen, L.; Peng, S.; Shuai, C. Mg bone implant: Features, developments and perspectives. *Mater. Des.* **2020**, *185*, 108259. [CrossRef]
2. Lee, S.; Yang, F.; Suh, J.; Yang, S.; Lee, Y.; Li, G.; Choe, H.S.; Suslu, A.; Chen, Y.; Ko, C.; et al. Anisotropic in-plane thermal conductivity of black phosphorus nanoribbons at temperatures higher than 100 K. *Nat. Commun.* **2015**, *6*, 8573. [CrossRef] [PubMed]
3. Mailoo, V.J.; Srinivas, V.; Turner, J.; Fraser, W.D. Beware of bone pain with bisphosphonates. *BMJ Case Rep.* **2019**, *12*, e225385. [CrossRef] [PubMed]
4. Hafezi, M.; Qin, L.; Mahmoodi, P.; Dong, M.; Dong, G. In vitro released characteristics of BSA lubricants from agarose hydrogel with tunable mechanical behaviors for artificial joint applications. *Biotribology* **2021**, *28*, 100200. [CrossRef]
5. Lu, H.; Ren, S.; Guo, J.; Li, Y.; Li, J.; Dong, G. Laser textured Co-Cr-Mo alloy stored chitosan/poly (ethylene glycol) composite applied on artificial joints lubrication. *Mater. Sci. Eng. C* **2017**, *78*, 239–245. [CrossRef] [PubMed]
6. Wang, X.; Han, X.; Li, C.; Chen, Z.; Huang, H.; Chen, J.; Wu, C.; Fan, T.; Li, T.; Huang, W.; et al. 2D materials for bone therapy. *Adv. Drug Delivery Rev.* **2021**, *178*, 113970. [CrossRef] [PubMed]
7. Wu, T.; Zhang, X.; Chen, K.; Chen, Q.; Yu, Z.; Feng, C.; Qi, J.; Zhang, D. The antibacterial and wear-resistant nano-ZnO/PEEK composites were constructed by a simple two-step method. *J. Mech. Behav. Biomed. Mater.* **2021**, *126*, 104986. [CrossRef] [PubMed]
8. Monich, P.R.; Henriques, B.; Novaes de Oliveira, A.P.; Souza, J.C.M.; Fredel, M.C. Mechanical and biological behavior of biomedical PEEK matrix composites: A focused review. *Mater. Lett.* **2016**, *185*, 593–597. [CrossRef]
9. Zhao, Y.; Wong, H.M.; Wang, W.; Li, P.; Xu, Z.; Chong, E.Y.; Yan, C.H.; Yeung, K.W.; Chu, P.K. Cytocompatibility, osseointegration, and bioactivity of three-dimensional porous and nanostructured network on polyetheretherketone. *Biomaterials* **2013**, *34*, 9264–9277. [CrossRef] [PubMed]
10. Xin, H.; Liu, R.; Zhang, L.; Jia, J.; He, N.; Gao, S.; Jin, Z. A comparative bio-tribological study of self-mated PEEK and its composites under bovine serum lubrication. *Biotribology* **2021**, *26*, 100171. [CrossRef]
11. Kurtz, S.M.; Devine, J.N. PEEK biomaterials in trauma, orthopedic, and spinal implants. *Biomaterials* **2007**, *28*, 4845–4869. [CrossRef] [PubMed]
12. He, M.; Huang, Y.; Xu, H.; Feng, G.; Liu, L.; Li, Y.; Sun, D.; Zhang, L. Modification of polyetheretherketone implants: From enhancing bone integration to enabling multi-modal therapeutics. *Acta Biomater.* **2021**, *129*, 18–32. [CrossRef] [PubMed]
13. Ma, H.; Suonan, A.; Zhou, J.; Yuan, Q.; Liu, L.; Zhao, X.; Lou, X.; Yang, C.; Li, D.; Zhang, Y.G. PEEK (Polyether-ether-ketone) and its composite materials in orthopedic implantation. *Arabian J. Chem.* **2021**, *14*, 102977. [CrossRef]
14. Zhao, X.; Xiong, D.; Wang, K.; Wang, N. Improved biotribological properties of PEEK by photo-induced graft polymerization of acrylic acid. *Mater. Sci. Eng. C* **2017**, *75*, 777–783. [CrossRef] [PubMed]
15. Wang, L.; He, H.; Yang, X.; Zhang, Y.; Xiong, S.; Wang, C.; Yang, X.; Chen, B.; Wang, Q. Bimetallic ions regulated PEEK of bone implantation for antibacterial and osteogenic activities. *Mater. Today Adv.* **2021**, *12*, 100162. [CrossRef]
16. Zhang, X.; Zhang, Y.; Jin, Z. A review of the bio-tribology of medical devices. *Friction* **2021**, *10*, 4–30. [CrossRef]
17. Ionescu, R.N.; Totan, A.R.; Imre, M.M.; Tancu, A.M.C.; Pantea, M.; Butucescu, M.; Farcasiu, A.T. Prosthetic materials used for implant-supported restorations and their biochemical oral interactions: A narrative review. *Materials* **2022**, *15*, 1016. [CrossRef] [PubMed]
18. Mishra, T.K.; Kumar, A.; Verma, V.; Pandey, K.N.; Kumar, V. PEEK composites reinforced with zirconia nanofiller. *Compos. Sci. Technol.* **2012**, *72*, 1627–1631. [CrossRef]
19. Nakonieczny, D.S.; Kern, F.; Dufner, L.; Antonowicz, M.; Matus, K. Alumina and zirconia-reinforced polyamide PA-12 composites for biomedical additive manufacturing. *Materials* **2021**, *14*, 6201. [CrossRef] [PubMed]
20. Soares, P.M.; Cadore-Rodrigues, A.C.; Souto Borges, A.L.; Valandro, L.F.; Pereira, G.K.R.; Rippe, M.P. Load-bearing capacity under fatigue and FEA analysis of simplified ceramic restorations supported by PEEK or zirconia polycrystals as foundation substrate for implant purposes. *J. Mech. Behav. Biomed. Mater.* **2021**, *123*, 104760. [CrossRef] [PubMed]

21. Belwanshi, M.; Jayaswal, P.; Aherwar, A. Mechanical behaviour investigation of PEEK coated titanium alloys for hip arthroplasty using finite element analysis. *Mater. Today Proc.* **2021**, in press. [CrossRef]
22. Priester, M.; Muller, W.D.; Beuer, F.; Schmidt, F.; Schwitalla, A.D. Performance of PEEK based telescopic crowns, a comparative study. *Dent. Mater.* **2021**, *37*, 1667–1675. [CrossRef] [PubMed]
23. Thiruchitrambalam, M.; Bubesh Kumar, D.; Shanmugam, D.; Jawaid, M. A review on PEEK composites—manufacturing methods, properties and applications. *Mater. Today Proc.* **2020**, *33*, 1085–1092. [CrossRef]
24. Hoskins, T.J.; Dearn, K.D.; Kukureka, S.N. Mechanical performance of PEEK produced by additive manufacturing. *Polym. Test.* **2018**, *70*, 511–519. [CrossRef]
25. Arif, M.F.; Kumar, S.; Varadarajan, K.M.; Cantwell, W.J. Performance of biocompatible PEEK processed by fused deposition additive manufacturing. *Mater. Des.* **2018**, *146*, 249–259. [CrossRef]
26. Verma, S.; Sharma, N.; Kango, S.; Sharma, S. Developments of PEEK (polyetheretherketone) as a biomedical material: A focused review. *Eur. Polym. J.* **2021**, *147*, 110295. [CrossRef]
27. Zhang, L.; Ren, Y.; Peng, S.; Guo, D.; Wen, S.; Luo, J.; Xie, G. Core-shell nanospheres to achieve ultralow friction polymer nanocomposites with superior mechanical properties. *Nanoscale* **2019**, *11*, 8237–8246. [CrossRef] [PubMed]
28. Bashandeh, K.; Lan, P.; Meyer, J.L.; Polycarpou, A.A. Tribological performance of graphene and PTFE solid lubricants for polymer coatings at elevated temperatures. *Tribol. Lett.* **2019**, *67*, 99. [CrossRef]
29. Shen, M.; Li, B.; Zhang, Z.; Zhao, L.; Xiong, G. Abrasive wear behavior of PTFE for seal applications under abrasive-atmosphere sliding condition. *Friction* **2019**, *8*, 755–767. [CrossRef]
30. Lin, Z.; Yue, H.; Gao, B. Enhancing tribological characteristics of PEEK by using PTFE composite as a sacrificial tribofilm-generating part in a novel dual-pins-on-disk tribometer. *Wear* **2020**, *460–461*, 203472. [CrossRef]
31. Haidar, D.R.; Alam, K.I.; Burris, D.L. Tribological insensitivity of an ultralow-wear poly (etheretherketone)–polytetrafluoroethylene polymer blend to changes in environmental moisture. *J. Phys. Chem. C* **2018**, *122*, 5518–5524. [CrossRef]
32. Peng, S.; Guo, Y.; Xie, G.; Luo, J. Tribological behavior of polytetrafluoroethylene coating reinforced with black phosphorus nanoparticles. *Appl. Surf. Sci.* **2018**, *441*, 670–677. [CrossRef]
33. Lv, Y.; Wang, W.; Xie, G.; Luo, J. Self-lubricating PTFE-based composites with black phosphorus nanosheets. *Tribol. Lett.* **2018**, *66*, 61. [CrossRef]
34. Zeng, Y.; Guo, Z. Synthesis and stabilization of black phosphorus and phosphorene: Recent progress and perspectives. *iScience* **2021**, *24*, 103116. [CrossRef] [PubMed]
35. Basov, D.N.; Fogler, M.M.; Garcia de Abajo, F.J. Polaritons in van der Waals materials. *Science* **2016**, *354*, aag1992. [CrossRef]
36. Sahoo, P.K.; Memaran, S.; Xin, Y.; Balicas, L.; Gutierrez, H.R. One-pot growth of two-dimensional lateral heterostructures via sequential edge-epitaxy. *Nature* **2018**, *553*, 63–67. [CrossRef] [PubMed]
37. Qu, G.; Xia, T.; Zhou, W.; Zhang, X.; Zhang, H.; Hu, L.; Shi, J.; Yu, X.F.; Jiang, G. Property-activity relationship of black phosphorus at the nano-bio interface: From molecules to organisms. *Chem. Rev.* **2020**, *120*, 2288–2346. [CrossRef] [PubMed]
38. Xiong, Z.; Zhang, X.; Zhang, S.; Lei, L.; Ma, W.; Li, D.; Wang, W.; Zhao, Q.; Xing, B. Bacterial toxicity of exfoliated black phosphorus nanosheets. *Ecotoxicol. Environ. Saf.* **2018**, *161*, 507–514. [CrossRef]
39. Naskar, A.; Kim, K.S. Black phosphorus nanomaterials as multi-potent and emerging platforms against bacterial infections. *Microb. Pathog.* **2019**, *137*, 103800. [CrossRef]
40. Shaw, Z.L.; Kuriakose, S.; Cheeseman, S.; Mayes, E.L.H.; Murali, A.; Oo, Z.Y.; Ahmed, T.; Tran, N.; Boyce, K.; Chapman, J.; et al. Broad-spectrum solvent-free layered black phosphorus as a rapid action antimicrobial. *ACS Appl. Mater. Interfaces* **2021**, *13*, 17340–17352. [CrossRef]
41. Anju, S.; Ashtami, J.; Mohanan, P.V. Black phosphorus, a prospective graphene substitute for biomedical applications. *Mater. Sci. Eng. C* **2019**, *97*, 978–993. [CrossRef] [PubMed]
42. Wang, Y.; Lu, X.; Yuan, N.; Ding, J. A novel nickel-copper alternating-deposition coating with excellent tribological and antibacterial property. *J. Alloys Compd.* **2020**, *849*, 156222. [CrossRef]
43. Fellah, M.; Hezil, N.; Touhami, M.Z.; AbdulSamad, M.; Obrosov, A.; Bokov, D.O.; Marchenko, E.; Montagne, A.; Alain, I.; Alhussein, A. Structural, tribological and antibacterial properties of (α + β) based ti-alloys for biomedical applications. *J. Mater. Res. Technol.* **2020**, *9*, 14061–14074. [CrossRef]
44. Yu, D.; Miao, K.; Li, Y.; Bao, X.; Hu, M.; Zhang, K. Sputter-deposited TaCuN films: Structure, tribological and biomedical properties. *Appl. Surf. Sci.* **2021**, *567*, 150796. [CrossRef]
45. Xu, J.; Ji, M.; Li, L.; Wu, Y.; Yu, Q.; Chen, M. Improving wettability, antibacterial and tribological behaviors of zirconia ceramics through surface texturing. *Ceram. Int.* **2022**, *48*, 3702–3710. [CrossRef]
46. Qin, W.; Ma, J.; Liang, Q.; Li, J.; Tang, B. Tribological, cytotoxicity and antibacterial properties of graphene oxide/carbon fibers/polyetheretherketone composite coatings on Ti-6Al-4V alloy as orthopedic/dental implants. *J. Mech. Behav. Biomed. Mater.* **2021**, *122*, 104659. [CrossRef] [PubMed]
47. Zhang, S.; Qin, Z.; Hou, Z.; Ye, J.; Xu, Z.; Qian, Y. Large-scale preparation of black phosphorus by molten salt method for energy storage. *Mater. Chem. Phys.* **2022**, *1*, 1–5. [CrossRef]
48. Liu, W.; Tao, Z.; Wang, D.; Liu, Q.; Wu, H.; Lan, S.; Dong, A. Engineering a black phosphorus-based magnetic nanosystem armed with antibacterial N-halamine polymer for recyclable blood disinfection. *Chem. Eng. J.* **2021**, *415*, 128888. [CrossRef]

49. Huang, S.; Xu, S.; Hu, Y.; Zhao, X.; Chang, L.; Chen, Z.; Mei, X. Preparation of NIR-responsive, ROS-generating and antibacterial black phosphorus quantum dots for promoting the MRSA-infected wound healing in diabetic rats. *Acta Biomater.* **2022**, *137*, 199–217. [CrossRef] [PubMed]
50. Luo, S.; Liu, R.; Zhang, X.; Chen, R.; Yan, M.; Huang, K.; Sun, J.; Wang, R.; Wang, J. Mechanism investigation for ultra-efficient photocatalytic water disinfection based on rational design of indirect Z-scheme heterojunction black phosphorus QDs/Cu$_2$O nanoparticles. *J. Hazard. Mater.* **2022**, *424*, 127281. [CrossRef]
51. Huang, B.; Tan, L.; Liu, X.; Li, J.; Wu, S. A facile fabrication of novel stuff with antibacterial property and osteogenic promotion utilizing red phosphorus and near-infrared light. *Bioact. Mater.* **2019**, *4*, 17–21. [CrossRef] [PubMed]

Article

Quaternized Amphiphilic Block Copolymers as Antimicrobial Agents

Chih-Hao Chang [1,2,*,†], Chih-Hung Chang [3,4,†], Ya-Wen Yang [5], Hsuan-Yu Chen [2], Shu-Jyuan Yang [6], Wei-Cheng Yao [7] and Chi-Yang Chao [8,*]

[1] Department of Orthopedics, National Taiwan University Hospital Jin-Shan Branch, No. 7, Yulu Rd., Wuhu Village, Jinshan Dist., New Taipei City 20844, Taiwan
[2] Department of Orthopedics, National Taiwan University College of Medicine, National Taiwan University Hospital, No. 1, Section 1, Jen-Ai Road, Taipei 100, Taiwan; hychen83@gmail.com
[3] Department of Orthopedic Surgery, Far Eastern Memorial Hospital, No. 21, Section 2, Nanya S. Road, Banciao Dist., New Taipei City 220, Taiwan; orthocch@mail.femh.org.tw
[4] Graduate School of Biotechnology and Bioengineering, Yuan Ze University, No. 135, Yuan-Tung Road, Chuang-Li Dist., Taoyuan 320, Taiwan
[5] Department of Surgery, National Taiwan University Hospital, No. 7, Chung Shan S. Rd., Taipei 10002, Taiwan; Yywivy@gmail.com
[6] Institute of Biomedical Engineering, College of Medicine and College of Engineering, National Taiwan University, No. 1, Section 1, Jen-Ai Road, Taipei 100, Taiwan; image0120@gmail.com
[7] Department of Anesthesiology and Pain Medicine, Min-Sheng General Hospital, No. 168, Ching-Kuo Rd., Taoyuan 330, Taiwan; m000924@e-ms.com.tw
[8] Department of Materials Science and Engineering, National Taiwan University, No. 1, Section 4, Roosevelt Road, Taipei 10617, Taiwan
* Correspondence: handchang@ntu.edu.tw (C.-H.C.); cychao138@ntu.edu.tw (C.-Y.C.)
† These authors contributed equally to this work and share first authorship.

Abstract: In this study, a novel polystyrene-block-quaternized polyisoprene amphipathic block copolymer (PS-*b*-PIN) is derived from anionic polymerization. Quaternized polymers are prepared through post-quaternization on a functionalized polymer side chain. Moreover, the antibacterial activity of quaternized polymers without red blood cell (RBCs) hemolysis can be controlled by block composition, side chain length, and polymer morphology. The solvent environment is highly related to the polymer morphology, forming micelles or other structures. The polymersome formation would decrease the hemolysis and increase the electron density or quaternized groups density as previous research and our experiment revealed. Herein, the PS-*b*-PIN with *N,N*-dimethyldodecylamine as side chain would form a polymersome structure in the aqueous solution to display the best inhibiting bacterial growth efficiency without hemolytic effect. Therefore, the different single-chain quaternized groups play an important role in the antibacterial action, and act as a controllable factor.

Keywords: antimicrobials agents; amphiphilic block copolymer; quaternized polymer; hemolysis; micelle

1. Introduction

Antibacterial agents are substances that can be added into materials to produce antibacterial properties in order to directly kill bacteria or inhibit the growth of bacteria for a long time. There are significant demands to develop new antibacterial agents while facing severe challenges. Since most antibiotics function by releasing low molecular weight biocide to kill bacteria or inhibit the growth of the bacteria, issues such as (1) rapid revolution of resistance mechanism to cause ineffectiveness, (2) limited capability due to chronic and repeating usage, and (3) environmental problems must be addressed. Consequently, the number of new antibiotics approved for marketing per year declines continuously. Instead of developing new antibiotics, the use of alternative antibacterial agents with different mechanisms to eliminate bacteria, such as silver nanocomposites, host-defense peptides, and synthetic polyelectrolytes, have drawn increasing attention recently [1–5]. Most silver nanocomposites have demonstrated good antibacterial ability; however, high nanosilver concentration may cause harmful environmental impacts, as well as cytotoxicity and genotoxicity in living organisms [6,7]. Host-defense peptides, usually found in the innate immune system, kill bacteria by damaging the cell membrane or invading the cell when approaching bacteria [8–10]. These antibacterial peptides generally have low cytotoxicity; however, the expensive production and susceptibility to proteolysis will challenge the success in further commercialization [11,12].

The synthetic polyelectrolytes bearing ionic groups, including phosphor-, sulfo- derivatives, and ammonium groups, have drawn increasing attention recently since they can be manufactured on a large scale with low cost, while killing bacteria in ways similar to peptides [13–15]. In addition, the antimicrobial or biological activity can be manipulated by tailoring the molecular structure, such as the molecular weight, density, and type of ionic groups, and even by building random or block copolymers [16,17]. Of the polyelectrolytes, quaternary ammonium groups are the mostly explored of late due to promising bacteria-killing performance and good environmental stability. Nevertheless, the positive antibacterial property is usually accompanied by high cytotoxicity (hemolysis), which becomes the most critical issue for the quaternized polyelectrolytes.

The balance between hydrophobicity and hydrophilicity of the polyelectrolyte is the key to achieving good anti-bacterial activity and low cytotoxicity concurrently, which may be achieved by molecular engineering on the polyelectrolytes. Homopolymers such as polynorbornene can be feasibly synthesized via homopolymerization of designated monomers [18]. However, tuning the hydrophobicity can only be achieved by changing the chemical structure of the polymer backbone, which requires specific molecular engineering on the monomer structure. Tew and colleagues have synthesized four polynorbornene homopolymers with different backbone structures to tailor hydrophobicity, one of which showed distinctive selectivity between anti-bacterial activity and hemolytic property [19]. In contrast to homopolymers, random copolymers can tailor amphiphilicity easily by adopting different monomers in a variety of compositions [20–22]. Mizutani et al. have prepared a series of degradable copolymers to examine the effect of polymer properties on their antimicrobial and hemolytic activities. The acrylate copolymer with quaternary ammonium groups and the acrylamide copolymers shows low or no antimicrobial and hemolytic activities [23]. Directly combining the useful monomers with functional groups selectivity is the advantage of random copolymer, but the lack of precise synthesis makes it difficult to control the antibacterial activity, hemolysis, and even cytotoxicity. Sen et al. have synthesized and compared series of amphiphilic pyridinium polymers, and observed that the spatial positioning of the charge and tail significantly influences the toxicity of these polymers. This result may be used as a guiding principle in the design of polymeric antimicrobial compounds with reduced toxicity [24]. Song et al. have tested four series of polymers with cationic and hydrophobic groups distributed along the backbone against six different bacterial species, and for host cytotoxicity. In their study, the antibacterial and hemolytic activities of polymers can be controlled by the exact distance of ammonium groups along the backbone [25].

Compared to homopolymers and random copolymers, block copolymers are more advantageous in controlling the microstructures, which are not well discussed in most homo or random copolymer literature. The antibacterial activity relates to the morphology in solution. By varying the lengths of the constituent blocks or ratio of the amphiphilic blocks, block copolymer could be transformed into a designated molecular structure [26]. Kenichi Kuroda et al. have reported that the block and random copolymers with similar polymer lengths and monomer compositions display the same level of bactericidal activity, but the block copolymers display selective activity against *E. coli* over red blood cells (RBCs) [27]. In a previous study, we synthesized a polystyrene-block-quaternized polyisoprene amphipathic block copolymer, denoted as PS-*b*-PIN for alkaline direct methanol fuel cells [28]. PS-*b*-PIN is amphiphilic block copolymer composed of a hydrophobic polystyrene (PS) segment and a quaternized polyisoprene (PIN) segment bearing pendant quaternary ammonium groups attached to the polyisoprene backbone via alkyl spacers. Since PS-*b*-PIN is amphiphilic block copolymer and contains many of quaternized side chains, it can be used as an antibacterial agent. Herein, we assembled the prepared PS-*b*-PIN with different long side chains into large polymersomes in the buffer solution to serve as antibacterial agents. Through precise synthesis of block copolymer, the antimicrobial activities can be controlled and polymer morphology manipulated to examine the relationship between them.

2. Materials and Methods

2.1. Materials

PS-*b*-PIN with different long side chains was provided by author Chi-Yang Chao from the Department of Materials Science and Engineering at National Taiwan University (NTU). Tetrahydrofuran (THF, Reagent Grade) was dried from a mixture of sodium under nitrogen atmosphere. Ethanol ($\geq$99.5%), dimethyl sulfoxide (DMSO), human hemoglobin, Triton X-100, and 3-[4,5-dimethylthiazol-2-yl]-2,5-diphenyl tetrazolium bromide (MTT) were purchased from Sigma-Aldrich (St. Louis, MO, USA).

2.2. Surface Charge Determination of PS-b-PIN

The PS-*b*-PIN stock solutions with different side chain length were diluted by ddH$_2$O to 1000 ppm. The zeta potential of the prepared PS-*b*-PIN solutions were determined using a Zetasizer Nano-ZS90 (MalvernInstruments Ltd., Malvern, UK), based on laser Doppler electrophoresis.

2.3. Antibacterial Analysis of PS-b-PIN

The antimicrobial efficacy of PS-*b*-PIN was investigated by using pathogenic bacterial strains of *Escherichia coli* (American Type Culture Collection (ATCC) 25922) and *Staphylococcus aureus* (ATCC 23235), obtained from the Department of Clinical Laboratory, Sciences and Medical Biotechnology at National Taiwan University (Taipei, Taiwan). Measurements of bacterial growth were obtained following the protocol of the ISO 22196 standard test methods. Typically, bacteria were cultivated in 3% Bacto tryptic-soy broth (TSB) (Becton Dickinson, Sparks, MD, USA) at 37 °C for 12 h. After serial dilution of the suspension, an aliquot of solution (2 mL) was spread on Luria-Bertani (LB) agars and incubated at 37 °C for 12 h. The bacterial concentration of the suspension was decided by colony counting assay with approximately 1×10^5 colony formation unit per milliliter (CFU/mL). The bacteria suspension was then diluted into 1×10^5 CFU/mL by TSB for later antibacterial tests.

To test the effect of side chain length of PS-*b*-PIN on antibacterial, the PS-*b*-PIN stock solutions with different side chain length were prepared at the concentration of 5000 ppm in 25% (*w/w*) ethanol aqueous solution and well dispersed by a sonication process for over 30 min before use. The TSB diluted PS-*b*-PIN solutions were then mixed with the *E. coli* with equal volume of 250 µL in separate micro tubes with final PS-*b*-PIN concentration of 0.4, 2, 10, 50 and 100 ppm. After orbital shaking incubation (37 °C, 30 rpm) for 16 h, the OD of TSB solutions containing *E. coli* were determined by ultraviolet-visible (UV-Vis) spectrophotometer (Cary 50 Conc; Varian, Palo Alto, CA, USA) at 600 nm. Sterile and inoculated culture media were used as a negative and positive control, respectively. The growth rate of *E. coli* was calculated according to the following equation:

% *E. coli* growth rate = 100 − (OD sample/OD positive control) × 100%

Moreover, the TSB diluted PS-*b*-PIN solutions were mixed with *E. coli* or *S. aureus* solution with equal volume of 250 µL in separate micro tubes with a final PS-*b*-PIN concentration of 100, 1000, and 2500 ppm. After orbital shaking incubation (37 °C, 30 rpm) for 3 and 5 h, bacterial suspensions were collected from the tube and then stained by BacLight LIVE/DEAD bacterial viability kit (Thermo Fisher Scientific, Waltham, MA, USA) following the manufacturer's instructions. The resulting samples were observed under a fluorescence microscope. Live and dead bacterial cells were counted through image analysis by ImageJ software (National Institutes of Health, Bethesda, MD, USA).

2.4. Cytotoxicity Test of PS-b-PIN

NIH 3T3 cells, derived from NIH Swiss mouse embryo cultures, were cultured in Dulbecco's modified Eagle's medium (DMEM) supplemented with 10% fetal bovine serum (FBS) and penicillin and streptomycin (100 μg/mL) at 37 °C in an atmosphere of 5% CO_2, and the culture medium was changed on alternate days.

NIH 3T3 cells were seeded into 48-well plates at the density of 1×10^3 cells/well and incubated for 24 h. The culture medium was pre-mixed with the prepared PS-b-PIN stock solution with final concentration of 1, 10, 100, 1000, and 2500 ppm for 24 h at 37 °C, and then followed by centrifugation at 2800 rpm for 5 min. The supernatant was then applied for culturing of pre-seeded NIH 3T3 cells for 24 h.

For cell viability assay, 0.5 mg/mL MTT solution was added to each well and incubated for 3 h at 37 °C. Subsequently, the MTT solution was aspirated, and the formed formazan crystals were dissolved in DMSO. The spectrophotometric absorbance at 570 nm was measured using a multi-well plate reader (PowerWave X, BioTek Instruments, Winooski, VT, USA).

2.5. Hemolysis Test of PS-b-PIN

Hemolytic activity was evaluated by examining hemoglobin release from rat red blood cells (RBCs) by direct contact to PS-b-PIN. Fresh rat whole blood was collected in spray-dried K2EDTA tube (Thermo Fischer Scientific, Waltham, MA, USA) and tested within three days. Next, 180 μL of hole blood was mixed with PS-b-PIN and 100 μL phosphate buffered saline buffer (PBS) with final concentration of 100, 1000, and 2500 ppm by totally volume of 300 μL in micro centrifuge tubes for 2 min. The RBCs in the mixed solution were then washed by repeatedly add of 1000 μL PBS, centrifugation (1200 rpm, 4 °C for 5 min) and supernatant was removed; this was repeated five times. The remaining RBCs were lysed to release of hemoglobin by adding of 800 μL H_2O and centrifuged under 3000 rpm at 4 °C for 5 min. Then, 10 μL of solution was mixed with 90 μL of Drabkin's reagent (Sigma-Aldrich, St. Louis, MO, USA) to oxidize the hemoglobin into cyanmethemoglobin and absorbance at 540 nm was measured using a microplate reader (PowerWave X, BioTek Instruments, Winooski, VT, USA). For standard calibration of the absorbance, human hemoglobin and Triton X-100 were used as negative and positive controls of the hemolysis test. Samples with a hemolysis ratio less than 5% were regarded as "no hemolysis" [27].

2.6. Morphology Determination by Transmission Electron Microscopy

PS-b-PIN were re-dissolved in ethanol/THF (v/v = 95/5) solution under ultrasonic oscillator for 24 h. A drop of the polymer solution was dipped on a copper mesh and vacuum dried for 24 h. The sample on the mesh was treated with ruthenium tetroxide (RuO_4) vapour for 3 min to stain the polystyrene rich domains. Transmission electron microscopy (TEM) images were obtained from Hitachi H-7100 Transmission Electron Microscope (Hitachi High-Technologies, Tokyo, Japan) equipped with a CCD camera using an accelerating voltage of 75 keV.

2.7. Statistical Analysis

All the data were presented as the mean ± standard deviation. Student's t-test was used to analyze the significance of the differences between groups. The value of $p < 0.05$ was considered statistically significant.

3. Results

3.1. Zeta Potential of PS-b-PIN

The previously developed PS-b-PIN diblock copolymer was synthesized according to the route depicted in Figure 1, provided by author Chi-Yang Chao from the Department of Materials Science and Engineering at National Taiwan University (NTU) [28]. The molecular weight of PS and PI in PS-*b*-PIBr was 6000 and 2000 g/mol, respectively, with the polydispersity index (PDI) of 1.20. Herein, the side chain length in PS-*b*-PIN could be adjusted by reacting PS-*b*-PIBr with trimethylamine (PS-*b*-PIN (C1-Br)), *N,N*-dimethylbutylamine (PS-*b*-PIN (C4-Br)), *N,N*-dimethyloctylamine (PS-*b*-PIN (C8-Br)), *N,N*-dimethyldodecylamine (PS-*b*-PIN (C12-Br)), *N,N*-dimethylhexadecylamine (PS-*b*-PIN (C16-Br)), or *N,N*-dimethylhexadecylamine (PS-*b*-PIN (C18-Br)). Figure 2 shows the zeta potential of PS-*b*-PIN polymer with different side chain lengths in an aqueous solution. Increasing the side chain length increased the zeta potential of PS-*b*-PIN particles, as the carbon side chain length was lower than 12. When the PS-*b*-PIN was conjugated with *N,N*-dimethylhexadecylamine (C16-Br) or *N,N*-dimethylhexadecylamine (C18-Br), there was no significant difference in zeta potential in comparison with that conjugated *N,N*-dimethyldodecylamine (C12-Br).

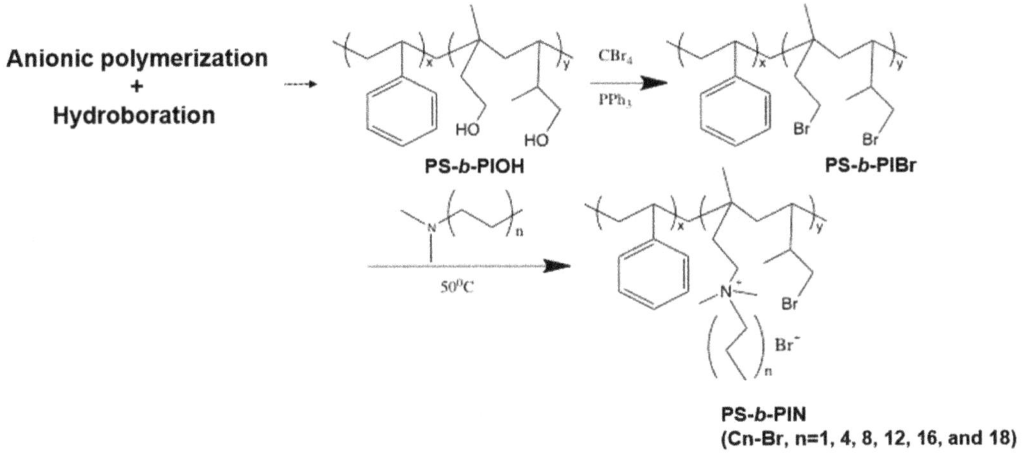

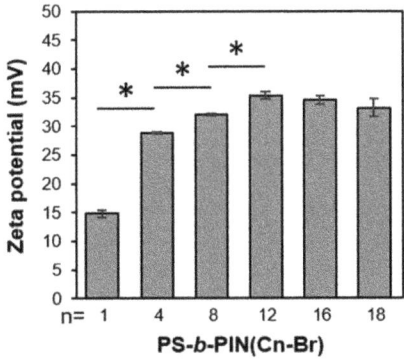

Figure 1. Synthesis of PS-*b*-PIN block copolymers with different side chain length.

Figure 2. The zeta potential of PS-*b*-PIN with different side chain length. *: $p < 0.05$.

3.2. Effect of Side Chain Length of PS-b-PIN on Antibacterial

In order to test the effect of side chain length of PS-*b*-PIN on inhibition of bacterial growth, the prepared PS-*b*-PIN (C1-Br), PS-*b*-PIN (C4-Br), PS-*b*-PIN (C8-Br), PS-*b*-PIN (C12-Br), PS-*b*-PIN (C16-Br), and PS-*b*-PIN (C18-Br) at the concentrations of 0.4, 2, 10, 50, and 100 ppm were incubated with *E. coli* for 16 h. The results of *E. coli* growth rate shown in Figure 3 reveal that there was a positive relationship between the bacterial growth rate and PS-*b*-PIN concentration. Moreover, the antibacterial effect of PS-*b*-PIN increased as the length side chain increased. However, the PS-*b*-PIN with *N,N*-dimethyldodecylamine as a side chain could achieve a better antibacterial effect due to the higher surface charge density, which could lead to ionic interaction with bacteria wall constituents and disruption of the cell bacteria membrane to cause bacterial death [29,30]. However, the PS-*b*-PIN with longer side chain length (C16-Br and C18-Br) could not exhibit an effective inhibition on the *E. coli* growth. It could be suggested that more flexible long carbon side chain would form intramolecular aggregates to reduce the active polymer chains, resulting in the lower antibacterial activity. Since the PS-*b*-PIN (C12-Br) had best effect on inhibiting bacterial growth, we decided to use this kind PS-*b*-PIN block copolymer sample for the following tests.

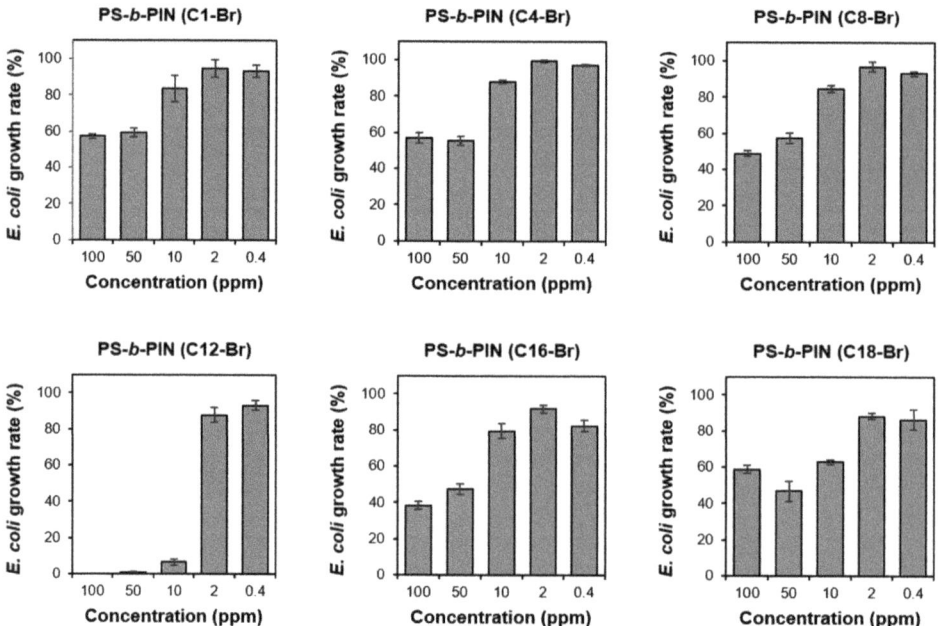

Figure 3. The *E. coli* growth rate after treatment with PS-*b*-PIN with different side chain lengths for 16 h.

To investigate the antibactericidal ability of PS-*b*-PIN against both Gram-negative and Gram-positive bacteria, 1×10^5 CFU/mL of bacteria were directly mixed with PS-*b*-PIN under orbital shaker incubator at 37 °C. In order to examine the antibacterial mechanism of PS-*b*-PIN, the bacteria number with different treatment were assayed at the incubation time of 3 and 5 h (Figure 4). Since PS-*b*-PIN has an absorption background at 600 nm under tested concentration, Live/Dead fluorescence assays were used. After 3 h of incubation, the number of *E. coli* was lower for control and PS-*b*-PIN treated groups. This was due to the lag phase of bacterial growth in which the number increase of bacteria was merely assayed. After 5 h incubation, *E. coli* started to grow in log phase and an obvious number increase was observed for the control group. In contrast, growth of *E. coli* was clearly inhibited by mixing with PS-*b*-PIN within a 5 h incubation. Similar growth tendency was also observed for *S. aureus*. At the concentration of 2500 ppm, the prepared PS-*b*-PIN against *E. coli* was more efficient than that against *S. aureus*. This greater antimicrobial efficiency might be attributed to the thickness of bacterial wall. The cell wall thickness of Gram-negative bacteria *E. coli* is around 7–8 nm, which was thinner and consequently more susceptible than that of Gram-positive bacteria *S. aureus* (20–80 nm) [31]. These results demonstrated that PS-*b*-PIN has antibacterial activity against not only Gram-negative but also Gram-positive bacteria in the short treating time.

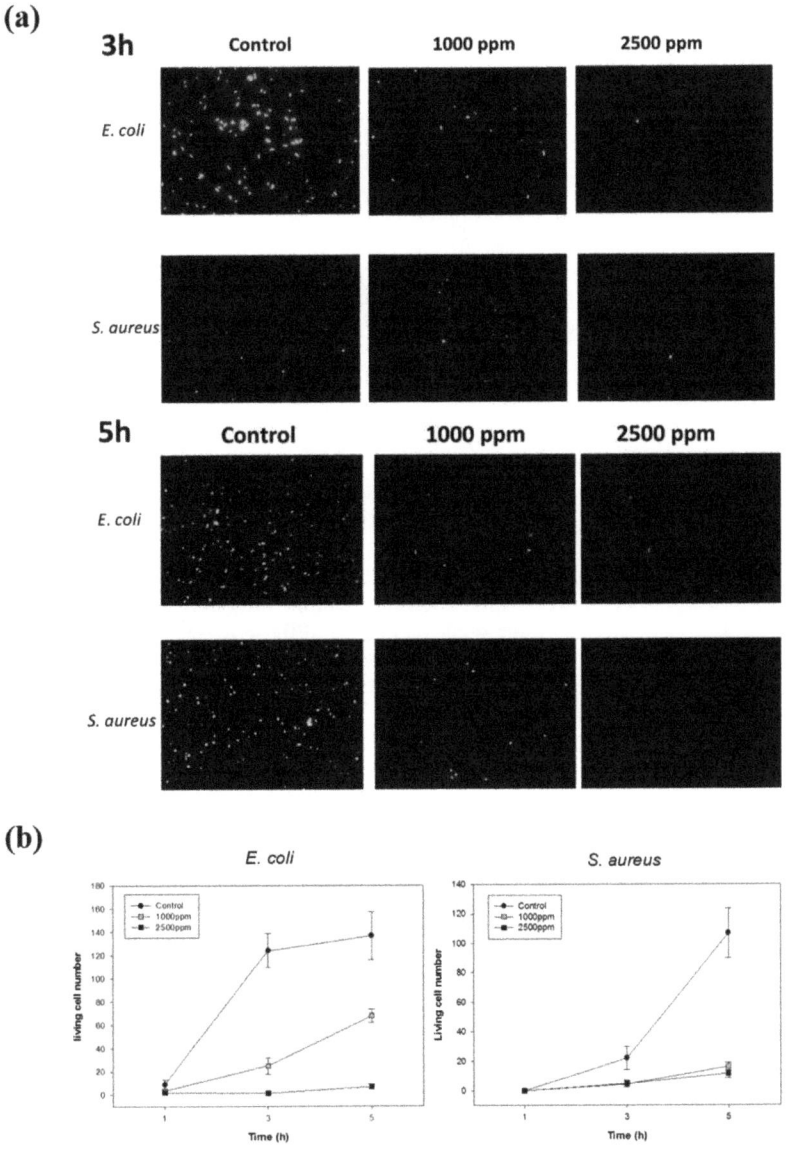

Figure 4. Inhibition of bacterial growth at different PS-*b*-PIN concentrations. (**a**) Fluorescence images and (**b**) the living number of *E. coli* and *S. aureus* with different PS-*b*-PIN concentrations treatment at 3 and 5 h.

3.3. Cytotoxicity of PS-b-PIN

As the pilot study of biocompatibility tests, cytotoxicity was investigated through MTT assays. Since PS-*b*-PIN can be well dispersed in aqueous solution, extract method was used [32]. Therefore, extract solution of PS-*b*-PIN was then used for the culturing of NIH 3T3 cells. Figure 5 shows the cell viability of NIH 3T3 cells incubated with PS-*b*-PIN extract solution for 24 h. It was observed that for PS-*b*-PIN with extracted concentration from 100 to 1000 ppm, over 80% of cell viability was reached. Since no cell toxicity dependency on PS-*b*-PIN concentration was observed, it was assumed that the copolymer chain of PS-*b*-PIN was stable in the aqueous phase and no toxic ingredients were dissolved in the solution.

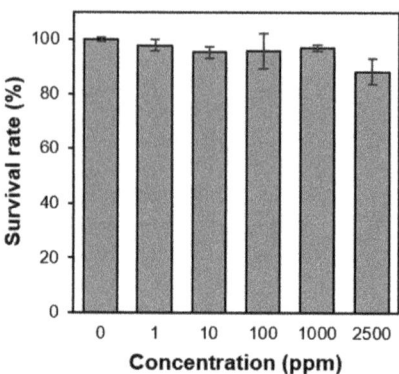

Figure 5. The cytotoxic effect toward 3T3 cells at different PS-*b*-PIN concentration.

3.4. Hemocompatibility of PS-b-PIN

The amphiphilic copolymer, PS-*b*-PIN, was further investigated with the breakage ability toward lipid bilayer of cell membrane. For that, lytic activity of the PS-*b*-PIN copolymer against RBCs was examined through hemolysis assays. By directly mixing copolymer with rat whole blood, PS-*b*-PIN showed no hemolysis (<5%) at all tested concentrations (100–2500 ppm). It is also noteworthy that the hemolysis does not show concentration dependence on copolymer (Figure 6). In contrast to PS-*b*-PIN, the non-ionic surfactant, Triton X-100, was observed to induce 95% hemolysis of RBCs. It is known that non-ionic surfactants such as Triton can exist as both micelles and monomers in solution and interfere with protein-lipid and lipid-lipid interaction of cell membrane as lysis agents [33]. On the other hand, the prepared PS-*b*-PIN, as an amphiphilic functional group, is theoretically stronger lysis agent for RBCs membranes, as it did for NIH 3T3 cell membranes [34]. The ability to destroy the RBCs and NIH 3T3 cell membranes might be attributed to the difference in the membrane structure and composition. It is well known that the RBCs has a double-layer phospholipid structure with a thickness of approximately 10 nm, in which many protein molecules intercalate on the membrane, and there is a strong net-like fiber tissue under the membrane to support the membrane structure [35]. Therefore, these intercalated proteins and net-like fiber tissue would provide RBCs with greater flexibility and deformability than NIH 3T3 cells against the PS-*b*-PIN destruction.

3.5. Morphologies of PS-b-PIN Polymersome

To examine the morphology of PS-*b*-PIN, 5000 ppm of PS-*b*-PIN was dissolved in ethanol/THF (v/v = 95/5) solution within an ultrasonic oscillator bath for one day. One drop of the polymer solution was dipped onto a copper mesh for TEM visualization observation. The PS-*b*-PIN was dried and treated with RuO_4 vapour to selectively stain the PS domains, which appeared dark in the TEM image, as shown in Figure 7. The hollow sphere with a diameter of 0.5–1 μm was clearly observed, and the thickness of a circle was around 80 nm for each sphere regardless the diameter. The image accordingly suggests that PS-*b*-PIN could form microstructure in solution. Since PS segment is hydrophobic and PIN segment is hydrophilic, PS-*b*-PIN would form large polymersomes in the buffer solution with PIN segments facing out by its hydrophilic interaction.

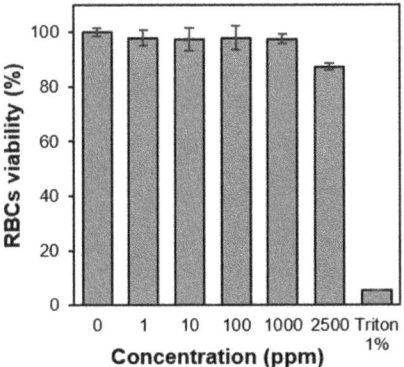

Figure 6. Hemolysis test of PS-*b*-PIN on RBC cells.

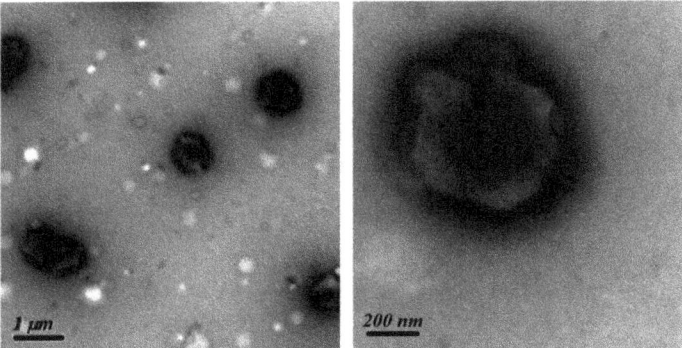

Figure 7. TEM images of PS-*b*-PIN with RuO$_4$ stained.

When a polymersome forms in the aqueous solution, the hydrophobic chains (PS segments) interfold inside, and positively charged quaternary ammonium chains (PIN segments) are exposed on the surface. On the other hand, the bacterial cell wall surface is negatively charged due to teichoic acids and lipopolysaccharides of Gram-positive and Gram-negative bacteria, respectively [36,37]. The polymersome then tends to interact and attach to the bacterial surface through charge-charge interaction. Since quaternary ammonium chains contain long alkyl structures, they can interrupt the bacterial cell membrane by inserting the alkyl chains into the lipid-lipid bilayer. As the size of the PS-*b*-PIN polymersome is approximately 500 nm, it can tear down the bacteria (about 3–4 µm) while maintaining its polymersome structure. After reacting with bacteria, the polymersome can disperse back into the solution again for another bactericidal function (Figure 8). Therefore, PS-*b*-PIN polymersome can kill bacteria while dispersed into the solution and also maintains a low bacterial number during a certain incubation time. However, PS-*b*-PIN may aggregate due to electrolytes and proteins in the physiological environment after a lengthy incubation time with bacteria.

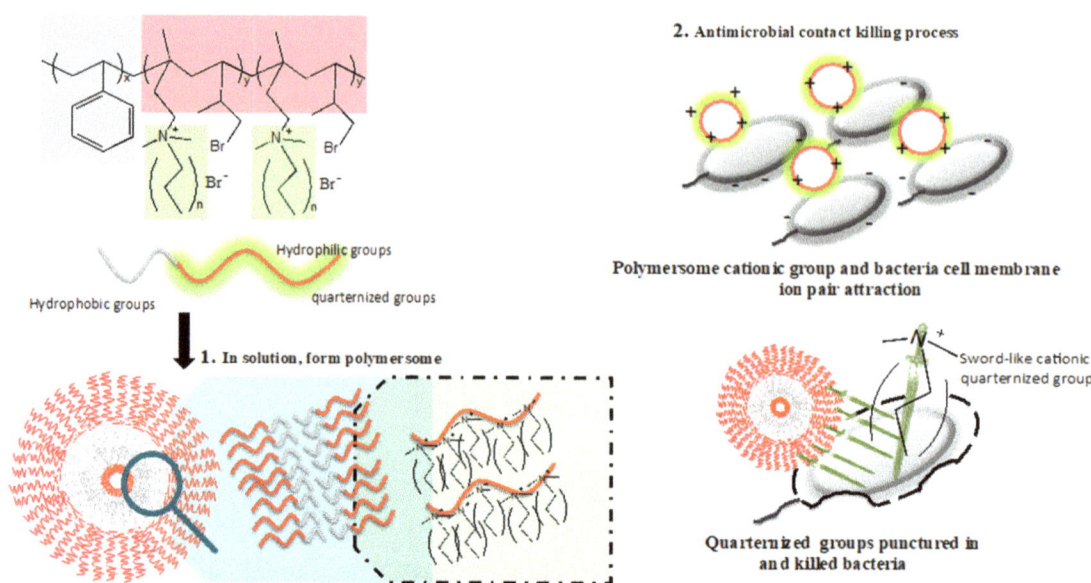

Figure 8. Schematic of proposed quaternized amphiphilic block copolymers PS-*b*-PIN polymersome formation and antibacterial activities.

4. Conclusions

In this study, we have investigated the effects of PS-block-PIN quaternized amphiphilic block copolymer on antibacterial activity. The amphiphilic block copolymer structure is a significant element in our research. Through molecular composition and morphology control, we were able to manipulate the morphology into polymersome in the solution. Due to the polymersome structure, it is possible to strike a balance among antibacterial, hemolytic, and cytotoxic activities. The block copolymer displayed high bactericidal activity against *E. coli*, while it displayed selective activity over RBCs and was not hemolytic. So far, we have manipulated the polymer structure to attain good antibacterial activity. This study highlights quaternized amphiphilic block copolymer structures as a new design agent to improve the activity by quaternization on side chains, as well as to understand the mechanism of antibacterial actions by the formation of morphology.

Author Contributions: Conceptualization, C.-H.C. (Chih-Hao Chang) and C.-Y.C.; funding acquisition, C.-H.C. (Chih-Hao Chang), C.-H.C. (Chih-Hung Chang), W.-C.Y. and C.-Y.C.; project administration, C.-H.C. (Chih-Hao Chang), C.-H.C. (Chih-Hao Chang) and C.-Y.C.; methodology and supervision, C.-H.C. (Chih-Hung Chang) and C.-Y.C.; investigation, C.-H.C. (Chih-Hung Chang), Y.-W.Y., H.-Y.C., S.-J.Y. and W.-C.Y.; writing—original draft preparation, C.-H.C. (Chih-Hung Chang), Y.-W.Y. and H.-Y.C.; writing—review and editing, C.-H.C. (Chih-Hung Chang), S.-J.Y. and C.-Y.C. All authors have read and agreed to the published version of the manuscript.

Funding: This work was supported by the Ministry of Science and Technology (MOST 107-2314-B-002-200-MY3), Far Eastern Memorial Hospital (107FTN17), Min-Sheng General Hospital (109F005-110-M) and National Taiwan University Hospital (106C101-41 and 107-S3948).

Data Availability Statement: The data presented in this study are available on request from the corresponding author.

Acknowledgments: The authors appreciate the help from the National Taiwan University College of Medicine and the National Taiwan University Hospital.

Conflicts of Interest: The authors declare no conflict of interest.

References

1. Motay, M.; Martel, D.; Vileno, B.; Soraru, C.; Ploux, L.; Méndez-Medrano, M.G.; Colbeau-Justin, C.; Decher, G.; Keller, N. Virtually Transparent TiO2/Polyelectrolyte Thin Multilayer Films as High-Efficiency Nanoporous Photocatalytic Coatings for Breaking Down Formic Acid and for Escherichia coli Removal. *ACS Appl. Mater. Interfaces* **2020**, *12*, 55766–55781. [CrossRef]
2. Albright, V.; Penarete-Acosta, D.; Stack, M.; Zheng, J.; Marin, A.; Hlushko, H.; Wang, H.; Jayaraman, A.; Andrianov, A.K.; Sukhishvili, S.A. Polyphosphazenes enable durable, hemocompatible, highly efficient antibacterial coatings. *Biomaterials* **2021**, *268*, 120586. [CrossRef]
3. Li, X.; Gui, R.; Li, J.; Huang, R.; Shang, Y.; Zhao, Q.; Liu, H.; Jiang, H.; Shang, X.; Wu, X.; et al. Novel Multifunctional Silver Nanocomposite Serves as a Resistance-Reversal Agent to Synergistically Combat Carbapenem-Resistant Acinetobacter baumannii. *ACS Appl. Mater. Interfaces* **2021**, *13*, 30434–30457. [CrossRef]
4. Duan, Y.; Wu, Y.; Yan, R.; Lin, M.; Sun, S.; Ma, H. Chitosan-sodium alginate-based coatings for self-strengthening anticorrosion and antibacterial protection of titanium substrate in artificial saliva. *Int. J. Biol. Macromol.* **2021**, *184*, 109–117. [CrossRef] [PubMed]
5. Cui, Z.; Luo, Q.; Bannon, M.S.; Gray, V.P.; Bloom, T.G.; Clore, M.F.; Hughes, M.A.; Crawford, M.A.; Letteri, R.A. Molecular engineering of antimicrobial peptide (AMP)-polymer conjugates. *Biomater. Sci.* **2021**, *9*, 5069–5091. [CrossRef]
6. Wahab, M.A.; Li, L.; Li, H.; Abdala, A. Silver Nanoparticle-Based Nanocomposites for Combating Infectious Pathogens: Recent Advances and Future Prospects. *Nanomaterials* **2021**, *11*, 581. [CrossRef] [PubMed]
7. Pachaiappan, R.; Rajendran, S.; Show, P.L.; Manavalan, K.; Naushad, M. Metal/metal oxide nanocomposites for bactericidal effect: A review. *Chemosphere* **2021**, *272*, 128607. [CrossRef] [PubMed]
8. Steinstraesser, L.; Kraneburg, U.; Jacobsen, F.; Al-Benna, S. Host defense peptides and their antimicrobial-immunomodulatory duality. *Immunobiology* **2011**, *216*, 322–333. [CrossRef] [PubMed]
9. Hancock, R.E.W.; Diamond, G. The role of cationic antimicrobial peptides in innate host defences. *Trends. Microbiol.* **2000**, *8*, 402–410. [CrossRef]
10. Hancock, R.E.W.; Lehrer, R. Cationic peptides: A new source of antibiotics. *Trends. Biotechnol.* **1998**, *16*, 82–88. [CrossRef]
11. Tamaki, M.; Kokuno, M.; Sasaki, I.; Suzuki, Y.; Iwama, M.; Saegusa, K.; Kikuchi, Y.; Shindo, M.; Kimura, M.; Uchida, Y. Syntheses of low-hemolytic antimicrobial gratisin peptides. *Bioorg. Med. Chem. Lett.* **2009**, *19*, 2856–2859. [CrossRef] [PubMed]
12. Marr, A.K.; Gooderham, W.J.; Hancock, R.E.W. Antibacterial peptides for therapeutic use: Obstacles and realistic outlook. *Curr. Opin. Pharmacol.* **2006**, *6*, 468–472. [CrossRef]
13. Xue, Y.; Pan, Y.F.; Xiao, H.N.; Zhao, Y. Novel quaternary phosphonium-type cationic polyacrylamide and elucidation of dual-functional antibacterial/antiviral activity. *RSC Adv.* **2014**, *4*, 46887–46895. [CrossRef]
14. Kanazawa, A.; Ikeda, T.; Endo, T. Antibacterial Activity of Polymeric Sulfonium Salts. *J. Polym. Sci. Pol. Chem.* **1993**, *31*, 2873–2876. [CrossRef]
15. Mowery, B.P.; Lee, S.E.; Kissounko, D.A.; Epand, R.F.; Epand, R.M.; Weisblum, B.; Stahl, S.S.; Gellman, S.H. Mimicry of antimicrobial host-defense peptides by random copolymers. *J. Am. Chem. Soc.* **2007**, *129*, 15474–15476. [CrossRef]
16. Palermo, E.F.; Kuroda, K. Structural determinants of antimicrobial activity in polymers which mimic host defense peptides. *Appl. Microbiol. Biotechnol.* **2010**, *87*, 1605–1615. [CrossRef]
17. Kuroda, K.; DeGrado, W.F. Amphiphilic polymethacrylate derivatives as antimicrobial agents. *J. Am. Chem. Soc.* **2005**, *127*, 4128–4129. [CrossRef]
18. Ilker, M.F.; Nusslein, K.; Tew, G.N.; Coughlin, E.B. Tuning the hemolytic and antibacterial activities of amphiphilic polynorbornene derivatives. *J. Am. Chem. Soc.* **2004**, *126*, 15870–15875. [CrossRef]
19. Tew, G.N.; Clements, D.; Tang, H.Z.; Arnt, L.; Scott, R.W. Antimicrobial activity of an abiotic host defense peptide mimic. *Biochim. Biophys. Acta* **2006**, *1758*, 1387–1392. [CrossRef]
20. Locock, K.E.S.; Michl, T.D.; Valentin, J.D.P.; Vasilev, K.; Hayball, J.D.; Qu, Y.; Traven, A.; Griesser, H.J.; Meagher, L.; Haeussler, M. Guanylated Polymethacrylates: A Class of Potent Antimicrobial Polymers with Low Hemolytic Activity. *Biomacromolecules* **2013**, *14*, 4021–4031. [CrossRef] [PubMed]
21. Locock, K.E.S.; Michl, T.D.; Stevens, N.; Hayball, J.D.; Vasilev, K.; Postma, A.; Griesser, H.J.; Meagher, L.; Haeussler, M. Antimicrobial Polymethacrylates Synthesized as Mimics of Tryptophan-Rich Cationic Peptides. *ACS Macro Lett.* **2014**, *3*, 319–323. [CrossRef]
22. Punia, A.; He, E.; Lee, K.; Banerjee, P.; Yang, N.L. Cationic amphiphilic non-hemolytic polyacrylates with superior antibacterial activity. *Chem. Commun.* **2014**, *50*, 7071–7074. [CrossRef]
23. Mizutani, M.; Palermo, E.F.; Thoma, L.M.; Satoh, K.; Kamigaito, M.; Kuroda, K. Design and Synthesis of Self-Degradable Antibacterial Polymers by Simultaneous Chain- and Step-Growth Radical Copolymerization. *Biomacromolecules* **2012**, *13*, 1554–1563. [CrossRef]
24. Sambhy, V.; Peterson, B.R.; Sen, A. Antibacterial and hemolytic activities of pyridinium polymers as a function of the spatial relationship between the positive charge and the pendant alkyl tail. *Angew. Chem. Int. Ed. Engl.* **2008**, *47*, 1250–1254. [CrossRef]
25. Song, A.R.; Walker, S.G.; Parker, K.A.; Sampson, N.S. Antibacterial Studies of Cationic Polymers with Alternating, Random, and Uniform Backbones. *ACS Chem. Biol.* **2011**, *6*, 590–599. [CrossRef] [PubMed]
26. Lee, H.C.; Lim, H.; Su, W.F.; Chao, C.Y. Novel Sulfonated Block Copolymer Containing Pendant Alkylsulfonic Acids: Syntheses, Unique Morphologies, and Applications in Proton Exchange Membrane. *J. Polym. Sci. Pol. Chem.* **2011**, *49*, 2325–2338. [CrossRef]

27. Oda, Y.; Kanaoka, S.; Sato, T.; Aoshima, S.; Kuroda, K. Block versus Random Amphiphilic Copolymers as Antibacterial Agents. *Biomacromolecules* **2011**, *12*, 3581–3591. [CrossRef] [PubMed]
28. Lee, H.C.; Liu, K.L.; Tsai, L.D.; Laic, J.Y.; Chao, C.Y. Anion exchange membranes based on novel quaternized block copolymers for alkaline direct methanol fuel cells. *RSC Adv.* **2014**, *4*, 10944–10954. [CrossRef]
29. Gottenbos, B.; Grijpma, D.W.; van der Mei, H.C.; Feijen, J.; Busscher, H.J. Antimicrobial effects of positively charged surfaces on adhering Gram-positive and Gram-negative bacteria. *J. Antimicrob. Chemother.* **2001**, *48*, 7–13. [CrossRef]
30. Rejane, C.G.; Sinara, T.B.M.; Odilio, B.G. Assis. Evaluation of the antimicrobial activity of chitosan and its quaternized derivative on E. coli and S. aureus growth. *Rev. Bras. Farmacogn.* **2016**, *26*, 122–127.
31. Eaton, P.; Fernandes, J.C.; Pereira, E.; Pintado, M.E.; Xavier Malcata, F. Atomic force microscopy study of the antibacterial effects of chitosans on Escherichia coli and Staphylococcus aureus. *Ultramicroscopy* **2008**, *108*, 1128–1134. [CrossRef]
32. Li, W.; Zhou, J.; Xu, Y. Study of the in vitro cytotoxicity testing of medical devices. *Biomed. Rep.* **2015**, *3*, 617–620. [CrossRef] [PubMed]
33. Schuck, S.; Honsho, M.; Ekroos, K.; Shevchenko, A.; Simons, K. Resistance of cell membranes to different detergents. *Proc. Natl. Acad. Sci. USA* **2003**, *100*, 5795–5800. [CrossRef] [PubMed]
34. Chrom, C.L.; Renn, L.M.; Caputo, G.A. Characterization and Antimicrobial Activity of Amphiphilic Peptide AP3 and Derivative Sequences. *Antibiotics* **2019**, *8*, 20. [CrossRef]
35. Mohandas, N.; Gallagher, P.G. Red cell membrane: Past, present, and future. *Blood* **2008**, *112*, 3939–3948. [CrossRef] [PubMed]
36. Brown, S.; Santa Maria, J.P., Jr.; Walker, S. Wall teichoic acids of gram-positive bacteria. *Annu. Rev. Microbiol.* **2013**, *67*, 313–336. [CrossRef] [PubMed]
37. Wang, J.E.; Dahle, M.K.; McDonald, M.; Foster, S.J.; Aasen, A.O.; Thiemermann, C. Peptidoglycan and lipoteichoic acid in gram-positive bacterial sepsis: Receptors, signal transduction, biological effects, and synergism. *Shock* **2003**, *20*, 402–414. [CrossRef]

Article

Non-Woven Sheet Containing Gemcitabine: Controlled Release Complex for Pancreatic Cancer Treatment

Kazuma Sakura [1,2,3,*], Masao Sasai [3], Takayuki Mino [4] and Hiroshi Uyama [4]

1. Respiratory Center, Osaka University Hospital, Osaka 565-0871, Japan
2. Department of Surgery, Graduate School of Medicine, Osaka University, Osaka 565-0871, Japan
3. Division of Translational Research, Osaka University Hospital, Osaka 565-0871, Japan; sasai-masao@tissue.med.osaka-u.ac.jp
4. Department of Applied Chemistry, Graduate School of Engineering, Osaka University, Osaka 565-0871, Japan; t.mino.OU2009@gmail.com (T.M.); uyama@chem.eng.osaka-u.ac.jp (H.U.)
* Correspondence: tg4c@surg1.med.osaka-u.ac.jp; Tel.: +81-6-6210-8289

Abstract: The 5-year survival rate for pancreatic cancer remains low, and the development of new methods for its treatment is actively underway. After the surgical treatment of pancreatic cancer, recurrence and peritoneal dissemination can be prevented by long-term local exposure to appropriate drug concentrations. We propose a novel treatment method using non-woven sheets to achieve this goal. Poly(L-lactic acid) non-woven sheets containing gemcitabine (GEM) were prepared, and GEM sustained release from this delivery system was investigated. Approximately 35% of the GEM dose was released within 30 d. For in vitro evaluation, we conducted a cell growth inhibition test using transwell assays, and significant inhibition of cell growth was observed. The antitumor effects of subcutaneously implanted GEM-containing non-woven sheets were evaluated in mice bearing subcutaneous Panc02 cells, and it was established that the sheets inhibited tumor growth for approximately 28 d. These results suggest the usefulness of GEM-containing non-woven sheets in pancreatic cancer treatment.

Keywords: pancreatic cancer; gemcitabine; controlled release; non-woven sheet; chemotherapy; antitumor efficacy; poly(L-lactic acid)

1. Introduction

Pancreatic cancer is difficult to manage, and the 5-year survival rate in patients suffering from the disease is less than 5% [1]. Because it is usually far advanced at the time of diagnosis, even after operative resection, relapse often occurs and is detected as multiple liver metastases, local recurrence, or peritoneal dissemination soon after surgery, resulting in a significantly poor disease prognosis. At present, surgical resection is the first and most effective therapeutic option for the treatment of localized pancreatic cancer without distant metastases; however, after pancreatectomy, 40.1% of patients have positive resection margins [2], 32.6% have local recurrence [3], and 13% develop peritoneal dissemination [4].

Therefore, it is important to control local recurrence after pancreatectomy to improve patient survival rates. Although intra-operative radiotherapy as an adjuvant treatment strategy has been used in a bid to reduce local recurrence, its effectiveness is still controversial as a recent randomized trial (ESPAC-1) showed that adjuvant chemoradiotherapy has a negative effect on patient survival [5]. Thus, it is thought that there is no effective local postoperative treatment option for pancreatic cancer. Trans-tissue local therapy has some advantages, such as a high local drug concentration at the target tissue and relatively low concentration in systemic organs, resulting in a significant enhancement of drug therapeutic effects and marked reduction in systemic adverse effects, in both frequency and severity.

Gemcitabine (GEM) is an anticancer drug classified as "pyrimidine antagonist" and is an antimetabolite. GEM enters the DNA of cancer cells, inhibits the synthesis of DNA required for cell division, eliminates the cancer cells, and suppresses the spreading and growth of the cancer. The cytotoxicity of gemcitabine is not limited to the S phase of the cell cycle and is equally effective in saturated and exponential cells. Toxicity appears to be the result of a combination of several effects on DNA synthesis [6]. In other words, GEM is an anticancer agent whose activity depends on the concentration and the time of contact with cells [7–9]. GEM has been used in pancreatic cancer, non-small cell lung cancer, biliary tract cancer, urinary epithelial cancer, unresectable breast cancer, and ovarian cancer that have worsened after chemotherapy [6]. Typical adverse events are myelosuppression, which is a decrease in white blood cells and platelets [6]. Clinical effectiveness of GEM in pancreatic cancer is limited by its rapid plasma metabolism and development of chemo-resistance [10]. To address this concern, we developed non-woven sheets containing GEM to be used in local therapy of pancreatic cancer, featuring a continuous release of GEM.

The treatment of local lesions by subcutaneously introducing an anticancer drug reservoir is a therapeutic strategy for local intraperitoneal cancer. It enables repeated and sustained drug release to a target lesion; however, several risks are associated with this method, such as catheter obliteration, infection via the catheter, and the development of mechanical ileus owing to the presence of the catheter in the abdominal cavity [11]. Some controlled release devices have also been developed for local cancer treatment. A study reported a local pancreatic cancer therapeutic strategy using a mixture of fibrin glue and GEM [12]. However, there are challenges to employing controlled release devices in clinical settings owing to the difficulty of maintaining a steady gradual supply of the drug and using these devices over an extended period of time. We focused on pancreatic cancer treatment and the prevention of its local recurrence and dissemination because chemotherapy for the disease is limited by the difficulty of drug administration after surgery. Anticancer drugs are often poured into the peritoneal cavity in hyperthermal conditions in cases in which cytological dissemination is detected by rapid pathological diagnosis or in those with a high risk of rapid recurrence [13]. However, maintaining the concentration of water-soluble anticancer drugs is difficult. Recently, electrospinning has received considerable scientific interest as a convenient and straightforward process for the fabrication of non-woven sheets of ultrafine fibrous biodegradable polymers [14]. Electrospinning is superior to conventional methods in that it can be conveniently used to fabricate various composite nanofibers that cannot be fabricated using conventional methods [15]. This implies that there is the possibility of fabricating novel local therapeutic devices for use in the human body.

The diameters of electrospun fibers often fall within the sub-micron range, whereas those of conventional polymer fibers are usually greater than the micron range [16]. The small diameter and non-woven morphology of these fibers result in a large specific surface area, which is advantageous for filter and biomedical applications.

2. Materials and Methods

2.1. Chemical Materials

Poly(L-lactic acid) (PLLA; H-900, M_n = 61.323, M_w = 123.284, M_w/M_n = 2.010) was purchased from Mitsui Chemicals (Tokyo, Japan). GEM was purchased from Eli Lilly Japan K.K. (Hyogo, Japan). The solvent 1,1,1,3,3-hexafluoro-2-propanol (HFIP) was purchased from Wako Chemicals (Osaka, Japan).

2.2. Cell Lines

Panc02, murine pancreatic cancer cells, were provided by Dr. Bunzo Nakata (Osaka City University Medical School, Osaka, Japan), and NIH-3T3 cells were purchased from RIKEN Bioresource Research Center (Ibaraki, Japan).

2.3. Animals

Female C57BL/6 mice (6–8-week-old), purchased from CLEA Japan (Tokyo, Japan), were kept under standard housing conditions and provided ad libitum access to food and water. Animals used in all experiments were acclimatized for at least one week in the breeding room of the animal experimentation unit of Osaka University Graduate School of Medicine. In vivo experiments were performed following a protocol approved by the Ethics Review Committee for Animal Experimentation of Osaka University Graduate School of Medicine (#21-055-0).

2.4. Preparation of GEM-Containing Non-Woven Sheets

HFIP was added to 1 g of PLLA and 0.1 g of GEM to make a total mass of 10 g. This solution was used as a PLLA solution with a GEM content of 10 wt%. Other solutions with different GEM contents used for the preparation of PLLA non-woven sheets were prepared in the same way. The solutions were dried for 3 h using a desiccator, with an applied voltage of 17 kV, an injection distance of 15 cm, an injection rate of 3 mL/h, a collector rotation speed of 100 rpm, and an injection volume of 4 mL, using electric field spinning equipment (IMC-164E, Imoto Machinery Co., Kyoto, Japan). The fabricated non-woven sheets were observed using a scanning electron microscope (S-3000N, Hitachi High-Tech Corp., Tokyo, Japan) and an ion-sputtering device (E-1010, Hitachi High-Tech Corp., Tokyo, Japan). The accelerating voltage was 15 kV, and the magnification was 4000 times.

2.5. Measurement of GEM Release In Vitro

Approximately 10 mg of 10 wt% GEM-containing PLLA non-woven sheet was cut into rectangles and immersed in phosphate-buffered saline (PBS) in a sample tube. The sample tube was placed on a bio shaker and shaken at 100 rpm at 37 °C. At a specific time point (0 h, 1 h, 4 h, 24 h, 7 d, 30 d, or 60 d), the non-woven sheet was removed and transferred to a new sample tube. Chloroform (1 mL) was added to the sample tube to dissolve the sheet, and then 1 mL of water was added to it. After proper stirring, the aqueous layer was drained off and high-performance liquid chromatography (HPLC) measurements were performed. The separation was performed on an TSKgel ODS-80Tm column (Tosoh, Osaka, Japan) using 0.1 M NaCl solution. The solution was passed through a 0.45-μm filter before being transferred to the sample tube. A GEM peak was observed at approximately 14 min of elution. The GEM concentration was calculated based on the peak area. To determine the quantity of GEM released, the measured GEM quantity was subtracted from the initial quantity.

2.6. Measurement of GEM Release In Vivo

Non-woven sheets were implanted into the backs of the mice. After a specified period of time (0 h, 1 h, 4 h, 24 h, 7 d, 30 d, or 60 d), the non-woven sheets were removed and dissolved, and the concentration of GEM in the water was measured using the same procedure used for measuring the in vitro GEM concentration.

Approximately 10 mg of 10 wt% GEM-containing PLLA non-woven sheet was cut out and implanted into the backs of the mice. After wound suturing, the sutures were cut again at 1 h, 4 h, 1 d, 7 d, 30 d, and 60 d, and the non-woven sheets were removed. These sheets were dried at room temperature (approximately 15–25 °C) to remove excess water. Then, 1 mL of chloroform was added to the sheets and stirred vigorously to dissolve them. After incubation at 37 °C for 1 d, 1 mL of water was added to the aqueous layer. GEM was extracted into the aqueous layer by stirring.

2.7. In Vitro Cytotoxicity Assay

Non-woven sheets with different GEM contents were cut into circular forms with a diameter of 14 mm and immersed in 70% ethanol before the cell culture experiments were conducted. The seeding densities of Panc02 cells and NIH-3T3 cells were standardized at 1×10^4 cells/well. Then, the plates were removed after every two days and washed with PBS. Floating cells were washed away, and the number of cells that adhered to the non-woven sheet was determined using a fluorescent plate reader (Synergy HT, Biotek Instruments, Inc., Winooski, VT, USA). For the experiment to evaluate growth inhibition through GEM slow release, cells were first seeded in a transwell plate (Transwell Permeable Supports, 0.4-μm with Size 24 Cluster Plate, PC Membrane, Costar®, Thermo Fisher Scientific, Waltham, MA, USA) and, after 6 h, the cut-out non-woven sheets were placed on a net.

The number of cells that adhered to the bottom of the plate after 1, 3, 5, and 7 d was determined using the DNA assay method. First, 1 mL of sodium lauryl sulfate (SDS) solution (SDS 20 mg, sodium chloride 0.90 g, trisodium citrate dehydrate 0.44 g, deionized water 100 mL) was added to each well, and the plates were incubated at room temperature (approximately 15–25 °C) for 1 h. Next, 100 μL from each well was dispensed into 96-well plates, and 100 μL of Hoechst solution (Hoechst 33258 20 μL, sodium chloride 0.18 g, trisodium citrate dehydrate 0.09 g, deionized water 20 mL) was added to each well, in order to measure the fluorescence intensity using a plate reader (setting parameters: excitation wavelength (nm) = 360, emission wavelength (nm) = 460, sensitivity = 80).

2.8. Antitumor Efficacy of the GEM-Containing Non-woven Sheets in a Subcutaneous Tumor-Bearing Mouse Model

Approximately 10 mg of 10 wt% GEM-containing non-woven sheets was cut and implanted into mouse back subcutaneous tissues. Then, 1 h, 4 h, 1 d, 2 d, and 7 d following wound suturing, tissues around the non-woven sheets (including subcutaneous fat) were excised. To verify that GEM sustained release leads to its transfer to surrounding tissues without being metabolized, we measured its concentrations in the tissues. As a control for the non-woven sheet, a GEM solution was injected into mouse subcutaneous tissue at a dose of 1 mg (20 mg/mL × 50 μL) per animal. GEM concentrations in the tissues of these control mice were measured at the same time points as those for the non-woven sheet-injected mice.

The weights of the excised tissues were measured, and saline was added to them at a weight several times greater so that the total volume was at least 0.6 mL. Then, the tissue suspension was homogenized using an ultrasonic homogenizer and stored in a refrigerator at −20 °C. GEM tissue concentration measurement was performed at SRL Inc. (Tokyo, Japan) using HPLC.

One million panc02 cells were injected into the dorsal region of C57BL/6 mice. One week after the cell injection, the tumor size reached approximately the size of a grain of rice (a length of approximately 5 mm). After confirmation of the tumor, 10 mg of 10 wt% GEM-containing PLLA non-woven sheet was implanted into mouse subcutaneous tissues. Similarly, as a control experiment, a GEM solution was injected into mouse subcutaneous tissues using a syringe (20 mg/mL × 50 μL). Then, tumor size was measured from the body surface in the days following wound suturing. For tumor size measurement, the long and orthogonal diameters, "a" and "b," respectively, of the tumors were measured using a ruler. Tumor size was determined using the formula:

$$\text{Tumor size (mm}^3\text{)} = ab^2/2 \qquad (1)$$

Tumor size was measured from the body surface after every 7 d, starting from the day of implantation of the non-woven sheet.

For the control group (n = 6), approximately 10 mg of GEM-free PLLA non-woven sheet was implanted into mouse subcutaneous tissue, whereas for the systemic group (n = 4), 50 μL of GEM aqueous solution was intraperitoneally administered to the mice at a concentration of 20 mg/mL; for the "GEM aqueous solution group" (n = 4), approximately 10 mg of GEM-free PLLA non-woven sheets and 50 μL of GEM aqueous solution at a concentration of 20 mg/mL were subcutaneously injected into the mice (GEM 1 mg/head). For the "GEM 0.5 mg-containing non-woven sheet group" (n = 4), approximately 5 mg of 10 wt% GEM-containing PLLA non-woven sheets was implanted into the mice (GEM 0.5 mg/head); for the "GEM 1.0 mg-containing non-woven sheet group" (n = 6), approximately 10 mg of 10 wt% GEM-containing PLLA non-woven sheets was implanted into the mice (GEM 1.0 mg/head).

2.9. Statistical Analyses

Student's t-test was used to determine statistical significance, and p-values less than 0.05 were considered statistically significant. All results are expressed as the mean ± standard error of the mean (SEM).

3. Results

3.1. Preparation of GEM-Containing Non-woven Sheets

Figure 1 shows the different non-woven sheets prepared at different GEM concentrations, as observed by scanning electron microscopy (Figure 1a). The degradability of the GEM-containing non-woven sheets was evaluated for three months under in vitro and in vivo conditions (Figure 1b).

(a)

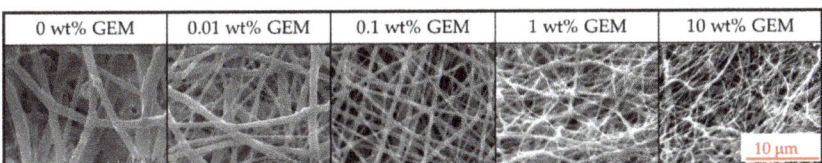

(b)

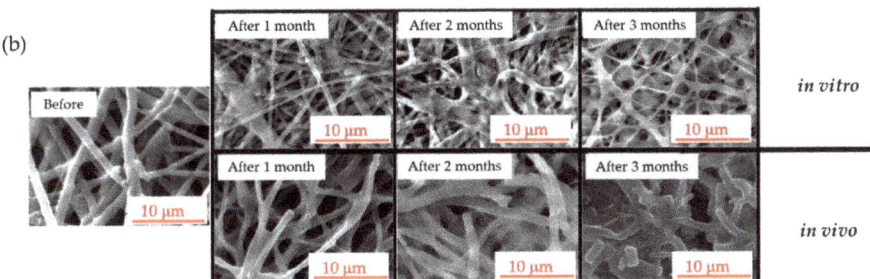

Figure 1. Scanning electron micrographs of GEM-containing non-woven sheets. (**a**) Scanning electron microscopy images of GEM-free and GEM-containing non-woven sheets at concentrations ranging from 0.01% to 10% content. Scale bar: 10 μm. (**b**) Scanning electron microscopy images of denatured GEM-free non-woven sheets after 1, 2, and 3 months of in vitro or in vivo evaluation. Scale bar: 10 μm.

3.2. Measurement of GEM Release from GEM-Containing Non-woven Sheets

Analysis of in vitro GEM release from the non-woven sheets showed an initial burst of 28.4 ± 2.0% on day 1, followed by a 29.8 ± 1.1% and 35.6 ± 2.1 % release by days 7 and 60, respectively (Figure 2a). For In vivo evaluation, the non-woven sheets implanted in the backs of mice were removed after some time, and the quantity of GEM released was determined by measuring GEM concentrations in the non-woven sheets. As in the in vitro evaluation, there was an initial burst of 30.1 ± 7.9 % on day 1, followed by a 30.5 ± 1.5% and 67.8 ± 1.5% release by days 7 and 60, respectively (Figure 2a).

(a)
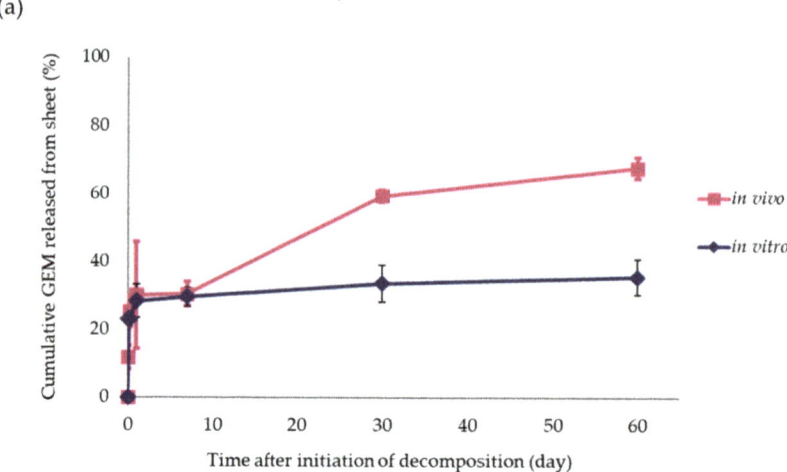

(b)
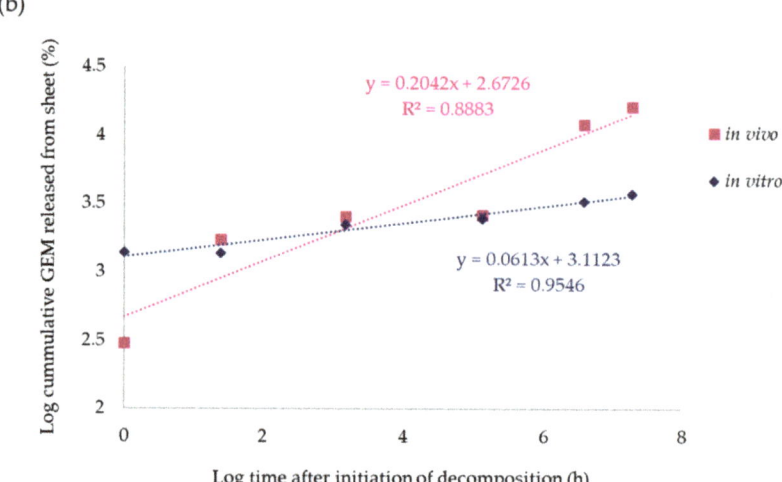

Figure 2. Percentage of GEM released from non-woven sheets with time in vitro and in vivo. (**a**) Cumulative release of GEM from 10 mg of 10 wt% GEM-containing non-woven sheets in vitro and in vivo was evaluated for a period of up to 60 d after non-woven sheet fabrication. After 60 d, 35.6 ± 2.1% and 67.8 ± 1.5% of GEM was cumulatively released in vitro and in vivo, respectively. (**b**) Korsmeyer–Peppas model: the approximate in vitro and in vivo expression levels are calculated as y = 0.0613x + 3.1123 (R^2 = 0.95) and y = 0.20427x + 2.6726 (R^2 = 0.89), respectively. Data are presented as the mean ± standard error of the mean (SEM).

3.3. In Vitro Cytotoxicity Assay

The evaluation of the cytotoxic activity of the non-woven sheets using a DNA assay showed that cells that were directly attached to the sheets exhibited high GEM concentration-dependent cytotoxicity (Figure 3a). The results of the DNA assay under non-contact conditions, using the transwell plate, showed that only non-woven sheets containing high GEM concentrations exhibited cytotoxic activity; sheets containing 0.01 wt% GEM showed no cytotoxic activity, even on day 7 (Figure 3b).

(a)
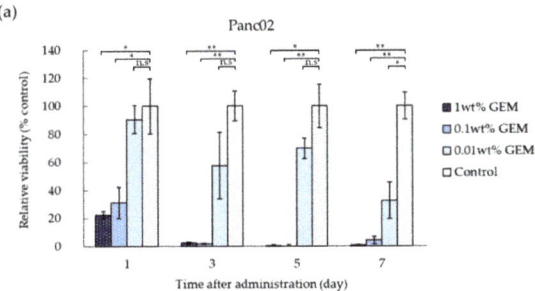

(b)
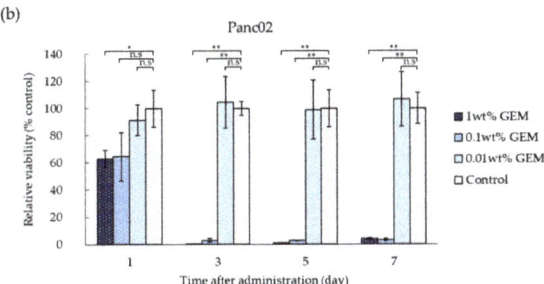

(c)
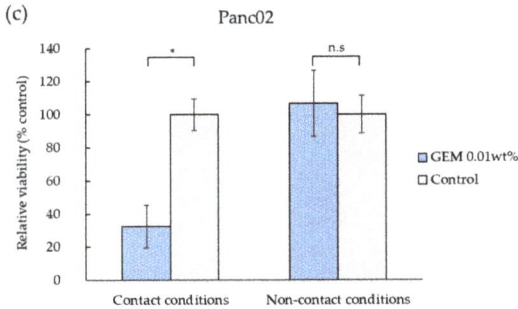

(d)
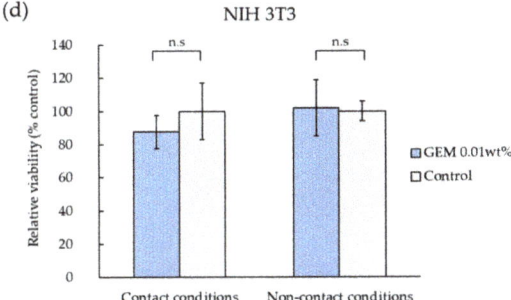

Figure 3. Cytotoxic effects of the non-woven sheets on the murine pancreatic cancer cell line (Panc02) and NIH-3T3 cells. (**a**) Relative cell viability in the treatment group over time (compared with the control group [GEM-free non-woven sheet]). Cell viability on days 1, 3, 5, and 7, respectively, was 22.3 ± 2.8%, 2.3 ± 0.9%, 0.05 ± 0.9%, and 0.6 ± 0.6% in the presence of 1 wt% GEM; 31.1 ± 9.9%, 1.8 ± 0.3%, −5.0 ± 5.8%, and 4.1 ± 2.6% in the presence of 0.1 wt% GEM; and 90.4 ± 9.9%, 57.7 ± 23.7%,

69.9 ± 7.2%, and 32.5 ± 13.0% in the presence of 0.01 wt% GEM (in decreasing GEM concentrations). (**b**) Relative cell viability in the treatment group over time (compared with the control group). Cell viability on days 1, 3, 5, and 7 was 62.9 ± 6.2%, 0.4 ± 0.1%, 1.0 ± 0.1%, and 4.0 ± 0.6% in the presence of 1 wt% GEM; 64.5 ± 17.8%, 3.0 ± 1.2%, 2.7 ± 0.1%, and 3.2 ± 0.6% in the presence of 0.1 wt% GEM; and 91.3 ± 11.4%, 104.6 ± 19.0%, 98.9 ± 21.9%, and 106.8 ± 19.9% in the presence of 0.01 wt% GEM, respectively (in decreasing GEM concentrations). (**c**) Cytotoxic effects of non-woven sheets containing 0.01 wt% GEM against Panc02 cells under contact and non-contact conditions. Non-woven sheets containing 0.01 wt% GEM exhibited significantly higher cytotoxic effects than the control under contact conditions. (**d**) Cytotoxic effects of non-woven sheets containing 0.01 wt% GEM against NIH-3T3 cells under contact and non-contact conditions. Data are presented as the mean ± standard error of the mean (SEM). Significant differences were determined using the Student's *t*-test (* $p < 0.05$; ** $p < 0.01$; n.s: Not significant).

3.4. Antitumor Efficacy of GEM-Containing Non-woven Sheets in Panc02 Tumor Cell-Bearing Mice

GEM tissue concentrations in areas surrounding the non-woven sheet were measured in both groups, namely, that subcutaneously administered the GEM solution and that subcutaneously implanted GEM-containing non-woven sheets (Figure 4a,b).

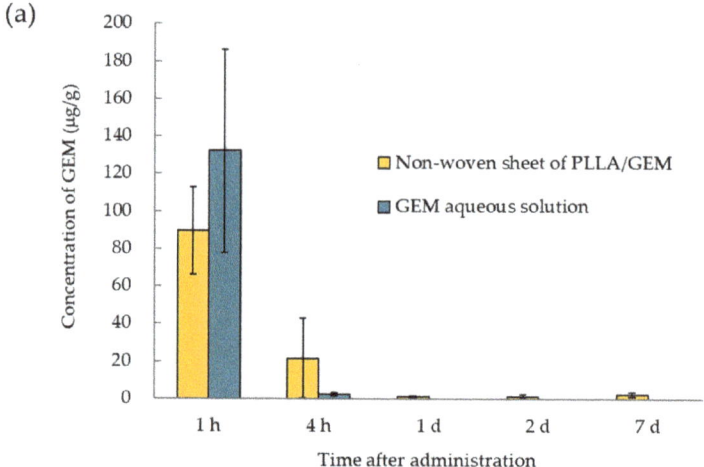

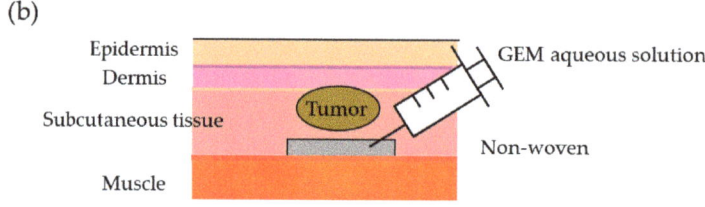

Figure 4. *Cont.*

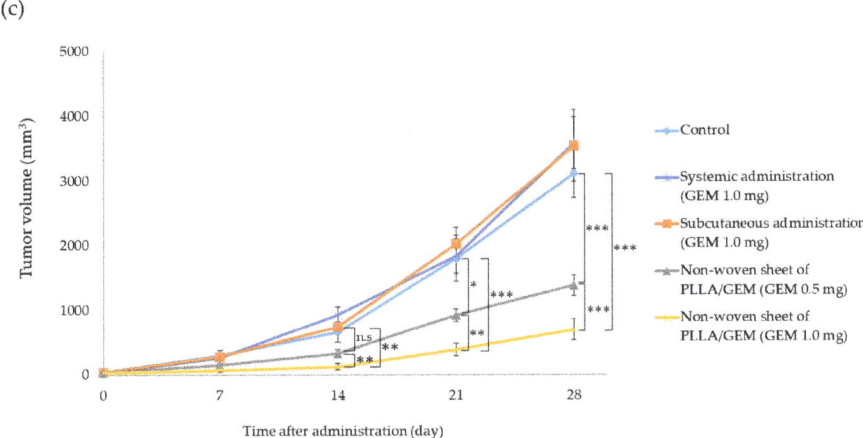

Figure 4. In vivo antitumor effects of the GEM-containing non-woven sheets. (**a**) GEM tissue concentrations. GEM concentrations in surrounding tissues following the implantation of the non-woven sheets in mice were measured at each time point. GEM concentrations released from the GEM-containing PLLA sheets at 1 h, 4 h, 1 d, 2 d, and 7 d were 89.6 ± 13.4, 21.5 ± 12.3, 1.1 ± 0.3, 1.3 ± 0.6, and 2.4 ± 0.8 μg/mL, respectively. In contrast, GEM concentrations released from the GEM-free PLLA sheets containing the GEM solution at 1 and 4 h were 132.4 ± 54.3 and 2.3 ± 0.8 μg/mL, respectively, but GEM was not detected after 1 d. (**b**) Diagram showing the transplantation process. Tumors were implanted intradermally, and non-woven sheets were implanted subcutaneously. GEM-free non-woven sheets were injected with GEM at the time of implantation. (**c**) Antitumor effects of the GEM-containing non-woven sheets. Mice were subcutaneously inoculated with Panc02 cells, and changes in tumor size were measured over time. Control group: Tumor size on days 0, 7, 14, 21, and 28 was 40.3 ± 12.1, 301.9 ± 78.2, 664.8 ± 157.4, 1801.2 ± 353.7, and 3118.3 ± 384.9 mm^3, respectively (n = 6). Systemic administration group: Tumor size on days 0, 7, 14, 21, and 28 was 36.0 ± 2.3, 260.4 ± 16.7, 929.0 ± 121.2, 1835.3 ± 269.8, and 3588.9 ± 397.7 mm^3, respectively (n = 4). Subcutaneous administration (GEM 1.0 mg) group: Tumor size on days 0, 7, 14, 21, and 28 was 38.0 ± 2.0, 283.5 ± 44.6, 744.4 ± 65.9, 2019.6 ± 257.3, and 3544.4 ± 556.9 mm^3, respectively (n = 4). Non-woven PLLA/GEM (GEM 0.5 mg) sheet group: Tumor size on days 0, 7, 14, 21, and 28 was 34.9 ± 9.7, 151.8 ± 34.7, 326.5 ± 62.1, 923.0 ± 96.8, and 1384.5 ± 158.4 mm^3, respectively (n = 4). Non-PLLA/GEM (GEM 1.0 mg) woven sheet group: Tumor size on days 0, 7, 14, 21, and 28 was 31.4 ± 7.6, 68.9 ± 24.0, 125.0 ± 45.5, 388.9 ± 98.6, and 702.2 ± 163.9 mm^3, respectively (n = 7). Results are presented as the mean ± SEM. Significant differences were determined using Student's t-test (* $p < 0.05$, ** $p < 0.001$, *** $p < 0.005$, n.s: Not significant).

After intradermally seeding tumor cells (Panc02) in mice, the preparations were administered and implanted to the mice, and tumor diameter was measured over a period of 28 d (Figure 4c). Each group consisted of four or more mice.

4. Discussion

We developed an electrospun biodegradable non-woven sheet for the controlled release of GEM. GEM is a standard anticancer drug used for pancreatic cancer treatment. It is a water-soluble antimetabolite; thus, it is not suitable for local administration in the peritoneal cavity [17]. PLLA was selected as the backing material polymer because of its biodegradability and biocompatibility [16,18]. In this study, we designed biocompatible GEM-containing nanofibers (GEM sheets) using electrospinning and evaluated their antitumor efficacy in murine pancreatic cancer models.

The scanning electron micrographs of non-woven fibers soaked in PBS showed that fiber diameter progressively decreased, but the shape of the fibers was preserved. Electrospinning of emulsions composed of a PLLA solution and GEM in the presence of HFIP as a catalyst yielded non-woven sheets with a mean fiber diameter of approximately 1–2 µm. As the GEM concentration increased, the fiber diameter progressively decreased because the dissolution of the water-soluble compound led to an increase in the ion concentration, which in turn increased the electrical conductivity. This increase in electrical conductivity resulted in a smaller fiber diameter, which in turn demonstrated similar characteristics to those reported previously [19]. There was a difference in fiber denaturation between the in vitro and in vivo experiments. In vivo, fiber diameter increased more than it did in vitro; in vivo, fiber fragmentation occurred over time, as it was blistered and cut into chunks. In vivo, PLLA fibers are degraded by proteinase K-catalyzed enzymes [20,21]. This difference in denaturation has a significant impact on the sustained release of GEM from the fibers.

In vitro, the cumulative release of GEM over 30 and 60 d from PLLA non-woven sheets containing 10 wt% GEM was approximately 35% and 36%, respectively, whereas that observed in vivo was 61% and 68%, respectively.

This long-term sustained release may be attributed to the fact that PLLA is a crystalline polymer. GEM was homogeneously dispersed in the solution and crystallized during fiber formation by field spinning. Differential scanning calorimetric measurements showed a GEM crystallinity of 35% (data not shown). In this crystallized state, GEM is suggested to exist in the amorphous region of the fiber [22,23]. GEM was initially released from the amorphous region close to the surface of the fiber (the initial burst), and that which was located in the amorphous region surrounded by the crystallized region was not released from the fiber within 60 d. In contrast, in vivo, hydrolysis of the crystallized region by proteinase K-catalyzed enzymes is thought to cause fiber disintegration, resulting in the sustained release of more GEM from the amorphous region surrounded by the crystallized region [24].

PLLA non-woven sheets containing more than 0.1 wt% GEM elicited antitumor effects against Panc02 cells (murine pancreatic cancer cell line) under both direct contact and non-contact conditions. The in vitro cytotoxicity evaluation showed that non-woven sheets containing GEM inhibited cell growth. As the non-woven sheets were set up as a scaffold material for the cells, to exclude the possibility of cytotoxicity being affected by contact between the cells and the non-woven sheets or GEM present on the surface, a cytotoxicity test with a transwell chamber was performed. It was established that cytotoxicity also occurred due to GEM release from the non-woven sheets. In direct contact conditions, cytotoxicity was high owing to the effect of direct exposure to the initial burst and the contact effect of the sheets, whereas under non-contact conditions, the effect of the initial burst was absent, and GEM cytotoxicity at 0.01 wt% was lower than that observed under contact conditions because its effects were due to diffusion alone. As a result, although cytotoxic activity was observed in non-woven sheets containing 0.01 wt% GEM under contact conditions, no cytotoxic activity was observed in these sheets under non-contact conditions. These results suggest that in vivo, non-woven sheets with low GEM concentrations only exhibit antitumor efficacy at their location of application. In addition, although NIH-3T3 has been shown to be sensitive to GEM [25], the cytotoxic effects of the sheets against NIH-3T3 cells were low even under contact conditions, suggesting that these effects do not affect normal cells at the location of application. These data show that PLLA non-woven sheets containing GEM possess a controlled release effect.

Next, we measured GEM concentrations in the subcutaneous tissues and blood of normal mice in which two pieces of non-woven sheets containing 10 wt% GEM were implanted or 2.0 mg of GEM solution was administered. The peak concentration of GEM in the blood of mice in the "non-woven sheet" group was approximately 1/100 that in the blood of mice in the aqueous group, with approximately 1 µg/mL GEM detected in mouse blood at least 7 d following sheet implantation (Supplementary Data; two pieces of 10 wt% GEM non-woven sheets and 100 µL of 20 mg/mL GEM solution).

Based on previous findings, we evaluated the effects of the GEM non-woven sheets on normal tissues. In the cytotoxicity assay, 13.5 mg of 0.01 wt% GEM was divided into 24 wells, with each well containing 1.35 μg of GEM. In the GEM release assay, approximately 30% of GEM was released within 24 h of sheet setting. Considering these data, in the cytotoxic assay, the quantity of GEM in each of the 24 wells was 1.35 μg × 30% = 0.405 μg/well. The volume of medium per well was 500 μL; therefore, GEM concentration per well was 0.81 μg/mL. There was a slight difference between the concentration of GEM in mouse blood 24 h after the implantation of the GEM-containing non-woven sheets and that calculated from 0.01 wt% GEM in the in vitro cytotoxic assay at 24 h. As in vivo and in vitro metabolisms are different, it is difficult to make a simple comparison between them, but we speculate that a GEM concentration of 1.3 μg/mL per well (when the release rate was calculated based on the in vitro concentration curve in Figure 2a) is not significantly different from the in vivo blood concentration. Therefore, the implantation of two pieces of 10 wt% GEM-containing sheets in vivo corresponded to a concentration of 0.01 wt% GEM in the cytotoxic assay. In the cytotoxic assay, 0.01 wt% GEM was found not to exhibit cytotoxic effects against NIH-3T3 cells and thus is considered to have negligible effects on normal tissues.

In the in vivo experiment, there was an initial burst in the tissues into which 10 mg of 10 wt% GEM (1 mg of GEM per head)-containing sheets was implanted 1–4 h following sheet implantation, followed by a sustained GEM concentration within the range of 1.2 to 2.4 μg/g from day 1 to at least 7 d after sheet implantation. The concentrations of GEM in the GEM-containing non-woven sheets were approximately the same as those in the blood of mice administered double the dose of the GEM-containing non-woven sheets. It was demonstrated that a constant GEM concentration was sustained in the tissues into which the GEM-containing non-woven sheets were implanted for a relatively long period of time. In addition, in vivo therapeutic experiments showed that local administration of the GEM solution (20 mg/mL × 50 μL) exerted no antitumor effects, whereas significant and consistent antitumor effects were observed for a relatively long period of time when the GEM-containing non-woven sheets were implanted close to the tumor site. When GEM is applied intraperitoneally for the treatment of post-surgical local pancreatic cancer recurrence, the released GEM is immediately metabolized owing to its water-soluble nature; however, GEM contained in the non-woven sheets is expected to remain localized and be gradually released over time. As GEM is a time- and concentration-dependent anticancer drug, prolonged contact with tumor cells is essential for its antitumor effects, and our non-woven formulation is expected to be an effective tool in that respect [7–9]. Thus, as it is possible to maintain local GEM concentrations without increasing its blood concentrations, the clinical use of this drug may follow its implantation in the resection site following pancreatic cancer surgery. However, the limited opportunity for use, arising from being restricted to a local site, may be a disadvantage, just as GIGLIADEL is currently used in clinical practice for the treatment of malignant gliomas [26,27]. Therefore, we would like to consider a form that can be used for occasions other than during surgery, such as ordinary anticancer drugs in the future.

5. Conclusions

We successfully prepared GEM-containing PLLA non-woven sheets and demonstrated that they exhibit antitumor effects against mouse pancreatic cancer cells in vitro and in vivo. In addition, we measured the tissue concentrations of GEM released from the GEM-containing non-woven sheets and demonstrated that these sheets could contribute to the inhibition of tumor cell growth. As our results were obtained in murine subcutaneous tumor models, the evaluation of PLLA non-woven sheet safety when attached to the remnant of the pancreatic stump and need to control drug release in the transplantation environment are issues that will be addressed in the future for the possible application of this formulation in clinical practice.

Supplementary Materials: The following are available online at https://www.mdpi.com/article/10.3390/polym14010168/s1, Figure S1: two pieces of 10 wt% GEM non-woven sheets and 100 µL of 20 mg/mL GEM solution.

Author Contributions: Conceptualization, H.U. and K.S.; data curation, M.S. and T.M.; investigation, M.S. and T.M.; writing—original draft preparation, M.S.; writing—review and editing, K.S.; project administration, H.U. and K.S.; supervision, K.S. All authors have read and agreed to the published version of the manuscript.

Funding: This research received no external funding.

Institutional Review Board Statement: In vivo experiments were performed following a protocol approved by the Ethics Review Committee for Animal Experimentation of Osaka University Graduate School of Medicine (#21-055-0).

Informed Consent Statement: Not applicable.

Data Availability Statement: The data presented in this study are available on reasonable request from the corresponding author.

Acknowledgments: We thank the staff of the Institute of Experimental Animal Sciences, Osaka University, and the members of the laboratories of the Graduate School of Medicine and Engineering.

Conflicts of Interest: The authors declare no conflict of interest.

References

1. Bengtsson, A.; Andersson, R.; Ansari, D. The actual 5-year survivors of pancreatic ductal adenocarcinoma based on real-world data. *Sci. Rep.* **2020**, *10*, 16425. [CrossRef] [PubMed]
2. Tummers, W.S.; Groen, J.V.; Sibinga Mulder, B.G.; Farina-Sarasqueta, A.; Morreau, J.; Putter, H.; van de Velde, C.J.; Vahrmeijer, A.L.; Bonsing, B.A.; Mieog, J.S.; et al. Impact of resection margin status on recurrence and survival in pancreatic cancer surgery. *Br. J. Surg.* **2019**, *106*, 1055–1065. [CrossRef] [PubMed]
3. Jones, R.P.; Psarelli, E.E.; Jackson, R.; Ghaneh, P.; Halloran, C.M.; Palmer, D.H.; Campbell, F.; Valle, J.W.; Faluyi, O.; O'Reilly, D.A.; et al. European Study Group for Pancreatic Cancer, Patterns of recurrence after resection of pancreatic ductal adenocarcinoma: A secondary analysis of the ESPAC-4 randomized adjuvant chemotherapy trial. *JAMA Surg.* **2019**, *154*, 1038–1048. [CrossRef] [PubMed]
4. Miyazaki, M.; Yoshitomi, H.; Shimizu, H.; Ohtsuka, M.; Yoshidome, H.; Furukawa, K.; Takayasiki, T.; Kuboki, S.; Okamura, D.; Suzuki, D.; et al. Repeat pancreatectomy for pancreatic ductal cancer recurrence in the remnant pancreas after initial pancreatectomy: Is it worthwhile? *Surgery* **2014**, *155*, 58–66. [CrossRef]
5. Neoptolemos, J.P.; Stocken, D.D.; Friess, H.; Bassi, C.; Dunn, J.A.; Hickey, H.; Beger, H.; Fernandez-Cruz, L.; Dervenis, C.; Lacaine, F.; et al. A randomized trial of chemoradiotherapy and chemotherapy after resection of pancreatic cancer. *N. Engl. J. Med.* **2004**, *350*, 1200–1210. [CrossRef]
6. Lucas, S.C.; Gisele, M. Gemcitabine: Metabolism and molecular mechanisms of action, sensitivity and chemoresistance in pancreatic cancer. *Eur. J. Pharmacol* **2014**, *741*, 8–16.
7. Ruiz van Haperen, V.W.; Veerman, G.; Noordhuis, P.; Vermorken, J.B.; Peters, G.J. Concentration and time dependent growth inhibition and metabolism in vitro by 2′,2′-difluoro-deoxycytidine (gemcitabine). *Adv. Exp. Med. Biol.* **1991**, *309a*, 57–60.
8. Hanauske, A.R.; Degen, D.; Marshall, M.H.; Hilsenbeck, S.G.; Grindey, G.B.; Von Hoff, D.D. Activity of 2′,2′-difluorodeoxycytidine (gemcitabine) against human tumor colony forming units. *Anticancer Drugs* **1992**, *3*, 143–146. [CrossRef]
9. Kornmann, M.; Butzer, U.; Blatter, J.; Beger, H.G.; Link, K.H. Pre-clinical evaluation of the activity of gemcitabine as a basis for regional chemotherapy of pancreatic and colorectal cancer. *Eur. J. Surg. Oncol.* **2000**, *26*, 583–587. [CrossRef]
10. Anupama, M.; Deepak, C.; Stephan, W.B.; Ram, I.M. Efficacy of gemcitabine conjugated and miRNA-205 complexed micelles for treatment of advanced pancreatic cancer. *Biomaterials* **2014**, *35*, 7077–7087.
11. Martin, R.C.; Robbins, K.; Tomalty, D.; O'Hara, R.; Bosnjakovic, P.; Padr, R.; Rocek, M.; Slauf, F.; Scupchenko, A.; Tatum, C. Transarterial chemoembolisation (TACE) using irinotecan-loaded beads for the treatment of unresectable metastases to the liver in patients with colorectal cancer: An interim report. *World J. Surg. Oncol.* **2009**, *7*, 80. [CrossRef]
12. Okino, H.; Maeyama, R.; Manabe, T.; Matsuda, T.; Tanaka, M. Trans-tissue, sustained release of gemcitabine from photocured gelatin gel inhibits the growth of heterotopic human pancreatic tumor in nude mice. *Clin. Cancer Res.* **2003**, *9*, 5786.
13. Ishikawa, T.; Kokura, S.; Sakamoto, N.; Ando, T.; Imamoto, E.; Hattori, T.; Oyamada, H.; Yoshinami, N.; Sakamoto, M.; Kitagawa, K.; et al. Phase II trial of combined regional hyperthermia and gemcitabine for locally advanced or metastatic pancreatic cancer. *Int. J. Hyperth.* **2012**, *28*, 597–604. [CrossRef] [PubMed]
14. Bognitzki, M.; Czado, W.; Frese, T.; Schaper, A.; Hellwig, M.; Steinhart, M.; Greiner, A.; Wendorff, J.H. Nanostructured fibers via electrospinning. *Adv. Mater.* **2001**, *13*, 70–72. [CrossRef]

15. Zeng, J.; Yang, L.; Liang, Q.; Zhang, X.; Guan, H.; Xu, X.; Chen, X.; Jing, X. Influence of the drug compatibility with polymer solution on the release kinetics of electrospun fiber formulation. *J. Control. Release* **2005**, *105*, 43–51. [CrossRef]
16. You, Y.; Min, B.-M.; Lee, S.J.; Lee, T.S.; Park, W.H. In vitro degradation behavior of electrospun polyglycolide, polylactide, and poly(lactide-co-glycolide). *J. Appl. Polym. Sci.* **2005**, *95*, 193–200. [CrossRef]
17. Gamblin, T.; Egorin, M.; Zuhowski, E.; Lagattuta, T.; Herscher, L.; Russo, A.; Libutti, S.; Alexander, H.; Dedrick, R.; Bartlett, D. Intraperitoneal gemcitabine pharmacokinetics: A pilot and pharmacokinetic study in patients with advanced adenocarcinoma of the pancreas. *Cancer Chemother. Pharmacol.* **2008**, *62*, 647–653. [CrossRef]
18. Yang, F.; Murugan, R.; Wang, S.; Ramakrishna, S. Electrospinning of nano/micro scale poly(l-lactic acid) aligned fibers and their potential in neural tissue engineering. *Biomaterials* **2005**, *26*, 2603–2610. [CrossRef] [PubMed]
19. Zeng, J.; Haoqing, H.; Schaper, A.; Wendorff, J.H.; Greiner, A. Poly-L-lactide nanofibers by electrospinning-Influence of solution viscosity and electrical conductivity on fiber diameter and fiber morphology. *e-Polymers* **2003**, *3*, 102–110. [CrossRef]
20. Tsuji, H.; Kidokoro, Y.; Mochizuki, M. Enzymatic degradation of biodegradable polyester composites of poly(L-lactic acid) and poly(ε-caprolactone). *Macromol. Mater. Eng.* **2006**, *291*, 1245–1254. [CrossRef]
21. Zeng, J.; Chen, X.; Liang, Q.; Xu, X.; Jing, X. Enzymatic degradation of poly(L-lactide) and poly(epsilon-caprolactone) electrospun fibers. *Macromol. Biosci.* **2004**, *4*, 1118–1125. [CrossRef] [PubMed]
22. Leuner, C.; Dressman, J. Improving drug solubility for oral delivery using solid dispersions. *Eur. J. Pharm. Biopharm.* **2000**, *50*, 47–60. [CrossRef]
23. Verreck, G.; Chun, I.; Rosenblatt, J.; Peeters, J.; Dijck, A.V.; Mensch, J.; Noppe, M.; Brewster, M.E. Incorporation of drugs in an amorphous state into electrospun nanofibers composed of a water-insoluble, nonbiodegradable polymer. *J. Control. Release* **2003**, *92*, 349–360. [CrossRef]
24. MacDonald, R.T.; McCarthy, S.P.; Gross, R.A. Enzymatic degradability of poly(lactide): Effects of chain stereochemistry and material crystallinity. *Macromolecules* **1996**, *29*, 7356–7361. [CrossRef]
25. Keyes, K.; Cox, K.; Treadway, P.; Mann, L.; Shih, C.; Faul, M.M.; Teicher, B.A. An in vitro tumor model: Analysis of angiogenic factor expression after chemotherapy. *Cancer Res.* **2002**, *62*, 5597–5602.
26. Bota, D.A.; Desjardins, A.; Quinn, J.A.; Affronti, M.L.; Friedman, H.S. Interstitial chemotherapy with biodegradable BCNU (Gliadel) wafers in the treatment of malignant gliomas. *Ther. Clin. Risk Manag.* **2007**, *3*, 707–715. [PubMed]
27. Ene, C.I.; Nerva, J.D.; Morton, R.P.; Barkley, A.S.; Barber, J.K.; Ko, A.L.; Silbergeld, D.L. Safety and efficacy of carmustine (BCNU) wafers for metastatic brain tumors. *Surg. Neurol. Int.* **2016**, *7* (Suppl. 11), S295–S299. [CrossRef]

Article

Composite Nanocellulose Fibers-Based Hydrogels Loading Clindamycin HCl with Ca²⁺ and Citric Acid as Crosslinking Agents for Pharmaceutical Applications

Pichapar O-chongpian [1], Mingkwan Na Takuathung [2], Chuda Chittasupho [1,3], Warintorn Ruksiriwanich [1,3], Tanpong Chaiwarit [1], Phornsawat Baipaywad [4] and Pensak Jantrawut [1,3,*]

1. Department of Pharmaceutical Sciences, Faculty of Pharmacy, Chiang Mai University, Chiang Mai 50200, Thailand; pichaparo@gmail.com (P.O.-c.); chuda.c@cmu.ac.th (C.C.); yammy109@gmail.com (W.R.); tanpong.c@gmail.com (T.C.)
2. Department of Pharmacology, Faculty of Medicine, Chiang Mai University, Chiang Mai 50200, Thailand; mingkwan.n@cmu.ac.th
3. Cluster of Research and Development of Pharmaceutical and Natural Products Innovation for Human or Animal, Chiang Mai University, Chiang Mai 50200, Thailand
4. Biomedical Engineering Institute, Chiang Mai University, Chiang Mai 50200, Thailand; phornsawat.b@cmu.ac.th
* Correspondence: pensak.j@cmu.ac.th; Tel.: +66-53944309

Abstract: Biocomposite hydrogels based on nanocellulose fibers (CNFs), low methoxy pectin (LMP), and sodium alginate (SA) were fabricated via the chemical crosslinking technique. The selected CNFs-based hydrogels were loaded with clindamycin hydrochloride (CM), an effective antibiotic as a model drug. The properties of the selected CNFs-based hydrogels loaded CM were characterized. The results showed that CNFs-based hydrogels composed of CNFs/LMP/SA at 1:1:1 and 2:0.5:0.5 mass ratios exhibited high drug content, suitable gel content, and high maximum swelling degree. In vitro assessment of cell viability revealed that the CM-incorporated composite CNFs-based hydrogels using calcium ion and citric acid as crosslinking agents exhibited high cytocompatibility with human keratinocytes cells. In vitro drug release experiment showed the prolonged release of CM and the hydrogel which has a greater CNFs portion ($C_2P_{0.5}A_{0.5}$/Ca + Ci/CM) demonstrated lower drug release than the hydrogel having a lesser CNFs portion ($C_1P_1A_1$/Ca + Ci/CM). The proportion of hydrophilic materials which were low methoxy pectin and sodium alginate in the matrix system influences drug release. In conclusion, biocomposite CNFs-based hydrogels composed of CNFs/LMP/SA at 1:1:1 and 2:0.5:0.5 mass ratios, loading CM with calcium ion and citric acid as crosslinking agents were successfully developed for the first time, suggesting their potential for pharmaceutical applications, such as a drug delivery system for healing infected wounds.

Keywords: nanocellulose fiber; hydrogel; low methoxyl pectin; sodium alginate; clindamycin

1. Introduction

Cellulose is the most abundant polymer produced by plants and microorganisms [1]. Normally, cellulose is fibrous with intermittent crystalline and amorphous sections. The separation of fibers results in nanoscale cellulose substances known as nanocellulose, which exists nanocrystals (CNCs) and cellulose nanofibers (CNFs) [2]. Nanocellulose has high number of hydroxyl groups, high mechanical strength, renewability, and low cost [3]. For these reasons, nanocellulose has been considered an ideal nanostructure for making new high-value materials in many fields and gained much attention and interest from researchers. Nanocellulose can be used alone or combined with other polymers or materials for various fields, such as wound dressing, food, cosmetics, tissue engineering, energy, electrospinning, bioprinting, and so on [1,4,5]. This study focuses on CNFs, which are micrometer-long entangled fibrils that contain both amorphous and crystalline cellulose domains. More specifically, CNFs have many unique properties, such as biodegradability, biocompatibility, high strength and modulus mechanical properties, large specific surface area, ability to form a strong entangled nanoporous network, as well as swelling in water and water absorptivity [6]. Due to their attractive properties, CNFs were selected to use as a biopolymer and investigated for their applications to be used as a nanostructure polymer to form a hydrogel.

Hydrogel fabrication from CNFs alone is an ongoing challenge due to the lack of common solvents for their dissolution. However, nanocellulose can disperse in some strong polar solvents (especially water) due to the strong interaction between the surface hydroxyls and solvent molecules. However, the hydrogen bonding between nanofibers still leads to aggregation at the micro-level [7,8]. So far, achieving a good dispersion of CNFs in aqueous media is still a major challenge in developing CNFs-based hydrogel. The addition of polyethylene glycol (PEG) in CNFs could improve the dispersity of CNFs in the aqueous phase. PEG could physically adsorb onto the surface of nanocelluloses via hydrogen bonding [9].

Furthermore, CNFs-based hydrogels are formed by mixing with other polymers to form new nanocomposite materials. Liu et al. fabricated CNF derived from the TEMPO-oxidised method in conjunction with different types of hemicellulose galactoglucomannan, xyloglucan, and xylan crosslinkers to produce a series of nanocellulose hydrogels [10]. Similarly, Yang et al. produced CNF-polyacrylamide composites hydrogels by forming ionic and covalent bonds with multivalent cations [11]. Orasugh et al. have prepared hydroxypropylmethyl cellulose-based nanocomposites with cellulose nanofibrils as a packaging and transdermal drug delivery system [12]. Carlström et al. have blended and crosslinked gelatin with CNFs to produce scaffolds with tuned degradation rates and enhanced mechanical properties [13]. Thus, our main goal of this study was to seek an optimal formulation of CNFs-based hydrogel for pharmaceutical applications, which could be served as a material analogue to use as a carrier for drugs or other biomolecules. In this report, we have focused on the development of hydrogels by using three biopolymers, including CNFs, low methoxy pectin (LMP), and sodium alginate (SA), for fabricating the hydrogels with three crosslinking solutions: citric acid or $CaCl_2$ or citric acid with $CaCl_2$. We hypothesized that the combination of these biopolymers and CNFs-based hydrogels' properties might lead to the invention of hydrogels with potential use as a drug carrier. After preparation, hydrogels were morphologically characterized, investigated for their gel content, swelling ratio, and mechanical strength, and tested for cytotoxicity in a human keratinocytes cell line. Then, clindamycin hydrochloride (CM), an active antibiotic against aerobic Gram-positive and anaerobic bacteria, mycoplasmas, and some protozoa [14], was used as a model drug and loaded into the selected CNFs-based hydrogel formulations. Assays of drug content and release profiles from these CM-loaded hydrogels were then performed.

2. Materials and Methods

2.1. Materials

Cellulose nanofibers (CNFs, white dry powder with nominal fiber width of 50 nm) were purchased from CelluloseLab, New Brunswick, Canada. Low methoxy pectin (LMP, degree of esterification = 29%) was purchased from Cargill™, Saint Germain, France. Sodium alginate (SA, white granule) was purchased from Qindao Bright Moon Seaweed Group Co., Ltd., Qingdao, China. Calcium chloride ($CaCl_2$) and citric acid monohydrate were purchased from RCI Labscan Ltd., Bangkok, Thailand. Distilled water was used as the solvent for preparing the hydrogels. All the other reagents were of analytical grade.

2.2. Preparation of CNFs-Based Hydrogels

CNFs-based hydrogels were prepared by varying the mass ratio between CNF and two polymeric precursors, LMP and SA. The compositions of the CNFs-based hydrogels are summarized in Table 1. Briefly, LMP and SA were dissolved in deionized water (DI) and heated to 60 ± 0.5 °C, hold for 1 h, and cooled to 35 ± 0.5 °C. The CNF powders were dispersed in 40% w/w PEG 1500 and mixed with the LMP and SA solution after which they were homogenized by a homogenizer (IKA T25 Ultra-Trurrax, IKA laboratory technology, Staufen, Germany) at 500 W. Homogeneous polymeric dispersions were obtained under homogenization at room temperature after 2 h. Each ten grams of a formulation were weighed and poured in 3 Petri dishes, envisioning three different crosslinking methods. The crosslinking solution, containing 0.5 M citric acid (Ci) or 3% w/w CaCl$_2$ (Ca) or 0.5 M citric acid, and 3% w/w CaCl$_2$ (Ci + Ca), was poured into each formulation. After 2 h, crosslinked hydrogels were taken out, and additionally washed with DI. Then, the excess surface DI was removed by gently blotting with a filter paper before being freeze-dried (Christ Beta 2-8 LD-plus, Osterode am Harz, Germany) for 24 h to obtain the crosslinked CNFs-based sponges. The CNFs-based hydrogels that demonstrated suitable absorption and tensile properties as well as low toxicity on human keratinocytes cell line were selected to load clindamycin hydrochloride (CM) as a model drug and investigated for drug content and in vitro drug release profile. Briefly, CM was simultaneously incorporated into the CNF dispersion to achieve a 1% w/w CM concentration then followed the steps described above.

Table 1. Composition of different CNFs-based hydrogels.

Mass Ratio	Polymer Composition/40 g			Sample Code		
	CNFs	LMP	SA	Calcium Crosslinking	Calcium and Citric Acid Crosslinking	Citric Acid Crosslinking
1:1:1	0.4	0.4	0.4	$C_1P_1A_1$/Ca	$C_1P_1A_1$/Ca + Ci	$C_1P_1A_1$/Ci
2:0.5:0.5	0.8	0.2	0.2	$C_2P_{0.5}A_{0.5}$/Ca	$C_2P_{0.5}A_{0.5}$/Ca + Ci	$C_2P_{0.5}A_{0.5}$/Ci
0.5:2:0.5	0.2	0.8	0.2	$C_{0.5}P_2A_{0.5}$/Ca	$C_{0.5}P_2A_{0.5}$/Ca + Ci	$C_{0.5}P_2A_{0.5}$/Ci
0.5:0.5:2	0.2	0.2	0.8	$C_{0.5}P_{0.5}A_2$/Ca	$C_{0.5}P_{0.5}A_2$/Ca + Ci	$C_{0.5}P_{0.5}A_2$/Ci

Note: CNFs = cellulose nanoafibers, LMP = low methoxyl pectin and SA = sodium algonate.

2.3. CNFs-Based Hydrogels Characterizations

2.3.1. Morphological Characterizations

After crosslinking, the digital images of the CNFs-based hydrogels were taken using a digital camera (Canon EOS 750D with an 18–55 mm lens, Canon, Inc., Tokyo, Japan). Scanning electron microscopy (SEM) images of coated CNFs-based hydrogels were acquired using JEOL JCM-7000 NeoScope™ Benchtop SEM (JEOL, Tokyo, Japan). Prior to imaging, uncoated hydrogels were mounted on aluminium stubs using double-sided carbon tape (NEM tape, Nisshin Co., Ltd., Tokyo, Japan) and coated with gold for 2 min then positioned on the stage in the imaging compartment of the device. Then, SEM images of all CNFs-based hydrogels were collected using a secondary electron detector at an acceleration voltage of 15 kV under low vacuum mode. Subsequently, a cross-section of CNFs-based hydrogels' morphology was conducted at magnifications of ×100.

2.3.2. CNFs-Based Hydrogels Thickness

The average thickness of hydrated CNFs-based hydrogels was determined using an outside micrometer (3203-25A, Insize Co, Ltd., Suzhou, China). Five random measurements were taken on each hydrogel. The average of the five values and their standard deviation (S.D.) of individual hydrogel was calculated. Thickness measurements were performed in triplicate.

2.3.3. Mechanical Strength Test of CNFs-Based Hydrogels

The mechanical strength of hydrated CNFs-based hydrogels was tested using a texture analyzer, TX. TA plus (Stable Micro Systems, Surrey, UK). An individual sample holder was constructed to facilitate the measurements of 2 cm × 2 cm hydrogel samples. The hydrogel was fixed on a plate with a cylindrical hole with a 9.0 mm diameter (the area of the sample holder hole was 63.56 mm^2). A cylindrical stainless probe (2 mm in diameter) with a plane flat-faced surface was used (with a probe contact area of 3.14 mm^2). The texture analyzer was adjusted for the probe's forward movement at a velocity of 1.0 mm/s. Measurement started when the probe had contacted the sample surface (triggering force). The probe moved on at a constant speed until the hydrogel was broken. The breakage of hydrogels was detected when the peak force of the texture profile analysis curve dropped. The applied force and the slope of the force–time curve were recorded. All the experiments were conducted at room temperature conditions (25 °C, 70% relative humidity). Five replicates were conducted for each hydrogel. The mechanical strength of the hydrogel was characterized by the puncture strength and Young's modulus [15].

2.3.4. Gel Content

The crosslinking efficiency was evaluated by gel content analysis in deionized water (DI) and phosphate buffer saline (PBS) pH 7.4. Freeze-dried CNFs-based hydrogels were cut into 2 cm × 2 cm dimension, and the initial dried weight (W_i) was measured. Then, the hydrogels were immersed in the testing media at room temperature for 48 h and dried for 48 h in a 40 °C oven (Memmert GmbH Co., KG, Schwabach, Germany) until they reached a constant weight (W_f). Gel content (GC) was calculated using the following Equation (1) [16]. The experiment was performed in triplicate under the same conditions and the average GC values were calculated.

$$GC\ (\%) = W_f/W_i \times 100 \quad (1)$$

2.3.5. Swelling Ratio

The swelling ratio of freeze-dried CNFs-based hydrogels was determined by using a gravimetric method [17]. Initially, freeze-dried CNFs-based hydrogels were cut into 2 cm × 2 cm pieces, and their dry weights (W_d) were measured. Each sample was immersed in a vial containing 20 mL deionized water (DI) and phosphate buffer saline (PBS) pH 7.4 at room temperature for 24 h. At certain intervals, the swollen hydrogels were withdrawn from DI or PBS. The wet weight of the swollen hydrogels (W_s) was measured after the removal of excess surface DI or PBS by gently blotting with a filter paper. The maximum swelling degree (%) was defined by the following Equation (2). These tests were carried out in triplicate under the same conditions, and the average values were reported.

$$\text{Maximum swelling degree (\%MSD)} = (W_s - W_d)/W_d \times 100. \quad (2)$$

2.4. Cell Culture

An immortalized human epidermal keratinocyte cell line, HaCaT cell, was obtained from Cell Lines Service GmbH (Eppelheim, Baden-Württemberg, Germany). The cells were cultured in Dulbecco's modified Eagle's media (DMEM) (Thermo Fisher Scientific, Waltham, MA, USA), supplemented with 10 % fetal bovine serum (Merck KGaA, Darmstadt, Germany) and antibiotics (100 U/mL penicillin and 100 µg/mL streptomycin) (Thermo Fisher Scientific, Waltham, MA, USA), and incubated under a humidified atmosphere of 37 °C, 5% CO_2.

2.5. Cell Viability Assay

Effects of the CNFs-based hydrogels on the viability of HaCaT cells were performed by using 3-(4,5-dimethylthiazol-2-yl)-2,5-diphenylte-trazolium bromide (MTT) (Sigma-Aldrich, Saint Louis, MO, USA). HaCaT cells were seeded in 96-well plates at a density of 2×10^5 cells per well and incubated for 24 h in complete DMEM. The indirect contact assay or elution test was performed by using the hydrogels extractable from the cell culture medium. The hydrogels were cut into a piece of 5 mm × 5 mm × 1 mm and sterilized by UV-irradiation for 30 min. Then, the hydrogel's extract was prepared by soaking the hydrogel in DMEM medium for 48 h and subsequently filtered through a 0.22-µm syringe filter [18]. After that, HaCaT cells were treated with 200 µL CNFs-based hydrogel extract for 24 h. The cell viability was performed by MTT assay for assessing cell metabolic activity. Briefly, 200 µL of MTT reagent (0.4 mg/mL) dissolved in culture medium was added into each conditioned well for 1–2 h. Then, the culture supernatants were discarded, and 100 µL of 100% dimethyl sulfoxide (DMSO) was added to solubilize the formazan complex. The plate was then measured for absorbance at 570 nm using a microplate reader (BioTek Instruments, Winooski, VT, USA).

The CNFs-based hydrogel formulations that provided suitable gel content and maximum swelling degree with low cytotoxicity on HaCaT cells were selected to incorporate clindamycin (CM). Then, drug content, gel content, maximum swelling degree, cytotoxicity on keratinocyte cell line, and in vitro drug release of CNFs-based hydrogels containing CM were investigated.

2.6. CM Loading Content

Three randomly taken freeze-dried CNFs-based hydrogels containing CM (about 0.1 g of hydrogel) were added into vials containing 20 mL of PBS buffer solution (pH 7.4) and set aside until the hydrogels dissolved completely by using a magnetic stirrer (50 rpm) at room temperature for 48 h. The solutions were filtered through a 0.45-µm membrane filter and then diluted. The average amount of drug-loaded into hydrogels was analyzed using a UV-spectrophotometer (UV2600i, Shimadzu Corporation, Kyoto, Japan) at 202 nm. The CM contents in hydrogel formulations were determined from the standard curve prepared with CM solution in a concentration range of 0.025–0.1 mg/mL with a high linear regression ($r^2 = 0.9962$). Each hydrogel sample was tested in triplicate. The CM content of the selected CNFs-based hydrogels was calculated with the following Equation (3):

$$\text{Drug loading content } (\%) = \frac{\text{Amount of drug in hydrogel}}{\text{Theorectical drug content}} \times 100 \quad (3)$$

2.7. In Vitro Drug Release Profile and Release Kinetic

CNF-based Hydrogels containing CM in a square shape (1 cm × 1 cm) were immersed in 20 mL PBS buffer pH 7.4 at 37 ± 0.5 °C. The medium was stirred continuously with a magnetic bar at 50 rpm. The dissolution media (3 mL) was taken at predetermined times (0.5, 1, 2, 3, 6, 12, 24, 30, 36, 48, and 72 h), and the phosphate buffer saline (3 mL) was replaced. The samples were analyzed for CM release using a UV-Vis spectrophotometer (UV2600i, Shimadzu, Kyoto, Japan) at 202 nm wavelength. All dissolution experiments were performed in triplicate. The amount of released drug was calculated with the following Equation (4):

$$\text{Amount of released drug } (\%) = \frac{\text{Amount of released drug at the specific time}}{\text{Amount of drug in hydrogel}} \times 100 \quad (4)$$

Drug release kinetics and mechanism were described by different drug release kinetic models. The first 60% of the data of the drug release was fixed with zero-order, first-order, Higuchi and Korsmeyer–Peppas models described in Equations (5)–(8), respectively [19].

The zero-order model describes that the dissolution rate is constant over the period of time and independent from drug concentration. The equation for the zero-model release is shown in Equation (5):

$$Q_t = Q_0 + k_0 t \quad (5)$$

where: Q_t is the cumulative amount of drug release at each predetermined time
Q_0 is the initial amount of drug
k_0 is zero-order kinetic constant
t is time

In first-order release kinetic, the drug release rate typically depends on its concentration. The first-order release kinetic was shown in Equation (6):

$$\log Q_0 - \log Q_t = \frac{k_1 t}{2.303} \quad (6)$$

where: Q_t is the cumulative amount of drug release at each predetermined time
Q_0 is the initial amount of drug
k_1 is first-order kinetic constant
t is time

The Higuchi model is developed to study the drug release of water-soluble and slightly water-soluble drugs incorporated in solid and/or semi-solid matrixes. The equation for the Higuchi model is shown in Equation (7):

$$Q_t = k_H t^{\frac{1}{2}} \qquad (7)$$

where: Q_t is the cumulative amount of drug release at each predetermined time
k_H is Higuchi kinetic constant
t is time

The Korsmeyer–Peppas model relates to the exponential drug release and the fractional drug release. It is used to explain the drug release from a polymer matrix. The Korsmeyer–Peppas model is shown Equation (8):

$$\frac{Q_t}{Q_0} = k t^n \qquad (8)$$

where: Q_t/Q_0 is the fraction of drug release
k is structural and geometrical constant
t is time

2.8. Statistical Analysis

The represented data are expressed as mean ± standard deviations (S.D.). The one-way ANOVA test was carried out using SPSS® statistics software version 17.0 (IBM Corporation, Armonk, NY, USA) to analyze the statistical significance of the results. The p level less than 0.01 were considered statistically different.

3. Results

3.1. Preparation and Morphological Characteristics of CNFs-Based Hydrogels

In the present study, we developed CNFs-based hydrogels as a material-analogue to use as a carrier for drugs or other therapeutic biomolecules. Hydrogel-forming biopolymers, including CNFs, LMP, and SA were used, and twelve different hydrogel formulations were prepared by chemically crosslinked using 3% w/w $CaCl_2$ solution (Ca), 3% w/w $CaCl_2$ in 0.5 M citric acid solution (Ca + Ci) or 0.5 M citric acid solution (Ci). The addition of PEG also made CNFs easily dispersed and helped to increase hydrogel's flexibility [9]. Therefore, following our optimization of CNFs-based hydrogel formulations using the appropriate ratio of each polymer and technique, flexible and non-brittle crosslinked hydrogels were successfully prepared. As shown in Figure 1, all formulations of $C_1P_1A_1$ and $C_2P_{0.5}A_{0.5}$ CNFs-based hydrogels were homogeneous and translucent. Meanwhile, the formulations with composition of CNFs less than 1 mass ratio, which was $C_{0.5}P_2A_{0.5}$ and $C_{0.5}P_{0.5}A_2$ formulations exhibited rough surface and an irregular shape which may be due to incomplete hydrogel formation (Figure S1). Moreover, a large standard deviation (>2 mm) of the thickness of $C_{0.5}P_2A_{0.5}$ and $C_{0.5}P_{0.5}A_2$ formulations (Table S1) indicates that these formulations were not suitable for further mechanical testing. This study found that the mass ratio of CNFs is a key parameter that influences the structure of the composite hydrogels that contained LMP and SA. Thus, $C_1P_1A0_1$ and $C_2P_{0.5}A_{0.5}$ CNFs-based hydrogels were focused. The SEM micrographs of $C_1P_1A0_1$ and $C_2P_{0.5}A_{0.5}$ CNFs-based hydrogels revealed the polymer matrix morphology and porosity with architectures strongly dependent on the composition of the hydrogels on the crosslinking method, as visible in Figure 1. CNFs appeared to be well-integrated with LMP and SA since no separation phase was noticed. The addition of crosslinking agents such as Ca, Ca + Ci, or Ci solutions was an important parameter that strongly influenced the morphology and porosity [17].

The morphological evaluation of the $CaCl_2$ crosslinked hydrogels revealed a significant difference in the morphology of the polymeric structuring when comparison with citric acid crosslinked samples. $CaCl_2$ crosslinked hydrogels revealed less porosity and dense polymer microstructures. For micrographs of citric acid crosslinked hydrogels, they revealed a highly porous morphology. In this study, we mixed 3% w/w $CaCl_2$ in 0.5M citric acid solution as crosslinking agents. We found that the morphology of the CNFs-based hydrogels crosslinked with Ca + Ci showed higher and homogeneous internal porosity than Ca crosslinked hydrogels.

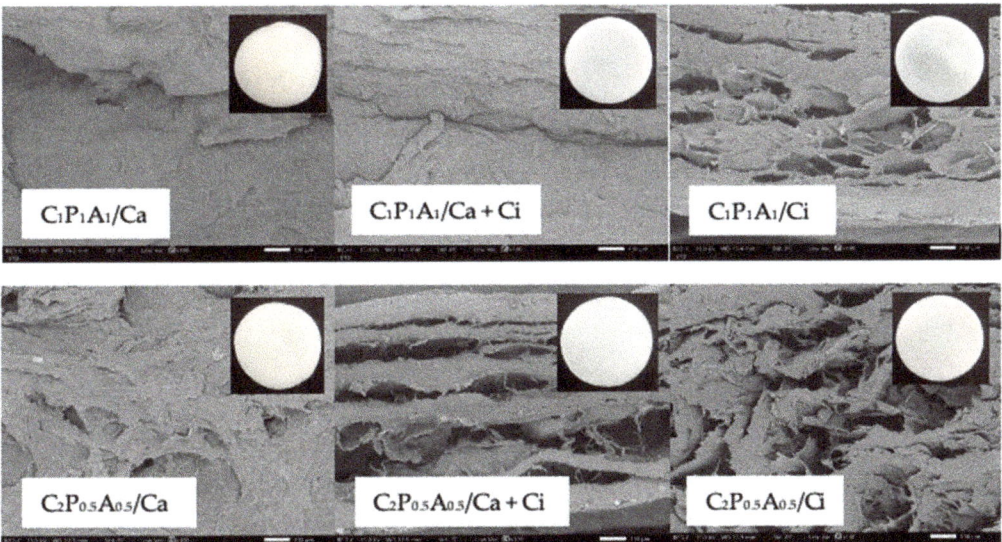

Figure 1. Influence of the composition and crosslinking agent of the hydrated CNFs-based hydrogels and the freeze-dried CNFs-based hydrogels composed of C:P:A with mass ratio 1:1:1 and 2:0.5:0.5 by SEM micrographs (cross-sections) at 100×.

3.2. Mechanical Properties of CNFs-Based Hydrogels

To investigate the influence of biopolymers ratio and type of crosslinking agent on the mechanical properties of CNFs-based hydrogels, puncture strength and Young's modulus were evaluated (Table 2). Compared with the same crosslinking agent, the puncture strength of CNFs-based hydrogels decreased with increasing CNFs portion. Despite the crosslinking method, LMP- and SA-richer hydrogels exhibited a higher puncture strength. This may be due to $C_1P_1A_1$ hydrogels having more crosslinked junctions within the polymeric matrix than $C_2P_{0.5}A_{0.5}$ hydrogels. Moreover, CNFs-based hydrogels using Ci as a crosslinking agent ($C_1P_1A_1$/Ci and $C_2P_{0.5}A_{0.5}$/Ci) exhibited lower puncture strength than other CNFs-based hydrogels using Ca and Ca + Ci as crosslinking agents. The decrease of the puncture strength was believed to be due to decreased crosslink density [20]. From Young's modulus values, it can be seen that the most elastic CNFs-based hydrogels were $C_2P_{0.5}A_{0.5}$/Ci. It can be observed that the citric acid-crosslinked hydrogels present higher elasticity compared to Ca and Ca + Ci-crosslinked hydrogels. However, Young's modulus values between CNFs-based hydrogels using Ca and Ca + Ci have no statistically significant difference. Our results suggested that the type of crosslinking agent and the ratio between the matrix components have a strong influence over the material strength.

Table 2. Thickness, tensile strength, and Young's modulus of CNFs-based hydrogel formulations.

Formulations	Thickness (mm)	Puncture Strength (N/mm^2)	Young's Modulus (N/mm^2)
$C_1P_1A_1$/Ca	6.896 ± 0.520 [a]	0.827 ± 0.058 [a]	0.159 ± 0.004 [a]
$C_1P_1A_1$/Ca + Ci	3.370 ± 0.187 [b]	0.602 ± 0.034 [b]	0.110 ± 0.007 [a]
$C_1P_1A_1$/Ci	3.464 ± 0.324 [b]	0.115 ± 0.006 [c]	0.053 ± 0.002 [b]
$C_2P_{0.5}A_{0.5}$/Ca	5.109 ± 0.382 [a]	0.384 ± 0.008 [d]	0.068 ± 0.014 [b]
$C_2P_{0.5}A_{0.5}$/Ca + Ci	3.470 ± 0.109 [b]	0.228 ± 0.008 [e]	0.066 ± 0.002 [b]
$C_2P_{0.5}A_{0.5}$/Ci	3.376 ± 0.075 [b]	0.063 ± 0.003 [f]	0.038 ± 0.002 [c]

For each test, average values with the same letter are not significantly different. Thus, average values with the different letters, e.g., 'a' or 'b' or 'c' or 'd' or 'e' or 'f' are statistically different ($p < 0.01$).

3.3. Gel Content and Swelling Properties of CNFs-Based Hydrogels

The structural stability of CNFs-based hydrogels was assessed in terms of GC (%) in two different media: DI and PBS in both physiological pH of 7.4, and the results were displayed in Figure 2. According to the gel fraction analysis, the formulations using Ca as a crosslinking agent ($C_1P_1A_1$/Ca and $C_2P_{0.5}A_{0.5}$/Ca) exhibited higher %GC (~45% to 55%) than the formulations crosslinked by Ca + Ci and Ci both in DI and PBS. The higher structural stability might be explained by a dense polymer matrix and strong microstructure due to its ability to generate a crosslinked network by hydrogen bonds and ionic bonds. The formulations using Ca + Ci and Ci as crosslinking agents exhibited lower %GC (~20% to 40%) because of the highly porous network and hydrogen bonds interaction with immersion media. The number of hydrophilic groups (–OH or –COOH) of the CNFs, pectin, and sodium alginate crosslinked by Ca + Ci or Ci was abundant. In the formulations containing citric acid, more than one hydrogen bond are formed by interactions with many functional groups per structural unit, while Ca can interact with only one group from each structural unit. The increase of the hydrophilic groups resulted in the increase of polarity of the composite and the enhancing solubility in water, respectively. Hence, the gel content is lowered [17,21]. Our results also found that the different ratios among the three polymers had revealed slightly significant differences in gel content.

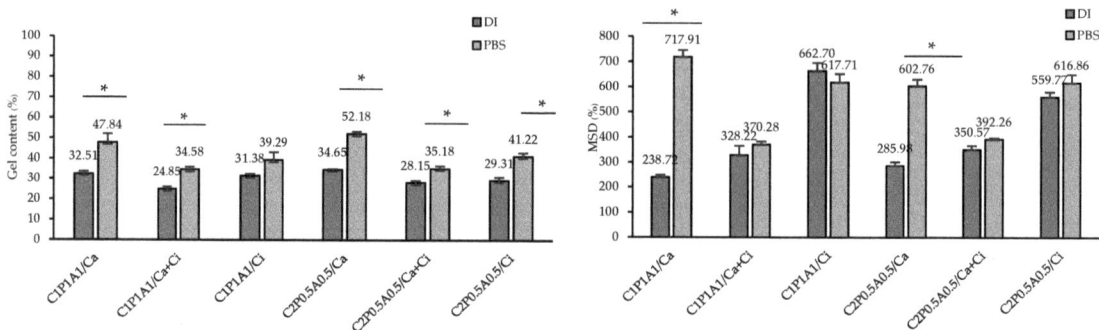

Figure 2. Gel content (**left**) and maximum swelling degree (MSD) (**right**) of CNFs-based hydrogels in DI and PBS at significant level of * $p < 0.01$.

The swelling properties of freeze-dried CNFs-based hydrogels were evaluated after 24 h, and the maximum swelling degree (%MSD) in two different media, DI and PBS, is presented in Figure 2. In DI, the formulations using Ci as a crosslinking agent ($C_1P_1A_1$/Ci and $C_2P_{0.5}A_{0.5}$/Ci) exhibited higher %MSD (~560% to 660%) than the formulations crosslinked by Ca and Ca + Ci due to the highly porous network and hydrogen interaction with water. Generally, hydrogels with a low degree of crosslinking showed higher water uptake ability since the highly crosslinked structure couldn't sustain much water within the gel structure [22,23]. In PBS, the %MSD of the hydrogels using Ci or Ca + Ci as a crosslinking agent revealed no substantial differences compared to the %MSD in DI. However, the formulations using Ca as a crosslinking agent ($C_1P_1A_1$/Ca and $C_2P_{0.5}A_{0.5}$/Ca) exhibited the lowest %MSD in DI (~235% to 285%) because of low hydrogen bonds in their network and dense polymer matrix reducing the water permeability. Nonetheless, the swelling capacity of the formulations using Ca as a crosslinking agent in PBS provided higher %MSD (~600% to 700%) because ion exchange promotes an additional relaxation of the network and enhances the swelling [17,24]. The formulations using Ca + Ci as crosslinking agents exhibited %MSD around 300% to 400% because they exhibited strong microstructure owing to generating a crosslinked network by hydrogen bonds and ionic bonds influencing swelling properties.

3.4. Effects of CNFs-Based Hydrogels on Cell Viability of HaCaT Cells

Considering the novel composite nanocellulose fiber-based hydrogels and crosslinking strategy, an evaluation of cellular response to the hydrogels was fundamentally required. We performed an MTT assay to measure the cell viability of keratinocytes, using HaCaT cells to specifically select the CNFs-based hydrogels that are not toxic to the cells for further experiments. The mitochondrial dehydrogenase performance measurement of CNFs-based hydrogels-treated cells cultured in complete media was performed. For CNFs-based hydrogels without CM, the results revealed good biocompatibility of the assessed materials in contact with HaCaT cells indicating that the combination between CNFs and LMP and SA using Ca + Ci as crosslinking agents was favorable for cell viability more than 80% (Figure 3). The same statistical difference was observed for CNFs-based hydrogels containing CM using Ca + Ci as crosslinking agents in comparison with CNFs-based hydrogels without CM. However, significantly decreased cell viability for Ca or Ci crosslinking agent was observed in $C_1P_1A_1$/Ca, $C_2P_{0.5}A_{0.5}$/Ca, $C_1P_1A_1$/Ci and $C_2P_{0.5}A_{0.5}$/Ci with and without CM (~40% to 60% cell viability). This may be due to calcium and citric in the hydrogel samples inducing alkaline (pH ~ 8) and acidic (pH ~ 4) media, respectively, which leads to diminishing cell viability and cell number. Meanwhile, the media pH of CNFs-based hydrogels using Ca + Ci ($C_1P_1A_1$/Ca + Ci and $C_2P_{0.5}A_{0.5}$/Ca + Ci) were around 6–7. Lönnqvist et al. studied the effect of alkaline and acidic pH on keratinocyte viability. Similar to our study, they showed that alkaline or acidic conditions are detrimental to cells even with a limited duration of exposure [25].

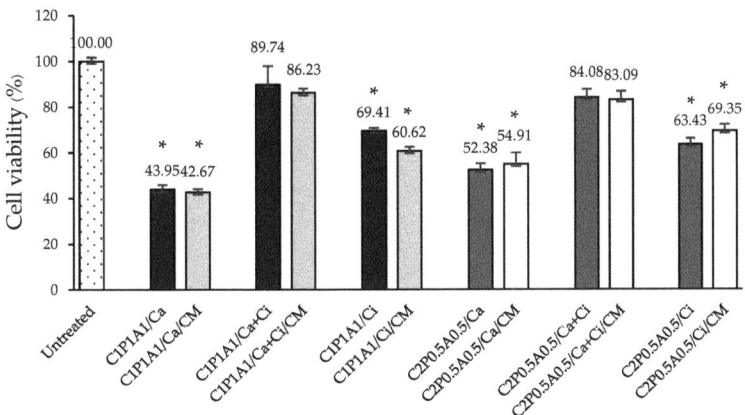

Figure 3. HaCaT cell viability of CNFs-based hydrogels and CNFs-based hydrogels containing CM at a significant level of * $p < 0.01$ in comparison with the untreated group.

Interestingly, the cell viability results were in accordance with the action of CNF-based hydrogels on HaCaT morphology as examined by phase-contrast microscopy. The morphology of HaCaT cells treated with CNFs-based hydrogels containing CM using Ca + Ci was similar to the untreated group, suggesting that these hydrogels showed a non-toxic effect to the cells (Figure 4b). In contrast, morphological changes of HaCaT cells treated with CNFs-based hydrogels using Ca or Ci were observed in a larger proportion of cells swelling and round-out, indicating necrosis. Some revealed shrinkage of cells with unclear nuclei, suggesting apoptosis (Figure 4a,c). Our results also found that the ratio among the three polymers had a minor effect on cell viability. The cells treated with DMSO as vehicle control (untreated group) at consistent concentrations existing in the CNFs-based hydrogels group did not cause any difference in cell viability. Thus, $C_1P_1A_1/Ca$, $C_2P_{0.5}A_{0.5}/Ca$, $C_1P_1A_1/Ci$ and $C_2P_{0.5}A_{0.5}/Ci$ hydrogels were not selected for our further study.

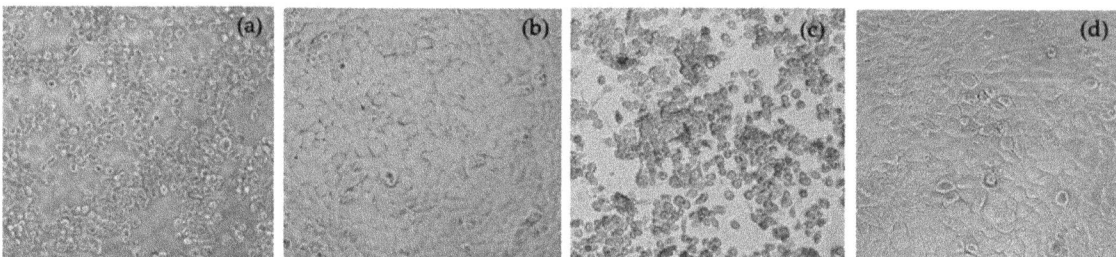

Figure 4. Morphology of HaCaT cells treated with CNFs-based hydrogels containing CM using Ca (**a**), Ca + Ci (**b**), Ci (**c**) as crosslinking agents and untreated (**d**) groups.

3.5. Drug Content and CNFs-Based Hydrogels Containing CM Characteristics

The clindamycin HCl (CM) contents in the selected CNFs-based hydrogels were determined and shown in Table 3. $C_1P_1A_1/Ca + Ci/CM$ and $C_2P_{0.5}A_{0.5}/Ca + Ci/CM$ hydrogels demonstrated more than 80% drug content, with $C_2P_{0.5}A_{0.5}/Ca + Ci/CM$ exhibited significantly higher drug contents than $C_1P_1A_1/Ca + Ci/CM$. This may be due to the greater porosity in the $C_2P_{0.5}A_{0.5}/Ca + Ci/CM$ structure than $C_1P_1A_1/Ca + Ci/CM$ hydrogel (Figure 5). The high porosity in the hydrogel structure is linked to a higher drug loading in the hydrogel matrix [26]. The percent gel content of both CNFs-based hydrogels containing CM was similar in DI and PBS with a slightly different in comparison with the $C_1P_1A_1/Ca + Ci$ and $C_2P_{0.5}A_{0.5}/Ca + Ci$ hydrogels without CM. However, %MSD in DI and PBS of $C_1P_1A_1/Ca + Ci/CM$ were significantly higher than $C_2P_{0.5}A_{0.5}/Ca + Ci/CM$. This may be due to the higher proportion of alginate and pectin. The hydrogen bonds formed by the interactions between alginate and pectin and water molecules could entrap media into the intramolecular space of the hydrogel structure. The maximum swollen occurs when the intracellular space is filled with media [27,28]. The gross appearance and SEM images of the selected formulation, $C_1P_1A_1/Ca + Ci/CM$ and $C_2P_{0.5}A_{0.5}/Ca + Ci/CM$, are shown in Figure 5. After the crosslinking procedure, $C_1P_1A_1/Ca + Ci/CM$ and $C_2P_{0.5}A_{0.5}/Ca + Ci/CM$ demonstrated homogeneity and translucency with smooth surfaces hydrogels. The cross-sectional SEM images of the freeze-dried $C_1P_1A_1/Ca + Ci/CM$ and $C_2P_{0.5}A_{0.5}/Ca + Ci/CM$ exhibited an entangled nanoporous network in the hydrogels' structure. The crystal structure of clindamycin HCl was not observed in the matrix of either hydrogel.

Table 3. Drug content, gel content and maximum swelling degree (MSD) of CNFs-based hydrogels containing CM.

Formulations	Drug Content (%)	Gel Content (%)		MSD (%)	
		DI	PBS	DI	PBS
$C_1P_1A_1$/Ca + Ci/CM	83.21 ± 2.42 [a]	24.94 ± 8.31 [a]	23.71 ± 2.02 [a]	522.22 ± 188.80 [a]	472.51 ± 30.98 [a]
$C_2P_{0.5}A_{0.5}$/Ca + Ci/CM	94.21 ± 4.05 [b]	21.54 ± 0.43 [a]	25.50 ± 0.79 [a]	345.04 ± 55.74 [b]	350.18 ± 18.48 [b]

For each test, average values with the same letter are not significantly different. Thus, average values with the different letter, e.g., 'a' or 'b' are statistically different ($p < 0.01$).

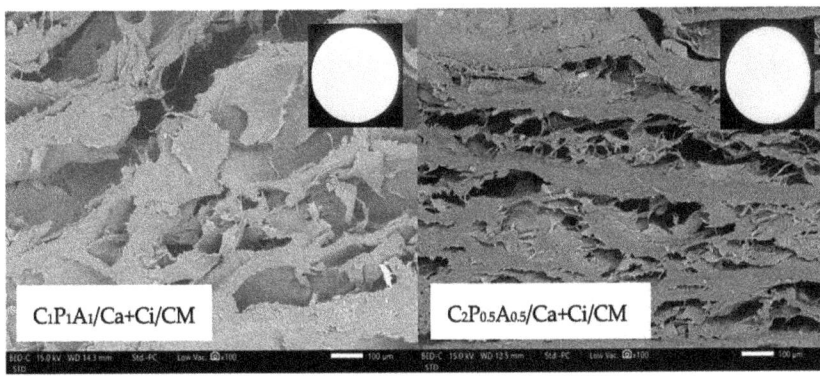

Figure 5. Gross appearance and SEM images of cross-section at 100× of $C_1P_1A_1$/Ca + Ci/CM (**left**) and $C_2P_{0.5}A_{0.5}$/Ca + Ci/CM (**right**).

3.6. CM Release Profile and Kinetic

The drug release of profiles of $C_1P_1A_1$/Ca + Ci and $C_2P_{0.5}A_{0.5}$/Ca + Ci hydrogels containing CM are shown in Figure 6. Both formulations exhibited a similar pattern. The drug release indicated that both $C_1P_1A_1$/Ca + Ci and $C_2P_{0.5}A_{0.5}$/Ca + Ci hydrogels showed prolonged drug release. Overall, the $C_1P_1A_1$/Ca + Ci and $C_2P_{0.5}A_{0.5}$/Ca + Ci gradually released CM in 3 h and then slightly released to about 100%. From 3 to 48 h, the $C_1P_1A_1$/Ca + Ci significantly released CM more than the $C_2P_{0.5}A_{0.5}$/Ca + Ci. For example, at 6 h, the $C_1P_1A_1$/Ca + Ci released CM of about 47%, whereas $C_2P_{0.5}A_{0.5}$/Ca + Ci displayed 34%. However, the $C_2P_{0.5}A_{0.5}$/Ca + Ci released CM around 100% within 48 h. The slower drug release of the $C_2P_{0.5}A_{0.5}$/Ca + Ci hydrogel formulation could be explained by polymer compositions. Both formulations were comprised of pectin and alginate, which are hydrophilic polymers. On the other hand, they also contained nanocellulose fiber, which is not soluble in water. The proportion of hydrophilic material in the matrix systems influences on drug release [29]. In another study, researchers have investigated tramadol release from hydroxypropyl methylcellulose (HPMC) matrices with and without ethyl cellulose (water-insoluble polymer). They concluded that the HPMC matrices with ethyl cellulose displayed slower drug release due to the presence of hydrophobic polymer reduced permeation of the solvent molecules and decreased diffusion of the drug from polymeric matrices, respectively [30]. Possibly, the hydrogel composition containing higher ratio of CNFs may slow down the dissolution of CM. Consequently, the $C_1P_1A_1$/Ca + Ci exhibited significantly faster CM release than that of $C_2P_{0.5}A_{0.5}$/Ca + Ci.

The release behavior of CM in this study was investigated by various models, for instance, zero-order, first-order, Higuchi, and Korsmeyer–Peppas models. The release rate constant and correlation coefficient (R^2) values are shown in Table 4. According to the models, the most appropriate model to evaluate CM release from the CNFs-based hydrogels was the Higuchi model due to higher R^2. In this study, the Higuchi model suggested that the CM release mechanism was a diffusion from the polymeric matrix [31]. Furthermore, the $C_1P_1A_1$/Ca + Ci showed a higher release rate constant (K) than the $C_2P_{0.5}A_{0.5}$/Ca + Ci. This corresponds to the drug release profile that displayed higher CM release from $C_1P_1A_1/Ca + Ci$ than $C_2P_{0.5}A_{0.5}/Ca + Ci$ hydrogel. The data were also fixed with the Korsmeyer–Peppas model to explain the possible release mechanism with n value. Theoretically, $n \leq 0.45$ corresponds to a Fickian diffusion mechanism, $0.45 < n < 0.89$ to a non-Fickian transport, $n = 0.89$ to Case II (relaxational) transport, and $n > 0.89$ to super case II transport. In this study, the n values of $C_1P_1A_1/Ca + Ci$ and $C_2P_{0.5}A_{0.5}/Ca + Ci$ were 0.8020 and 0.8466, respectively, which corresponded to non-Fickian transport referring that the solute transport process in which the polymer relaxation time is approximate the characteristic solvent diffusion time. This means that the solvent absorption and active compound release depend on polymer swelling and polymer/solvent couple viscoelastic properties (Mario and Gabriele, 2005) [32].

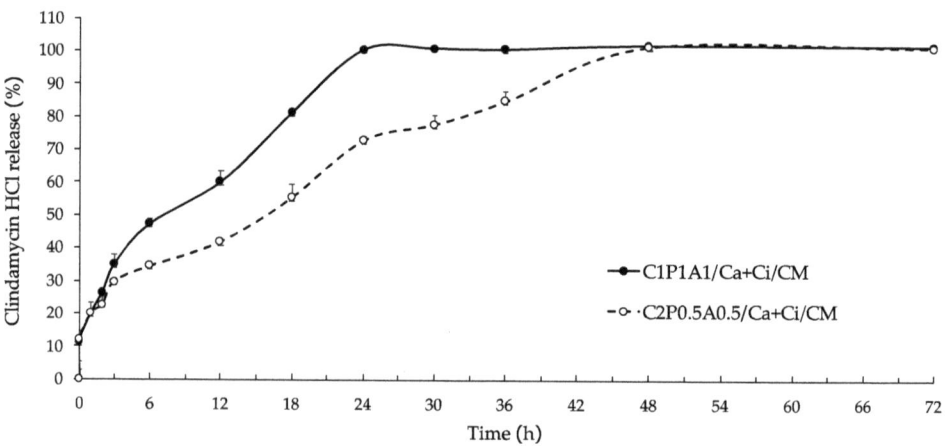

Figure 6. Clindamycin hydrochloride release profile of $C_1P_1A_1/Ca + Ci/CM$ and $C_2P_{0.5}A_{0.5}/Ca + Ci/CM$ hydrogels.

Table 4. Release kinetic data of the CNFs-based hydrogel containing CM in various drug release models.

Kinetic Models	Parameters	Samples	
		$C_1P_1A_1/Ca + Ci/CM$	$C_2P_{0.5}A_{0.5}/Ca + Ci/CM$
Zero-oder	R^2	0.9004	0.9247
	k_0 (h^{-1})	17.1810	17.8590
First order	R^2	0.7244	0.7673
	k_1 (h^{-1})	7.0787	13.6510
Higuchi	R^2	0.9960	0.9839
	k_H (h$^{1/2}$)	18.3460	13.550
Korsemeyer-Peppas	R^2	0.8981	0.9423
	k(h^{-n})	0.0598	0.0361
	n	0.8020	0.8466

4. Conclusions

CNFs-based composite hydrogels were successfully developed with low methoxyl pectin and sodium alginate using calcium and citric acid as crosslinking agents. Among 12 hydrogel formulations, $C_1P_1A_1/Ca + Ci$ and $C_2P_{0.5}A_{0.5}/Ca + Ci$ demonstrated suitable gel content and maximum swelling degree with low cytotoxicity on HaCaT cells, and were selected to load CM and evaluate for drug content and in vitro drug release profile.

$C_1P_1A_1/Ca + Ci/CM$ and $C_2P_{0.5}A_{0.5}/Ca + Ci/CM$ showed CM contents greater than 80%. The in vitro drug release data revealed that the cumulative drug release percentage decreased with the increase of CNFs portions in the composite CNFs-based hydrogels with $C_2P_{0.5}A_{0.5}/Ca + Ci/CM$ as the prolonged CM release profile up to 3 days. Based on the present study, we concluded that CNFs/LMP/SA using Ca + Ci composite hydrogels can be used for pharmaceutical applications as a prolonged release system in transdermal drug delivery. However, the developed CNFs-based hydrogels also require further studies to evaluate their stability at different pH (acidic, neutral and basic), in vitro antibacterial activity, and/or in vivo treatment of skin infection.

Supplementary Materials: The following are available online at https://www.mdpi.com/article/10.3390/polym13244423/s1, Figure S1: Gross appearance and SEM images of the cross-section at 100× of CNFs-based hydrogels that composed of C:P:A with mass ratio 0.5:2:0.5 and 0.5:0.5:2, Table S1: Thickness of CNFs-based hydrogels that composed of C:P:A with mass ratio 0.5:2:0.5 and 0.5:0.5:2 formulations.

Author Contributions: Investigation, P.O.-c. and P.J.; methodology, P.O.-c., M.N.T., W.R., P.B., and P.J.; supervision, P.J.; validation, P.O.-c., M.N.T., T.C., and P.J.; writing—original draft, P.O.-c., T.C., and P.J.; writing—review & editing, M.N.T., C.C., and P.J. All authors have read and agreed to the published version of the manuscript.

Funding: This study received financial support from the National Research Council of Thailand (NRCT), and a partial funding from Chiang Mai University.

Institutional Review Board Statement: Not applicable.

Informed Consent Statement: Not applicable.

Data Availability Statement: The study did not report any data.

Acknowledgments: The authors acknowledge the Faculty of Pharmacy, Chiang Mai University and National Research Council of Thailand (NRCT): N41A640346.

Conflicts of Interest: The authors declare no conflict of interest.

References

1. Ullaha, W.; Ul-Islam, M.; Khan, S.; Kim, Y.; Park, J.K. Structural and physico-mechanical characterization of bio-cellulose produced by a cell-free system. *Carbohydr. Polym.* **2016**, *136*, 908–916. [CrossRef]
2. Klemm, D.; Kramer, F.; Moritz, S.; Lindström, T.; Ankerfors, M.; Gray, D.; Dorris, A. Nanocelluloses: A new family of nature-based materials. *Angew. Chem. Int. Ed.* **2011**, *50*, 5438–5466. [CrossRef] [PubMed]
3. Eyley, S.; Thielemans, W. Surface modification of cellulose nanocrystals. *Nanoscale* **2014**, *6*, 7764–7779. [CrossRef]
4. Lin, N.; Dufresne, A. Nanocellulose in biomedicine: Current status and future prospect. *Eur. Polym. J.* **2014**, *59*, 302–325. [CrossRef]
5. Shi, Z.; Gao, X.; Ullah, M.W.; Li, S.; Wang, Q.; Yang, G. Electroconductive natural polymer-based hydrogels. *Biomaterials* **2016**, *111*, 40–54. [CrossRef] [PubMed]
6. Tejado, A.; Alam, M.N.; Antal, M.; Yang, H.; van de Ven, T.G.M. Energy requirements for the disintegration of cellulose fibers into cellulose nanofibers. *Cellulose* **2012**, *19*, 831–842. [CrossRef]
7. Chu, Y.; Sun, Y.; Wu, W.; Xiao, H. Dispersion Properties of Nanocellulose: A Review. *Carbohydr. Polym.* **2020**, *250*, 116892. [CrossRef]
8. Foster, E.J.; Moon, R.J.; Agarwal, U.P.; Bortner, M.J.; Bras, J.; Camarero-Espinosa, S.; Youngblood, J. Current characterization methods for cellulose nanomaterials. *Chem. Soc. Rev.* **2018**, *47*, 2609–2679. [CrossRef] [PubMed]
9. Chen, L.H.; Wang, Q.Q.; Hirth, K.; Baez, C.; Agarwal, U.P.; Zhu, J.Y. Tailoring the yield and characteristics of wood cellulose nanocrystals (CNC) using concentrated acid hydrolysis. *Cellulose* **2015**, *22*, 1753–1762. [CrossRef]
10. Liu, J.; Chinga-Carrasco, G.; Cheng, F.; Xu, W.; Willför, S.; Syverud, K.; Xu, C. Hemicellulose-reinforced nanocellulose hydrogels for wound healing application. *Cellulose* **2016**, *23*, 3129–3143. [CrossRef]
11. Yang, J.; Xu, F.; Han, C.R. Metal ion mediated cellulose nanofibrils transient network in covalently crosslinked hydrogels: Mechanistic insight into morphology and dynamics. *Biomacromolecules* **2017**, *18*, 1019–1028. [CrossRef]
12. Orasugha, J.T.; Saha, N.R.; Rana, D.; Sarkar, G.; Mollick, M.R.; Chattoapadhyay, A.; Mitra, B.C.; Mondal, D.; Ghosh, S.K.; Chattopadhyay, D. Jute cellulose nano-fibrils/hydroxypropylmethylcellulose nanocomposite: A novel material with potential for application in packaging and transdermal drug delivery system. *Ind. Crops Prod.* **2018**, *112*, 633–643. [CrossRef]

13. Carlström, I.E.; Rashad, A.; Campodoni, E.; Sandri, M.; Syverud, K.; Bolstad, A.I.; Mustafa, K. Crosslinked gelatin-nanocellulose scaffolds for bone tissue engineering. *Mater. Lett.* **2020**, *264*, 127326. [CrossRef]
14. Kim, J.O.; Choi, J.Y.; Park, J.K.; Kim, J.H.; Jin, S.G.; Chang, S.W.; Li, D.X.; Hwang, M.-R.; Woo, J.S.; Kim, J.-A.; et al. Development of Clindamycin-Loaded Wound Dressing with Polyvinyl Alcohol and Sodium Alginate. *Biol. Pharm. Bull.* **2008**, *31*, 2277–2282. [CrossRef]
15. Chaiwarit, T.; Rachtanapun, P.; Kantrong, N.; Jantrawut, P. Preparation of clindamycin hydrochloride loaded de-esterified low-methoxyl mango peel pectin film used as a topical drug delivery system. *Polymers* **2020**, *12*, 1006. [CrossRef] [PubMed]
16. Gulrez, S.K.H.; Al-Assaf, S.; Phillips, G.O. Hydrogels: Methods of preparation, characterisation and applications. In *Progress in Molecular and Environmental Bioengineering from Analysis and Modeling to Technology Applications*; Carpi, A., Ed.; Janeza Trdine: Rijeka, Croatia, 2011; Volume 9, pp. 117–151.
17. Lungu, A.; Cernencu, A.I.; Dinescu, S.; Balahura, R.; Mereuta, P.; Costache, M.; Syverud, K.; Stancu, I.C.; Iovu, H. Nanocellulose-enriched hydrocolloid-based hydrogels designed using a Ca^{2+} free strategy based on citric acid. *Mater. Des.* **2021**, *197*, 109200. [CrossRef]
18. Siqueira, P.; Siqueira, É.; de Lima, A.; Siqueira, G.; Pinzón-Garcia, A.; Lopes, A.; Segura, M.; Isaac, A.; Pereira, F.; Botaro, V. Three-Dimensional Stable Alginate-Nanocellulose Gels for Biomedical Applications: Towards Tunable Mechanical Properties and Cell Growing. *Nanomaterials* **2019**, *9*, 78. [CrossRef]
19. Kheawfu, K.; Kaewpinta, A.; Chanmahasathien, W.; Rachtanapun, P.; Jantrawut, P. Extraction of nicotine from tobacco leaves and development of fast dissolving nicotine extract film. *Membranes* **2021**, *11*, 403. [CrossRef]
20. Rosiak, M.T.; Darmawan, D.; Zainuddin, S. Irradiation of polyvinyl alcohol and polyvinyl pyrrolidone blended hydrogel for wound dressing. *Radiat. Phys. Chem.* **2001**, *62*, 107–113.
21. Mok, C.F.; Ching, Y.C.; Abu Osman, N.A.; Muhamad, F.; Mohd Junaidi, M.U.; Choo, J.H. Preparation and characterization study on maleic acid crosslinked poly(vinyl alcohol)/chitin/nanocellulose composites. *J. Appl. Polym. Sci.* **2020**, *137*, 49044. [CrossRef]
22. Balakrishnan, B.; Mohanty, M.; Umashankar, P.; Jayakrishnan, A. Evaluation of an in situ forming hydrogel wound dressing based on oxidized alginate and gelatin. *Biomaterials* **2005**, *26*, 6335–6342. [CrossRef] [PubMed]
23. Choi, Y.S.; Hong, S.R.; Lee, Y.M.; Song, K.W.; Park, M.H.; Nam, Y.S. Study on gelatin-containing artificial skin: I. Preparation and characteristics of novel gelatin-alginate sponge. *Biomaterials* **1999**, *20*, 409–417. [CrossRef]
24. Bajpai, S.K.; Sharma, S. Investigation of swelling/degradation behaviour of alginate beads crosslinked with Ca^{2+} and Ba^{2+} ions. *React. Funct. Polym.* **2004**, *59*, 129–140. [CrossRef]
25. Lönnqvist, S.; Emanuelsson, P.; Kratz, G. Influence of acidic pH on keratinocyte function and re-epithelialisation of human in vitro wounds. *J. Plast. Surg. Hand Surg.* **2015**, *49*, 346–352. [CrossRef]
26. Hoare, T.R.; Kohane, D.S. Hydrogels in drug delivery: Progress and challenges. *Polymer* **2008**, *49*, 1993–2007. [CrossRef]
27. Hasan, N.; Cao, J.; Lee, J.; Kim, H.; Yoo, J.-W. Development of clindamycin-loaded alginate/pectin/hyaluronic acid composite hydrogel film for the treatment of MRSA-infected wounds. *J. Pharm. Investig.* **2021**, *51*, 597–610. [CrossRef]
28. Lee, J.; Hlaing, S.P.; Cao, J.; Hasan, N.; Ahn, H.-J.; Song, K.-W.; Yoo, J.-W. In situ hydrogel-forming/nitric oxide-releasing wound dressing for enhanced antibacterial activity and healing in mice with infected wounds. *Pharmaceutics* **2019**, *11*, 496. [CrossRef] [PubMed]
29. Maderuelo, C.; Zarzuelo, A.; Lanao, J.M. Critical factors in the release of drugs from sustained release hydrophilic matrices. *J. Control. Release.* **2011**, *154*, 2–19. [CrossRef]
30. Tiwari, S.B.; Murthy, T.K.; Raveendra Pai, M.; Mehta, P.R.; Chowdary, P.B. Controlled release formulation of tramadol hydrochloride using hydrophilic and hydrophobic matrix system. *AAPS PharmSciTech* **2003**, *4*, 18–23. [CrossRef]
31. Maheswari, K.M.; Devineni, P.K.; Deekonda, S.; Shaik, S.; Uppala, N.P.; Nalluri, B.N. Development and Evaluation of Mouth Dissolving Films of Amlodipine Besylate for Enhanced Therapeutic Efficacy. *J. Pharm.* **2014**, *2014*, 520949. [CrossRef]
32. Mario, G.; Gabriele, G. Mathematical Modelling and Controlled Drug Delivery: Matrix Systems. *Curr. Drug Deliv.* **2005**, *2*, 97–116.

Article

WSG, a Glucose-Rich Polysaccharide from *Ganoderma lucidum*, Combined with Cisplatin Potentiates Inhibition of Lung Cancer *In Vitro* and *In Vivo*

Wei-Lun Qiu [1,†], Wei-Hung Hsu [1,2,3,†], Shu-Ming Tsao [4], Ai-Jung Tseng [1], Zhi-Hu Lin [1], Wei-Jyun Hua [1,5], Hsin Yeh [1], Tzu-En Lin [6], Chien-Chang Chen [7], Li-Sheng Chen [3,*] and Tung-Yi Lin [1,5,8,*]

1. Institute of Traditional Medicine, National Yang Ming Chiao Tung University, Taipei 11221, Taiwan; qgdfms@gm.ym.edu.tw (W.-L.Q.); seanhsu5817@gmail.com (W.-H.H.); aurora.tseng@gmail.com (A.-J.T.); tiger77749@gmail.com (Z.-H.L.); karta603@gm.ym.edu.tw (W.-J.H.); shin28jenna@gmail.com (H.Y.)
2. LO-Sheng Hospital Ministry of Health and Welfare, New Taipei 242, Taiwan
3. School of Oral Hygiene, College of Oral Medicine, Taipei Medical University, Taipei 110, Taiwan
4. Department of Biotechnology and Laboratory Science in Medicine, National Yang Ming Chiao Tung University, Taipei 11221, Taiwan; tsaosm@gmail.com
5. Program in Molecular Medicine, National Yang Ming Chiao Tung University, Taipei 11221, Taiwan
6. Institute of Biomedical Engineering, National Yang Ming Chiao Tung University, Hsinchu 30010, Taiwan; telin@nctu.edu.tw
7. The General Education Center, Ming Chi University of Technology, New Taipei 243, Taiwan; chang@mail.mcut.edu.tw
8. Biomedical Industry Ph.D. Program, National Yang Ming Chiao Tung University, Taipei 11221, Taiwan
* Correspondence: samchen@tmu.edu.tw (L.-S.C.); biotungyi@gmail.com or tylin99@nycu.edu.tw (T.-Y.L.)
† These authors contributed equally to this work.

Abstract: Lung cancer has the highest global mortality rate of any cancer. Although targeted therapeutic drugs are commercially available, the common drug resistance and insensitivity to cisplatin-based chemotherapy, a common clinical treatment for lung cancer, have prompted active research on alternative lung cancer therapies and methods for mitigating cisplatin-related complications. In this study, we investigated the effect of WSG, a glucose-rich, water soluble polysaccharide derived from *Ganoderma lucidum*, on cisplatin-based treatment for lung cancer. Murine Lewis lung carcinoma (LLC1) cells were injected into C57BL/6 mice subcutaneously and through the tail vein. The combined administration of WSG and cisplatin effectively inhibited tumor growth and the formation of metastatic nodules in the lung tissue of the mice. Moreover, WSG increased the survival rate of mice receiving cisplatin. Co-treatment with WSG and cisplatin induced a synergistic inhibitory effect on the growth of lung cancer cells, enhancing the apoptotic responses mediated by cisplatin. WSG also reduced the cytotoxic effect of cisplatin in both macrophages and normal lung fibroblasts. Our findings suggest that WSG can increase the therapeutic effectiveness of cisplatin. In clinical settings, WSG may be used as an adjuvant or supplementary agent.

Keywords: *Ganoderma lucidum*; polysaccharides; cisplatin; synergistic effect; anti-lung cancer

1. Introduction

Ganoderma lucidum is a well-known ingredient in traditional Chinese medicine. *G. lucidum* has long been used in Asian countries, in particular in China, where it is regarded as an elixir that promotes health and longevity [1–3]. Modern pharmaceutical studies have reported various biological properties of *G. lucidum* (e.g., anticancer, antioxidant, antimicrobial, immunomodulatory, and anti-inflammatory) [4–8]. The three main components of *G. lucidum*, polysaccharides, triterpenoids, and proteins, have been extensively examined [9,10]. Notably, *G. lucidum* polysaccharides (GLPs) have been commercially tested as clinical drugs and supplements for the treatment of several inflammatory diseases because of the anti-inflammatory, antioxidant, and immunomodulatory activities [11,12].

In addition, one study effectively used GLPs as a material for nanoparticle conjugation because they mediated immunocyte activation and increased radio-sensitivity in the tumor microenvironment [13].

In the past several decades, numerous studies have demonstrated that GLPs play a pivotal role in immunomodulatory and anticancer activities. For example, in one study, GLPs-RF3, a kind of GLP, activates toll-like receptor 4 (TLR4)-mediated extracellular signal-regulated kinase (ERK), c-Jun N-terminal kinase, and p38, leading to induce interleukin-1 expression in mouse spleen cells [14]. GLPs induce various types of cytokine expression to activate macrophages and lymphocytes [6] and accelerate the recovery of cyclophosphamide-disrupted immune functions [15]. In a mouse model, β-D-glucan from GLPs effectively inhibit sarcoma-180 solid tumors [16]. GLPs-RF3 attenuates breast cancer tumorigenesis by enhancing the ubiquitination-dependent degradation of transforming growth factor beta (TGF-β) receptors [17]. Other investigations of GLPs have focused on the association between their immunocyte-activating and anticancer activities. For example, GLPs exhibit antitumor activities through the TLR4-dependent activation of B cells and macrophages in a BALB/c mouse model [18]. FMS, a fucose-enriched polysaccharide purified from GLPs-RF3, suppresses lung tumorigenesis through the induction of anti-GloboH-series epitopes antibodies [5]. Taken together, these studies indicate that treatment with GLPs can effectively suppress tumor growth by enhancing the activities of tumor-associated immune effector cells, suggesting that GLPs may constitute an efficacious immunopotentiating adjuvant to anticancer drugs.

Cisplatin, a platinum-based agent, is a well-known anticancer drug that has long been used as an effective clinical chemotherapeutic agent [19]. Its interference with DNA repair mechanisms leads to DNA damage and subsequently triggers apoptotic responses [20]. Clinically, cisplatin has advantages in survival and symptom control in lung cancer therapy because many patients are usually diagnosed with locally advanced or metastatic disease when they are diagnosed with lung cancer. Therefore, cisplatin and its derivatives constitute the first-line drugs for clinical chemotherapy of lung cancer. However, patients undergoing cisplatin-based therapy often experience side effects, such as immunosuppression [21,22]. Studies have reported that combination treatments of cisplatin and herbal medicines exhibited greater efficacy than cisplatin alone did [23,24]. These studies suggest that herbal extracts combined with cisplatin may be a novel cancer treatment strategy. Currently, increasing studies have indicated that GLPs could be considered for use as supplements for patients with cancer [25]. However, the efficacy of combination treatment with GLPs and chemotherapy for lung cancer has yet to be established. Therefore, the synergistic effects of GLPs and cisplatin in lung cancer therapy must be explored.

We previously identified the chemical characterization of WSG, a water soluble glucose-rich polysaccharide derived from GLPs-RF3, and demonstrated it presents anti-lung cancer activity via downregulation of EGFR and TGFβR [26]. In this study, we aimed to investigate whether WSG in combination with cisplatin could overcome lung cancer progression *in vitro* and *in vivo*. Furthermore, we assessed the benefits of co-treatment with WSG and cisplatin on cell viability of macrophages and lung fibroblast *in vitro*.

2. Materials and Methods

2.1. Materials and Antibodies

WSG, a water soluble glucose-rich polysaccharide with an average molecular mass of approximately 1000 kDa, was isolated from GLPs-RF3 (Wynlife Healthcare, Inc., San Diego, CA, USA. Cat: 500T; Lot: WT20010703) as described previously [26]. Cisplatin (P4394) was purchased from Sigma-Aldrich (Saint Louis, MO, USA). An Annexin V/PI apoptosis kit was purchased from Life Technologies (Carlsbad, CA, USA). Anti-β-actin and anti-EGFR antibodies were acquired from GeneTex (Hsinchu, Taiwan). Anti-phosphorylated AKT (S473) antibody was purchased from Santa Cruz Biotechnology (Santa Cruz, CA, USA), anti-phosphorylated ERK1/2 antibody was purchased from Sigma-Aldrich (Steinheim, Germany).

2.2. Mouse Allogeneic Tumor Model

Six-week-old C57BL/6 male mice were purchased from the National Laboratory Animal Center (Taipei, Taiwan) and kept in specific pathogen-free conditions in the Laboratory Animal Center of National Yang Ming Chao Tung University. The experimental procedures and study design were approved by the Institutional Animal Care and Use Committee (approval No: 1090215). After one week of adaptation, the mice were randomly assigned to one of four groups, with five individuals in each group. In the control group, the mice were intraperitoneally injected with phosphate-buffered saline (PBS). In the WSG group, the mice were intraperitoneally injected with 75 mg/kg WSG every 2 days [5,26]. In the cisplatin group, the mice daily received intraperitoneal injections of 2.3 mg/kg cisplatin from day 1 to day 5. In the combination treatment group, the mice were co-treated with WSG and cisplatin.

To create a tumor growth mouse model, LLC1 cells (2×10^5 cells in 100 µL of serum-free medium) were subcutaneously injected into the right flank. Tumors were calculated at 2-day intervals with digital vernier calipers, and tumor volume was calculated as follows: $V = (L \times W^2)/2$, where V is volume in cubic millimeters, L is tumor length in millimeters, and W is tumor width in millimeters [27]. The mice were sacrificed using carbon dioxide (CO_2) on day 21. To create a tumor homing mouse model, LLC1 cells (2×10^5 cells in 100 µL of serum-free medium) were injected into the lateral tail vein. The mice were sacrificed using CO_2 after 28 days. The lung tissue of each mouse was harvested and fixed in paraformaldehyde (10%) for the examination of tumor homing and growth. Hematoxylin and eosin (H&E) staining was applied to the tumor colonies, and the number and tumor volume of the colonies were determined through microscopy.

2.3. Cell Lines and Cell Culture

Human lung adenocarcinoma A549 cells, murine LLC1 cells, human lung fibroblast MRC-5, and murine macrophages Raw264.7 were acquired from the Bioresource Collection and Research Center (BCRC; Hsinchu, Taiwan). Human lung adenocarcinoma CL1-5 cells were obtained from Dr. P.-C. Yang of National Taiwan University. The lung cancer cell lines were cultured as previously described [28,29]. The MRC-5 cells were cultured in minimum essential medium supplemented with 10% heat-inactivated fetal bovine serum (FBS; VWR Life Science Seradigm, Radnor, PA, USA), 2 mM L-glutamine, 0.1 mM nonessential amino acids, 1.5 g/L sodium bicarbonate, and 1.0 mM sodium pyruvate at 37 °C in an environment of 95% air and 5% CO_2. The RAW 264.7 cells were kept in Dulbecco's modified Eagle medium (GIBCO-Life Technologies) that was supplemented with 3.7 g/L $NaHCO_3$, 5% heat-inactivated FBS, and 100 units/mL each of penicillin and streptomycin (Biological Industries, Cromwell, CT, USA).

2.4. Cell Viability and Synergistic Analysis

Cell viability was evaluated using a crystal violet assay. For each experiment, 5×10^4 cells were seeded into each well of 12-well plates. After incubation overnight to allow for cell attachment, the cells were subjected to treatment with various concentrations of WSG or cisplatin or a combination of the two for 24 or 48 h. Subsequently, the cells were rinsed with PBS and stained with 1% crystal violet solution for 1 h. The specifics of the procedure were described previously [23]. The percentages of cell viability were normalized for comparison with the control group. Next, as conducted in our previous study [23], the synergistic effects of the combination treatment, as indicated by Chou-Talalay combination indexes (CIs), were analyzed using the CompuSyn program (Paramus, NJ, USA) [30]. The CI was <1, confirming the synergistic effects between the tested drugs.

2.5. Cell Protein Extraction for Western Blotting

The cancer cells were rinsed two times with 1 mL of cold PBS supplemented with 1% Na_3VO_4 and then scraped into a lysis buffer containing a mixture of proteinase (Sigma-Aldrich, Saint Louis, MO, USA) and phosphatase inhibitors (MedChem Express, Monmouth Junction, NJ, USA) [31]. Next, the cell lysates were centrifuged at 4 °C and 13,000× g for 10 min to obtain the supernatant as the cellular proteins or extracts. Protein concentrations were determined using a Bradford assay (Bio-Rad, Hercules, CA, USA). For Western blotting, 30 μg cell extracts were separated through 10% sodium dodecyl sulfate–polyacrylamide gel electrophoresis (SDS-PAGE) and transferred to polyvinylidene difluoride (PVDF) membranes to detect the indicated molecules, with β-actin used as the loading control. The Western blotting assay was performed as described previously [32].

2.6. Sample Preparation for Apoptosis Analysis

The cancer cells were seeded in 6-cm plates at a density of 2×10^5 cells/plate. After overnight incubation, the cells were treated with WSG (120 μg/mL) and/or cisplatin (10 μM) for 48 h and subsequently washed with PBS and harvested after trypsinization. The cell pellets were washed with cold PBS, centrifuged, and then collected. Next, the cell pellets were re-suspended in annexin-binding buffer and stained with propidium iodide (PI) and annexin V–fluorescein isothiocyanate (FITC) using an apoptosis detection kit (BD Pharmingen, USA) in accordance with the manufacturer's instructions. The samples were examined using a flow cytometer (BD FACSCalibur, Franklin lakes, NJ, USA), and the raw data were analyzed for apoptosis by using FlowJo software (National Institute of Mental Health, Bethesda, MD, USA).

2.7. Statistical Analysis

All presented results are representative of three independent experiments unless otherwise indicated. All data are expressed as means ± standard deviations of the indicated number of experiments. Between-group differences were analyzed through a t test using GraphPad Prism8 software (San Diego, CA, USA). Differences were considered significant at $p < 0.05$.

3. Results and Discussion

3.1. WSG Synergistically Enhances Cisplatin-Induced Cytotoxicity in Lung Cancer Cells

Previously, we have demonstrated that WSG effectively inhibited the viability of A549 and LLC1 cells [26]. Herein, we investigated the impact of WSG combined with cisplatin treatment on the viability of lung cancer cells *in vitro*. A crystal violet assay was performed to examine the efficacy of the combination treatment in suppressing lung cancer A549 and LLC1 cells. The cells were co-treated with cisplatin (0–20 μM) and low-dose WSG (60 or 120 μg/mL) for 24 h. The combination treatment exhibited stronger inhibitory effects than WSG or cisplatin treatment alone (Figure 1A). The interactions between WSG and cisplatin were evaluated with Chou-Talalay CIs [33]. The CIs for all concentrations were less than 1 (Figure 1B), suggesting that the combination treatment exerted a reliable, synergistic, cytotoxic effect on the A549 and LLC1 cells. Regarding the dose-reduction indexes (DRIs) of the combination treatment [33], the DRIs exceeded 1 for both WSG and cisplatin (Figure 1C). Moreover, the IC_{50} of cisplatin in combination with WSG were calculated. As shown in Figure 1D, WSG dramatically decreased the IC_{50} values of cisplatin in A549 and LLC1 cells in a concentration-dependent manner, with WSG of 120 μg/mL reducing significantly the IC_{50} of cisplatin from 10 to 1 μM and 0.5 μM, respectively. These results indicate that the combination of WSG and cisplatin may promote anticancer activity and attenuate the side effects of the individual treatments, supporting a previous finding that GLPs may reduce the occurrence of chemotherapy side effects in patients with cancer [34]. In short, WSG and cisplatin exhibited synergistic cytotoxic effects in the lung cancer cells.

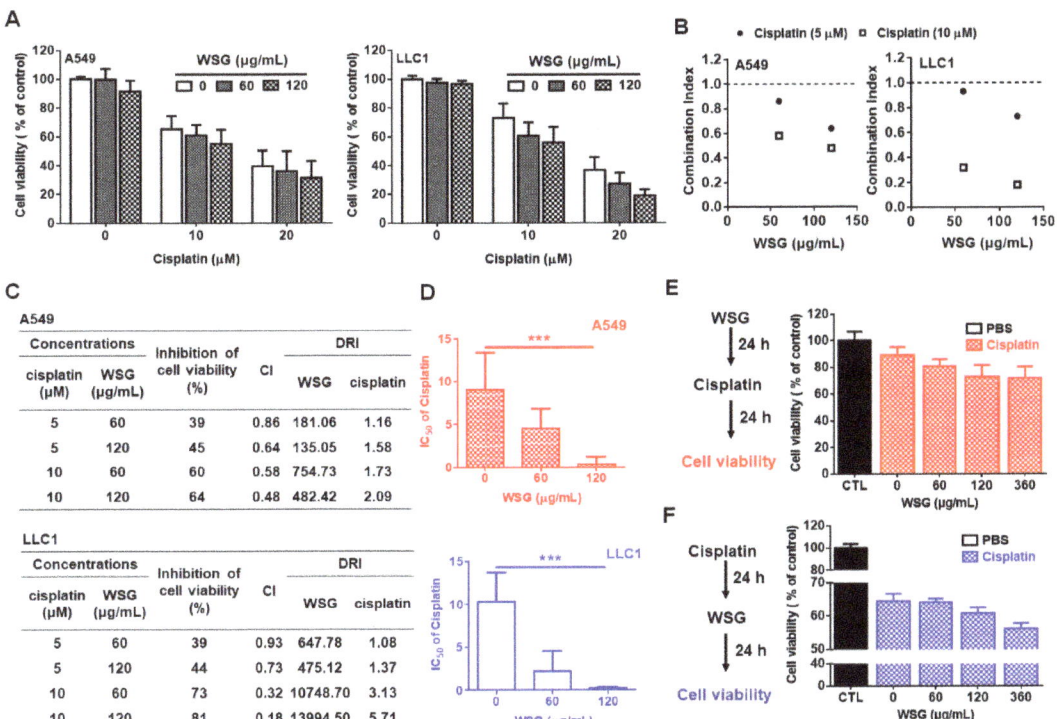

Figure 1. WSG enhances the cytotoxic effects of cisplatin on A549 and LLC1 cells. (**A**) A549 and LLC1 cells were subjected to 24 h combination treatment using different concentrations of WSG and cisplatin. A crystal violet staining assay was conducted for cell viability examination. Each sample in the combination treatment group was normalized against each untreated control. (**B,C**) CI and DRI of the combination treatment in the A549 and LLC1 cells, determined using CompuSyn software. (**D**) WSG reduced the IC_{50} of cisplatin in A549 and LLC1 cells. The IC_{50} values of cisplatin were measured using CompuSyn Software. Significant differences between the treatment and control groups are presented (*** $p < 0.001$). (**E**) Left: the schematic design for the in vitro cell viability assay of WSG (0–360 µg/mL)→cisplatin (5 µM) sequential treatment. Right: The viability of the cells was determined by using the crystal violet assay. (**F**) Left: the schematic design for the in vitro cell viability assay of cisplatin (5 µM)→WSG (0–360 µg/mL) sequential treatment. Right: The viability of the cells was determined by using the crystal violet assay. Each sample in the combination treatment group was normalized against each untreated control. Error bars indicated SD.

To mimic the potential clinical complementary strategy, we further examined the effect of a sequential treatment of WSG and cisplatin on lung cancer A549 cells. Two types of sequential treatments were performed. Type one (type 1): cancer cells were treated with WSG, followed by cisplatin incubation (Figure 1E; WSG→Cisplatin). By contrast, in the treatment of type two (type 2), cells were treated with cisplatin, followed by WSG incubation (Figure 1F; Cisplatin→WSG). As expected, we found that no matter whether WSG or cisplatin was given first, the sequential treatment presented a more effective inhibition rate of the cell viability (Figure 1E,F). Together, we demonstrated that WSG could enhance the cisplatin-inhibited viability of lung cancer cells *in vitro*.

3.2. WSG Enhances Cisplatin-Induced Apoptotic Response in Lung Cancer Cells

To further examine these synergistic cytotoxic effects, we used an annexin V–FITC/PI staining kit to test whether WSG enhanced cisplatin-induced cell death. As shown in Figure 2, apoptosis occurred in 32.7–45.7% of the A549 cells co-treated with cisplatin (10 µM) and WSG (120 µg/mL). The corresponding rate among the LLC1 cells ranged from 27.6% to 51.8%. These results suggest that the combination of cisplatin and WSG was much more effective in inducing cell death than cisplatin alone. In essence, the *in vitro* experiments demonstrated that the combination of cisplatin and WSG exerted reliable, synergistic, and cytotoxic effects, inducing cell death in the lung cancer cells.

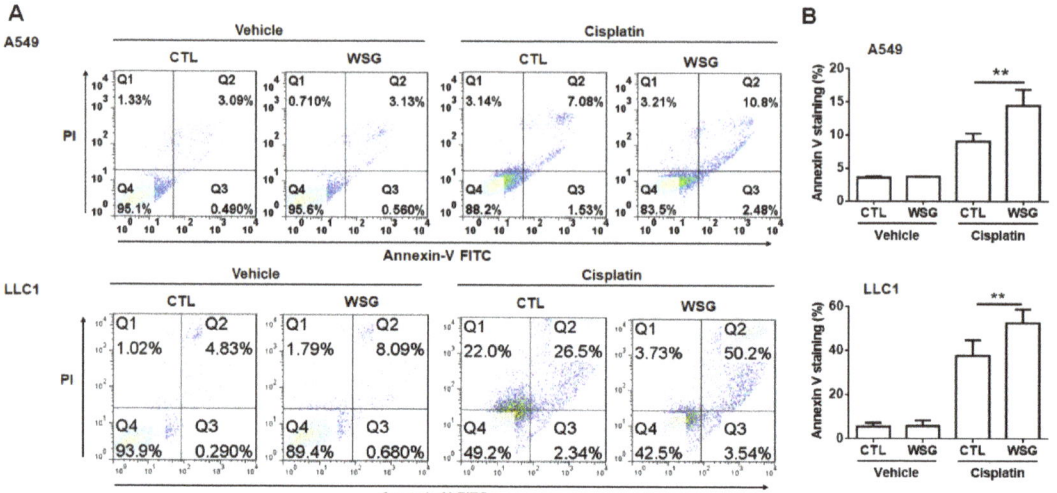

Figure 2. WSG enhances cisplatin-induced apoptotic responses. (**A**) After 48 h exposure to WSG (120 μg/mL) and/or cisplatin (10 μM), the cells were subjected to co-staining with the annexin V–FITC/PI kit. Flow cytometry was performed for apoptosis analysis. The percentages of apoptotic cells in early and late apoptosis were determined using FlowJo software. (**B**) The data, representative of three separate experiments, are presented as means ± standard deviations; error bars reflect standard deviations. Significant differences between the treatment and control groups are presented (** $p < 0.01$).

3.3. WSG Retains Viability in RAW 264.7 and MRC-5 Cells Treated with Cisplatin

Patients receiving cisplatin-based therapy often experience complications, including low blood cell counts, irreversible hearing loss (ototoxicity), and permanent neuronal and renal damage [21,22]. According to cisplatin manufacturers, the medication reduces the synthesis of blood cells, specifically white blood cells, leaving patients susceptible to infection. Therefore, we examined the effects of WSG on cisplatin-treated RAW 264.7 macrophages. Low concentrations of WSG (20–150 μg/mL) did not affect viability, but a high concentration (300 μg/mL) increased viability by approximately 30% (Figure 3A). We also examined the cytotoxic effect of cisplatin on RAW264.7 cells. As expected, cisplatin substantially and dose-dependently reduced the viability of the RAW 264.7 cells (Figure 3B), consistent with results of a previous study [35]. Notably, WSG increased the viability of RAW 264.7 cells upon cisplatin treatment (Figure 3C). Previous study shows that polysaccharides from *Cudrania tricuspidata* fruit mediated with a reduction in reactive oxygen species (ROS) production and mitochondrial transmembrane potential loss contributes to cytoprotective action in cisplatin-treated RAW264.7 cells [36]. Whether WSG could mediate cisplatin-induced ROS production and mitochondria damage needs to be dissected in the future. In addition, cisplatin exerted cytotoxic effects on the MRC-5 cells, in line with previous results [37]. We found that WSG did not exhibit cytotoxic effects on the MCR-5 cells (Figure 3D) but increased cell viability of the MRC-5 cells upon cisplatin treatment (Figure 3E). Increasing evidence shows that GLPs might attenuate the cytotoxicity induced by cisplatin to improve chemotherapy-induced fatigue [38]. Herein, we demonstrated that WSG protects white blood cells and lung fibroblasts from the adverse side effects of cisplatin. The results collectively indicate that WSG may be a safe anticancer agent for inhibiting the survival of lung cancer cells.

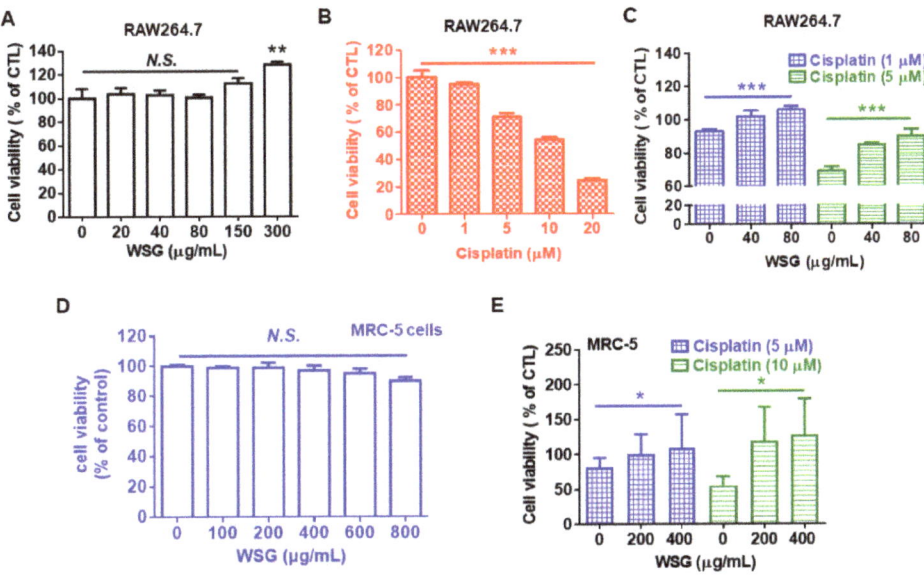

Figure 3. WSG reduces cisplatin-induced cytotoxic effects on RAW 264.7 and MRC-5 cells. (**A**) RAW 264.7 murine macrophages were subjected to 24 h treatment with various concentrations (0–300 μg/mL) of WSG. (**B**) RAW 264.7 cells were subjected to treatment with various concentrations (0–20 μM) of cisplatin for 24 h. (**C**) RAW 264.7 cells were subjected to 24 h treatment with both WSG (40 and 80 μg/mL) and cisplatin (1 and 5 μM). (**D**) MCR-5 cells were subjected to 48 h treatment with various concentrations (0–800 μg/mL) of WSG, and their viability was assessed through crystal violet staining. WSG treatment group data were normalized for comparison with an untreated control. (**E**) MRC-5 fibroblasts were subjected to 48 h treatment with a combination of WSG (200 and 400 μg/mL) and cisplatin (5 and 10 μM). A crystal violet assay was performed for cell viability examination. Treatment groups were normalized for comparison with an untreated control. Data are presented as means ± standard deviations. Significant differences between the treatment and control groups are presented (* $p < 0.05$, ** $p < 0.01$, *** $p < 0.001$).

3.4. Co-Treatment with WSG and Cisplatin Effectively Suppresses Lung Tumor Growth in LLC1-Bearing Mice

As the above results show, co-treatment with WSG and cisplatin effectively inhibits lung cancer cells *in vitro*. We further explored the impact of WSG co-treatment with cisplatin on lung tumor growth *in vivo*. Initially, LLC1 cells were hypodermically injected into the hypodermic dorsum of mice that were randomly divided into four groups: control (PBS), WSG, cisplatin, and WSG combined with cisplatin. The tumor growth rates were monitored over 21 days (Figure 4A). As expected, compared with the control group, both WSG and cisplatin effectively inhibited tumor growth and reduced tumor weight in the LLC1-bearing mice (Figure 4B,C), and stronger anticancer activity was achieved with WSG than with cisplatin. The tumor-suppressive effects of co-treatment with cisplatin and WSG in the LLC1-bearing mice were significantly stronger than those of WSG or cisplatin alone (Figure 4B,C). In addition, the combination treatment did not affect the body weight of the mice, indicating the nontoxicity of this therapeutic strategy (data not shown). These results suggest that WSG may function both as an antitumor agent and as a chemotherapeutic adjuvant enhancing the anticancer effects of cisplatin.

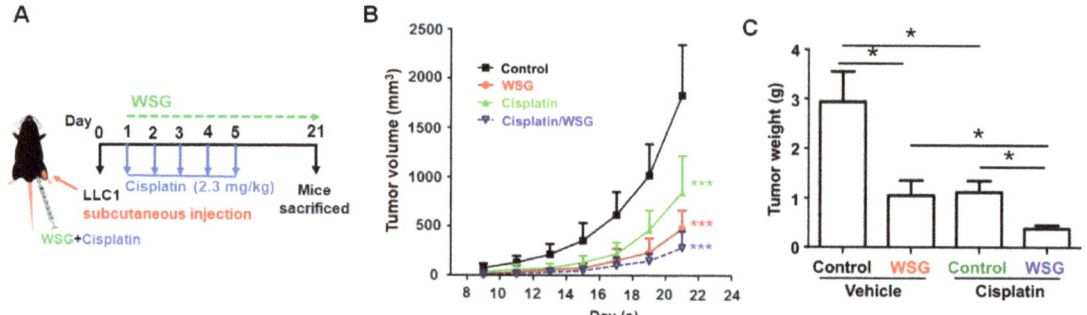

Figure 4. Co-treatment with WSG and cisplatin suppresses lung tumor growth in mice bearing LLC1 cells. (**A**) Experimental schematic. LLC1-bearing mice (n = 5 in each group) were intraperitoneally injected with cisplatin and WSG. (**B**) Tumor volume was monitored for the indicated times. (**C**) Tumor weights were determined after each intervention. The bars represent means ± standard deviations. Significant differences between the treatment and control groups are presented (* $p < 0.05$, *** $p < 0.001$).

3.5. Combination of WSG and Cisplatin Effectively Suppresses Tumor Growth in Lung Tissues of LLC1 Cells-Bearing Mice

We further investigated the antitumor effects of the combination treatment on the lung tissue of the mice (Figure 5A). To mimic tumor progression in the lung tissue, we injected LLC1 cells into the mice through the lateral tail vein to trap them and allow them to develop into tumor lesions [39]. We initially counted the tumor nodules in the lung tissue of mice receiving the combination treatment. As shown in Figure 5B,C, WSG significantly suppressed lung cancer homing and growth, a result observed in our previous study [26]. Specifically, the combination treatment reduced the number of tumor nodules considerably more than cisplatin treatment alone (Figure 5B,C). H&E staining of the tumor nodules revealed fewer instances of tumor homing (nodules) in the lung tissue of the mice receiving the combination treatment than in the tissue of the mice receiving WSG or cisplatin alone (Figure 5B). However, it did not exert stronger tumor-suppressive effects than WSG treatment did alone (Figure 5C). Notably, the combination treatment reduced tumor volume substantially more than WSG or cisplatin alone (Figure 5D), suggesting that the combination strategy not only inhibited tumor nodules formation but also suppressed tumor growth in lung tissues. In addition, the survival rate of the mice receiving the combination treatment was significantly higher than that of the mice receiving the control treatment or cisplatin alone (Figure 5E). However, the mice treated with WSG alone had the highest overall survival rate. Our results indicate that WSG has potential as a therapeutic intervention for controlling disease progression and extending life in lung cancer.

In this study, our results showed that the combination with WSG potentiated the cell-killing efficacy (apoptosis) of cisplatin in lung cancer cells in vitro. More remarkably, we found that the antitumor effect of cisplatin was significantly enhanced with the combination with WSG in the lung cancer-bearing mouse model. Notably, the combination of WSG with cisplatin showed selectivity in enhancing the efficacy of cisplatin between cancer cells and normal cells, suggesting that WSG may play different roles for contributing to the cellular responses in cisplatin-treated lung cancer and normal cells. Increasing evidence shows that multiple intracellular regulators may contribute to cisplatin resistance [40]. For instance, the excision repair cross-complementation group 1 (ERCC1) enzyme plays a pivotal role in cisplatin-induced DNA damage and recombination. Clinically, patients with lung cancer and ERCC1-positive tumors appear to worsen from adjuvant cisplatin-based chemotherapy [41]. Expression of copper-transporting P-type adenosine triphosphatase A (ATP7A) is associated with cisplatin resistance in lung cancer cells [42]. However, WSG did not modulate the expression of ERCC1 and ATP7A (data not shown), suggesting that WSG may not be involved in the cisplatin-associated DNA repair system and copper-transporting.

Notably, ERK signaling contributes to cisplatin resistance [43]. We previously found WSG significantly reduced phosphorylation of ERK1/2 [26], suggesting that WSG improved the activity of cisplatin against lung cancer cells via the targeting of ERK signaling.

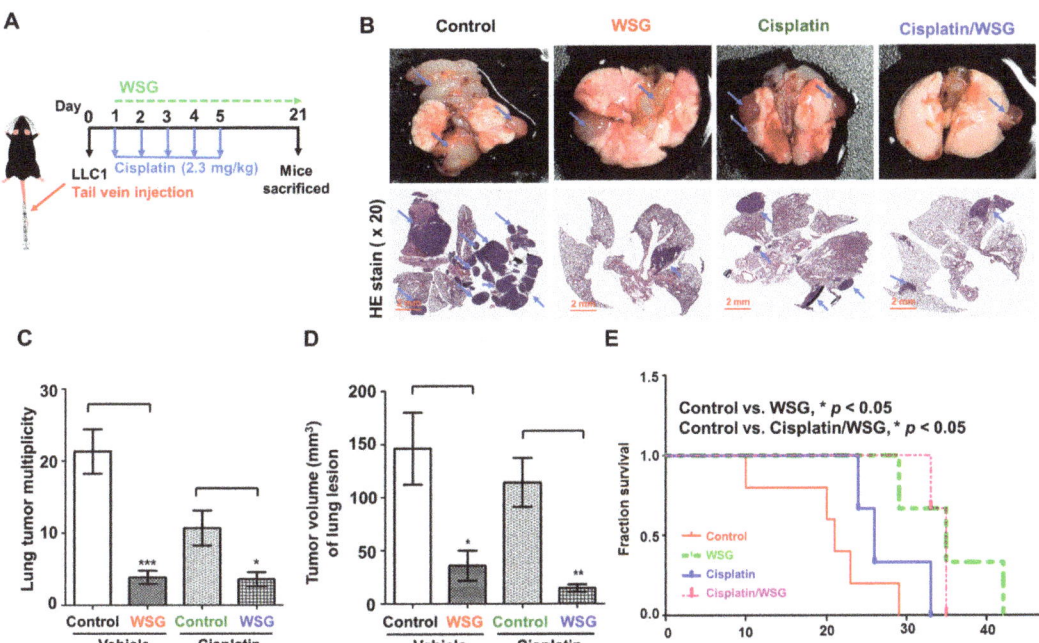

Figure 5. Co-treatment with WSG and cisplatin suppresses tumorigenesis in the lung tissues of mice injected with LLC1 cells. (**A**) Experimental schematic. LLC1 cells were injected through the lateral tail veins of the mice ($n = 5$). The LLC1-bearing mice were intraperitoneally injected with cisplatin and WSG. (**B**) At the end of the treatment, lung tissue was collected. One of five tissue samples is presented. As indicated by the blue arrows, tumor lesions were observed on the surface of the lung tissues. Lower panels of the lung sections under H&E staining (20× magnification). Bar scale: 2 mm. (**C**) Tumors (nodules) in the lung lesions. (**D**) Tumor volume (mm^3) of the lung lesions. Each bar represents the mean ± standard deviation. Significant differences between the treatment and control groups are presented (* $p < 0.05$, ** $p < 0.01$, *** $p < 0.001$). (**E**) Combination treatment increased the survival rate of LLC1-bearing mice ($n = 8$).

4. Conclusions

WSG is a novel glucose-rich polysaccharide isolated from commercially available GLPs-RF3. We previously identified the potential anticancer mechanism of WSG [26]. Cisplatin-based chemotherapy is one of the most common and effective treatments for lung cancer [44]. To investigate the clinical applications of WSG, we conducted a preclinical study of WSG on tumor-bearing mice receiving chemotherapy. WSG enhanced the tumor-suppressive effects of cisplatin *in vivo* as well as *in vitro*. In addition, WSG enhanced the cytotoxic and tumor-suppressive effects of cisplatin and increased the survival rates of LLC1-bearing mice. Furthermore, in macrophages and lung fibroblasts, WSG attenuated the cytotoxic effects of cisplatin. Thus, WSG may be considered for use as an adjuvant or dietary supplement to improve outcomes in patients receiving cisplatin-based chemotherapy.

Author Contributions: Conceptualization, T.-Y.L.; methodology, W.-L.Q. and S.-M.T.; validation, W.-L.Q.; formal analysis, W.-L.Q. and S.-M.T.; investigation, W.-L.Q., S.-M.T., A.-J.T., Z.-H.L., W.-J.H. and H.Y.; data curation, W.-L.Q., W.-H.H. and S.-M.T.; writing—original draft preparation, W.-H.H. and T.-Y.L.; writing—review and editing, T.-Y.L.; visualization, T.-E.L., C.-C.C. and L.-S.C.; supervision, T.-Y.L.; project administration, T.-Y.L.; funding acquisition, W.-H.H. and T.-Y.L. All authors have read and agreed to the published version of the manuscript.

Funding: This study was supported by grants from the Ministry of Science and Technology, Taiwan (The Young Scholar Fellowship Program: MOST 110-2636-B-A49A-501) and New Taipei City Chinese Medical Doctors Association (NTCCM-10801).

Institutional Review Board Statement: The experimental procedures and study design were approved by the Institutional Animal Care and Use Committee (approval No: 1090215).

Informed Consent Statement: Not applicable.

Data Availability Statement: The data presented in this study are available on request from the corresponding author.

Acknowledgments: We thank H.-Y.H. for his insightful commentary as well as S.-T.W. and M.-H.L. for their technical assistance.

Conflicts of Interest: The authors declare no conflict of interest.

References

1. Shiao, M.S. Natural products of the medicinal fungus Ganoderma lucidum: Occurrence, biological activities, and pharmacological functions. *Chem. Rec.* **2003**, *3*, 172–180. [CrossRef] [PubMed]
2. Seweryn, E.; Ziala, A.; Gamian, A. Health-Promoting of Polysaccharides Extracted from Ganoderma lucidum. *Nutrients* **2021**, *13*, 2725. [CrossRef] [PubMed]
3. Peng, H.H.; Wu, C.Y.; Hsiao, Y.C.; Martel, J.; Ke, P.Y.; Chiu, C.Y.; Liau, J.C.; Chang, I.T.; Su, Y.H.; Ko, Y.F.; et al. Ganoderma lucidum stimulates autophagy-dependent longevity pathways in Caenorhabditis elegans and human cells. *Aging* **2021**, *13*, 13474–13495. [CrossRef]
4. Liu, W.; Wang, H.; Pang, X.; Yao, W.; Gao, X. Characterization and antioxidant activity of two low-molecular-weight polysaccharides purified from the fruiting bodies of Ganoderma lucidum. *Int. J. Biol. Macromol.* **2010**, *46*, 451–457. [CrossRef] [PubMed]
5. Liao, S.F.; Liang, C.H.; Ho, M.Y.; Hsu, T.L.; Tsai, T.I.; Hsieh, Y.S.; Tsai, C.M.; Li, S.T.; Cheng, Y.Y.; Tsao, S.M.; et al. Immunization of fucose-containing polysaccharides from Reishi mushroom induces antibodies to tumor-associated Globo H-series epitopes. *Proc. Natl. Acad. Sci. USA* **2013**, *110*, 13809–13814. [CrossRef]
6. Lai, C.Y.; Hung, J.T.; Lin, H.H.; Yu, A.L.; Chen, S.H.; Tsai, Y.C.; Shao, L.E.; Yang, W.B.; Yu, J. Immunomodulatory and adjuvant activities of a polysaccharide extract of Ganoderma lucidum in vivo and in vitro. *Vaccine* **2010**, *28*, 4945–4954. [CrossRef]
7. Wang, Y.Y.; Khoo, K.H.; Chen, S.T.; Lin, C.C.; Wong, C.H.; Lin, C.H. Studies on the immuno-modulating and antitumor activities of Ganoderma lucidum (Reishi) polysaccharides: Functional and proteomic analyses of a fucose-containing glycoprotein fraction responsible for the activities. *Bioorganic Med. Chem.* **2002**, *10*, 1057–1062. [CrossRef]
8. Cor, D.; Knez, Z.; Knez Hrncic, M. Antitumour, Antimicrobial, Antioxidant and Antiacetylcholinesterase Effect of Ganoderma Lucidum Terpenoids and Polysaccharides: A Review. *Molecules* **2018**, *23*, 649. [CrossRef] [PubMed]
9. Boh, B.; Berovic, M.; Zhang, J.; Zhi-Bin, L. Ganoderma lucidum and its pharmaceutically active compounds. *Biotechnol. Annu. Rev.* **2007**, *13*, 265–301. [CrossRef]
10. Gill, B.S.; Navgeet; Kumar, S. Ganoderma lucidum targeting lung cancer signaling: A review. *Tumour. Biol.* **2017**, *39*, 1010428317707437. [CrossRef]
11. Chen, Y.; Zhang, H.; Wang, Y.; Nie, S.; Li, C.; Xie, M. Sulfated modification of the polysaccharides from Ganoderma atrum and their antioxidant and immunomodulating activities. *Food Chem.* **2015**, *186*, 231–238. [CrossRef] [PubMed]
12. Gao, Y.; Tang, W.; Dai, X.; Gao, H.; Chen, G.; Ye, J.; Chan, E.; Koh, H.L.; Li, X.; Zhou, S. Effects of water-soluble Ganoderma lucidum polysaccharides on the immune functions of patients with advanced lung cancer. *J. Med. Food* **2005**, *8*, 159–168. [CrossRef] [PubMed]
13. Yu, H.; Yang, Y.; Jiang, T.; Zhang, X.; Zhao, Y.; Pang, G.; Feng, Y.; Zhang, S.; Wang, F.; Wang, Y.; et al. Effective Radiotherapy in Tumor Assisted by Ganoderma lucidum Polysaccharide-Conjugated Bismuth Sulfide Nanoparticles through Radiosensitization and Dendritic Cell Activation. *ACS Appl. Mater. Interfaces* **2019**, *11*, 27536–27547. [CrossRef] [PubMed]
14. Chen, H.S.; Tsai, Y.F.; Lin, S.; Lin, C.C.; Khoo, K.H.; Lin, C.H.; Wong, C.H. Studies on the immuno-modulating and anti-tumor activities of Ganoderma lucidum (Reishi) polysaccharides. *Bioorganic Med. Chem.* **2004**, *12*, 5595–5601. [CrossRef]
15. Zhu, X.L.; Chen, A.F.; Lin, Z.B. Ganoderma lucidum polysaccharides enhance the function of immunological effector cells in immunosuppressed mice. *J. Ethnopharmacol.* **2007**, *111*, 219–226. [CrossRef]
16. Sone, Y.; Okuda, R.; Wada, N.; Kishida, E.; Misaki, A. Structures and Antitumor Activities of the Polysaccharides Isolated from Fruiting Body and the Growing Culture of Mycelium of Ganoderma lucidum. *Agric. Biol. Chem.* **1985**, *49*, 2641–2653. [CrossRef]

17. Tsao, S.M.; Hsu, H.Y. Fucose-containing fraction of Ling-Zhi enhances lipid rafts-dependent ubiquitination of TGFbeta receptor degradation and attenuates breast cancer tumorigenesis. *Sci. Rep.* **2016**, *6*, 36563. [CrossRef]
18. Shao, B.M.; Dai, H.; Xu, W.; Lin, Z.B.; Gao, X.M. Immune receptors for polysaccharides from Ganoderma lucidum. *Biochem. Biophys. Res. Commun.* **2004**, *323*, 133–141. [CrossRef]
19. Barabas, K.; Milner, R.; Lurie, D.; Adin, C. Cisplatin: A review of toxicities and therapeutic applications. *Vet. Comp. Oncol.* **2008**, *6*, 1–18. [CrossRef]
20. Dasari, S.; Tchounwou, P.B. Cisplatin in cancer therapy: Molecular mechanisms of action. *Eur. J. Pharmacol.* **2014**, *740*, 364–378. [CrossRef]
21. Astolfi, L.; Ghiselli, S.; Guaran, V.; Chicca, M.; Simoni, E.; Olivetto, E.; Lelli, G.; Martini, A. Correlation of adverse effects of cisplatin administration in patients affected by solid tumours: A retrospective evaluation. *Oncol. Rep.* **2013**, *29*, 1285–1292. [CrossRef] [PubMed]
22. Benkafadar, N.; Menardo, J.; Bourien, J.; Nouvian, R.; François, F.; Decaudin, D.; Maiorano, D.; Puel, J.-L.; Wang, J. Reversible p53 inhibition prevents cisplatin ototoxicity without blocking chemotherapeutic efficacy. *EMBO Mol. Med.* **2017**, *9*, 7–26. [CrossRef] [PubMed]
23. Hsu, H.Y.; Lin, T.Y.; Hu, C.H.; Shu, D.T.F.; Lu, M.K. Fucoidan upregulates TLR4/CHOP-mediated caspase-3 and PARP activation to enhance cisplatin-induced cytotoxicity in human lung cancer cells. *Cancer Lett.* **2018**, *432*, 112–120. [CrossRef]
24. Baharuddin, P.; Satar, N.; Fakiruddin, K.S.; Zakaria, N.; Lim, M.N.; Yusoff, N.M.; Zakaria, Z.; Yahaya, B.H. Curcumin improves the efficacy of cisplatin by targeting cancer stem-like cells through p21 and cyclin D1-mediated tumour cell inhibition in non-small cell lung cancer cell lines. *Oncol. Rep.* **2016**, *35*, 13–25. [CrossRef] [PubMed]
25. Didem, S.; Shile, H. Ganoderma lucidum Polysaccharides as An Anti-cancer Agent. *Anti-Cancer Agents Med. Chem.* **2018**, *18*, 667–674.
26. Hsu, W.H.; Qiu, W.L.; Tsao, S.M.; Tseng, A.J.; Lu, M.K.; Hua, W.J.; Cheng, H.C.; Hsu, H.Y.; Lin, T.Y. Effects of WSG, a polysaccharide from Ganoderma lucidum, on suppressing cell growth and mobility of lung cancer. *Int. J. Biol. Macromol.* **2020**, *165*, 1604–1613. [CrossRef]
27. Euhus, D.M.; Hudd, C.; LaRegina, M.C.; Johnson, F.E. Tumor measurement in the nude mouse. *J. Surg. Oncol.* **1986**, *31*, 229–234. [CrossRef]
28. Wu, C.T.; Lin, T.Y.; Hsu, H.Y.; Sheu, F.; Ho, C.M.; Chen, E.I. Ling Zhi-8 mediates p53-dependent growth arrest of lung cancer cells proliferation via the ribosomal protein S7-MDM2-p53 pathway. *Carcinogenesis* **2011**, *32*, 1890–1896. [CrossRef]
29. Qiu, W.L.; Tseng, A.J.; Hsu, H.Y.; Hsu, W.H.; Lin, Z.H.; Hua, W.J.; Lin, T.Y. Fucoidan increased the sensitivity to gefitinib in lung cancer cells correlates with reduction of TGFbeta-mediated Slug expression. *Int. J. Biol. Macromol.* **2020**, *153*, 796–805. [CrossRef]
30. Chou, T.-C. Drug Combination Studies and Their Synergy Quantification Using the Chou-Talalay Method. *Cancer Res.* **2010**, *70*, 440–446. [CrossRef]
31. Lin, T.Y.; Hsu, H.Y. Ling Zhi-8 reduces lung cancer mobility and metastasis through disruption of focal adhesion and induction of MDM2-mediated Slug degradation. *Cancer Lett.* **2016**, *375*, 340–348. [CrossRef]
32. Hsu, H.Y.; Hua, K.F.; Lin, C.C.; Lin, C.H.; Hsu, J.; Wong, C.H. Extract of Reishi polysaccharides induces cytokine expression via TLR4-modulated protein kinase signaling pathways. *J. Immunol.* **2004**, *173*, 5989–5999. [CrossRef]
33. Chou, T.C. Theoretical basis, experimental design, and computerized simulation of synergism and antagonism in drug combination studies. *Pharmacol. Rev.* **2006**, *58*, 621–681. [CrossRef]
34. Zeng, P.; Guo, Z.; Zeng, X.; Hao, C.; Zhang, Y.; Zhang, M.; Liu, Y.; Li, H.; Li, J.; Zhang, L. Chemical, biochemical, preclinical and clinical studies of Ganoderma lucidum polysaccharide as an approved drug for treating myopathy and other diseases in China. *J. Cell. Mol. Med.* **2018**, *22*, 3278–3297. [CrossRef]
35. Liu, Y.; Ni, X.Y.; Chen, R.L.; Li, J.; Gao, F.G. TIPE attenuates the apoptotic effect of radiation and cisplatin and promotes tumor growth via JNK and p38 activation in Raw264.7 and EL4 cells. *Oncol. Rep.* **2018**, *39*, 2688–2694. [CrossRef] [PubMed]
36. Byun, E.B.; Song, H.Y.; Kim, W.S.; Han, J.M.; Seo, H.S.; Park, S.H.; Kim, K.; Byun, E.H. Protective Effect of Polysaccharides Extracted from Cudrania tricuspidata Fruit against Cisplatin-Induced Cytotoxicity in Macrophages and a Mouse Model. *Int. J. Mol. Sci.* **2021**, *22*, 7512. [CrossRef] [PubMed]
37. Yamaguchi, M.; Tomihara, K.; Heshiki, W.; Sakurai, K.; Sekido, K.; Tachinami, H.; Moniruzzaman, R.; Inoue, S.; Fujiwara, K.; Noguchi, M. Astaxanthin ameliorates cisplatin-induced damage in normal human fibroblasts. *Oral Sci. Int.* **2019**, *16*, 171–177. [CrossRef]
38. Ouyang, M.Z.; Lin, L.Z.; Lv, W.J.; Zuo, Q.; Lv, Z.; Guan, J.S.; Wang, S.T.; Sun, L.L.; Chen, H.R.; Xiao, Z.W. Effects of the polysaccharides extracted from Ganoderma lucidum on chemotherapy-related fatigue in mice. *Int. J. Biol. Macromol.* **2016**, *91*, 905–910. [CrossRef]
39. Gomez-Cuadrado, L.; Tracey, N.; Ma, R.; Qian, B.; Brunton, V.G. Mouse models of metastasis: Progress and prospects. *Dis. Model. Mech.* **2017**, *10*, 1061–1074. [CrossRef]
40. Chen, S.-H.; Chang, J.-Y. New Insights into Mechanisms of Cisplatin Resistance: From Tumor Cell to Microenvironment. *Int. J. Mol. Sci.* **2019**, *20*, 4136. [CrossRef] [PubMed]
41. Olaussen, K.A.; Dunant, A.; Fouret, P.; Brambilla, E.; Andre, F.; Haddad, V.; Taranchon, E.; Filipits, M.; Pirker, R.; Popper, H.H.; et al. DNA repair by ERCC1 in non-small-cell lung cancer and cisplatin-based adjuvant chemotherapy. *N. Engl. J. Med.* **2006**, *355*, 983–991. [CrossRef] [PubMed]

42. Inoue, Y.; Matsumoto, H.; Yamada, S.; Kawai, K.; Suemizu, H.; Gika, M.; Takanami, I.; Iwazaki, M.; Nakamura, M. Association of ATP7A expression and in vitro sensitivity to cisplatin in non-small cell lung cancer. *Oncol. Lett.* **2010**, *1*, 837–840. [CrossRef] [PubMed]
43. Salaroglio, I.C.; Mungo, E.; Gazzano, E.; Kopecka, J.; Riganti, C. ERK is a Pivotal Player of Chemo-Immune-Resistance in Cancer. *Int. J. Mol. Sci.* **2019**, *20*, 2505. [CrossRef] [PubMed]
44. Wang, G.; Reed, E.; Li, Q.Q. Molecular basis of cellular response to cisplatin chemotherapy in non-small cell lung cancer (Review). *Oncol. Rep.* **2004**, *12*, 955–965. [CrossRef] [PubMed]

Article

Wrinkling on Stimuli-Responsive Functional Polymer Surfaces as a Promising Strategy for the Preparation of Effective Antibacterial/Antibiofouling Surfaces

Carmen M. González-Henríquez [1,2,*], Fernando E. Rodríguez-Umanzor [1,3], Matías N. Alegría-Gómez [1,3], Claudio A. Terraza-Inostroza [4], Enrique Martínez-Campos [5], Raquel Cue-López [5], Mauricio A. Sarabia-Vallejos [6], Claudio García-Herrera [6] and Juan Rodríguez-Hernández [7]

1. Departamento de Química, Facultad de Ciencias Naturales, Matemáticas y del Medio Ambiente, Universidad Tecnológica Metropolitana, Santiago 7800003, Chile; guezu@utem.cl (F.E.-R.-U.); matias.alegriag@utem.cl (M.N.A.-G.)
2. Programa Institucional de Fomento a la Investigación, Desarrollo e Innovación, Universidad Tecnológica Metropolitana, Santiago 8940000, Chile
3. Programa PhD en Ciencia de Materiales e Ingeniería de Procesos, Universidad Tecnológica Metropolitana, Santiago 8940000, Chile
4. Research Laboratory for Organic Polymer (RLOP), Facultad de Química y Farmacia, Pontificia Universidad Católica de Chile, Santiago 7810000, Chile; cterraza@uc.cl
5. Group of Organic Synthesis and Bioevaluation, Instituto Pluridisciplinar, Universidad Complutense de Madrid, Associated Unit to the ICTP-IQM-CSIC, 28040 Madrid, Spain; emartinezcampos79@gmail.com (E.M.-C.); raquelcuelopez1x@gmail.com (R.C.-L.)
6. Departamento de Ingeniería Mecánica, Facultad de Ingeniería, Universidad de Santiago de Chile, Santiago 9170022, Chile; mauricio.sarabia@usach.cl (M.A.S.-V.); Claudio.garcia@usach.cl (C.G.-H.)
7. Polymer Functionalization Group, Departamento de Química Macromolecular Aplicada, Instituto de Ciencia y Tecnología de Polímeros-Consejo Superior de Investigaciones Científicas (ICTP-CSIC), 28006 Madrid, Spain; jrodriguez@ictp.csic.es

* Correspondence: carmen.gonzalez@utem.cl

Abstract: Biocompatible smart interfaces play a crucial role in biomedical or tissue engineering applications, where their ability to actively change their conformation or physico-chemical properties permits finely tuning their surface attributes. Polyelectrolytes, such as acrylic acid, are a particular type of smart polymers that present pH responsiveness. This work aims to fabricate stable hydrogel films with reversible pH responsiveness that could spontaneously form wrinkled surface patterns. For this purpose, the photosensitive reaction mixtures were deposited via spin-coating over functionalized glasses. Following vacuum, UV, or either plasma treatments, it is possible to spontaneously form wrinkles, which could increase cell adherence. The pH responsiveness of the material was evaluated, observing an abrupt variation in the film thickness as a function of the environmental pH. Moreover, the presence of the carboxylic acid functional groups at the interface was evidenced by analyzing the adsorption/desorption capacity using methylene blue as a cationic dye model. The results demonstrated that increasing the acrylic acid in the microwrinkled hydrogel effectively improved the adsorption and release capacity and the ability of the carboxylic groups to establish ionic interactions with methylene blue. Finally, the role of the acrylic acid groups and the surface topography (smooth or wrinkled) on the final antibacterial properties were investigated, demonstrating their efficacy against both gram-positive and gram-negative bacteria model strains (*E. coli* and *S. Aureus*). According to our findings, microwrinkled hydrogels presented excellent antibacterial properties improving the results obtained for planar (smooth) hydrogels.

Keywords: pH-sensitive hydrogel; poly(acrylic acid); quartz crystal microbalance (QCM); ellipsometric measurement; antibacterial activity

1. Introduction

Smart polymers are defined as materials that can actively alter their properties (such as volume or shape) in response to external changes [1]. For example, pH-sensitive materials can pass from open fully solvated coils to desolvated globular conformations over a small range of pH. These compounds generally contain some acid groups (–COOH, –SO$_3$H) or basic groups (–NH$_2$) in their chain. They have been utilized in different medical applications, mainly as drug vehicles or for gene and protein delivery applications [2], pH-sensitive materials for monitoring tissue acidosis [3], adsorbent materials for the remotion of cationic and anionic dyes [4], surface modification for contact lens [5], wound healing hydrogels with antibacterial activity [6–8], or as a pH-sensitive indicator to detect bacterial growth [9].

Polyelectrolytes are a particular type of smart polymers whose repeating units bear electrolyte groups, such as –COOH or –NH$_2$ groups. Generally, the weak polyelectrolyte has an average dissociation constant (pKa or pKb) in the range of ~2 to ~10, which means that it would be partially dissociated at intermediate pH. Poly(acrylic acid) (PAAc) is a type of weak reversible polyelectrolyte that could accept or donate protons in response to pH variations. Interestingly, PAAc is widely used in applications for biomedical engineering, drug release, agriculture, and environmental protection [10]. Several methods have tested its pH-sensitiveness in the literature; for example, Sun et al. [11] synthesize microgels of poly(acrylic acid-*co*-vinylamine) (P(AAc-*co*-VAm)) for drug release applications. They studied the microgels' zeta potential and swelling ratio, demonstrating that using P(AAc-*co*-VAm) makes it possible to generate a sustainable drug release system.

Similarly, Roghani-Mamaqani et al. [12] synthesized cellulose nanocrystals (CNCs)-grafted block copolymers of acrylic acid, N-isopropylacrylamide, and poly(N-isopropylacrylamide) (PNIPAAm). Dynamic light scattering (DLS) was used to measure the compounds' hydrodynamic radius to find their lower critical solution temperature (LCST) and pH responsiveness. Interestingly, the block copolymers reversibly form a core-shell structure with PNIPAAm as core and PAAc as shell above LCST at higher pH values. Another example of PAAc used as a pH-responsive carrier for drug release applications was reported by Gupta and Purwar [13], who prepared hydrogel nanofibrous mats via electrospinning methods using PAAc and polyethylene glycol (PEG) as crosslinker. The nanofibrous mats were loaded with amoxicillin for antimicrobial activity. The results demonstrate that drug-loaded mats form a clear inhibition zone with gram-positive and gram-negative bacteria.

In some applications related to adsorption mechanisms, drug release, tissue engineering, soft machines, or soft robotics, the material should present on-demand switching from strong to weak surface adhesion [14]. Indeed, Yang et al. [15] reported an approach for switchable adhesion between hydrogels (P(PAAc-*co*-AAm)) based on a mechanical wrinkling process. Thus, the formation and characterization of wrinkles are tailored by regulating the prestretch of adherend and optimizing adhesive ingredients synergistically. Furthermore, the topographic characteristics of the wrinkled patterns of different scales of length and amplitude in polymeric materials have been used to selectively control the adherence and proliferation of various cells [16]. These micro-nano structures allow the development of surfaces with effective antibacterial activity by direct contact (bactericidal) or preventing bacterial fouling at the surface (antifouling) [17]. Our research group worked with hydrogels with wrinkled surface patterns, based on the monomers 2-hydroxyethyl methacrylate (HEMA) and 2,2,2-trifluoroethyl methacrylate (TFMA), with poly(ethylene glycol) diacrylate (PEGDA$_{575}$) as a crosslinking agent [18]. These studies demonstrated that the wrinkled patterns based on hydrogels with an amphiphilic balance in their components could act as selective membranes for cell growth and have properties such as biocompatibility and antibacterial activity.

These characteristics are essential because the transfer of microorganisms can quickly contaminate sterile polymeric materials in the medical area. By regulating the surface's microstructure and adherence strength via pH-responsive changes, it is possible to fabricate switchable surfaces suitable for variable applications, like tissue engineering and drug/contaminant release/adsorption. Moreover, it has been demonstrated that PAAc presents antimicrobial properties against some antibiotic-resistant bacterial strains, often related to nosocomial infections [19,20]. Therefore, innovative material design approaches to achieve biologically active surfaces that prevent bacteria colonization are valuable in several biomedical-related fields [21].

This work aims to fabricate stable hydrogel films with reversible pH responsiveness that could spontaneously form wrinkled surface patterns and present antimicrobial properties against both gram-positive and gram-negative bacteria strains. The films comprise both AAc and PEGDA$_{575}$ as monomer and crosslinking agent, respectively. Incorporating AAc in the reaction mixture affects the material's swelling degree and the micelle's dimensions, thus inducing changes in their hydrophilic nature and mechanical properties. More importantly, as will be discussed, these functional and wrinkled hydrogel films present important novel features compared to previously reported wrinkled surfaces. First of all, the presence of carboxylic acid groups will enable us to carry out chemical modification reactions. In this sense, the adsorption/desorption tests using methylene blue (MB) as a cationic dye will be described demonstrating that the wrinkled micropattern presence improves the material's capacity to adsorb or release the MB. Moreover, the antibacterial performance of these materials, as evidenced by the LIVE/DEAD assays, was, on the one hand, significantly improved in the case of wrinkled surfaces in comparison to planar hydrogels. On the other hand, these materials are effective against both gram-positive (*Staphylococcus aureus*) and gram-negative (*E. coli*) bacteria.

2. Materials, Equipment, and Methods

2.1. Materials

The hydrogel composites were synthesized using different mole ratios of acrylic acid (AAc, 99.0%) and poly(ethylene glycol) diacrylate (PEGDA), with an average molecular weight of 575 g/mol. From now on, the composites *net-poly*(AAc-*co*-PEGDA$_{575}$) will be designated according to their relative composition between AAc and PEGDA$_{575}$ (0:1, 1:1, 5:1, 10:1, and 15:1). 2-hydroxy-4'-(2-hydroxyethoxy)-2-methylpropiophenone (Irgacure 2959, 98.0%) was used as a photo-initiator and methylene blue as a dye. All these reactives were acquired from Sigma–Aldrich (St. Louise, MO, USA) and were utilized as received without further purification.

3-hydroxytyraminium chloride (dopamine) was used to generate self-assembled monolayers (SAMs) on the glass substrates with the finality of favoring the adhesion with the hydrogel film. These reactives were acquired from Merck (Darmstadt, Germany). The tris(hydroxymethyl) aminomethane and acid chloride were purchased from Winkler Ltda. (Santiago, Chile). Round glass coverslips (nominal thickness: 0.13–0.16 mm) from Ted Pella, Inc., (Redding, CA, USA) were employed to deposit the hydrogel films via spin-coating. Quartz crystals (6 MHz, crystal area ø 6 mm) were acquired from Ted Pella Inc. (Redding, CA, USA).

2.2. Equipment

^{1}H-NMR spectra were recorded on a Bruker 400 Ultrashield^{TM} spectrometer (Bruker Corp.,Billerica, MA, USA). The compounds were dissolved in dimethyl sulfoxide (DMSO-d$_6$) using TMS (tetramethylsilane) as an internal standard at 60 °C.

A spin coater model KW-4A from Chemat Scientific (Northridge, CA, USA), coupled with a Rocker Chemker 410 oil-free vacuum pump from Rocker Scientific Co. (New Taipei, Taiwan), was used to deposit the reaction mixture over the pre-treated round glass coverslips. Photopolymerizations were carried out using a UV lamp (9 W) with an emission peak centered at λ = 365 nm from Vilber Lourmat Inc. (Marne-la-Vallée, France).

The polymeric films' chemical composition and depth profiles were determined using a confocal Raman model CRM-Alpha 300 RA (WITec, Ulm, Germany) equipped with Nd:YAG dye laser (maximum power output of 50 mW at 532 nm). The Raman spectra were taken point by point with a resolution step of 100 nm. Cross-section images were acquired using this methodology. Attenuated total reflection-Fourier transform infrared (ATR-FTIR) was used to determine the chemical composition of the compounds in a Nicolet IS 5 FT-IR (Thermo Scientific, Waltham, MA, USA).

The hydrogels' thermal stability was acquired by thermogravimetric analysis TGA-Q500 system (TA Instruments, New Castle, DE, USA). The data were obtained under a nitrogen atmosphere with a heating rate of 10 °C min^{-1}, from ambient temperature to 650 °C. Additionally, enthalpy changes were analyzed with a differential scanning calorimeter, DSC-7 (Perkin Elmer Inc., Wellesley, MA, USA), under a nitrogen atmosphere at a heating rate of 10 °C/min. The glass transition (Tg) was obtained from the second heating cycle.

Bresser Trino Researcher II (4–100×) trinocular microscope from Bresser GmbH (Rhede, Germany), coupled with a 5 Mp CCD color camera (Bresser GmbH, Rhede, Germany), was used as the first approach for visualizing the topography of hydrogel wrinkled surface patterns. Additionally, sample topographies were obtained at room temperature using an atomic force microscope (AFM) model NaioAFM, from Nanosurf AG (Liestal, Switzerland) in intermittent contact mode at different scan ranges (25 × 25 µm^2 and 50 × 50 µm^2). Images were treated using the offline software Gwyddion [22,23].

The particle size (hydrodynamic radius) at different concentrations and pH of the compounds, namely *net-poly*(AAc-*co*-PEGDA$_{575}$), were obtained by DLS using a Zetasizer, model NanoS90 (Malvern Instruments Ltd., Malvern, UK), equipped with a 4 mW He-Ne laser (λ = 633 nm) at an angle of 90°. AR4 Abbe-refractometer from A. KRÜSS Optronic (Hamburg, Germany) was used to measure hydrogels' refractive index. The contact angle using a ThetaLite optical tensiometer from Attension-Beloin Scientific (Gothenburg, Sweden) was measured. Experimentally, a droplet of 4 µL was gently deposited on the top of the films. This measurement was performed in three to five different film sectors to ensure reproducibility; three samples were also analyzed in each case.

A multi-angle laser ellipsometer model SE400Adv, from SENTECH Instrument GmbH (Berlin, Germany), measured film thickness under different pHs (3.3 and 6.5). A Quartz Crystal Microbalance (QCM) was utilized to check the film's thickness changes when the copolymer was exposed to different pHs. The QCM corresponds to a high-resolution thickness monitor system MTM-20 (Cressington Scientific Instruments Ltd., Watford, UK) designed for a sputter coater system. A Plasma Prep III system from SPI (Structure Probe Inc., West Chester, PA, USA) coupled to a process plasma controller and a vacuum pump model DUO 3 from Pfeiffer Vacuum GmbH (Asslar, Germany) was used for argon plasma exposure. Additionally, a V-730 UV-Vis spectrophotometer from Jasco Inc. (Easton, MD, USA) was used to evaluate the dye adsorption/desorption process.

For biological evaluation, *Staphylococcus aureus* (25293) and *Escherichia coli* (10536) bacteria strains were purchased from American Type Culture Collection (ATCC, Manassas, VA, USA). Luria-Bertani (LB) broth media (12780052) was purchased from ThermoFisher (Waltham, MA, USA). A spectrophotometer Specord 205 (Analytik Jena, Jena, Germany) was used to measure the optical density of bacteria culture. An inverted fluorescence microscope IX51 from Olympus Co. (Tokyo, Japan) was used to evaluate bacteria viability over wrinkled films.

2.3. Methods

2.3.1. Step 1. Substrate Functionalization

First, the substrates were washed with a soap solution and rinsed with distilled water and then acetone and isopropyl alcohol to remove grease traces. Then, polydopamine was deposited in the glass slides' surface according to the protocol reported by Nirasay et al. [24]. In this procedure, the glass was immersed in a 2 g/L dopamine solution in 10 mM Tris-HCl phosphate buffer (pH 8.5) and gently mixed on an orbital shaker for 3 h at room temperature to generate a SAM of polydopamine on the glass surface. The protocol followed in this case was not the same as Nirasay et al.; therefore, the film thickness (monitored via ellipsometry) resulted in a mean measured thickness of ~10 nm, slightly different than that obtained by the authors. Afterward, these substrates were rinsed with abundant deionized water, dried with ultra-pure nitrogen gas, and stored for further use.

2.3.2. Step 2. Preparation of the Photosensitive Reaction Mixture

All the hydrogels were prepared in transparent flasks (with a septum) covered from light. These reaction mixtures were deposited over polydopamine-treated glasses. The amount of reactive used for the synthesis is shown in Table 1. The proposed chemical structure is depicted in Figure S1 (Supplementary Material Section).

Table 1. Reactive amounts used to synthesize the hydrogels net-poly(AAc-co-PEGDA$_{575}$).

Mole Composition Ratio	AAc (g)	PEGDA$_{575}$ (g)	Irgacure 2959 * (µL)	MilliQ Water (µL)
0:1	0	0.5	6.63	41.7
1:1	0.063		13.3	83.4
5:1	0.313		40.0	250.0
10:1	0.626		72.9	458.3
15:1	0.939		106.0	666.0

* 60 mg of Irgacure 2959 in 300 µL MeOH.

2.3.3. Step 3. Deposition and Irradiation

The reaction mixture was deposited using the spin-coating technique to form homogenous hydrogel thin films. These homogeneous layers were obtained by applying subsequent velocity steps from 500 rpm, 1000 rpm, and 1500 rpm for 18 s. Nonuplets of each sample were deposited to ensure reproducibility and obtain essential data for the analysis, thus achieving statistically reliable results.

2.3.4. Step 4. Deswelling, UV Light Irradiation, and Plasma Exposure Processes

Afterward, the hydrogel films were exposed to UV light irradiation (λ = 365 nm) for 5 min. Later, the samples were exposed to rough vacuum (1×10^{-3} torr) for 3 h to extract the material's remnant solvent. This deswelling process generates two different zones on the material, a top rigid layer of more deswelled material and a soft hydrated foundation. The samples were then exposed to argon plasma (5–7 s), which serves as an external stimulus to spontaneously generate wrinkled patterns on the film's surface due to the mechanical interaction of the deswelled layer and the hydrogel's soft foundation. This process also probably oxidizes and polymerizes the top layer of the film due to the generation of free radicals on the surface via frontal vitrification [25]. Finally, the sample is exposed to UV light for 30 min to fully polymerize the film and fix the hydrogel surface's wrinkled pattern (Figure 1).

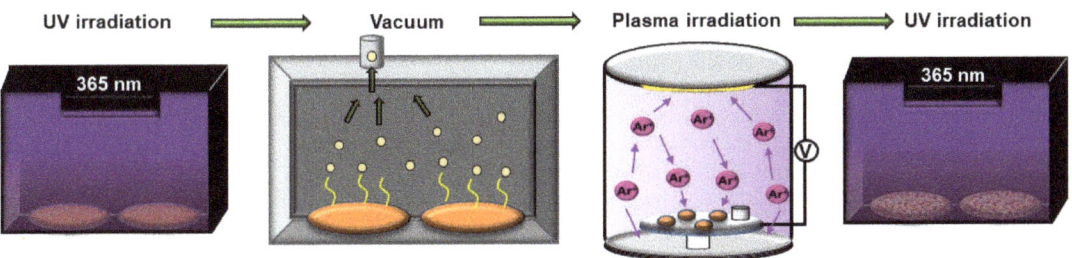

Figure 1. Schematic representation of the subsequent steps used to generate the wrinkle patterns.

To generate smooth (flat) films, the reaction mixture previously deposited by spin-coating was exposed to UV-light for 35 min to generate a total polymerization of the material. Afterward, the films were exposed to vacuum for 2 days and finally irradiated with argon plasma.

2.4. Evaluation Tests

2.4.1. Dye Adsorption/Desorption Tests

Adsorption/desorption experiments were carried out using two hydrogel films (with and without wrinkles). Following a typical experimental procedure [26], a dye solution of MB with a concentration of 5×10^{-3} mM was prepared at pH 7 to test its adsorption/desorption kinetics. Subsequently, on a glass coverslip (1 cm^2), with the hydrogel film on the surface, the MB dye mixture was added upon stirring at a speed of 400 rpm for 300 min. Approximately 600 µL of the dye solution was extracted at different times. The decrease in absorbance of the typical MB peaks was monitored using a UV-Vis spectrophotometer. The dye concentration was calculated at each time using the Beer–Lambert law [27]. Finally, desorption tests were carried out by immersing each MB-loaded hydrogel in an aqueous HCl solution at pH 2 for 60 min. All tests were carried out at room temperature and atmospheric pressure.

2.4.2. Antibacterial Evaluation

Samples were first sterilized in a 12-well plate, using two initial washes with 1 mL of phosphate-buffered saline (PBS). Subsequently, surfaces were exposed to 40 min of UV light before a final wash with PBS. For bacteria inoculum, 1:10 dilutions of previously obtained 0.8 O.D. bacterial suspensions (*Staphylococcus aureus* and *Escherichia coli*) stored at 4 °C were prepared in LB media and left at 37 °C under constant shaking of 125 rpm for 18–24 h. Next, it was verified that the optical density of bacteria cultures was 0.8 at 600 nm using a spectrophotometer previous to cell seeding over films. Then, 1 mL of this inoculum was added to each sample and incubated for 1 h at 37 °C. Once the inoculum was removed, bacteria adhered to the surfaces to grow in PBS for 48 h. LIVE/DEAD™ BacLight™ (ThermoFisher Scientific) kit test was used to evaluate biofilm formation and viability. Staining was performed for 15 m in dark conditions at room temperature, followed by rinsing twice with PBS. Samples were photographed using a fluorescence microscope (Olympus IX 51). This essay analyzes the membrane integrity of the cell, staining green (SYTO® 9) if they are intact or red (propidium iodide) if the membrane is compromised. Green fluorescent bacteria were photographed using FITC filter ($\lambda_{ex}/\lambda_{em}$ = 490/525 nm), and red fluorescent-labeled cells were observed with a TRICT filter ($\lambda_{ex}/\lambda_{em}$ = 550/600 nm). Images were acquired by triplicate using 200× magnification, and bacterial coverage was calculated using the software ImageJ.

3. Results and Discussion

3.1. Preparation of the Wrinkled Functional Surfaces

The preparation of the stimuli-responsive wrinkles was carried out following the scheme depicted in Figure 1. The hydrogel films were firstly formed via spin-coating technique over polydopamine-functionalized round glass coverslips. The films were then exposed to UV, vacuum, and argon plasma to trigger the wrinkled patterns on the hydrogel surface. Finally, the films were exposed again to UV irradiation.

It has been reported that the plasma treatment induces surface oxidation and an increase in the polymerization degree of the material [28]. To confirm this affirmation, two representative samples (5:1 and 15:1) were selected to perform ATR-FTIR. These results show an increase in the carbonyl group vibration at 1722 cm^{-1} after argon plasma irradiation. The percentage increment in the peak integrated area was 64.9% and 99.4% for the samples 5:1 and 15:1, respectively. In Figure S2a (Supplementary Material Section), is possible to observe the spectra for both mentioned samples, normalized to the highest intensity peak at 903 cm^{-1}, attributed to C–O–C symmetric stretching band.

As expected, both FT-IR Spectra showed an increase in three signals (2866 cm^{-1}, 1728 cm^{-1}, and 1093 cm^{-1}). The signal intensity increase confirmed that the oxygen-containing groups were incorporated onto the surface via the plasma treatment. Similar behavior was observed on the signals at 2944 cm^{-1} and 2866 cm^{-1}, related to the antisymmetric/symmetric vibration of –CH$_3$ and –CH$_2$– stretching groups. At ~1728 cm^{-1} and ~1093 cm^{-1}, the C=O vibration and C–O–C asymmetric stretching band were shown, respectively.

3.2. Characterization of the Hydrogels before Film Deposition

^{1}H-NMR, DLS, and thermal studies (TGA and DSC analyses) were performed using the wrinkled hydrogel films deposited over a glass substrate (non-functionalized). The solid film was detached from the substrate by immersing them in nanopure water in all the cases. Afterward, the films were dried under vacuum at 60 °C. Then, the samples were grounded, and the obtained powder (dry hydrogel) was utilized to perform chemical and physical characterization.

Information about the monomer and crosslinking agent conversion and therefore the wrinkled hydrogels' chemical composition was obtained via ^{1}H-NMR and Raman confocal spectroscopy. To carry out an exhaustive analysis of the signals observed by ^{1}H-NMR, the data were compared to the polymeric form of the two isolate ingredients (polyAAc and polyPEGDA$_{575}$). The polyPEGDA$_{575}$ was synthesized by following the same copolymerization methodology and was coined as 0:1. In the case of polyAAc, it was synthesized using the procedure proposed by Rafi Shaik et al. [29], which was corroborated by ATR-FTIR (Figure S2b, Supplementary Material Section). By analyzing the ^{1}H-NMR signals and the area under each peak (Figure S3, Supplementary Material Section), it is possible to demonstrate the formation of a crosslinked structure and determine the monomer and crosslinking agent conversion [18]. Detailed identification of ^{1}H-NMR signals is performed in the Supplementary Material Section. To determine the concentration of AAc that reacted (Table 2), two signals from the NMR spectrum were analyzed: from the bare monomer (AAc) at 6.11 and 6.04 ppm (CH$_2$=CH–C(O)–OH) and from their polymeric backbone chain at 4.10 and 4.08 ppm (–CH(COOH)–CH$_2$)$_n$–). The second signal is related to the total monomer present in the sample, both from the monomer that reacts and the one that does not. Thus, comparing these values, it is possible to obtain the amount of monomer that effectively reacts. Thus, the conversion degree can be calculated by dividing the integral of the respective polymer peak by the sum of integral monomer peak+integral polymer peak. Additionally, the signals selected to determine the concentration of PEGDA$_{575}$ that reacted were two: from the bare monomer (PEGDA$_{575}$) at 6.21 and 6.16 ppm (CH$_2$=CH–C(O)–OR) and their polymeric chain at 3.68 and 3.63 ppm (–CH$_2$–CH$_2$–C(O)–O–CH$_2$–).

Table 2. Real composition of net-poly(AAc-co-PEGDA$_{575}$).

Mole Ratio (Expected Composition)	Conversion Degree of Each Monomer within the Wrinkled Hydrogel Formed		Real Composition
	AAc	PEGDA$_{575}$	
0:1	0%	81.4%	0:1
1:1	98.7%	94.2%	1.0:0.95
5:1	91.3%	95.0%	4.72:0.95
10:1	99.7%	92.3%	9.93:0.92
15:1	94.3%	89.3%	14.1:0.95

The results of polymeric percentage conversion and the expected composition are shown in Table 2. Compared to the expected feed composition, the copolymer formed presents a slightly lower concentration of PEGDA$_{575}$ than expected according to the amount added in the reacting mixture. All the reaction samples present a real composition remarkably similar to the expected one according to the amounts of reactives introduced in the mixture. Similar results have also been found in previous studies performed by our research group [14,18].

In addition to the ^{1}H-NMR, the chemical analysis of the wrinkled hydrogels synthesized using different feed mole ratios was carried out via Raman spectroscopy (Figure 2). These results show the characteristic bands for this kind of hydrogels, such as the antisymmetric/symmetric vibration of –CH$_3$ and –CH$_2$– groups, located at 2941 cm^{-1} and 2880 cm^{-1}, respectively. In addition, it is possible to detect that the peak at 1723 cm^{-1} corresponds to the carbonyl group from PEGDA$_{575}$ or AAc (>C=O or –COOH) and has a significant intensity increase when the inclusion of AAc in the mixture occurs, which is expectable. A similar situation occurs with the band frequency at 1634 cm^{-1} and 1601 cm^{-1}, corresponding to the C=C stretching mode, which could also be utilized to measure the degree of crosslinking of the polymer meshwork. Thus, when increasing the mole ratio (from 0:1 to 15:1), the copolymer presents more free double bonds due to PEGDA$_{575}$ as a crosslinking agent. Similar behavior is observed for the bands in the range 1460–1480 cm^{-1}, associated with the antisymmetric –CH$_3$ bending and –CH$_2$– scissoring.

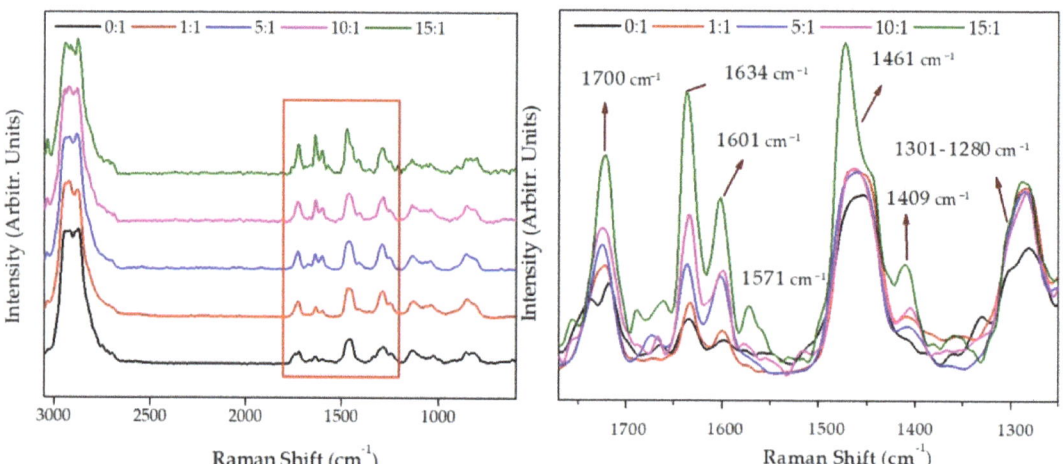

Figure 2. Raman spectra of *net-poly*(AAc-co-PEGDA$_{575}$) at a different mole ratio, i.e., AAc:PEGDA$_{575}$ of 0:1, 1:1, 5:1, 10:1, and 15:1.

An intense signal is observed near 1571 cm^{-1} due to the stretching vibration of the conjugated system of –C=C– and >C=O groups. This band suddenly appears in the samples 15:1 and 10:1, behavior related to the AAc concentration utilized in the reaction mixture (copolymer). On the other hand, the characteristic Raman signals from carboxylic groups can be observed at 1409 cm^{-1}, which corresponds to the interaction of oxygen-hydrogen in-plane bending (C–OH in-plane bend). From 1301 cm^{-1} to 1280 cm^{-1}, two coupled signals could be observed: skeletal vibration –(CH$_2$)$_n$– in-phase twist and C–O stretching. These bands' intensities also increased according to carboxylic acid present in the hydrogels' surfaces. These variations could also be related to the polymerization degree of each composite. Finally, four signals with strong-medium intensities are shown at 1128 cm^{-1}, 1035 cm^{-1}, 852 cm^{-1}, and 815 cm^{-1}, which should be related to C–C skeletal stretching.

In Figure 3 are depicted the TGA and the DSC traces for the different hydrogels, and in Table 3 are summarized the most relevant results obtained from the TGA and DSC curves. From TGA data, it is possible to determine the weight loss at different temperatures (260 °C, 340 °C, and 460 °C). At 260 °C is shown a slight weight loss increase from 0.8% to 14.0% and at 340 °C from 6.0% to 27.2% when the amount of AAc is increased in the mixture from 0 to 15 mole (0:1 to 15:1). Similarly, it is possible to observe a residual mass increase from 3.3% to 9.1% at 460 °C. These results show that the thermal stability of the compound slightly decreases with the AAc inclusion, leaving a more significant residual mass at high temperatures. Jeong et al. [30] obtained similar results using *poly*AAc hydrogels, which contain different metronidazole concentrations. According to the obtained thermograms, three main steps of thermal decomposition are shown: the first one is related to the evaporation of adsorbed water (weight loss about 0.2–2.7%); the second could be attributed to the destruction of *poly*AAc lateral groups, leading to decarboxylation or anhydride formation (weight loss of ~2.6–11.9% between 200 °C and 300 °C); and the third step is associated to main chain scission and composite depolymerization above 300 °C.

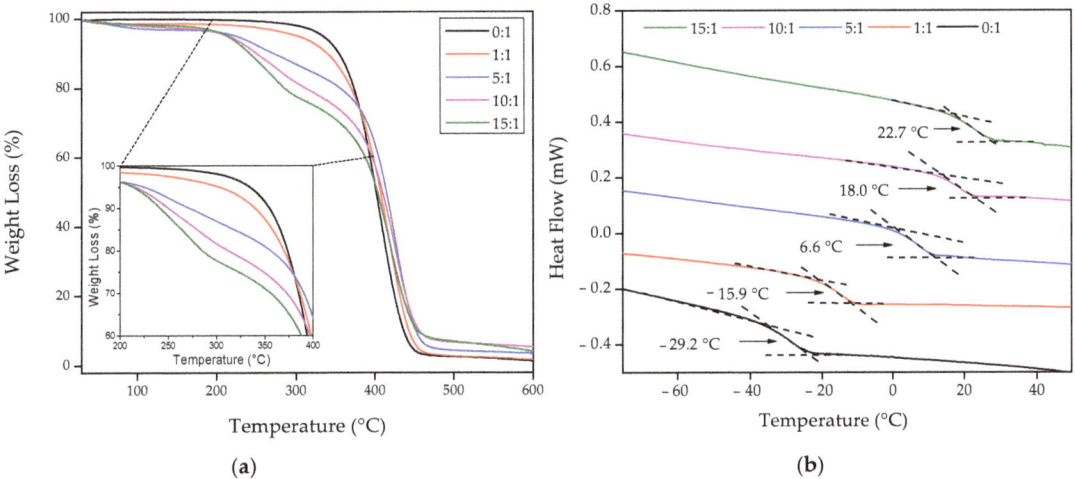

Figure 3. Thermal analysis of the wrinkled hydrogel surfaces with different chemical compositions: (**a**) TGA and (**b**) DSC.

Table 3. Thermal studies of the hydrogels *net-poly*(AAc-*co*-PEGDA$_{575}$) at different mole ratios for two temperatures (260 °C and 340 °C), with residual mass at 460 °C, and their respective Tg.

Samples	TGA (Weight Loss)			DSC
	260 °C	340 °C	Residual Mass, 460 °C	Tg (°C)
0:1	0.80%	6.00%	3.30%	−29.2
1:1	2.70%	10.00%	4.20%	−15.9
5:1	8.60%	17.50%	6.90%	6.6
10:1	11.50%	23.60%	9.20%	18
15:1	14.00%	27.20%	9.10%	22.7

By analyzing the DSC curves, the glass transition temperature (Tg) was determined as the temperature at the mid-point of the endothermic rise, measured from the post-transition baselines' extension [31]. Thus, the five hydrogels' DSC curves possess only one Tg each, which increased from −29.2 °C to 22.7 °C with the increasing inclusion of AAc in the mixture (from 0 to 15 mole). These Tg values are associated with the concentration of monomer or crosslinking agent used for synthesizing the hydrogel. According to the literature, the Tg of *poly*AAc is 123 °C [32,33], and for *poly*PEGDA$_{575}$, the Tg was found from −40 to −30 °C [34]. It is possible to conclude that the Tg values obtained for the composites are in concordance with each hydrogel's compositional ratio, showing an increase when AAc is included in the mixture, as was expectable.

3.3. Wrinkling on Solid Supports to Produce Stimuli-Responsive Microstructured Hydrogels

The successive vacuum and UV irradiation steps lead to wrinkled hydrogel films with variable wrinkle characteristics (wavelength and amplitude). In addition to the preliminary visualization of the wrinkle formation by optical microscopy (Figure S3, Supplementary

Material), AFM microscopy was used to obtain more accurate information. Figure 4a shows a set of AFM images for the samples at different mole ratios (from 0:1 to 15:1). Additionally, the top right corner inset represents the two-dimensional Fast Fourier Transform (2D-FFT) of the respective AFM micrographs. Information about the wavelength, amplitude, roughness, area increase percentage, and aspect ratio (wrinkle height/width) can be obtained by analyzing these images. Finally, the samples' thicknesses can also be determined by measuring the border's depth profile (see Figure 4b). For all the examples, the film thicknesses were found in the range of 1.5–1.7 µm, thus indicating that the spin-coating technique generates homogenous films with similar thicknesses independently of the sample composition used in each case, as expected. These results were later corroborated via ellipsometric techniques.

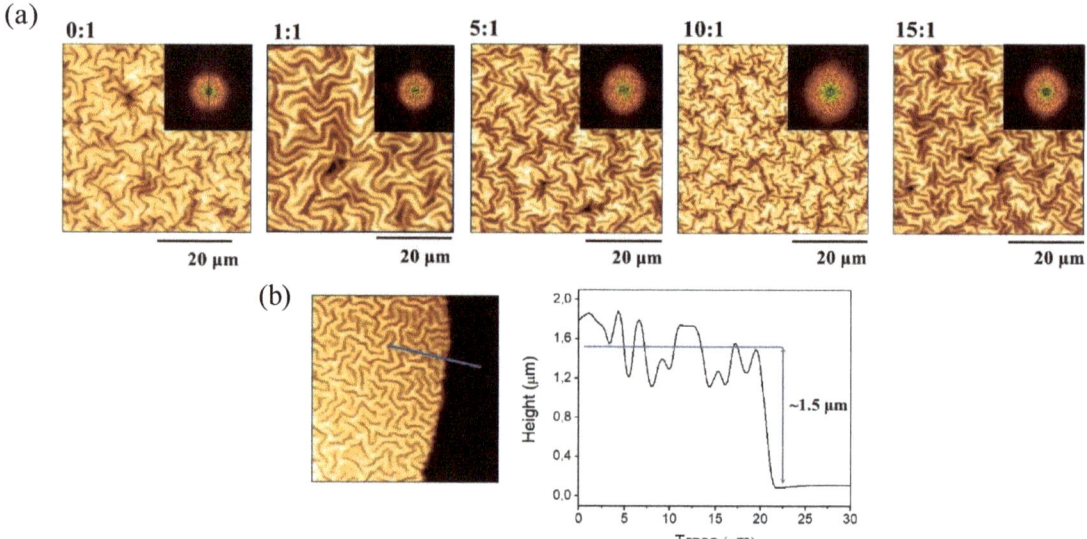

Figure 4. (a) AFM topography images of the wrinkled patterns and their corresponding 2D-FFT and (b) film thickness measurement using AFM images.

The distribution of the wrinkled patterns is not significantly affected by the variation of hydrogel composition; in all cases, the distribution is homogenous in all directions, forming herringbone-like structures [35]. The 2D-FFT results allow us to demonstrate this affirmation due to concentrical rings with different intensities, which is a typical pattern characteristic of homogeneous wrinkled distributions that do not have any preferred ordering direction [36–38]. This kind of homogenous herringbone-like structure is typical of free equi-biaxial contractions [39,40], which, in this case, was produced by a mechanical contraction as a result of the deswelling process when the sample is dried under vacuum.

As it is possible to observe from the data, the width (wavelength) of the wrinkles increases dramatically when the mole ratio concentration changes from 0:1 to 1:1 (Figure 5a); this could be related to the inclusion of AAc in the mixture, which could alter the mechanical properties of the material, thus affecting the wrinkling process mediated by surface instabilities. Then, for the samples 5:1, 10:1, and 15:1, is possible to detect a slight decrease in wrinkle width with the inclusion of AAc in the mixture. Interestingly, the wrinkle height (amplitude) and wrinkle width (wavelength) possess a similar behavior for these three samples (Figure 5a). This tendency is also reflected in the wrinkles' aspect ratio (amplitude/wavelength), whose values are between 0.26 ± 0.13 and 0.51 ± 0.21 (Figure 5c). Aspects ratios between those ranges indicate that these samples' wrinkled patterns are considered ripples or small ridge patterns [41]. A similar situation occurs for the roughness and area increase percentage measured for each sample; an increase from sample 0:1 to 1:1 is detected, probably due to the AAc insertion, and then a slight decrease is observed for the samples 5:1, 10:1, and 15:1. The area increase percentage was obtained from the AFM micrographies via the determination of the real surface area and the projected area of the analyzed sector.

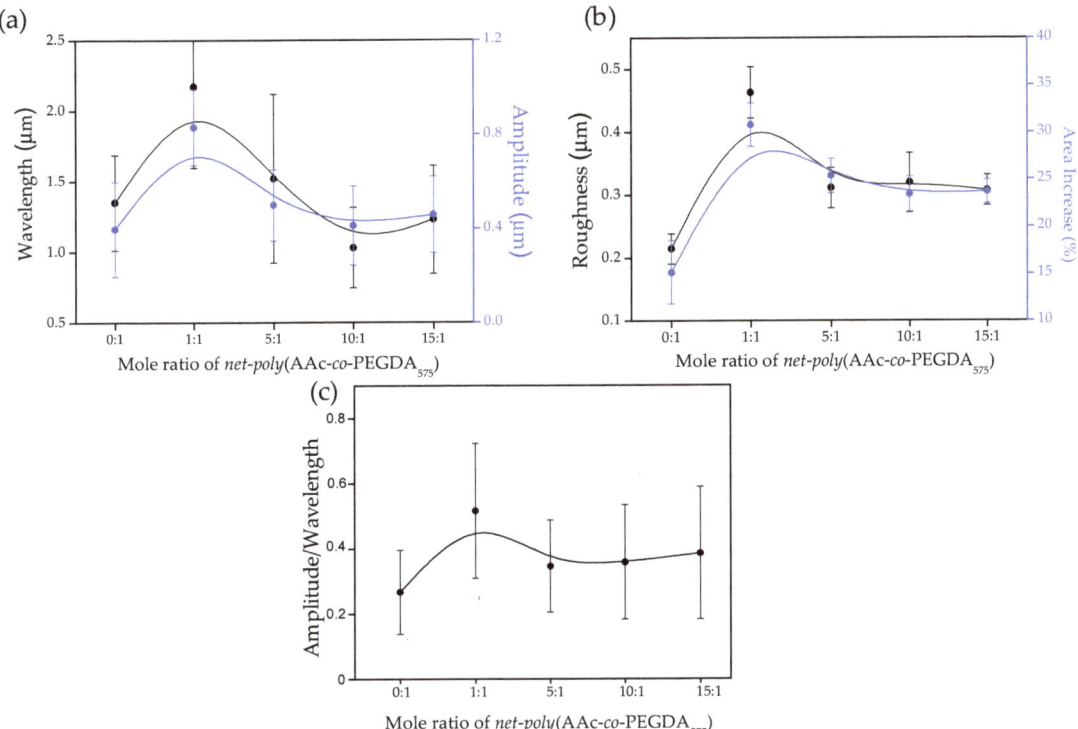

Figure 5. Morphological characteristics obtained from AFM analysis. (**a**) Wrinkle wavelength and amplitude, (**b**) roughness and area increase, and (**c**) aspect ratio of the hydrogel films of *net-poly*(AAc-*co*-PEGDA$_{575}$) at a different mole ratio (0:1, 1:1, 5:1, 10:1, and 15:1).

As demonstrated by several research groups [42–44] and also reported by our group [18,45,46], the variation of microstructural patterns, particularly wrinkled patterns, could affect material biocompatibility and cell adherence due to some microstructural cues present on the surface of the material. A priori, the AAc-based hydrogel films fabricated in this work are expected to improve their compatibility with biological tissues by altering their wrinkled microstructural patterns.

3.4. Evaluation of the pH Response of the Wrinkled Surfaces

Several complementary techniques were employed to characterize the wrinkled hydrogel surfaces' pH sensitiveness, i.e., DLS, water contact angle, QCM (Quart crystal microbalance), and ellipsometry measurements.

DLS was carried out for aqueous solutions of the copolymers. The hydrogels in the buffer (3 mg of the copolymer in 5 mL of pH) were agitated in small flasks for 72 h and sonicated to ensure a complete hydrogel dissociation. Hydrodynamic radius, using distilled water and pHs 3.3 and 6.5 buffers, were obtained through this method. The introduction of -COOH groups in the hydrogel alters the interaction with water at a determined pH. Thus, at pH lower than 4.5, most -COOH groups from the AAc are protonated, while at higher pHs, they are ionized, leading to broken hydrogen bonds and swollen microgels [47]. Figure 6a shows the microparticles' hydrodynamic radius as a function of pH. In general, as is possible to observe, at pH 1.0, the hydrodynamic radius is lower than at pH 9.0. This behavior is related to the stretching of the molecules and their hydration. At low pH, *poly*AAc adopts a compact (but not fully collapsed) globular conformation. However, ionization occurs when the pH increases above its pKa, and the polymers expand into a fully solvated, open-coil conformation [48]. With this, it is possible to observe that the hydrodynamic radius increases with AAc concentration in the composite. Moreover, the difference between the hydrodynamic radius at pHs 3.3 and 6.5 becomes larger when AAc increases. Interestingly, at concentrations of 10:1 and 15:1, the difference in the hydrodynamic radius for the different pH stops growing, indicating that a plateau is reached over 10:1 concentration of AAc. Additionally, from the DLS-Titration curve, it is possible to conclude an increase in hydrodynamic radius upon an increase in pH, behavior related to a possible molecular association started at pH ≥ pKa (located at 4.5). These results show the familiar expansion of the polymeric chain upon the pKa; similar behavior was studied by Swift et al. [49] in the polymers *poly*AAc and *poly*(AAc-*co*-ACE) (ACE, Acenaphthylene). Whereas the DLS analysis provides information about the capability of the size and polydispersity of the swollen hydrogels in a dispersed state, with the finality of characterizing the hydrophilic or hydrophobic nature of the materials, static contact angles were measured using the sessile drop methodology.

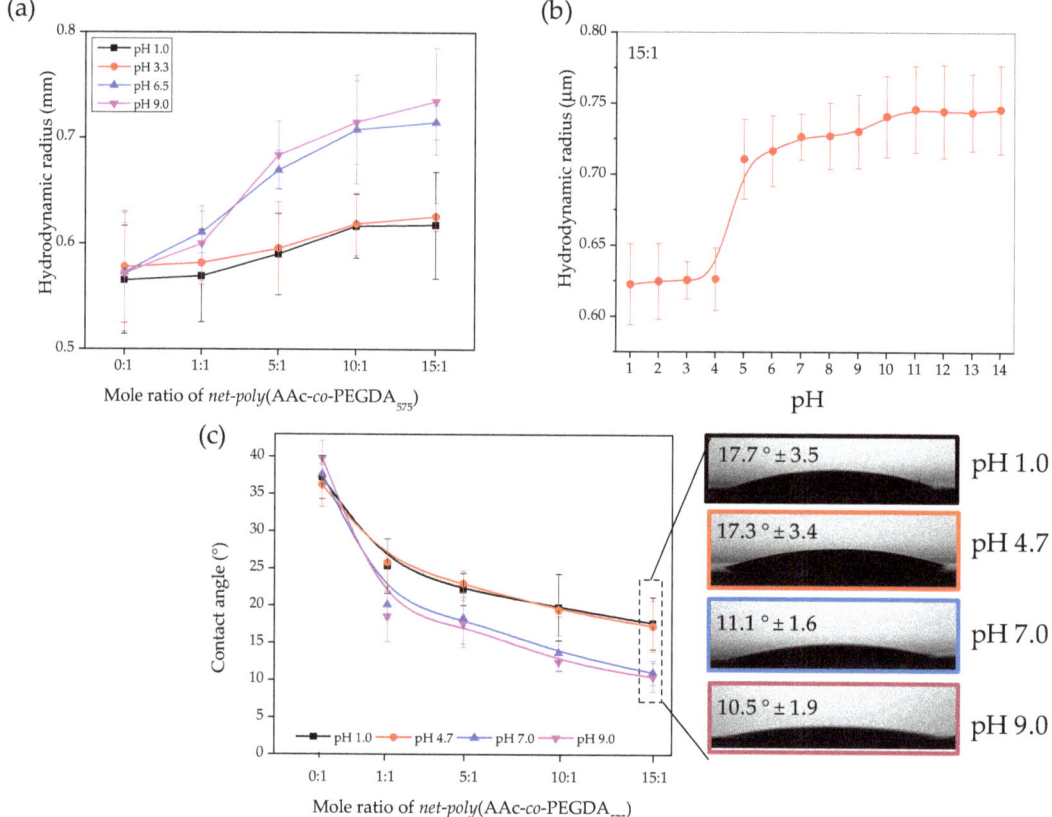

Figure 6. (a) Dependence of the hydrodynamic radius at four different pHs (1.0, 3.3, 6.5, and 9.0) for the samples *poly*(AAc-*co*-PEGDA$_{575}$), (b) DLS-Titration curve of the for the sample 15:1 using several pHs as analysis medium, and (c) contact angle of a small drop on the hydrogel surface with different pH, together with some representative images.

As depicted in Figure 6c, is possible to conclude that copolymers with the highest amount of acrylic acid in their main chain show the lowest contact angle due to the incorporation of polar groups (–COOH), which increase the hydrophilicity of the hydrogels. Additionally, the pH-sensitive capacity of copolymers was studied through contact angle measurement at different pH (1.0 to 9.0). Thus, at increasing the pH values ($\geq$pKa [50,51]), a slight decrease in the contact angle for all the samples that contain acrylic acid segments (1:1 to 15:1) is observed. These low values are related to –COOH functional groups in the polymer chains, predominantly undissociated (at pH 3, these groups' dissociation degree (α) is equal to 0.03). At pH 4.5, the number of –COOH groups is the same as –COO− (α = 0.5). As the pH rises above 4.5, polyelectrolyte dissociation increases rapidly; at pH 6, it equals 0.97, and at pH 9, reaches values close to 1 [52–54].

Palacio-Cuesta et al. [55] studied the variation of the contact angle as a function of the wt% of copolymer *poly*(MMA-*co*-AA) in flat and wrinkled surfaces. Thus, the planar surface slightly decreases the contact angle in comparison with wrinkles surfaces.

According to the results obtained via DLS analysis, the samples 10:1 and 15:1 presented the larger changes with the pH. These two were selected to address the hydrogel films' physical changes when exposed to different pH mediums using QCM and ellipsometric measurements (Figure 7).

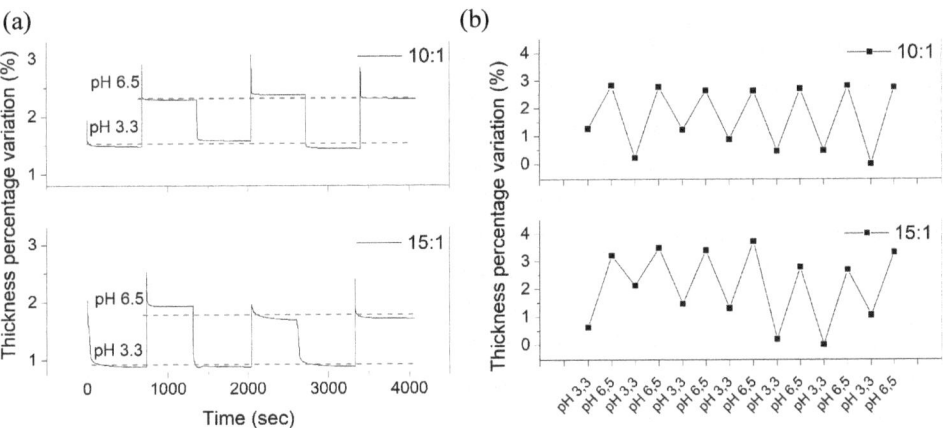

Figure 7. Time dependence of the changes concerning (**a**) QCM thickness and (**b**) ellipsometric measurement upon cycling exposure; Table 3. 3 and 6.5 for two hydrogels (10:1 and 15:1).

A QCM consists of a thin quartz disc placed between a pair of gold electrodes. Due to quartz's piezoelectricity, it is possible to excite a shear mechanical oscillation of the crystal by applying an AC voltage across the substrate. The sensor's resonance frequency depends on the total oscillating mass, including any medium or material coupled to the substrate. The resonance frequency follows the Sauerbrey relation, which is inversely proportional to the film's areal mass density (including the solvent within the film) [56]. Indeed, QCM measures changes in the material's mass density, which could be related in some cases to thickness increase. In this case, the hydrogel thin films absorb or release solvent depending on the medium's pH, which variates the sample's mass density and possibly the film thickness. Similar studies were performed by Borisov et al., which utilized the QCM to analyze the effect of pH and electrical stimuli on brushed grafted polymers [57].

In our case, two buffers based on citric acid and sodium hydroxide with pH 3.3 and 6.5 were used to test the copolymer's pH response. A small drop of the reaction mixture dissolved in methanol (200 μL of the mixture in 4 mL of dissolvent) was deposited on the quartz sensor and irradiated with UV light for their polymerization. The drop was allowed to evaporate at ambient conditions for few minutes, and the frequency measurement was carried out. Then, the process was repeated using a different buffer. Figure 7a shows the time dependence vs. thickness percentage variation of the samples (10:1 and 15:1) with three pH cycles. Thus, when the pH is increased from 3.3. to 6.5, the carboxylic groups become ionized, the AAc chains are expected to stretch, and the brush becomes thicker and increasingly hydrated, thus increasing its mass density and probably increasing its thickness. The QCM response was reversible and fast (less than five minutes) when the pH was decreased from 6.5 to 3.3, reaching a similar value as the measured initially. The percentage thickness difference is almost the same for the two concentrations of AAc analyzed (10:1 and 15:1), similar to the hydrodynamic radius results obtained from the DLS analysis. The compounds that present lower concentrations of AAc, like 1:1 or 5:1, do not show any significant thickness variation.

Ellipsometry measurements were also performed on the thin, solid hydrogel films to detect physical changes and corroborate the QCM results. Biesalski et al. [58] described a weak polyacid brush's synthesis and swelling behavior attached to a solid surface. Thus, pH values (with and without monovalent salts) affect the layers' thickness, monitored by multiple-angle null-ellipsometry. In our case, ellipsometry measurements were performed to corroborate the results obtained by QCM. Thus, the reaction mixture composed of AAc and PEGDA$_{575}$ was dissolved in methanol (200 µL in 800 µL of dissolvent) and deposited via spin-coating on SAMs-silicon wafer (SAM thickness was previously studied by this technique).

Afterward, the films were polymerized using UV light and immersed in water; then, the thickness was studied in dry and swelling states at different pHs. Figure 7b shows the swollen thickness of two different hydrogels (10:1 and 15:1) as a pH (6.5 and 3.3) function. By increasing the pH of the medium (from 3.3 to 6.5), the hydrogel film thickness shows significant conformational changes related to the number of dissociated carboxylic acid groups, i.e., a higher degree of dissociation implies a higher charge density, a higher osmotic pressure of the counterions, and therefore an increased thickness of the film. The thickness measurement was repeated cyclically in the same spot by varying the pH on the medium (seven pH cycles were performed for each sample), showing similar results in each cycle, i.e., a thickness increase when the pH increases from 3.3 to 6.5. Finally, and similarly to the results obtained from QCM and DLS data, the film thickness difference between pH 3.3 and pH 6.5 did not vary significantly between the two concentrations (15:1 and 10:1). This effect could probably be related to the molecular weight of the obtained compounds after deposition. According to Borisova et al. [57], the variation of micelle hydrodynamic radius depends on the compounds' molecular weight and not on the compound's chemical conformation. According to the NMR spectra, the nominal molecular formula weight of the repetitive units for each sample (10:1 and 15:1) are similar (~1220 g/mol and ~1510 g/mol, respectively); therefore, it is expectable that the film thickness variation between pH 3.3 and pH 6.5 is comparable. The film thickness variation could be an exciting capability for materials intended to be used in biomedical or tissue engineering due to its capacity to actively change its conformation under different pH mediums, useful as drug carriers or biosensors. Finally, for films with lower concentrations of AAc (1:1 and 5:1), the thickness percentage variation was small enough to be considered as measurement errors, so it was not included in the study. This effect could be related to the swelled state of the polymeric films. Small, conformational changes are easily detectable when the compound is in solution due to the molecules' freedom degrees. In a swelled state, the molecule is restricted; thus, the conformational changes produced due to pH variations are difficult to detect.

Finally, the pH response has not only an effect on the film thickness but also on the wrinkle characteristics, i.e., wavelength and amplitude. As depicted in Figure 8, both amplitude and wavelength increased with the pH. The changes observed are reversible upon changing the environmental pH. Interestingly, this system based on acrylic acid complements previous works in which the use of diethylaminoethyl methacrylate (DEAEMA) (thermal-pH responsive) allowed the preparation of wrinkles with lower wavelength and amplitude upon increasing the pH [14]. It is interesting to note that, in those previous works, the wavelength of the wrinkles formed remains clearly above 5 µm, while with this current approach, the wavelength has been significantly reduced to the range of 1.2–1.7 µm.

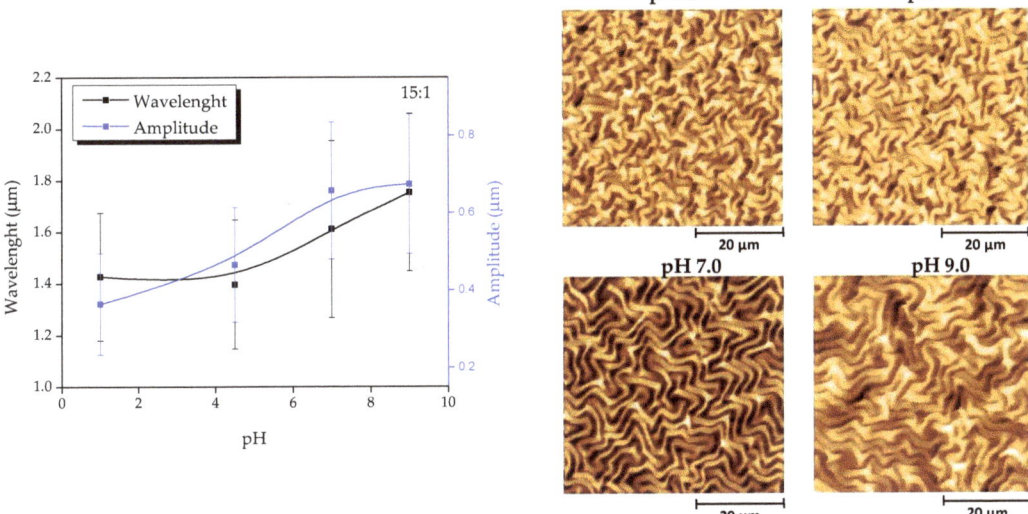

Figure 8. Right: Evolution of the wrinkle characteristics (wavelength and amplitude) for the wrinkled hydrogel 15:1 as a function of Table 1. Obtained at different pH values.

Finally, the availability of the carboxylic acid groups to establish interactions with ionic molecules was investigated by the adsorption of methylene blue [26]. Hydrogel-based biopolymers are one of the materials most used as adsorbents due to their ability to absorb and retain large amounts of water in their structure. This behavior is mainly due to the presence of polar functional groups that strongly interact with water molecules, such as amino (–NH$_2$), hydroxyl (–OH), and carboxyl (–COOH) groups [59]. These hydrogels' acidic or basic groups can accept or release protons due to pH variations, generating charges on the polymeric structure. In carboxylic acid, at a pH above pKa, the –COOH groups dissociate to form carboxylate ions (–COO–), which strongly interact with cationic dyes, like MB [60]. Accordingly, methylene blue (MB) adsorption and desorption studies were carried out in smooth and wrinkled hydrogel samples to confirm the hydrogel's response capacity at pH above and below pKa and the influence of wrinkled morphologies on adsorption. It is expected that samples with micro-patterns on their surface should present a higher surface area for contact with the medium, thus increasing its adsorption/desorption capacities.

Accordingly, Figure 9a shows the results obtained by UV-Vis spectroscopy at pH 7. The changes in the concentration of MB during the removal process are monitored by following the most intense adsorption band (664 nm). It was observed that after 120 min of magnetic stirring, the adsorption rate remained constant, managing to remove 61.4%, 23.9%, 82.9%, and 42.3% in the samples 0:1 (wrinkled), 0:1 (smooth), 15:1 (wrinkled), and 15:1 (smooth), respectively. The results show that the incorporation of the AAc in the polymer produces a considerable increase in the adsorption capacity of MB, as was expectable. These results are also observable in Figure 9c, where the hydrogels with a 15:1 concentration present an intense pigmentation compared to the 0:1 concentration. Additionally, it is possible to observe that the presence of a microwrinkled pattern on the hydrogel films generates a significant increase in adsorption capacity of the material (an increase of 40.7% for sample 0:1 and 41.3% for sample 15:1), which is directly related to the rise in the contact area when forming micro-morphologies to surface level (Figure 5b). With pH values below the pKa, the adsorption percentage of the dye is controlled mainly by the –COOH groups of the adsorbent. Under this condition, the –COOH dissociate to form –COO–, where the negatively charged oxygen atom generates an electrostatic attraction with the positive charges of the =N$^+$ fractions of the cationic dye [61].

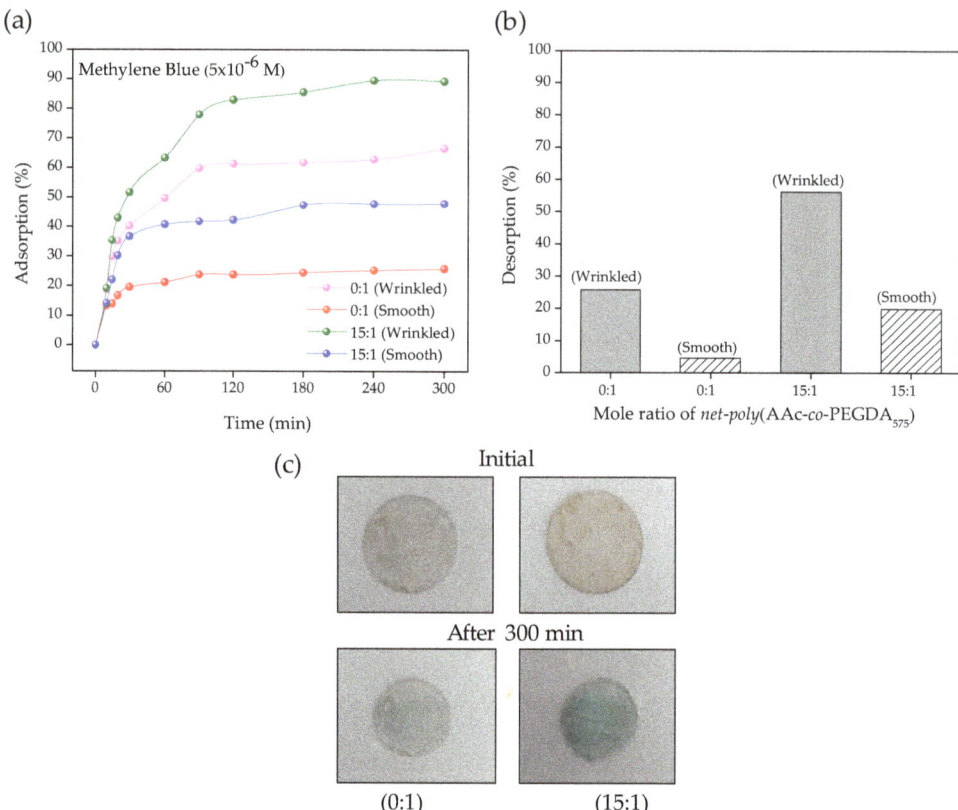

Figure 9. (a) MB adsorption percentage, (b) MB desorption percentage from smooth and wrinkled hydrogels with a concentration of 0:1 and 15:1, and (c) photographs of glasses with different hydrogel concentrations before and after MB adsorption.

The hydrogel samples loaded with MB were immersed in an acidified aqueous solution (pH 2) for 60 min to conduct the dye desorption tests. In these processes, a 25.8%, 4.5%, 56.2%, and 19.9% desorption was achieved for the samples 0:1 (wrinkled), 0:1 (smooth), 15:1 (wrinkled), and 15:1 (smooth), respectively (Figure 9b). Although both concentrations manage to desorb part of the adsorbed dye, the values for the 15:1 sample are higher than in hydrogels without acrylic acid (0:1). This behavior confirms the pH responsiveness of hydrogels with carboxylate functional groups, which lose their charge at pH below pKa. Therefore, electrostatic interactions between adsorbate and dye decrease, increasing the desorption percentage, as was expected.

3.5. Antibacterial Evaluation of the PAA Functionalized Hydrogel Wrinkled Films

Provided the capacity of the carboxylic acid groups to establish interactions, the capacity of these functional groups to act as antibacterial materials for biomedical purposes was explored. For this purpose, smooth and wrinkled films with increasing acrylic acid content (1:1, 5:1, and 15:1) were evaluated as antibacterial surfaces. In this evaluation, *Staphylococcus aureus* was chosen as a model due to their implication in nosocomial infections, especially with antibiotic-resistant strains. In addition, the *Escherichia coli* bacterial strain has also been evaluated as a gram-negative model. Thus, a solution with the bacteria was allowed to adhere to the wrinkled films and then proliferate for 48 h. After that, the films were removed from the incubator, and the viability of the bacterial cells at the film surface was evaluated using a LIVE/DEAD fluorometric staining.

Figure 10a shows live (green) and dead (red) *S. aureus* bacteria over films (wrinkled, smooth, and control (C)). The samples designated as control (C) correspond to a planar surface formed exclusively by PEGDA$_{575}$, i.e., without acrylic acid. Over smooth control surface, viable bacteria colonies were observed. However, the presence of acrylic acid induces the apparition of red-colored bacteria, i.e., dead bacteria. Interestingly, when the acrylic acid content increases on smooth surfaces, the number of dead bacteria becomes higher, evidencing the antimicrobial behavior of this surface functionalization.

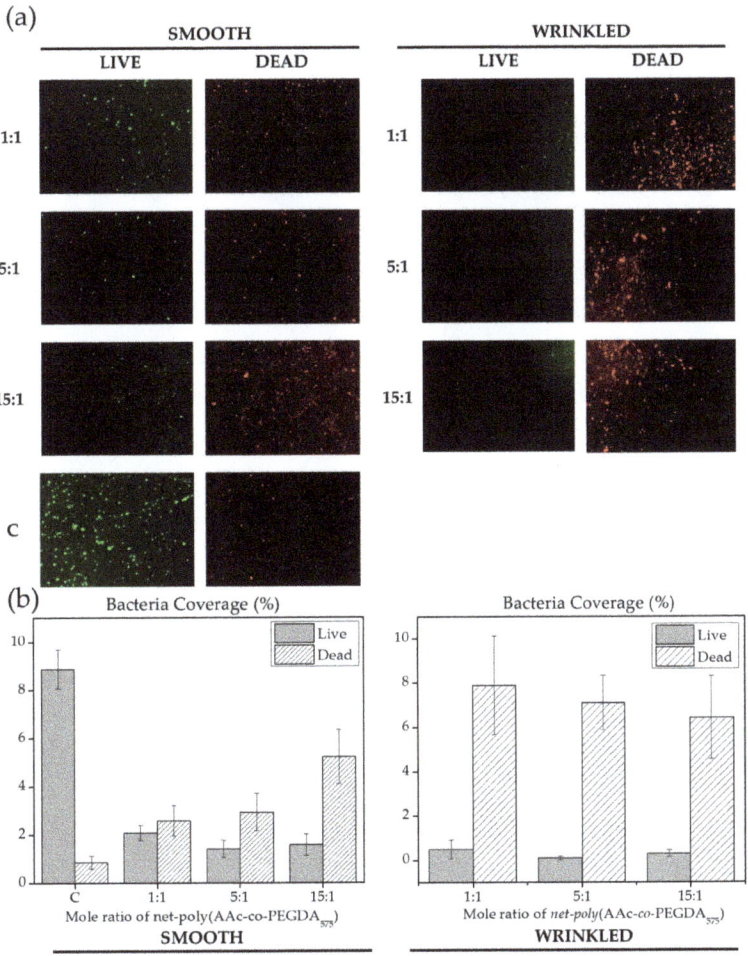

Figure 10. LIVE/DEAD assay for smooth and wrinkled AAc films incubated with *Staphylococcus aureus* bacteria strain: (**a**) live (green) and dead (red) bacteria over wrinkled, smooth, and control (C) films after 48 h of incubation and (**b**) surface area coverage (%) of bacteria culture after ImageJ analysis.

Moreover, even fewer isolated living bacteria were photographed over wrinkled samples, whereas a high amount of dead clusters were detected on the three different wrinkled surfaces. In all wrinkled samples, dead bacteria's percentage of area occupied is significantly higher than occupied by live bacteria, demonstrating an enhancing action compared to their smooth counterparts (Figure 10b). This observation can be explained by considering that the microwrinkled surface structure favors the bacterial attachment to the surface. Once attached (most probably on the valleys of the wrinkles), they are exposed to a larger amount of antimicrobial acrylic acid groups.

Furthermore, the total *S. aureus* coverage of wrinkled films (Figure 10b), especially dead bacteria's percentage of the occupied area, shows a descending trend as the proportion of AAc augments. This fact could be a consequence of the first initial cell adhesion steps, where wrinkled surfaces with a high proportion of AAc could prevent *S. aureus* colonization showing an additional antifouling effect. After that, bacteria proliferation over wrinkled films was prevented, as viability was compromised due to the AAc action in the bacteria lipid membrane and peptidoglycan layer.

In addition, a similar evaluation was carried out using *E. coli* as an experimental model. Figure 11 shows the antimicrobial activity for smooth and wrinkled samples, with a dramatic decrease of bacteria viability in all films containing acrylic acid. In this case, higher proportions of AAc provoked an increase of dead *E. coli* surface coverage (Figure 11b), showing a remarkable efficacy for 15:1 AAc wrinkled films. Therefore, the antifouling effect was not detected using *E. coli* as an experimental model, suggesting a different early interaction between AAc films and gram-negative bacteria. However, *E. coli* contact killing was also improved using wrinkled surfaces, demonstrating the potentiality of this strategy. Different polymeric supports have previously discussed this synergic behavior between surface composition and topography [16], showing results similar to the obtained but without the pH responsiveness capacity of the AAc-based wrinkled hydrogel film presented in this study. Finally, these results demonstrate an apparent antibacterial effect for all evaluated wrinkled samples, suggesting their application as coatings for clinical devices or biomedical applications.

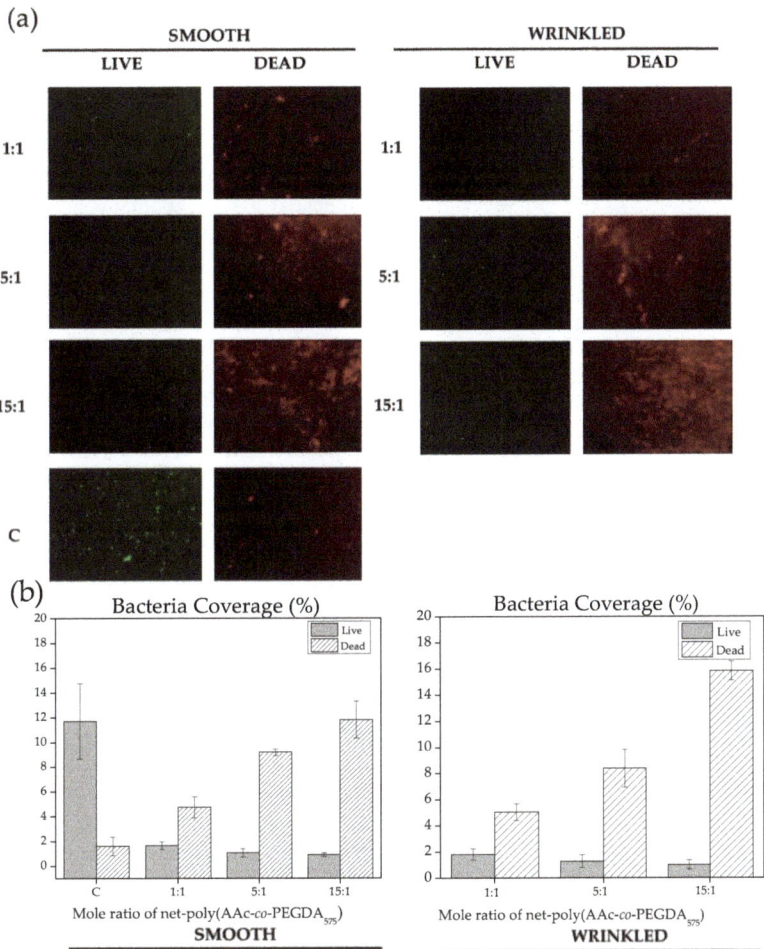

Figure 11. LIVE/DEAD assay for smooth and wrinkled AAc films incubated with *Escherichia coli* bacteria strain: (**a**) live (green) and dead (red) bacteria over wrinkled, smooth, and control (C) films after 48 h of incubation and (**b**) surface area coverage (%) of bacteria culture after ImageJ analysis.

4. Conclusions

Hydrogel composites were synthesized using different mole ratios of AAc and PEGDA$_{575}$. ^{1}H-NMR studies demonstrate that the procedure followed to synthesize the polymers results in almost the same composition as theoretically expected. Together with this, TGA and DSC were also carried out to investigate the samples' thermal behavior. These results show that the weight loss increases with the amount of AAc included in the mixture. A similar situation occurs for the residual mass at 460 °C and the Tg values, increasing from −29.2 to 22.7 °C with an AAc increase, as expected.

Once the compounds' chemical and thermal characterization were finished, the samples were spin-coated spin over SAM-dopamine functionalized glasses to form thin films. Then, wrinkled patterns were formed spontaneously at the top of the films by exposing the samples to UV radiation, vacuum, and argon plasma. Raman spectroscopy was used to corroborate the deposited samples' chemical composition, showing that characteristic bands of the AAc and PEGDA$_{575}$ could be detected in the appropriate proportions for each compound, indicating that the sample deposited maintains the chemical structure of the synthesized composite. Using AFM, the wrinkled patterns' dimension was characterized, showing a stabilization of the wrinkle wavelength and amplitude for the samples 5:1, 10:1, and 15:1.

Contact angle measurements were carried out to analyze the samples' polar nature at different pHs. These results indicate that at larger pHs, the contact angle of the wrinkled surfaces increases for all the composites, including AAc, resulting in more hydrophobic samples when the pH increases over AAc pKa. This study aimed to fabricate hydrogel films with reversible pH responsiveness that spontaneously form wrinkled surface patterns; to this end, QCM and ellipsometry measurements were carried out on the thin films, resulting in notable thickness variations of the films depending on the pH medium. The hydrogels' response to pH changes was fast and could be entirely rationalized based on the theory of conformational changes. The adsorption/desorption kinetics of the cationic dye (MB) at different pH allow to confirm the response capacity of hydrogels against pH and determine the influence of surface modifications on the interaction of external agents comparing the wrinkled and smooth hydrogel samples. An increase in the contact surface area produces a considerable increase in the adsorption of dyes, which is of great importance to address the current environmental problems associated with colored industrial waste.

Interestingly, the inclusion of AAc in the hydrogel mixture affects the dimensions of the wrinkled patterns, theoretically altering the material capacity to interact with biological matter and dyes. On the other hand, despite these physical changes, the pH responsiveness of the material (QCM and ellipsometry) or their polar nature (contact angle) does not seem affected, thus becoming a proper smart material for biomedical and dyes adsorption applications. Furthermore, an antibacterial effect has been detected for all AAc wrinkled films, showing an improved efficacy for *S. aureus and E. coli* strains compared to their smooth counterparts. In this scenario, AAc hydrogel coatings could efficiently avoid bacterial proliferation and prevent nosocomial infections in clinical scenarios.

Supplementary Materials: The following are available online at https://www.mdpi.com/article/10.3390/polym13234262/s1, Figure S1: Schematic representation of the proposed chemical structure of *net-poly*(AAc-co-PEGDA$_{575}$); Figure S2: ATR-FTIR spectra of (a) *net-poly*(AAc-co-PEGDA$_{575}$) with a mole ratio of 5:1 and 15:1 before and after argon plasma irradiation and (b) polyAAc (PAAc); Figure S3: ^{1}H-NMR spectra of *net-poly*(AAc-co-PEGDA$_{575}$) at different mole ratio from 0:1 to 15:1 and poly-AAc; Figure S4: Optical microscopy images of the samples *net-poly*(AAc-co-PEGDA$_{575}$) at different mole ratios, NMR data: Signals chemical identification.

Author Contributions: C.M.G.-H.: Conceptualization, methodology, investigation, resources, writing—original draft, supervision, project administration, funding acquisition. F.E.R.-U.: Formal analysis, investigation, visualization. M.N.A.-G.: Formal analysis, investigation, visualization. C.A.T.-I.: Conceptualization, validation, review and editing. E.M.-C.: Methodology, formal analysis, investigation, visualization. R.C.-L.: Formal analysis, investigation, visualization. M.A.S.-V.: Conceptualization, methodology, formal analysis, investigation, writing—original draft, visualization. C.G.-H.: Conceptualization, validation, review and editing. J.R.-H.: Conceptualization, methodology, validation, resources, writing—original draft, project administration, funding acquisition. All authors have read and agreed to the published version of the manuscript.

Funding: This research was funded by FONDECYT Grant N° 1170209. Thanks to FONDEQUIP N° EQM150101 for the use of FE-SEM facilities. J. Rodriguez-Hernandez acknowledges financial support from Ministerio de Ciencia, Innovación y Universidades (Project RTI2018-096328-B-I00). C.M. Gonzalez-Henriquez also acknowledges project L318-02 (UTEM) "Concurso de Fortalecimiento para Equipamiento Científico y Tecnológico, Convocatoria 2018" for plasma exposure system acquisition. M.A. Sarabia and C. García-Herrera acknowledge the financial support given by Universidad de Santiago de Chile through Project DICYT 052116GH_POSTDOC. Finally, this study was funded by VRAC Grant Number LPR20-03 and L318-02 of Universidad Tecnológica Metropolitana.

Institutional Review Board Statement: Not applicable.

Informed Consent Statement: Not applicable.

Data Availability Statement: The data presented in this study are available on request from the corresponding author.

Conflicts of Interest: The authors declare no conflict of interest.

References

1. Yang, P.; Zhu, F.; Zhang, Z.; Cheng, Y.; Wang, Z.; Li, Y. Stimuli-responsive polydopamine-based smart materials. *Chem. Soc. Rev.* **2021**, *50*, 8319–8343. [CrossRef] [PubMed]
2. Mane, S.R.; Sathyan, A.; Shunmugam, R. Biomedical applications of pH-responsive amphiphilic polymer nanoassemblies. *ACS Appl. Nano Mater.* **2020**, *3*, 2104–2117. [CrossRef]
3. Bhat, A.; Amanor-Boadu, J.M.; Guiseppi-Elie, A. Toward impedimetric measurement of acidosis with a pH-responsive hydrogel sensor. *ACS Sens.* **2020**, *5*, 500–509. [CrossRef] [PubMed]
4. Liu, W.; Hu, R.; Li, Y.; Huang, Y.; Wang, Y.; Wei, Z.; Yu, E.; Guo, X. Cross-linking of poly(dimethylaminoethyl methacrylate) by phytic acid: pH-responsive adsorbent for high-efficiency removal of cationic and anionic dyes. *RSC Adv.* **2020**, *10*, 4232–4242. [CrossRef]
5. Guo, Y.; Qian, S.; Wang, L.; Zeng, J.; Miao, R.; Meng, Y.; Jin, Y.; Chen, H.; Wang, B. Reversible antibiotic loading and pH-responsive release from polymer brushes on contact lenses for therapy and prevention of corneal infections. *J. Mater. Chem. B* **2020**, *8*, 10087–10092. [CrossRef]
6. Ma, M.; Zhong, Y.; Jiang, X. Thermosensitive and pH-responsive tannin-containing hydroxypropyl chitin hydrogel with long-lasting antibacterial activity for wound healing. *Carbohydr. Polym.* **2020**, *236*, 116096. [CrossRef] [PubMed]
7. Fu, Y.; Yang, L.; Zhang, J.; Hu, J.; Duan, G.; Liu, X.; Li, Y.; Gu, Z. Polydopamine antibacterial materials. *Mater. Horizons* **2021**, *8*, 1618–1633. [CrossRef] [PubMed]
8. Li, X.; Wu, B.; Chen, H.; Nan, K.; Jin, Y.; Sun, L.; Wang, B. Recent developments in smart antibacterial surfaces to inhibit biofilm formation and bacterial infections. *J. Mater. Chem. B* **2018**, *6*, 4274–4292. [CrossRef] [PubMed]
9. Ko, Y.; Jeong, H.Y.; Kwon, G.; Kim, D.; Lee, C.; You, J. pH-responsive polyaniline/polyethylene glycol composite arrays for colorimetric sensor application. *Sens. Actuators B Chem.* **2020**, *305*, 127447. [CrossRef]
10. Lim, L.S.; Rosli, N.A.; Ahmad, I.; Mat Lazim, A.; Mohd Amin, M.C.I. Synthesis and swelling behavior of pH-sensitive semi-IPN superabsorbent hydrogels based on poly(acrylic acid) reinforced with cellulose nanocrystals. *Nanomaterials* **2017**, *7*, 399. [CrossRef]
11. Chen, Y.; Sun, P. pH-sensitive polyampholyte microgels of poly(acrylic acid-co-vinylamine) as injectable hydrogel for controlled drug release. *Polymers* **2019**, *11*, 285. [CrossRef]
12. Zeinali, E.; Haddadi-Asl, V.; Roghani-Mamaqani, H. Synthesis of dual thermo- and pH-sensitive poly(N-isopropylacrylamide-co-acrylic acid)-grafted cellulose nanocrystals by reversible addition-fragmentation chain transfer polymerization. *J. Biomed. Mater. Res. Part A* **2018**, *106*, 231–243. [CrossRef] [PubMed]
13. Gupta, P.; Purwar, R. Electrospun pH responsive poly (acrylic acid-co- acrylamide) hydrogel nanofibrous mats for drug delivery. *J. Polym. Res.* **2020**, *27*, 296. [CrossRef]
14. González-Henríquez, C.M.; Alfaro-Cerda, P.A.; Veliz-Silva, D.F.; Sarabia-Vallejos, M.A.; Terraza, C.A.; Rodriguez-Hernandez, J. Micro-wrinkled hydrogel patterned surfaces using pH-sensitive monomers. *Appl. Surf. Sci.* **2018**, *457*, 902–913. [CrossRef]
15. Li, Q.; Zhang, P.; Yang, C.; Duan, H.; Hong, W. Switchable adhesion between hydrogels by wrinkling. *Extreme Mech. Lett.* **2021**, *43*, 101193. [CrossRef]
16. Nguyen, D.H.K.; Bazaka, O.; Bazaka, K.; Crawford, R.J.; Ivanova, E.P. Three-dimensional hierarchical wrinkles on polymer films: From chaotic to ordered antimicrobial topographies. *Trends Biotechnol.* **2020**, *38*, 558–571. [CrossRef] [PubMed]
17. Modaresifar, K.; Azizian, S.; Ganjian, M.; Fratila-Apachitei, L.E.; Zadpoor, A.A. Bactericidal effects of nanopatterns: A systematic review. *Acta Biomater.* **2019**, *83*, 29–36. [CrossRef]
18. González-Henríquez, C.M.; Sarabia-Vallejos, M.A.; Terraza, C.A.; del Campo-García, A.; Lopez-Martinez, E.; Cortajarena, A.L.; Casado-Losada, I.; Martínez-Campos, E.; Rodríguez-Hernández, J. Design and fabrication of biocompatible wrinkled hydrogel films with selective antibiofouling properties. *Mater. Sci. Eng. C* **2019**, *97*, 803–812. [CrossRef]
19. Gratzl, G.; Paulik, C.; Hild, S.; Guggenbichler, J.P.; Lackner, M. Antimicrobial activity of poly(acrylic acid) block copolymers. *Mater. Sci. Eng. C* **2014**, *38*, 94–100. [CrossRef]
20. Vargas-Alfredo, N.; Martínez-Campos, E.; Santos-Coquillat, A.; Dorronsoro, A.; Cortajarena, A.L.; del Campo, A.; Rodríguez-Hernández, J. Fabrication of biocompatible and efficient antimicrobial porous polymer surfaces by the Breath Figures approach. *J. Colloid Interface Sci.* **2018**, *513*, 820–830. [CrossRef] [PubMed]
21. Cosgrove, S.E.; Qi, Y.; Kaye, K.S.; Harbarth, S.; Karchmer, A.W.; Carmeli, Y. The Impact of Methicillin Resistance in Staphylococcus aureus Bacteremia on Patient Outcomes: Mortality, Length of Stay, and Hospital Charges. *Infect. Control Hosp. Epidemiol.* **2005**, *26*, 166–174. [CrossRef]
22. Klapetek, P.; Valtr, M.; Nečas, D.; Salyk, O.; Dzik, P. Atomic force microscopy analysis of nanoparticles in non-ideal conditions. *Nanoscale Res. Lett.* **2011**, *6*, 514. [CrossRef]
23. Nečas, D.; Klapetek, P. Gwyddion: An open-source software for SPM data analysis. *Open Phys.* **2012**, *10*, 181–188. [CrossRef]
24. Nirasay, S.; Badia, A.; Leclair, G.; Claverie, J.; Marcotte, I. Polydopamine-Supported Lipid Bilayers. *Materials* **2012**, *5*, 2621–2636. [CrossRef]

25. González-Henríquez, C.M.; Sarabia Vallejos, M.A.; Rodríguez-Hernández, J. Wrinkles obtained by frontal polymerization/vitrification. In *Wrinkled Polymer Surfaces*; Springer International Publishing: Cham, Switzerland, 2019; pp. 63–84.
26. Yuan, Z.; Wang, J.; Wang, Y.; Liu, Q.; Zhong, Y.; Wang, Y.; Li, L.; Lincoln, S.F.; Guo, X. Preparation of a poly(acrylic acid) based hydrogel with fast adsorption rate and high adsorption capacity for the removal of cationic dyes. *RSC Adv.* **2019**, *9*, 21075–21085. [CrossRef]
27. Bhuyan, M.M.; Chandra Dafader, N.; Hara, K.; Okabe, H.; Hidaka, Y.; Rahman, M.M.; Mizanur Rahman Khan, M.; Rahman, N. Synthesis of potato starch-acrylic-acid hydrogels by gamma radiation and their application in dye adsorption. *Int. J. Polym. Sci.* **2016**, *2016*, 1–11. [CrossRef]
28. Tan, G.; Chen, R.; Ning, C.; Zhang, L.; Ruan, X.; Liao, J. Effects of argon plasma treatment on surface characteristic of photopolymerization PEGDA-HEMA hydrogels. *J. Appl. Polym. Sci.* **2012**, *124*, 459–465. [CrossRef]
29. Shaik, M.; Kuniyil, M.; Khan, M.; Ahmad, N.; Al-Warthan, A.; Siddiqui, M.; Adil, S. Modified polyacrylic acid-zinc composites: Synthesis, characterization and biological activity. *Molecules* **2016**, *21*, 292. [CrossRef] [PubMed]
30. Jeon, J.-O.; Baik, J.; An, S.-J.; Jeon, S.-I.; Lee, J.-Y.; Lim, Y.-M.; Park, J.-S. Development and characterization of cross-linked poly(acrylic acid) hydrogel containing drug by radiation-based techniques. *Preprints* **2018**, 2018010028. [CrossRef]
31. Song, P.; Zhang, Y.; Kuang, J. Preparation and characterization of hydrophobically modified polyacrylamide hydrogels by grafting glycidyl methacrylate. *J. Mater. Sci.* **2007**, *42*, 2775–2781. [CrossRef]
32. Huang, Y.; Yu, H.; Xiao, C. pH-sensitive cationic guar gum/poly (acrylic acid) polyelectrolyte hydrogels: Swelling and in vitro drug release. *Carbohydr. Polym.* **2007**, *69*, 774–783. [CrossRef]
33. Maurer, J.J.; Eustace, D.J.; Ratcliffe, C.T. Thermal characterization of poly(acrylic acid). *Macromolecules* **1987**, *20*, 196–202. [CrossRef]
34. Keim, T.; Gall, K. Synthesis, characterization, and cyclic stress-influenced degradation of a poly(ethylene glycol)-based poly(beta-amino ester). *J. Biomed. Mater. Res. Part A* **2010**, *92A*, 702–711. [CrossRef] [PubMed]
35. Yin, J.; Yagüe, J.L.; Eggenspieler, D.; Gleason, K.K.; Boyce, M.C. Deterministic order in surface micro-topologies through sequential wrinkling. *Adv. Mater.* **2012**, *24*, 5441–5446. [CrossRef]
36. Peng, S.; Li, W.; Zhang, J. Diffraction-pattern based on spontaneous wrinkled thin films. *Mater. Trans.* **2017**, *58*, 1–5. [CrossRef]
37. Park, H.-G.; Jeong, H.-C.; Jung, Y.H.; Seo, D.-S. Control of the wrinkle structure on surface-reformed poly(dimethylsiloxane) via ion-beam bombardment. *Sci. Rep.* **2015**, *5*, 12356. [CrossRef] [PubMed]
38. Jeong, H.-C.C.; Park, H.-G.G.; Lee, J.H.; Seo, D.-S.S. Localized ion-beam irradiation-induced wrinkle patterns. *ACS Appl. Mater. Interfaces* **2015**, *7*, 23216–23222. [CrossRef]
39. Takei, A.; Jin, L.; Hutchinson, J.W.; Fujita, H. Ridge localizations and networks in thin films compressed by the incremental release of a large equi-biaxial pre-stretch in the substrate. *Adv. Mater.* **2014**, *26*, 4061–4067. [CrossRef]
40. Huang, X.; Li, B.; Hong, W.; Cao, Y.-P.; Feng, X.-Q. Effects of tension–compression asymmetry on the surface wrinkling of film–substrate systems. *J. Mech. Phys. Solids* **2016**, *94*, 88–104. [CrossRef]
41. Wang, Q.; Zhao, X. A three-dimensional phase diagram of growth-induced surface instabilities. *Sci. Rep.* **2015**, *5*, 8887. [CrossRef]
42. Guvendiren, M.; Burdick, J.A. Stem cell response to spatially and temporally displayed and reversible surface topography. *Adv. Healthc. Mater.* **2013**, *2*, 155–164. [CrossRef]
43. Viswanathan, P.; Guvendiren, M.; Chua, W.; Telerman, S.B.; Liakath-Ali, K.; Burdick, J.A.; Watt, F.M. Mimicking the topography of the epidermal-dermal interface with elastomer substrates. *Integr. Biol.* **2016**, *8*, 21–29. [CrossRef] [PubMed]
44. Guvendiren, M.; Burdick, J.A. The control of stem cell morphology and differentiation by hydrogel surface wrinkles. *Biomaterials* **2010**, *31*, 6511–6518. [CrossRef] [PubMed]
45. González-Henríquez, C.M.; Rodriguez-Umanzor, F.E.; Almagro-Correa, J.; Sarabia-Vallejos, M.A.; Martínez-Campos, E.; Esteban-Lucía, M.; del Campo-García, A.; Rodríguez-Hernández, J. Biocompatible fluorinated wrinkled hydrogel films with antimicrobial activity. *Mater. Sci. Eng. C* **2020**, *114*, 111031. [CrossRef] [PubMed]
46. González-Henríquez, C.M.; Galleguillos-Guzmán, S.C.; Sarabia-Vallejos, M.A.; Santos-Coquillat, A.; Martínez-Campos, E.; Rodríguez-Hernández, J. Microwrinkled pH-sensitive hydrogel films and their role on the cell adhesion/proliferation. *Mater. Sci. Eng. C* **2019**, *103*, 109872. [CrossRef]
47. Quan, C.-Y.; Wei, H.; Sun, Y.-X.; Cheng, S.-X.; Shen, K.; Gu, Z.-W.; Zhang, X.-Z.; Zhuo, R.-X. Polyethyleneimine modified biocompatible poly(N-isopropylacrylamide)-based nanogels for drug delivery. *J. Nanosci. Nanotechnol.* **2008**, *8*, 2377–2384. [CrossRef]
48. Katchalsky, A.; Eisenberg, H. Molecular weight of polyacrylic and polymethacrylic acid. *J. Polym. Sci.* **1951**, *6*, 145–154. [CrossRef]
49. Swift, T.; Paul, N.; Swanson, L.; Katsikogianni, M.; Rimmer, S. Förster resonance energy transfer across interpolymer complexes of poly(acrylic acid) and poly(acrylamide). *Polymer* **2017**, *123*, 10–20. [CrossRef]
50. Michaels, A.S.; Morelos, O. Polyelectrolyte adsorption by kaolinite. *Ind. Eng. Chem.* **1955**, *47*, 1801–1809. [CrossRef]
51. Borozenko, O.; Ou, C.; Skene, W.G.; Giasson, S. Polystyrene-block-poly(acrylic acid) brushes grafted from silica surfaces: pH- and salt-dependent switching studies. *Polym. Chem.* **2014**, *5*, 2242. [CrossRef]
52. Wiśniewska, M.; Chibowski, S. Influence of temperature and purity of polyacrylic acid on its adsorption and surface structures at the ZrO_2/polymer solution interface. *Adsorpt. Sci. Technol.* **2005**, *23*, 655–667. [CrossRef]
53. Aureau, D.; Ozanam, F.; Allongue, P.; Chazalviel, J.-N. The titration of carboxyl-terminated monolayers revisited: In situ calibrated fourier transform infrared study of well-defined monolayers on silicon. *Langmuir* **2008**, *24*, 9440–9448. [CrossRef]

54. Aulich, D.; Hoy, O.; Luzinov, I.; Brücher, M.; Hergenröder, R.; Bittrich, E.; Eichhorn, K.-J.; Uhlmann, P.; Stamm, M.; Esser, N.; et al. In situ studies on the switching behavior of ultrathin poly(acrylic acid) polyelectrolyte brushes in different aqueous environments. *Langmuir* **2010**, *26*, 12926–12932. [CrossRef]
55. Palacios-Cuesta, M.; Cortajarena, A.L.; García, O.; Rodríguez-Hernández, J. Fabrication of functional wrinkled interfaces from polymer blends: Role of the surface functionality on the bacterial adhesion. *Polymers* **2014**, *6*, 2845–2861. [CrossRef]
56. Richter, R.P.; Brisson, A.R. Following the formation of supported lipid bilayers on mica: A study combining AFM, QCM-D, and ellipsometry. *Biophys. J.* **2005**, *88*, 3422–3433. [CrossRef] [PubMed]
57. Borisova, O.V.; Billon, L.; Richter, R.P.; Reimhult, E.; Borisov, O.V. PH- and electro-responsive properties of poly(acrylic acid) and poly(acrylic acid)-block-poly(acrylic acid-grad-styrene) brushes studied by quartz crystal microbalance with dissipation monitoring. *Langmuir* **2015**, *31*, 7684–7694. [CrossRef] [PubMed]
58. Biesalski, M.; Johannsmann, D.; Rühe, J. Synthesis and swelling behavior of a weak polyacid brush. *J. Chem. Phys.* **2002**, *117*, 4988–4994. [CrossRef]
59. Bahram, M.; Mohseni, N.; Moghtader, M. An introduction to hydrogels and some recent applications. In *Emerging Concepts in Analysis and Applications of Hydrogels*; InTech: London, UK, 2016.
60. Lv, Q.; Shen, Y.; Qiu, Y.; Wu, M.; Wang, L. Poly(acrylic acid)/poly(acrylamide) hydrogel adsorbent for removing methylene blue. *J. Appl. Polym. Sci.* **2020**, *137*, 49322. [CrossRef]
61. Jana, S.; Ray, J.; Mondal, B.; Pradhan, S.S.; Tripathy, T. pH responsive adsorption/desorption studies of organic dyes from their aqueous solutions by katira gum-cl-poly(acrylic acid-co-N-vinyl imidazole) hydrogel. *Colloids Surf. A Physicochem. Eng. Asp.* **2018**, *553*, 472–486. [CrossRef]

Article

Characterisation and Modelling of an Artificial Lens Capsule Mimicking Accommodation of Human Eyes

Huidong Wei [1,2,*], James S. Wolffsohn [1,*], Otavio Gomes de Oliveira [2] and Leon N. Davies [1]

- [1] College of Health and Life Sciences, Aston University, Birmingham B4 7ET, UK; l.n.davies@aston.ac.uk
- [2] Rayner Intraocular Lenses Limited, Worthing BN14 8AQ, UK; otaviogomes@rayner.com
- * Correspondence: h.wei1@aston.ac.uk (H.W.); j.s.w.wolffsohn@aston.ac.uk (J.S.W.); Tel.: +44-(0)-121-204-4140 (J.S.W.)

Abstract: A synthetic material of silicone rubber was used to construct an artificial lens capsule (ALC) in order to replicate the biomechanical behaviour of human lens capsule. The silicone rubber was characterised by monotonic and cyclic mechanical tests to reveal its hyper-elastic behaviour under uniaxial tension and simple shear as well as the rate independence. A hyper-elastic constitutive model was calibrated by the testing data and incorporated into finite element analysis (FEA). An experimental setup to simulate eye focusing (accommodation) of ALC was performed to validate the FEA model by evaluating the shape change and reaction force. The characterisation and modelling approach provided an insight into the intrinsic behaviour of materials, addressing the inflating pressure and effective stretch of ALC under the focusing process. The proposed methodology offers a virtual testing environment mimicking human capsules for the variability of dimension and stiffness, which will facilitate the verification of new ophthalmic prototype such as accommodating intraocular lenses (AIOLs).

Keywords: silicone rubber; biomechanical; hyper-elastic; constitutive model; FEA

1. Introduction

The eye's crystalline lens capsule is a membrane with a thickness ranging from 5 to 30 µm [1–3], forming the capsular bag, which encompasses the lens substance. A primary biomechanical function of the in vivo lens capsule is to facilitate the mechanism of ocular accommodation [4]. According to Helmholtz [5], the ciliary muscle contracts when the eye focuses on a near target, relaxing the zonules attached to the lens equator, and enabling the lens surfaces to become more prolate (curved) for increased optical power (accommodation). Conversely, relaxation of the ciliary muscle, to view distant objects, causes centrifugal tensioning of the zonules and a corresponding flattening the lens, leading to decreased optical power (dis-accommodation). During the accommodating process, given its high Young's Modulus (of 1 MPa) compared to the lens substance (of 1 Pa), the capsular bag is able to mould the shape of the internal lens [6,7]. It has been found that the biomechanical properties of lens capsule are relatively less affected by age, except the geometric inhomogeneity of the thickness distribution [3,8].

The subjective visual difficulties experienced with near vision by increasing age, known as presbyopia, is primarily attributed to the reduction in accommodation due to the rising stiffness of lens substance [9,10]. A potential treatment of presbyopia is the replacement of lens substance by functional artificial lenses, such as mechanically accommodating intraocular lenses or lenses filled with polymers [11–13]. These techniques, however, rely on the retained biomechanical function of the capsular bag, which is compromised during cataract removal and lens implantation surgery, to restore some or all of the accommodative process. When the lens substance is removed from the capsular bag similar to the surgery, it shows minor difference of the reaction force with the lens containing substance, addressing the primary biomechanical function of lens capsule [14].

The accommodative function of the crystalline lens has been studied with ex vivo testing and numerical modelling [15]. The shape change of lenses during accommodation in vivo can be replicated by inducing radial stretch using a lens stretching device [16–18]. The deformation and reaction force of lenses by such devices have been used to investigate the aetiology of presbyopia by combining optical power and biometry measurements [19,20]. Further, the biomechanical behaviour of lenses can be better understood by constructing finite element models [10,15,21–24]. Although the mechanical behaviour of the lens capsule itself has been studied by performing uniaxial or inflation tests [1,6], few studies directly investigate the behaviour of the capsule under accommodative forces. For the development of ophthalmic products treating presbyopia, it is essential to test any prototype inside the capsule to characterise how it will perform under different conditions; however, the access to human tissues is limited and animal capsules have different biomechanical properties compared to that of human capsules [6,14].

Thus, this study aimed to generate an alternative product able to mimic the biomechanical performance of human capsules, where a silicone rubber can be used to replicate the hyper-elastic response of human capsule [25]. The base material was characterised experimentally by mechanical tests at different conditions. A constitutive model was subsequently calibrated using the testing data and incorporated into a finite element analysis (FEA) model to simulate an accommodating test of an artificial capsule made from the same material. The characterisation together with the FEA modelling provided an insight into the intrinsic behaviour of materials and the mechanism of accommodation.

2. Materials and Methods

2.1. Material Preparation

An industrial-grade, room temperature vulcanised (RTV) silicone (Model: ZA13, Zhermack, Rovigo, Italy) was selected to be the base material. The silicone rubber had a low Shore A Hardness (S) (of 13) after curing with an equivalent Young's modulus (E) of 0.5 MPa by Equation (1) [26,27], which was close to the lowest boundary of stiffness of human lens capsules ex vivo [1,3]. The material had a base fluid A and catalyst fluid B, which was mixed thoroughly at a ratio of 1:1. The catalyst fluid activated the condensation of the base fluid, allowing for a working time of 40 to 50 min at 23 °C. According to the operating instructions from the supplier, a vacuum was applied to the mixed fluid for 3 min to eliminate any air pockets. The degassed fluid was carefully poured into a mould and set for 24 h at room temperature (of 23 °C) to form a solid. Two types of strips were prepared for the mechanical characterisation, with dimensions of 6 mm × 25 mm × 10 mm (strip A) and 100 mm × 25 mm × 2 mm (strip B), respectively (Figure 1a). The moulded strips had stable shapes (Figure 1b) with a tolerance thickness of ± 0.08 mm.

$$E = \frac{0.0981(56 + 7.66S)}{0.137505(254 - 2.54S)}, \quad (1)$$

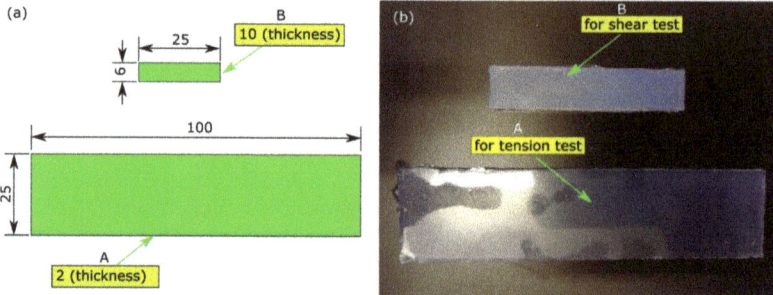

Figure 1. Testing specimens for characterisation: (**a**) dimensions (unit: mm); (**b**) moulded strips.

2.2. Mechanical Characterisation

A universal mechanical testing machine (Model: 5942, Instron, Norwood, MA, USA) was used for material characterisation, which had a maximum loading capacity of 500 N, maximum extension of 488 mm, and speed between 0.05 and 2500 mm/min. The strip A was installed onto the machine for the uniaxial tension test, which was fixed at the bottom and elongated on the top by two respective grips (Figure 2a). An initial separation of 75 mm was assigned between the two grips as the gauge length. The testing protocols were defined by an integrated software Bullhill (Version: 3.0, Instron, Norwood, MA, USA). A monotonic tension (MT) was defined by applying a displacement of 90 mm at 100 mm/min, resulting in a nominal strain of 1.2 at $0.02\ s^{-1}$. After the first primary loading (PL), a sagging state (of strip A) was observed when the sample returned to the initial separation with zero displacement (Figure 2b). This implied a negative force of the specimen under displacement-controlled deformation, recognised as the property change of rubber materials at repeated loading (Mullins effect). The process was repeated by performing reloading (RL) on the same specimen five more times.

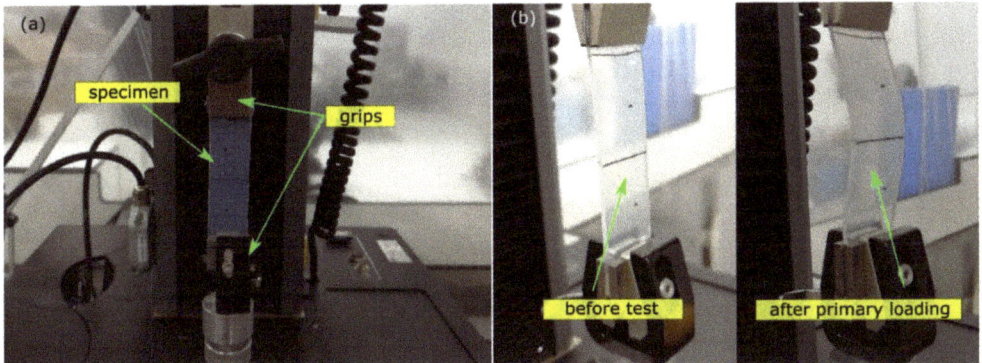

Figure 2. Uniaxial tension test: (**a**) testing setup; (**b**) shapes of specimen before test and after primary loading.

After the monotonic test, a cyclic tension (CT) testing was performed on a new specimen by six cycles of loading and reloading operation at 100 mm/min. Four amplitudes of displacement (15, 37.5, 75, and 90 mm) were defined during loading, aiming for a final nominal strain of 0.2, 0.5, 1.0 and 1.2, respectively (at $0.02\ s^{-1}$). A force-controlled unloading was defined to avoid sagging by halting the top grip when the reaction force was zero. In addition to the cyclic test at low speed (of 100 mm/min), another cyclic test was performed on a new specimen by increasing the speed to 1000 mm/min, which provided a condition with 10 times the strain rate (of $0.2\ s^{-1}$). The repeatability of the material behaviour was checked by comparing the results at the different testing conditions.

Strip B was used for a simple shear test on the testing machine by designing customised fixtures (Figure 3a). Two long aluminium base bars were fixed by two clamps on the machine, offset by a gap of 6 mm, to allow the specimen to be installed along the elongation axis. It was bonded onto the surfaces of the bases by adhesive made of Ethyl 2-cyanoacrylate (Brand: Loctite control, Henkel, Winsford, UK). This application introduced a single lapped in-plane shear test (Figure 3b). By using the software (Bullhill, Version: 3.0), a monotonic shear (MS) test was defined by assigning a displacement of 8.4 mm at 8 mm/min (with gauge length of 6 mm), creating a maximum shear strain of 1.4 at $0.02\ s^{-1}$. Similar to the tension test, one primary loading (PL) and five reloading (RL) operations were applied on the same specimen for MS. The cyclic shear (CS) test by force-controlled protocol was conducted on new specimens by applying four amplitudes of displacement (1.2, 3.0, 6.0, and 8.4 mm) at 8 mm/min to achieve different levels of maximum shear strain (of 0.2, 0.5, 1.0 and 1.4).

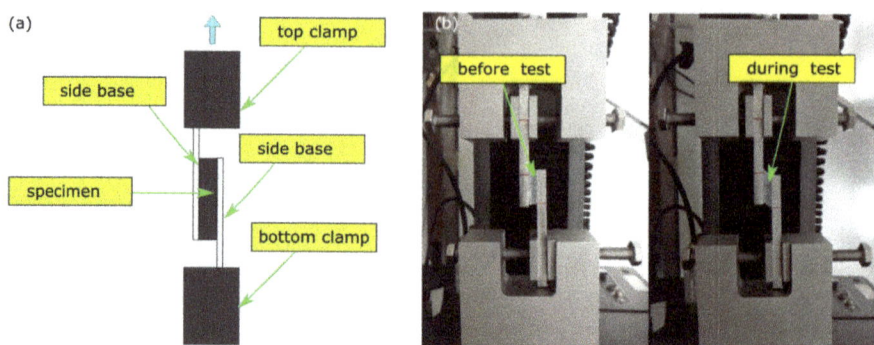

Figure 3. Simple shear test: (**a**) schematic testing setup; (**b**) shapes of specimen before and during test.

2.3. Accommodating Test

An artificial lens capsule (ALC) was manufactured by using the same batch of silicone material and moulding process (Figure 4a) [28]. The ALC had an anterior and posterior half with an average thickness of 150 ± 40 μm measured by calliper. It had an ellipsoidal shape with a radius of 4.5 mm, 1.6 mm, and 2.55 mm along the equator, anterior sagitta and posterior sagitta, respectively. An extension ring with a width of 0.5 mm and thickness of 1 mm was fitted around the equator to allow the ALC to be fixed onto a support structure, which has eight branches equally distributed (45° angle) and mounted on a lens radial stretching system (LRSS) (Figure 4b) [28]. Each branch had independent radial motion by performing a radial cut of the joint region. The LRSS provided stretch and release of each branch simultaneously radially to mimic the action of the ciliary muscles and adjoining zonules on the capsule. The equatorial stretch of the ALC was calibrated at two linear nominal speeds (NS) of 0.5 mm/s and 0.05 mm/s, driven by a stepper motor. A duration of 2.7 s and 27 s was needed, respectively, to offer a diameter change of 0.9 mm, which was similar to the displacement observed in vivo to a 10 D accommodation stimulus [29,30]. The ALC was filled with ophthalmic viscosurgical devices (OVDs) after it was mounted to the LRSS, which helped to recover and maintain the initial shape of the ALC under pressurisation. Preliminary cyclic testing was performed on the ALC at a high speed (NS = 0.5 mm/s) and then on the same ALC at a low speed (NS = 0.05 mm/s). The shape change of the ALC was monitored from the side view by a digital microscope camera (Model: Cmex18pro, Euromex, Arnhem, The Netherland) with a resolution of 20-million pixels. A load cell was installed on one of the arms of the LRSS to measure the reaction force of 3 tests at each condition. The stretching force was derived by extracting the nonlinear force contributed other than the ALC.

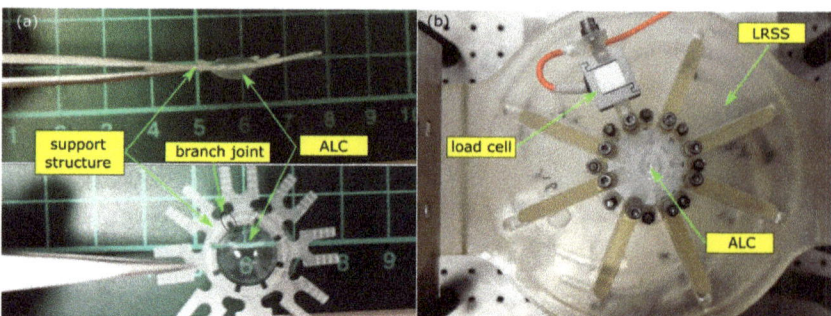

Figure 4. Accommodating test: (**a**) an artificial lens capsule (ALC) with support structure; (**b**) ALC on the lens radial stretching system (LRSS).

2.4. Constitutive Modelling

An isotropic polynomial incompressible strain function was used to model the constitutive behaviour of material (Equation (2)). The initial shear modulus (μ) can be estimated (Equation (3)) and the equivalent Young's modulus (E) can be deduced based on the incompressibility of materials (Equation (4)).

$$U = \sum_{i+j=1}^{N} C_{ij}(\bar{I}_1 - 3)^i (\bar{I}_2 - 3)^j + \sum_{i=1}^{N} \frac{1}{D_i}(J-1)^{2i}, \quad (2)$$

$$\mu = 2(C_{10} + C_{01}), \quad (3)$$

$$E \approx 3\mu, \quad (4)$$

where J is the determinant of the deformation gradient (**F**) denoting the volume change (Equation (5)), $\bar{I}_1$, $\bar{I}_2$ is the deviatoric first and second invariant of the right Cauchy-Green tensor (**C**) under finite deformation (Equations (6)–(8)).

$$J = \det(\mathbf{F}), \quad (5)$$

$$\mathbf{C} = \mathbf{F}^T \mathbf{F}, \quad (6)$$

$$\bar{I}_1 = J^{-1/3} \operatorname{tr}(\mathbf{C}), \quad (7)$$

$$\bar{I}_2 = \frac{1}{2} J^{-1/3} \left[(\operatorname{tr}(\mathbf{C}))^2 - \operatorname{tr}(\mathbf{C}^2) \right], \quad (8)$$

Three parameters (C_{10}, C_{01}, C_{20}) were calibrated in the constitutive model by tension and shear test data with the assumption of incompressibility ($J \equiv 1$). The calibration was based on a theoretical modelling of the deformation [31–33]. A least square approach (Equation (9)) was used to acquire the best fit of the stress-strain relationship by minimising the stress difference (R^2) between test data (σ_i^t) and modelling result (σ_i^m) under the same strain. The data from PL were used to fit the model across the whole strain level. To incorporate the Mullins effect, the cyclic RL data were used to fit another group of material parameters. As a small strain (of less than 0.14) was expected during accommodation [34], the RL data within the strain of 0.2 was employed and the residual strain was shifted to zero during calibration.

$$R^2 = \sum_{i=1}^{N} (\sigma_i^t - \sigma_i^m)^2, \quad (9)$$

A one-fourth geometric model of the artificial capsule surface was constructed for FEA by 3-node shell elements (S3) with an open-source meshing software (Salome, Version: 8.3.0, CEA & EDF, Paris, France) (Figure 5a). To represent the radial cutting of the joints, two groups of repeated nodes sharing identical coordinates were built, belonging to different elements of each side, which allowed the separation of each side under circumferential tension. Symmetric boundary conditions were defined at the edges of the plane along X = 0 and Z = 0. Loads were applied by controlling the relative amplitudes to replicate the accommodating test (Figure 5b). Within a time-amplitude of 0.1, a linear pressure load was applied on the interior surface of the capsule to simulate the application of OVDs. A trial-and-error of pressure value was conducted to yield a similar profile of the ALC to that measured in the experiment. A linear displacement from time-amplitude of 0.1 to 1 was provided along the outer equatorial edge to simulate the radial stretch. A static analysis was performed by an open-source FEA solver (CalculiX, Version: 2.17, Friedrichshafen, Germany) [35]. The total reaction force of the nodes between two radials cutting positions (of 45° angle) was exported and the stretching force was derived by the relative values between the time amplitude of 0.1 and 1.

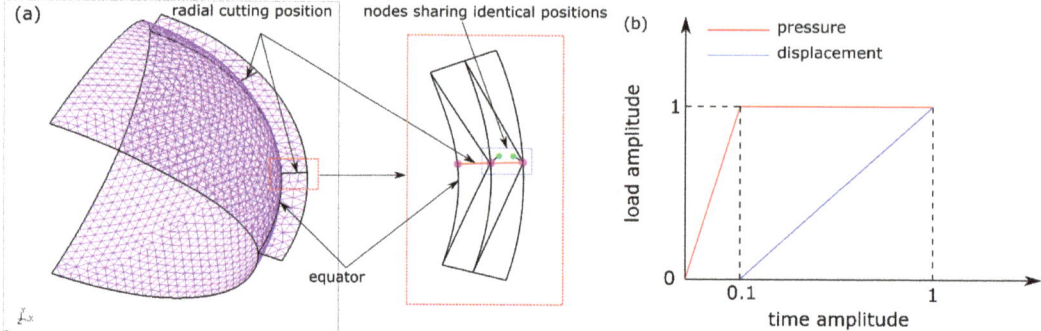

Figure 5. Modelling on accommodating test of ALC: (**a**) FE model; (**b**) loading step.

3. Results

3.1. Monotonic Test

The mechanical behaviour of the silicone material was displayed by the stress-strain relationship under MT and MS test at 0.02 s^{-1} (Figure 6). At primary loading (MT_PL) (Figure 6a), the virgin material showed an evident hyper-elastic behaviour by a linear curve at low strain regime (< 0.6) and nonlinear curve beyond. Among the reloading process (MT_RL), there was an offset of stress-strain path with that of MT_PL, which is known as the Mullins effect of rubbers, implying the stress softening due to the internal damage of materials [36]. This corresponded to the sagging of the specimen where the initial motion of grip was to overcome the sagging with zero reaction force. Under MS at 0.02 s^{-1} (Figure 6b), the nonlinearity of stress-strain relationship was more evident during RL than that of PL. The offset between PL and RL indicated the existence of Mullins effect, whilst the residual strain was observed to be negligible.

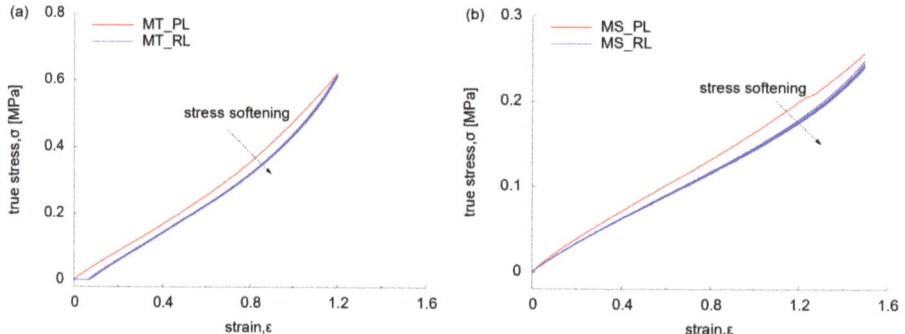

Figure 6. Stress-strain relationship under monotonic test: (**a**) at uniaxial tension (0.02 s^{-1}); (**b**) at simple shear (0.02 s^{-1}).

3.2. Cyclic Test

The Mullins effect was further displayed by the result of cyclic tension (CT) and shear (CS) tests, with variable softening behaviour at different strain amplitude (Figure 7). It was found that the stress-strain relationship of MT_PL and MS_PL was along the response of the first primary loading (PL) path of each amplitude for both CT (Figure 7a) and CS tests (Figure 7b), by minor deviations. By increasing the strain amplitude, the stress softening behaviour became more evident, with elevated residual strain at zero stress. At high strain amplitude (of over 1.0), there was a slightly different mechanical response of materials along the reloading and unloading paths. The influence of strain rate was revealed by comparing the behaviour of material at two strain rates (of 0.02 and 0.2 s^{-1}) under CT

(Figure 7c). There was no marked difference on the stress-strain relationship along the PL and RL path by similar overall hyper-elasticity and stress softening behaviour, which indicated the irrelevance of mechanical behaviour to strain rate.

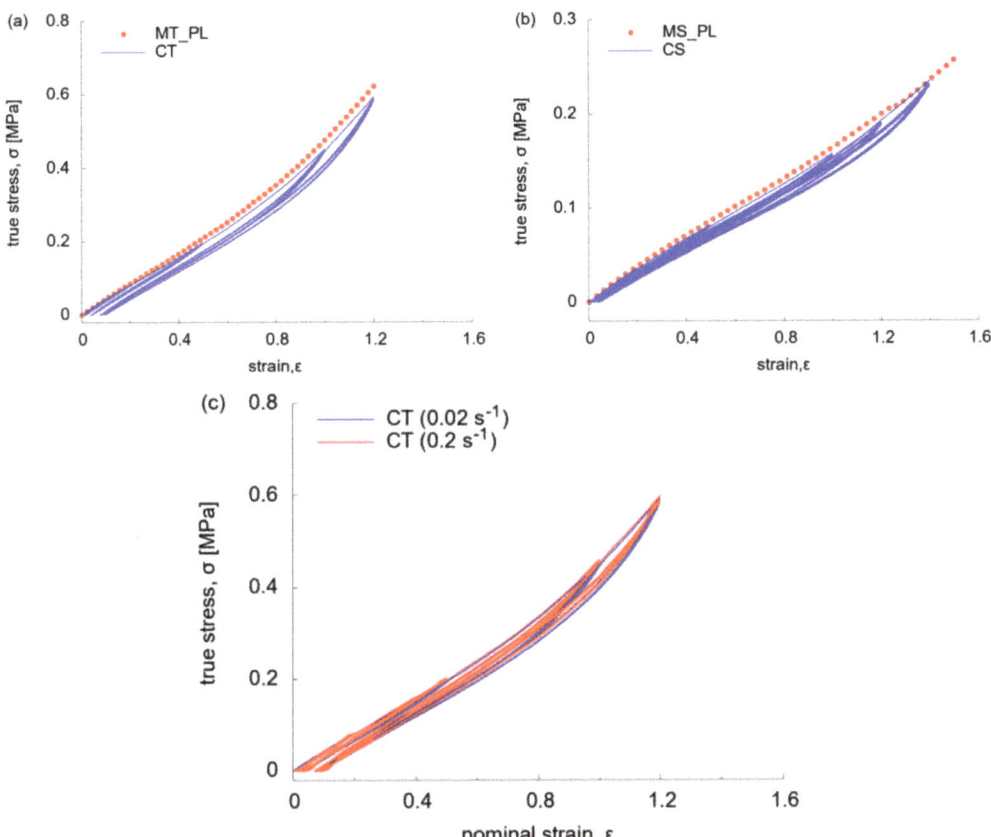

Figure 7. Stress-strain relationship under cyclic test: (**a**) at uniaxial tension (0.02 s^{-1}); (**b**) at simple shear (0.02 s^{-1}); (**c**) at uniaxial tension (0.02 s^{-1} vs. 0.2 s^{-1}).

3.3. Modelling and Simulation

The material parameters of the constitutive model (C_{10}, C_{01} and C_{20} in Equation (2)) were calibrated by the data from monotonic and cyclic test along primary-loading (PL) and reloading (RL) paths (Table 1). The shear moduli (μ) and equivalent Young's moduli (E) were calculated to obtain the initial stiffness of materials (Equations (3) and (4)). It was observed that the material at RL exhibited higher stiffness than that of PL; this was due to Mullins effect and more accurate fitting of the data points at small strain regime (of less than 0.2). The Young's modulus at PL and RL was around 0.45 and 0.49 MPa, respectively, which was in accordance with the value (of 0.5 MPa) estimated by the hardness of material (of Shore A Hardness 13).

Table 1. Material parameters of the constitutive model (unit: MPa).

	C_{10}	C_{01}	C_{20}	μ	E
PL	2.426×10^{-2}	5.015×10^{-2}	3.791×10^{-3}	1.488×10^{-1}	4.464×10^{-1}
RL	5.000×10^{-8}	8.160×10^{-2}	5.000×10^{-6}	1.632×10^{-1}	4.896×10^{-1}

The performance of the constitutive model was displayed by comparing it to the experimental test (Figure 8). Along the primary loading path of monotonic uniaxial tension (MT_PL) and simple shear (MS_PL) at 0.02 s^{-1} (Figure 8a), there was a good consistency between experiment and modelling over the whole strain regime; this indicated the uniqueness of material parameters and their applicability for different modes of deformation. By using the reloading data under cyclic testing, the modelling results with two groups of material parameters (PL and RL) were compared with the experimental test (Figure 8b). Within an initial small strain (of 0.02), there was no evident diversity between the two models with PL and RL. As the strain increased, a slightly steeper stress-strain relationship was identified by the modelling with RL, which captured the material behaviour better. The contrast indicated the suitability of using modelling with PL only for virgin material, but with RL for the material after the initial stretch.

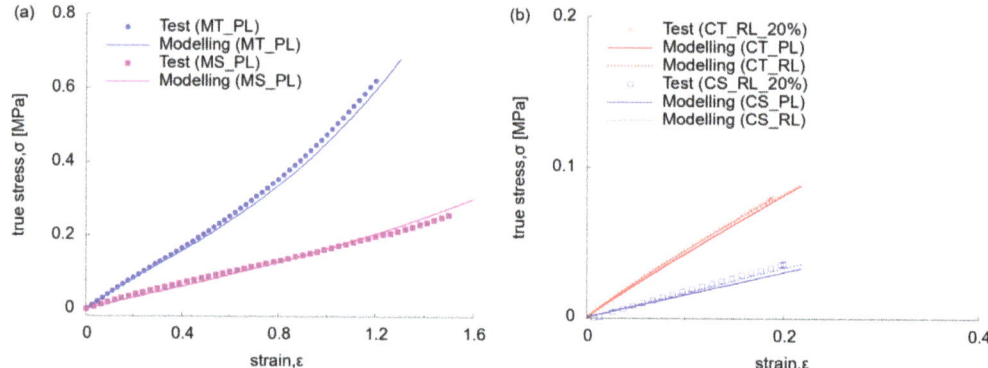

Figure 8. Comparison of stress-strain relationship between experiment and modelling: (**a**) primary loading under tension and shear test (at 0.02 s^{-1}); (**b**) reloading under tension and shear (strain < 0.2 at 0.02 s^{-1}).

In FEA, a large bulk modulus (K = 150 MPa) was employed to satisfy the incompressibility, which was 1000 times the shear modulus (of 0.15 MPa), resulting in an equivalent Poisson's ratio (of 0.4995) close to 0.5. An internal pressure of 2.0 kPa was found by trial and error to achieve a similar profile of the ALC pressurised with OVDs. The shape changes of ALC by FEA_PL and FEA_RL were illustrated by comparing the initial profile (mesh) and final profile (surface) (Figure 9a,b). By applying radial stretch, an increase of equatorial diameter was found, resulting in a decrease of sagittal distance at anterior and posterior half. A homogeneous strain area covering a big region of the anterior and posterior surface, indicated an approximate strain magnitude of 0.15 (of anterior half) and 0.065 (of posterior half). There was a marked stress-intensive region near the cutting line of the joints, with a strain magnitude of 0.2. There was no evident difference in the deformation between FEA_PL and FEA_RL, which was primarily due to the incompressibility of the material under displacement-controlled deformation.

As there was similar shape change by FEA, the historic profile of the ALC by FEA_RL was compared to the experiment tests (Figure 9c,d). During the preliminary test of the ALC at high speed (NS = 0.5 mm/s) (Figure 9c), the initial shape of the ALC was very well represented by FEA, implying the pressure load applied to the modelling was suitable to simulate the application of OVDs. Comparatively, there was less inflating state of the ALC with secondary testing at low speed (NS = 0.05 mm/s) than with FEA (Figure 9d), which could be attributed to the pressure loss after the preliminary cyclic tests. Under the displacement history of the LRSS, there was a minor change of profile at the low displacement (of 0.1 mm) at the initial stretch (t < 1.5 s at NS = 0.5 mm/s and t < 15 s at NS = 0.05 mm/s) between the experiment and FEA. The subsequent stretching process from the experiment was well predicted by FEA under equivalent stretch, implying similar anterior and posterior curvatures and sagittal distance.

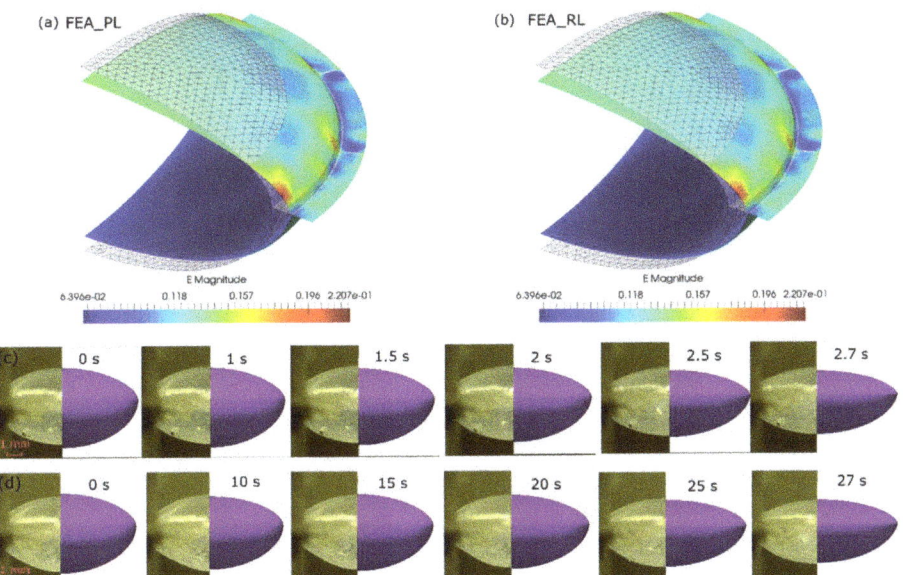

Figure 9. Shape change of ALC under accommodating: (**a**) initial shape and strain profile of deformed shape by FEA_PL; (**b**) initial shape and strain profile of deformed shape by FEA_RL; (**c**) comparison between experiment and FEA_RL (NS = 0.5 mm/s); (**d**) comparison between experiment and FEA_RL (NS = 0.05 mm/s).

The shape change of the ALC was further compared by fitting the initial and final profile with circular arcs to acquire the curvature radius of anterior (R_A) and posterior (R_P) surfaces and sagittal distance (Figure 10a,b). In the FEA modelling, the curvature radius before stretch was found to be R_A = 6.9 mm and R_P = −5.3 mm with a sagittal distance of 4.9 mm. Compared to the experimental test at NS = 0.5 mm/s (Figure 10a), a curvature radius of R_A = 7.3 mm and R_P = −5.4 mm with a sagittal distance of 4.7 mm was observed, implying a good correspondence of curvatures and minor over-inflation at the posterior half (Figure 10a). The pressure loss of the ALC during experimental testing at NS = 0.05 mm/s was exhibited by the initially lower inflation compared to FEA, revealing a result of R_A = 7.0 mm and R_P = −4.8 mm with a sagittal distance of 4.6 mm (Figure 10b). After stretch, FEA predicted a profile of R_A = 10.7 mm (ΔR_A = 3.8 mm) and R_P = −6.9 mm (ΔR_P = 1.6 mm), with a sagittal distance of 3.9 mm (Δ = 1 mm). The experimental result after stretch at NS = 0.5 mm/s was ΔR_A = 2.1 mm, ΔR_P = 1.0 mm and decrease of sagittal distance of 0.6 mm, indicating a smaller change than with FEA modelling (Figure 10a). At the secondary test conducted at NS = 0.05 mm/s (Figure 10b), the stretch introduced a profile change of ΔR_A = 2.3 mm, ΔR_P = 0.9 mm and a decrease of sagittal distance of 0.7 mm, resulting in a profile closer to the FEA modelling.

Despite the similar change of profile of the ALC by FEA_PL and FEA_RL, there were different reaction forces from FEA simulation. At high speed (NS = 0.5 mm/s) (Figure 10c), both FEA modelling with PL and RL indicated a similar linear reaction force versus the displacement to the testing data from three repeats. Within a displacement of 0.2 mm, the deviations between FEA_PL and FEA_RL were very small. At higher stretch, there was more correspondence with the testing data by FEA_RL, implying better applicability of the material parameters. In the secondary test conducted at NS = 0.05 mm/s (Figure 10d), a small deviation was observed between test and modelling, within a small displacement of 0.1 mm, due to a higher reaction force from the experimental test. The steeper curve indicated a stronger force than that at a high speed and FEA modelling, regardless of the rate independence of the material. This was probably attributable to the status of the ALC with pressure loss, introducing bigger turbulence force from the shape change under radial stretch. As the displacement reached 0.2, a decaying reaction force was observed during the testing process, indicating the existence of relaxation that was not incorporated in FEA, leading to an apparently higher reaction force than in the experimental test.

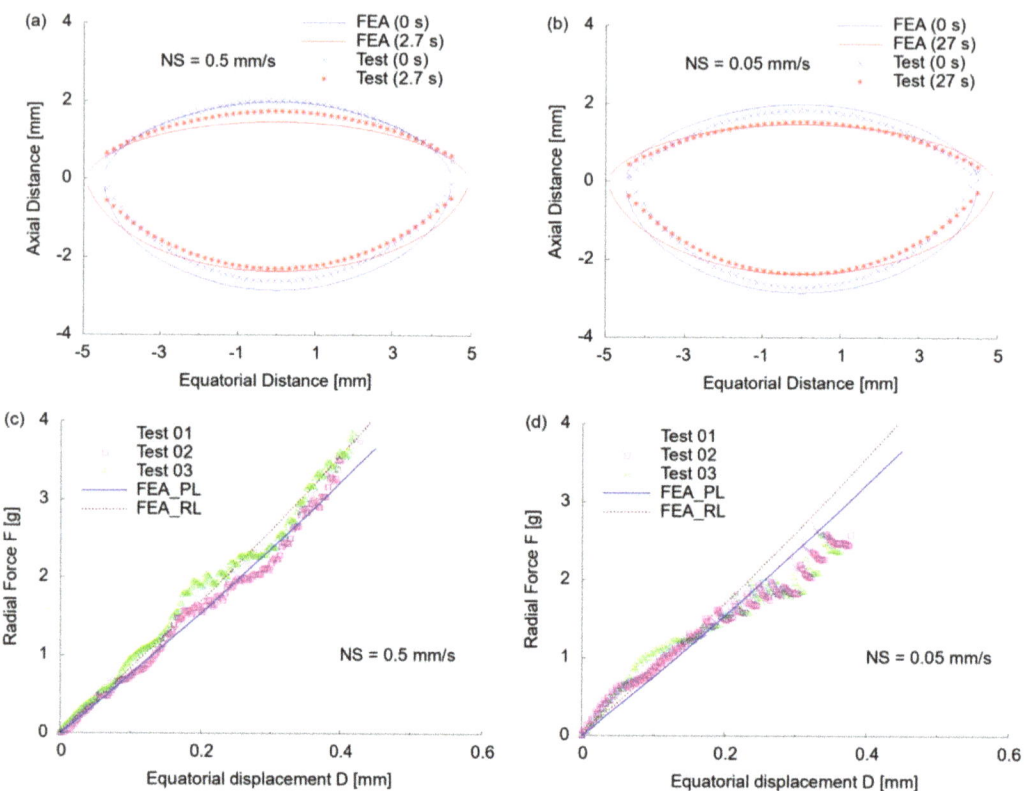

Figure 10. Comparison of profile and reaction force of ALC between experiment and FEA modelling: (**a**) change of profile at high speed (NS = 0.5 mm/s); (**b**) change of profile at low speed (NS = 0.05 mm/s); (**c**) radial stretching force at high speed (NS = 0.5 mm/s); (**d**) radial stretching force at low speed (NS = 0.05 mm/s).

4. Discussion

Under small strain regime (of less than 10%), the mechanical response of the human capsule is usually assumed to be linear-elastic with a Young's modulus of 0.7 to 1.5 MPa [37,38]. This assumption conformed to the results (of 0.4 to 1.5 MPa) under uniaxial tension of the human's capsule ex vivo [1,3,39]. The silicone rubber for the artificial capsule was selected with a stiffness equivalent to a modulus (of 0.5 MPa), close to the lowest limit of human capsule under uniaxial tension. Different values beyond this range have been reported with higher values (of 2.0 to 8.0 MPa) and lower values (of 0.03 to 0.3 MPa) [2,40]. Although the liner-elastic model provided similar results to the hyper-elastic model for the inflation test of capsule [37,38], there was a large disparity with the modelling of uniaxial tension [41]. This highlighted the need for characterising a material by using different modes of deformation. The mechanical nonlinearity of the human capsule turned more significant with a high strain regime [1,3,39], where hyper-elastic models became more appropriate [15,24,42]. More complex behaviours of the biological capsules, such as stress softening, viscoelasticity (creep, relaxation) and anisotropy, exist at different loading conditions [1,43–49]. The stress softening, i.e., Mullins effect, was observed for silicone rubber under cyclic loading, and was more evident at higher strain levels [36,50]. A complete understanding of this behaviour relies on more comprehensive testing and modelling [51,52], not conducted in the current study but was simplified by using two groups of material parameters to take account of the reloading data. The viscoelasticity of silicone rubber, such as the rate dependence, was reported to be negligible [53,54], which was proved by the uniaxial tension test at different speeds.

The ALC in free state had an equatorial radius (of 4.5 mm) close to the biometry of the human capsule at accommodated state (of 4.5 to 5.0 mm) [10,14,19,55,56]. There was a definite sagittal distance of the FEA model (of 4.9 mm) similar to an aged biological capsule (of 4.8 mm at 45 years old) [14,19,55]. These dimensions have been widely employed for FEA modelling of accommodation [37,38,57]. A big difference of spatial thickness existed between the ALC and the human capsule. A homogenous membrane with an average thickness (of 150 µm) was employed in the FEA model, whilst the human capsule has a much smaller thickness (of 5 to 30 µm) and large inhomogeneity [1,8,25]. This relatively large thickness of the ALC was beneficial for the moulding operation and prevention of collapse [28]. In the accommodating simulation, Helmholtz's theory was observed to change the shape of capsule of the zonules, which was driven by the ciliary muscle [58]. The zonules have been found to locate on the anterior, equatorial, and posterior region of the capsule by inhomogeneous distribution and implied the dependence on age [59,60]. These properties can be defined with different arrangement and stiffness in the modelling [15,34,37,38,61,62], where the influence was found to be significant for displacement-controlled condition [37,61], but negligible for force-controlled condition [38]. Compared to the zonule distribution of a biological model, the ALC model was simplified significantly by placing the attachments around the equatorial region with a similar width to that found in the human lens (of 0.2 to 0.4 mm) [15,37,38]. The stiffness was strongly enhanced by using a relatively large thickness (of 1 mm) to transfer the radial stretch more effectively whilst the circumferential force was eliminated by performing radial cuts as suggested by other studies [17,18,63].

The accommodation of eyes can be modelled numerically by using the capsule only with the application of pressure from lens substance [7,24,42,62]. The relative pressure from the lens substance was defined to be 4 to 6 mmHg (of 0.5 to 0.8 kPa) under accommodative stimulation [24,42], whilst the resisting pressure of the lens capsule can reach as high as 40 mmHg (of 5.0 kPa) [48]. A high pressure value (of 2.0 kPa) of the ALC was deduced by a reverse approach with FEA to be compensated for its high thickness (of 150 µm) in contrast to the human capsule (of 5 to 30 µm), providing the insight into the over-inflation status in a previous study [28]. Compared to the studies using a flat surface as the initial shape of the human capsule, the reference state of the ALC was defined to be the moulded shape, significantly diminishing the pretension effect of materials before exerting stretch. The curvature radius of the pressurised ALC in the FEA model (R_A = 6.9 mm, R_P = −5.3 mm) was in accordance to the human capsule of young age (of 19 to 29 years) [10,21,22]. The modelling resulted in the surface of posterior capsule after stretch showing a curvature radius (R_P = −6.9 mm) consistent with the in vivo study and simulation [10,15,21–24]. The deformed surface of the anterior capsule (R_A = 10.7 mm) had a curvature radius between that reported for young (of 12.0 mm) and old (of 9.0 to 9.8 mm) lenses [19,64]. To achieve the change on curvature of radius aforementioned and sagittal distance (of 1 mm) for the ALC, a radial displacement (of 0.5 mm) was exerted onto the equator in the experimental test and FEA simulation. In contrast, a higher radial stretch (of 2 mm) was employed to the rigid part in the ex vivo stretching test of human lenses [14,17,20,55,65]. An evident accommodative power change has been observed by radial stretch (of 0.8 to 1.0 mm) on the ciliary body [66]. MRI studies have shown the adjustment of lens diameter is between 0.3 and 0.6 mm and lens thickness between 0.1 and 0.6 mm under binocular accommodative stimulus conditions (of 0.1 to 8.0 D) [19,21,29,64]. These findings have been replicated by the modelling of accommodation with FEA [15,37,38,57,62,67,68], the approach of which was employed in the presented FEA modelling and further advanced by the accompanied validation from experimental testing.

The contraction force was a resulting parameter for the displacement-controlled deformation driven by the radial stretch [37,57,62,68], which can also be explicitly defined as the stimulus for load-controlled deformation [15,24,38,67]. Experimental and numerical studies have shown a single or multiple linear relationship between the stretching force and radial stretch [14,17–19,37,68]. The multilinear correlation was attributed to the influence of testing environmental factors (mobilisation, pretension, and dynamic effects) during the stretching process by a previous study [28], which was further validated by the linearity and accordance of FEA by extracting the force contribution only from the stretch of the ALC. The overall contraction force of the human lens ranged from 3 g to 13 g as assessed by ex vivo stretching tests [14,17–19,69], which is slightly reduced by the extraction of lens substance [14,18]. Most numerical studies showed a similar stretching force over a wide range (of 2 to 14 g) [15,24,37,38,57,62,67,68]. A larger stretching force (of 20 to 30 g) required for the ALC from experiment and FEA, was primarily due to the higher thickness of ALC (of 150 μm) compared to the human lens capsule (of 5 to 30 μm) [1,2,6], which needs to be further reduced in order to provide a smaller average overall stretching force.

5. Conclusions

An artificial lens capsule (ALC) based on soft silicone has been developed to replicate the mechanical behaviour of lens capsule during the accommodating process. A hyperelastic behaviour of the base material together with stress softening (Mullins effect) and rate independence were demonstrated by mechanical characterisation under different modes of deformation. The constitutive model was calibrated using experiments, which captured the mechanical behaviour by numerical modelling. The accommodating test of the ALC was successfully modelled by using FEA incorporating the constitutive model. The deformation of the ALC showed convincing similarity to the human capsule by both experimental testing and modelling. The applicability of the FEA modelling was highlighted by presenting the correspondence of the deformation; this can be used to adjust the thickness to correct for the discrepancy of the contraction force of the ALC due to its manufactured thickness profile. The development of the ALC with relevant mechanical behaviour will facilitate the examination of ophthalmic implants for overcoming presbyopia implanted within capsular bag, such as accommodating intraocular lenses (AIOLs). The FEA modelling provides a convenient approach to examine and validate the function of implants within the human capsular bag of variable stiffness and geometry.

Author Contributions: Conceptualization, H.W., J.S.W., O.G.d.O. and L.N.D.; methodology, H.W., J.S.W., O.G.d.O. and L.N.D.; software, H.W.; validation, H.W., O.G.d.O. and J.S.W.; formal analysis, H.W.; investigation, H.W.; resources, J.S.W. and O.G.d.O.; data curation, H.W.; writing—original draft preparation, H.W.; writing—review and editing, J.S.W., O.G.d.O. and L.N.D.; visualization, L.N.D.; supervision, J.S.W.; project administration, L.N.D.; funding acquisition, J.S.W. and L.N.D. All authors have read and agreed to the published version of the manuscript.

Funding: This research was funded by INNOVATE UK under the Knowledge Transfer Partnership Project between Aston University and Rayner Intraocular Lenses Limited, grant number 11718. L.N.D. was funded, in part, by a research grant (27616) from the College of Optometrists, UK.

Institutional Review Board Statement: Not applicable.

Informed Consent Statement: Not applicable.

Acknowledgments: Luiz Melk, Matt Clayton, and Dario Vecchi from Rayner are acknowledged for the support in sample preparation and testing.

Conflicts of Interest: The authors declare no conflict of interest.

References

1. Krag, S.; Andreassen, T.T. Mechanical properties of the human lens capsule. *Prog. Retin. Eye Res.* **2003**, *22*, 749–767. [CrossRef]
2. Fisher, R.F. Elastic constants of the human lens capsule. *J. Physiol.* **1969**, *201*, 1–19. [CrossRef] [PubMed]
3. Krag, S.; Olsen, T.; Andreassen, T.T. Biomechanical characteristics of the human anterior lens capsule in relation to age. *Investig. Ophthalmol. Vis. Sci.* **1997**, *38*, 357–363.
4. Huang, D.; Xu, C.; Guo, R.; Ji, J.; Liu, W. Anterior lens capsule: Biomechanical properties and biomedical engineering perspectives. *Acta Ophthalmol.* **2021**, *99*, 1–8. [CrossRef] [PubMed]
5. Ovenseri-Ogbomo, G.O.; Oduntan, O.A. Mechanism of accommodation: A review of theoretical propositions. *Afr. Vis. Eye Health* **2015**, *74*, 1–6. [CrossRef]
6. Fisher, R.F. The significance of the shape of the lens and capsular energy changes in accommodation. *J. Physiol.* **1969**, *201*, 21–47. [CrossRef] [PubMed]
7. Schachar, R.A.; Bax, A.J. Mechanism of human accommodation as analyzed by nonlinear finite element analysis. *Compr. Ther.* **2001**, *27*, 122–132. [CrossRef]
8. Fisher, R.F.; Pettet, B.E. The postnatal growth of the capsule of the human crystalline lens. *J. Anat.* **1972**, *112*, 207–214.
9. Heys, K.R.; Cram, S.L.; Truscott, R.J.W. Massive increase in the stiffness of the human lens nucleus with age: The basis for presbyopia? *Mol. Vis.* **2004**, *10*, 956–963. [PubMed]
10. Dubbelman, M.; Van der Heijde, G.; Weeber, H. Change in shape of the aging human crystalline lens with accommodation. *Vis. Res.* **2005**, *45*, 117–132. [CrossRef] [PubMed]
11. Weikert, M.P. Update on bimanual microincisional cataract surgery. *Curr. Opin. Ophthalmol.* **2006**, *17*, 62–67. [CrossRef] [PubMed]
12. Wolffsohn, J.S.; Davies, L.N. Presbyopia: Effectiveness of correction strategies. *Prog. Retin. Eye Res.* **2019**, *68*, 124–143. [CrossRef] [PubMed]
13. Pepose, J.S.; Burke, J.S.; Qazi, M.A. Accommodating Intraocular Lenses. *Asia-Pac. J. Ophthalmol.* **2017**, *6*, 350–357. [CrossRef]
14. Ziebarth, N.; Borja, D.; Arrieta, E.; Aly, M.; Manns, F.; Dortonne, I.; Nankivil, D.; Jain, R.; Parel, J.-M. Role of the Lens Capsule on the Mechanical Accommodative Response in a Lens Stretcher. *Investig. Opthalmol. Vis. Sci.* **2008**, *49*, 4490–4496. [CrossRef] [PubMed]
15. Lanchares, E.; Navarro, R.; Calvo, B. Hyperelastic modelling of the crystalline lens: Accommodation and presbyopia. *J. Optom.* **2012**, *5*, 110–120. [CrossRef]
16. Webb, J.N.; Dong, C.; Bernal, A.; Scarcelli, G. Simulating the Mechanics of Lens Accommodation via a Manual Lens Stretcher. *J. Vis. Exp.* **2018**, e57162. [CrossRef] [PubMed]
17. Ehrmann, K.; Ho, A.; Parel, J. Biomechanical analysis of the accommodative apparatus in primates. *Clin. Exp. Optom.* **2008**, *91*, 302–312. [CrossRef] [PubMed]
18. Cortés, L.P.; Burd, H.J.; Montenegro, G.A.; D'Antin, J.C.; Mikielewicz, M.; Barraquer, R.I.; Michael, R. Experimental Protocols for Ex Vivo Lens Stretching Tests to Investigate the Biomechanics of the Human Accommodation Apparatus. *Investig. Ophthalmol. Vis. Sci.* **2015**, *56*, 2926–2932. [CrossRef] [PubMed]
19. Manns, F.; Parel, J.-M.; Denham, D.; Billotte, C.; Ziebarth, N.; Borja, D.; Fernandez, V.; Aly, M.; Arrieta, E.; Ho, A.; et al. Optomechanical Response of Human and Monkey Lenses in a Lens Stretcher. *Investig. Ophthalmol. Vis. Sci.* **2007**, *48*, 3260–3268. [CrossRef] [PubMed]
20. Pierscionek, B.K. Age-related response of human lenses to stretching forces. *Exp. Eye Res.* **1995**, *60*, 325–332. [CrossRef]
21. Doi, T.; Tanabe, S.; Fujita, I. Matching and correlation computations in stereoscopic depth perception. *J. Vis.* **2011**, *11*, 1–16. [CrossRef]
22. Reilly, M.A. A quantitative geometric mechanics lens model: Insights into the mechanisms of accommodation and presbyopia. *Vis. Res.* **2014**, *103*, 20–31. [CrossRef]
23. Ljubimova, D.; Eriksson, A.; Bauer, S. Aspects of eye accommodation evaluated by finite elements. *Biomech. Model. Mechanobiol.* **2007**, *7*, 139–150. [CrossRef] [PubMed]
24. David, G.; Pedrigi, R.M.; Humphrey, J.D. Accommodation of the human lens capsule using a finite element model based on nonlinear regionally anisotropic biomembranes. *Comput. Methods Biomech. Biomed. Eng.* **2017**, *20*, 302–307. [CrossRef]
25. Fisher, R.F.; Wakely, J. The elastic constants and ultrastructural organization of a basement membrane (lens capsule). *Proc. R. Soc. London. Ser. B Biol. Sci.* **1976**, *193*, 335–358.
26. Gent, A.J. On the relation between indentation hardness and Young's modulus. *Inst. Rubber Ind. Trans.* **1958**, *34*, 46–57. [CrossRef]
27. Meththananda, I.M.; Parker, S.; Patel, M.P.; Braden, M. The relationship between Shore hardness of elastomeric dental materials and Young's modulus. *Dent. Mater.* **2009**, *25*, 956–959. [CrossRef]
28. Wei, H.; Wolffsohn, J.S.; de Oliveira, O.G.; Davies, L.N. An Artificial Lens Capsule with a Lens Radial Stretching System Mimicking Dynamic Eye Focusing. *Polymers* **2021**, *13*, 3552. [CrossRef] [PubMed]
29. Sheppard, A.L.; Evans, C.J.; Singh, K.D.; Wolffsohn, J.S.; Dunne, M.C.M.; Davies, L.N. Three-Dimensional Magnetic Resonance Imaging of the Phakic Crystalline Lens during Accommodation. *Investig. Ophthalmol. Vis. Sci.* **2011**, *52*, 3689–3697. [CrossRef] [PubMed]
30. Kasthurirangan, S.; Markwell, E.L.; Atchison, D.; Pope, J.M. In Vivo Study of Changes in Refractive Index Distribution in the Human Crystalline Lens with Age and Accommodation. *Investig. Ophthalmol. Vis. Sci.* **2008**, *49*, 2531–2540. [CrossRef] [PubMed]

31. Rashid, B.; Destrade, M.; Gilchrist, M.D. Mechanical characterization of brain tissue in tension at dynamic strain rates. *J. Mech. Behav. Biomed. Mater.* **2014**, *33*, 43–54. [CrossRef]
32. Upadhyay, K.; Subhash, G.; Spearot, D. Visco-hyperelastic constitutive modeling of strain rate sensitive soft materials. *J. Mech. Phys. Solids* **2020**, *135*, 103777. [CrossRef]
33. Wu, Y.; Wang, H.; Li, A. Parameter Identification Methods for Hyperelastic and Hyper-Viscoelastic Models. *Appl. Sci.* **2016**, *6*, 386. [CrossRef]
34. Wilkes, R.P.; Reilly, M.A. A pre-tensioned finite element model of ocular accommodation and presbyopia. *Int. J. Adv. Eng. Sci. Appl. Math.* **2016**, *8*, 25–38. [CrossRef]
35. Dhondt, G. *The Finite Element Method for Three-Dimensional Thermomechanical Applications*; John Wiley & Sons: Chichester, UK, 2004.
36. Harwood, J.A.C.; Mullins, L.; Payne, A.R. Stress softening in natural rubber vulcanizates. Part II. Stress softening effects in pure gum and filler loaded rubbers. *J. Appl. Polym. Sci.* **1965**, *9*, 3011–3021. [CrossRef]
37. Burd, H.; Judge, S.; Cross, J. Numerical modelling of the accommodating lens. *Vis. Res.* **2002**, *42*, 2235–2251. [CrossRef]
38. Hermans, E.; Dubbelman, M.; van der Heijde, G.; Heethaar, R. Estimating the external force acting on the human eye lens during accommodation by finite element modelling. *Vis. Res.* **2006**, *46*, 3642–3650. [CrossRef] [PubMed]
39. Krag, S.; Andreassen, T.T. Mechanical Properties of the Human Posterior Lens Capsule. *Investig. Ophtalmol. Vis. Sci.* **2003**, *44*, 691–696. [CrossRef]
40. Efremov, Y.M.; Bakhchieva, N.A.; Shavkuta, B.S.; Frolova, A.A.; Kotova, S.L.; Novikov, I.; Akovantseva, A.A.; Avetisov, K.S.; Avetisov, S.E.; Timashev, P.S. Mechanical properties of anterior lens capsule assessed with AFM and nanoindenter in relation to human aging, pseudoexfoliation syndrome, and trypan blue staining. *J. Mech. Behav. Biomed. Mater.* **2020**, *112*, 104081. [CrossRef]
41. Burd, H.J. A structural constitutive model for the human lens capsule. *Biomech. Model. Mechanobiol.* **2009**, *8*, 217–231. [CrossRef]
42. David, G.; Humphrey, J.D. Finite element model of stresses in the anterior lens capsule of the eye. *Comput. Methods Biomech. Biomed. Eng.* **2007**, *10*, 237–243. [CrossRef]
43. Rebouah, M.; Machado, G.; Chagnon, G.; Favier, D. Anisotropic Mullins stress softening of a deformed silicone holey plate. *Mech. Res. Commun.* **2013**, *49*, 36–43. [CrossRef]
44. Reilly, M.A.; Hamilton, P.D.; Perry, G.; Ravi, N. Comparison of the behavior of natural and refilled porcine lenses in a robotic lens stretcher. *Exp. Eye Res.* **2009**, *88*, 483–494. [CrossRef] [PubMed]
45. Mordi, J.A.; Ciuffreda, K.J. Dynamic aspects of accommodation: Age and presbyopia. *Vis. Res.* **2004**, *44*, 591–601. [CrossRef]
46. Glasser, A.; Campbell, M.C. Biometric, optical and physical changes in the isolated human crystalline lens with age in relation to presbyopia. *Vis. Res.* **1999**, *39*, 1991–2015. [CrossRef]
47. Beers, A.; Van Der Heijde, G. In vivo determination of the biomechanical properties of the component elements of the accommodation mechanism. *Vis. Res.* **1994**, *34*, 2897–2905. [CrossRef]
48. Heistand, M.R.; Pedrigi, R.M.; Delange, S.L.; Dziezyc, J.; Humphrey, J.D. Multiaxial mechanical behavior of the porcine anterior lens capsule. *Biomech. Model. Mechanobiol.* **2005**, *4*, 168–177. [CrossRef] [PubMed]
49. Reilly, M.A.; Cleaver, A. Inverse elastographic method for analyzing the ocular lens compression test. *J. Innov. Opt. Health Sci.* **2017**, *10*, 1–12. [CrossRef]
50. Diani, J.; Fayolle, B.; Gilormini, P.; Diani, J.; Fayolle, B.; Gilormini, P. A review on the Mullins effect. *Eur. Polym. J.* **2009**, *45*, 601–612. [CrossRef]
51. Dorfmann, A.; Ogden, R. A constitutive model for the Mullins effect with permanent set in particle-reinforced rubber. *Int. J. Solids Struct.* **2004**, *41*, 1855–1878. [CrossRef]
52. Liao, Z.; Hossain, M.; Yao, X.; Navaratne, R.; Chagnon, G. A comprehensive thermo-viscoelastic experimental investigation of Ecoflex polymer. *Polym. Test.* **2020**, *86*, 106478. [CrossRef]
53. Guo, L.; Lv, Y.; Deng, Z.; Wang, Y.; Zan, X. Tension testing of silicone rubber at high strain rates. *Polym. Test.* **2016**, *50*, 270–275. [CrossRef]
54. Liao, Z.; Hossain, M.; Yao, X. Ecoflex polymer of different Shore hardnesses: Experimental investigations and constitutive modelling. *Mech. Mater.* **2020**, *144*, 103366. [CrossRef]
55. Pierscionek, B.K. In Vitro Alteration of Human Lens Curvatures by Radial Stretching. *Exp. Eye Res.* **1993**, *57*, 629–635. [CrossRef] [PubMed]
56. Dubbelman, M.; Van der Heijde, G. The shape of the aging human lens: Curvature, equivalent refractive index and the lens paradox. *Vis. Res.* **2001**, *41*, 1867–1877. [CrossRef]
57. Martin, H.; Guthoff, R.; Terwee, T.; Schmitz, K.-P. Comparison of the accommodation theories of Coleman and of Helmholtz by finite element simulations. *Vis. Res.* **2005**, *45*, 2910–2915. [CrossRef] [PubMed]
58. Helmholtz, H. Ueber die Accommodation des Auges. *Graefe Arch. Clin. Exp. Ophthalmol.* **1855**, *2*, 1–74. [CrossRef]
59. Glasser, A.; Kaufman, P.L. The mechanism of accommodation in primates. *Ophthalmology* **1999**, *106*, 863–872. [CrossRef]
60. Farnsworth, P.; Shyne, S. Anterior zonular shifts with age. *Exp. Eye Res.* **1979**, *28*, 291–297. [CrossRef]
61. Wang, K.; Venetsanos, D.T.; Hoshino, M.; Uesugi, K.; Yagi, N.; Pierscionek, B.K. A Modeling Approach for Investigating Opto-Mechanical Relationships in the Human Eye Lens. *IEEE Trans. Biomed. Eng.* **2020**, *67*, 999–1006. [CrossRef] [PubMed]
62. Liu, Z.; Wang, B.; Xu, X.; Wang, C. A study for accommodating the human crystalline lens by finite element simulation. *Comput. Med. Imaging Graph.* **2006**, *30*, 371–376. [CrossRef] [PubMed]

63. Burd, H.J.; Wilde, G.S. Finite element modelling of radial lentotomy cuts to improve the accommodation performance of the human lens. *Graefe Arch. Clin. Exp. Ophthalmol.* **2016**, *254*, 727–737. [CrossRef]
64. Strenk, S.A.; Semmlow, J.L.; Strenk, L.M.; Munoz, P.; Gronlund-Jacob, J.; Demarco, J.K. Age-related changes in human ciliary muscle and lens: A magnetic resonance imaging study. *Investig. Ophthalmol. Vis. Sci.* **1999**, *40*, 1162–1169.
65. Nankivil, D.; Heilman, B.M.; Durkee, H.; Manns, F.; Ehrmann, K.; Kelly, S.; Arrieta-Quintero, E.; Parel, J.-M. The Zonules Selectively Alter the Shape of the Lens During Accommodation Based on the Location of Their Anchorage Points. *Investig. Ophthalmol. Vis. Sci.* **2015**, *56*, 1751–1760. [CrossRef] [PubMed]
66. Fisher, R.F. The force of contraction of the human ciliary muscle during accommodation. *J. Physiol.* **1977**, *270*, 51–74. [CrossRef]
67. Wang, K.; Venetsanos, D.; Wang, J.; Pierscionek, B.K. Gradient moduli lens models: How material properties and application of forces can affect deformation and distributions of stress. *Sci. Rep.* **2016**, *6*, 31171. [CrossRef] [PubMed]
68. Pour, H.M.; Kanapathipillai, S.; Zarrabi, K.; Manns, F.; Ho, A. Stretch-dependent changes in surface profiles of the human crystalline lens during accommodation: A finite element study. *Clin. Exp. Optom.* **2015**, *98*, 126–137. [CrossRef] [PubMed]
69. Michael, R.; Mikielewicz, M.; Gordillo, C.; Montenegro, G.A.; Cortés, L.P.; Barraquer, R.I. Elastic Properties of Human Lens Zonules as a Function of Age in Presbyopes. *Investig. Ophthalmol. Vis. Sci.* **2012**, *53*, 6109–6114. [CrossRef] [PubMed]

Article

Biopolymer Hydrogel Scaffolds Containing Doxorubicin as A Localized Drug Delivery System for Inhibiting Lung Cancer Cell Proliferation

Chuda Chittasupho [1,2,*], Jakrapong Angklomklew [3], Thanu Thongnopkoon [4], Wongwit Senavongse [3], Pensak Jantrawut [1,2] and Warintorn Ruksiriwanich [1,2]

1. Department of Pharmaceutical Sciences, Faculty of Pharmacy, Chiang Mai University, Chiang Mai 50200, Thailand; pensak.amuamu@gmail.com (P.J.); warintorn.ruksiri@cmu.ac.th (W.R.)
2. Cluster of Research and Development of Pharmaceutical and Natural Products Innovation for Human or Animal, Chiang Mai University, Chiang Mai 50200, Thailand
3. Department of Biomedical Engineering, Faculty of Engineering, Srinakharinwirot University, Nakhon Nayok 26120, Thailand; jakrapong.a@hotmail.com (J.A.); wongwit@g.swu.ac.th (W.S.)
4. Department of Pharmaceutical Technology, Faculty of Pharmacy, Srinakharinwirot University, Nakhon Nayok 26120, Thailand; thanu@g.swu.ac.th
* Correspondence: chuda.c@cmu.ac.th; Tel.: +66-53944342; Fax: +66-53944390

Abstract: A hydrogel scaffold is a localized drug delivery system that can maintain the therapeutic level of drug concentration at the tumor site. In this study, the biopolymer hydrogel scaffold encapsulating doxorubicin was fabricated from gelatin, sodium carboxymethyl cellulose, and gelatin/sodium carboxymethyl cellulose mixture using a lyophilization technique. The effects of a crosslinker on scaffold morphology and pore size were determined using scanning electron microscopy. The encapsulation efficiency and the release profile of doxorubicin from the hydrogel scaffolds were determined using UV-Vis spectrophotometry. The anti-proliferative effect of the scaffolds against the lung cancer cell line was investigated using an MTT assay. The results showed that scaffolds made from different types of natural polymer had different pore configurations and pore sizes. All scaffolds had high encapsulation efficiency and drug-controlled release profiles. The viability and proliferation of A549 cells, treated with gelatin, gelatin/SCMC, and SCMC scaffolds containing doxorubicin significantly decreased compared with control. These hydrogel scaffolds might provide a promising approach for developing a superior localized drug delivery system to kill lung cancer cells.

Keywords: gelatin; sodium carboxymethyl cellulose; scaffold; A549 cells; freeze drying

1. Introduction

Lung cancer is the leading cause of death in men and women, responsible for almost 25% of all cancer deaths [1]. Chemotherapy is the first-line therapy for small-cell lung cancer (SCLC) pre-operation and post-operation, and stage IV non-small cell lung cancer (NSCLC) [2]. Chemotherapy is also used as an adjuvant chemotherapy to kill remaining cancer cells and to reduce the chance of recurrence. In addition, adjuvant chemotherapy shrinks the tumor before surgery and reduces the spread of cancer, making surgery less invasive and more effective. Chemotherapy is commonly administered to patients intravenously at the maximum tolerable doses, which may cause severe toxicity in healthy tissues [3]. Although patient survival rates from chemotherapy treatments are high, drug resistance and drug-related toxicity severely limit clinical outcomes [4]. Doxorubicin (DOX) is an anthracycline type of chemotherapy, widely used for the treatment of solid tumors, including lung tumor [5]. A DOX HCl injection causes cumulative and dose-dependent cardiotoxicity. Cardiotoxicity, which develops at later stages of therapy or after the treatment, can cause severe cardiomyopathy and congestive heart failure, leading to patient death [6]. The administration of a DOX HCl injection does not efficiently deliver the drug to the target site at a therapeutic concentration and fails to maintain sufficient drug accumulation in the tumor. Some fractions of the drug administered to the patients reach tumor cells and some may transport to other tissues, leading to a reduced efficacy and an increased toxicity in normal tissues [7].

Scaffold, a localized drug delivery system, is a promising platform for maintaining the drug concentration in the blood circulation at low levels while increasing the drug level at the therapeutic concentration that reaches tumor cells. The scaffolds can release the drug to kill the remaining tumors that remain due to an incomplete surgical removal of the tumor cells from the lung, the main cause of recurrence and metastasis [8]. The localized drug delivery systems include a tumor proximity implant and an intra-tumoral injection. The retention of the chemotherapeutic drug at the tumor site decreases drug concentration in the blood circulation, hence limiting drug exposure to other normal tissues [9,10]. The drug retained at the tumor ensures the therapeutic efficacy of the drug. Hydrogel scaffolds have many advantages, including a high drug encapsulation efficiency, a controlled and sustained release of chemotherapeutic drugs, and a high water uptake capability with both solid and liquid properties [11]. The properties of the hydrogel scaffold such as morphology, pore size and the drug release profile can be modulated through a ratio of the polymer mixture [12,13].

Gelatin is a natural biomaterial with the advantages of biocompatibility, biodegradability, low immunogenicity, low cost, and includes many functional groups that allow for structure modification [14]. However, gelatin is sensitive to enzymatic degradation. SCMC is a chemically modified water-soluble polysaccharide widely used as a drug delivery system and drug delivery matrix. The advantages of SCMC are its high biocompatibility, biodegradability, and low immunogenicity [15]. Crosslinked SCMC has a high water absorption capacity and a suitable swelling degree to form hydrogel with a dynamic viscoelastic property [16]. In this study, the hydrogel scaffold encapsulating the DOX was constructed using a varying gelatin and SCMC ratio. The effects of a crosslinker on the scaffold morphology and pore size were observed. The water swelling capacity and drug release kinetic profile of the hydrogel scaffold loaded with DOX were characterized. The anti-proliferative effect of the scaffolds against the lung cancer cell line was investigated. We hypothesized that the addition of SCMC to gelatin may present a promising approach for developing a superior localized drug delivery hydrogel scaffold by improving morphology, pore size, drug loading efficiency, the swelling property, and drug release profile.

2. Materials and Methods

2.1. Materials

Gelatin, sodium carboxymethyl cellulose (SCMC) (viscosity = 400–1000 mPa.s, degree of substitution = 0.60–0.95), and glutaraldehyde 25% aqueous solution (Sigma–Aldrich Inc., St. Louis, MO, USA). The DOX HCl injection solution (Pfizer Inc. New York, NY, USA). A549 cells, a human adenocarcinoma cell line derived from lung cancer cell of a male patient aged 58 (Japanese Collection of Research Bioresources (JRCB) Cell Bank, Tokyo, Japan). Dulbecco's Modified Eagle Media (DMEM), fetal bovine serum (FBS), penicillin–streptomycin (10,000 U/mL of penicillin and 10,000 U/mL of streptomycin), and MTT (3-(4,5-dimethylthiazol-2-yl)-2,5-diphenyltetrazolium bromide) (Life Technologies Inc., Carlsbad, CA, USA).

2.2. Preparation of Gelatin-SCMC Porous Scaffold

The porous scaffolds, generated from a gelatin and SCMC mixture at different ratios, were fabricated using the swelling and freeze drying processes. Gelatin powder was dissolved in distilled water at 70 °C to obtain a 3% w/v gelatin solution. Sodium carboxymethyl cellulose (SCMC) was dissolved in distilled water at a concentration of 2% w/v. The two polymer solutions were mixed at different ratios, i.e., 10:0, 9:1, 8:2, 7:3, 6:4, and 5:5 under constant stirring. The polymer mixture was poured into 24-well plates. The glutaraldehyde solution (2% w/v) was added dropwise to provide the final concentration at 0.2% w/v. The resulting hydrogel was washed three times with glycine (0.1 M) and distilled water to inactivate unreacted aldehyde groups. The hydrogels were swollen at 4 °C for 24 h, and were then kept at −80 °C for 18 h. Then, the frozen hydrogel was freeze-dried for 48 h to form a porous scaffold. For the scaffold containing the anti-cancer drug, DOX (167 µg/mL) was added into the polymer mixture before crosslinking. The dried scaffolds were stored at 4 °C until use.

2.3. Scaffold Morphological Investigation by Scanning Electron Microscopy

The morphology of the scaffolds was analyzed using the HITACHI S-3400N scanning electron microscope (Hitachi High-Tech in America Inc., Schaumburg, IL, USA) to evaluate their structure and porosity. Scaffold samples were fixed on an aluminum stub with conductive double-sided adhesive tape and coated with gold in an argon atmosphere prior to the photograph [17]. The images were captured at 100× magnification (20 kV). The average pore sizes for each scaffold were randomly measured for Martin's diameter and were recorded to obtain an average with a standard deviation.

2.4. Scaffold Swelling Capacity Determination

The swelling capacity of scaffolds was tested by measuring their water uptake capacity. The scaffolds were immersed in a phosphate-buffered saline (PBS), pH 7.4. Portions of the scaffolds were cut and weighed, measuring at around 2.5 mg. The initial weight of the scaffold was recorded (W_0). The scaffolds were then immersed in 20 mL of PBS, pre-incubated at 37 °C for 45 min and at 37 °C for 5 h. The scaffolds were removed, and the surface adsorbed PBS was removed using filter paper. The scaffolds were weighed again (W_1). The swelling degree percentage of the porous scaffold was calculated using Equation (1), and the average value was obtained from the three experiments.

$$Swelling\ degree\ (\%) = \frac{W_1 - W_0}{W_0} \times 100\% \qquad (1)$$

where W_0 is the weight of the scaffold before swelling and W_1 is the weight of scaffold after swelling.

2.5. Drug Encapsulation and Loading Efficiency in Scaffold

The scaffold containing DOX was cut and weighed, measuring at around 2.5 mg. The scaffolds were dissolved in deionized water (500 µL) under sonication until they were completely dissolved. The solution was analyzed to detect the amount of encapsulated the DOX using a fluorescence microplate reader (Spectramax M3, Molecular Devices Inc., San Jose, CA, USA) at Ex. 482 nm and Em. 590 nm [18]. A calibration curve was constructed by plotting the mean fluorescent intensity of DOX in deionized water versus concentrations ranging from 0.5–500 µg/mL. The linear regression method was used for drug quantitation. The DOX concentration in the scaffold samples was calculated according to a linear equation of a standard curve. The absorbance of the blank scaffolds was subtracted from that of the samples. The encapsulation efficiency (%) and loading efficiency (%) were calculated using Equations (2) and (3), respectively.

$$Encapsulation\ efficiency\ (\%) = \frac{Amount\ of\ DOX\ in\ the\ scaffold}{Amount\ of\ DOX\ initially\ added} \times 100\% \qquad (2)$$

$$Loading\ efficiency\ (\%) = \frac{Amount\ of\ DOX\ in\ the\ scaffold}{Weight\ of\ Scaffold} \times 100\% \qquad (3)$$

2.6. In Vitro Drug Release of Scaffold

The scaffolds containing doxorubicin were cut to provide a 2.5 mg mass, containing 16.7 µg of DOX. Each sample was immersed in 500 µL of PBS, pH 7.4 and incubated at 37 °C for 15, 30, 60 min and 24 h. At pre-determined intervals, samples were centrifuged at

13,000 rpm for 10 min. The amount of DOX released into the supernatant was measured using a fluorescence microplate reader at Ex. 482 and Em. 590 nm (Spectramax M3, Molecular Devices Inc., San Jose, CA, USA). A calibration curve was constructed by plotting mean fluorescent intensity of the DOX in PBS versus the DOX concentrations ranging from 0.5–500 μg/mL. The cumulative release of DOX (%) was calculated using Equation (4).

$$Cumulative\ release\ of\ DOX\ (\%) = \frac{Amount\ of\ DOX\ released\ from\ the\ scaffold}{Amount\ of\ DOX\ loaded\ in\ the\ scaffold} \times 100\% \quad (4)$$

2.7. Cell Culture

The A549 cells were cultured in Dulbecco's Modified Eagle's Medium (DMEM), supplemented with 10% FBS and 1% penicillin–streptomycin, and maintained in a humidified incubator with an atmosphere of 5% CO_2, at 37 °C (19).

2.8. In Vitro Cell Viability Study

The A549 cells (2.5×10^3 cells/mL) were developed in a 24-well plate for 24 h. Both scaffolds containing DOX (42 μg) and scaffolds without DOX were added to the cells. The DMEM, without the serum, (500 μL/well) was added to the scaffold. A free DOX solution (250 μg/mL) in DMEM was added to the cells as a positive control [19,20]. The cells without samples incubated in DMEM without the serum were used as a negative control. Cells were incubated with scaffold samples for 24 and 48 h at 37 °C, 5% CO_2. The samples were removed, and cells were washed three times with PBS and were incubated in a culture medium containing 0.5 mg/mL MTT (500 μL/well) for 2 h at 37 °C, 5% CO_2. After incubation, the MTT solution was removed, and DMSO (350 μL/well) was added to solubilize formazan products. The absorbance was measured at 550 and 650 nm for reference (Spectramax M3, Molecular Devices Inc., San Jose, CA, USA) [19]. The percentage of the cell viability was calculated as a ratio of mean absorbance with respect to the mean absorbance of negative control wells (Equation (5)).

$$Cell\ viability\ (\%) = \frac{(A550 - A650 control) - (A550 - A650 sample)}{(A550 - A650 control)} \times 100\% \quad (5)$$

2.9. Statistical Analysis

A statistical evaluation of data was performed using an analysis of variance (one-way ANOVA). The Newman–Keuls test was used as a post hoc test to assess the significance of differences. In all cases, a value of $p < 0.05$ was accepted as significant.

3. Results and Discussion

3.1. Hydrogel Scaffold Formation

Polymeric hydrogels can be prepared as three-dimensional hydrophilic networks capable of releasing drugs at the controlled rates. In this study, hydrogel scaffolds were fabricated by modulating the composition of the polymer and the crosslinker to provide properties according to specific applications. The hydrogels were successfully formed from 3% w/v gelatin, 2% w/v SCMC, and a 9:1 gelatin/SCMC mixture without glutaraldehyde (Table 1). The hydrogel formation was affected by the addition of glutaraldehyde. The hydrogel did not form when the glutaraldehyde was added to a 2% w/v SCMC solution, but this effect was not observed when glutaraldehyde was added to a 3% w/v gelatin solution. The hydrogel formed when a 9:1 and an 8:2 gelatin/SCMC were used to composite the hydrogel in the presence of glutaraldehyde. Increasing the ratio of the SCMC in the gelatin/SCMC polymer solution resulted in an unsuccessful formation of hydrogel in the presence of glutaraldehyde. In the 7:3, 6:4, and 5:5 gelatin/SCMC ratios with the presence of the crosslinker, the hydrogels were not formed.

Table 1. Formation of scaffold prepared from gelatin, sodium carboxymethyl cellulose, and gelatin/SCMC mixture.

Sample Name	3% w/v Gelatin/2% w/v SCMC Ratio	Glutaraldehyde 0.2% w/v	Scaffold Formation
Gelatin	10:0	-	✓
Gelatin + Glutaraldehyde	10:0	✓	✓
SCMC	0:10	-	✓
SCMC + Glutaraldehyde	0:10	✓	✗
Gelatin/SCMC	9:1	-	✓
Gelatin/SCMC + Glutaraldehyde	9:1	✓	✓
Gelatin/SCMC + Glutaraldehyde	8:2	✓	✓
Gelatin/SCMC + Glutaraldehyde	7:3	✓	✗
Gelatin/SCMC + Glutaraldehyde	6:4	✓	✗
Gelatin/SCMC + Glutaraldehyde	5:5	✓	✗

Hydrogels can be formed through various methods, including free radical polymerization, irradiation crosslinking, chemical crosslinking and physical crosslinking. The physical crosslinking method involves interactions such as polyelectrolyte complexation (ionic interaction), hydrogen bonding and hydrophobic association. A gelatin-based hydrogel can be prepared through both chemical crosslinking and physical crosslinking techniques. In the case of chemical crosslinking, gelatin-based hydrogel can be developed using dialdehyde (glutaraldehyde) or formaldehyde as a crosslinking agent. With regard to physical crosslinking, a gelatin-based hydrogel can be formed via hydrogen bonding between an electron deficient hydrogen atom and a functional group of high electronegativity [21]. An SCMC is a water soluble polysaccharide which can be fabricated as hydrogel scaffolds [21–23]. The hydrogen bonding between polymer chains, i.e., intramolecular hydrogen bonding, of SCMC is hypothesized to be the mechanism of hydrogel formation [24]. With regard to the hydrogel of the gelatin/SCMC mixture formed without using glutaraldehyde, the molecular interaction between the positively charged gelatin and the negatively charged carboxymethyl groups of SCMC have been reported in the preparation of a controlled delivery of microparticles [25]. This might provide one of the mechanisms through which to form hydrogel besides the intramolecular hydrogen bonding formation of each polymer.

Glutaraldehyde can react with various functional groups of proteins, e.g., amine, thiol, phenol and imidazole, since the reactive amino acid sided-chains are nucleophiles [26]. The crosslinking of the gelatin through the addition of glutaraldehyde was largely attributed to the Schiff's base between the aldehyde and the two free amino groups in lysine or the hydroxylysine of gelatin [27]. SCMC provides some of the hydroxyl groups that react with glutaraldehyde to form hemiacetal or acetal rings [28]. However, the reaction that occurs between the hydroxyl of SCMC and the aldehyde that forms hemiacetals is reversible. The equilibrium may not favor the formation of hemiacetals [29]. In addition, hemiacetals are prone to hydrolysis, leading to an unsuccessful hydrogel formation of SCMC with the addition of glutaraldehyde [28].

The SEM images showing the morphology and porosity of blank scaffolds and scaffolds containing DOX are presented in Figures 1–3. The pore structures of the scaffolds produced by gelatin, SCMC, and the 9:1 ratio gelatin/SCMC mixture, in the presence and absence of glutaraldehyde, are different. The gelatin scaffolds contained regular shapes of pores and uniform pore size (Figure 1). The gelatin/SCMC scaffolds contained a less uniform pore size and shape (Figure 2). The morphology of the pores in the scaffolds prepared using 100% SCMC exhibited an undefined morphology (Figure 3). The average pore size of each type of scaffold is presented in Figure 4. The scaffolds that were made using gelatin possessed the largest pore sizes with the narrowest pore size distribution. The average pore size of scaffolds made from gelatin was in the range of 101–197 μm. The SCMC scaffolds contained the smallest pore size with a wide distribution. The average pore sizes of SCMC scaffolds were in a range of 55–81 μm. The pore sizes of the gelatin/SCMC scaffolds were between those of the gelatin and SCMC scaffolds, at 58–141 μm. All the scaffolds were prepared through the addition of a crosslinker and had some interconnectivity between pores, confirming the use of the crosslinking strategy via glutaraldehyde. However, the morphology and the structure of the scaffolds without the crosslinker addition proved better than those of the scaffolds formed by crosslinking. In addition, to avoid using a chemical crosslinker, all the scaffolds that were used for further investigation were prepared in the absence of glutaraldehyde. The incorporation of DOX into the hydrogel did not affect the pore size and shape of any of the scaffolds.

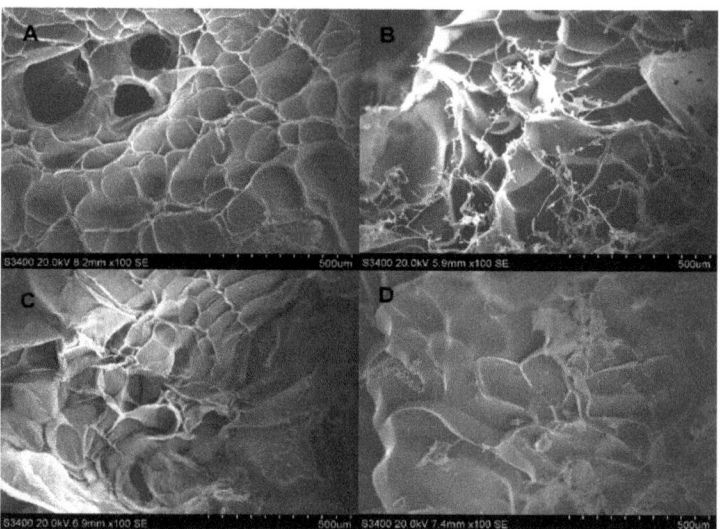

Figure 1. SEM images of hydrogel porous scaffolds made from solution of (**A**) 3% w/w gelatin (100×); (**B**) 3% w/w gelatin and 2% glutaraldehyde (100×); (**C**) 3% w/w gelatin and DOX (100×) and (**D**) 3% w/w gelatin, DOX, and 2% glutaraldehyde (100×).

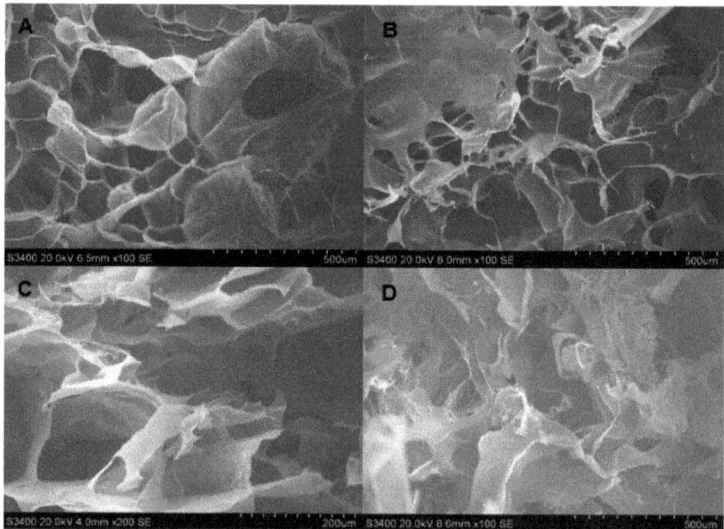

Figure 2. SEM images of hydrogel porous scaffolds made from solution of (**A**) 9:1 gelatin/SCMC solution (100×); (**B**) 9:1 gelatin/SCMC solution and 2% glutaraldehyde (100×); (**C**) 9:1 gelatin/SCMC solution and DOX (100×) and (**D**) 9:1 gelatin/SCMC solution, DOX, and 2% glutaraldehyde (100×).

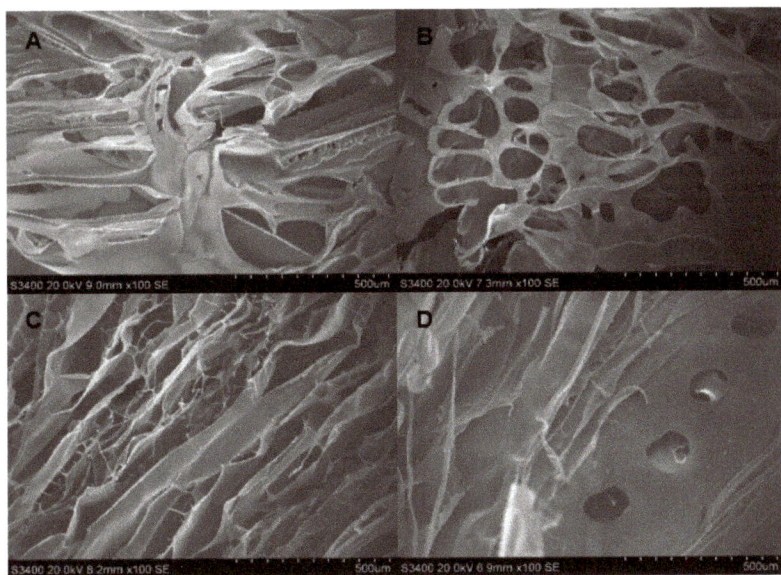

Figure 3. SEM images of hydrogel porous scaffolds made from solution of (**A**) 2% w/v SCMC solution (100×); (**B**) 2% w/v SCMC and 2% glutaraldehyde (100×); (**C**) 2% w/v SCMC and DOX (100×) and (**D**) 2% w/v SCMC solution, DOX, and 2% glutaraldehyde (100×).

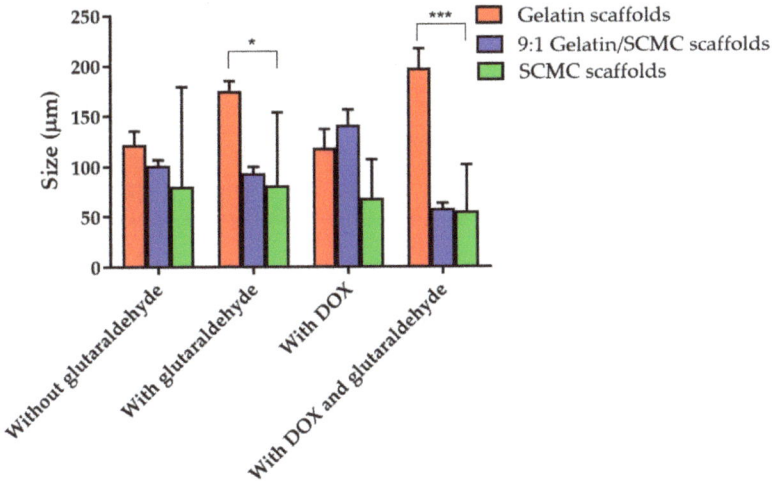

Figure 4. Pore sizes of gelatin, 9:1 gelatin/SCMC, and SCMC scaffolds with and without 2% w/v glutaraldehyde. Data are expressed as mean ± SEM., * and *** indicate $p < 0.05$ and $p < 0.001$, respectively.

The pore size and pore porosity depend on the density of the hydrogel and on the presence of a crosslinker. As shown in Figure 4, the pore size of the 9:1 gelatin/SCMC hydrogel scaffold was smaller than that of the gelatin scaffold and larger than that of the SCMC scaffold. However, in the presence of doxorubicin and glutaraldehyde, the pore size was significantly decreased ($p < 0.0001$). These results were consistent with the SEM images presented in Figure 2D. The crosslinker resulted in the formation of an interconnected pore and a smaller pore size [30]. Yang et al. reported that gelatin scaffolds crosslinked with glutaraldehyde had the smallest pore size compared to other crosslinkers [31].

3.2. Degree of Hydrogel Scaffold Swelling

The scaffolds composed of blank gelatin, 9:1 gelatin/SCMC, and SCMC and scaffolds containing DOX were subjected to swelling in PBS, pH 7.4 at 37 °C. Each scaffold showed a different degree of swelling. The results of the swelling study revealed that the SCMC scaffold had the highest swelling degree compared with the 9:1 gelatin/SCMC mixture, and the gelatin scaffolds (Figure 5). The swelling degrees of the gelatin, the 9:1 gelatin/SCMC, and the SCMC were 652.4 ± 26.9%, 1214.0 ± 173.9%, and 1655.8 ± 366.7%, respectively. The addition of DOX to the scaffold did not affect the swelling property of the scaffold. The swelling degree indicated the excellent absorption and retention properties of the scaffolds. The water absorption property depends on the ratio of SCMC in the scaffold. The greater swelling degree of the SCMC scaffolds may be a result of to the higher number of –OH and –COOH groups available for hydrogen bonding with water molecules compared with gelatin [32]. The swelling property of hydrogel is based on water absorption through an open porous structure by the capillary force, the type and the degree of the polymer crosslinking, and the polymer chain length [22,33,34]. DOX is a freely water soluble drug and the amount of DOX added to the scaffold was relatively low compared to the polymer mass. Therefore, the incorporation of DOX in all scaffolds did not influence the swelling properties of all types of scaffolds.

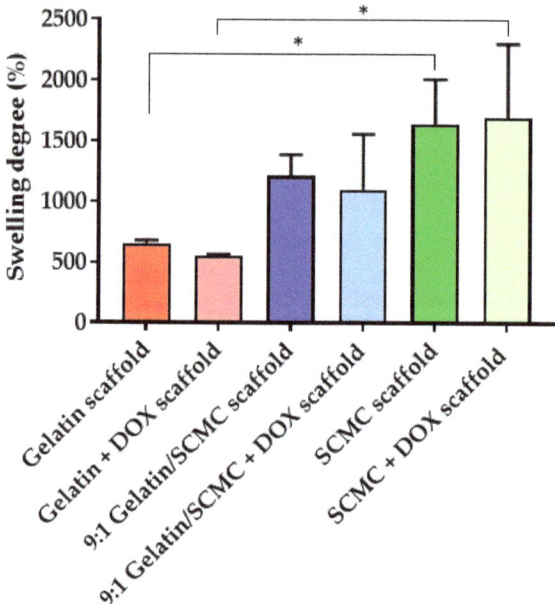

Figure 5. Swelling index (%) of gelatin, 9:1 gelatin/SCMC, and SCMC scaffolds with and without DOX. Data are expressed as mean ± S.D., * indicated $p < 0.05$.

3.3. Drug Encapsulation and Loading Efficiency of Hydrogel Scaffolds

The encapsulation efficiency of the DOX in the gelatin, the 9:1 gelatin/SCMC, and the SCMC scaffolds was 100.7 ± 0.8, 100.9 ± 0.9, and 99.3 ± 7.2% (Figure 6). The amount of DOX embedded in the scaffolds was 167 µg/30 mg of the scaffold, calculated based on the encapsulation efficiency. The high encapsulation efficiency of DOX might be a result of to the incorporation of the DOX, which was performed through the diffusion and absorption processes of the scaffolds. The loading efficiency of the gelatin, the 9:1 gelatin/SCMC, and the SCMC scaffolds were 1.38 ± 0.15%, 0.78 ± 0.07%, and 1.05 ± 0.09%, respectively.

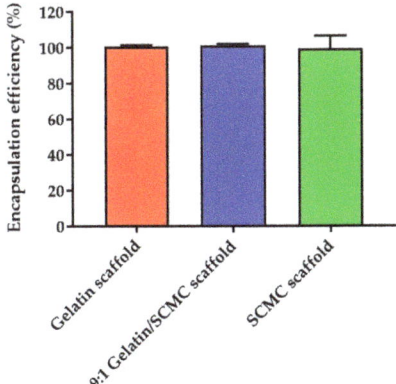

Figure 6. DOX encapsulation efficiency (%) of gelatin, 9:1 gelatin/SCMC, and SCMC scaffolds. Data are expressed as mean ± S.D.

The drug loading efficiency is usually controlled by the polymer composition and can be estimated relative to the swelling degree in the loading solvent [35]. The drug encapsulation efficiency was around 100% for all of the types of scaffolds, which may be a result of three primary reasons. The first reason is that the amount of DOX loaded into the scaffolds was relatively low. The second reason is that interactions may occur between the DOX and the gelatin or the SCMC, such as an electrostatic interaction, hydrogen bonding, and an intermolecular hydrophobic interaction [36–40]. The third reason is the high swelling degree of the polymer. Both gelatin and SCMC are hydrophilic polymers, therefore, the drug loading capacity in these hydrogels relies on the water solubility of the drug. A DOX is a freely water soluble drug, possessing a water solubility of 50 mg/mL. It was previously shown that drug loading increases relative to the hydrogel swelling [35]. We assumed that the maximum loading efficiency could be higher than the results in this study. The loading of the drug may be designed based on the investigation. In our study, the loading capacity was optimal for the cytotoxicity in the A549 cell line. The drug loading efficiency can be modulated for animal or clinical studies.

3.4. In Vitro Drug Release Profiles of Hydrogel Scaffolds

The release profiles of DOX from the gelatin, the 9:1 gelatin/SCMC, and the SCMC scaffolds were investigated at 37 °C for 24 h. Figure 7 shows the comparison of drug release profiles from three different types of scaffolds. The gelatin scaffold displayed the highest rate of drug release followed by the 9:1 gelatin/SCMC scaffold, and the SCMC scaffold, respectively. As shown in Figure 7, the maximum values of the DOX cumulative release from the gelatin, the 9:1 gelatin/SCMC, and the SCMC scaffolds were 31.75 ± 3.24%, 27.16 ± 4.95%, and 26.91 ± 5.15%, respectively. The burst release of the DOX from hydrogel scaffolds was a result of DOX diffusion from the surpassing surface area in the porous scaffolds. The pores of the hydrogel scaffolds increased the potential surface area for drug adsorption. After the hydrogel scaffolds were briefly immersed in PBS, the polymers became hydrated and swelled to form a gel layer on the surface of the system, which could retard the penetration of water into the scaffold core. It was reported that a higher degree of swelling increased the thickness of the gel layer, and reduced the drug release rate [41,42]. The gelatin had a lower swelling degree, compared with the 9:1 gelatin/SCMC, and the SCMC hydrogel scaffolds, resulting in a faster release rate in the first few hours. The incomplete release of drug from the gelatin and the SCMC scaffolds was previously reported [32,43]. The degradation of scaffolds upon biodegradation might induce further release of the drug from the scaffolds [31]. Gelatin scaffolds are susceptible to degradation in the presence of metalloproteinase enzymes within the interstitial space of a tumor [44]. The SCMC scaffolds are biodegradable and can be hydrolyzed through cleaving at the glycosidic linkages [45].

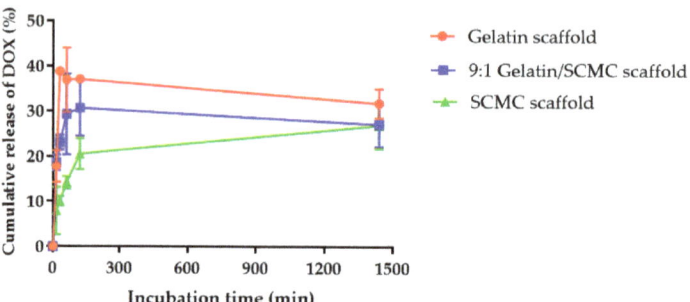

Figure 7. Release profiles of DOX from gelatin, 9:1 gelatin/SCMC, and SCMC scaffolds. Data are expressed as mean ± S.D.

The localized drug delivery system was designed to achieve a high therapeutic concentration of chemotherapeutic drugs in tumor sites and to minimize the systemic side effects of the drug [46]. The chemotherapeutic drugs administered locally usually expose a tumor for a short period of time, leading to an insufficient therapeutic level. The gelatin, the 9:1 gelatin/SCMC, and the SCMC hydrogel scaffolds containing DOX have been developed to provide a biphasic drug release profile which had a burst release of around 37, 30, and 20%, respectively. The gelatin and the SCMC hydrogel scaffolds were previously shown to release higher amounts of the drug due to the biodegradable properties of the polymers, providing a sustained release of the remaining dose over a period of time [46,47]. This could assist in avoiding the repeated administration of the drugs to the patient. Thus, the biphasic drug delivery system, combining an immediate drug release followed by a sustained release, may be favored for local tumor treatment.

The release profile of all the samples suggested a diffusion-controlled model. The release of the drug was usually controlled by the rate of hydrogel scaffold degradation [48]. However, the release profile was not influenced by scaffold degradation during the period of this investigation. In addition to the morphology of the porous scaffold, the degree of swelling may lead to the entrapment of the drug in the hydrogel scaffold. Since the DOX was water-soluble, the increased ability of the scaffold to swell in water resulted in a greater amount of the drug being retained in the hydrogel scaffold. In this study, the amount of DOX loaded into the hydrogel scaffold was relatively low (167 µg/30 mg) to ensure the uniformity of drug distribution throughout the scaffold and to maintain the sink conditions during the release experiment [49]. Therefore, the release profiles of DOX of the three different scaffolds were not different.

3.5. Effects of Hydrogel Scaffolds on A549 Cell Viability

The DOX was incorporated into the porous hydrogel scaffolds to control the drug release to tumors as a localized drug delivery system. Porous hydrogel scaffolds encapsulating the DOX were formed by mixing a 3% *w/v* gelatin solution, a 2% *w/v* SCMC solution, and a gelatin and SCMC mixture solution at a 9:1 ratio. The DOX exhibited a slow release profile from the scaffold, with around 30% being released within 24 h, indicating the sustained release of DOX to tumors. The A549 lung cancer cell viability, after the application of the scaffold on 24 and 48 h, was analyzed using an MTT assay. The A549 cells significantly decreased in cell number upon exposure to DOX scaffolds (Figure 8). The blank scaffolds prepared from the gelatin, SCMC, and the gelatin/SCMC mixture also reduced cell viability to some extent. Berg et al. reported that the viability of A549 cells exposed to 3D printed alginate/gelatin scaffolds decreased to below 70%, and a substantial loss in their cell number was observed, despite the high content of proteins and growth factors, which was most likely a result of the lower degree of porosity [50]. In our study, A549 cell viability significantly decreased after the treatment with blank gelatin, the gelatin/SCMC mixture, and the SCMC scaffolds from 82%, 85%, to 58%, respectively. This might also result from the smaller porosity and lower degree of porosity observed in the SEM images, resulting in an inadequate supply of nutrients to the cells. The viability of A549 cells, treated with gelatin, gelatin/SCMC, and SCMC scaffolds containing DOX for 24 h were 17, 13, and 22%, respectively. The greater cell viability of A549 after exposure to the SCMC scaffold might result from a lower drug release rate compared with the gelatin and gelatin/SCMC scaffolds. The viability of cells incubated with the DOX solution was comparable to that of cells incubated with the DOX scaffolds. The DOX solution, gelatin scaffold, 9:1 gelatin/SCMC scaffold, the SCMC scaffolds containing DOX, and blank scaffolds decreased A549 cell proliferation in a time-dependent manner. Specifically, the cell viability of A549 cells treated with gelatin, 9:1 gelatin/SCMC, and SCMC scaffolds containing DOX for 48 h was reduced to 1.8, 1.6, and 2.9 fold, respectively, compared with 24-h incubation.

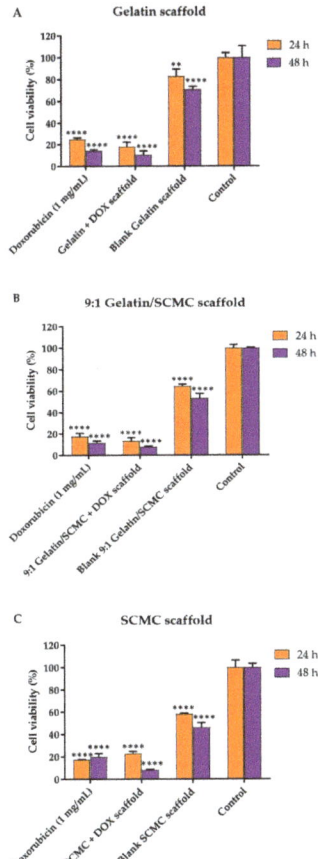

Figure 8. Viability of A549 cells treated with (**A**) gelatin (**B**) 9:1 gelatin/SCMC, and (**C**) SCMC scaffolds containing DOX. Data are expressed as mean ± S.D., ** indicated $p < 0.01$, and **** indicated $p < 0.0001$.

4. Conclusions

Porous hydrogel gelatin, SCMC, and gelatin/SCMC mixture scaffolds were developed as a localized drug delivery system for lung cancer treatment. The DOX was loaded within the scaffolds at a high encapsulation efficiency. The ratio of the gelatin and of the SCMC affected the pore size and swelling capacity of the scaffolds. However, it did not affect the drug encapsulation efficiency. All of the scaffolds sustained the release of the drug and effectively killed lung cancer cells. The results suggested these scaffolds to be a promising localized drug delivery system for lung cancer treatment prior to, or after surgery.

Author Contributions: Conceptualization, C.C. and J.A.; methodology, C.C. and J.A.; formal analysis, C.C.; investigation, J.A.; resources, C.C.; writing—original draft preparation, C.C. and T.T.; writing—review and editing, C.C. and T.T.; supervision, W.S.; project administration, C.C.; funding acquisition, P.J. and W.R. All authors have read and agreed to the published version of the manuscript.

Funding: This research was funded by Srinakharinwirot University and was partially supported by Chiang Mai University.

Institutional Review Board Statement: Not applicable.

Informed Consent Statement: Not applicable.

Data Availability Statement: The data presented in this study are available on request from the corresponding author.

Conflicts of Interest: The authors declare no conflict of interest.

References

1. Ferlay, J.; Colombet, M.; Soerjomataram, I.; Parkin, D.M.; Piñeros, M.; Znaor, A.; Bray, F. Cancer statistics for the year 2020: An overview. *Int. J. Cancer* **2021**. [CrossRef] [PubMed]
2. Zappa, C.; Mousa, S.A. Non-small cell lung cancer: Current treatment and future advances. *Transl. Lung Cancer Res.* **2016**, *5*, 288–300. [CrossRef]
3. Duan, X.; Li, Y. Physicochemical characteristics of nanoparticles affect circulation, biodistribution, cellular internalization, and trafficking. *Small* **2013**, *9*, 1521–1532. [CrossRef] [PubMed]
4. Pilkington, G.; Boland, A.; Brown, T.; Oyee, J.; Bagust, A.; Dickson, R. A systematic review of the clinical effectiveness of first-line chemotherapy for adult patients with locally advanced or metastatic non-small cell lung cancer. *Thorax* **2015**, *70*, 359–367. [CrossRef]
5. Melguizo, C.; Cabeza, L.; Prados, J.; Ortiz, R.; Caba, O.; Rama, A.R.; Delgado, A.V.; Arias, J.L. Enhanced antitumoral activity of doxorubicin against lung cancer cells using biodegradable poly(butylcyanoacrylate) nanoparticles. *Drug Des. Dev. Ther.* **2015**, *9*, 6433–6444.
6. Zhao, L.; Zhang, B. Doxorubicin induces cardiotoxicity through upregulation of death receptors mediated apoptosis in cardiomyocytes. *Sci. Rep.* **2017**, *7*, 44735. [CrossRef]
7. Wu, W.; Luo, L.; Wang, Y.; Wu, Q.; Dai, H.-B.; Li, J.-S.; Durkan, C.; Wang, N.; Wang, G.-X. Endogenous pH-responsive nanoparticles with programmable size changes for targeted tumor therapy and imaging applications. *Theranostics* **2018**, *8*, 3038–3058. [CrossRef]
8. Shi, X.; Cheng, Y.; Wang, J.; Chen, H.; Wang, X.; Li, X.; Tan, W.; Tan, Z. 3D printed intelligent scaffold prevents recurrence and distal metastasis of breast cancer. *Theranostics* **2020**, *10*, 10652–10664. [CrossRef]
9. Yang, Y.; Qiao, X.; Huang, R.; Chen, H.; Shi, X.; Wang, J.; Tan, W.; Tan, Z. E-jet 3D printed drug delivery implants to inhibit growth and metastasis of orthotopic breast cancer. *Biomaterials* **2020**, *230*, 119618. [CrossRef]
10. Dang, H.P.; Shafiee, A.; Lahr, C.A.; Dargaville, T.R.; Tran, P.A. Local Doxorubicin Delivery via 3D-Printed Porous Scaffolds Reduces Systemic Cytotoxicity and Breast Cancer Recurrence in Mice. *Adv. Ther.* **2020**, *3*, 2000056. [CrossRef]
11. Chittasupho, C.; Thongnopkoon, T.; Burapapisut, S.; Charoensukkho, C.; Shuwisitkul, D.; Samee, W. Stability, permeation, and cytotoxicity reduction of capsicum extract nanoparticles loaded hydrogel containing wax gourd extract. *Saudi Pharm. J.* **2020**, *28*, 1538–1547. [CrossRef]
12. Chaiwarit, T.; Ruksiriwanich, W.; Jantanasakulwong, K.; Jantrawut, P. Use of Orange Oil Loaded Pectin Films as Antibacterial Material for Food Packaging. *Polymers* **2018**, *10*, 1144. [CrossRef] [PubMed]
13. Panyamao, P.; Ruksiriwanich, W.; Sirisa-ard, P.; Charumanee, S. Injectable Thermosensitive Chitosan/Pullulan-Based Hydrogels with Improved Mechanical Properties and Swelling Capacity. *Polymers* **2020**, *12*, 2514. [CrossRef] [PubMed]
14. Song, R.; Murphy, M.; Li, C.; Ting, K.; Soo, C.; Zheng, Z. Current development of biodegradable polymeric materials for biomedical applications. *Drug Des. Dev. Ther.* **2018**, *12*, 3117–3145. [CrossRef] [PubMed]
15. Rahman, M.S.; Hasan, M.S.; Nitai, A.S.; Nam, S.; Karmakar, A.K.; Ahsan, M.S.; Shiddiky, M.J.A.; Ahmed, M.B. Recent Developments of Carboxymethyl Cellulose. *Polymers* **2021**, *13*, 1345. [CrossRef]

16. Nerurkar, N.L.; Elliott, D.M.; Mauck, R.L. Mechanical design criteria for intervertebral disc tissue engineering. *J. Biomech.* **2010**, *43*, 1017–1030. [CrossRef]
17. Thongnopkoon, T.; Chittasupho, C. Curcumin composite particles prepared by spray drying and in vitro anti-cancer activity on lung cancer cell line. *J. Drug Deliv. Sci. Technol.* **2018**, *45*, 397–407. [CrossRef]
18. Chittasupho, C.; Kewsuwan, P.; Murakami, T. CXCR4-targeted Nanoparticles Reduce Cell Viability, Induce Apoptosis and Inhibit SDF-1α Induced BT-549-Luc Cell Migration In Vitro. *Curr. Drug Deliv.* **2017**, *14*, 1060–1070. [CrossRef] [PubMed]
19. Chittasupho, C.; Athikomkulchai, S. Nanoparticles of Combretum quadrangulare leaf extract induce cytotoxicity, apoptosis, cell cycle arrest and anti-migration in lung cancer cells. *J. Drug Deliv. Sci. Technol.* **2018**, *45*, 378–387. [CrossRef]
20. Chittasupho, C.; Lirdprapamongkol, K.; Kewsuwan, P.; Sarisuta, N. Targeted delivery of doxorubicin to A549 lung cancer cells by CXCR4 antagonist conjugated PLGA nanoparticles. *Eur. J. Pharm. Biopharm.* **2014**, *88*, 529–538. [CrossRef]
21. El-Sherbiny, I.M.; Yacoub, M.H. Hydrogel scaffolds for tissue engineering: Progress and challenges. *Glob. Cardiol. Sci. Pract.* **2013**, *2013*, 316–342. [CrossRef]
22. Zhu, J.; Marchant, R.E. Design properties of hydrogel tissue-engineering scaffolds. *Expert Rev. Med. Devices.* **2011**, *8*, 607–626. [CrossRef] [PubMed]
23. Yang, Y.; Lu, Y.-T.; Zeng, K.; Heinze, T.; Groth, T.; Zhang, K. Recent Progress on Cellulose-Based Ionic Compounds for Biomaterials. *Adv. Mater.* **2021**, *33*, 2000717. [CrossRef] [PubMed]
24. Kashfipour, M.A.; Mehra, N.; Dent, R.S.; Zhu, J. Regulating Intermolecular Chain Interaction of Biopolymer with Natural Polyol for Flexible, Optically Transparent and Thermally Conductive Hybrids. *Eng. Sci.* **2019**, *8*, 11–18. [CrossRef]
25. Devi, N.; Maji, T.K. Preparation and evaluation of gelatin/sodium carboxymethyl cellulose polyelectrolyte complex microparticles for controlled delivery of isoniazid. *AAPS PharmSciTech* **2009**, *10*, 1412–1419. [CrossRef] [PubMed]
26. Migneault, I.; Dartiguenave, C.; Bertrand, M.J.; Waldron, K.C. Glutaraldehyde: Behavior in aqueous solution, reaction with proteins, and application to enzyme crosslinking. *BioTechniques* **2004**, *37*, 790–802. [CrossRef] [PubMed]
27. Lin, J.; Pan, D.; Sun, Y.; Ou, C.; Wang, Y.; Cao, J. The modification of gelatin films: Based on various cross-linking mechanism of glutaraldehyde at acidic and alkaline conditions. *Food Sci. Nutr.* **2019**, *7*, 4140–4146. [CrossRef] [PubMed]
28. Buhus, G.; Popa, M.; Desbrieres, J. Hydrogels based on carboxymethylcellulose and gelatin for inclusion and release of chloramphenicol. *J. Bioact. Compat. Polym.* **2009**, *24*, 525–545. [CrossRef]
29. Xiao, Z.; Xie, Y.; Militz, H.; Mai, C. Effect of glutaraldehyde on water related properties of solid wood. *Holzforschung* **2010**, *64*, 483–488. [CrossRef]
30. McKegney, M.; Taggart, I.; Grant, M.H. The influence of crosslinking agents and diamines on the pore size, morphology and the biological stability of collagen sponges and their effect on cell penetration through the sponge matrix. *J. Mater. Sci. Mater. Med.* **2001**, *12*, 833–844. [CrossRef] [PubMed]
31. Yang, G.; Xiao, Z.; Long, H.; Ma, K.; Zhang, J.; Ren, X.; Zhang, J. Assessment of the characteristics and biocompatibility of gelatin sponge scaffolds prepared by various crosslinking methods. *Sci. Rep.* **2018**, *8*, 1616. [CrossRef]
32. Akalin, G.O.; Pulat, M. Preparation and Characterization of Nanoporous Sodium Carboxymethyl Cellulose Hydrogel Beads. *J. Nanomater.* **2018**, *2018*, 9676949. [CrossRef]
33. Chai, Q.; Jiao, Y.; Yu, X. Hydrogels for Biomedical Applications: Their Characteristics and the Mechanisms behind Them. *Gels* **2017**, *3*, 6. [CrossRef] [PubMed]
34. Omidian, H.; Rocca, J.G.; Park, K. Advances in superporous hydrogels. *J. Control. Release* **2005**, *102*, 3–12. [CrossRef] [PubMed]
35. Kim, S.W.; Bae, Y.H.; Okano, T. Hydrogels: Swelling, drug loading, and release. *Pharm. Res.* **1992**, *9*, 283–290. [CrossRef] [PubMed]
36. Huang, C.-H.; Chuang, T.-J.; Ke, C.-J.; Yao, C.-H. Doxorubicin–Gelatin/Fe3O4–Alginate Dual-Layer Magnetic Nanoparticles as Targeted Anticancer Drug Delivery Vehicles. *Polymers* **2020**, *12*, 1747. [CrossRef]
37. Long, J.T.; Cheang, T.Y.; Zhuo, S.Y.; Zeng, R.F.; Dai, Q.S.; Li, H.P.; Fang, S. Anticancer drug-loaded multifunctional nanoparticles to enhance the chemotherapeutic efficacy in lung cancer metastasis. *J. Nanobiotechnol.* **2014**, *12*, 37. [CrossRef]
38. Wang, A.; Cui, Y.; Li, J.; van Hest, J.C.M. Fabrication of Gelatin Microgels by a "Cast" Strategy for Controlled Drug Release. *Adv. Funct. Mater.* **2012**, *22*, 2673–2681. [CrossRef]
39. Capanema, N.S.V.; Mansur, A.A.P.; Carvalho, S.M.; Carvalho, I.C.; Chagas, P.; de Oliveira, L.C.A.; Mansur, H.S. Bioengineered carboxymethyl cellulose-doxorubicin prodrug hydrogels for topical chemotherapy of melanoma skin cancer. *Carbohydr. Polym.* **2018**, *195*, 401–412. [CrossRef]
40. Li, M.; Tang, Z.; Lin, J.; Zhang, Y.; Lv, S.; Song, W.; Huang, Y.; Chen, X. Synergistic Antitumor Effects of Doxorubicin-Loaded Carboxymethyl Cellulose Nanoparticle in Combination with Endostar for Effective Treatment of Non-Small-Cell Lung Cancer. *Adv. Healthc. Mater.* **2014**, *3*, 1877–1888. [CrossRef]
41. Sujja-areevath, J.; Munday, D.L.; Cox, P.J.; Khan, K.A. Relationship between swelling, erosion and drug release in hydrophillic natural gum mini-matrix formulations. *Eur. J. Pharm. Sci.* **1998**, *6*, 207–217. [CrossRef]
42. Skoug, J.W.; Mikelsons, M.V.; Vigneron, C.N.; Stemm, N.L. Qualitative evaluation of the mechanism of release of matrix sustained release dosage forms by measurement of polymer release. *J. Control. Release* **1993**, *27*, 227–245. [CrossRef]
43. Kimura, Y.; Tabata, Y. Controlled release of stromal-cell-derived factor-1 from gelatin hydrogels enhances angiogenesis. *J. Biomater. Sci. Polym. Ed.* **2010**, *21*, 37–51. [CrossRef]
44. Wu, D.C.; Cammarata, C.R.; Park, H.J.; Rhodes, B.T.; Ofner, C.M., 3rd. Preparation, drug release, and cell growth inhibition of a gelatin: Doxorubicin conjugate. *Pharm. Res.* **2013**, *30*, 2087–2096. [CrossRef] [PubMed]

45. Leo, E.; Angela Vandelli, M.; Cameroni, R.; Forni, F. Doxorubicin-loaded gelatin nanoparticles stabilized by glutaraldehyde: Involvement of the drug in the cross-linking process. *Int. J. Pharm.* **1997**, *155*, 75–82. [CrossRef]
46. Kuang, G.; Zhang, Z.; Liu, S.; Zhou, D.; Lu, X.; Jing, X.; Huang, Y. Biphasic drug release from electrospun polyblend nanofibers for optimized local cancer treatment. *Biomater. Sci.* **2018**, *6*, 324–331. [CrossRef]
47. Sun, X.; Zhao, X.; Zhao, L.; Li, Q.; D'Ortenzio, M.; Nguyen, B.; Xu, X.; Wen, Y. Development of a hybrid gelatin hydrogel platform for tissue engineering and protein delivery applications. *J. Mater. Chem. B* **2015**, *3*, 6368–6376. [CrossRef]
48. Siepmann, J.; Göpferich, A. Mathematical modeling of bioerodible, polymeric drug delivery systems. *Adv. Drug Deliv. Rev.* **2001**, *48*, 229–247. [CrossRef]
49. Ong, Y.X.J.; Lee, L.Y.; Davoodi, P.; Wang, C.-H. Production of drug-releasing biodegradable microporous scaffold using a two-step micro-encapsulation/supercritical foaming process. *J. Supercrit. Fluids* **2018**, *133*, 263–269. [CrossRef]
50. Berg, J.; Hiller, T.; Kissner, M.S.; Qazi, T.H.; Duda, G.N.; Hocke, A.C.; Hippenstiel, S.; Elomaa, L.; Weinhart, M.; Fahrenson, C.; et al. Optimization of cell-laden bioinks for 3D bioprinting and efficient infection with influenza A virus. *Sci. Rep.* **2018**, *8*, 13877. [CrossRef]

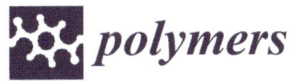

Article

pH-Responsive Succinoglycan-Carboxymethyl Cellulose Hydrogels with Highly Improved Mechanical Strength for Controlled Drug Delivery Systems

Younghyun Shin [1,†], Dajung Kim [1,†], Yiluo Hu [1], Yohan Kim [1], In Ki Hong [2], Moo Sung Kim [3] and Seunho Jung [1,4,*]

[1] Center for Biotechnology Research in UBITA (CBRU), Department of Bioscience and Biotechnology, Konkuk University, Seoul 05029, Korea; syh4969@naver.com (Y.S.); dajung903@naver.com (D.K.); lannyhu0806@hotmail.com (Y.H.); shsks1@daum.net (Y.K.)
[2] Covergence Technology Laboratory, Kolmar Korea, 61, Heolleung-ro-8-gil, Seocho-gu, Seoul 06800, Korea; inkiaaa@kolmar.co.kr
[3] Macrocare, 32 Gangni 1-gil, Cheongju 28126, Korea; rnd@macrocare.net
[4] Center for Biotechnology Research in UBITA (CBRU), Department of Systems Biotechnology & Institute for Ubiquitous Information Technology and Applications (UBITA), Konkuk University, Seoul 05029, Korea
* Correspondence: shjung@konkuk.ac.kr; Tel.: +82-2-450-3520
† These authors contributed equally to this work.

Abstract: Carboxymethyl cellulose (CMC)-based hydrogels are generally superabsorbent and biocompatible, but their low mechanical strength limits their application. To overcome these drawbacks, we used bacterial succinoglycan (SG), a biocompatible natural polysaccharide, as a double crosslinking strategy to produce novel interpenetrating polymer network (IPN) hydrogels in a non-bead form. These new SG/CMC-based IPN hydrogels significantly increased the mechanical strength while maintaining the characteristic superabsorbent property of CMC-based hydrogels. The SG/CMC gels exhibited an 8.5-fold improvement in compressive stress and up to a 6.5-fold higher storage modulus (G′) at the same strain compared to the CMC alone gels. Furthermore, SG/CMC gels not only showed pH-controlled drug release for 5-fluorouracil but also did not show any cytotoxicity to HEK-293 cells. This suggests that SG/CMC hydrogels could be used as future biomedical biomaterials for drug delivery.

Keywords: hydrogels; carboxymethyl cellulose; succinoglycan; metal coordination; drug delivery; swelling properties

1. Introduction

Hydrogels are a three-dimensional hydrophilic polymer capable of absorbing large amounts of water [1]. Due to their unique properties, hydrogels are widely used in pharmaceutical, tissue engineering, biomedical, cosmetic, and drug delivery systems [2–6]. Natural polymers and synthetic polymers are used to manufacture hydrogels. Compared to synthetic-based polymers, natural polymers tend to be environmentally friendly, reproducible, and biocompatible [7]. Among them, natural polysaccharides derived from microorganisms are very ideal candidates for the fabrication of new hydrogels because they have biodegradability, versatility, and biocompatibility [8]. In many studies, bacterial polysaccharides such as chitosan [9], starch [10], xanthan gum [11], gellan gum [12], and alginate [7] have been reported as hydrogel components. Particularly, acidic bacterial polysaccharide-based hydrogels have a very wide application field because they can successfully perform drug delivery depending on the pH [13].

Succinoglycan is an acidic exopolysaccharide (EPS) derived from soil microorganisms *Sinorhizobium* and *Agrobacterium* [14,15]. It plays an important role in the development of the root nodule between bacteria and the Alfalfa legume [14]. Succinoglycan is a polysaccharide with repeating octasaccharides composed of seven glucose residues and one galactose residue. They have succinate, pyruvate, and acetate groups as non-carbohydrate substituents [16]. Due to the functional groups with a carboxyl group, it is easy to crosslink with metal cations such as Fe^{3+} and Cr^{3+} [17], and is very sensitive to pH [18]. In addition, succinoglycan can maintain the physical consistency of its physical properties even under extreme conditions such as high temperature, high shear rate, high salinity, or ionic concentration [19,20]. In particular, the high heat stability of succinoglycan showed mass stability of about 60%, even at 600 °C, as measured using thermogravimetric analysis (TGA) [17].

Carboxmethyl cellulose (CMC), a derivative of cellulose, is synthesized by the reaction between cellulose and chloroacetic acid [21]. Unlike cellulose, CMC is relatively soluble in water and can absorb large amounts of water. Due to these characteristics, CMC has numerous potentials as a superabsorbent hydrogel, such as controlled fertilizer and agrochemicals, drug delivery, wound dressing, and tissue engineering [22–24]. However, despite these advantages, CMC hydrogels are limited in their application due to the critical disadvantage of weak mechanical strength. Therefore, various multi-component CMC hydrogel systems have been reported for the purpose of improving mechanical strength [25]. For example, the addition of polyacrylamide, acrylic acid, and carboxymethyl β-cyclodextrin increased the mechanical strength of CMC-based hydrogels [26–28]. However, these synthetic-based polymers have disadvantages of low biodegradability, versatility, and biocompatibility compared to natural polysaccharides derived from microorganisms [8]. In addition, since they are not acidic polysaccharides, there is a limit to pH-dependent drug delivery. Therefore, CMC-based hydrogels using acidic polysaccharides have been reported, but there is a disadvantage of being in the form of beads or modified forms [29–32].

To our knowledge, there have been no reports of non-beaded CMC-based hydrogels capable of pH-responsive drug delivery using succinoglycan, an unmodified bacterial polysaccharide to significantly improve the mechanical strength of hydrogels while maintaining characteristic superabsorbency. We hypothesized that bacterial succinoglycan could increase the mechanical strength of CMC-based hydrogels because succinoglycan can maintain a stable consistency of physical properties even in extreme environments [7,17].

Here, succinoglycan (SG), an unmodified natural polysaccharide, was successfully used to increase the mechanical strength of the CMC hydrogel. Since polysaccharide-based IPN (interpenetrating polymer network) hydrogels provide excellent biocompatibility, mechanical strength, and excellent phase stability [33,34], an SG/CMC IPN hydrogel (SG/CMC gel) was fabricated by cross-linking the hydroxyl groups and carboxyl groups present in both SG and CMC through Fe^{3+} ions. These structures were characterized using Fourier Transform Infrared (FTIR) Spectroscopy, Thermogravimetric Analysis (TGA), Field Emission Scanning Electron Microscopy (FE-SEM), a rheology test, and a compressive test. We also investigated the pH-responsive drug release properties using 5-fluorouracil as a model drug.

2. Materials and Methods

2.1. Materials

The bacterial strain (*Sinorhizobium meliloti* Rm1021) was supplied by the Microbial Carbohydrate Resource Bank (MCRB) at Konkuk University (Seoul, Korea). CMC (Mw = 250,000 g/mol with degree of substitution 0.7–based on manufacturer's data) was obtained from Sigma Aldrich (St. Louis, MO, USA). Iron (III) chloride hexahydrate (97.5%) was purchased from Daejung Chemicals & Metals Co., Ltd. (Siheung-si, Korea) and 5-fluorouracil (5-FU) purchased from Sigma Aldrich (St. Louis, MO, USA). All other chemicals were of analytical grade and used without further purification.

2.2. Growth Conditions and Production and Preparation of Succinoglycan

The isolation and purification of succinoglycan from *S. meliloti* Rm1021 was performed as previously described [18]. Bacteria were cultured in medium comprised of d-mannitol

(10 g/L), glutamic acid (1.5 g/L), K_2HPO_4 (5 g/L), KH_2PO_4 (5 g/L), $MgSO_4 \cdot 7H_2O$ (0.2 g/L), and $CaCl_2 \cdot 2H_2O$ (0.04 g/L), which was adjusted to a pH of 7.00 at 30 °C for 7 days with shaking (180 rpm). After, cells were centrifuged at 8000× g for 15 min at 4 °C and the supernatant was collected. To obtain succinoglycan, three volumes of ethanol were added to the supernatant. Furthermore, the precipitated succinoglycan was dissolved in distilled water and dialyzed (MWCO 12–14 kDa, distilled water for 3 days). After collection, succinoglycan purified via dialysis was lyophilized for later use. The molecular weights of succinoglycan were estimated via gel permeation chromatography (GPC) analysis. GPC was performed using a Waters Breeze System equipped with a Waters 1525 Binary pump and a Waters 2414 refractive index detector and was performed at 30 °C with a flow rate of 0.8 mL min^{-1} using 0.02 N sodium nitrate as a solvent. The molecular weight (Mw) of succinoglycan, as estimated via GPC, is 1.8×10^5 Da.

2.3. Preparation of Fe^{3+}-Crosslinked Layered-SG/CMC IPN Hydrogels

CMC and lyophilized SG were dissolved in distilled water at room temperature and stirred to obtain a clear polymer solution, respectively. As it is difficult to obtain a uniform hydrogel during the manufacturing process by adding Fe^{3+} solution to the polymer solution, we performed with reference to the preparation of a layered hydrogel [35]. To prepare a hydrogel, the prepared SG solution was added to the CMC solution and stirred for 30 min. The resultant polymer mixture solution was poured into a mold. After that, the aqueous polymer mixture solution was poured into special molds with circular grooves (20 cm in diameter, 10 cm in height) and soaked in aqueous iron (III) chloride ($FeCl_3$) solution (30 mM) for 18 h. These samples were completely immersed in deionized water and the deionized water was sufficiently changed to remove the nomadic trivalent iron ions from the hydrogel via diffusion of ion. SG/CMC IPN hydrogels obtained from circular groove were labeled as SxCx gels, where x = 1, 2, and 3 represent the relative proportions occupied by the polymer solution, respectively. The composition of SG, CMC, and SG/CMC IPN hydrogels are shown in Table 1.

Table 1. Composition of Composition of SG/CMC IPN gel.

Sample Name	Succinoglycan (%)	CMC (%)	Concentration of $FeCl_3$ (mM)
SG gel	100	-	30
S3C1 gel	75	25	30
S1C1 gel	50	50	30
S1C3 gel	25	75	30
CMC gel	-	100	30

2.4. Characterization of Fe^{3+}-Crosslinked Layered SG/CMC IPN Hydrogels

The FTIR spectra of each sample were taken using an FTIR spectrometer (Spectrum Two FTIR, Perkin Elmer), and obtained with a resolution of 0.5 cm^{-1} using 8 scans and a wavenumber range of 4000–600 cm^{-1}. All samples were mixed with KBr powder in a certain ratio (1:100) and then compressed to form pellets.

2.5. Thermogravimetric Analysis (TGA)

The hydrogels were washed by soaking in an excess of 70% ethanol for 2 days and were then dried at 60 °C and lyophilized. Lyophilized hydrogels were grinded in a blender. Thermogravimetric Analysis (TGA) was performed using a Perkin Elmer Pyris1 thermogravimetric analyzer. The thermal analyzer was operated under nitrogen atmospheric pressure and regulated by compatible PC commands. The dried sample (10 mg) was placed in a crucible and heated in an enclosed system with a linear temperature increase at a rate of 10 °C/min over a temperature range of 25 to 600 °C.

2.6. Field Emission Scanning Electron Microscopy (FE-SEM) Analysis

The cross-sectional morphology of SG/CMC IPN hydrogels were observed using FE-SEM (JSM-7800F Prime, JEOL Ltd., Akishima, Japan). The samples were quickly frozen and then lyophilized for 24 h. For observation, the surface of cross-sectioned hydrogel was coated with a thin layer of platinum at 10 mA for 60 s in a vacuum. SEM/EDS analysis of hydrogels was performed with FE-SEM (AURIGA, Carl Zeiss, Oberkochen, Germany) with an Energy selective Backscattered (EsB) detector. All hydrogels were platinum-coated at 10 mA, 120 s before examination using field emission scanning electron microscope.

2.7. Compression Test

Compression test was performed by using Instron E3000LT (Instron Inc., Norwood, MA, USA) by preparing a hydrogel disk with a height of 15 mm and a diameter of 20 mm. The sample was placed on a plate and compressed at a rate of 5 mm/min. Compressive stress was recorded when the sample was compressed with a strain of 70%. The measurement was performed in triplicate. All the hydrogels used in the measurements were prepared in the same manner to the rheological analysis.

2.8. Rheological Experiments

The rheological properties of the hydrogel were analyzed by oscillating angular frequency sweep and temperature ramp tests using a DHR-2 rheometer (TA Instruments, New Castle, DE, USA) equipped with a 20-millimeter parallel plate. The angular frequency was swept from 0.1 to 100 rad/s at a strain of 0.5% at 25 °C. The hydrogel samples were prepared as disks and were analyzed with a parallel plate measuring 20 mm in diameter, and the gap between plates was adjusted to 1.3 mm. A stress strain amplitude sweep test was conducted on samples from 0.1% to a maximum strain of 100% at 1.0 Hz to determine the limit of the linear viscoelastic region. All hydrogel samples used for the measurement were prepared with the same mold and the same amount of water. Each measurement was performed in triplicate. The temperature ramp test was conducted at a constant angular frequency of 10 rad/s and a constant strain of 1.0%. It was also carried out at 10 to 70 °C with an insulating cover to maintain the temperature.

2.9. Equilibrium Swelling Ratio and pH Sensitivity Measurements

The swelling properties of SG/CMC IPN hydrogels were investigated in various pH buffer solutions (pH 2, 4, 6, 8, 10). Each of the dried SG/CMC gels were immersed in pH buffer solutions at 37 °C to induce an equilibrium state. Then, the mass of the swollen hydrogel was weighed at various times after removing excess surface water. The swelling ratio was determined using following equation:

$$\text{Swelling ratio } (\%) = \frac{w_s - w_d}{w_s} \quad (1)$$

where w_s is the weight of the swollen hydrogel and w_d is the weight of the dried hydrogel in a vacuum oven prior to PBS immersion. After separating the hydrogel from the solution for accurate weight measurement, the solution remaining on the surface was gently wiped off using a laboratory tissue. Each measurement was performed in triplicate.

2.10. Drug Loading and Drug Release

5-Fluorouracil (5-FU) was used as the model drug to analyze drug release properties of SG/CMC IPN gels in different pH conditions. To put it simply, 1 mg of 5-FU was dissolved in 1 mL of SG/CMC polymer solution. The concentration of 5-FU in the hydrogels was fixed at 1 mg/mL. The obtained hydrogels were incubated in 40 mL of various pH buffer solutions (pH = 1.2, 7.4) at 37 °C with 50 rpm of constant stirring. At each interval time points, 500 µL of released medium was taken out for UV–Vis spectroscopic analysis and then same volume of fresh solution was added. The concentrations of 5-FU were analyzed using a spectrophotometer (UV2450, Shimadzu Corporation, Kyoto, Japan) at a wavelength of 266 nm. The cumulative amount of the drug was calculated by the following equation:

$$\text{Cumulative amount of the drug} = C_n V + \sum_{i=1}^{i=n-1} C_i V_i \qquad (2)$$

where V is the release of medium volume, V_i is the sampling volume, and C_n and C_i are the 5-FU concentrations in the release medium and extraction sample. All measurements were conducted in triplicate.

2.11. In Vitro Cytotoxicity

The cytotoxicity of the hydrogels was evaluated using WST-8 assays using human embryonic kidney 239 cells (HEK-293, Korean Cell Line Bank, Korea). For direct cytotoxicity tests, HEK-293 cells were seeded into 24-well culture plates at a concentration of 3×10^4 cells per well with minimum essential medium (MEM, WELGENE, Gyeongsan-si, Korea) containing 10% fetal bovine serum and 1% penicillin/streptomycin; then, 5 mg of hydrogel sample was added to the wells and plates were incubated at 37 °C in an atmosphere containing 5% CO_2. MEM medium was used as a negative control, and 10% (v/v) dimethylsulfoxide (DMSO) dissolved in the MEM medium was used as a positive control. After a 48-hour incubation, WST-8 assay reagent (QuantiMax, BIOMAX, Seoul, Korea) was added to each well and absorbance was measured at 450 nm. Cell viability was determined using the following formula:

$$\text{Cell viability } (\%) = \frac{\text{Absorbance of cells with hydrogel}}{\text{Absorbance of negative control cells}} \qquad (3)$$

All assays were repeated in triplicate for each sample.

3. Result and Discussion

3.1. Characterization of SG/CMC IPN Hydrogels

The coordination of the Fe^{3+}-crosslinked layered SG/CMC IPN hydrogels were confirmed through FTIR analysis. Figure 1 shows the FTIR spectra of SG, CMC, and S1C1 gels, and Table 2 shows the shift of the characteristic absorption peaks of the FTIR spectra. Here, the coordination of SxCx with Fe^{3+} ions was explained by S1C1 gel, and similar peak shifts occurred in the other SxCx gels. The FTIR spectrum of succinoglycan exhibited absorption peaks at 3326 cm^{-1}, corresponding to the –OH stretching bands, and the C=O stretching carbonyl esters of the acetate group showed absorption peaks at 1728 cm^{-1} [36–38]. Additionally, the absorption peaks at 1629, 1382, and 1074 cm^{-1} were attributed to the asymmetric C=O stretching vibration of the succinate and pyruvate functional groups, the symmetric stretching vibration of the carboxylate –COO– group in the acid residue, and the asymmetric C-O-C stretching vibration, respectively [36,37]. The FTIR spectrum of CMC showed a broad absorption band at 3408 cm^{-1}, which was related to the stretching frequency of the –OH group, and the peaks at 1625, 1422, and 1076 cm^{-1} were related to the asymmetric, symmetric stretching vibrations of the carboxylate groups, and C–C bending of CMC, respectively [39,40].

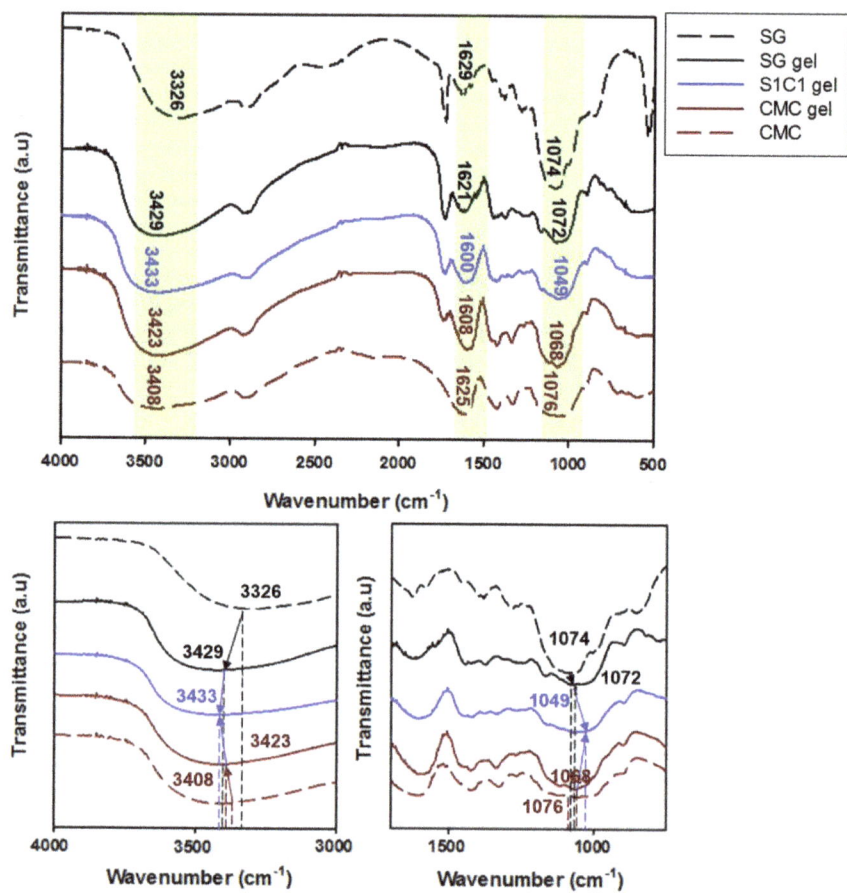

Figure 1. FTIR spectra of SG gel, S1C1 gel, and CMC gel.

However, when the Fe^{3+} ions are crosslinked with the polymer, the S1C1 gel absorption peaks shifted to 3433, 1600, and 1049 cm^{-1}, respectively. When the polymer coordinated with the Fe^{3+} ions, the –OH stretching band of the polymer mixture significantly red shifted to 3433 cm^{-1}. This proves that the –OH group and Fe^{3+} ion of the polymer mixture are coordinated [41]. In addition, the specific peak associated with the carboxyl group was shifted to 1600 cm^{-1}, and the asymmetric C-O-C stretching peak by backbone sugar shifted to 1049 cm^{-1}. The shift of these peaks is also due to the coordination of the carboxyl group of the polymer mixture with the Fe^{3+} ions [42,43]. In conclusion, the Fe^{3+} ion was strongly coordinated by hydrogen bonds between the hydroxyl groups and carboxyl groups of SG and CMC. A schematic representation of the presumable coordination mechanism of the SG and CMC polymer solutions and $FeCl_3$ based on the above findings is shown in Figure 2.

Table 2. Shift of the main peaks in the FTIR spectra of SG gel, S1C1 gel, and CMC gel.

Peak Assignment		Peak Wavenumber (cm^{-1})			Reference
		SG	CMC	S1C1	
OH stretch		3326	3408	3433	[36,37,39,41]
C=O stretch	Acetate	1728	-	1600	[36,37,39,40]
	Succinate/pyruvate	1629	-		
	Carboxymethyl ether	-	1625		[40]
asymmetric C-O-C stretch		1074	1076	1049	[36,37,44]

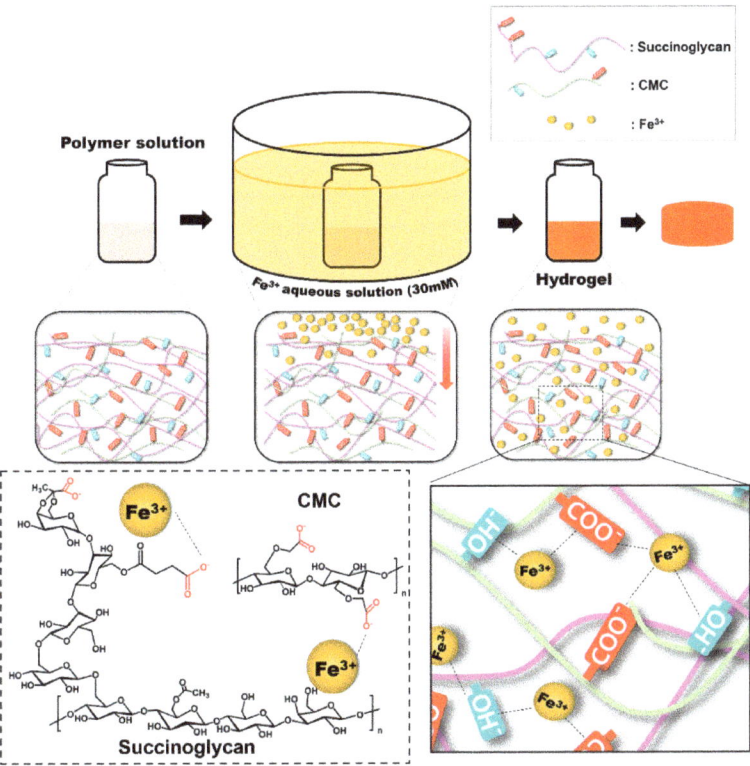

Figure 2. Schematic illustration of model of SG/CMC IPN hydrogels formation mechanism.

3.2. Thermogravimetric Analysis (TGA) and Derivative Thermogravimetry (DTG)

Figure 3 illustrates thermogravimetric analysis (TGA) and the derivative thermogravimetry (DTG) curves of SG gel, S1C1 gel, and CMC gel. As observed, the mass loss seen near the initial 120 °C in all the hydrogels was due to the water evaporation in the hydrogel [26,45]. The SG gel exhibited a 43.4% mass loss in the temperature range of 211–397 °C, and the major DTG picks appeared twice at 264 and 509 °C. The second phase was related with the degradation of the thermally stable structure formed by cross linkage and strong bonds in succinoglycan [17]. Additionally, the weight of the CMC gel decreased significantly between 220 and 495 °C. Thereafter, the maximum DTG peak was observed at 278 °C in the DTG curve, and the degradation steps in this range of CMC gel occurred due to the cellulose backbone cleavage and fragmentation [46]. In comparison, the weight loss in the S1C1 gel was observed at 210 and 392 °C, with maximum DTG peaks at 247 and 523 °C. The weight loss in this range was 38.9%, showing a more moderate weight loss at high temperatures, and it was shown that the formation of a hydrogel with metals reduced the skeletal decomposition of SG and CMC. In addition, a 50% weight loss was observed for SG gel and CMC gel at 386 and 323 °C, respectively, whereas a 50% weight loss was observed for S1C1 gel at 398 °C. Therefore, it was found that the thermal stability of the hydrogel was increased through IPN formation in the process of coordinating SG and CMC with Fe^{3+}. This is the first report to increase the thermal stability of a non-beaded CMC-based hydrogel using succinoglycan, an unmodified natural polysaccharide.

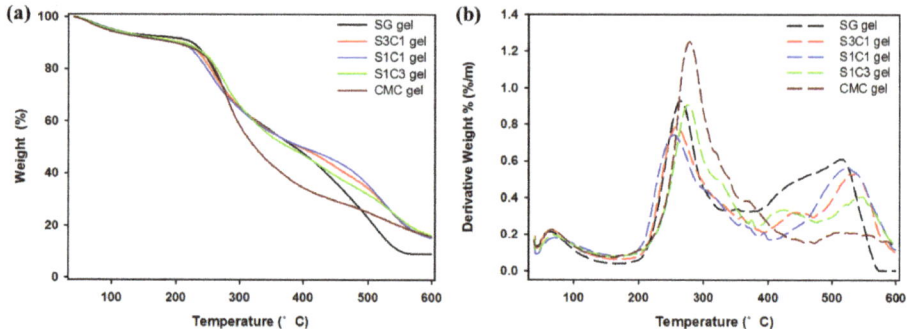

Figure 3. (a) TGA and (b) DTG curves of SG/CMC IPN gel.

3.3. Compression Mechanical Properties of SG/CMC IPN Hydrogels

Compression tests were performed to investigate the effect of SG on the mechanical properties of Fe^{3+} crosslinked layered SG/CMC IPN gels. Figure 4 shows the compressive strain curves. Compared to CMC only gels, SG/CMC IPN gels exhibit a higher compressive stress. As the ratio of SG content from S1C3 to S3C1 increased, the compressive stress of SG/CMC IPN gels improved (S3C1 gel: 0.0483 MPa > S1C1 gel: 0.0412 MPa > S1C3 gel: 0.0128 MPa > CMC gel: 0.0057 MPa). This result is exhibited because SG is able to maintain its physical consistency, even under extreme conditions [19,20]. It has been reported that the addition of SG or modified-SG increases the mechanical strength of the hydrogel [17,47]. Therefore, these results indicate that SG effectively enhanced the mechanical strength of the Fe^{3+} crosslinked layered SG/CMC IPN gels. It has been reported to increase the mechanical strength of CMC-based hydrogels using synthetic polymers or modified polymers [26,27,29]. However, it is significant that the mechanical strength of CMC-based hydrogel as a non-bead form was increased by using unmodified natural polysaccharide.

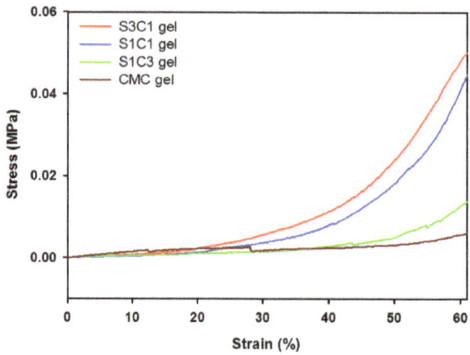

Figure 4. Compressive stress–strain curves of hydrogels with different SG concentrations.

3.4. Rheological Behavior of SG/CMC IPN Hydrogels

As shown in Figure 5a, to obtain information on the mechanical properties of the hydrogel, the oscillation angular frequency sweep test was performed. The oscillatory angle frequency sweep experiment was conducted from 0.1 to 100 rad/s at a fixed strain of 0.5. As can be seen in Figure 5, the storage modulus value G′ of all the hydrogels is larger than the loss modulus value G″. This indicates that the hydrogel is not in a fluid sol state, but a stable solid gel state. Additionally, compared with CMC gel, hydrogel containing SG has a larger storage modulus G′ at all frequencies. Comparing the actual G′ value, the CMC gel is about 1010 Pa. However, the G′ value of the S1C1 gel is 6060 Pa, and the G′ value of the S3C1 gel with the highest SG ratio is 6560 Pa. This is an increase of about 6.0 times and 6.5 times, respectively, compared to the CMC gel. This means that the addition of SG to the CMC-based hydrogel improves the mechanical strength, and the mechanical properties can be adjusted according to the amount of SG.

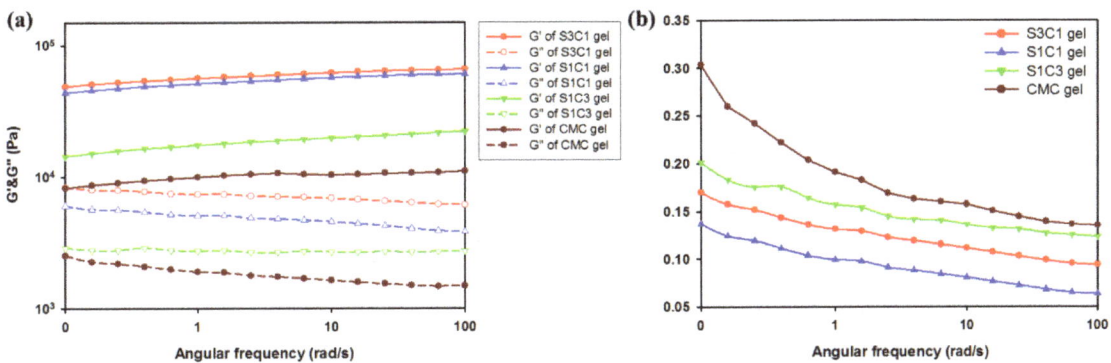

Figure 5. Rheological observations from the oscillation angular frequency sweep test: (**a**) Storage modulus (G′, filled symbols) and loss modulus (G″, empty symbols) of hydrogels; (**b**) The loss tangent (tan δ) of hydrogels.

The loss tangent (tan δ) was shown to measure the viscoelastic properties of CMC and SxCx gels (Figure 5b). Figure 5b shows that the SG/CMC IPN gels have low tan δ values at all frequencies. This shows that the gel was stably formed. In particular, the S1C1 gel prepared with a 1:1 ratio of SG and CMC had the lowest tan δ value, indicating the formation of the most mechanically strong hydrogel. Therefore, in subsequent experiments, S1C1 was set as the general SxCx.

Temperature ramp rheological studies of hydrogels were conducted at a constant frequency and strain. In Figure 6, the Tm value where the G″ value is higher than the G′ value did not occur in all the hydrogels. However, it can be seen that the G′ value of the CMC gel became rapidly insignificant. Therefore, the application in various fields is limited. However, in the case of the S1C1 gel with 2% of SG added to the CMC hydrogel, it can be seen that the G′ value is clearly maintained. This means that due to the strong mechanical strength of SG, it did not deteriorate during heating to 70 °C.

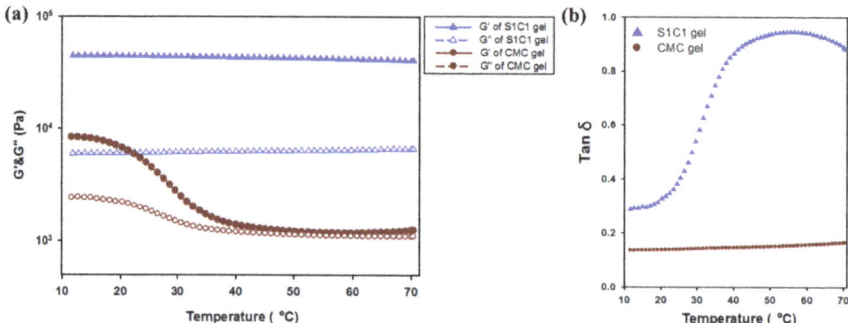

Figure 6. Temperature dependence of hydrogel through rheological observation of the oscillation temperature ramp test. (**a**) Storage modulus (G′, filled symbols) and loss modulus (G″, empty symbols) of hydrogels. (**b**) The loss tangent (tan δ) of hydrogels.

3.5. Field Emission Scanning Electron Microscopy (FE-SEM)

The FE-SEM image of the SG/CMC gel cross-section is shown in Figure 7. Figure 7a–c were each hydrogel observed at the same magnification and Figure 7d was the S1C1 gel observed at high magnification. A three-dimensional interconnected porous structure was observed on the cross-section surface of the SG gel. These pores can be related to the degree of cross-linking and mechanical strength [47]. The SEM image of the CMC gel shows an increase in the surface area of the hydrogel due to the coarse, small, and frequent porous structure. This large surface area supports the superabsorbency of CMC-based hydrogels by facilitating water molecules to diffuse into the polymer network [48]. On the other hand, the S1C1 gel shows both the interconnected porous structure of the SG gel and the frequent porous structure of the CMC gel. Figure 7d shows the structure of the same S1Cl gel when viewed at a relatively high magnification. As shown in Figure 7d, the layered structure was shown, which may be due to the diffused Fe^{3+} ions cross-linked with the polymer in each layer during SG/CMC IPN gel formation. Fe^{3+} ions form a first cross-link at the interface between $FeCl_3$ and SG/CMC, and then permeate from the top to the bottom of the SG/CMC polymer mixture solution to form a layered hydrogel.

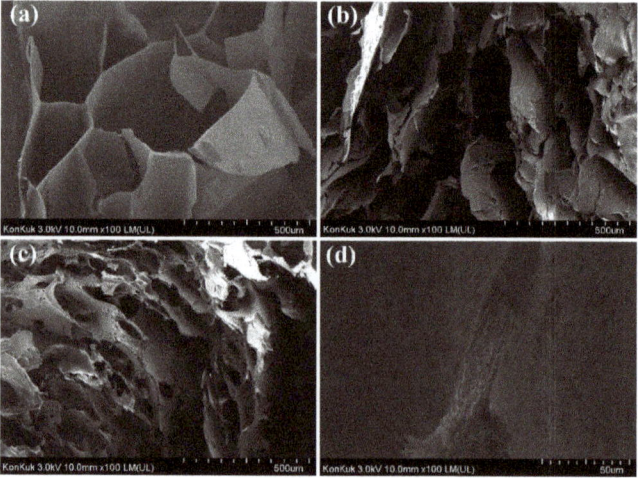

Figure 7. FE-SEM images illustrating the microstructures of cross-sectioned hydrogels: (**a**) SG gel; (**b**) S1C1 gel (Low magnification); (**c**) CMC gel; (**d**) S1C1 gel (High magnification).

EDS confirmed that Fe^{3+} ions were uniformly distributed in the polymer mixture (Figure S1). As shown in Figure S1d, when the S1C1 gel is cut into cross-sections, Fe^{3+} ions are evenly distributed from the top to the bottom of the hydrogel. This means that the SG/CMC IPN hydrogel was formed through the uniform penetration of Fe^{3+} ions.

3.6. Swelling Behavior of SG/CMC IPN Hydrogels

Figure 8 shows the pH-sensitive swelling behavior of SG/CMC IPN gel at an acidic and alkaline pH. The swelling equilibrium appeared on an average of 6 h, and the hydrogel was immersed in a pH buffer of five points for 2 days for sufficient swelling. SG and CMC-based hydrogels are known to have a large influence on swelling depending on the pH. This pH-dependent swelling behavior is due to the carboxyl groups of SG and CMC. At an acidic pH (pH 2, 4), most carboxylic acid groups are protonated. In addition, the hydrogen bonding interaction between the carboxylate and hydroxyl groups is enhanced. Therefore, the SG/CMC IPN tends to become denser and, consequently, the swelling value decreases [49]. At a higher pH (pH 6, 8), some carboxylic acid groups are ionized; therefore, the hydrogen bonds are broken [49,50]. Therefore, the electrostatic repulsion between the carboxylic acid groups is increased compared to an acidic pH. Therefore, the SG/CMC IPN tends to swell more. The reason that the swelling ratio is reduced in a basic solution (pH 10) is that Fe^{3+} ions prevent the anion–anion repulsion of the carboxylic acid group [50]. Similar results have been reported in many previous reports [49–52].

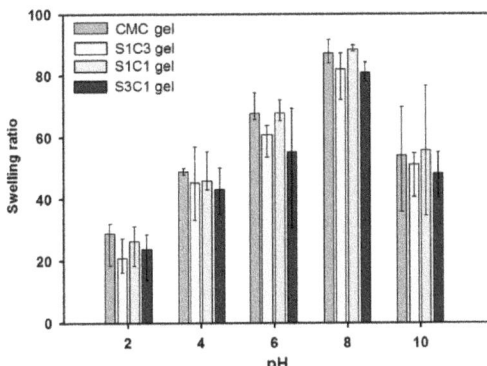

Figure 8. The influence of different pH conditions (pH 2, 4, 6, 8, and 10) on the equilibrium swelling ratio of SG, CMC gel, and SG/CMC IPN hydrogel at 37 °C.

Additionally, then, the hydrogels showed a difference in the swelling ratio according to the SG/CMC ratio, but all showed a similar swelling behavior. The three-dimensional structure of the hydrogel prepared using the IPN crosslinking method is expected to provide a wide space for expansion according to the pH because SG and CMC are formed based on hydrogen bonding. These swelling test results suggest that SG/CMC IPN hydrogels using a double-crosslinking strategy can effectively maintain the superabsorbency of CMC while maintaining the swelling changes with pH changes.

3.7. Drug Release Profiles of 5-FU

Figure 9 shows the drug release behavior for 16 h in pH 1.2 and pH 7.4 buffer solutions. 5-FU was released within 7 h at pH 1.2, of which 54% was released from the S1C1 gel. In contrast, at pH 7.4, 5-FU was released more rapidly than at pH 1.2. The S1C1 gel completely released the loaded drug 7 h after the start of the release. In Figure 9c, 5-FU was released from the S1C1 gel to clearly show the difference in release according to pH conditions. The buffer was changed from pH 1.2 to pH 7.4 3 h after the start of the release experiment. After changing to the pH 7.4 condition, 5-FU was rapidly released and then completely released within 7 h. These results clearly show the difference in the release pattern of 5-FU according to the pH.

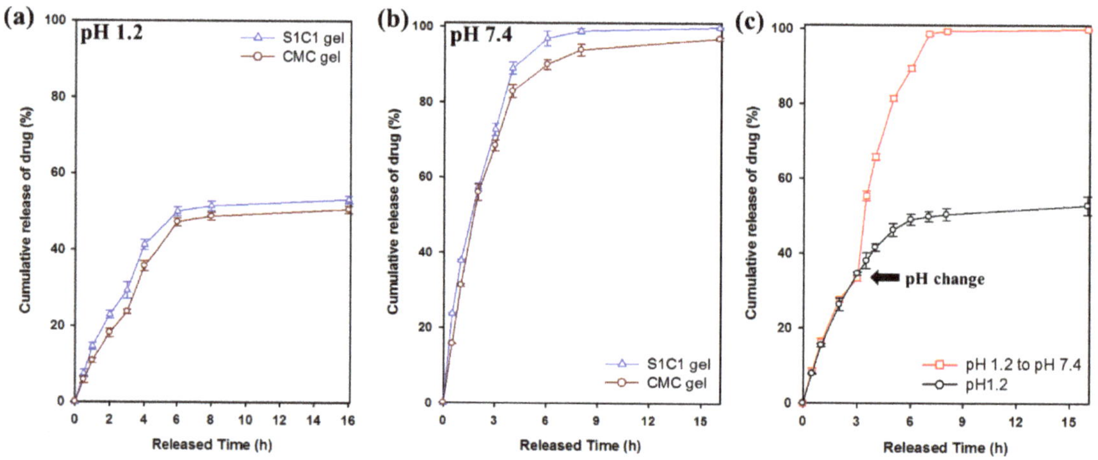

Figure 9. Cumulative release of 5-FU from of SG, CMC gel, and SG/CMC IPN hydrogels at 37 °C: (**a**) pH 1.2 condition; (**b**) pH 7.4 condition; (**c**) Change the pH from 1.2 to 7.4 after 3 h of experiment.

This pattern occurs because of the carboxylic acid groups present in SG and CMC. At pH 1.2, the hydrogel shrinks because the carboxyl group is protonated, and hydrogen bonding is strengthened. Therefore, it can slow the release of the drug. On the other hand, at pH 7.4, the release rate is increased by increasing the hydrophilicity of the polymer because the carboxylic acid group is ionized. Therefore, these results clearly suggest that SG/CMC IPN gels with higher drug release under neutral conditions can be used to deliver drugs to carriers [53]. This is the first report of pH-dependent drug release in gel form instead of bead form using unmodified natural polysaccharide. Table S1 summarizes the results of other previously published papers.

3.8. Cytotoxicity of Hydrogels

The manufactured hydrogel must be non-toxic for use in drug delivery and biomedical applications. The in vitro cytotoxicity of SG/CMC IPN gels was assessed using a WST-8 assay. Cell viability was evaluated and investigated by adding 5 mg of hydrogel to the cultured HEK-293 cells, followed by further culturing [54]. As shown in Figure 10, the cell viability of HEK-293 cells after 2 days of additional culture was more than 99.5% for all the hydrogels, showing no significant difference from the negative control group. In particular, in the case of the three SG/CMC IPN gels, cell viability was similar to that of CMC gels, which is known to be non-toxic. It has been reported that hydrogels that form IPN through Fe^{3+} coordination do not have cytotoxicity [55–59]. On the other hand, the positive control treated with DMSO showed a cell viability of 25%. The results indicate that the SG/CMC IPN gel does not cause negative effects on HEK-293 cells and can be used for biomedical applications such as drug delivery.

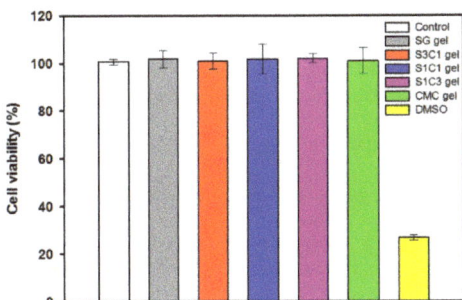

Figure 10. Cytotoxicity of SG/CMC IPN hydrogels against HEK 293 cells.

4. Conclusions

In this study, SG/CMC hydrogels were prepared by applying the IPN polymerization of SG to CMC gels via a double cross-linking strategy of two polymers using ionic cross-linking between carboxyl groups and Fe^{3+}. This hydrogel enables an effective pH-responsive drug delivery with significantly improved mechanical strength while maintaining the super absorbency, which is the unique advantage of CMC gel. The resulting SG/CMC IPN gel could control the rheological properties, the cross section, and pore size of the gels, depending on the relative ratio of SG. In addition, the prepared hydrogel exhibited a pH-responsive swelling property, and the drug release of 5-FU where the release was significantly increased at a physiological pH (pH 7.4) compared to an acidic condition (pH 1.2). In addition, the SG/CMC IPN hydrogel was not cytotoxic. Overall, our results mentioned above suggest that the prepared SG/CMC IPN hydrogels have potential for use in a variety of biomedical applications such as cosmetics, food engineering, and controlled drug delivery systems.

Supplementary Materials: The following are available online at https://www.mdpi.com/article/10.3390/polym13183197/s1, Figure S1: (a) SEM image of S1C1 gel, (b–d) EDS mapping image for C, O, and Fe, Table S1: Summary of gel characteristics, compressive test, TGA/DTG, and pH-dependent drug release results obtained in this study with other previously published papers.

Author Contributions: S.J. designed and supervised the experiments and edited the paper; Y.S. performed the experiments and wrote the paper; D.K. performed the experiments and wrote the initial manuscript; Y.H. and Y.K. performed the methodology; I.K.H. and M.S.K. validated and reviewed the paper. All authors have read and agreed to the published version of the manuscript.

Funding: This work was supported by the Technology Innovation Program (20016324, Development of biomaterials based on novel microbial exopolysacchrides) funded by the Ministry of Trade, Industry & Energy (MOTIE, Korea) and partially supported by the National Research Foundation of Korea (NRF) grants funded by the Korea government (MSIT) (NRF-2021R1A2C1013120), SDG.

Institutional Review Board Statement: Not applicable.

Informed Consent Statement: Not applicable.

Data Availability Statement: Not applicable.

Conflicts of Interest: The authors declare no conflict of interest.

References

1. Qi, X.; Wei, W.; Shen, J.; Dong, W. Salecan polysaccharide-based hydrogels and their applications: A review. *J. Mater. Chem. B* **2019**, *7*, 2577–2587. [CrossRef]
2. Hoffman, A.S. Hydrogels for biomedical applications. *Adv. Drug Deliv. Rev.* **2012**, *64*, 18–23. [CrossRef]
3. Van Vlierberghe, S.; Dubruel, P.; Schacht, E. Biopolymer-based hydrogels as scaffolds for tissue engineering applications: A review. *Biomacromolecules* **2011**, *12*, 1387–1408. [CrossRef] [PubMed]

4. Sun, C.; Jia, H.; Lei, K.; Zhu, D.; Gao, Y.; Zheng, Z.; Wang, X. Self-healing hydrogels with stimuli responsiveness based on acylhydrazone bonds. *Polymer* **2019**, *160*, 246–253. [CrossRef]
5. Feng, Q.; Wei, K.; Lin, S.; Xu, Z.; Sun, Y.; Shi, P.; Li, G.; Bian, L. Mechanically resilient, injectable, and bioadhesive supramolecular gelatin hydrogels crosslinked by weak host-guest interactions assist cell infiltration and in situ tissue regeneration. *Biomaterials* **2016**, *101*, 217–228. [CrossRef]
6. Balakrishnan, B.; Banerjee, R. Biopolymer-based hydrogels for cartilage tissue engineering. *Chem. Rev.* **2011**, *111*, 4453–4474. [CrossRef]
7. Kim, S.; Jung, S. Biocompatible and self-recoverable succinoglycan dialdehyde-crosslinked alginate hydrogels for pH-controlled drug delivery. *Carbohydr. Polym.* **2020**, *250*, 116934. [CrossRef]
8. Kumar, V.; Mittal, H.; Alhassan, S.M. Biodegradable hydrogels of tragacanth gum polysaccharide to improve water retention capacity of soil and environment-friendly controlled release of agrochemicals. *Int. J. Biol. Macromol.* **2019**, *132*, 1252–1261.
9. Crescenzi, V.; Francescangeli, A.; Taglienti, A.; Capitani, D.; Mannina, L. Synthesis and partial characterization of hydrogels obtained via glutaraldehyde crosslinking of acetylated chitosan and of hyaluronan derivatives. *Biomacromolecules* **2003**, *4*, 1045–1054. [CrossRef]
10. Ismail, H.; Irani, M.; Ahmad, Z. Starch-based hydrogels: Present status and applications. *Int. J. Polym. Mater. Polym. Biomater.* **2013**, *62*, 411–420. [CrossRef]
11. Gils, P.S.; Ray, D.; Sahoo, P.K. Characteristics of xanthan gum-based biodegradable superporous hydrogel. *Int. J. Biol. Macromol.* **2009**, *45*, 364–371. [CrossRef] [PubMed]
12. Koivisto, J.T.; Joki, T.; Parraga, J.E.; Pääkkönen, R.; Ylä-Outinen, L.; Salonen, L.; Jönkkäri, I.; Peltola, M.; Ihalainen, T.O.; Narkilahti, S. Bioamine-crosslinked gellan gum hydrogel for neural tissue engineering. *Biomed. Mater.* **2017**, *12*, 025014. [CrossRef] [PubMed]
13. Buwalda, S.J.; Vermonden, T.; Hennink, W.E. Hydrogels for therapeutic delivery: Current developments and future directions. *Biomacromolecules* **2017**, *18*, 316–330. [CrossRef] [PubMed]
14. Halder, U.; Banerjee, A.; Bandopadhyay, R. Structural and functional properties, biosynthesis, and patenting trends of Bacterial succinoglycan: A review. *Indian J. Microbiol.* **2017**, *57*, 278–284. [CrossRef]
15. Stredansky, M.; Conti, E.; Bertocchi, C.; Matulova, M.; Zanetti, F. Succinoglycan production by Agrobacterium tumefaciens. *J. Ferment. Bioeng.* **1998**, *85*, 398–403. [CrossRef]
16. Simsek, S.; Wood, K.; Reuhs, B.L. Structural analysis of succinoglycan oligosaccharides from Sinorhizobium meliloti strains with different host compatibility phenotypes. *J. Bacteriol.* **2013**, *195*, 2032–2038. [CrossRef]
17. Kim, D.; Kim, S.; Jung, S. Fabrication and Characterization of Polysaccharide Metallohydrogel Obtained from Succinoglycan and Trivalent Chromium. *Polymers* **2021**, *13*, 202. [CrossRef]
18. Hu, Y.; Jeong, D.; Kim, Y.; Kim, S.; Jung, S. Preparation of Succinoglycan Hydrogel Coordinated With Fe^{3+} Ions for Controlled Drug Delivery. *Polymers* **2020**, *12*, 977. [CrossRef]
19. Jahanbin, K.; Moini, S.; Gohari, A.R.; Emam-Djomeh, Z.; Masi, P. Isolation, purification and characterization of a new gum from Acanthophyllum bracteatum roots. *Food Hydrocoll.* **2012**, *27*, 14–21. [CrossRef]
20. McKellar, R.; Van Geest, J.; Cui, W. Influence of culture and environmental conditions on the composition of exopolysaccharide produced by Agrobacterium radiobacter. *Food Hydrocoll.* **2003**, *17*, 429–437. [CrossRef]
21. Barkhordari, S.; Yadollahi, M.; Namazi, H. pH sensitive nanocomposite hydrogel beads based on carboxymethyl cellulose/layered double hydroxide as drug delivery systems. *J. Polym. Res.* **2014**, *21*, 454. [CrossRef]
22. Nnadi, F.; Brave, C. Environmentally friendly superabsorbent polymers for water conservation in agricultural lands. *J. Soil Sci. Environ. Manag.* **2011**, *2*, 206–211.
23. Elbarbary, A.M.; Ghobashy, M.M. Controlled release fertilizers using superabsorbent hydrogel prepared by gamma radiation. *Radiochim. Acta* **2017**, *105*, 865–876. [CrossRef]
24. Zohourian, M.M.; Kabiri, K. Superabsorbent polymer materials: A review. *Iran. Polym. J.* **2008**, *17*, 451–477.
25. Lin, F.; Lu, X.; Wang, Z.; Lu, Q.; Lin, G.; Huang, B.; Lu, B. In situ polymerization approach to cellulose–polyacrylamide interpenetrating network hydrogel with high strength and pH-responsive properties. *Cellulose* **2019**, *26*, 1825–1839. [CrossRef]
26. Jeong, D.; Kim, C.; Kim, Y.; Jung, S. Dual crosslinked carboxymethyl cellulose/polyacrylamide interpenetrating hydrogels with highly enhanced mechanical strength and superabsorbent properties. *Eur. Polym. J.* **2020**, *127*, 109586. [CrossRef]
27. Chen, W.; Bu, Y.; Li, D.; Liu, C.; Chen, G.; Wan, X.; Li, N. High-strength, tough, and self-healing hydrogel based on carboxymethyl cellulose. *Cellulose* **2020**, *27*, 853–865. [CrossRef]
28. Jeong, D.; Joo, S.-W.; Hu, Y.; Shinde, V.V.; Cho, E.; Jung, S. Carboxymethyl cellulose-based superabsorbent hydrogels containing carboxymehtyl β-cyclodextrin for enhanced mechanical strength and effective drug delivery. *Eur. Polym. J.* **2018**, *105*, 17–25. [CrossRef]
29. Janarthanan, G.; Shin, H.S.; Kim, I.-G.; Ji, P.; Chung, E.-J.; Lee, C.; Noh, I. Self-crosslinking hyaluronic acid–carboxymethylcellulose hydrogel enhances multilayered 3D-printed construct shape integrity and mechanical stability for soft tissue engineering. *Biofabrication* **2020**, *12*, 045026. [CrossRef]
30. Kim, M.S.; Park, S.J.; Gu, B.K.; Kim, C.-H. Ionically crosslinked alginate–carboxymethyl cellulose beads for the delivery of protein therapeutics. *Appl. Surf. Sci.* **2012**, *262*, 28–33. [CrossRef]
31. Agarwal, T.; Narayana, S.G.H.; Pal, K.; Pramanik, K.; Giri, S.; Banerjee, I. Calcium alginate-carboxymethyl cellulose beads for colon-targeted drug delivery. *Int. J. Biol. Macromol.* **2015**, *75*, 409–417. [CrossRef] [PubMed]

32. Bhattacharya, S.S.; Shukla, S.; Banerjee, S.; Chowdhury, P.; Chakraborty, P.; Ghosh, A. Tailored IPN hydrogel bead of sodium carboxymethyl cellulose and sodium carboxymethyl xanthan gum for controlled delivery of diclofenac sodium. *Polym. Plast. Technol. Eng.* **2013**, *52*, 795–805. [CrossRef]
33. Park, J.S.; Yeo, J.H. Swelling Properties of Chitosan/CMC/PEGDA based Semi-IPN Hydrogel. In Proceedings of the 4th World Congress on New Technologies, Madrid, Spain, 19–21 August 2018.
34. Aalaie, J.; Vasheghani-Farahani, E.; Rahmatpour, A.; Semsarzadeh, M.A. Gelation rheology and water absorption behavior of semi-interpenetrating polymer networks of polyacrylamide and carboxymethyl cellulose. *J. Macromol. Sci. Part B* **2013**, *52*, 604–613. [CrossRef]
35. Kang, M.; Oderinde, O.; Liu, S.; Huang, Q.; Ma, W.; Yao, F.; Fu, G. Characterization of Xanthan gum-based hydrogel with Fe^{3+} ions coordination and its reversible sol-gel conversion. *Carbohydr. Polym.* **2019**, *203*, 139–147. [CrossRef]
36. Cho, E.; Choi, J.M.; Kim, H.; Tahir, M.N.; Choi, Y.; Jung, S. Ferrous iron chelating property of low-molecular weight succinoglycans isolated from Sinorhizobium meliloti. *Biometals* **2013**, *26*, 321–328. [CrossRef]
37. Moosavi-Nasab, M.; Taherian, A.R.; Bakhtiyari, M.; Farahnaky, A.; Askari, H. Structural and rheological properties of succinoglycan biogums made from low-quality date syrup or sucrose using agrobacterium radiobacter inoculation. *Food Bioprocess Technol.* **2012**, *5*, 638–647. [CrossRef]
38. Yang, X.; Xu, G. The influence of xanthan on the crystallization of calcium carbonate. *J. Cryst. Growth* **2011**, *314*, 231–238. [CrossRef]
39. Pushpamalar, V.; Langford, S.J.; Ahmad, M.; Lim, Y.Y. Optimization of reaction conditions for preparing carboxymethyl cellulose from sago waste. *Carbohydr. Polym.* **2006**, *64*, 312–318. [CrossRef]
40. Yadollahi, M.; Namazi, H. Synthesis and characterization of carboxymethyl cellulose/layered double hydroxide nanocomposites. *J. Nanoparticle Res.* **2013**, *15*, 1563. [CrossRef]
41. Singh, T.; Trivedi, T.J.; Kumar, A. Dissolution, regeneration and ion-gel formation of agarose in room-temperature ionic liquids. *Green Chem.* **2010**, *12*, 1029–1035. [CrossRef]
42. Wang, B.; Liao, L.; Huang, Q.; Cheng, Y. Adsorption behaviors of benzonic acid by carboxyl methyl konjac glucomannan gel microspheres cross-linked with Fe^{3+}. *J. Chem. Eng. Data* **2012**, *57*, 72–77. [CrossRef]
43. Swamy, B.Y.; Yun, Y.-S. In vitro release of metformin from iron (III) cross-linked alginate–carboxymethyl cellulose hydrogel beads. *Int. J. Biol. Macromol.* **2015**, *77*, 114–119. [CrossRef]
44. El-Sakhawy, M.; Kamel, S.; Salama, A.; Tohamy, H.-A.S. Preparation and infrared study of cellulose based amphiphilic materials. *Cell. Chem. Technol.* **2018**, *52*, 193–200.
45. Chang, C.; Duan, B.; Cai, J.; Zhang, L. Superabsorbent hydrogels based on cellulose for smart swelling and controllable delivery. *Eur. Polym. J.* **2010**, *46*, 92–100. [CrossRef]
46. Shen, D.; Gu, S.; Bridgwater, A.V. Study on the pyrolytic behaviour of xylan-based hemicellulose using TG–FTIR and Py–GC–FTIR. *J. Anal. Appl. Pyrolysis* **2010**, *87*, 199–206. [CrossRef]
47. Kim, S.; Jeong, D.; Lee, H.; Kim, D.; Jung, S. Succinoglycan dialdehyde-reinforced gelatin hydrogels with toughness and thermal stability. *Int. J. Biol. Macromol.* **2020**, *149*, 281–289. [CrossRef]
48. El-Mohdy, H.A. Radiation initiated synthesis of 2-acrylamidoglycolic acid grafted carboxymethyl cellulose as pH-sensitive hydrogel. *Polym. Eng. Sci.* **2014**, *54*, 2753–2761. [CrossRef]
49. Betancourt, T.; Pardo, J.; Soo, K.; Peppas, N.A. Characterization of pH-responsive hydrogels of poly (itaconic acid-g-ethylene glycol) prepared by UV-initiated free radical polymerization as biomaterials for oral delivery of bioactive agents. *J. Biomed. Mater. Res. Part A Off. J. Soc. Biomater. Jpn. Soc. Biomater. Aust. Soc. Biomater. Korean Soc. Biomater.* **2010**, *93*, 175–188. [CrossRef] [PubMed]
50. Bao, Y.; Ma, J.; Li, N. Synthesis and swelling behaviors of sodium carboxymethyl cellulose-g-poly (AA-co-AM-co-AMPS)/MMT superabsorbent hydrogel. *Carbohydr. Polym.* **2011**, *84*, 76–82. [CrossRef]
51. Bukhari, S.M.H.; Khan, S.; Rehanullah, M.; Ranjha, N.M. Synthesis and characterization of chemically cross-linked acrylic acid/gelatin hydrogels: Effect of pH and composition on swelling and drug release. *Int. J. Polym. Sci.* **2015**, *2015*, 187961. [CrossRef]
52. Reis, A.V.; Guilherme, M.R.; Cavalcanti, O.A.; Rubira, A.F.; Muniz, E.C. Synthesis and characterization of pH-responsive hydrogels based on chemically modified Arabic gum polysaccharide. *Polymer* **2006**, *47*, 2023–2029. [CrossRef]
53. Bashir, S.; Hina, M.; Iqbal, J.; Rajpar, A.; Mujtaba, M.; Alghamdi, N.; Wageh, S.; Ramesh, K.; Ramesh, S. Fundamental concepts of hydrogels: Synthesis, properties, and their applications. *Polymers* **2020**, *12*, 2702. [CrossRef] [PubMed]
54. Capella, V.; Rivero, R.E.; Liaudat, A.C.; Ibarra, L.E.; Roma, D.A.; Alustiza, F.; Mañas, F.; Barbero, C.A.; Bosch, P.; Rivarola, C.R. Cytotoxicity and bioadhesive properties of poly-N-isopropylacrylamide hydrogel. *Heliyon* **2019**, *5*, e01474. [CrossRef] [PubMed]
55. Lu, L.; Tian, T.; Wu, S.; Xiang, T.; Zhou, S. A pH-induced self-healable shape memory hydrogel with metal-coordination cross-links. *Polym. Chem.* **2019**, *10*, 1920–1929. [CrossRef]
56. Han, N.; Xu, Z.; Cui, C.; Li, Y.; Zhang, D.; Xiao, M.; Fan, C.; Wu, T.; Yang, J.; Liu, W. A Fe^{3+}-crosslinked pyrogallol-tethered gelatin adhesive hydrogel with antibacterial activity for wound healing. *Biomater. Sci.* **2020**, *8*, 3164–3172. [CrossRef]
57. Wang, Q.; Pan, X.; Lin, C.; Ma, X.; Cao, S.; Ni, Y. Ultrafast gelling using sulfonated lignin-Fe^{3+} chelates to produce dynamic crosslinked hydrogel/coating with charming stretchable, conductive, self-healing, and ultraviolet-blocking properties. *Chem. Eng. J.* **2020**, *396*, 125341. [CrossRef]

58. Thakur, N.; Sharma, B.; Bishnoi, S.; Jain, S.; Nayak, D.; Sarma, T.K. Biocompatible Fe^{3+} and Ca^{2+} dual cross-linked G-quadruplex hydrogels as effective drug delivery system for pH-responsive sustained zero-order release of doxorubicin. *ACS Appl. Bio Mater.* **2019**, *2*, 3300–3311. [CrossRef]
59. Patwa, R.; Zandraa, O.; Capáková, Z.; Saha, N.; Sáha, P. Effect of iron-oxide nanoparticles impregnated bacterial cellulose on overall properties of alginate/casein hydrogels: Potential injectable biomaterial for wound healing applications. *Polymers* **2020**, *12*, 2690. [CrossRef]

Article

pH and Reduction Dual-Responsive Bi-Drugs Conjugated Dextran Assemblies for Combination Chemotherapy and In Vitro Evaluation

Xiukun Xue [1], Yanjuan Wu [1,*], Xiao Xu [2], Ben Xu [1], Zhaowei Chen [2,*] and Tianduo Li [1,*]

[1] Shandong Provincial Key Laboratory of Molecular Engineering, School of Chemistry and Chemical Engineering, Qilu University of Technology (Shandong Academy of Science), Jinan 250353, China; Jasonxue@yeah.net (X.X.); xuben2019@126.com (B.X.)

[2] Institute of Food Safety and Environment Monitoring, College of Chemistry, Fuzhou University, Fuzhou 350108, China; 201310023@fzu.edu.cn

* Correspondence: wuyanjuan5@qlu.edu.cn (Y.W.); chenzw@fzu.edu.cn (Z.C.); ylpt6296@vip.163.com (T.L.)

Citation: Xue, X.; Wu, Y.; Xu, X.; Xu, B.; Chen, Z.; Li, T. pH and Reduction Dual-Responsive Bi-Drugs for Combination Chemotherapy and In Vitro Evaluation. *Polymers* **2021**, *13*, 1515. https://doi.org/10.3390/polym13091515

Academic Editor: Luis García-Fernández

Received: 22 March 2021
Accepted: 28 April 2021
Published: 8 May 2021

Publisher's Note: MDPI stays neutral with regard to jurisdictional claims in published maps and institutional affiliations.

Copyright: © 2021 by the authors. Licensee MDPI, Basel, Switzerland. This article is an open access article distributed under the terms and conditions of the Creative Commons Attribution (CC BY) license (https://creativecommons.org/licenses/by/4.0/).

Abstract: Polymeric prodrugs, synthesized by conjugating chemotherapeutic agents to functional polymers, have been extensively investigated and employed for safer and more efficacious cancer therapy. By rational design, a pH and reduction dual-sensitive dextran-di-drugs conjugate (oDex-g-Pt+DOX) was synthesized by the covalent conjugation of Pt (IV) prodrug and doxorubicin (DOX) to an oxidized dextran (oDex). Pt (IV) prodrug and DOX were linked by the versatile efficient esterification reactions and Schiff base reaction, respectively. oDex-g-Pt+DOX could self-assemble into nanoparticles with an average diameter at around 180 nm. The acidic and reductive (GSH) environment induced degradation and drug release behavior of the resulting nanoparticles (oDex-g-Pt+DOX NPs) were systematically investigated by optical experiment, DLS analysis, TEM measurement, and in vitro drugs release experiment. Effective cellular uptake of the oDex-g-Pt+DOX NPs was identified by the human cervical carcinoma HeLa cells via confocal laser scanning microscopy. Furthermore, oDex-g-Pt+DOX NPs displayed a comparable antiproliferative activity than the simple combination of free cisplatin and DOX (Cis+DOX) as the extension of time. More importantly, oDex-g-Pt+DOX NPs exhibited remarkable reversal ability of tumor resistance compared to the cisplatin in cisplatin-resistant lung carcinoma A549 cells. Take advantage of the acidic and reductive microenvironment of tumors, this smart polymer-dual-drugs conjugate could serve as a promising and effective nanomedicine for combination chemotherapy.

Keywords: polymeric prodrug; dual-sensitive; combination chemotherapy; drug conjugation; dextran

1. Introduction

Nanosized drug delivery systems (nDDSs) with an ability to load multiple chemotherapeutic agents simultaneously have gained considerable interest for precise and efficient cancer therapy [1–6]. The nDDSs currently developed for combination chemotherapy include inorganic nanoparticles (NPs), liposomes, organic-inorganic hybrid materials, and polymeric NPS, etc. [7–9]. Among various nDDSs, polymeric drug delivery systems have played an integral role due to several additional advantageous properties, such as biocompatibility/biodegradability, synthetic flexibility, and tailorable properties [10–13]. Encapsulation of dual drugs into polymeric NPs or conjugation of one drug onto the polymer backbones while loading another drug into the formed nanosystems are among the most pervasive strategies [14–18]. As an example, recently, a novel dual-drugs delivery system composed of nanocrystalline cellulose and L-lysine were developed for efficient co-encapsulation of model chemotherapeutic curcumin and methotrexate [19]. In another study, disulfide-containing camptothecin (CPT) was conjugated to poly (L-glutamic acid)-graft-methoxy poly (ethylene glycol) (PLG-g-mPEG) via esterification reaction for the preparation of an amphiphilic biodegradable prodrug (PLG-g-mPEG/CPT) [18]. The prodrug could co-assemble with DOX into micellar NPs for combination cancer therapy. Previous reports demonstrated the application of these polymeric nDDSs for combination chemotherapy could simply conquer the problem of poor water solubility of most hydrophobic chemotherapeutic agents, improve their biodistribution, enhance treatment efficiency, and reduce the systemic toxicity. Moreover, an effective way to further enhance the chemotherapeutic efficiency is to adopt tumor site-specific responsive linkage for drug conjugation. Specifically, it is well-known that the corresponding pH value of tumor extracellular microenvironment, endosomes, and lysosomes is approximately 6.8, 5.5–6.0, and 4.5–5.0, which is lower than the physiological pH 7.4 [20–22]. Thus, various acidic pH-cleavable linkages, such as hydrazone, orthoester, acetal, and carbamate have been successfully designed and synthesized for drug conjugation to realize efficient intracellular drug release [23]. In addition, the high concentration gradient of glutathione (GSH), a thiol-containing tripeptide, between the intracellular (1–10 mM) and extracellular areas (2–20 µM) is frequently used as a reductive stimuli to boost the drug release from nanocarriers [24]. A series of reduction-responsive linkages have been developed for drug conjugation. For example, CPT was conjugated to the side chains of poly (methacrylate) via thioether bond, which could be cleaved by GSH via thiolysis [24]. Notably, a prominent hallmark of tumor cell is the heterogeneous coexistence of lower pH environment and overproduced intracellular GSH compared with those in blood and normal cells [24,25]. Accordingly, the pH and reduction dual-sensitive prodrug nDDSs have recently received increasing attention.

Dextran, with α-1,6-glycosidically linked glucose chains, is a natural and highly water-soluble polymer, which has great utility in biomedical applications. Dextran has many hydroxyl groups that can be readily chemically modified. Importantly, dextran is a kind of FDA-approved natural polymer. Several studies have reported the use of dextran and its derivatives as nanocarriers for cancer imaging and therapy [26–30]. Herein, we prepare a novel pH and reduction dual-sensitive polymeric prodrug oDex-g-Pt+DOX, in which Pt (IV) prodrug was conjugated to the oxidized dextran (oDex) as reduction-sensitive segment, and then DOX was conjugated via acid-cleavable hydrazone bond (Scheme 1). The amphiphilic oDex-g-Pt+DOX could further self-assemble into well-defined NPs in aqueous solution with the average diameters of ~180 nm. Considering that DOX can restrain the DNA remodeling, and further inhibit the repairment of cisplatin-damaged duplex DNA by suppressing the activity of topoisomerase II, simultaneous delivery of Pt (IV) and DOX may work in a synergistic way to overcome drug resistance. Our results revealed that this novel polymeric prodrug oDex-g-Pt+DOX NPs presented pH and reduction dual-triggered drug release behavior, could effectively kill cancer cells after intracellular internalization, and reverse cisplatin resistance in cisplatin-resistant lung carcinoma A549 cells (A549/DDP cells).

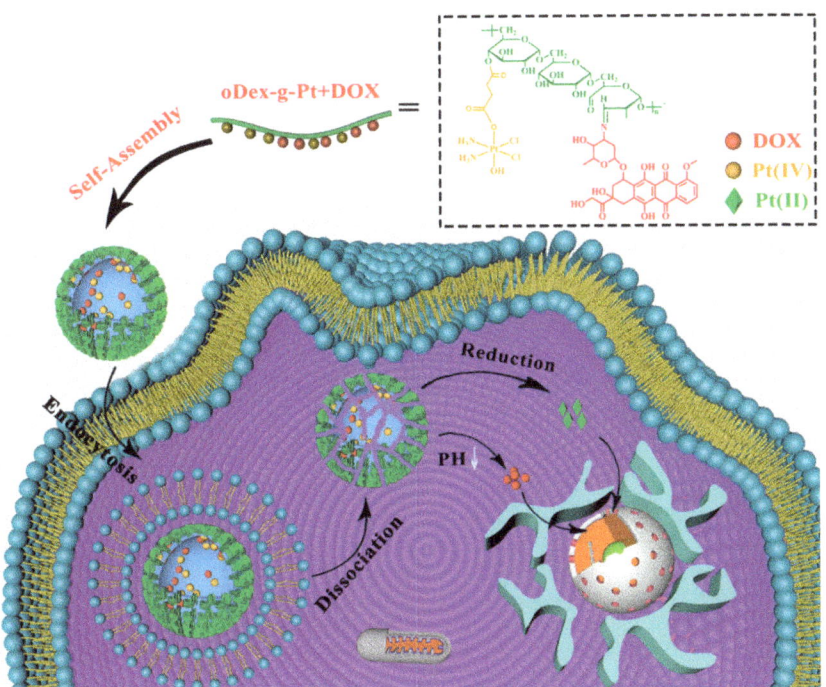

Scheme 1. Schematic representation of the self-assembly of oDex-g-Pt+DOX (platinum plus DOX conjugated oxidized dextran), and its intracellular action after endocytosis for combinational chemotherapy.

2. Materials and Methods

2.1. Materials

Dextran10 k, 3-(4,5-dimethylthiazol-2-yl)-2,5-diphenyltetrazolium bromide (MTT, 98%), sodium metaperiodate (NaIO$_4$, ≥99%), doxorubicin hydrochloride (DOX·HCl, 98%), and Hoechst 33,258 (≥98%) was obtained from Sigma-Aldrich (Shanghai, China). Cisplatin (99.8%) was bought from Shandong Boyuan Pharmaceuti-cal Co., Ltd. (Jinan, Shandong, China). Hydrogen peroxide solution (H$_2$O$_2$, 30 wt% in water), succinic anhydride (99%), triethylamine (TEA, ≥99.5%), and 1-ethyl-3-(3-dimethylaminopropyl) carbodiimide hydrochloride (EDC·HCl, ≥98%) were purchased from Aladdin Chemistry Co. Ltd. (Shanghai, China). Phosphate-buffered saline (PBS), dulbecco's modified eagle's medium (DMEM)/high-glucose medium, and trypsin were obtained from GE Healthcare Life Sciences (Beijing, China). Gibco Fetal Bovine Serum (FBS) was purchased from Thermo Fisher Scientific Inc. (Shanghai, China). Lyso-tracker was purchased from Shanghai Biyuntian Biological Co., Ltd. (Shanghai, China). All chemical reagents were used as received without purification if not mentioned otherwise. Dimethyl sulfoxide (DMSO, ≥99%) was purchased from Sinopharm Chemical Reagent Co., Ltd. (Shanghai, China), and distilled followed by dried with calcium hydride (CaH$_2$) for 2 weeks.

2.2. Measurements

Proton nuclear magnetic resonance (^{1}H NMR) spectra were collected at room temperature (rt) on a Bruker AVANCE II 400 NMR spectrometer using deuterated dimethyl sulfoxide (DMSO-d$_6$) as solvent. Fourier transform infrared (FT-IR) spectra were measured on a Thermo Scientific Nicolet IS10 instrument. The platinum content was recorded on an inductively coupled plasma mass spectrometer (ICP-MS) and inductively coupled plasma optical emission spectrometer (ICP-OES, Thermoscientific, USA). UV-visible absorption spectra were used for quantitative determination of the DOX concentration by a Cary 5000 UV-Vis-NIR spectrometer. The diameters and polydispersity index (PDI) measurements were performed at rt on a Malvern Zetasizer Nano ZS90 instrument with a vertically polarized He-Ne laser. The morphology of the NPs was investigated by transmission electron microscopy (TEM) using a JEOL JEM 2100 electron microscope. Confocal laser scanning microscopy (CLSM) micrographs were visualized with a Leica SP8 CLSM image system. Quantitative DOX fluorescent analysis of cellular uptake was investigated by BD FACSCalibur flow cytometry imaging system.

2.3. Synthesis of Oxidized Dextran (oDex)

Firstly, Dex (2.0 g) was dissolved in 20 mL distilled water under 50 °C for 20 min. Then, $NaIO_4$ solution (53 mg·mL^{-1}, 20 mL) was added to the above transparent solution for oxidation process. After stirring for 6 h at rt, 2 mL ethylene glycol was added in order to stop the oxidation reaction. The mixture was purified by dialysis (MWCO: 7000 Da) against distilled water for 3 days and finally lyophilized.

2.4. General Procedure for Synthesis of Platinum Conjugated oDex (oDex-g-Pt)

The synthesis of diamminedichloro-dihydroxyplatinum (c,c,t-$[Pt(NH_3)_2Cl_2(OH)_2]$, DHP) and succinic anhydride modified DHP (c,c,t-$[Pt(NH_3)_2Cl_2(OH)(O_2CCH_2CH_2CH_2CO_2H)]$, Pt (IV) prodrug) were presented in the Supporting Information according to the previous reports (Scheme S1) [31]. In general, oDex (1.0 g), and Pt (IV) (1085 mg, 2.5 mmol) were dissolved in dried DMSO (40 mL) under argon, then EDC·HCl (4793 mg, 25 mmol) and DMAP (305 mg, 2.5 mmol) were added. The mixture was vigorously stirred at rt for 72 h while the argon inert atmosphere was maintained. The solution was dialyzed against distilled water for 48 h using a pre-swelled Spectra/Por Regenerated Cellulose membrane (MWCO: 3500 Da) to remove DMSO, unreacted Pt (IV), and condensation reagents. Finally, the light yellow solution was freeze-dried to obtain oDex-g-Pt conjugates.

2.5. Conjugation of DOX with oDex-g-Pt

DOX was incorporated onto the pendants of oDex-g-Pt through a hydrazone bond. DOX·HCl (10 mg) was dissolved in dried DMSO (1 mL) with the help of TEA (20 µL) to remove HCl. The oDex-g-Pt (100 mg) was also dissolved in 10 mL of dried DMSO. Then, those two solutions were mixed and reacted at rt in dark for another 72 h. Afterwards, the mixture was dialyzed (MWCO: 3500 Da) against DMSO for 24 h to remove unconjugated DOX, and then against the phosphate buffer saline (PBS, pH 7.4) for 48 h to remove DMSO. Finally, the DOX conjugated oDex-g-Pt product (oDex-g-Pt+DOX) was obtained as dark red powder after lyophilization. oDex-g-DOX, a control, was similarly prepared by covalent conjugation of DOX to the oDex (Scheme S2).

2.6. Preparation and Characterization of oDex-g-Pt and oDex-g-Pt+DOX NPs

Briefly, oDex-g-Pt+DOX (10 mg) was dissolved in DMSO (2 mL), and stirred at rt in dark for 12 h. The solution was dropwise injected into Milli-Q water (8 mL) within 10 min and continuously stirred for another 4 h. Then the suspension was dialyzed (MWCO: 3500 Da) against Milli-Q water for 48 h to prepare the oDex-g-Pt+DOX NPs. The oDex-g-Pt NPs and oDex-g-DOX NPs were also prepared similarly. The morphology, diameters and PDI of oDex-g-DOX NPs, oDex-g-Pt NPs and oDex-g-Pt+DOX NPs in Milli-Q water were determined by TEM and DLS measurements. The content of Pt in the oDex-g-Pt NPs and oDex-g-Pt+DOX NPs were measured by ICP-OES. UV–vis spectra of DOX and oDex-g-Pt+DOX NPs were recorded. The content of DOX in the oDex-g-Pt+DOX NPs was then calculated by using a DOX standard curve. The Pt and DOX loading content (DLC) were calculated as following equations:

$$DLC\ (\%) = [\text{drug in NPs/total weight of NPs}] \times 100\% \tag{1}$$

2.7. pH- and Reduction-Sensitivity of oDex-g-Pt+DOX NPs

Changes in the particle sizes and size distributions of oDex-g-Pt+DOX NPs were observed using DLS and TEM measurements after incubation in PBS (pH = 7.4, 0.1 M) buffer containing GSH at 0 mM and 10 mM. The colloidal stability of oDex-g-Pt+DOX NPs in acetate buffer solution (ABS, pH = 5.0) containing GSH at 0 mM and 10 mM was also characterized. In a typical experiment, oDex-g-Pt+DOX NPs (1 mg/mL) were prepared in PBS buffer (pH = 7.4) as above, GSH was added at specified concentrations with continuous stirring. At predetermined time intervals, 3 mL of sample was used for DLS characterization, the average diameter and PDI were recorded. Moreover, the morphology changes of oDex-g-Pt+DOX NPs were analyzed by TEM. Tyndall effect of oDex-g-Pt+DOX NPs in pH 7.4 and pH 5.0 with 10mM GSH for 24 h were also measured.

2.8. pH- and Reduction-Activated Drugs Release from oDex-g-Pt+DOX NPs

The DOX and Pt release profiles from oDex-g-Pt+DOX NPs were investigated using a dialysis technique. In general, the lyophilized oDex-g-Pt+DOX NPs (2 mg) was resuspended in 2 mL of PBS (0.1 M, pH = 7.4), PBS (0.1 M, pH = 7.4) containing 10 mM GSH, ABS (0.1 M, pH = 5.0), or ABS (0.1 M, pH = 5.0) containing 10 mM GSH. Samples were placed into Spectra/Por Regenerated Cellulose membrane (MWCO: 3500 Da), and immersed into the corresponding buffer solutions (18 mL). Dialysis was continued in a shaking culture incubator at 37 °C, and 2 mL aliquots were sampled and replenished with equivalent volumes of fresh buffer at desired time points. Concentrations of DOX released into the aqueous solution from oDex-g-Pt+DOX NPs in acidic and reductive environments were quantitatively measured using UV−vis spectroscopy, and the percentages of DOX release were calculated by interpolation from its standard calibration curve. Simultaneously, the Pt concentration was detected via ICP-MS after nitrolysis. The drugs released from oDex-g-Pt+DOX NPs were expressed as percentage of cumulative drugs in the solution to the total drugs in the oDex-g-Pt+DOX NPs as a function of time.

2.9. Cell Culture

Human cervical carcinoma HeLa cells, normal fibroblasts L929 cells, lung carcinoma A549 cells, and A549/DDP cells were obtained from the institute of biochemistry and cell biology, Chinese academy of sciences (Shanghai, China). The cell culture medium was DMEM containing 10% FBS and 100 IU/mL penicillin-streptomycin, and was replaced every 2 days to keep the exponential growth of cells. The cells were incubated under 5% CO_2 at 37 °C.

2.10. In Vitro Cytotoxicity

The MTT assay was used to assess the biocompatibility of dextran and oDex against L929 and HeLa cells according to the previous reports [10]. For L929 test, cells at 5×10^3 cells/well were placed into 96-plates and incubated with DMEM (100 μL) for 24 h. Then, the culture medium was replenished with 200 μL of fresh DMEM medium containing different amounts of dextran or oDex (31.25, 62.5, 125, 250, 500 and 1000 μg/mL) for another 48 h incubation, respectively. MTT solution (20 μL, 5 mg/mL in PBS) was added to each well for another 4 h co-incubation at 37 °C, followed by removal of solution containing MTT. DMSO (150 μL) was added to each well to dissolve the purple formazan crystals formed by live cells. Finally, after shaking for 10 min, the absorbance of purple formazan product was recorded at 490 nm by a microplate reader. The cellular viability was calculated according to the following equation:

$$\text{Cellular viability (\%)} = [\text{OD of the treated cells}/\text{OD of control cells}] \times 100\% \qquad (2)$$

(optical density: OD).

The nano formulations oDex-g-DOX NPs, oDex-g-Pt NPs, oDex-g-Pt+DOX NPs, free cisplatin (Cis), free DOX, Pt (IV) and the combinational forms were also evaluated for in vitro cytotoxicity utilizing HeLa, A549, and A549/DDP cell lines by the standard MTT assay. Briefly, after HeLa cells were cultivated as above, the culture medium was replaced by Cis, DOX, Pt (IV), Cis + DOX, Pt (IV) + DOX, oDex-g-Pt NPs, and oDex-g-Pt+DOX NPs at a final DOX concentration from 2.437–156 µM or Pt concentration from 3.375–216 µM for 48 h or 72 h of incubation. Then, MTT assay was completed as above.

The half maximal inhibitory concentration (IC_{50} value) is the drug concentration that can inhibit or kill 50% of the cells.

2.11. In Vitro Cellular Uptake and Intracellular Distribution

The in vitro cellular uptake efficiency and intracellular drugs distribution of oDex-g-Pt+DOX NPs was investigated in HeLa cells using the CLSM technique. HeLa cells were seeded on the microscope slides in a 6-well plate with 3×10^5 cells/well with DMEM medium (2 mL), and cultured overnight at 37 °C in a humidified atmosphere containing 5% CO_2. Then, the medium was removed, 2 mL of oDex-g-Pt+DOX NPs solution (5 µg DOX/mL) was added. After further 0.5 h or 4 h incubation, cells were gently rinsed with cold PBS. Subsequently, cells were co-incubated with Lyso-tracker (1 mL, 50 nM) medium for another 30 min. After that, the cells were washed twice with cold PBS (10 mM, pH = 7.4) and fixed with paraformaldehyde solution (4% in 10 mM PBS, pH = 7.4) for 20 min. Then, the cell nuclei were dyed with Hoechst 33,258 (10 µg/mL, 1 mL) for 10 min, and the in vitro intracellular fluorescence was recorded for Hoechst 33,258 (405 nm), DOX (480 nm), and Lyso-tracker (555 nm) on a Leica SP8 CLSM image system.

Flow cytometric analysis was also used to evaluate the in vitro cellular uptake efficiency of free DOX and oDex-g-Pt+DOX NPs. For flow cytometry measurements (FCM), HeLa cells were planted into 12-well plates at a density of 1×10^5 cells/well and cultured for 24 h. Subsequently, the DMEM culture medium was discarded, and cells were treated with free DOX or oDex-g-Pt+DOX NPs a fixed DOX-equivalent concentration, followed by 0.5 h or 4 h incubation. Afterwards, the cells were washed with PBS (10 mM, pH = 7.4) and detached by trypsin. After centrifugation, the cells were resuspended in cold PBS (10 mM, pH = 7.4, 0.5 mL) for flow cytometry measurements.

2.12. Statistical Analysis

All the experiments had three replicates ($n = 3$) at least. Data were presented as mean standard deviation (SD). The one-way ANOVA analysis with a Tukey post-hoc test was carried out to analyze the statistical significance: * $p < 0.5$, ** $p < 0.01$, *** $p < 0.001$, **** $p < 0.0001$.

3. Results

3.1. Design, Synthesis and Characterization of Drug, Drug-Polymer Conjugates

Dextran, a kind of hydrophilic and biocompatible polymer with massive side hydroxyl groups, has been extensively studied for biomedical applications, such as antibacterial biofilm, drug delivery particles, bioink for 3D printing, and so on [21,32]. DOX is one of the most commonly used and effective antineoplastic drugs with outstanding activity against various solid tumors. DOX can inhibit the synthesis of DNA by intercalating into DNA duplex and further inhibiting nucleic acid biosynthesis. Cisplatin, a leading platinum-based chemotherapeutic drug, has been widely applied for the treatment of various cancers, including breast, ovarian, head and neck, bladder, thyroid, prostate, and non small cell lung cancer [33]. Although cisplatin has gained great success in clinical cancer treatment, there are still some crucial concerns associated with the use of cisplatin, on account of cisplatin resistance and significant side effects. Many cancer cells have intrinsic and/or quickly acquired resistance, leading to unsatisfactory therapeutic efficacy. Meanwhile, cisplatin has severe side effects to the normal tissues, particularly chronic neurotoxicity and acute nephrotoxicity, which is mainly due to the lack of specificity. More importantly, previous reports have indicated that the combination of DOX and cisplatin could enhance the anticancer efficiency, decrease the side effects, and further reverse the cisplatin resistance [34,35]. As shown in Scheme 2, in our design, dextran was oxidized by NaIO$_4$ to prepare the oDex with abundant of side hydroxyl and aldehyde groups. Subsequently, Pt (IV) prodrug, a cisplatin derivative, and DOX were gradually grafted to oDex through coupling reactions to prepare oDex-g-Pt+DOX conjugates. The synthesized molecules and polymers were listed in Table 1.

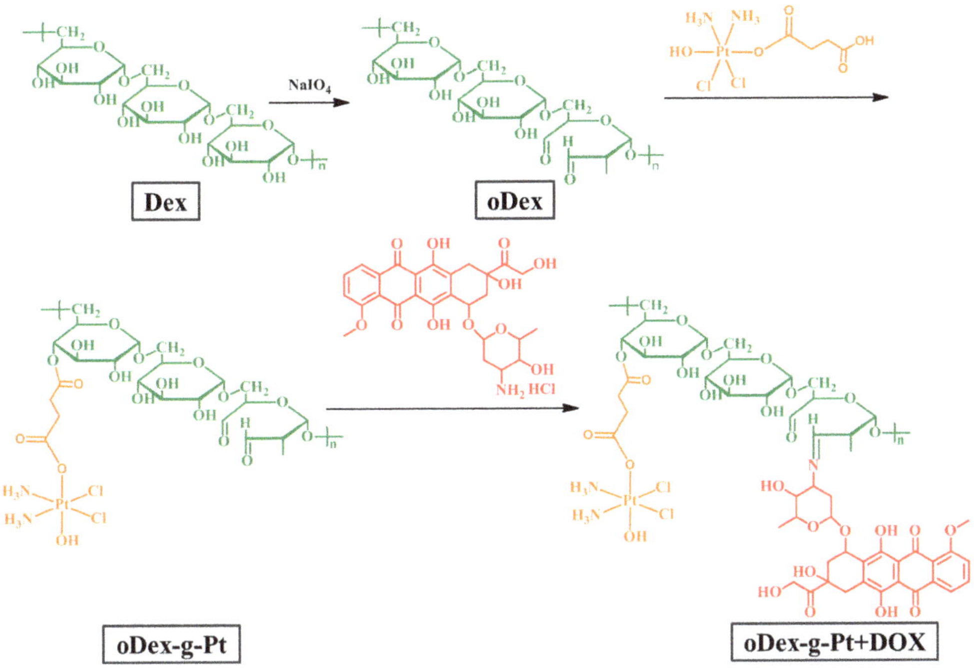

Scheme 2. Synthetic scheme of the oDex (oxidized dextran), oDex-g-Pt (platinum conjugated oxidized dextran), and oDex-g-Pt+DOX (platinum plus DOX conjugated oxidized dextran).

Table 1. The full name and corresponding abbreviation for the synthesized molecules and polymers.

Full Name	Abbreviation
diamminedichloro-dihydroxyplatinum	DHP
succinic anhydride modified DHP	Pt (IV)
oxidized dextran	oDex
platinum conjugated oxidized dextran	oDex-g-Pt
DOX conjugated oxidized dextran	oDex-g-DOX
platinum plus DOX conjugated oxidized dextran	oDex-g-Pt+DOX

Initially, DHP was functionalized with succinic anhydride to synthesize a Pt (IV) prodrug with one carboxyl group as axial ligand. The structures of DHP, Pt (IV) prodrug, and oDex-g-DOX were characterized using ^{1}H NMR and FT-IR spectra (Figures S1–S3). The chemical structures of oDex, oDex-g-Pt, and oDex-g-Pt+DOX were also confirmed by both ^{1}H NMR (Figure 1) and FT-IR spectra (Figure S3). As shown in Figure 1, ^{1}H NMR spectrum of oDex exhibited a new peak at δ 9.6 ppm, which was the characteristic resonance signal of the aldehyde group (-CHO). The successful synthesis of oDex was further confirmed by FT-IR spectrum, a new absorption band at 1734 cm^{-1} assigned to the aldehyde group in oDex (Figure S3). In oDex-g-Pt, a new proton resonance peak that arose at δ 2.4 ppm was be ascribed to methylene group adjacent to Pt, which confirmed the graft of succinic anhydride modified DHP (Pt (IV) prodrug) to oDex. For oDex-g-Pt+DOX, the peaks at δ 7.0–8.0 ppm were the signals of benzene protons from DOX segment, and the new stretching vibration located at 1635 cm^{-1} (ν−C=N−) was attributed to imine bond, demonstrated the successful synthesis. Moreover, the DLC of DOX and Pt in oDex-g-Pt+DOX were calculated to be 11.26% and 5.89%, respectively. The precise chemical structures of drugs and drugs-oDex conjugates ensure the possibility of the clinical applications. Furthermore, the UV-Vis spectra of free DOX and oDex-g-Pt+DOX were also measured. As shown in Figure S4, the absorption peak of DOX could be clearly observed in oDex-g-DOX and oDex-g-Pt+DOX system, which indicated the existence of DOX.

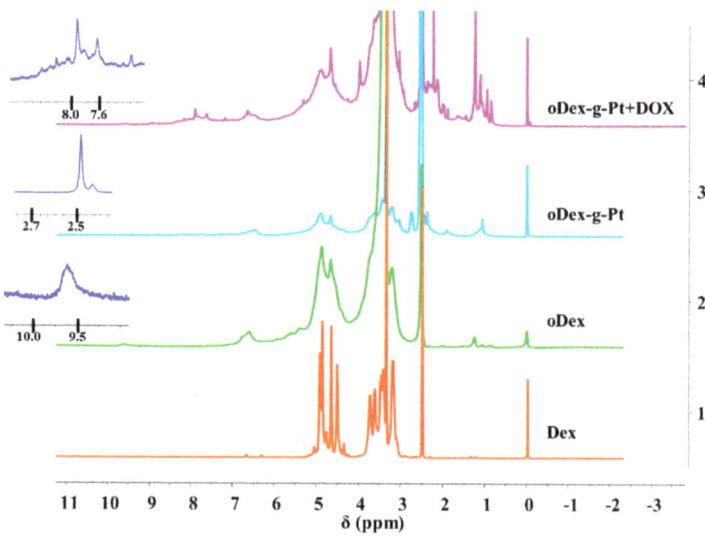

Figure 1. The ^{1}H NMR spectra of Dex (dextran), oDex (oxidized dextran), oDex-g-Pt (platinum conjugated oxidized dextran), and oDex-g-Pt+DOX (platinum plus DOX conjugated oxidized dextran) in DMSO-d$_6$.

3.2. Preparation and Characterization of oDex-g-Pt NPs and oDex-g-Pt+DOX NPs

In our work, Pt (IV) and DOX acted not only as chemotherapeutic drugs, but also as hydrophobic segment in oDex-g-Pt, oDex-g-DOX and oDex-g-Pt+DOX. Meanwhile, oDex had a great quantity of hydroxyl groups, which were hydrophilic. It is reasonable that oDex-g-Pt, oDex-g-DOX and oDex-g-Pt+DOX could self-assemble into NPs in aqueous medium. oDex-g-Pt NPs, oDex-g-DOX and oDex-g-Pt+DOX NPs were prepared using a simple dialysis method. Particle size and size distribution play significant roles in intravenous activities. Figure 2A,B showed the DLS results of oDex-g-Pt NPs and oDex-g-Pt+DOX NPs. oDex-g-Pt NPs had an average particle size of 95 nm and a low PDI of 0.17 in Milli-Q water. Under same condition, oDex-g-Pt+DOX NPs had a mean particle size of 183 nm and a PDI of 0.22. The size and size distribution of control group, oDex-g-DOX, were displayed in Figure S5. In addition, TEM micrographs indicated that both oDex-g-Pt NPs and oDex-g-Pt+DOX NPs had clear spherical morphology (Figure 2C,D). Notably, the size of oDex-g-Pt NPs and oDex-g-Pt+DOX NPs determined by TEM was smaller than DLS results due to the swelling of NPs in aqueous medium for DLS measurements. All the results demonstrated that both oDex-g-Pt and oDex-g-Pt+DOX could successfully assemble into small and uniform NPs. The appropriate sizes of oDex-g-Pt+DOX NPs benefited the increased intratumoral accumulation via the enhanced permeability and retention effect [26]. Moreover, the zeta potential values of oDex-g-Pt and oDex-g-Pt+DOX were

−3.18 mV and 5.80 mV, respectively (Figure S6), which also demonstrated the successful conjugation of DOX.

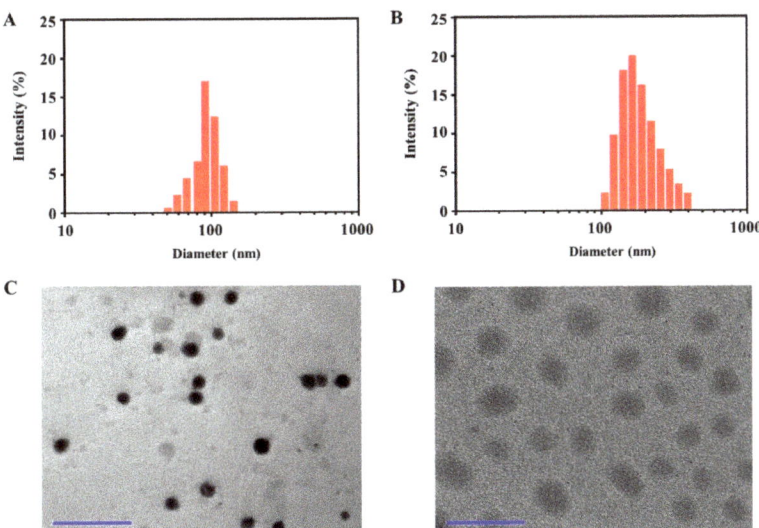

Figure 2. Characterization of oDex-g-Pt NPs (platinum conjugated oxidized dextran nanoparticles) and oDex-g-Pt+DOX NPs (platinum plus DOX conjugated oxidized dextran nanoparticles). Particle size and size distributions of oDex-g-Pt NPs (**A**) and oDex-g-Pt+DOX NPs (**B**) in Milli-Q water. TEM images of the oDex-g-Pt NPs (**C**) and oDex-g-Pt+DOX NPs (**D**) (Scale bars: 500 nm).

3.3. Stability of oDex-g-Pt+DOX NPs

For oDex-g-Pt+DOX NPs, Pt (IV) prodrug was conjugated to the oDex via the esterification reaction, and DOX was introduced via the schiff base reaction. As is well known, Pt (IV) prodrug can be simply reduced into active Pt (II) drugs by reductive agents, such as sodium ascorbic, DTT and GSH, and the imine linkage is sensitive to the acidic environment. Consequently, the oDex-g-Pt+DOX NPs can respond well to the tumor acidic and reductive microenvironment, and then quickly dissociate to release active DOX and Pt (II) drugs.

Initially, the average size and size distribution changes of oDex-g-Pt+DOX NPs in response to pH and GSH were followed by DLS measurements. DOX was conjugated onto the oDex-g-Pt via Schiff base reaction, the formed hydrazone linkages are acidity-labile. Then, two pH values, pH 5.0 and pH 7.4, were selected for the stability evaluation of the prepared oDex-g-Pt+DOX NPs. pH 5.0 and pH 7.4 were used to mimic the tumor intracellular components and normal physiological conditions, respectively. In Figure 3A,B, no significant variations in mean size and PDI were observed for oDex-g-Pt+DOX NPs in PBS at physiological pH (pH = 7.4) over 48 h, which indicated that oDex-g-Pt+DOX NPs had high stability under such conditions. However, oDex-g-Pt+DOX NPs underwent gradual swelling at pH 5.0, in which the average diameter increased from ca. 204 to 372 nm, and the PDI value was in a constant state of change. These changes might be ascribed to the acid-triggered release of DOX, which induced the instability of the NPs. To evaluate reduction-sensitivity of oDex-g-Pt+DOX NPs, the NPs were incubated in PBS (pH = 7.4) with 10 mM GSH. The results showed that the size and size distribution were constantly changing with increase in incubation time. Generally, both the average diameter and PDI increased. These changes suggested the initial reduction of the Pt (IV) groups increasing the hydrophilicity of oDex-g-Pt+DOX NPs, and consequent disassembly and re-assembly of the NPs. Notably, DOX was still conjugated to oDex as hydrophobic segment. Certainly, the disassembly and re-assembly could also be detected in PBS (pH = 5.0) with 10 mM GSH. As shown in Figure 3A,B, the average size increased from ca. 251 to 703 nm within 15 h, and then progressively decreased to 289 nm in 48 h. Simultaneously, the PDI value showed a similar phenomenon. In particular, size and size distribution of the oDex-g-Pt+DOX NPs incubated in PBS (pH = 5.0), PBS (pH = 7.4) with 10 mM GSH, and PBS (pH = 5.0) with 10 mM GSH for 24 h are shown in Figure 3C–E, respectively.

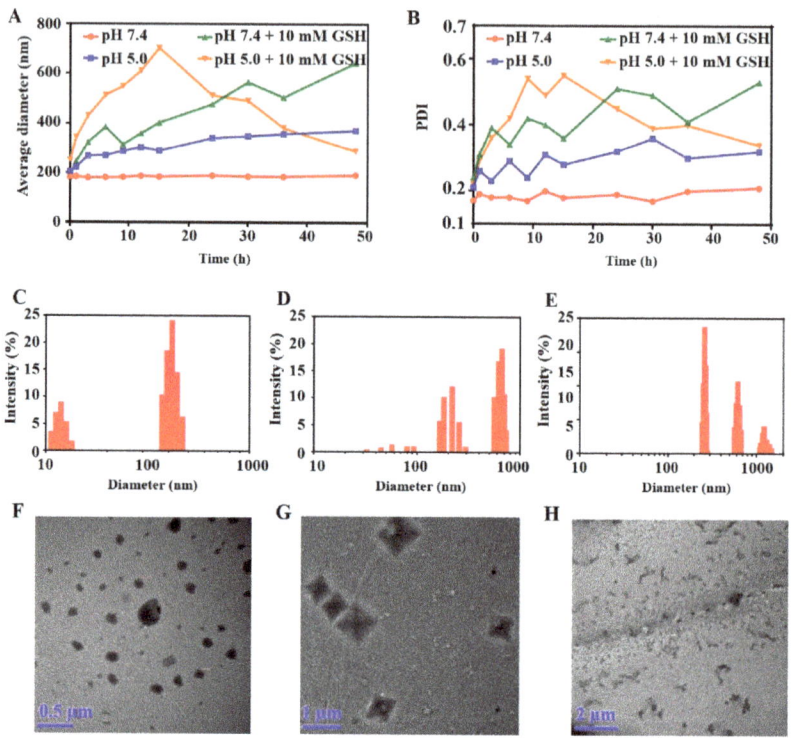

Figure 3. Stability evaluation of oDex-g-Pt+DOX NPs (platinum plus DOX conjugated oxidized dextran nanoparticles) monitored by DLS and TEM. (**A**) Average particle size of oDex-g-Pt+DOX NPs after incubation under different conditions. (**B**) Size distributions of oDex-g-Pt+DOX NPs after incubation under different conditions. Size and size distribution of oDex-g-Pt+DOX NPs incubated in pH 5.0 (**C**), pH = 7.4 with 10 mM GSH (**D**), and pH = 5.0 with 10 mM GSH (**E**) for 24 h. TEM images of oDex-g-Pt+DOX NPs incubated in pH 5.0 (**F**), pH = 7.4 with 10 mM GSH (**G**), and pH = 5.0 with 10 mM GSH (**H**) for 24 h.

Importantly, the acid- and reduction-induced instability could be further supported by TEM images. As shown in Figure 3F, after incubated in pH 5.0 for 24 h, the morphology of NPs was irregular, size distribution increased, and obvious aggregation could be observed. Interestingly, polyhedron appeared in the TEM image after 24 h incubation reductive condition (Figure 3G). As expected, size, size distribution, and morphology of the oDex-g-Pt+DOX NPs coexisted in pH = 5.0 with 10 mM GSH were sharply changed, which was ascribed to the quick release of dual drugs (DOX and Pt) (Figure 3H). Tyndall effect was further used to investigate the disassembly of oDex-g-Pt+DOX. As shown in Figure S7, oDex-g-Pt+DOX NPs had obvious tyndall effect. However, after treated in pH 5.0 with 10 mM GSH for 24 h, tyndall effect decreased sharply. The above results indicated the dissociations of oDex-g-Pt+DOX NPs in the acid and reductive environment.

3.4. In Vitro Drugs Release

DOX and Pt were conjugated to oDex via stimuli-responsive linkages and drug release behaviors of the oDex-g-Pt+DOX NPs were pH and reduction dual-sensitive. The in vitro drug release was detected at pH 5.0 or 7.4 to mimic acidic microenvironment of tumor intracellular components and normal pH of physiological conditions, respectively. Moreover, different pH with 10 mM GSH was used to mimic the reductive microenvironment of tumor intracellular components. As shown in Figure 4, no significant burst release of DOX and Pt was observed from oDex-g-Pt+DOX NPs. The cumulative DOX release from oDex-g-Pt+DOX NPs increased apparently as the pH decreased from 7.4 to 5.0 (Figure 4A). At physiological pH, there was only approximately 18.9% of DOX releasing from oDex-g-Pt+DOX NPs after 48 h. However, 51.4% of DOX was released at pH 5.0. The accelerated DOX release at pH 5.0 was attributed to the acid-cleavable oxime bond in oDex-g-Pt+DOX NPs. Incubation of the oDex-g-Pt+DOX NPs at pH 7.4 with 10 mM GSH lead to a slightly increased cumulative release of DOX (25.5%) after 48 h, and more than 79% of DOX was released at pH 5.0 with 10 mM GSH within the same period. This might be ascribable to the release of Pt, which induced the hydrophilic conversion processes of the NPs. During the Pt release procession, oDex-g-Pt+DOX NPs released more Pt as the addition of 10 mM GSH and decrease of pH. As depicted in Figure 4B, the release proportions of Pt from oDex-g-Pt+DOX NPs was faster and reached 56.4% in pH 7.4 with 10 mM GSH in 48 h, which was much higher than those in pH 7.4 (21.9%), and final release of Pt anticancer drug was approximately 77.3% after incubation with 20 mM GSH at the same time point (Figure S8), further indicating the reduction-responsive degradation of oDex-g-Pt+DOX NPs. As expected, the cumulative Pt release profiles indicated that the highest contents of released Pt from oDex-g-Pt+DOX NPs was 76.6% at pH 5.5 with 10 mM GSH after 48 h. The acidity-accelerated Pt release profiles might be owing to the pH-responsive DOX release and a more hydrophilic shell of NPs. These results proved that oDex-g-Pt+DOX NPs were comparatively stable under normal physiological conditions, while the chemical conjugated DOX and Pt were released by breaking the pH-sensitive hydrozone bond and reduction-sensitive Pt (IV) group, respectively. The pH and reduction dual sensitivity will endow the oDex-g-Pt+DOX NPs with high-selective intracellular release of DOX and Pt for combination chemotherapy.

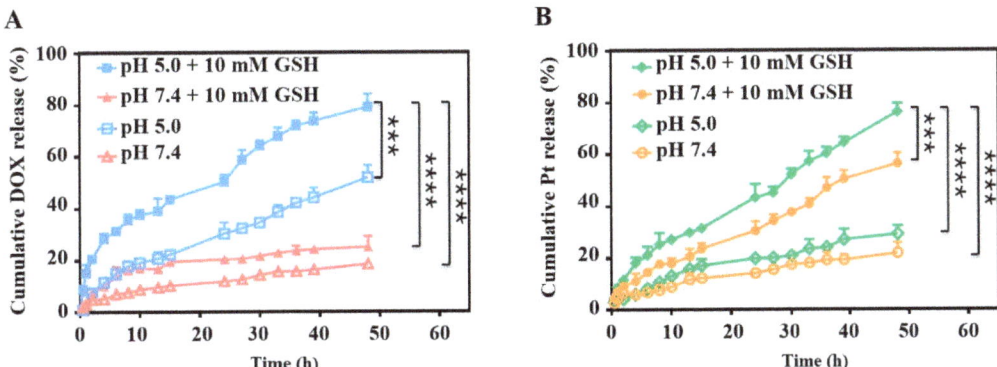

Figure 4. Drugs release profiles. (**A**) DOX and (**B**) Pt release behaviors from oDex-g-Pt+DOX NPs (platinum plus DOX conjugated oxidized dextran nanoparticles) at different conditions. Data points represent mean ± s.d. ($n = 3$) from three independent experiments. Statistical significance was calculated via one-way ANOVA analysis with a Tukey post-hoc test. *** $p < 0.001$, **** $p < 0.0001$.

3.5. In Vitro Cytotoxicity and Cellular Uptake

The promising and desirable result of a nDDS for chemotherapy is the maintenance (or reinforcement) of the drugs' efficacy and, at the same time, the safety of carrier on healthy cells. Initially, the biocompatibility of Dex and oDex were investigated via MTT assays in L929 cells and HeLa cells. The results revealed that both Dex and oDex displayed no obvious cytotoxicity toward L929 cells and HeLa cells even at a high concentration of 1.0 mg/mL (Figure 5). These results indicated that both Dex and oDex had good biosafety.

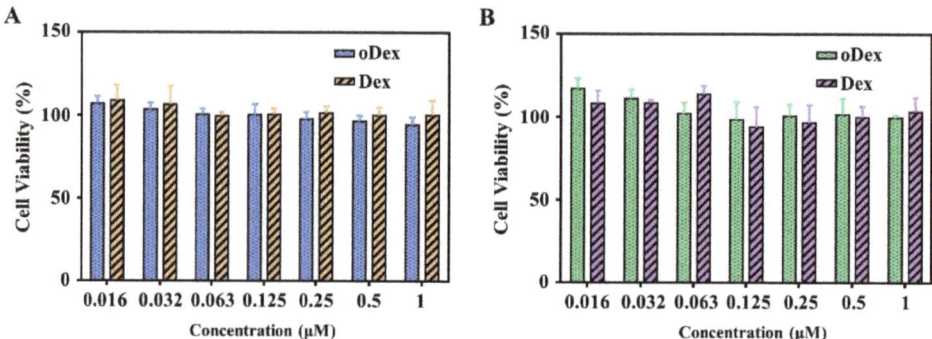

Figure 5. Cytotoxicity assay. Cell viability of L929 (**A**) and HeLa (**B**) cells after 48 h incubation with Dex (dextran) or oDex (oxidized dextran) at different concentrations.

The in vitro cell proliferation inhibition activity of various drugs and their nanoformulations were further evaluated against HeLa and A549 cells by MTT assay. The results are shown in Figure 6 and Figure S9. Obviously, with the increase of drug concentration and the evolution of time, the enhanced anti-proliferative activities were displayed toward all the test formulations. For instance, as shown in Figure 6A,B and Figure S9A,B, after 48 h or 72 h co-incubation with HeLa cells, cytotoxicity order of various DOX and Pt formulations at dosages of 13.5 µM was as follows: Cis+DOX > Pt (IV)+DOX ≈ DOX > Cis > oDex-g-Pt+DOX > Pt (IV) > oDex-g-Pt. The free drugs (Cis, DOX and Cis+DOX) showed better in vitro anti-cancer efficacy than nanoformulations. This might be caused by the rapid passive diffusion of free drugs and the prolonged drug release from nanoformulations via endocytosis process. Cis+DOX indicated the highest cellular cytotoxicity, owing to the combinational effect between DOX and Pt. More fascinatingly, oDex-g-Pt+DOX NPs showed better inhibition efficacy compared with oDex-g-Pt NPs. In addition, the IC_{50} values of all drugs and nanoformulations decreased along with the extension of co-incubation time from 48 to 72 h. The oDex-g-Pt+DOX NPs displayed enhanced cellular cytotoxicity with IC_{50} values of 14.5 µM for 48 h and <3.375 µM for 72 h. Then, the in vitro cellular cytotoxicity of drugs and nanoformulations toward A549 cells were also detected, and similar results were obtained (Figure 6C and Figure S9C). The findings demonstrated that the acid and reduction dual-sensitive oDex-g-Pt+DOX prodrugs showed great potential for combination chemotherapy.

Cisplatin is a common chemotherapeutic drug in clinic. However, its direct administration via intravenous injection causes serious side effects and severe drug resistance [36]. Previous reports demonstrated that synergetic delivery with DOX and cisplatin could work in concerted way or overcome drug resistance [34]. The cellular cytotoxicity of drugs and nanoformulations was further assayed using cisplatin-resistant cancer cells (A549/DDP) (Figure 6D and Figure S9D). Notably, the IC_{50} value of cisplatin against A549 cells was 7.04 µM (Figure S9D). In Figure 6D and Figure S9D, A549/DDP cells showed obvious resistance to cisplatin and Pt (IV), the IC50 value of cisplatin and Pt (IV) against A549/DDP cells was both higher than 216 µM. As expected, DOX itself was more susceptive than cisplatin in A549/DDP cells, when cisplatin was used with DOX, the cell death was dramatically increased with significantly reduced IC50 values (15.5 µM). Meanwhile, Pt (IV)+DOX groups displayed high cytotoxicity towards A549/DDP cells. It was exciting to find that oDex-g-Pt+DOX NPs exhibited a closer in vitro cellular cytotoxicity to the Cis+DOX group. These results suggested that DOX and cisplatin had synergistic therapeutic effect, and oDex-g-Pt+DOX NPs might overcome the acquired drug resistance to cisplatin to some extent.

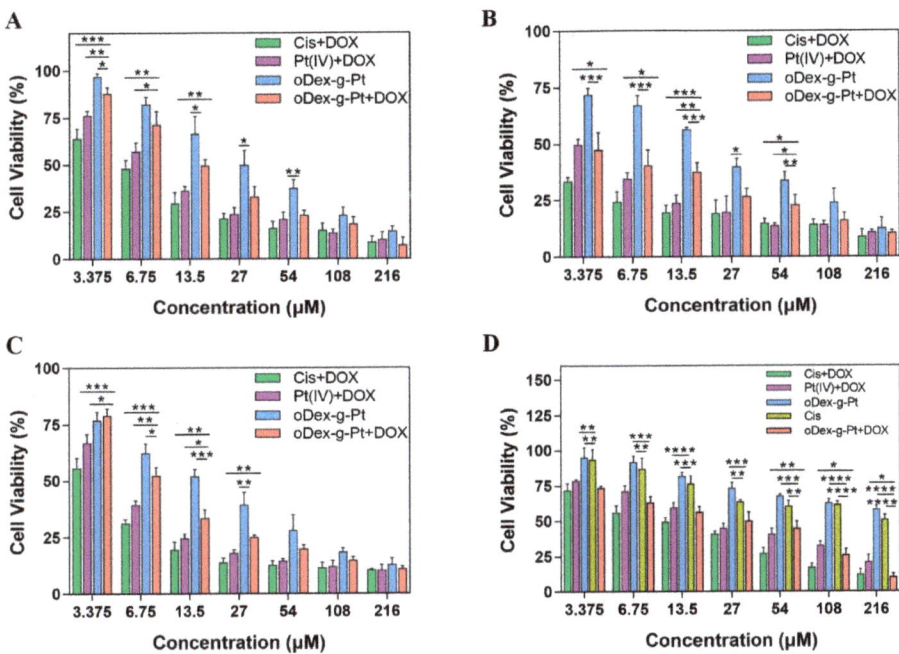

Figure 6. In vitro anti-cancer efficacy of drugs and nanodrugs. Cell viability curves of HeLa cells incubated with drugs and nanodrugs for 48 h (**A**) or 72 h (**B**). Cell viability curves of A549 cells (**C**) and A549/DDP cells (**D**) after incubation with drugs and nanodrugs for 48 h. Data points represent mean ± s.d. (n = 3) from three independent experiments. Statistical significance was calculated via one-way ANOVA analysis with a Tukey post-hoc test. * $p < 0.5$, ** $p < 0.01$, *** $p < 0.001$, **** $p < 0.0001$.

Subsequently, the in vitro cellular internalization and intracellular drug release behaviors of prodrugs drugs release performance of oDex-g-Pt+DOX NPs were examined in HeLa cells by CLSM. oDex-g-Pt+DOX NPs were incubated with HeLa cells for 0.5 h and 4 h, respectively. Hoechst 33,258 and red lyso-tracker were applied to stain the nucleus (blue imaging) and the lysosomes (green imaging), respectively. The subcellular localization of DOX was observed by the red fluorescence imaging. As shown in Figure 7, a time-dependent cellular internalization was clearly observed for oDex-g-Pt+DOX NPs, much higher DOX fluorescence intensities were exhibited at 4 h than at 0.5 h. A much weaker DOX fluorescence was detected after co-incubation with oDex-g-Pt+DOX NPs prodrugs for 0.5 h, and the DOX fluorescence was mainly co-localized within the lysosomes/endosomes in the cytoplasm. As the culture time was prolonged to 4 h, fluorescence signal of DOX in HeLa cells was significantly intensified, indicating the enhanced cellular uptake efficiency. Moreover, fluorescence of DOX in both cytoplasm and nuclei could be observed, and significant fluorescence of DOX was displayed at the perinuclear region, which indicated that oDex-g-Pt+DOX NPs were continuously transferred into the cells, and DOX was gradually released. The results proved that oDex-g-Pt+DOX NPs entered HeLa cells slowly via the endocytosis pathway, and DOX was sustainably released from the nanoformations.

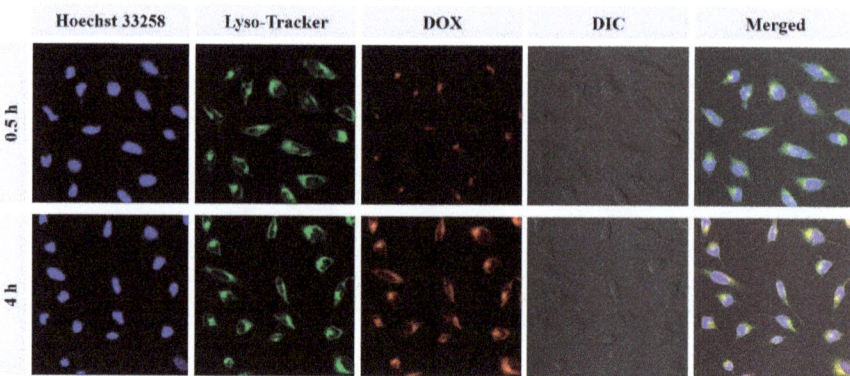

Figure 7. Cellular uptake of oDex-g-Pt+DOX NPs (platinum plus DOX conjugated oxidized dextran nanoparticles) was determined by confocal laser scanning microscopy.

The cell uptake of free DOX and oDex-g-Pt+DOX NPs toward HeLa cells for both 0.5 and 4 h was further analyzed by FCM. The fluorescent intensities of DOX were in the following order: oDex-g-Pt+DOX NPs group with 0.5 h incubation < DOX group with 0.5 h incubation < DOX group with 4 h incubation < oDex-g-Pt+DOX NPs group with 4 h incubation, as shown in Figure 8. After 0.5 h co-incubation, free DOX exhibited the stronger fluorescent intensity compared with oDex-g-Pt+DOX NPs, owing to quick diffusion pathway of free DOX. However, with the extension of culture time from 0.5 to 4 h, the oDex-g-Pt+DOX NPs exhibited upregulated DOX fluorescent intensity, which might be due to the intracellular drug accumulation from the sustained drug release of oDex-g-Pt+DOX NPs. The results indicated that the smart oDex-g-Pt+DOX NPs conjugates showed efficient endocytosis and intracellular drugs release.

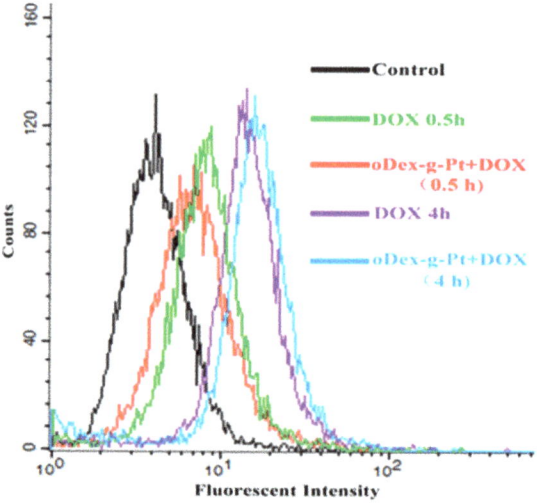

Figure 8. Cellular uptake. FCM of HeLa cells after incubation with free DOX and oDex-g-Pt+DOX NPs (platinum plus DOX conjugated oxidized dextran nanoparticles) for 0.5 h or 4 h.

4. Conclusions

In this work, an amphiphilic oDex-g-Pt+DOX prodrug was designed and synthesized as a smart drug delivery platform for combination chemotherapy application. oDex-g-

Pt+DOX was developed by the versatile esterification reaction between the hydroxyl group of oDex and carboxyl group in Pt (IV) prodrug, and then the prepared polymer was conveniently functioned via the facile schiff base reaction between aldehyde groups of oDex and amino group in DOX. The obtained oDex-g-Pt+DOX conjugates spontaneously formed into NPs in aqueous medium with an average diameter of 183 nm. The smart oDex-g-Pt+DOX contained several fascinating advantages: (i) facile synthesis of the dual drugs delivery system with an esterification and schiff base reaction; (ii) plenty stability in the normal physiological environment without initial burst release; (iii) acid- and reduction-triggered dissociation, DOX and Pt release; (iv) besides, the multifunctional oDex-g-Pt+DOX NPs with its important merits including enhanced and combinational anti-tumor efficacy in vitro, especially in the cisplatin-resistant cell line. Therefore, the exploited acid and reduction dual-responsive oDex-g-Pt+DOX NPs exhibited great potential for clinical combination chemotherapy.

Supplementary Materials: The following are available online at https://www.mdpi.com/article/10.3390/polym13091515/s1, Scheme S1: Synthetic scheme of the diamminedichloro-dihydroxyplatinum (DHP) and succinic anhydride modified DHP (Pt (IV) prodrug). Scheme S2: Synthetic scheme of the oDex-g-DOX (DOX conjugated oxidized dextran). Figure S1: The ^{1}H NMR spectra of diamminedichloro-dihydroxyplatinum (DHP) (A), and succinic anhydride modified DHP (Pt (IV) prodrug) (B) in DMSO-d_6. Figure S2: Fourier-transform infrared spectra (FT-IR) spectra of the cisplatin, diamminedichloro-dihydroxyplatinum (DHP) and succinic anhydride modified DHP (Pt (IV) prodrug). Figure S3: Fourier-transform infrared spectra (FT-IR) spectra of the Dex, oDex, oDex-g-Pt, and oDex-g-Pt+DOX. Figure S4: UV–visible spectrum of Dex, oDex, oDex-g-Pt, DOX and oDex-g-Pt+DOX in aqueous solution (A) and DMSO (B). Figure S5: Zeta potential measurement of oDex-g-Pt (A) and oDex-g-Pt+DOX (B). Figure S6: Tyndall effect seen for oDex-g-Pt NPs (A), oDex-g-Pt NPs cultured in pH 7.4 with 10 mM GSH for 24 h (B), oDex-g-Pt+DOX NPs (C), and oDex-g-Pt+DOX NPs cultured in pH 5.0 with 10 mM GSH for 24 h (D). Figure S7: In vitro anti-cancer efficacy of drugs and nanodrugs. Cell viability curves of HeLa cells incubated with drugs and nanodrugs for 48 h (A) or 72 h (B). Cell viability curves of A549 cells (C) and A549/DDP cells (D) after incubation with drugs and nanodrugs for 48 h. Figure S8: Drugs release profiles. Pt release behaviors from oDex-g-Pt+DOX NPs (platinum plus DOX conjugated oxidized dextran nanoparticles) in the presence of 20 mM GSH. Figure S9: In vitro anti-cancer efficacy of drugs and nanodrugs. Cell viability curves of HeLa cells incubated with drugs and nanodrugs for 48 h (A) or 72 h (B). Cell viability curves of A549 cells (C) and A549/DDP cells (D) after incubation with drugs and nanodrugs for 48 h.

Author Contributions: Conceptualization, Y.W. and Z.C.; methodology, X.X. (Xiukun Xue); software, X.X. (Xiukun Xue) and X.X. (Xiao Xu); supervision, Y.W. and T.L.; formal analysis, Z.C. and B.X.; investigation, Y.W.; data curation, Y.W.; writing—original draft preparation, Y.W. and Z.C.; writing—review and editing, Y.W. and Z.C.; funding acquisition, Y.W. and Z.C. All authors have read and agreed to the published version of the manuscript.

Funding: This work was financially supported by Natural Science Foundation of Shandong Province (No. ZR2018BB050), the National Natural Science Foundation of China (No. 51803097), Program for Scientific Research Innovation Team in Colleges and Universities of Jinan (No. 2019GXRC021), and Minjiang Scholar Startup Package of Fuzhou University.

Institutional Review Board Statement: Not applicable.

Informed Consent Statement: Not applicable.

Data Availability Statement: Data is contained within the article or supplementary material.

Acknowledgments: We would like to express our deepest gratitude to Zhijian Luo at Integrated Hospital of Traditional Chinese Medicine, Southern Medical University, Guangzhou. Thanks for his assistance in the cytotoxicity assay against A549/DDP cells.

Conflicts of Interest: The authors declare no conflict of interest.

References

1. Bhattacharyya, J.; Weitzhandler, I.; Ho, S.B.; McDaniel, J.R.; Li, X.; Tang, L.; Liu, J.; Dewhirst, M.; Chilkoti, A. Encapsulating a Hydrophilic Chemotherapeutic into Rod-like Nanoparticles of a Genetically Encoded Asymmetric Triblock Polypeptide Improves its Efficacy. *Adv. Funct. Mater.* **2017**, *27*, 1605421. [CrossRef]
2. Cheng, C.; Sui, B.; Wang, M.; Hu, X.; Shi, S.; Xu, P. Carrier-Free Nanoassembly of Curcumin-Erlotinib Conjugate for Cancer Targeted Therapy. *Adv. Healthc. Mater.* **2020**, *9*, e2001128. [CrossRef] [PubMed]
3. Cui, X.; Sun, Y.; Shen, M.; Song, K.; Yin, X.; Di, W.; Duan, Y. Enhanced Chemotherapeutic Efficacy of Paclitaxel Nanoparticles Co-delivered with MicroRNA-7 by Inhibiting Paclitaxel-Induced EGFR/ERK pathway Activation for Ovarian Cancer Therapy. *ACS Appl. Mater. Interfaces* **2018**, *10*, 7821–7831. [CrossRef]
4. Park, N.H.; Cheng, W.; Lai, F.; Yang, C.; Florez de Sessions, P.; Periaswamy, B.; Wenhan Chu, C.; Bianco, S.; Liu, S.; Venkataraman, S.; et al. Addressing Drug Resistance in Cancer with Macromolecular Chemotherapeutic Agents. *J. Am. Chem. Soc.* **2018**, *140*, 4244–4252. [CrossRef] [PubMed]
5. Chen, Z.; Wang, Z.; Gu, Z. Bioinspired and Biomimetic Nanomedicines. *Acc. Chem. Res.* **2019**, *52*, 1255–1264. [CrossRef]
6. Chen, Z.; Li, Z.; Lin, Y.; Yin, M.; Ren, J.; Qu, X. Biomineralization inspired surface engineering of nanocarriers for pH-responsive, targeted drug delivery. *Biomaterials* **2013**, *34*, 1364–1371. [CrossRef] [PubMed]
7. Wang, T.; Wang, D.; Yu, H.; Wang, M.; Liu, J.; Feng, B.; Zhou, F.; Yin, Q.; Zhang, Z.; Huang, Y.; et al. Intracellularly Acid-Switchable Multifunctional Micelles for Combinational Photo/Chemotherapy of the Drug-Resistant Tumor. *ACS Nano* **2016**, *10*, 3496–3508. [CrossRef]
8. Li, L.; Gu, W.; Chen, J.; Chen, W.; Xu, Z.P. Co-delivery of siRNAs and anti-cancer drugs using layered double hydroxide nanoparticles. *Biomaterials* **2014**, *35*, 3331–3339. [CrossRef]
9. Wang, H.; Zhu, W.; Liu, J.; Dong, Z.; Liu, Z. pH-Responsive Nanoscale Covalent Organic Polymers as a Biodegradable Drug Carrier for Combined Photodynamic Chemotherapy of Cancer. *ACS Appl. Mater. Interfaces* **2018**, *10*, 14475–14482. [CrossRef]
10. Wu, Y.; Kuang, H.; Xie, Z.; Chen, X.; Jing, X.; Huang, Y. Novel hydroxyl-containing reduction-responsive pseudo-poly(aminoacid) via click polymerization as an efficient drug carrier. *Polym. Chem.* **2014**, *5*, 4488–4498. [CrossRef]
11. Hartl, N.; Adams, F.; Merkel, O.M. From Adsorption to Covalent Bonding: Apolipoprotein E Functionalization of Polymeric Nanoparticles for Drug Delivery across the Blood–Brain Barrier. *Adv. Therap.* **2020**, *4*, 2000092. [CrossRef]
12. Karlsson, J.; Vaughan, H.J.; Green, J.J. Biodegradable Polymeric Nanoparticles for Therapeutic Cancer Treatments. *Annu. Rev. Chem. Biomol. Eng.* **2018**, *9*, 105–127. [CrossRef]
13. Feng, J.; Wen, W.; Jia, Y.G.; Liu, S.; Guo, J. pH-Responsive Micelles Assembled by Three-Armed Degradable Block Copolymers with a Cholic Acid Core for Drug Controlled-Release. *Polymers* **2019**, *11*, 511. [CrossRef] [PubMed]
14. Zhu, D.; Wu, S.; Hu, C.; Chen, Z.; Wang, H.; Fan, F.; Qin, Y.; Wang, C.; Sun, H.; Leng, X.; et al. Folate-targeted polymersomes loaded with both paclitaxel and doxorubicin for the combination chemotherapy of hepatocellular carcinoma. *Acta Biomater.* **2017**, *58*, 399–412. [CrossRef] [PubMed]
15. Wu, H.; Jin, H.; Wang, C.; Zhang, Z.; Ruan, H.; Sun, L.; Yang, C.; Li, Y.; Qin, W. Synergistic Cisplatin/Doxorubicin Combination Chemotherapy for Multidrug-Resistant Cancer via Polymeric Nanogels Targeting Delivery. *ACS Appl. Mater. Interfaces* **2017**, *9*, 9426–9436. [CrossRef]
16. Wu, P.; Wang, X.; Wang, Z.; Ma, W.; Guo, J.; Chen, J.; Yu, Z.; Li, J.; Zhou, D. Light-Activatable Prodrug and AIEgen Copolymer Nanoparticle for Dual-Drug Monitoring and Combination Therapy. *ACS Appl. Mater. Interfaces* **2019**, *11*, 18691–18700. [CrossRef]
17. Zhu, C.; Xiao, J.; Tang, M.; Feng, H.; Chen, W.; Du, M. Platinum covalent shell cross-linked micelles designed to deliver doxorubicin for synergistic combination cancer therapy. *Int. J. Nanomed.* **2017**, *12*, 3697–3710. [CrossRef] [PubMed]
18. Li, Y.; Yang, H.; Yao, J.; Yu, H.; Chen, X.; Zhang, P.; Xiao, C. Glutathione-triggered dual release of doxorubicin and camptothecin for highly efficient synergistic anticancer therapy. *Colloids. Surf. B Biointerfaces* **2018**, *169*, 273–279. [CrossRef]
19. Moghaddam, S.V.; Abedi, F.; Alizadeh, E.; Baradaran, B.; Annabi, N.; Akbarzadeh, A.; Davaran, S. Lysine-embedded cellulose-based nanosystem for efficient dual-delivery of chemotherapeutics in combination cancer therapy. *Carbohydr. Polym.* **2020**, *250*, 116861. [CrossRef]
20. Sharma, P.K.; Taneja, S.; Singh, Y. Hydrazone-Linkage-Based Self-Healing and Injectable Xanthan-Poly (ethylene glycol) Hydrogels for Controlled Drug Release and 3D Cell Culture. *ACS Appl. Mater. Interfaces* **2018**, *10*, 30936–30945. [CrossRef]
21. Xu, W.; Ding, J.; Xiao, C.; Li, L.; Zhuang, X.; Chen, X. Versatile preparation of intracellular-acidity-sensitive oxime-linked polysaccharide-doxorubicin conjugate for malignancy therapeutic. *Biomaterials* **2015**, *54*, 72–86. [CrossRef]
22. Cai, Z.; Zhang, H.; Wei, Y.; Xie, Y.; Cong, F. Reduction- and pH-Sensitive Hyaluronan Nanoparticles for Delivery of Iridium(III) Anticancer Drugs. *Biomacromolecules* **2017**, *18*, 2102–2117. [CrossRef]
23. Zheng, L.; Zhang, X.; Wang, Y.; Liu, F.; Peng, J.; Zhao, X.; Yang, H.; Ma, L.; Wang, B.; Chang, C.; et al. Fabrication of Acidic pH-Cleavable Polymer for Anticancer Drug Delivery Using a Dual Functional Monomer. *Biomacromolecules* **2018**, *19*, 3874–3882. [CrossRef] [PubMed]
24. Yin, W.; Ke, W.; Lu, N.; Wang, Y.; Japir, A.; Mohammed, F.; Pan, Y.; Ge, Z. Glutathione and Reactive Oxygen Species Dual-Responsive Block Copolymer Prodrugs for Boosting Tumor Site-Specific Drug Release and Enhanced Antitumor Efficacy. *Biomacromolecules* **2020**, *21*, 921–929. [CrossRef] [PubMed]
25. Dai, Y.; Xu, C.; Sun, X.; Chen, X. Nanoparticle design strategies for enhanced anticancer therapy by exploiting the tumour microenvironment. *Chem. Soc. Rev.* **2017**, *45*, 3830–3852. [CrossRef]

26. Li, D.; Su, T.; Ma, L.; Yin, F.; Xu, W.; Ding, J.; Li, Z. Dual-acidity-labile polysaccharide-di-drugs conjugate for targeted cancer chemotherapy. *Eur. J. Med. Chem.* **2020**, *199*, 112367. [CrossRef]
27. Braga, C.B.; Perli, G.; Becher, T.B.; Ornelas, C. Biodegradable and pH-Responsive Acetalated Dextran (Ac-Dex) Nanoparticles for NIR Imaging and Controlled Delivery of a Platinum-Based Prodrug into Cancer Cells. *Mol. Pharm.* **2019**, *16*, 2083–2094. [CrossRef] [PubMed]
28. Curcio, M.; Cirillo, G.; Paoli, A.; Naimo, G.D.; Mauro, L.; Amantea, D.; Leggio, A.; Nicoletta, F.P.; Iemma, F. Self-assembling Dextran prodrug for redox- and pH-responsive co-delivery of therapeutics in cancer cells. *Colloids. Surf. B Biointerfaces* **2020**, *185*, 110537. [CrossRef] [PubMed]
29. Atanase, L.; Desbrieresc, J.; Riess, G. Micellization of synthetic and polysaccharides-based graft copolymers in aqueous media. *Prog. Polym. Sci.* **2017**, *73*, 32–60. [CrossRef]
30. Atanase, L. Micellar Drug Delivery Systems Based on Natural Biopolymers. *Polymers* **2021**, *13*, 477. [CrossRef]
31. He, S.; Cong, Y.; Zhou, D.; Li, J.; Xie, Z.; Chen, X.; Jing, X.; Huang, Y. A dextran–platinum(IV) conjugate as a reduction-responsive carrier for triggered drug release. *J. Mater. Chem. B* **2015**, *3*, 8203–8211. [CrossRef] [PubMed]
32. Turner, P.R.; Murray, E.; McAdam, C.J.; McConnell, M.A.; Cabral, J.D. Peptide Chitosan/Dextran Core/Shell Vascularized 3D Constructs for Wound Healing. *ACS Appl. Mater. Interfaces* **2020**, *12*, 32328–32339. [CrossRef]
33. Daraba, O.; Cadinoiu, A.; Rata, D.; Atanase, L.; Vochita, G. Antitumoral Drug-Loaded Biocompatible Polymeric Nanoparticles Obtained by Non-Aqueous Emulsion Polymerization. *Polymers* **2020**, *12*, 1018. [CrossRef] [PubMed]
34. He, L.; Sun, M.; Cheng, X.; Xu, Y.; Lv, X.; Wang, X.; Tang, R. pH/redox dual-sensitive platinum (IV)-based micelles with greatly enhanced antitumor effect for combination chemotherapy. *J. Colloid Interface Sci.* **2019**, *541*, 30–41. [CrossRef]
35. Xu, C.; Wang, Y.; Guo, Z.; Chen, J.; Lin, L.; Wu, J.; Tian, H.; Chen, X. Pulmonary delivery by exploiting doxorubicin and cisplatin co-loaded nanoparticles for metastatic lung cancer therapy. *J. Control Release* **2019**, *295*, 153–163. [CrossRef] [PubMed]
36. Guo, D.; Xu, S.; Huang, Y.; Jiang, H.; Yasen, W.; Wang, N.; Su, Y.; Qian, J.; Li, J.; Zhang, C.; et al. Platinum(IV) complex-based two-in-one polyprodrug for a combinatorial chemo-photodynamic therapy. *Biomaterials* **2018**, *177*, 67–77. [CrossRef] [PubMed]

Article

Gram Scale Synthesis of Dual-Responsive Dendritic Polyglycerol Sulfate as Drug Delivery System

Felix Reisbeck, Alexander Ozimkovski, Mariam Cherri, Mathias Dimde, Elisa Quaas, Ehsan Mohammadifar *, Katharina Achazi * and Rainer Haag *

Institute of Chemistry and Biochemistry, Freie Universität Berlin, Takustr. 3, 14195 Berlin, Germany; freisbeck@zedat.fu-berlin.de (F.R.); saschaoz@zedat.fu-berlin.de (A.O.); mcherri6@zedat.fu-berlin.de (M.C.); mathias.dimde@fu-berlin.de (M.D.); equaas@zedat.fu-berlin.de (E.Q.)
* Correspondence: ehsan@zedat.fu-berlin.de (E.M.); katharina.achazi@fu-berlin.de (K.A.); haag@chemie.fu-berlin.de (R.H.); Tel.: +49-30-838-52632 (E.M.); +49-30-838-59815 (K.A.); +49-30-838-52633 (R.H.)

Citation: Reisbeck, F.; Ozimkovski, A.; Cherri, M.; Dimde, M.; Quaas, E.; Mohammadifar, E.; Achazi, K.; Haag, R. Gram Scale Synthesis of Dual-Responsive Dendritic Polyglycerol Sulfate as Drug Delivery System. *Polymers* 2021, *13*, 982. https://doi.org/10.3390/polym13060982

Academic Editor: Luis García-Fernández

Received: 26 February 2021
Accepted: 17 March 2021
Published: 23 March 2021

Publisher's Note: MDPI stays neutral with regard to jurisdictional claims in published maps and institutional affiliations.

Copyright: © 2021 by the authors. Licensee MDPI, Basel, Switzerland. This article is an open access article distributed under the terms and conditions of the Creative Commons Attribution (CC BY) license (https://creativecommons.org/licenses/by/4.0/).

Abstract: Biocompatible polymers with the ability to load and release a cargo at the site of action in a smart response to stimuli have attracted great attention in the field of drug delivery and cancer therapy. In this work, we synthesize a dual-responsive dendritic polyglycerol sulfate (DR-dPGS) drug delivery system by copolymerization of glycidol, ε-caprolactone and an epoxide monomer bearing a disulfide bond (SSG), followed by sulfation of terminal hydroxyl groups of the copolymer. The effect of different catalysts, including Lewis acids and organic bases, on the molecular weight, monomer content and polymer structure was investigated. The degradation of the polymer backbone was proven in presence of reducing agents and *candida antarctica* Lipase B (CALB) enzyme, which results in the cleavage of the disulfides and ester bonds, respectively. The hydrophobic anticancer drug Doxorubicin (DOX) was loaded in the polymer and the kinetic assessment showed an enhanced drug release with glutathione (GSH) or CALB as compared to controls and a synergistic effect of a combination of both stimuli. Cell uptake was studied by using confocal laser scanning microscopy with HeLa cells and showed the uptake of the Dox-loaded carriers and the release of the drug into the nucleus. Cytotoxicity tests with three different cancer cell lines showed good tolerability of the polymers of as high concentrations as 1 mg mL^{-1}, while cancer cell growth was efficiently inhibited by DR-dPGS@Dox.

Keywords: drug delivery system; dual-responsiveness; dendritic polyglycerol sulfates

1. Introduction

The development of novel drug delivery systems (DDS) for application in chemotherapy has been widely studied. A major drawback of established chemotherapeutic agents such as Doxorubicin (DOX) is their poor solubility in water and unspecific toxicity due to lack of targeting [1]. One method to overcome these drawbacks is the use of polymeric DDS for the encapsulation and targeted release of the drug at the cancer site. Prominent examples for the basis of DDS include natural polymers such as chitosan and hyaluronic acid (HA) and synthetic polymers such as polyglutamic acid (PGA), poly(lactic-co-glycolic acid) (PLGA), polyethylene glycol (PEG) and dendritic polymers [2–5]. The utilization of DDS allows for prolonging circulation time of drug conjugates and therefore, enhanced permeation and retention can be achieved as passive targeting due to the leaky vasculature of tumors. However, in recent years research revealed that this effect displays heterogeneity depending on the tumor and that the uptake of nanoparticles in cancerous tissue is due to an active mechanism rather than passive accumulation [6,7].

In order to enable site-specific release of the drug, inherent properties of cancerous tissue can be exploited. Stimuli such as the decreased pH regime, overexpression of enzymes, presence of reactive oxygen species (ROS) as well as an increased redox potential due to the elevated concentration of the tripeptide glutathione (GSH) are possible rationales for the design of DDS and the incorporation of stimuli-responsive motifs into the polymer structures [8].

Dendritic polyglycerol sulfate (dPGS) has been vastly investigated in the past years as a potential candidate for a variety of medical applications [9]. It exhibits features, such as well-established synthesis, tunable molecular weight, size, surface charge, and flexible globular shape. Furthermore, the polyether backbone as well as its overall negative surface charge lead to biocompatibility up to high concentrations and high water solubility, while maintaining an inertness of the polymer to biological components such as proteins or phospholipid membranes [10]. Originally developed as an alternative for heparin [11], in vivo experiments showed a very high anti-inflammatory activity of dPGS [12]. This led to the investigation of dPGS as potential candidate in different inflammation-related applications, including bone targeting and treatment of osteoarthritis and neurological disorders [13–15]. As inflammation (and thus a higher L/P-selectin concentration) is concomitant with tumor growth, dPGS-based systems exhibited an enhanced tumor targetability, which could be shown by in vivo experiments [16–18].

Taking all these advantages into account, dPGS would be a potential candidate for the development of a drug delivery system. However, one drawback for such applications is its extreme hydrophilicity, hampering the encapsulation of (mostly hydrophobic) anticancer drugs as well as lack of biodegradability in physiological medium. Thus far, several different works have attempted to improve these drawbacks by incorporating hydrophobic and cleavable segments in the dPG backbone [19–22]. However, these attempts have always been challenging due to the laborious procedures and low synthesis scale.

In this work, we present a new method for gram scale synthesis of a DDS based on a sulfated dendritic terpolymer of glycidol, ε-caprolactone and 2-((2-(oxiran-2-ylmethoxy)ethyl)disulfanyl)ethan-1-ol (SSG) for the potential targeted delivery of Doxorubicin (Figure 1). First, the optimal reaction conditions with different catalysts for the copolymerization were screened. The optimized copolymer was then further sulfated and loaded with the hydrophobic anticancer drug Doxorubicin (DOX). As noted, the sulfation of these carriers leads to an increased tumor targetability, whereas the incorporation of ester groups and disulfide bonds allows for the degradation and drug release at the tumor site. Although numerous publications focusing on the controlled and targeted drug delivery of anticancer drugs have been reported, developing new methods for the synthesis of low-cost biocompatible drug delivery systems in gram scale as well as scalable methods for drug loading are highly demanded for further biomedical applications. This novel DDS can be synthesized in gram scale within two facile reaction steps, shows high tumor targeting and its degradation under enzymatic and/or reductive conditions renders the drug release of DOX into cells.

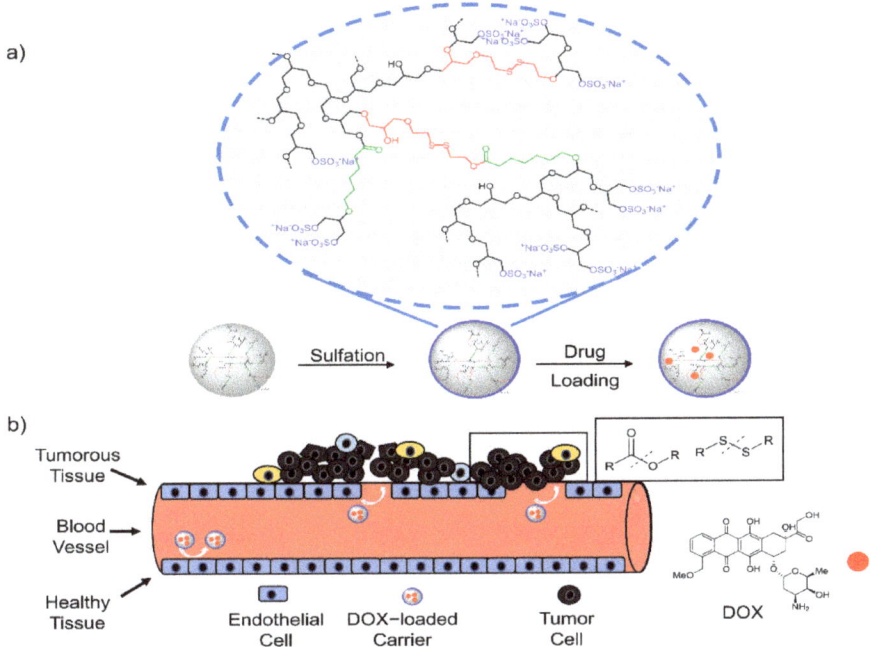

Figure 1. Schematic illustration of synthesis and DOX-loading of the polymeric DDS with subsequent drug release. (**a**) Synthesis of copolymer, sulfation and Dox-loading; (**b**) tumor targeting and underlying mechanism of drug release.

2. Materials and Methods

Potassium tert-butoxide was purchased from ABCR (Karlsruhe, Germany). Anhydrous Toluene and DTT were bought from Acros (Geel, Belgium). Diphenyl phosphate was provided by TCI Chemicals (Tokyo, Japan). 2-Hydroxyethyl disulfide, DCM, Sn(Oct)$_2$, Strontium ispopropoxide, Mg(HMDS)$_2$, TBD, DBU, TCEP, and GSH were purchased from Sigma Aldrich (Merck KGaA, Darmstadt, Germany) and used without further purification. *Tert*-Butanol was bought from Grüssing GmbH (Filsum, Germany). Glycidol (Acros) and ε-CL (TCI) were distilled prior to use and stored in a Schlenk flask over a molecular sieve (4 Å). Cyanine-5-amine was purchased from Lumiprobe (Hannover, Germany). Pur-A-Lyzer Maxi 6000 Dialysis Kit was bought from Sigma Aldrich. Doxorubicin hydrochloride was purchased from ABCR GmbH.

2.1. Instrumentation

NMR spectra were recorded on a Jeol Eclipse 500 MHz (Tokyo, Japan) or a Bruker AVANCE III 700 MHz spectrometer (Billerica, MA, USA). Chemical shifts δ were reported in ppm and the deuterated solvent peak was used as internal standard. All spectra were recorded at 300 K. NMR data were reported including: chemical shift, multiplicity (s = singlet, d = doublet, t = triplet, m = multiplet), integration and coupling constants (s) in Hertz (Hz). Multiplets were recorded over the range (ppm) in which they appear in the spectrum.

Elemental analysis was performed with a VARIO EL III (Elementar).

GPC measurements in DMF were performed with on a customized chromatography system (PSS Polymer Standards GmbH, Mainz, Germany). A 5 cm precolumn (PSS-SDV in DMF, 5 μm particle size) coupled with a 30 cm column (PSS SDV linear M in DMF, 5 μm particle size) and a differential refractometer was used to separate and analyze the samples. The mobile phase was DMF (10 mM LiBr) at a flow rate of 1.0 mL min^{-1}. The columns were heated at 50 °C, while the differential refractometer detector was kept at 35 °C. For each measurement, 50 µL of a prefiltered (PTFE 0.2 µm) 1.5 mg mL^{-1} sample solution was injected. The data were processed using the WinGPC unichrome software from PSS. Molecular weights and molecular weight distribution were obtained relative to a poly(methyl methacrylate) standard.

For the purification of the copolymer, tangential flow filtration was performed using a 10 kDa regenerated cellulose cassette (Merck) in a cassette holder (Sartorius). The flow of the solution through the system was induced by a peristaltic pump (Gibson). The rotor speed was kept at the maximal operating one. The sulfated copolymers were purified by dialysis in brine to water. Benzoylated cellulose membrane (Sigma Aldrich, MWCO = 2 kDa) was chosen for this step.

UV/Vis measurements were conducted on an Agilent Cary 8454 UV-visible spectrophotometer, using half-micro quartz cuvettes.

Fourier transform infrared spectroscopy (FTIR) spectra were recorded on a Nicolet AVATAR 320 FT-IR 5 SXC (Thermo Fisher Scientific, Waltham, MA, USA) with a DTGS detector from 4000 to 650 cm^{-1}. Sample measurement was performed by dropping a solution of the compound and letting the solvent evaporate for a few seconds.

Raman spectra were recorded on a Bruker (Karlsruhe, Germany) MultiRAM II equipped with a low-temperature Ge detector (1064 nm, 100–180 mW) with 256 scans at a resolution of 2 cm^{-1}.

2.2. Synthesis

2.2.1. 2-((2-(oxiran-2-ylmethoxy)ethyl)disulfanyl)ethan-1-ol (SSG Monomer)

SSG Monomer was synthesized according to a modified published procedure [23]. Briefly, a solution of 2-Hyroxyethyl disulfide (13.9 g; 90 mmol; 1.0 equiv.) in *tert*-Butanol (150 mL) was added dropwise via dropping funnel to a solution of KO^{t}Bu (10.1 g; 90 mmol; 1.0 equiv.) in *tert*-Butanol (225 mL). After 4 h, excess epichlorohydrin (64.4 mL; 600 mmol; 6.7 equiv.) was added dropwise and the reaction mixture was stirred overnight at room temperature. The formed salt was filtered off and the solvent removed under reduced pressure. The crude product was dissolved in DCM (150 mL), extracted with water (3 × 50 mL) and dried over Na$_2$SO$_4$. After purification by column chromatography (eluent EtOAc 4:1 Hexane v/v), the product (5.9 g; 28 mmol; 31%) was obtained as a yellow oil. ^{1}H NMR spectra were in accordance with reported data. The successful synthesis of the monomer was confirmed by ^{1}H and ^{13}C NMR spectroscopy (see Supplementary Information).

2.2.2. Polymerization Procedure

The SSG monomer (3.15 g, 15 mmol, 1.0 equiv.), glycidol (10 mL; 150 mmol; 10 equiv.) and ε-CL (1.66 mL; 15 mmol; 1.0 equiv.) were mixed and formed a clear solution. The catalyst (1.5 mmol; 0.1 equiv.) was dissolved in anhydrous toluene (7.5 mL) and divided into 5 equal portions. To initiate the reaction, the monomer solution (1.0 mL) and the catalyst solution (2.5 mL) were added to a 2-necked flask with installed mechanical stirrer at 70 °C with a stirring speed of 120 rpm. The remaining monomer solution was added via a syringe pump over the next 8 h, whereas a portion of the catalyst solution was added to the reaction mixture every 2 h. The reaction was kept at 70 °C under constant stirring for 48 h and terminated by the addition of water (30 mL). Crude products were purified by tangential flow filtration (MWCO 10 kDa), dried via lyophilization, and analyzed by means of GPC, NMR, IR and Raman spectroscopy as well as elemental analysis. For Sn(Oct)$_2$ and DBU catalysts, no toluene was used as solvent, since these are liquids in pure form.

2.2.3. Sulfation of Polymers

This reaction was performed according to an established protocol [24]. The unsulfated polymer was dried at 60 °C under high vacuum overnight and then dissolved in anhydrous

DMF (concentration 0.029 g/mL). Then, a solution of sulfur trioxide pyridine complex (1.5 eq/OH group) in anhydrous DMF (concentration 0.1 g/mL) was added via syringe pump over 2 h and the reaction was kept at 60 °C for 24 h. The pH was adjusted to 7 with 1 M NaOH, solvent was removed under reduced pressure and the polymer dissolved in brine. Finally, dialysis (MWCO 2 kDa) was performed with brine solution, slowly decreasing the salt content to pure water over 96 h. After lyophilization, the product was obtained as a white to yellowish solid. The products were analyzed by means of NMR, GPC and IR spectroscopy. The degree of sulfation was calculated based on elemental analysis.

2.2.4. Preparation of Doxorubicin (Free Base)

DOX·HCl (100 mg; 0.172 mmol, 1.0 equiv.) was dissolved in water (200 mL) and added with DCM (200 mL) to a separation funnel. Then, NEt$_3$ (1 mL; 7.2 mmol; 42.0 equiv.) was added, the organic layer was separated, and the aqueous phase extracted with DCM (5 × 50 mL). The organic phase was dried over Na$_2$SO$_4$, the solvent removed under reduced pressure and the product dried under high vacuum. The product was obtained in quantitative yield and stored in the freezer. ESI-MS: m/z. Calculated 544.1819 g/mol. Found 544.1908 g/mol ([M+H]+).

2.3. Drug Encapsulation

To 10 mg of Doxorubicin (free base), 1 mL of a solution of the sulfated polymer in Milli-Q water (100 mg/mL) was added dropwise at room temperature at 1200 rpm. In order to separate conjugates from the free drug, the solution was transferred into a Falcon tube, centrifuged at 4000 rpm for 5 min and finally purified by column chromatography with a Sephadex G-25. Conjugates were partially lyophilized and the drug-loading content was determined by UV/Vis measurements.

2.4. Degradation Study with Reducing Agents

For degradation studies, 10 mM solutions of DTT, TCEP and GSH were prepared. For the latter, the pH was adjusted to 7.4 with 1 M NaOH. Solutions were degassed meticulously for 2 h. Polymers (40 mg) were dissolved in the respective solutions (10 mg mL^{-1}) and incubated at 37 °C for 24 h. The solutions were dialyzed (MWCO 1 kDa) against water and the dialysate was replaced every 6 h. Finally, solutions were lyophilized and analyzed by means of ^{1}H NMR. As a control, polymer solutions were prepared in PBS and processed in the same manner.

2.5. Degradation Study with Lipase B

This degradation study was performed according to an established protocol [25]. Briefly, polymers (30 mg) were dissolved in PBS (60 mL). Then, *Candida antarctica* Lipase B (60 mg; 200 wt% of polymer) and a few drops of n-Butanol were added and incubated at 37 °C for 72 h. The resin was filtered off, the solution lyophilized and analyzed by means of ^{1}H NMR. As a control, the same solution was prepared and processed without the addition of resin-immobilized enzyme.

2.6. Release Study of Doxorubicin

DR-dPGS@DOX conjugates with an overall DOX content of 0.2 mg were dissolved in solutions (2 mL) of the reducing agents and/or CALB resin. Prior to use, all solutions were treated exactly as for the degradation study. As controls, DR-dPGS@DOX in PBS without any reducing agent and pure DOX in PBS were used. All solutions were transferred into a Float-a-lyzer dialysis kit (MW 6–8 kDa) inside a Falcon tube filled with 30 mL of the respective medium and incubated at 37 °C. The DO content inside the dialysis kit was measured at constant time points (0 h, 0.5 h, 1 h, 2 h, 3 h, 4 h, 5 h, 6 h, 8 h, 24 h, 26 h, 30 h, 48 h, 54 h, and 72 h) in order to measure the released drug. For normalization, the initial absorbance at 0 h was set to 100% content and the cumulative released was calculated based on that.

2.7. Cytotoxicity Studies

The effect of the compounds on three cancer cell lines, A549, HeLa and MCF-7, was determined using the cell viability assay Cell Counting Kit 8 (CCK-8) from Sigma Aldrich Chemie GmbH (Taufkirchen, Germany) according to the manufacturer's instructions. A549 (DSMZ no.: ACC 107), HeLa (DSMZ no.: ACC 57) and MCF-7 (DSMZ no.: ACC 115) cells were obtained from Leibniz-Institute DSMZ—Deutsche Sammlung von Mikroorganismen und Zellkulturen GmbH and cultured in Dulbecco's Modified Eagle's Medium (DMEM) supplemented with 10% (v/v) FBS, 100 U/mL penicillin and 100 µg/mL Streptomycin (all from Gibco BRL, Eggenstein, Germany). Cells were regularly subcultured at least twice a week when they reached 70% to 90% confluency. For the cytotoxicity assay, 90 µL of a cell suspension in DMEM containing 5×10^4 cells per mL were seeded in each inner well of a 96-well plate and incubated over night at 37 °C and 5% CO2. In the outer wells, 90 µL DMEM without cells were added. The next day, serial dilutions of all the samples were prepared and 10 µL each were added to the cells in triplicates and in addition to one outer well for background correction. SDS (1%) and nontreated cells served as controls. After another 48 h at 37 °C and 5% CO2, the CCK-8 solution was added (10 µL/well) and absorbance at a measurement wavelength of 450 nm and a reference wavelength of 650 nm was measured after approximately 3 h incubation using a Tecan plate reader (infinite pro200, TECAN-reader Tecan Group Ltd. Männedorf, Switzerland). Measurements were performed in triplicate. The cell viability was calculated by setting the nontreated control to 100% after subtracting the background using Excel software. All cell experiments were conducted according to German genetic engineering laws and German biosafety guidelines in the laboratory (safety level 1).

IC50 values were calculated with GraphPad Prism 6.01 using the log(inhibitor) vs. normalized response equation.

2.8. Confocal Laser Scanning Microscopy (CLSM)

The uptake of Cy5-labeled DR-dPGS and DR-dPGS@DOX in HeLa cells was analyzed using Confocal Laser Scanning Microscopy (CLSM). The cells were propagated as described above. For CLSM, HeLa cells were seeded in 8-well ibidi slides (ibidi treat) in 270 µL DMEM. After cell attachment 4 h to 24 h, a postseeding 30 µL of solution containing compound were added for 3 h or 20 h. Before imaging, the cells were stained with Hoechst 33342 (1 µg/mL), washed with PBS and covered with fresh cell culture medium (DMEM). Confocal images were taken with an inverted confocal laser scanning microscope Leica DMI6000CSB SP8 (Leica, Wetzlar, Germany) with a 63x/1.4 HC PL APO CS2 oil immersion objective using the manufacture given LAS X software in sequential mode with the following channel settings: Transmission Ch (grey intensity values), excitation laser line 405 nm, detection of transmitted light (photomultiplier); Ch1 (Hoechst 33342): excitation laser line 405 nm, detection range 410 nm–485 nm (hybrid detector); Ch2 (Doxorubicin): excitation laser line 488 nm, detection range 496 nm–629 nm (hybrid detector); Ch3 (Cy5 dye): excitation laser line 488 nm, detection range 638 nm–797 nm (photomultiplier).

3. Results

3.1. Polymer Synthesis and Analysis

All polymers were synthesized by ring-opening copolymerization (ROP) of glycidol, ε-caprolactone (ε-CL) and 2-((2-(oxiran-2-ylmethoxy)ethyl)disulfanyl)ethan-1-ol (SSG), using Lewis acids and organic bases as catalysts in a one-pot gram-scale reaction. Due to the fact that a higher PCL content can lead to water solubility issues, we decided to keep the ε-CL content fixed at 10 mol% in the feed. We sought the optimal conditions for a gram-scale batch size polymerization at 70 °C with a catalyst that renders reasonable yields, molecular weights above 10 kDa, low polydispersity (Đ), a sufficient degree of branching (DB) and incorporation of intact ester structures of the CL and disulfide bonds of the SSG, respectively. As has been stated by Kizhakkedathu et al. before, typical temperatures for the ring-opening polymerization (ROP) of such monomers is around 95 °C; however, the SSG monomer is not stable and disulfide linkages undergo decomposition into thiols and thioethers [26]. Similar observations were made, and therefore, the polymerization temperature was kept strictly at 70 °C.

In order to find optimal conditions for the synthesis of the copolymers, a wide scope of catalysts were screened. Thus far, a variety of catalysts has been studied for the ROP of caprolactones and oxiranes including Lewis acids such as Sn(oct)$_2$, strontium isporopoxide (Sr(OiPR)$_2$), Mg(HMDS)$_2$ and organocatalysts such as 1,8-diaza-[5.4.0]undec-7-ene (DBU) and 1,5,7-triazabicyclo[4.4.0]dec-5-ene (TBD) and diphenyl phosphate (DPP) [27–29]. In our group, the synthesis of dPG-PCL copolymers was reported recently, utilizing Sn(oct)$_2$ as a suitable catalyst [30]. All aforementioned catalysts were used for ROP of glycidol (Gly), ε-CL and SSG with the feed molar ratio of [Gly]/[ε-CL]/[SSG]:[10]/[1]/[1] (Table 1).

Table 1. Polymer synthesis and catalyst screening for polymerizations.

Catalyst	Mn [a] [kDa]	Mw [a] [kDa]	Đ	SS Content [b] [%]	SS Content [c] [%]	CL Content [b] [%]	DB [b] [%]	Yield [g]
Sn(Oct)$_2$	12.8	17.1	1.3	2.3	2.6	5.8	27	1.2
Sr(OiPr)$_2$	23.4	38.8	1.6	4.2	3.5	5.2	56	1.8
DBU	1.2	3.4	2.8	4.2	3.6	3.5	56	1.0
TBD	0.4	3.1	7.8	3.7	3.9	6.8	50	2.0
Mg(HMDS)$_2$	<0.5	N/D	N/D	N/D	3.6	N/D	N/D	2.9
DPP	<0.5	N/D	N/D	N/D	N/D	N/D	N/D	0

(a) measured by GPC in DMF; (b) measured by ^{1}H NMR; (c) measured by EA.

All polymers were analyzed by means of ^{1}H and ^{13}C NMR, GPC, IR and Raman spectroscopy. Typical spectra of products can be seen in Figure 2 for the example of the Sr(OiPr)$_2$-catalyzed polymer.

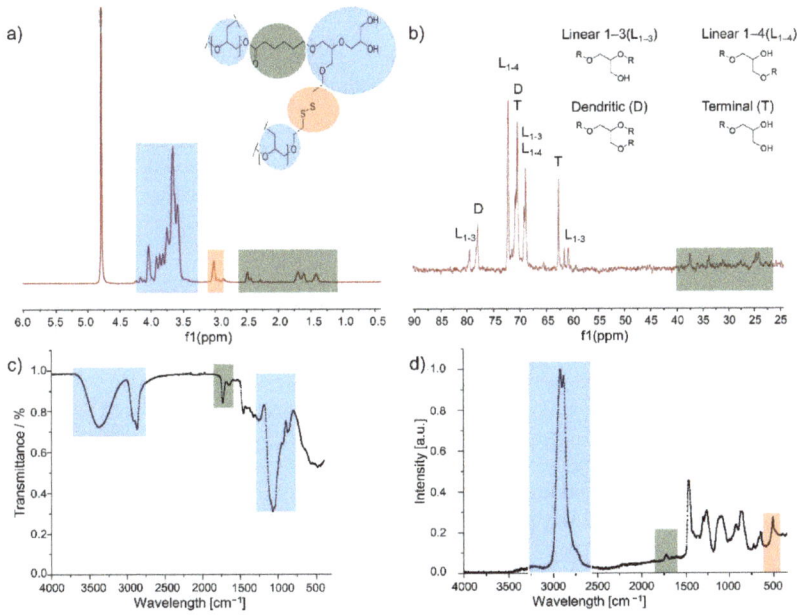

Figure 2. Analysis of the synthesized copolymers with color-coded structural units of glycidol (blue), SSG (red) and ε-CL (grey). (**a**) ^{1}H NMR spectrum with assignment of specific peaks of the three different monomer species; (**b**) inverse gated ^{13}C NMR spectrum with assignment of the signals of linear, dendritic and terminal carbon signals for the calculation of the degree of branching; (**c**) IR spectrum; (**d**) Raman spectrum.

As the ^{1}H NMR spectrum (Figure 2a) displays quantifiable signals of the dPG backbone (4.2–3.5 ppm), the signals of the protons of the methylene groups next to the disulfide bonds (3.0 ppm) and the aliphatic units of ε-CL, particularly the methylene bridge next to the carbonyl group (2.5 ppm), it is the proof of the incorporation of the three monomers into the polymer backbone and the basis for the determination of disulfide and ε-CL content in Table 1. The monomer content was calculated based on NMR and elemental analysis (Table 1). The details of the calculations are explained in the Supplementary Information. As opposed to the tedious synthesis of perfect dendrimers, DR-dPGS can be synthesized in a large batch size within two reaction steps and maintain the advantages of its dendrimer analogue; however, the degree of branching is only 56% (Table 1).

Besides the composition of the polymers, their structure is of high importance for the application as a DDS. Therefore, the degree of branching was calculated based on the definition by Frey [31], by using integrals obtained by overnight IG ^{13}C NMR spectra:

$$DB = \frac{2\,D}{2\,D + L} \tag{1}$$

where D is the relative integral of dendritic units and L is the relative integral of linear units of type L_{1-3} and L_{1-4} as indicated in Figure 2b.

Furthermore, IR spectroscopy (Figure 2c) was used to show the presence of the structural units in the copolymer. The broad peaks around 3400 cm^{-1} and 3000 cm^{-1} can be assigned to the hydroxy groups of the polyether and its C-H bonds, respectively. The presence of ε-CL can be determined with this method due to the strong carbonyl bond in IR spectroscopy at 1725 cm^{-1}. Moreover, the signal induced by the C-O bonds is strongly present at 1100 cm^{-1}. As disulfide bonds cannot be detected easily with IR spectroscopy, Raman spectroscopy (Figure 2d) is a useful method to determine their presence and generate a more complete image of the polymer structure. The absorbance band at 500 cm^{-1} in the Raman spectrum is assigned to the disulfide bond of the SSG monomer [32]. The results lead to the conclusion that all monomers were incorporated into the polymer structure.

As noted, ideally catalysts render polymers with molecular weights above 10 kDa and narrow molecular weight distribution in good yields, a sufficient degree of branching (DB) and the incorporation of intact ester and disulfide bonds in the polymer. Taking the results from Table 1 into account, considering molecular weight, degree of branching, monomer content and potential catalyst toxicity, the copolymer synthesized with Sr(OiPr)$_2$ was selected for further experiments and will be further referred to as Dual-Responsive dendritic Polyglycerol Sulfate DR-dPGS.

3.2. Degradation Study

In order to prove the dual-responsiveness of the system, it is necessary to show its degradation when being exposed to a reducing agent or an ester-cleaving enzyme. For this a degradation study using the reducing agents DTT, TCEP and the physiologically relevant GSH as well as the enzyme *Candida antarctica* Lipase B (CALB, commercially known as Novozyme-435) were used. Degradation under reductive and enzymatic conditions was monitored for 24 h and 72 h by using NMR spectroscopy (Figure 3).

The degree of degradation was deduced from the comparison of the relevant signals to the integral at 4.0 ppm, which stays constant during the entire process (Figure S1). As can be seen, the incorporated disulfides are cleaved as the integral decreases, whereas the thioether-related integral at 2.80 ppm and the thiol-related integral at 2.70 ppm increase (Figure 3a). As proof of principle, the reducing agents DTT and TCEP show a decrease in disulfide content of 79% and 92%, respectively. The physiologically more relevant GSH reduces the disulfide content in the copolymer of 31% within 24 h. Even though it is not as efficient as the other reducing agents, the ratio of GSH to its dimer GSSG is kept constant by enzymes in the cells and therefore is renewed more frequently than in a degradation study [33]. The incubation of the copolymer with CALB leads to a decrease in the α-carbonyl signal and an increase in its counterpart next to the carboxyl group (Figure 3b) and shows 74% ester degradation within 72 h.

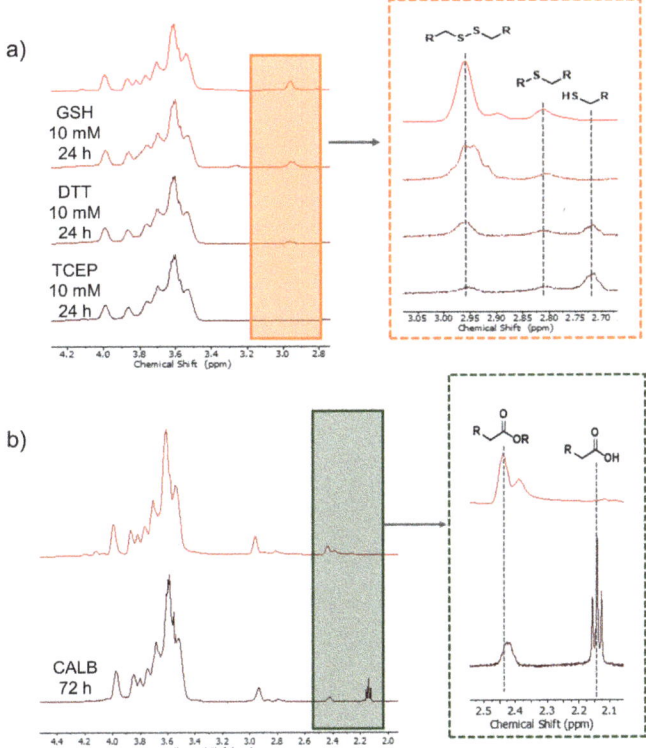

Figure 3. Degradation Study of DR-dPG. (**a**) ^{1}H NMR spectra after 24 h with incubation at reductive conditions with GSH, DTT, and TCEP. The decrease in the integral of methylene group next to the disulfide and increase in its thiol counterpart are highlighted; (**b**) degradation under enzymatic conditions with CALB. The decrease in the methylene bridge in α-position to the ester and increase in its respective acid counterpart are highlighted.

3.3. Sulfation

DR-dPG was then further sulfated by an established procedure, using sulfur trioxide pyridine complex in DMF at 60 °C overnight [24]. The reaction was followed by dialysis against brine, slowly decreasing the salt concentration to pure water. The sulfated copolymer was then analyzed by means of zeta potential measurements and elemental analysis (EA). The latter is the basis for the calculation of the degree of sulfation (DS). The DS was calculated based on the results obtained in Table 1. A detailed calculation is given in the supporting information.

For dPG, sulfation leads to roughly an increase of double the molecular weight, as statistically each glycidol unit leads to the formation of one hydroxyl group which can be converted into a sulfate group. Here, the M_n for DR-dPGS was calculated as 45.5 kDa (a detailed calculation is given in the Supplementary Information).

3.4. Drug-Loading and Release Study

Moreover, the efficacy of DR-dPGS in encapsulating Doxorubicin (free base) was investigated. For this purpose, the polymer was loaded with a targeted amount of 10 wt% and purified prior to analysis by UV/Vis measurements (Figure 4a).

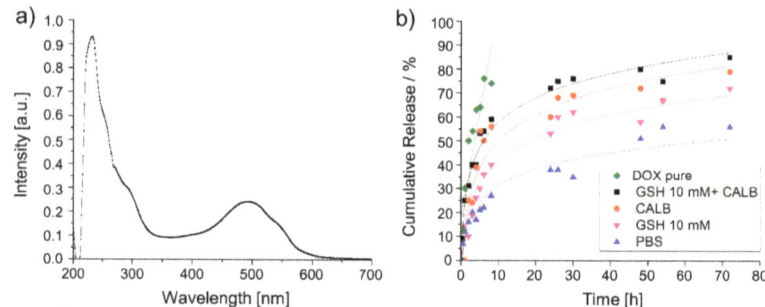

Figure 4. Determination and of drug loading content and subsequent release of DOX under reductive and/or enzymatic conditions. (**a**) UV/Vis spectrum of DR-dPGS@DOX; (**b**) release curve of DOX (linear fit), GSH+CALB, CALB, GSH, and PBS (Logarithmic Log3P1 fit).

The amount of loaded drug was calculated based on a calibration curve measured at different concentrations (Figure S2). The drug-loading capacity (DLC) was determined according to the following equation:

$$DLC = \frac{m\,(Drug)}{m\,(total) - m\,(Drug)} \times 100 \qquad (2)$$

As successful loading of Doxorubicin and the degradation of the copolymer could be proven (Figure 4a), we next developed a set-up to monitor the triggered release of DOX under reductive or enzymatic degradation. As GSH concentrations of up to 10 mM can be reached in the cytosol, [34] we chose this for proof-of-concept experiments for the reductive environment. DOX release under enzymatic conditions was screened accordingly to the degradation study. Furthermore, we chose a combination of GSH and CALB in order to investigate the dual-responsiveness of the system. As controls we chose pure doxorubicin in PBS as well as DR-dPGS@DOX in PBS without any stimulus. The results from the release study can be seen in Figure 4b.

The experiment renders information about the on-going release during degradation. The free DOX was released completely within the first 8 h while the release of DR-dPGS@DOX in PBS occurred in a sustained manner. Furthermore, the dual-responsiveness of the polymer conjugate can be shown with reductive and/or enzymatic conditions. Incubation with 10 mM GSH renders a release of 72% of drug over time, enzymatic degradation with CALB leads to 79% drug release. Interestingly, a synergistic effect of the combination of both stimuli can be detected, with the combined conditions leading to 85% release of the loaded drug within 72 h.

3.5. Cytotoxicity and Cell Uptake

In order to investigate the effect of this novel copolymer on cells and to verify the cytotoxic behavior of drug-loaded carriers, we performed cytotoxicity studies using three different cancer cell lines: HeLa cervix carcinoma cells, A549 lung carcinoma cells and MCF-7 breast cancer cells (Figure S3). The experiments show that the copolymers DR-dPGS without the drug doxorubicin were well tolerated until the highest tested concentration of 1 mg mL^{-1}. DOX encapsulated in DR-dPGS, as well as free doxorubicin, caused a similar concentration dependent decrease in the cells' viability in all three cell lines. The effect of free and encapsulated doxorubicin was best in HeLa and A549 cells; MCF-7 cells seemed to be more robust against the drug. The results indicate that the cytotoxic properties of the drug are amplified by the copolymer, which can be further proven by the IC50 values obtained from these measurements and can be seen in Table 2. Interestingly, DR-DPGS@DOX outperforms the free drug by roughly an order of magnitude.

Table 2. Calculated IC50 values for DR-dPGS@DOX and free DOX.

	IC50 DR-dPGS@DOX (µg/mL)	IC50 DOX (µg/mL)
HeLa	0.036	0.117
A549	0.039	0.2549
MCF-7	0.444	6.163

In order to visualize the cellular uptake and fate of the DR-dPGS and DOX, we used confocal laser scanning microscopy (CLSM). For this study we monitored living HeLa cells treated with DR-dPGS and DR-dPGS@DOX for 3 h and 20 h. DR-dPGS was labelled before imaging and loading with a hydrophobic cyanine dye (Cy5). The images displayed in Figure 5 clearly show that DR-dPGS and DOX are taken up by the cells and their signals accumulate over time. The DR-dPGS is mainly located in the cytosol outside the nucleus in distinct areas. These spots are most probably lysosomes and point to an endocytotic uptake pathway of DR-dPGS as shown for other sulfated polymers [35]. In contrast DOX is mainly located in the nucleus, which is in line with the literature and proves that the drug is released from DR-dPGS [36]. Furthermore, this can also be seen by the drastically reduced Hoechst signal, as DOX also interacts with DNA and thus replaces Hoechst [37]. Moreover, after 20 h of incubation the toxic effect of DOX is clearly seen as fewer cells can be observed and the remaining cells show morphological changes compared to nontreated cells or cells only treated with empty DR-dPGS. Therefore, it can be concluded that DR-dPGS can successfully transport and release doxorubicin in cancer cells.

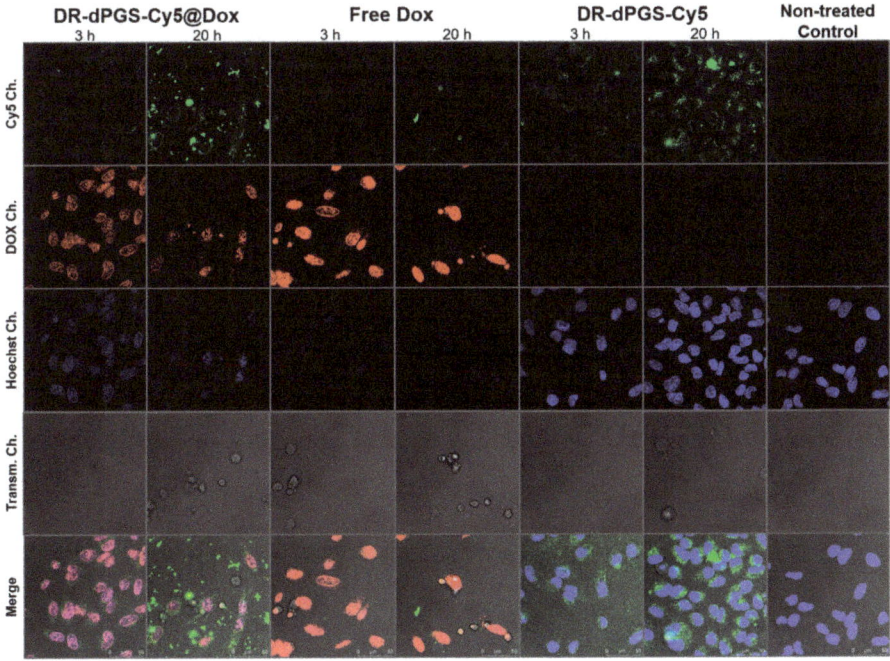

Figure 5. Confocal Microscope images of HeLa cells treated with DR-dPGS-Cy5 with and without DOX, free DOX for 3 h and 20 h, respectively, and nontreated cells. Blue: Hoechst (nuclei); red: DOX, green: Cy5-labelled carrier. Scale bar indicates 50 µm.

4. Discussion

The copolymerization of glycidol, SSG and ε- caprolactone was investigated with regard to the impact of different catalysts on molecular weight, disulfide and ester content as well as degree of branching. After analysis by means of GPC, ^{1}H and ^{13}C NMR, IR and Raman spectroscopy, the Sr(OiPr)$_2$-catalyzed polymer DR-dPG was selected and its degradation under reductive or enzymatic conditions could be proven by monitoring the signals of protons next to the disulfide and ester bond, respectively. The polymer was then further sulfated and DR-dPGS was loaded with the hydrophobic drug DOX. A DLC of 2.7% was calculated based on UV/Vis measurements. A release study with 10 mM GSH and CALB revealed that the drug can be released via reductive and enzymatic stimuli, as well as a synergistic effect of both, leading to 85% drug release over 72 h. Cytotoxicity measurements with three different cancer cell lines show that the polymer itself is well-tolerated, whereas DR-dPGS@DOX showed enhanced cytotoxic effects compared to the free drug, indicating that the cytotoxic behavior of the drug is maintained after encapsulation. This could also be shown with calculated IC50 values. The cellular uptake of dye-labeled DR-DPGS@DOX was examined with confocal laser scanning microscopy. Observations indicate that the carriers are taken up efficiently by HeLa cells. The polymer itself accumulates in the cytosol and releases the drug, which accumulates in the cell nucleus. These properties together with simple and gram scale synthesis make the synthesized copolymer a promising candidate for future biomedical applications.

Supplementary Materials: The following are available online at https://www.mdpi.com/2073-4360/13/6/982/s1, Figure S1:1H NMR spectra obtained by the degradation study; Figure S2: DOX calibration; Figure S3: Cell viability results obtained from the cell viability assay (CCK-8) upon incubation of A549, HeLa and MCF-7 cells with (a) DR-dPGS, (b–d) DR-DPGS@DOX (DLC = 2.7%) and free DOX after 48 h of treatment; Figure S4: 1H NMR spectrum of the SSG monomer; Figure S5: 13C spectrum of the SSG monomer; Figure S6: 1H NMR spectrum of the Sn(Oct)2-catalyzed polymer; Figure S7: 13C NMR spectrum of the Sn(Oct)2-catalyzed polymer; Figure S8: 1H NMR spectrum of the Sr(OiPr)2-catalyzed polymer; Figure S9: 13C NMR spectrum of the Sr(OiPr)2-catalyzed polymer; Figure S10: 1H NMR spectrum of the DBU-catalyzed polymer; Figure S11: 13C NMR spectrum of the DBU-catalyzed polymer; Figure S12: 1H NMR spectrum of the TBD-catalyzed polymer; Figure S13: 13C NMR spectrum of the TBD-catalyzed polymer; Figure S14: GPC elugram of DR-dPG.

Author Contributions: Conceptualization, methodology, synthesis in vitro studies, data interpretation, writing—original draft preparation, visualization: F.R. and E.M. Conceptualization, data interpretation, writing—original draft preparation, synthesis, in vitro studies: A.O. Visualization, data interpretation: M.C. Conceptualization, data interpretation: M.D. In vitro studies: E.Q. Data interpretation, visualization, writing—original draft preparation, in vitro studies: K.A. Conceptualization, guidance, resources, funding acquisition: R.H. All authors have read and agreed to the published version of the manuscript.

Funding: This study was supported by the Dahlem Research School (DRS) as well as SFB 765 and the core facility BioSupraMol (www.biosupramol.de, accessed on 22 March 2021), both funded by the Deutsche Forschungsgemeinschaft (DFG).

Institutional Review Board Statement: Not applicable.

Informed Consent Statement: Not applicable.

Data Availability Statement: Data are contained within the article or Supplementary Materials.

Acknowledgments: We thank Gene Senges for measuring Raman spectroscopy and Thi Mai Phuong Neumann-Tran for measuring GPC. We are grateful to Kathleen Hernandez for proofreading this manuscript.

Conflicts of Interest: The authors declare no conflict of interest.

References

1. Karabasz, A.; Bzowska, M.; Szczepanowicz, K. Biomedical Applications of Multifunctional Polymeric Nanocarriers: A Review of Current Literature. *Int. J. Nanomed.* **2020**, *15*, 8673–8696. [CrossRef]
2. Zhang, Y.; Sun, T.; Jiang, C. Biomacromolecules as carriers in drug delivery and tissue engineering. *Acta Pharma. Sin. B* **2018**, *8*, 34–50. [CrossRef] [PubMed]
3. Abbina, S.; Vappala, S.; Kumar, P.; Siren, E.M.J.; La, C.C.; Abbasi, U.; Brooks, D.E.; Kizhakkedathu, J.N. Hyperbranched polyglycerols: Recent advances in synthesis, biocompatibility and biomedical applications. *J. Mater. Chem. B* **2017**, *5*, 9249–9277. [CrossRef] [PubMed]
4. Gagliardi, A.; Giuliano, E.; Venkateswararao, E.; Fresta, M.; Bulotta, S.; Awasthi, V.; Cosco, D. Biodegradable Polymeric Nanoparticles for Drug Delivery to Solid Tumors. *Front. Pharmacol.* **2021**, *12*, 601626. [CrossRef] [PubMed]
5. Vlakh, E.; Ananyan, A.; Zashikhina, N.; Hubina, A.; Pogodaev, A.; Volokitina, M.; Sharoyko, V.; Tennikova, T. Preparation, Characterization, and Biological Evaluation of Poly(Glutamic Acid)-b-Polyphenylalanine Polymersomes. *Polymers* **2016**, *8*, 212. [CrossRef] [PubMed]
6. Golombek, S.K.; May, J.-N.; Theek, B.; Appold, L.; Drude, N.; Kiessling, F.; Lammers, T. Tumor targeting via EPR: Strategies to enhance patient responses. *Adv. Drug Deliv. Rev.* **2018**, *130*, 17–38. [CrossRef] [PubMed]
7. Sindhwani, S.; Syed, A.M.; Ngai, J.; Kingston, B.R.; Maiorino, L.; Rothschild, J.; MacMillan, P.; Zhang, Y.; Rajesh, N.U.; Hoang, T.; et al. The entry of nanoparticles into solid tumours. *Nat. Mater.* **2020**, *19*, 566–575. [CrossRef] [PubMed]
8. Dong, H.; Pang, L.; Cong, H.; Shen, Y.; Yu, B. Application and design of esterase-responsive nanoparticles for cancer therapy. *Drug Deliv.* **2019**, *26*, 416–432. [CrossRef]
9. Rades, N.; Licha, K.; Haag, R. Dendritic Polyglycerol Sulfate for Therapy and Diagnostics. *Polymers* **2018**, *10*, 595. [CrossRef]
10. Khandare, J.; Calderón, M.; Dagia, N.M.; Haag, R. Multifunctional dendritic polymers in nanomedicine: Opportunities and challenges. *Chem. Soc. Rev.* **2012**, *41*, 2824–2848. [CrossRef]
11. Türk, H.; Haag, R.; Alban, S. Dendritic Polyglycerol Sulfates as New Heparin Analogues and Potent Inhibitors of the Complement System. *Bioconjug. Chem.* **2004**, *15*, 162–167. [CrossRef] [PubMed]
12. Dernedde, J.; Rausch, A.; Weinhart, M.; Enders, S.; Tauber, R.; Licha, K.; Schirner, M.; Zügel, U.; von Bonin, A.; Haag, R. Dendritic polyglycerol sulfates as multivalent inhibitors of inflammation. *Proc. Natl. Acad. Sci. USA* **2010**, *107*, 19679–19684. [CrossRef] [PubMed]
13. Schneider, T.; Welker, P.; Licha, K.; Haag, R.; Schulze-Tanzil, G. Influence of dendritic polyglycerol sulfates on knee osteoarthritis: An experimental study in the rat osteoarthritis model. *BMC Musculoskelet. Disord.* **2015**, *16*, 387. [CrossRef]
14. Maysinger, D.; Ji, J.; Moquin, A.; Hossain, S.; Hancock, M.A.; Zhang, I.; Chang, P.K.Y.; Rigby, M.; Anthonisen, M.; Grütter, P.; et al. Dendritic Polyglycerol Sulfates in the Prevention of Synaptic Loss and Mechanism of Action on Glia. *ACS Chem. Neurosci.* **2018**, *9*, 260–271. [CrossRef]
15. Maysinger, D.; Gröger, D.; Lake, A.; Licha, K.; Weinhart, M.; Chang, P.K.Y.; Mulvey, R.; Haag, R.; McKinney, R.A. Dendritic Polyglycerol Sulfate Inhibits Microglial Activation and Reduces Hippocampal CA1 Dendritic Spine Morphology Deficits. *Biomacromolecules* **2015**, *16*, 3073–3082. [CrossRef]
16. Zhong, Y.; Dimde, M.; Stöbener, D.; Meng, F.; Deng, C.; Zhong, Z.; Haag, R. Micelles with Sheddable Dendritic Polyglycerol Sulfate Shells Show Extraordinary Tumor Targetability and Chemotherapy in Vivo. *ACS Appl. Mater. Interfaces* **2016**, *8*, 27530–27538. [CrossRef]
17. Ferber, S.; Tiram, G.; Sousa-Herves, A.; Eldar-Boock, A.; Krivitsky, A.; Scomparin, A.; Yeini, E.; Ofek, P.; Ben-Shushan, D.; Vossen, L.I.; et al. Co-targeting the tumor endothelium and P-selectin-expressing glioblastoma cells leads to a remarkable therapeutic outcome. *eLife* **2017**, *6*, e25281. [CrossRef] [PubMed]
18. Weinhart, M.; Gröger, D.; Enders, S.; Riese, S.B.; Dernedde, J.; Kainthan, R.K.; Brooks, D.E.; Haag, R. The Role of Dimension in Multivalent Binding Events: Structure–Activity Relationship of Dendritic Polyglycerol Sulfate Binding to L-Selectin in Correlation with Size and Surface Charge Density. *Macromol. Biosci.* **2011**, *11*, 1088–1098. [CrossRef]
19. Zabihi, F.; Graff, P.; Schumacher, F.; Kleuser, B.; Hedtrich, S.; Haag, R. Synthesis of poly(lactide-co-glycerol) as a biodegradable and biocompatible polymer with high loading capacity for dermal drug delivery. *Nanoscale* **2018**, *10*, 16848–16856. [CrossRef]
20. Adeli, M.; Namazi, H.; Du, F.; Hönzke, S.; Hedtrich, S.; Keilitz, J.; Haag, R. Synthesis of multiarm star copolymers based on polyglycerol cores with polylactide arms and their application as nanocarriers. *RSC Adv.* **2015**, *5*, 14958–14966. [CrossRef]
21. Kainthan, R.K.; Janzen, J.; Kizhakkedathu, J.N.; Devine, D.V.; Brooks, D.E. Hydrophobically derivatized hyperbranched polyglycerol as a human serum albumin substitute. *Biomaterials* **2008**, *29*, 1693–1704. [CrossRef]
22. Kurniasih, I.N.; Liang, H.; Rabe, J.P.; Haag, R. Supramolecular Aggregates of Water Soluble Dendritic Polyglycerol Architectures for the Solubilization of Hydrophobic Compounds. *Macromol. Rapid Commun.* **2010**, *31*, 1516–1520. [CrossRef] [PubMed]
23. Son, S.; Shin, E.; Kim, B.-S. Redox-Degradable Biocompatible Hyperbranched Polyglycerols: Synthesis, Copolymerization Kinetics, Degradation, and Biocompatibility. *Macromolecules* **2015**, *48*, 600–609. [CrossRef]
24. Ferraro, M.; Silbereis, K.; Mohammadifar, E.; Neumann, F.; Dernedde, J.; Haag, R. Biodegradable Polyglycerol Sulfates Exhibit Promising Features for Anti-inflammatory Applications. *Biomacromolecules* **2018**, *19*, 4524–4533. [CrossRef]
25. Stefani, S.; Sharma, S.K.; Haag, R.; Servin, P. Core-shell nanocarriers based on PEGylated hydrophobic hyperbranched polyesters. *Eur. Polym. J.* **2016**, *80*, 158–168. [CrossRef]

26. Shenoi, R.A.; Chafeeva, I.; Lai, B.F.L.; Horte, S.; Kizhakkedathu, J.N. Bioreducible hyperbranched polyglycerols with disulfide linkages: Synthesis and biocompatibility evaluation. *J. Polym. Sci. Part A Polym. Chem.* **2015**, *53*, 2104–2115. [CrossRef]
27. Kamber, N.E.; Jeong, W.; Waymouth, R.M.; Pratt, R.C.; Lohmeijer, B.G.G.; Hedrick, J.L. Organocatalytic Ring-Opening Polymerization. *Chem. Rev.* **2007**, *107*, 5813–5840. [CrossRef] [PubMed]
28. Kim, S.; Wittek, K.I.; Lee, Y. Synthesis of poly(disulfide)s with narrow molecular weight distributions via lactone ring-opening polymerization. *Chem. Sci.* **2020**, *11*, 4882–4886. [CrossRef]
29. Bandelli, D.; Weber, C.; Schubert, U.S. Strontium Isopropoxide: A Highly Active Catalyst for the Ring-Opening Polymerization of Lactide and Various Lactones. *Macromol. Rapid Commun.* **2019**, *40*, 1900306. [CrossRef]
30. Mohammadifar, E.; Zabihi, F.; Tu, Z.; Hedtrich, S.; Nemati Kharat, A.; Adeli, M.; Haag, R. One-pot and gram-scale synthesis of biodegradable polyglycerols under ambient conditions: Nanocarriers for intradermal drug delivery. *Polym. Chem.* **2017**, *8*, 7375–7383. [CrossRef]
31. Hölter, D.; Burgath, A.; Frey, H. Degree of branching in hyperbranched polymers. *Acta Polym.* **1997**, *48*, 30–35. [CrossRef]
32. Hernández, B.; Pflüger, F.; López-Tobar, E.; Kruglik, S.G.; Garcia-Ramos, J.V.; Sanchez-Cortes, S.; Ghomi, M. Disulfide linkage Raman markers: A reconsideration attempt. *J. Raman Spectrosc.* **2014**, *45*, 657–664. [CrossRef]
33. Schafer, F.Q.; Buettner, G.R. Redox environment of the cell as viewed through the redox state of the glutathione disulfide/glutathione couple. *Free Radic. Biol. Med.* **2001**, *30*, 1191–1212. [CrossRef]
34. Cheng, R.; Feng, F.; Meng, F.; Deng, C.; Feijen, J.; Zhong, Z. Glutathione-responsive nano-vehicles as a promising platform for targeted intracellular drug and gene delivery. *J. Control. Release* **2011**, *152*, 2–12. [CrossRef]
35. Kepsutlu, B.; Wycisk, V.; Achazi, K.; Kapishnikov, S.; Pérez-Berná, A.J.; Guttmann, P.; Cossmer, A.; Pereiro, E.; Ewers, H.; Ballauff, M.; et al. Cells Undergo Major Changes in the Quantity of Cytoplasmic Organelles after Uptake of Gold Nanoparticles with Biologically Relevant Surface Coatings. *ACS Nano* **2020**, *14*, 2248–2264. [CrossRef]
36. Tacar, O.; Sriamornsak, P.; Dass, C.R. Doxorubicin: An update on anticancer molecular action, toxicity and novel drug delivery systems. *J. Pharm. Pharmacol.* **2013**, *65*, 157–170. [CrossRef] [PubMed]
37. Elgart, V.; Lin, J.-R.; Loscalzo, J. Determinants of drug-target interactions at the single cell level. *PLoS Comput. Biol.* **2018**, *14*, e1006601. [CrossRef] [PubMed]

Article

Assessing the Influence of Dyes Physico-Chemical Properties on Incorporation and Release Kinetics in Silk Fibroin Matrices

Bruno Thorihara Tomoda, Murilo Santos Pacheco, Yasmin Broso Abranches, Juliane Viganó, Fabiana Perrechil and Mariana Agostini De Moraes *

Department of Chemical Engineering, Institute of Environmental, Chemical and Pharmaceutical Sciences, Universidade Federal de São Paulo, Rua São Nicolau 210, Diadema 09913-030, Brazil; btomoda@gmail.com (B.T.T.); murilopch@outlook.com (M.S.P.); yasbroso@gmail.com (Y.B.A.); julianevigano@gmail.com (J.V.); fabiana.perrechil@unifesp.br (F.P.)
* Correspondence: mamoraes@unifesp.br

Abstract: Silk fibroin (SF) is a promising and versatile biodegradable protein for biomedical applications. This study aimed to develop a prolonged release device by incorporating SF microparticles containing dyes into SF hydrogels. The influence of dyes on incorporation and release kinetics in SF based devices were evaluated regarding their hydrophilicity, molar mass, and cationic/anionic character. Hydrophobic and cationic dyes presented high encapsulation efficiency, probably related to electrostatic and hydrophobic interactions with SF. The addition of SF microparticles in SF hydrogels was an effective method to prolong the release, increasing the release time by 10-fold.

Keywords: biopolymer; hydrogel; microparticles; dye release

1. Introduction

Silk is a natural protein produced in the form of fibers by insect larvae, such as silkworms, spiders, scorpions, mites, and flies, varying in amino acid composition, chemical structure and properties according to its origin [1–3]. Two proteins compose the silk obtained from the silkworm cocoons: fibroin and sericin, the former is used for the development of biomaterials and the latter is removed due to reports of adverse events and hypersensitivity [1]. Several studies were carried in the production of silk fibroin (SF) based materials, showing different processes for the incorporation of active compounds, as well as the effects on the delivery system, highlighting the versatility of this protein [2,4–6]. SF matrices are mostly hydrophobic and amphoteric ion exchangers, highly dependent on pH, due to the presence of weak acid groups and bases. Thus, SF behaves as an anion exchanger at a pH lower than the isoelectric point (pI = 4.5) and as a cation exchanger at a pH above the pI [7]. SF structure consists of two protein chains, a "heavy" and a "light" chain. The "heavy" chain is mainly formed by ordered hydrophobic macromolecules and "light" chain is formed by polar amino acid residues. Due to these different blocks, SF is capable to perform hydrophilic and hydrophobic interactions, the latter being the main interaction of SF structure [8]. SF has mild processing conditions and can be used to produce hydrogels, microparticles or films [6,9].

Hydrogels are three-dimensional and hydrophilic polymeric structures, capable of absorbing large amounts of water or biological fluids, with potential applications in the biomedical area [10]. The physical properties of hydrogels, more than any other class of synthetic biomaterials, are the ones that most resemble living tissues, due to their high water content and their soft, rubbery consistency [11]. SF hydrogels' formation is a kinetic process that occurs spontaneously from the metastable aqueous solution due to SF molecules' aggregation tendency, moving from an amorphous conformation (random coil) to a stable conformation of the β-sheet type [12]. Although spontaneous, the formation can be accelerated by adding organic solvents, such as ethanol, which dehydrates SF molecules [12,13]. Moraes et al. [14] developed SF hydrogels capable to sustain diclofenac release for 10 h, showing a high release rate in the first three hours, releasing approximately 80% of the maximum release capacity. Other studies also show a fast release of drugs incorporated into SF hydrogels [14,15], emphasizing the need for strategies to achieve a prolonged and sustained release.

Encapsulation of drugs or bioactive compounds into biopolymeric microparticles is a suitable approach to prolong the release. Besides, as delivery systems, microparticles must protect the therapeutic agent from degradation and denaturation, in addition to controlling the release profile of the compound into the medium [16,17]. SF microparticles are an innovative tool for encapsulating drugs and developing drug delivery systems. Due to the formation of β-sheet structures, SF release devices are formed by crystalline regions, more stable in water, and exhibiting high rates of *in vivo* degradation [18,19]. Thus, aiming to prolong the release profile from SF hydrogels, a combined device of microparticles and hydrogel might have a high potential, allowing to modulate the system to the desired application and to obtain a more prolonged release rate [15,20,21].

Numata, Yamazaki, and Naga [15] developed a double release system of fluorescent dyes from SF hydrogels containing SF nanoparticles, aiming at a slow and prolonged release of the dye incorporated in the nanoparticles and a quick release of the dye added in the hydrogel. The release kinetics showed the efficiency of incorporating nanoparticles in the hydrogel, sustaining the release for up to 5 days. However, the influence of the dyes' properties on the dye-matrix interaction was not evaluated.

In this context, we proposed the synthesis of SF hydrogels containing SF microparticles to release model dyes, aiming at applications as release devices. The novelty of this work is the incorporation of model dyes with different charge, hydrophilicity and molar mass, for mapping the influence of the physico-chemical properties of the compound on the SF matrix and the interaction during incorporation and release, allowing to expand for applications in drugs and other bioactive molecules. The model dyes used were methylene blue (MB), rose bengal (RB), rhodamine B (RhB), and neutral red (NR).

2. Materials and Methods

2.1. Materials

Silkworm cocoons from *Bombyx mori* were supplied by silk company Bratac (Londrina, Paraná, Brazil). MB, RB, RhB, and NR were purchased from Sigma-Aldrich (St. Louis, MO, USA). All other chemicals were purchased from Synth (Diadema, São Paulo, Brazil). Ultrapure water from the Milli-Q system (Millipore, Burlington, MA, USA) was used.

The nomenclature used to identify the samples was MP for microparticles, HG for hydrogels, SF for silk fibroin, MB for methylene blue, RB for rose bengal, RhB for rhodamine B, and NR for neutral red. The microparticles incorporated with the dyes are identified by 'MP' + 'acronym of the dye or silk fibroin', and the hydrogels containing the microparticles with the dyes are identified by 'HG' + 'MP' + 'acronym of the dye or silk fibroin'.

2.2. Preparation of Silk Fibroin Aqueous Solution

Sericin was removed from the silkworm cocoons with 1 g/L aqueous sodium carbonate solution (Na_2CO_3) for 30 min in a thermostatic bath at 85 °C. This procedure was repeated three times, and then the SF threads were washed with plenty of distilled water. SF was dried at room temperature for 48 h and cut to an average size of 2 mm to facilitate dissolution. For each 10 g of SF, 100 mL of a ternary solution of calcium chloride, ethanol, and water ($CaCl_2$: CH_3CH_2OH: H_2O, 1:2:8 molar) were added and maintained at 85 °C in a thermostatic bath TE-184 (Tecnal, Piracicaba, São Paulo, Brazil) until complete dissolution of SF for a maximum of 90 min.

The SF saline solution (0.1 g/mL) was dialyzed in distilled water for 3 days, at 10 °C, to remove the calcium from the solution and obtain an SF aqueous solution. The dialysis water was changed every 24 h, in the volumetric ratio of 1:15 SF solution:water, with SF aqueous solution concentration of ca. 0.074 g/mL.

2.3. Preparation of Silk Fibroin Microparticles

SF microparticles were prepared by the atomization method, in which SF aqueous solution was sprayed into an absolute ethanol coagulation bath. The atomization system consists of a nozzle with a diameter of 0.5 mm (Labmaq, Ribeirão Preto, São Paulo, Brazil), a peristaltic pump TE-BP-01 Mini (Tecnal, Piracicaba, São Paulo, Brazil), and a compressor model OP8.1/30II (Pressure, Maringá, Paraná, Brazil). The SF solution was pumped (flow rate around 0.045 L/h) to the atomizing nozzle, where the solution was broken up into smaller droplets (compressed air pressure at 1 bar). The droplets came into contact with the absolute ethanol under stirring. After forming the microparticles, they were kept at rest for 30 min and then stirred for 10 min.

Each dye solution (1×10^{-4} mol/L) was incorporated into the microparticles by adsorption, by immersing 0.1 g of SF microparticles into 10 mL of dye solution and stirring at 30 rpm by a tube revolver (Thermo Scientific, Waltham, MA, USA) until the equilibrium (3 h) at room temperature (24 °C). The dyes used were MB, RB, RhB, and NR. These dyes were selected due to their different properties showed in Table 1. The rational selection was made to verify the influence of the charge, molar mass, and hydrophilicity of the dye in the SF matrices. MB, RB, and RhB are anionic dyes, and NR is a slightly cationic dye at neutral pH.

Table 1. Properties of the dyes used to load the SF microparticles.

Dye	Hydrophilicity	Solubility in Water (mg/mL) *	Molar Mass (g/mol) *	pKa **
MB	Hydrophilic	50	319.85	3.14
RB	Hydrophilic	100	973.67	4.7
RhB	Hydrophobic	7.8 [22]	479.02	3.7
NR	Hydrophobic	10	288.78	6.8

* Data obtained from supplier safety datasheet. ** Data obtained from PubChem.

2.4. Microparticles Characterization

The encapsulation efficiency (EE%) (Equation (1)) was calculated by the ratio between the mass of loaded dye in the microparticles (obtained by mass balance from the residual mass of dye quantified in the supernatant) and the mass of compound added in the fresh dye solution. The quantification of the dyes was performed by spectroscopy in a Genesys 10S UV/Vis spectrophotometer (ThermoScientific, Waltham, MA, USA), using wavelength of 664 nm for MB, 530 nm for RB, 542 nm for RhB, and 520 nm for NR. The calibration curves and fitting parameters are shown in Supplementary Materials, Figure S1.

$$Encapsulation\ efficiency\ (\%) = 100 \times \frac{mass\ of\ compound\ incorporated}{mass\ of\ compound\ added}. \quad (1)$$

The microparticles' average size was assessed by optical microscopy using a Primo Star microscope (Zeiss, Oberkochen, Baden-Württemberg, Germany) with a 10× objective. Mean particle diameter and log-normal frequency distribution were calculated from image analysis of at least five (5) different images containing 70–100 microparticles using the ImageJ software.

The morphology of microparticles was observed in a scanning electron microscope (SEM). The samples were frozen with liquid nitrogen, freeze-dried, and covered with gold in the Sputter Coater, Model K450 (Emitech, Kent, UK), with an Au layer thickness estimated at 200 Å. Samples were observed in a Leo 440i-6070, (LEO Electron Microscopy, Oxford, UK) with an accelerating voltage of 10 kV and a current of 50 mA.

The microparticles containing the dyes were analyzed with a LEICA DMi8 Confocal Microscope (Leica Microsystems, Wetzlar, Hessen, Germany) with 10× objective at 580 nm emission wavelength, coupled to the LAS X software to observe the dye's homogeneity through the microparticle matrix.

Fourier transform infrared spectroscopy (FTIR) was performed on SF microparticles before and after the dyes loading to verify possible SF-dye interactions. An Agilent Cary 630 FTIR spectrophotometer (Agilent, Santa Clara, CA, USA) was used in ATR mode, and the samples were analyzed in the wavelength range from 1200 to 2000 cm^{-1}, with a resolution of 4 cm^{-1} and 128 scans.

2.5. Preparation of Silk Fibroin Hydrogels

The SF aqueous solution (0.074 g/mL) was used to produce the hydrogels and a 50% ethanol solution was slowly added to the SF aqueous solution (10 mL) in a 1:1 ratio [14]. To produce hydrogels with microparticles, 0.1 mg (wet mass) of SF microparticles loaded with dye were added to the SF aqueous solution containing 50% ethanol solution. The mixtures were then added to molds (diameter 4 cm). The hydrogel formation time was measured just after the ethanol addition into the SF solution until the hydrogel' full gelation, where there is no movement of the liquid solution verified by the tube inversion method. The tube inversion method consists in turning the mold downwards to note whether solution stops to flow due to its own weight.

2.6. Hydrogel Characterization

The hydrogel was characterized by a scanning electron microscope (SEM) to observe the morphology. The samples were prepared by supercritical fluid drying [23] to maintain their original structure, fractured in liquid nitrogen, covered with gold and then were observed as previously described in Section 2.4.

Fourier transform infrared spectroscopy (FTIR) was performed on SF hydrogels with and without microparticles loaded with dyes to verify possible structural changes and SF-dye interactions. The same methodology described at Section 2.4 was used.

The rheological behavior was evaluated by dynamic oscillatory tests on an MCR 92 rheometer (Anton Paar, Graz, Austria) equipped with a parallel plate geometry with a diameter of 49.96 mm and a gap of 1 mm. For the analysis, the SF solution was mixed with the 50% (v/v) ethanol solution, with and without the addition of the microparticles, and immediately transferred to the rheometer plate preheated at 37 °C. A time sweep up to 150 min was carried out, with temperature, strain, and frequency fixed at 37 °C, 1%, and 0.1 Hz, respectively. After 150 min, the gelled samples were subjected to a frequency sweep from 0.01 Hz to 100 Hz at 37 °C and fixed deformation at 1%.

Thermogravimetric analysis (TGA) was performed to verify the behavior, thermostability, and hydrogel degradation peaks under temperature variation. A TGA/differential scanning calorimetry (DSC)1 equipment, coupled with an MX5 microanalytical balance (Mettler Toledo, Columbus, OH, USA) was used. The analysis was carried out in the temperature range of 25 to 600 °C, at a heating rate of 10 °C/min and nitrogen flow rate of 50 mL/min.

The heat flow related to the chemical transitions in hydrogels and microparticles due to temperature changes was obtained from differential scanning calorimetry (DSC) analysis. A Shimadzu DSC-50 (Shimadzu, Nakagyo-ku, Kyoto, Japan) was used with a nitrogen flow rate of 50 mL/min and with a temperature range from 25 to 250 °C, at a heating rate of 10 °C/min.

2.7. Release Kinetics of the Dyes

The release kinetics study of SF hydrogels and microparticles loaded with dyes was carried out at 37 °C under constant stirring for up to 24 h in a phosphate-buffered saline solution (PBS) release medium. At certain time intervals up to the equilibrium, a release medium aliquot was collected and analyzed by UV/Vis spectroscopy Genesys 10S (ThermoScientific, Waltham, MA, USA). The medium's partial renewal was performed and consisted of replacing the sampled aliquot by the same volume of fresh release medium.

In the release of the loaded dyes from the SF hydrogels, it was necessary to add Protease XIV at ca. 300 µL/mL to the release medium [15,19], since the compounds did not release in preliminary tests, persisting the retention in microparticles and hydrogels for at least 72 h. Protease XIV works by degrading the crystalline structure β-sheet of SF hydrogels in smaller fibroin filaments of smaller β-sheet and random coil structures. These smaller structures are again fragmented until reaching a simple SF molecule [19]. For the release from the microparticles, the release medium was only the phosphate-buffered saline solution.

The mathematical models of Higuchi [24], Peppas [25], Peppas–Sahlin [26], Burst release [27], and Hopfenberg [28] were used to fit the release kinetics data. The fitting parameters were used to determine the dye release mechanism from the SF matrix with Origin software (Originlab).

3. Results and Discussion

3.1. Silk Fibroin Microparticles

The dyes were incorporated at neutral pH and, depending on each pKa (Table 1), the dyes showed different charges. MB, RB, and RhB have an anionic character (negative charge), and the NR has a weak cationic character (slightly positive charge) at neutral pH. Besides, MB and RB are hydrophilic dyes but with very different molar mass, while RhB and NR are hydrophobic dyes. Thus, we proceeded with the comparison between RhB (anionic) and NR (cationic) to assess the influence of the charge; MB (319.85 g/mol) and RB (973.67 g/mol) to assess the effect of molar mass; and MB (hydrophilic) and RhB (hydrophobic) to assess the hydrophilicity influence.

SF microparticles prepared by the atomization method presented spherical shape (Figure 1A, Figure S2), and no visible change in morphology was observed regarding the dye incorporation into SF microparticle. The microparticle size was controlled by the opening of the atomizer nozzle, liquid and compressed air flow rates [29]. The particle size plays an important role in the release control and the performance of microparticles during drug administration. Particle size generally ranges from 1 to 1000 µm and, together with the structure, has to be defined depending on the desired application [30]. The diameter of SF microparticles prepared in our study was comparable to those prepared by different methods, such as emulsification-diffusion [31] and spray-drying [32]. Nevertheless, it was bigger than those from the coacervation method [33]. The results show an increase of 15% to 65% in the SF microparticles mean diameter when the dyes were incorporated. However, the Tukey-Kramer test analysis at 95% confidence indicates no significant difference between the mean diameters of the loaded microparticles regarding the raw SF microparticle. Figure 1B presents the Log-normal particle size distribution of SF microparticles, which confirmed the homogeneity of microparticles. In addition to the increase of particle size, the encapsulation of dyes also led to an increase in the polidispersity of the samples.

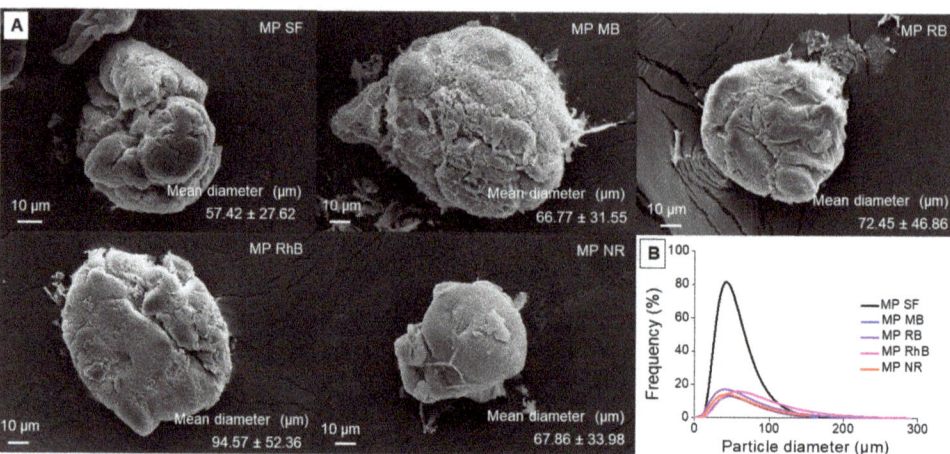

Figure 1. (**A**) Scanning electron microscopy micrograph of silk fibroin (SF) microparticle produced by the atomization method with and without dye loading. The inset values show the mean particle diameter. (**B**) Log-Normal of particle size distribution of SF microparticles.

From the confocal micrographs (Figure 2), it is possible to infer that the dyes were incorporated and evenly dispersed in the SF microparticles, since the fluorescence associated with the dyes is uniform and without flaws in the microparticles (Figure S3).

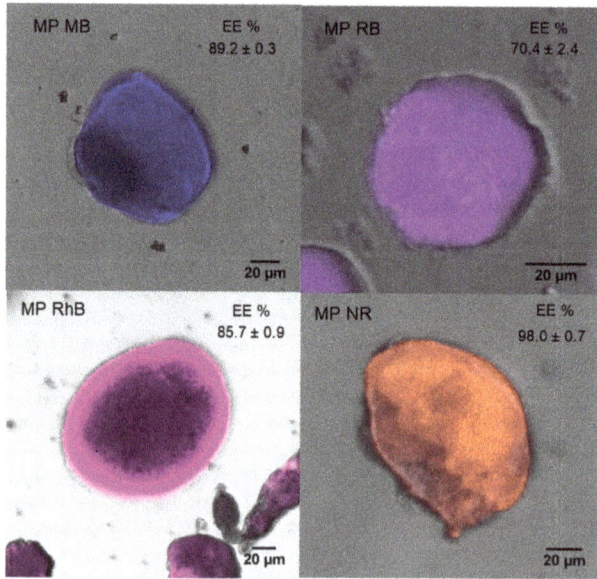

Figure 2. Confocal microscopy images of SF microparticles produced by the atomization method and loaded with dyes. The inset values represent the encapsulation efficiency of the dyes.

The dye encapsulation efficiency on SF microparticles is presented in Figure 2. The NR, which has a positive net charge, showed the highest encapsulation efficiency, due to an electrostatic interaction between the dye and SF, which has a negative net charge after dialysis. Besides, the high encapsulation efficiency of NR can be related to the capacity of SF to make hydrophobic interactions, having great interaction with this compound.

In contrast, RB has a hydrophilic and anionic character, having less interaction with SF and, consequently, less adsorption capacity in SF microparticles. Although MB has a hydrophilic and anionic character, the MB molar mass is smaller than the RB, which may have influenced the incorporation due to a greater barrier to adsorption on the SF microparticle surface. Moreover, the results in Figure 2 reveal that all dyes have high encapsulation efficiency (>70%) in the SF microparticles, even for the RB dye.

Numata, Yamazaki, and Naga [15] showed an encapsulation efficiency of 35% for the texas red dye and 55% for the RhB and fluorescein isothiocyanate in the SF nanoparticles for particle diameter ranging from 175 to 209 nm. The results presented by these authors can be justified by the different dyes characteristics and the incorporation methodology, in which the dye was added to the fibroin solution before the preparation of the microparticles. The texas red is a hydrophilic and anionic dye, with fewer interactions and less incorporation in the microparticles. The RhB, on the other hand, is a hydrophobic and cationic dye and the fluorescein isothiocyanate is a hydrophobic and anionic dye. Therefore, due to hydrophobic interactions, it was expected that RhB and fluorescein isothiocyanate could be incorporated in greater quantities in the SF nanoparticles. Additionally, fluorescein isothiocyanate has a greater hydrophobic character, presenting a greater microparticle and dye interaction (mainly governed by hydrophobic interactions), and a dye release for up to 24 h [15].

The FTIR spectra of the SF microparticles with and without dyes are presented in Figure 3. The amide I bands are located close to 1620 cm^{-1}, and the amide II bands are located between 1516 and 1520 cm^{-1}. The stable β-sheet structure (silk II) is mostly found when the wavelengths of amide I, II, and III are presented in 1630, 1515, and 1260 cm^{-1}, respectively, while the α-helix (silk I) is found in 1660, 1540, and 1230 cm^{-1} [34]. Thus, it is possible to confirm that the SF microparticles with and without the model dyes are in the most stable conformation, i.e., β-sheet structure. This result was expected for SF microparticles because the microparticles were produced by the atomization method, where the SF solution was atomized in the coagulation bath of ethanol, an organic solvent that induces β-sheet formation in SF [35]. The same trend was observed for the SF microparticles loaded with dyes, since the microparticles were produced and stabilized in ethanol before the incorporation of the dyes.

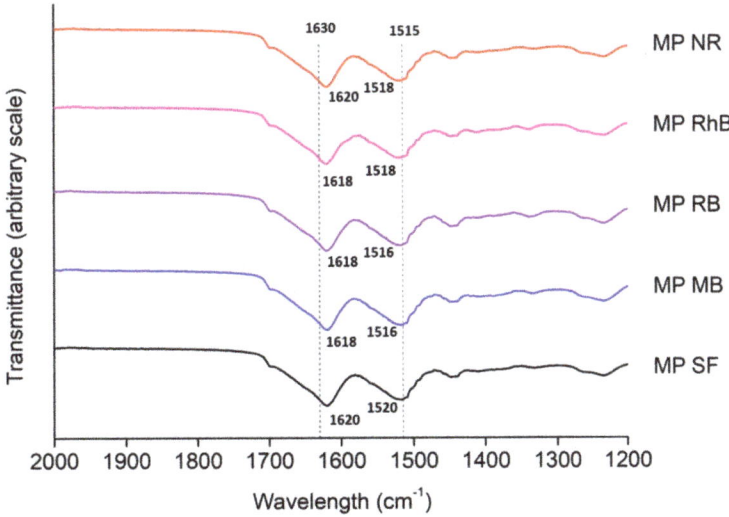

Figure 3. Fourier transform infrared spectroscopy (FTIR) spectra of SF microparticles with and without dyes.

3.2. Silk Fibroin Hydrogels

Figure 4 presents the photographs of the SF hydrogels (4 cm diameter). The microparticles were dispersed in the hydrogel, which acquired the color of the dye (Figure 4). Moraes et al. [14] prepared SF hydrogels using the same methodology (mixing with ethanol) in approximately 27 min, which is in accordance with the gelation time of SF hydrogel in the present study. Matsumoto et al. [36] verified that the gelation process of SF depends on several variables, such as protein concentration, temperature, and pH. Kim et al. [37] achieved shorter gelation time and better mechanical properties of SF hydrogels by increasing SF concentration. In this study, we chose to keep SF concentration constant and decrease of gelation time was achieved by addition of ethanol. In addition, it was observed that the incorporation of loaded microparticles increased the gelation time of the SF hydrogels. The gelation time of SF with the loaded microparticles estimated by the tube inversion method ranged from 80 to 85 min, except for the HG + MP NR, in which gelation time was 110 min.

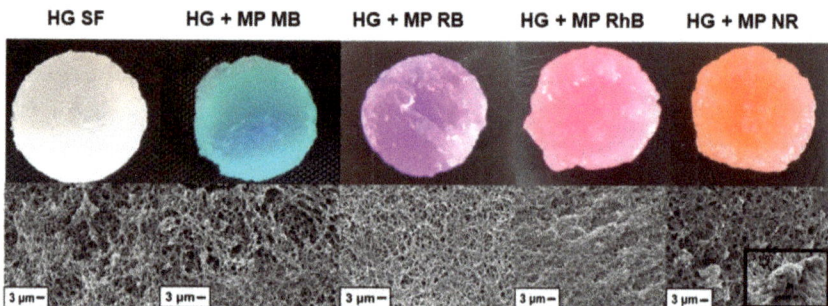

Figure 4. Photographs and scanning electron microscope (SEM) fracture micrographs of SF hydrogels with and without microparticles loaded with dye.

The SEM micrographs (Figure 4) indicate that all SF hydrogels (with or without microparticles) have an interconnected porous structure forming the hydrogel matrix. Figure 4 (HG + MP NR inset) shows a micrograph emphasizing the presence of microparticles in the hydrogel network, indicating the effectiveness of incorporating microparticles into hydrogels.

In order to understand the effect of time on the formation of hydrogels, rheological tests were performed with time sweep measurements, to verify the behavior of the SF hydrogel through the observation of the storage (G') and loss moduli (G") at a constant frequency (Figure 5). For the hydrogels without microparticles, at the beginning of the gelation process, G" is greater than G', in which both G" and G' are increasing linearly, and, at around 600 s, there is a crossover point and G' becomes higher than G", increasing the viscoelasticity due to the gelation process. The linear increase in G' after the crossover point of HG SF in Figure 5 indicates an increase in viscoelasticity that occurs due to the formation of β-sheet structures [38]. Hydrogels containing the microparticles have G' greater than G" from the beginning of the process, but there was a delay in the curve, indicating a longer time required to start the gelation, ranging from 1500 to 2000 s. Besides, at the end of the gelation process, there is a plateau in which the G' value for hydrogels without the addition of microparticles was about 10-fold greater than G' for hydrogels containing microparticles. These results indicate that the addition of microparticles containing dyes reduced the interactions between the SF molecules, taking a long time to start the hydrogel formation and resulting in a gel with less elastic character. The comparison between samples containing microparticles with different dyes showed that the dye type had little influence on gelation time and viscoelastic properties. These results are in agreement with the gelation time observed by the tube inversion method.

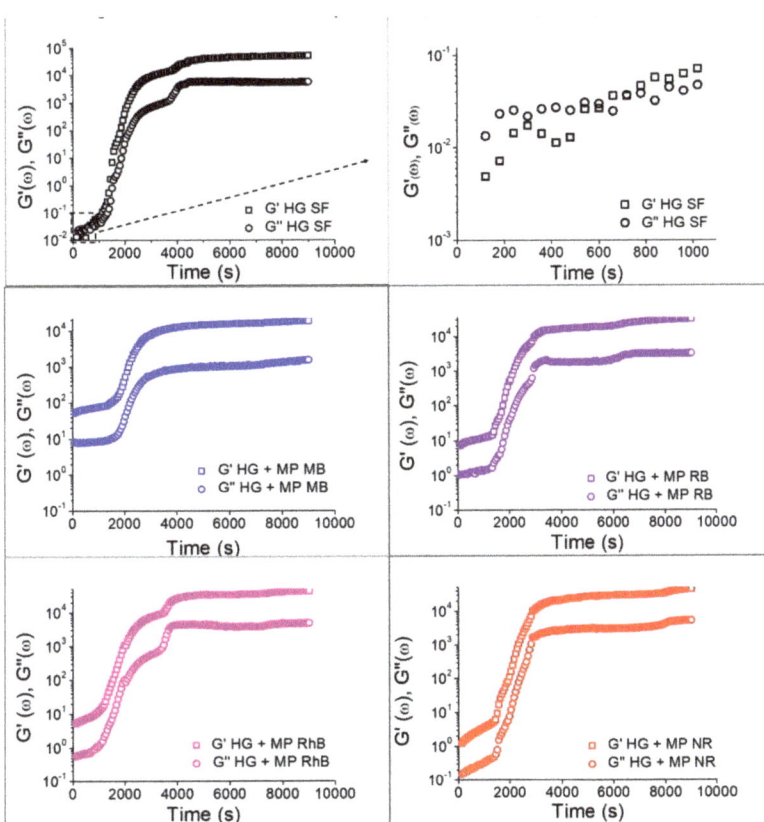

Figure 5. Storage modulus (G′) and Loss modulus (G″) as a function of time of the SF hydrogel and SF hydrogels containing the microparticles loaded with dyes.

Frequency sweep measurements were also performed to evaluate the effect of the microparticles incorporation on the SF hydrogels' structure (Figure 6). All samples showed G′ higher than G″ and both moduli practically independent on frequency, which is a mechanical spectrum characteristic of gel [39]. Figure 6 also shows that both G′ and G″ decreased due to the microparticles addition in the hydrogel. Thus, the reduction of G′ and G″ indicates that the SF hydrogels structure was affected by the presence of the microparticles. Nejadnik et al. [40] found the same influence due to incorporating calcium phosphate nanoparticles in hyaluronic acid hydrogels. The storage modulus (G′) is directly proportional to the gel behavior [41], indicating that the SF hydrogel without microparticles showed a greater gel behavior. From Figure 6 and the gelation time results (Figure 5), it appears that the hydrogels containing microparticles with the largest gelation times also showed a greater G′, indicating hydrogels with a more structured gel network. These results can be related to the hydrogel microstructure (Figure 4), since the HG + MP RhB and HG + MP NR, which showed a more compact structure by SEM, also exhibit a more structured gel network by rheological tests.

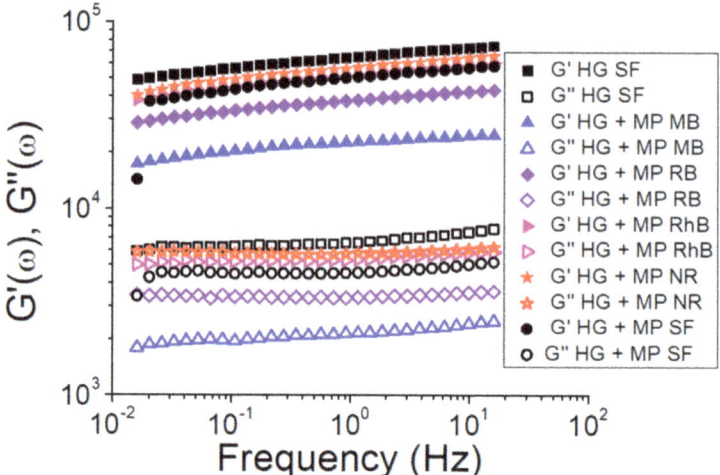

Figure 6. Storage (G′) and loss moduli (G″) as a function of the frequency of SF hydrogels with and without microparticles loaded with dyes. Closed points refer to the storage modulus (G′) and open points refer to the loss modulus (G″).

Figure 7A shows the FTIR spectra of the SF hydrogels with and without the microparticles loaded with the dyes. The amide I bands are located around 1620 to 1637 cm^{-1} and the amide II bands are located between 1523 and 1525 cm^{-1}. The SF hydrogels with and without the microparticles containing the dyes show β-sheet conformation, similar to the SF microparticles (Figure 3). just like in microparticles, β-sheet conformation was expected in hydrogels due to the gelation and aggregation of SF molecules. Comparing the SF hydrogel's spectra with the hydrogels containing the microparticles, it was observed that there was no chemical change in the hydrogel structure by incorporating the microparticles. Additionally, new bands related to the dyes could not be observed, which is expected because the point analysis method (FTIR-ATR) was used. Therefore, peaks related to the dyes would only be observed if the FTIR diamond tip was placed exactly on top of a microparticle containing the dye.

Regarding TGA (Figure 7B), a peak of degradation around 100 °C and another peak of degradation around 300 °C were observed. The first peak is related to the water loss from the hydrogel, while the second peak is related to the side chains of the residual amino acid groups [42], which is in accordance with the literature [12]. The degradation temperature did not change when microparticles were incorporated into the SF hydrogel. Thus, there is no significant change in the hydrogel's thermal stability due to the microparticles' presence.

The DSC thermogram (Figure 7C) indicates a peak between 118 °C for the SF hydrogel, at 115 °C for the hydrogel containing microparticles with MB, RB, and RhB, and at 130 °C for NR, probably related to SF glass transition temperature [43]. The rise in temperature at this peak indicates that there is a strong interaction between SF and NR. This interaction is related to the SF microparticles' capacity to adsorb the NR dye in large quantities (Figure 2), due to the dye hydrophobic and cationic character.

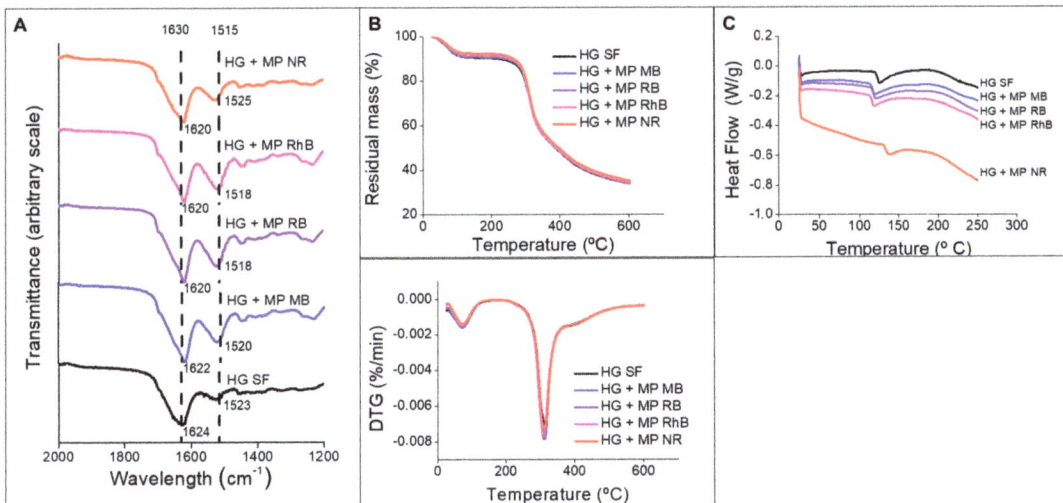

Figure 7. (**A**) FTIR spectra of SF hydrogels with and without microparticles loaded with dyes. (**B**) Thermogravimetric analysis (TGA) and derivative thermogravimetry (DTG) thermograms of SF hydrogels with and without microparticles loaded with dyes. (**C**) DSC thermogram of SF hydrogels with and without microparticles loaded with dyes.

3.3. Release Kinetics of the Dyes

The dyes release from SF microparticles exhibited a rapid release profile (Figure 8A), reaching equilibrium in approximately 90 min. MB had the highest mass fraction released, followed by RhB, RB, and NR. The dye with a hydrophobic character and positive charge, i.e., NR, showed a lower mass fraction released, due to the strong electrostatic and hydrophobic interactions with fibroin. For the MB dye, a higher release into the medium was already expected due to the dye's anionic and hydrophilic characteristics, which indicates a weak interaction with SF. RB is an anionic and hydrophilic dye, just like the MB dye; however, the lower release of this dye is probably influenced by the RB molecule's size, which has a molar mass three times greater than MB, which can hinder the diffusion and release of the dye in the release medium.

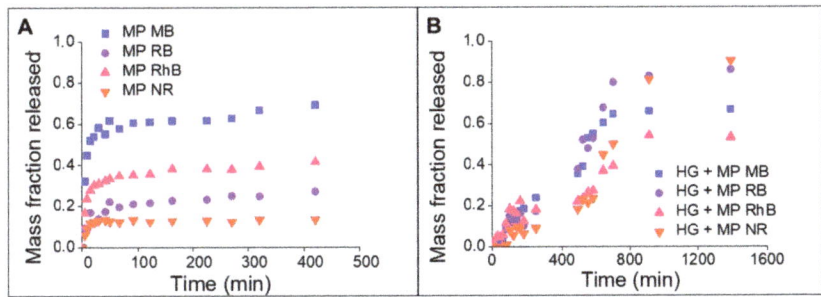

Figure 8. Release kinetics. (**A**) SF microparticles loaded with dyes. (**B**) SF hydrogels containing the microparticles loaded with dyes.

Table 2 shows the models' parameters obtained by the mathematical fitting. All models presented a satisfactory fit ($R^2 > 0.90$) to the experimental data. The parameter 'n' of the Peppas and Peppas–Sahlin models is equal to 0.45, indicating that Fickian Diffusion for spherical geometry is the predominant mechanism in releasing dyes from SF microparticles [25,44].

Table 2. Parameters obtained by fitting the Peppas, Peppas–Sahlin, Higuchi, and burst release models to the SF microparticles' dye release data.

Model	Equation	Parameters	MP MB	MP RB	MP RhB	MP NR
Peppas	$\frac{M_L}{M_\infty} = Kt^n$	K (1/s^n)	0.0238	0.0054	0.0128	0.0050
		n	0.4500	0.4504	0.4500	0.4500
		R^2	0.9009	0.9250	0.9799	0.9656
Peppas–Sahlin	$\frac{M_L}{M_\infty} = K_1 t^n + K_2 t^{2n}$	K_1 (1/s^n)	0.0224	0.0050	0.0129	0.0050
		K_2 (1/s^{2n})	0.0000	1.48×10^{-5}	0.0000	0.0000
		n	0.4500	0.4500	0.4500	0.4500
		R^2	0.9067	0.9083	0.9688	0.9531
Higuchi	$\frac{M_L}{M_\infty} = K_h \sqrt{t}$	K_h (1/s$^{0.5}$)	0.0157	0.0037	0.0091	0.0035
		R^2	0.9154	0.9346	0.9739	0.9593
Burst release	$\frac{M_L}{M_\infty} = Kt^n + B$	K (1/s^n)	0.0206	0.0043	0.0124	0.0048
		n	0.4500	0.4748	0.4500	0.4500
		B	0.0422	0.0055	0.0071	0.0035
		R^2	0.9235	0.9076	0.9718	0.9564

The release profile of the SF hydrogels containing the microparticles loaded with dyes showed a slow and prolonged release, persisting for approximately 900 min (15 h) (Figure 8B). Numata, Yamazaki, and Naga [15] observed the release of approximately 90% of RhB incorporated in SF hydrogels within 1 h. On the other hand, FITC, the other model drug used in the study, was loaded on SF nanoparticles incorporated in the hydrogels and was totally released just after 5 days.

Comparing the time to release the dyes from the loaded microparticles and SF hydrogel containing the loaded microparticles, there is a 10-fold increase in the release time, showing that the strategy of incorporating the microparticles in the hydrogel is effective in reducing the dyes' release time.

For the release in the hydrogels, it was necessary to add Protease XIV, also used by Numata, Cebe, and Kaplan [15,19] since the dyes did not release in preliminary tests, in which the retention of dyes in the SF hydrogels containing the microparticles persisted for at least 72 h. The addition of Protease XIV was not necessary for dyes release from SF microparticles, since dyes were incorporated on SF microparticles by adsorption, being retained on SF microparticles surface, which allowed dyes release to the medium more easily. The different behaviors on dye release from SF microparticles and SF hydrogels are probably related to the dye availability to the medium: in the microparticles, the dyes are adsorbed to the surface, while in the hydrogels the dyes are located inside the SF hydrogel structured network, retained in the microparticles surface. Protease XIV was employed to degrade the stable structure crystalline β-sheet of SF hydrogels, leading the SF filaments to break into smaller β-sheet and random coil structures.

The release of RB dye showed different behaviors: a small dye fraction was released from the SF microparticles loaded with dyes and a high fraction from the SF hydrogel containing the microparticles. This behavior can be related to the fact that RB has a bigger molar mass than the other dyes, influencing its diffusion from the microparticles surface to the release medium. On the other hand, the addition of Protease XIV in the hydrogel release assays led to a higher release of RB, since restriction of RB diffusion was broken up by the smaller SF structures.

Table 3 shows the parameters obtained by fitting the Peppas, Peppas–Sahlin, Hopfenberg, and Higuchi models to the experimental data of dyes released from the SF hydrogels containing microparticles loaded with dyes. MB had the best fitting ($0.85 < R^2 < 0.96$), and the Peppas model was the one that best described the behavior of the dyes' release from SF hydrogel matrices ($R^2 > 0.84$). The Peppas–Sahlin, Hopfenberg, and Higuchi models did not fit well the release data; thus, they were not adequate to describe the release behavior of all dyes.

Table 3. Parameters obtained by fitting the Peppas, Peppas–Sahlin, Hopfenberg, and Higuchi to the release data of the dyes loaded in the microparticles incorporated in the SF hydrogels.

Model	Equation	Parameters	HG + MP MB	HG + MP RB	HG + MP RhB	HG + MP NR
Peppas	$\frac{M_L}{M_\infty} = Kt^n$	K (1/s^n) n R^2	0.002 0.813 0.960	0.002 0.890 0.928	0.012 0.483 0.914	0.001 0.890 0.844
Peppas–Sahlin	$\frac{M_L}{M_\infty} = K_1 t^n + K_2 t^{2n}$	K_1 (1/s^n) K_2 (1/s^{2n}) n R^2	0.002 0.001 0.450 0.949	- - - -	- - - -	0.000 3.81×10^{-4} 0.527 0.863
Hopfenberg	$\frac{M_L}{M_\infty} = 1 - \left(1 - \frac{k_0 t}{c_0 r}\right)^n$	k_0 (cm^2/mol.s) R^2	0.000 0.953	0.000 0.943	- -	- -
Higuchi	$\frac{M_L}{M_\infty} = K_h \sqrt{t}$	K_h (1/s$^{0.5}$) R^2	0.013 0.845	0.016 0.768	0.011 0.920	0.011 0.690

Considering only the Peppas model, the parameter 'n' indicates that the Case-II Transport mechanism [25] is associated with a release dependence on matrix degradation. This behavior was predominant in the release of the RB and NR. The anomalous transport mechanism was predominant in the MB and RhB release, having a greater influence of the Fickian Diffusion mechanism in the RhB release and a greater influence of the Case-II Transport mechanism in the MB release. As the releases of the dyes from SF hydrogels incorporated with loaded SF microparticles are controlled mainly by the Anomalous Transport and Case-II mechanisms, there is little influence of the Fickian Diffusion mechanism [25]. The influence of the Case-II Transport mechanism, that is, the release dependence on matrix degradation, can be noticed in all the release curves since the fraction of released compound presents a lag time behavior and slower release behavior. Such behavior was expected because it was necessary the addition of Protease XIV to induce the SF hydrogel matrices degradation [19]. In this way, the fraction of released dye can remain stable or increase, as the hydrogel matrix is degraded, allowing the dye to leave its network.

The incorporation of the microparticles loaded with dyes in the SF hydrogels allowed a slow and prolonged dyes release, although accelerated by the action of the enzyme Protease XIV. The release of dyes loaded in the microparticles incorporated in the hydrogels proved to be dependent on the degradation of the hydrogel and microparticles by the Protease XIV, which, by degrading the β-sheet structure of fibroin, facilitates the dyes release into the medium.

4. Conclusions

SF microparticles and hydrogels were produced aiming the development of a prolonged release device. The SF microparticles showed high adsorption capacity of dyes, ranging from 70 to 98%. Positive charged dye (NR) showed the high adsorption capacity, which was related to the electrostatic interaction with the SF microparticles (negatively charged), in addition to hydrophobic interactions. The MB, RB, and RhB have negative charges under the study conditions (neutral pH), even though they exhibited adsorption capacity of at least 70%, indicating the presence of minor hydrophilic interactions (MB and RB) and hydrophobic interactions (RhB) with the SF microparticles. Thus, although fibroin mostly performs hydrophobic interactions, the influence of hydrophilic and electrostatic interactions was observed in the loading of dyes into SF microparticles. These interactions allow a more effective incorporation of the dyes in the microparticles.

The SEM images showed that incorporating the loaded microparticles in the SF hydrogels was carried out effectively. The SF hydrogels chemical structure was not affected by the addition of the microparticles. The rheological and thermal properties, on the other hand, were affected by the microparticles, probably due to changes in the hydrogel network organization.

The dyes release from the microparticles reached equilibrium in approximately 90 min, while from the hydrogel containing the SF microparticles persisted for approximately

900 min, exhibiting a release 10 times longer. Besides, the dyes release from the SF microparticles showed Fickian Diffusion as the predominant mechanism. In contrast, SF hydrogel containing SF microparticles presented the Anomalous Mechanism and the Transport of Case-II as the predominant mechanisms, indicating that the dye's release is dependent on the matrix degradation.

The strategy used to incorporate microparticles in hydrogels was effective in prolonging the release. Moreover, it was possible to better understand the influence of the charge, hydrophilicity, and size of the molecules on the adsorption and the release of dyes from SF matrices. Thus, this study opens new possibilities for the development of prolonged-release devices, and it can also be expanded to applications with other molecules, such as drugs and bioactive compounds.

Supplementary Materials: The following are available online at https://www.mdpi.com/2073-4360/13/5/798/s1. Figure S1: Calibration curve and fitting parameters for MB, RB, RhB and NR dyes, Figure S2: Optical microscopy image obtained from SF microparticles loaded with dyes, Figure S3: Confocal microscopy images of SF microparticles loaded with dyes.

Author Contributions: Investigation, B.T.T.; Methodology, M.S.P., Y.B.A., B.T.T., and J.V.; Project administration, M.A.D.M.; Supervision, F.P. All authors have read and agreed to the published version of the manuscript.

Funding: This research was funded FAPESP (2018/15539-6 and 2019/08975-7), Coordination of Superior Level Personnel Improvement (CAPES, finance code 001), and CNPq (Brazilian National Council of Technological and Scientific Development).

Institutional Review Board Statement: Not applicable.

Informed Consent Statement: Not applicable.

Data Availability Statement: The data presented in this study are available on request from the corresponding author.

Acknowledgments: The authors would like to thank FAPESP (2018/15539-6 and 2019/08975-7), Coordination of Superior Level Personnel Improvement (CAPES, finance code 001), and CNPq (Brazilian National Council of Technological and Scientific Development) for the financial support, and silk company Bratac for gently supplying silkworm cocoons.

Conflicts of Interest: The authors declare no conflict of interest.

References

1. Altman, G.H.; Diaz, F.; Jakuba, C.; Calabro, T.; Horan, R.L.; Chen, J.; Lu, H.; Richmond, J.; Kaplan, D.L. Silk-based biomaterials. *Biomaterials* **2003**, *24*, 401–416. [CrossRef]
2. Qi, Y.; Wang, H.; Wei, K.; Yang, Y.; Zheng, R.-Y.; Kim, I.S.; Zhang, K.-Q. A Review of Structure Construction of Silk Fibroin Biomaterials from Single Structures to Multi-Level Structures. *Int. J. Mol. Sci.* **2017**, *18*, 237. [CrossRef] [PubMed]
3. Vepari, C.; Kaplan, D.L. Silk as a biomaterial. *Prog. Polym. Sci.* **2007**, *32*, 991–1007. [CrossRef]
4. Koh, L.-D.; Cheng, Y.; Teng, C.-P.; Khin, Y.-W.; Loh, X.-J.; Tee, S.-Y.; Low, M.; Ye, E.; Yu, H.-D.; Zhang, Y.-W.; et al. Structures, mechanical properties and applications of silk fibroin materials. *Prog. Polym. Sci.* **2015**, *46*, 86–110. [CrossRef]
5. Rockwood, D.N.; Preda, R.C.; Yücel, T.; Wang, X.; Lovett, M.L.; Kaplan, D.L. Materials fabrication from Bombyx mori silk fibroin. *Nat. Protoc.* **2011**, *6*, 1612–1631. [CrossRef]
6. Wenk, E.; Merkle, H.P.; Meinel, L. Silk fibroin as a vehicle for drug delivery applications. *J. Control. Release* **2011**, *150*, 128–141. [CrossRef] [PubMed]
7. Chen, J.; Minoura, N.; Tanioka, A. Transport of pharmaceuticals through silk fibroin membrane. *Polymer* **1994**, *35*, 2853–2856. [CrossRef]
8. Sashina, E.S.; Bochek, A.M.; Novoselov, N.P.; Kirichenko, D.A. Structure and Solubility of Natural Silk Fibroin. *Russ. J. Appl. Chem.* **2006**, *79*, 869–876. [CrossRef]
9. Mandal, B.B.; Mann, J.K.; Kundu, S.C. Silk fibroin/gelatin multilayered films as a model system for controlled drug release. *Eur. J. Pharm. Sci.* **2009**, *37*, 160–171. [CrossRef] [PubMed]
10. Peppas, N.A.; Bures, P.; Leobandung, W.; Ichikawa, H. Hydrogels in pharmaceutical formulations. *Eur. J. Pharm. Biopharm.* **2000**, *50*, 27–46. [CrossRef]
11. Ratner, B.D.; Hoffman, A.S. Synthetic Hydrogels for Biomedical Applications. In *Hydrogels for Medical and Related Applications*; American Chemical Society: Washington, DC, USA, 1976; pp. 1–36.

12. Nogueira, G.M.; De Moraes, M.A.; Rodas, A.C.D.; Higa, O.Z.; Beppu, M.M. Hydrogels from silk fibroin metastable solution: Formation and characterization from a biomaterial perspective. *Mater. Sci. Eng. C* **2011**, *31*, 997–1001. [CrossRef]
13. Kasoju, N.; Hawkins, N.; Pop-Georgievski, O.; Kubies, D.; Vollrath, F. Silk fibroin gelation via non-solvent induced phase separation. *Biomater. Sci.* **2016**, *4*, 460–473. [CrossRef]
14. De Moraes, M.A.; Mahl, C.R.A.; Silva, M.F.; Beppu, M.M. Formation of silk fibroin hydrogel and evaluation of its drug release profile. *J. Appl. Polym. Sci.* **2015**, *132*, 1–6. [CrossRef]
15. Numata, K.; Yamazaki, S.; Naga, N. Biocompatible and Biodegradable Dual-Drug Release System Based on Silk Hydrogel Containing Silk Nanoparticles. *Biomacromolecules* **2012**, *13*, 1383–1389. [CrossRef]
16. Suri, S.; Ruan, G.; Winter, J.; Schmidt, C.E. Microparticles and Nanoparticles. In *Biomaterials Science*; Elsevier: Amsterdam, The Netherlands, 2013; pp. 360–388.
17. Wenk, E.; Wandrey, A.J.; Merkle, H.P.; Meinel, L. Silk fibroin spheres as a platform for controlled drug delivery. *J. Control. Release* **2008**, *132*, 26–34. [CrossRef] [PubMed]
18. Mwangi, T.K.; Bowles, R.D.; Tainter, D.M.; Bell, R.D.; Kaplan, D.L.; Setton, L.A. Synthesis and characterization of silk fibroin microparticles for intra-articular drug delivery. *Int. J. Pharm.* **2015**, *485*, 7–14. [CrossRef] [PubMed]
19. Numata, K.; Cebe, P.; Kaplan, D.L. Mechanism of enzymatic degradation of beta-sheet crystals. *Biomaterials* **2010**, *31*, 2926–2933. [CrossRef] [PubMed]
20. Li, H.; Zhu, J.; Chen, S.; Jia, L.; Ma, Y. Fabrication of aqueous-based dual drug loaded silk fibroin electrospun nanofibers embedded with curcumin-loaded RSF nanospheres for drugs controlled release. *RSC Adv.* **2017**, *7*, 56550–56558. [CrossRef]
21. Wu, P.; Liu, Q.; Wang, Q.; Qian, H.; Yu, L.; Liu, B.; Li, R. Novel silk fibroin nanoparticles incorporated silk fibroin hydrogel for inhibition of cancer stem cells and tumor growth. *Int. J. Nanomed.* **2018**, *13*, 5405–5418. [CrossRef]
22. Gad, H.M.H.; El-Sayed, A.A. Activated carbon from agricultural by-products for the removal of Rhodamine-B from aqueous solution. *J. Hazard. Mater.* **2009**, *168*, 1070–1081. [CrossRef]
23. Viganó, J.; Meirelles, A.A.D.; Náthia-Neves, G.; Baseggio, A.M.; Cunha, R.L.; Maróstica Junior, M.R.; Meireles, M.A.A.; Gurikov, P.; Smirnova, I.; Martínez, J. Impregnation of passion fruit bagasse extract in alginate aerogel microparticles. *Int. J. Biol. Macromol.* **2019**, *155*, 1060–1068. [CrossRef] [PubMed]
24. Higuchi, T. Mechanism of Sustained-Action Medication. Theoretical Analysis of Rate of Release of Solid Drugs Dispersed in Solid Matrices. *J. Pharm. Sci.* **1963**, *52*, 1145–1149. [CrossRef]
25. Ritger, P.L.; Peppas, N.A. A simple equation for description of solute release II. Fickian and anomalous release from swellable devices. *J. Control. Release* **1987**, *5*, 37–42. [CrossRef]
26. Peppas, N.A.; Sahlin, J.J. A simple equation for the description of solute release. III. Coupling of diffusion and relaxation. *Int. J. Pharm.* **1989**, *57*, 169–172. [CrossRef]
27. Kim, H.; Fassihi, R. Application of Binary Polymer System in Drug Release Rate Modulation. 2. Influence of Formulation Variables and Hydrodynamic Conditions on Release Kinetics. *J. Pharm. Sci.* **1997**, *86*, 323–328. [CrossRef]
28. Hopfenberg, H.B. Controlled Release from Erodible Slabs, Cylinders, and Spheres. *Control. Release Polym. Formul.* **1976**, *33*, 26–32. [CrossRef]
29. Perrechil, F.A.; Sato, A.C.K.; Cunha, R.L. κ-Carrageenan-sodium caseinate microgel production by atomization: Critical analysis of the experimental procedure. *J. Food Eng.* **2011**, *104*, 123–133. [CrossRef]
30. Lengyel, M.; Kállai-Szabó, N.; Antal, V.; Laki, A.J.; Antal, I. Microparticles, Microspheres, and Microcapsules for Advanced Drug Delivery. *Sci. Pharm.* **2019**, *87*, 20. [CrossRef]
31. Baimark, Y.; Srihanam, P.; Srisuwan, Y.; Phinyocheep, P. Preparation of Porous Silk Fibroin Microparticles by a Water-in-Oil Emulsification-Diffusion Method. *J. Appl. Polym. Sci.* **2010**, *118*, 1127–1133. [CrossRef]
32. Dyakonov, T.; Yang, C.H.; Bush, D.; Gosangari, S.; Majuru, S.; Fatmi, A. Design and Characterization of a Silk-Fibroin-Based Drug Delivery Platform Using Naproxen as a Model Drug. *J. Drug Deliv.* **2012**, *2012*, 490514. [CrossRef]
33. Lammel, A.S.; Hu, X.; Park, S.H.; Kaplan, D.L.; Scheibel, T.R. Controlling silk fibroin particle features for drug delivery. *Biomaterials* **2010**, *31*, 4583–4591. [CrossRef] [PubMed]
34. Rusa, C.C.; Bridges, C.; Ha, S.; Tonelli, A.E. Conformational Changes Induced in Bombyx mori Silk Fibroin by Cyclodextrin Inclusion Complexation. *Macromolecules* **2005**, *38*, 5640–5646. [CrossRef]
35. Silva, M.F.; De Moraes, M.A.; Nogueira, G.M.; Rodas, A.C.D.; Higa, O.Z.; Beppu, M.M. Glycerin and Ethanol as Additives on Silk Fibroin Films: Insoluble and Malleable Films. *J. Appl. Pol.* **2012**, *128*, 115–122. [CrossRef]
36. Matsumoto, A.; Chen, J.; Collette, A.L.; Kim, U.J.; Altman, G.H.; Cebe, P.; Kaplan, D.L. Mechanisms of silk fibroin sol-gel transitions. *J. Phys. Chem. B* **2006**, *110*, 21630–21638. [CrossRef]
37. Kim, U.J.; Park, J.; Li, C.; Jin, H.J.; Valluzzi, R.; Kaplan, D.L. Structure and properties of silk hydrogels. *Biomacromolecules* **2004**, *5*, 786–792. [CrossRef] [PubMed]
38. Yucel, T.; Cebe, P.; Kaplan, D.L. Vortex-Induced Injectable Silk Fibroin Hydrogels. *Biophys. J.* **2009**, *97*, 2044–2050. [CrossRef]
39. Steffe, J.F. *Rheological Methods in Food Process Engineering*, 2nd ed.; Freeman Press: Walled Lake, MI, USA, 1996; ISBN 0963203614.
40. Nejadnik, M.R.; Yang, X.; Bongio, M.; Alghamdi, H.S.; Van den Beucken, J.J.J.P.; Huysmans, M.C.; Jansen, J.A.; Hilborn, J.; Ossipov, D.; Leeuwenburgh, S.C.G. Self-healing hybrid nanocomposites consisting of bisphosphonated hyaluronan and calcium phosphate nanoparticles. *Biomaterials* **2014**, *35*, 6918–6929. [CrossRef] [PubMed]

41. Braga, A.L.M.; Cunha, R.L. The effects of xanthan conformation and sucrose concentration on the rheological properties of acidified sodium caseinate-xanthan gels. *Food Hydrocoll.* **2004**, *18*, 977–986. [CrossRef]
42. Ribeiro, M.; de Moraes, M.A.; Beppu, M.M.; Monteiro, F.J.; Ferraz, M.P. The role of dialysis and freezing on structural conformation, thermal properties and morphology of silk fibroin hydrogels. *Biomatter* **2014**, *4*, e28536. [CrossRef] [PubMed]
43. Magoshi, J.; Nakamura, S. Studies on Physical Properties and Structure of Silk. Glass Transition and Crystallization of Silk Fibroin. *J. Appl. Polym. Sci.* **1975**, *19*, 1013–1015. [CrossRef]
44. Siepmann, J.; Siepmann, F. Mathematical modeling of drug delivery. *Int. J. Pharm.* **2008**, *364*, 328–343. [CrossRef] [PubMed]

Article

Effects of Two Melt Extrusion Based Additive Manufacturing Technologies and Common Sterilization Methods on the Properties of a Medical Grade PLGA Copolymer

Marion Gradwohl [1,2,3], Feng Chai [1], Julien Payen [3], Pierre Guerreschi [1,4], Philippe Marchetti [2,5] and Nicolas Blanchemain [1,*]

[1] U1008 Controlled Drug Delivery Systems and Biomaterials, Institut National de la Santé et de la Recherche Médicale (INSERM), Centre Hospitalier Régional Universitaire de Lille (CHU Lille), University of Lille, F-59000 Lille, France; marion.gradwohl@gmail.com (M.G.); feng.hildebrand@univ-lille.fr (F.C.); pierre.guerreschi@chru-lille.fr (P.G.)

[2] UMR 9020–UMR-S 1277–Canther–Cancer Heterogeneity, Plasticity and Resistance to Therapies, Institut de Recherche contre le Cancer de Lille, University Lille, CNRS, Inserm, CHU Lille, F-59000 Lille, France; philippe.marchetti@inserm.fr

[3] LATTICE MEDICAL, F-59120 Loos, France; julien.payen@lattice-medical.com

[4] Service de Chirurgie Plastique Reconstructrice et Esthétique, CHU de Lille, F-59037 Lille, France

[5] Banque de Tissus, Centre de Biologie-Pathologie, CHU Lille, F-59000 Lille, France

* Correspondence: nicolas.blanchemain@univ-lille.fr; Tel.: + 33-3-2096-4975

Citation: Gradwohl, M.; Chai, F.; Payen, J.; Guerreschi, P.; Marchetti, P.; Blanchemain, N. Effects of Two Melt Extrusion Based Additive Manufacturing Technologies and Common Sterilization Methods on the Properties of a Medical Grade PLGA Copolymer. *Polymers* **2021**, *13*, 572. https://doi.org/10.3390/polym13040572

Academic Editor: Luis García-Fernández

Received: 30 January 2021
Accepted: 12 February 2021
Published: 14 February 2021

Publisher's Note: MDPI stays neutral with regard to jurisdictional claims in published maps and institutional affiliations.

Copyright: © 2021 by the authors. Licensee MDPI, Basel, Switzerland. This article is an open access article distributed under the terms and conditions of the Creative Commons Attribution (CC BY) license (https://creativecommons.org/licenses/by/4.0/).

Abstract: Although bioabsorbable polymers have garnered increasing attention because of their potential in tissue engineering applications, to our knowledge there are only a few bioabsorbable 3D printed medical devices on the market thus far. In this study, we assessed the processability of medical grade Poly(lactic-*co*-glycolic) Acid (PLGA)85:15 via two additive manufacturing technologies: Fused Filament Fabrication (FFF) and Direct Pellet Printing (DPP) to highlight the least destructive technology towards PLGA. To quantify PLGA degradation, its molecular weight (gel permeation chromatography (GPC)) as well as its thermal properties (differential scanning calorimetry (DSC)) were evaluated at each processing step, including sterilization with conventional methods (ethylene oxide, gamma, and beta irradiation). Results show that 3D printing of PLGA on a DPP printer significantly decreased the number-average molecular weight (M_n) to the greatest extent (26% M_n loss, $p < 0.0001$) as it applies a longer residence time and higher shear stress compared to classic FFF (19% M_n loss, $p < 0.0001$). Among all sterilization methods tested, ethylene oxide seems to be the most appropriate, as it leads to no significant changes in PLGA properties. After sterilization, all samples were considered to be non-toxic, as cell viability was above 70% compared to the control, indicating that this manufacturing route could be used for the development of bioabsorbable medical devices. Based on our observations, we recommend using FFF printing and ethylene oxide sterilization to produce PLGA medical devices.

Keywords: additive manufacturing; sterilization; medical devices; bioabsorbable; polymer

1. Introduction

In recent years, additive manufacturing (AM), also known as 3D printing, has penetrated the healthcare industry, as it enables producing patient specific implants that are customized to the patient's anatomy using complex geometry. Among AM technologies, melt-extrusion based methods are notably attractive, as they do not require an additional crosslinker or solvents which could lead to a biocompatibility issue. AM was initially used for prototyping; however, it has rapidly become a real production tool because it allows reducing costs and product development times compared to traditional processing methods such as injection molding [1].

Bioabsorbable polymers—including polylactid acid, polyglycolid acid, polycaprolactone, and their copolymers—have been widely used for the manufacturing of medical devices such as orthopedics screws, plate, sutures, or absorbable stents [2–6]. As they are fully absorbed by the body and present excellent biocompatibility properties, they have generated immense interest for the manufacturing of scaffolds for tissue engineering and regenerative applications [7]. Indeed, using a bioabsorbable polymer improves patient post-healing, as it avoids needing a second intervention to remove the implant [8–10].

Poly(lactic-*co*-glycolic) Acid (PLGA) has been extremely commonly used in biomedical and pharmaceutical applications as a polymer. Its tunable degradation rate and mechanical characteristics can be obtained by playing with its lactide/glycolide ratio to fit desirable properties [4]. PLGA degrades mainly by hydrolysis of its ester bonds in physiological conditions, causing random chain scission [11]. Hydrolytic degradation of PLGA is known to be autocatalytic and the presence of acidic bioproducts may accelerate its degradation rate [12]. Thanks to its melt behavior and thermal properties, PLGA can be processed via melt-extrusion based additive manufacturing technologies such as fused filament fabrication (FFF) and direct extrusion-based 3D printing (DPP) [13]. Fused filament fabrication is the most commonly used AM technology where a filament is melted into a heating nozzle and deposed layer-by-layer to form the 3D construct. This process requires the manufacturing of a filament in an early stage at high temperatures and by applying mechanical stress that can alter the polymer properties before printing [14]. Direct extrusion-based 3D printing using pneumatic or screw extrusion-based systems enables using polymer pellets as a raw material and hence avoids one manufacturing step compared to FFF [15].

Polymer-based implantable medical devices should be sterilized to eliminate any living microorganisms to prevent infections. The sterilization method must be chosen in order to be the least destructive to the polymer properties and implant morphology. Moist heat is one of the most widely used sterilization methods, as it presents the advantages of being fast, simple, effective and absent of any toxic residues [16,17]. Nevertheless, the high temperatures and humidity required to eliminate viable microorganisms causes excessive degradation of the polymer material that makes it unsuitable for the sterilization of bioabsorbable polymers such as PLGA [18]. Ethylene oxide (EO) is the sterilization technique generally used for polymer-based medical devices, as it can be performed at low-temperatures, which leads to minimal changes in molecular weight compared to other conventional methods [19–21]. The major concern with EO sterilization is the presence of toxic residues that can reside on the implant after the process [22–24]. In order to remove EO residues such as ethylene glycol and ethylene hydrochloride, an aeration of EO sterilized medical devices must be performed which leads to a lengthening of the final process [25,26]. The Food and Drug Administration (FDA) has recently opened an innovation challenge to identify new sterilization techniques as alternatives to EO and this sterilization technique therefore might disappear in the next few decades [27]. Irradiation sterilization methods such as gamma or beta radiation also offer the possibility to sterilize at low temperatures [16,18,19]. Moreover, these methods are rapid and effective but are known to result in changes in material properties such as the molecular weight [28–30].

When dealing with the medical device industry, manufacturing process control has to be taken into consideration to ensure safety and performance of products [31]. Impacts of the whole manufacturing process must be evaluated and controlled to ensure reliable and reproducible clinical performance of devices. As a matter of fact, additive manufacturing of bioabsorbable polymer and sterilization of the final products can significantly impact the polymer features, so the whole process must be chosen in order to reduce changes in the physico-chemical properties of materials.

In this work, we assessed the impact of the full additive manufacturing route from medical-grade PLGA granules to sterile 3D constructs by comparing FFF and DPP technologies. A focus has been made on determining the impact of the whole manufacturing process by evaluating changes in crystallinity, thermal properties, and molecular weight.

Different sterilization techniques such as EO, gamma radiation, and beta radiation were also investigated to determine the most appropriate method for sterilizing PLGA.

2. Materials and Methods

2.1. Material

Medical Grade 85:15 poly(L-lactide-*co*-glycolide) PURASORAB PLG 8523 (inherent viscosity 2,3 dl/g) was obtained from PURAC (CORBION, Gorinchem, The Netherlands). PLGA pellets were stored in a freezer at −15 degrees to minimize degradation and were dried at 40 °C before processing.

2.2. Filament and Scaffold Fabrication

Circular porous disc of 20 (diameter) × 1 (width) mm were designed using Solidworks Software (Figure 1c) and fabricated via FFF and DPP printing technologies for further experiments. For both printers, a nozzle of 0.40 mm was chosen to produce bioabsorbable samples.

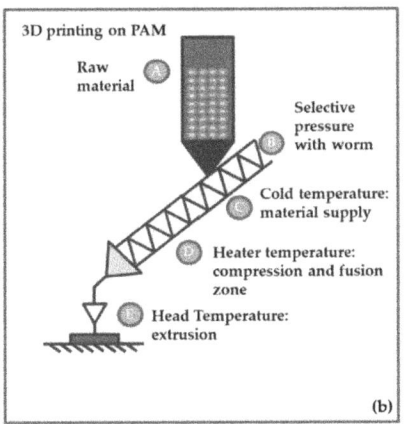
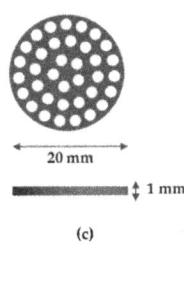

Figure 1. (a) Description of the manufacturing process n°1, including extrusion of the Poly(Lactide-*co*-Glycolide) (PLGA) filament and Fused Filament Fabrication; (b) Description of the manufacturing process n°2: Direct Pellet Printing; (c) Dimensions of the 3D printed sample.

SFor Fused Filament Fabrication (Process n°1, Figure 1a), PLGA 85:15 was first melt-spun into 2.85 mm filaments using a Composer Series 350 extruder (3DEVO, Utrecht, The Netherlands). PLGA was extruded with a speed of 7.4 rpm and the filaments diameter was controlled with an optical sensor with a precision of +/− 0.05 mm. PLGA filament was thereafter used to manufacture samples on an Ultimaker 3 printer (Ultimaker, Utrecht, The Netherlands).

A PAM (Pellet Additive Manufacturing) printer from POLLEN (Ivry-sur-Seine, France) was used in this study as a DPP printer (Process n°2, Figure 1b). This technology used pellets as a raw material and the polymer was heated progressively thanks to three heating zones: a cold extruder, main extruder, and extruder head (Figure 1b). The machine has an extrusion system inspired by the principles of an injection molding machine. A heated cylinder (zone C, Figure 1), fitted with a proprietary extrusion screw (A, Figure 1b), deposits the material evenly. Unlike standard FFF printers, the print bed moves to allow the part to be manufactured. Extrusion and printing profiles are shown in Table 1, where only parameters that may affect PLGA properties are presented (temperature, speed, flow).

Table 1. Manufacturing process parameters.

			PROCESS n°1		
Filament Extrusion Parameters	Zone 1 Temperature (°C)	Zone 2 Temperature (°C)	Zone 3 Temperature (°C)	Zone 4 Temperature (°C)	Extrusion speed (rpm)
	205 °C	210 °C	210 °C	200 °C	5 rpm
Printing Profile on FFF Printer		Printing Temperature	Bed Temperature	Printing speed	Flow (%)
		200 °C	65 °C	60 mm/s	100%
			PROCESS n°2		
Printing Profile on DPP Printer	Cold Temperature (°C)	Extruder Temperature (D)	Head Temperature (E)	Bed Temperature	Printing speed	Flow
	65 °C	170 °C	210 °C	65 °C	10 mm/s	100%

2.3. Scaffold Sterilization

Samples were first packed in TYVEK sterilization pouches (STERICLIN, Feuchtwangen, Germany) and sent for sterilization by ethylene oxide (STERYLENE, Civrieux d'Azergues, France), gamma radiation, and beta radiation (IONISOS, Dagneux, France). The EO oxide sterilization treatment cycle was not validated according to ISO 1135 certification. Briefly, samples were first preconditioned in a separate chamber and were thereafter exposed to EO for 2 h. An aeration was then performed during 48 h to remove ethylene oxide residue from samples. For both gamma and beta sterilization, three radiation doses were evaluated: 15 kGy, 25 kGy, and 50 kGy. Gamma and beta sterilization treatment cycles were not validated according to ISO 1137 certification.

2.4. DSC

The thermal properties of PLGA at each step of the manufacturing process were determined by differential scanning calorimetry (DSC) under a nitrogen atmosphere with a Mettler Toledo DSC 1 apparatus (Greifensee, Switzerland). Approximatively 10 mg of samples were placed in individually aluminum pans and were heated from 20 °C to 200 °C with a heating rate of 10 °C min^{-1}. The Glass Transition temperature (T_g) was taken as the midpoint temperature and the melting temperature as the maximum of the endothermic peak; all the data were taken from the first heating run. Crystallinity was calculated according to the following equation: % Cristallinity= $\Delta H_m / \Delta H_m°$, where ΔH_m is the melting enthalpy of the sample determined from DSC and $\Delta H_m°$ is the enthalpy of melting for 100% crystalline PLA 93 J·g^{-1} [32].

2.5. TGA

Thermogravimetric Analysis (TGA) was carried out on pellets to determine the degradation temperature of PLGA. The degradation temperature was defined as the temperature where PLGA has lost 5% of its initial mass. TGA analysis was used to define the temperature limit to avoid material degradation and therefore helped to define the appropriate printing temperature. TGA measurements were performed with a TGA Q50 apparatus (TA Instruments, New Castle, DE, USA) under nitrogen from 20 to 400 °C with a 10 °C/min heating rate.

2.6. Gel Permeation Chromatography (GPC)

Molecular weight distribution and weight-average (M_w) and number-average molecular weight (M_n) of PLGA were evaluated using a WATERS E2695 separations module system (Waters Corp., Milford, MA, USA) equipped with a Wyatt T-REX refractive index detector (Wyat Technology, Santa Barbara, CA, USA) and three STYRAGEL columns (HR1, HR3, and HR4) calibrated with polystyrene standards. Samples (30 mg) were first dissolved in Chloroform for 3 h and then dissolved in THF. Each sample was filtered through a 0.45 μm PTFE filter. Samples were eluted with THF with a flow rate of 1.0 mL·min^{-1}. Each sample was run in triplicate, and the data were reported as averages.

2.7. Cytotoxicity Assay

The cytotoxicity of samples was further evaluated using the extraction method (ISO 10993-5), with the NIH3T3 mouse embryo Fibroblast (ATCC®CRL-1658™) by LGC Standards SARL, Molsheim, France. Sample extracts were prepared under sterile conditions by adding 200 mg into 1 mL of MEM–α culture medium supplemented with 10% fetal bovine serum (FBS, Gibco®, Thermo Fisher Scientific, Illkirch-Graffenstaden, France), which was incubated at 37 °C and with agitation at 80 rpm for 72 h to reach the requirements of the FDA for an implantable medical device. The complete culture medium was also incubated under the same conditions as the negative control. On the same day, NIH3T3 cells were plated at 4×10^3 cells/well in a 96−well tissue culture polystyrenes plate and grown in a 100 µL/well MEM–α medium supplement with 10% FBS at 37 °C and 5% CO_2 for 24 h. The 96-well plate was partitioned into columns: the culture medium only (no cells); cells incubated in the culture medium (control); and cells incubated in an extraction medium. The medium for the monolayer cell culture was then replaced by the 100 µL/well sterile original sample extracts (filtered through a 0.2 µm sterile syringe filter) or the control medium. After 24 h of exposure of the cells to the sample extracts or control at 37 °C and 5% CO_2, the cell viability was determined using AlamarBlue®(Gibco®, Thermo Fisher Scientific, Illkirch-Graffenstaden, France) assay. Briefly, the medium in each well was replaced with a 10% AlamarBlue®solution in the culture medium (200 µL/well), and the plate was incubated at 37 °C and 5% CO_2 for 2 h. One hundred and fifty microliters of reacted solution per well were transferred into a 96-well plate (Fluoro–LumiNunc™, ThermoScientific, Illkirch-Graffenstaden, France). The intensity of fluorescence was determined using a Twinkle LB 970 Microplate Fluorometer (Berthold, Bad Wildbad, Germany) with an excitation wavelength of 530 nm and an emission wavelength of 590 nm. The cell survival rate was expressed by the percentage of cell viability with respect to the value of the control.

3. Results

3.1. Repetability and Printing Quality

PLGA samples were successively printed for direct pellet printing (DPP) and fused filament fabrication (FFF). However, due to melt behavior of PLGA in DPP extrusion process (Figure 2c), it was more difficult to obtain repeatability compared to for the conventional FFF process (Figure 2c). Printing profile parameters such as temperature or flow were always slightly adjusted to avoid under extrusion issues, whereas with FFF the same printing profile was used for the whole study. Moreover, this can be observed by dispersity of weight control results after 3D printing (Figure 2): DPP printed 3D samples reveals high variability (m_{min} = 0.204 g, m_{max} = 0.486 g, mean = 0.346 ± 0.07 g, n = 33) whereas mass values of FFF printed ones were less dispersed (m_{min} = 0.284 g, m_{max} = 0.333 g, mean = 0.311 ± 0.01 g, n = 56).

Repeatability of the two additive manufacturing process was investigated by successively printing five PLGA circular discs and by evaluating their molecular weight. For each sample, molecular weight analysis was performed at three different positions, and data were represented as the mean and standard deviation of these values (Figure 3).

Firstly, the final molecular weight of PLGA was lower for DPP printed samples than for FFF samples. Secondly, PLGA molecular weight of 3D perforated discs seemed to be relatively constant after successive printing, whereas for DPP the molecular weight of PLGA decreased after each print. This molecular weight decrease during successive print could be explained by the longer and uncontrolled residence time of PLGA at elevated temperatures in the extrusion mechanism of a DPP printer.

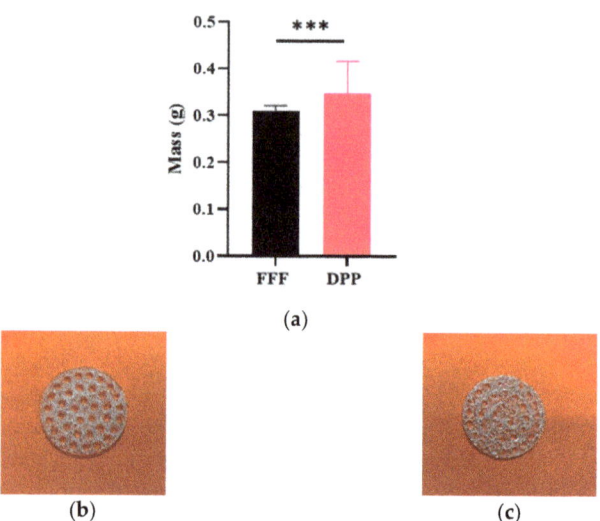

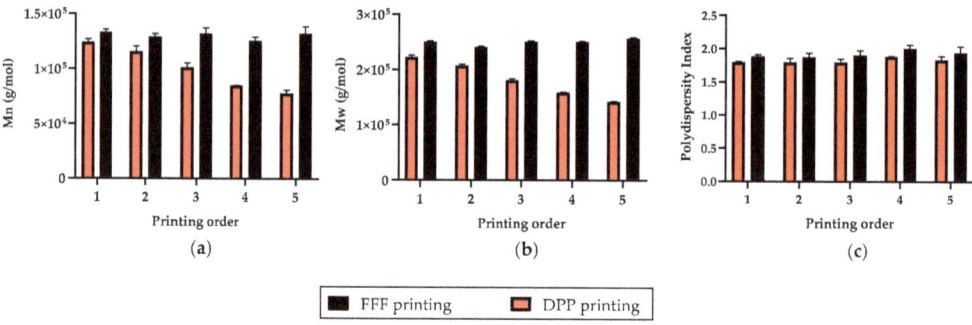

Figure 2. (a) Comparison of mass of PLGA samples printed on FFF and PAM printer. All data were analyzed using a non parametric student t-test *** $p < 0.001$, (b) FFF and (c) DPP printed samples.

Figure 3. (a) Evaluation of (a) the number-average molecular weight (M_n), (b) weight-average (M_w), and (c) polydispersity index (IP) of PLGA 85:15 from successively printed samples; All data are expressed as mean ± SD ($n = 3$).

3.2. Impact of the Manufacturing Process

3.2.1. Molecular Weight

Molecular weight analysis was performed by GPC to evaluate the impact of each step of the manufacturing process on the degradation of PLGA 85:15 and to evaluate the less impacting process between FFF and DPP (Figure 4). The weight average molecular weight and average molecular weight of PLGA decreased to about 10% of the original value after the filament-extrusion process. Size exclusion chromatography profiles indicated a decrease respectively by 19% and 26% of the PLGA M_n from 1.53×10^5 g/mol for pristine material to 1.24×10^5 g/mol after process n°1 and 1.13×10^5 g/mol for process n°2. Even though Fused Filament Fabrication requires two thermal steps, its molecular weight drop is less important than for the DPP printer.

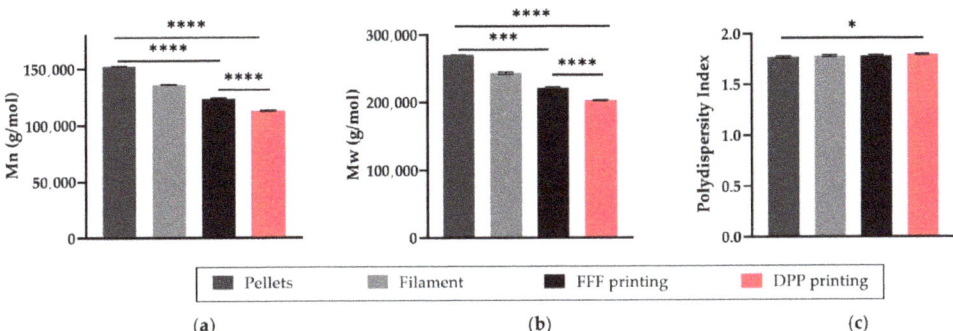

Figure 4. (a) Impact of additive manufacturing steps on the (a) M_n, (b) M_w and (c) polydispersity index of PLGA 85:15; All data are expressed as mean ± SD (n = 3) and were analyzed using Anova one-way test and compared to the control * $p < 0.05$, *** $p < 0.001$, **** $p < 0.0001$.

3.2.2. Thermal Analysis

Thermal characterization of PLGA pellets by DSC and ATG was used to determine the optimal temperatures for extrusion and 3D printing; the melting point and degradation temperature were T_m = 149 °C and $T_{d5\%}$ =304 °C, respectively (Figure 5). The raw pellet displayed a large crystallinity pic compared to processed PLGA, indicating that prior to filament fabrication or 3D printing, the material was semi-crystalline with a glass transition temperature of T_g = 63 °C. The results showed a very low crystallinity rate of PLGA 3D printed constructs for both process due to a rapid cooling rate, which limits crystallization. Nevertheless, a slight endothermic pic corresponding to a crystallinity value of 3.8% was observed for DPP printed constructs, probably due to a slower cooling time inside the 3D printing machine chamber.

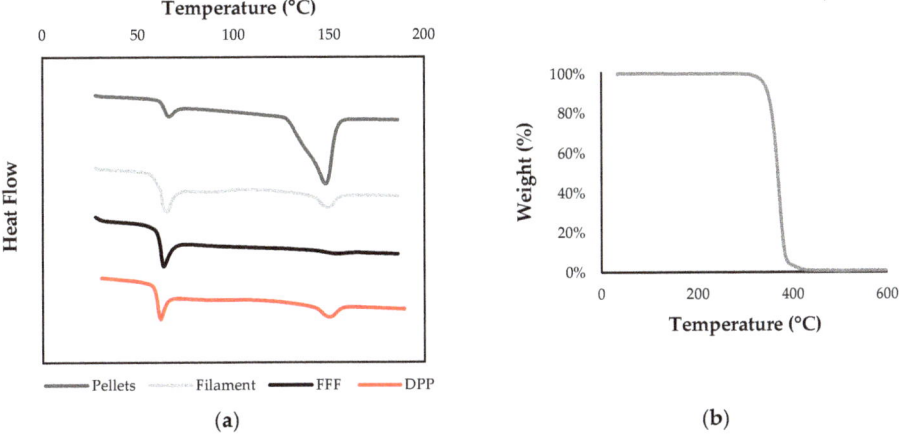

Figure 5. (a) Differential scanning calorimetry (DSC) thermograms of PLGA 85:15 raw material, filament, PAM, and Ultimaker 3 3D printed constructs with a heating rate of 10 °C/min for the first heating run. (b) TGA thermogram of PLGA 85:15 pellets.

3.3. Impact of Sterilization

As demonstrated by the microbiological assay, sterilization of the samples was achieved effectively with each treatment gamma radiation, beta radiation, and ethylene oxide of methods because no clouding of the media was observed after 48 h of incubation time (data not shown). PLGA 85:15 samples sterilized by ethylene oxide, gamma, and beta irradiation were compared to those that were non-sterile in terms of changes in the weight average and number average molecular weight (Figure 6). For gamma and beta irradiation, all samples changed from their initial of M_n 1.062 × 10^5 g/mol as a result of sterilization. The molecular weight data indicate that the EO sterilization processes has no significant influence on the molecular weight of PLGA 85:15 material. Gamma or beta irradiation lead to a significant drop in the initial PLGA molecular weight in a dose-dependent manner. Changes in the polydispersity index compared to the control were only statistically significant for the β50kGy group ($p < 0.05$). However, beta irradiation sterilization seems to have less impacts on PLGA molecular mass properties.

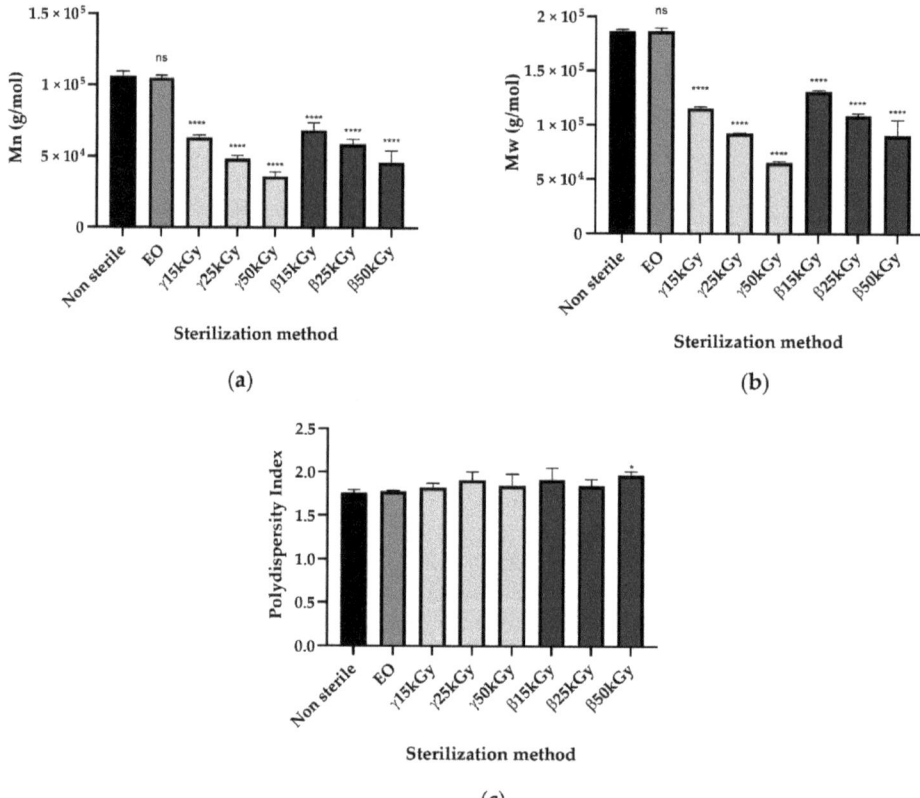

Figure 6. Impact of ethylene oxide as well as beta, rand, and gamma irradiation sterilization methods on the (a) M_n, (b) M_w, and (c) polydispersity index of PLGA 85:15; All data are expressed as mean ± SD (n = 3) and were analyzed using Anova one-way test and compared to the control * $p < 0.05$, ** $p < 0.01$, *** $p < 0.001$, **** $p < 0.0001$.

Scanning Differential Calorimetry (DSC) was performed to determine whether sterilization by EO, gamma, and beta irradiation could influence the crystallinity and glass transition temperature (T_g) of PLGA material compared to the unsterilized starting 3D constructs (Figure 7). After EO sterilization, the T_g value of PLGA slightly increased compared to the control but the difference was not statistically different ($p > 0.05$). For gamma and beta irradiation, only the γ15kGy and γ25kGy groups showed statistically significant differences in T_g values compared to the control ($p < 0.05$), though T_g values remained far above human body temperatures (Figure 7a).

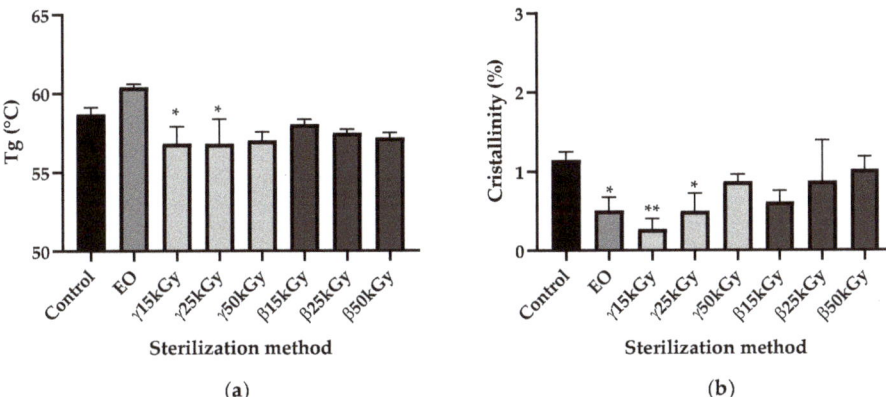

Figure 7. Impact of ethylene oxide, as well as beta, rand, and gamma irradiation sterilization methods on the (**a**) Glass transition temperature (T_g), and (**b**) the crystallinity of PLGA 85:15; All data are expressed as mean ± SD (n = 3) and were analyzed using Anova one-way test and compared to control * $p < 0.05$, ** $p < 0.01$.

EO sterilization leads to a significant decrease in PLGA crystallinity from 1.15% to 0.50% ($p < 0.05$). For gamma and beta irradiation, only γ15kGy and γ25kGy groups show a statistically difference in crystallinity compared to the control (respectively $p < 0.01$ and $p < 0.05$); however crystallinity rates remained relatively low, so we could not conclude that ethylene oxide, gamma, and beta irradiations sterilizations have a real impact on PLGA crystallinity (Figure 7b).

3.4. Cytotoxicity

Regarding the cytotoxicity assay, media extracts of PLGA samples did not affect cell viability of NIH3T3 fibroblast cells, which remained higher than 70% (Figure 8). Thus, these results demonstrated that no cytotoxic compounds are generated during EO, gamma, and beta sterilization methods.

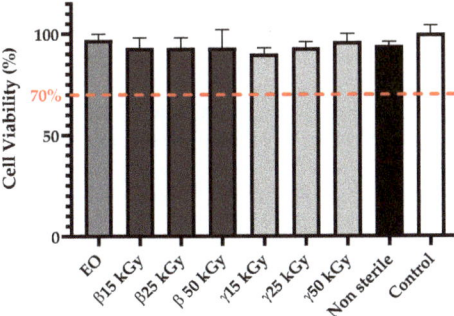

Figure 8. Impact of ethylene oxide, as well as beta, rand, and gamma irradiation methods on cytotoxicity of PLGA constructs after sterilization; All data are expressed as mean ± SD (n = 8).

4. Discussion

The aim of this study was to quantify PLGA 85:15 degradation when processed during all the production steps from additive manufacturing to terminal sterilization in order to select the production route that would have the least impact on its properties. Two additive manufacturing technologies, FFF and DPP, were compared as well as three sterilization methods: ethylene oxide, gamma irradiation, and beta irradiation. The design of the samples was chosen for the purpose of challenging 3D printers, as they require a lot of retraction and travel of print heads, thus resulting in bad printing quality if these tasks are not well managed by the device. As PLGA is sensitive to elevated heat, it likely degrades while printing; this means that printing parameters such as temperature must be adjusted to minimize degradation. Indeed, Widmer et al. have shown that PLGA's molecular weight was decreased when the extrusion temperature was increasing [33]. These optimized printing temperatures were determined thanks to DSC and TGA analysis on pristine materials by testing different temperatures above its melting point and below its degradation temperature to achieve appropriate melt behavior. Information obtained by GPC analysis reflects how PLGA degrades during the manufacturing process. Samples printed via FFF were successively manufactured with adequate quality, but this process took longer than for DPP, as it first requires the manufacturing of the filament to occur. Furthermore, based on our experience, a large amount of polymer pellets is needed for filament extrusion, which can be an issue for research applications in particular when designing in-house material. One might assume that adding a thermal step such as filament extrusion to the manufacturing process will degrade more PLGA; however, GPC results reveals that the opposite occurred. In fact, for the PAM process, due to longer residence time of PLGA in the extrusion mechanism at elevated temperatures, raw material that wait to be processed begin to degrade progressively. Molecular weight analysis of successive printing has confirmed this hypothesis, as the M_n of PLGA decreased after each print. In a study, Shim et al. have demonstrated the impact of the residence time of PLGA at elevated temperatures on its thermal degradation by using solid free form fabrication [34]. Furthermore, to reach good printing quality by printing PLGA on PAM, printing speeds have been drastically reduced, which also increases this elongated heating period and therefore degrades PLGA. Working with a higher printing speed on DPP printer results in the under extrusion phenomenon. The Ultimaker printer enables us to increase the printing speed compared to PAM, furthermore with the FFF process, the PLGA filament is only heated punctually and during a relatively short time. Degradation generated during the manufacturing process must be considered as a bioabsorbable polymer choice for tissue engineering applications, as it impacts the resorption time announced by suppliers. Regarding printing quality, the FFF process leads to a relative smooth surface of perforated discs, whereas with the DPP process, PLGA samples show a rough top surface that could lead to undesired inflammation and discomfort after implantation in vivo. In the case of medical devices, quality control is primordial and to reach ISO 13485 certification, production must be carried out to demonstrate the ability of the process to achieve the expected results reproducibly and repeatedly. Information obtained by mass control after printing reveals that DPP printed samples has shown a large dispersity in term of mass compared to FDM printed ones and this would be an issue during process qualification because differences in mass would lead to differences in degradation behavior. Thermally-induced PLGA degradation observed in DPP printing leads to a lack of repeatability of 3D samples [35]. After an additive manufacturing step, PLGA 85:15 shows a low value of crystallinity, which is relatively interesting for bioabsorption during degradation. Indeed, as water penetration is more difficult in the crystalline area, having an amorphous polymer might reduce the chances of retaining a remaining crystalline fraction that take a longer time to degrade after the reconstruction of living tissues. As the aim of the study was to choose a manufacturing process that would minimize PLGA degradation, we would recommend for medical devices production to proceed to classic Fused Filament Fabrication in order to reach required quality specifications to be externally certified under ISO 13485. However, the DPP process should be used with more thermal stable bioabsorbable polymers, as it enables skipping the filament extrusion step and thus simplifying manufacturing process, thereby reducing the costs and qualification times that are essential for medical device development. Polycaprolactone could be a promising polymer to use with the DPP process thanks to its low melting point, thermal stability, and good melt viscosity properties [36–38]. Ahlinder et al. have demonstrated that the PCL molecular weight was not significantly altered after 3D printing, which makes 3D printing suitable for an additive manufacturing process that requires a longer residence time above the melting point, such as using a PAM printer. Every implantable medical device must be sterile, and sterilization is performed to eliminate viable microorganisms and thus minimize the risk of complications such as infection after implantation within the human body. For a medical device to be labeled sterile, the security assurance level (SAL) defined by the European Pharmacopeia must be SAL $\leq 10^{-6}$, meaning that the theoretical probability that a unit is nonsterile is less than one in one million. In this study, sterility testing was performed to demonstrate the absence of viable contaminating microorganisms remaining after the sterilization process. For each investigated sterilization process, a simple microbiological assay was conducted and was achieved effectively with each treatment, as no clouding of the Mueller Hinton was observed after 48 h of incubation (data not shown). However, this test only proves the sterility of the surface of the samples, as PLGA's bioabsorbable sterility must be maintained during degradation to avoid undesirable infection complications after surgery. Indeed, even if processing PLGA 85:15 via additive manufacturing requires high temperatures that would probably kill microorganisms, the possibility cannot be excluded that still some of them could be entrapped within the layer of molten polymer and thus be released during its degradation. Furthermore, tiny crevices are present between the layer structure of a 3D printed samples where microorganisms can hide, and these parts will be more difficult to sterilize. As a key characteristic of gamma irradiation is its high penetration capacity within the bulk material, remaining viable microorganisms should be killed after performing sterilization, whereas this could be an issue for beta and ethylene oxide sterilization. Due to the limited penetration depth of electrons, sterilization via beta irradiation could be compromised for thick and complex devices; sterilization via beta irradiation should be suitable for the sterilization of 1 mm thick PLGA discs, though further study must be done to confirm our hypothesis. For ethylene oxide however, the presence of viable microorganisms within the bulk material could be a potential issue, as EO sterilization is known to be a surface sterilization. Nevertheless, the EO can penetrate some materials and can possibly sterilize the germs present in the interlayers of 3D constructs. A sterility test must be carried out during process validation to verify the destruction of the bioburden, which will allow the effectiveness of EO between layers to be confirmed. Sterilization treatments were investigated to find an effective method which is less destructive for PLGA. Gamma sterilization and beta sterilization were found to significantly decrease PLGA molecular weight, whereas after ethylene oxide treatment, PLGA's molecular weight remains unchanged. Cleaning medical devices in order to reduce the initial bioburden before sterilization could be a solution to apply a minimal radiation dose for gamma and beta sterilization to be efficient. However, even with the smallest radiation dose tested of 15 kGy, M_n loss of PLGA was still statistically significant with $40 \pm 2\%$ and $35 \pm 5\%$ loss rates for gamma and beta irradiations, respectively ($p < 0.0001$). Faisant et al. have shown that the decrease in PLGA drug-free microparticles M_w as a function of radiation dose is almost linear and even at 4.3 kGy, degradation occurs [39]. The main issue for radiation sterilization is the polymer compatibility; indeed, these methods can degrade polymer structure because cross-linking, chain scission, or a combination of the two can occur. In our case, gamma and beta radiation induced PLGA chain scission, which result in shorter chains of macromolecules, thus leading to a significant decrease in its molecular weight. In some tissue engineering applications, this undesirable drop in PLGA molecular weight can be very problematic regarding polymer degradation, as it could lead to early loss of its properties such as its scaffold mechanical integrity and then can compromise new tissue formation. Furthermore, as beta radiation is known to have a limited depth penetration, it can cause an issue for thick devices. Ionizing rays will have more difficulties penetrating to the bulk of the device and thus can result in a heterogeneity of molecular weight within its structure. For PLGA 85:15, we would recommend using ethylene oxide sterilization, as it does not lead to chain-scission; moreover, a low-temperature ethylene oxide process enables us to sterilize PLGA devices under its glass transition temperatures, thus avoiding thermal transitions that can strongly affect mechanical properties. One of the drawbacks of the ethylene oxide method is the possible presence of EO toxic residues after sterilization such as ethylene chlorohydrin or ethylene glycol. The factors influencing the presence of these residues can be the material nature and geometry of the device to be sterilized, non-optimal aeration time, or the packaging choice. Furthermore, the inactivation of microorganisms leads to the presence of endotoxins which are components of the bacteria's membranes and are known to be pyrogenic molecules. The use of ethylene oxide sterilization therefore requires checking the complete safety of the device during process validation according to the ISO 1135. Among all sterilization methods, the tested EO process seems to be the most suitable for PLGA 85:15; however, additional work must be done to evaluate if EO sterilization leads to changes in PLGA's mechanical properties and morphology.

5. Conclusions

This work focused on finding the appropriate melt additive manufacturing route as well as sterilization process that minimize PLGA degradation. Analysis of data reveals that direct pellet printing causes high polymer degradation due to the uncontrolled residence time of PLGA in a molten state in the extrusion mechanism. This study has also shown that beta and gamma irradiation cause damages to the PLGA scaffold, while ethylene oxide sterilization is likely to have a destructive effect on PLGA molecular integrity. Based on our observations, we recommend using classic FFF printing and ethylene oxide sterilization to produce PLGA medical devices. However Direct Pellet Printing could be a promising additive manufacturing technology for more thermally stable aliphatic polyesters such as PCL. The methodology proposed in this study can be used as a guideline for the development of adequate production processes from additive manufacturing to terminal sterilization of any bioabsorbable thermoplastic polymer for medical device applications.

Author Contributions: Conceptualization, M.G. and N.B.; Data curation, M.G.; Formal analysis, M.G.; Investigation, M.G. and F.C.; Methodology, M.G., F.C., and N.B.; Project administration, N.B., J.P. and P.M.; Supervision, N.B., P.M., J.P., and P.G.; Writing—original draft, M.G.; Writing—review & editing, N.B., P.M., P.G., and J.P. All authors have read and agreed to the published version of the manuscript.

Funding: MG is a recipient of CIFRE doctoral fellowship attributed by the ANRT (Association Nationale Recherche Technologie). This research was also funded by financial support from the Région Hauts-de France (POPSTER).

Institutional Review Board Statement: Not applicable.

Informed Consent Statement: Not applicable.

Data Availability Statement: No new data were created or analyzed in this study. Data sharing is not applicable to this article.

Acknowledgments: The authors thank Aurelie Malfait (from Université de Lille, Institut Chevreul, UMET, UMR 8207, 59000 lille), for the SEC analysis and for her help with the measurements. The authors thank Mickaël Maton for his help with the DSC analysis, microbiological assay, and with the measurements (from University of Lille, INSERM U1008).

Conflicts of Interest: The authors declare no conflict of interest.

References

1. Franchetti, M.; Kress, C. An Economic Analysis Comparing the Cost Feasibility of Replacing Injection Molding Processes with Emerging Additive Manufacturing Techniques. *Int. J. Adv. Manuf. Technol.* **2017**, *88*, 2573–2579. [CrossRef]
2. Nikoubashman, O.; Heringer, S.; Feher, K.; Brockmann, M.-A.; Sellhaus, B.; Dreser, A.; Kurtenbach, K.; Pjontek, R.; Jockenhövel, S.; Weis, J.; et al. Development of a Polymer-Based Biodegradable Neurovascular Stent Prototype: A Preliminary In Vitro and In Vivo Study. *Macromol. Biosci.* **2018**, *18*, 1700292. [CrossRef]
3. Liao, L.; Peng, C.; Li, S.; Lu, Z.; Fan, Z. Evaluation of Bioresorbable Polymers as Potential Stent Material—In Vivo Degradation Behavior and Histocompatibility. *J. Appl. Polym. Sci.* **2017**, *134*. [CrossRef]

4. de Melo, L.P.; Salmoria, G.V.; Fancello, E.A.; de Mello Roesler, C.R. Effect of Injection Molding Melt Temperatures on PLGA Craniofacial Plate Properties during In Vitro Degradation. *Int. J. Biomater.* **2017**, *2017*, 1256537. [CrossRef]
5. Maurus, P.B.; Kaeding, C.C. Bioabsorbable Implant Material Review. *Oper. Tech. Sports Med.* **2004**, *12*, 158–160. [CrossRef]
6. Konan, S.; Haddad, F.S. A Clinical Review of Bioabsorbable Interference Screws and Their Adverse Effects in Anterior Cruciate Ligament Reconstruction Surgery. *Knee* **2009**, *16*, 6–13. [CrossRef]
7. Ramot, Y.; Haim-Zada, M.; Domb, A.J.; Nyska, A. Biocompatibility and Safety of PLA and Its Copolymers. *Adv. Drug Deliv. Rev.* **2016**, *107*, 153–162. [CrossRef]
8. Mohseni, M.; Hutmacher, D.W.; Castro, N.J. Independent Evaluation of Medical-Grade Bioresorbable Filaments for Fused Deposition Modelling/Fused Filament Fabrication of Tissue Engineered Constructs. *Polymers* **2018**, *10*, 40. [CrossRef]
9. Prabhu, B.; Karau, A.; Wood, A.; Dadsetan, M.; Liedtke, H.; DeWitt, T. Bioresorbable Materials for Orthopedic Applications (Lactide and Glycolide Based). In *Orthopedic Biomaterials: Progress in Biology, Manufacturing, and Industry Perspectives*; Li, B., Webster, T., Eds.; Springer International Publishing: Cham, Switzerland, 2018; pp. 287–344. ISBN 978-3-319-89542-0.
10. Gunatillake, P.A.; Adhikari, R. Biodegradable Synthetic Polymers for Tissue Engineering. *Eur. Cell Mater.* **2003**, *5*, 1–16. [CrossRef]
11. Chye Joachim Loo, S.; Ping Ooi, C.; Chiang Freddy Boey, Y. Influence of Electron-Beam Radiation on the Hydrolytic Degradation Behaviour of Poly(Lactide-Co-Glycolide) (PLGA). *Biomaterials* **2005**, *26*, 3809–3817. [CrossRef]
12. Woodard, L.N.; Grunlan, M.A. Hydrolytic Degradation and Erosion of Polyester Biomaterials. *ACS Macro Lett.* **2018**, *7*, 976–982. [CrossRef]
13. Ahlinder, A.; Fuoco, T.; Finne-Wistrand, A. Medical Grade Polylactide, Copolyesters and Polydioxanone: Rheological Properties and Melt Stability. *Polym. Test.* **2018**, *72*, 214–222. [CrossRef]
14. Ahlinder, A.; Fuoco, T.; Morales-López, Á.; Yassin, M.A.; Mustafa, K.; Finne-Wistrand, A. Nondegradative Additive Manufacturing of Medical Grade Copolyesters of High Molecular Weight and with Varied Elastic Response. *J. Appl. Polym. Sci.* **2020**, *137*, 48550. [CrossRef]
15. Jain, S.; Fuoco, T.; Yassin, M.A.; Mustafa, K.; Wistrand, A.F. Printability and Critical Insight into Polymer Properties during Direct-Extrusion Based 3D Printing of Medical Grade Polylactide and Copolyesters. *Biomacromolecules* **2019**, *21*, 388–396. [CrossRef]
16. Sterilization of Implantable Polymer-Based Medical Devices: A Review. ScienceDirect. Available online: https://www-sciencedirect-com.ressources-electroniques.univ-lille.fr/science/article/pii/S0378517317311304?via%3Dihub (accessed on 30 July 2019).
17. Sterilization Techniques for Biodegradable Scaffolds in Tissue Engineering Applications. Available online: https://www.ncbi.nlm.nih.gov/pmc/articles/PMC4874054/ (accessed on 30 July 2019).
18. Holy, C.E.; Cheng, C.; Davies, J.E.; Shoichet, M.S. Optimizing the Sterilization of PLGA Scaffolds for Use in Tissue Engineering. *Biomaterials* **2000**, *22*, 25–31. [CrossRef]
19. Haim Zada, M.; Kumar, A.; Elmalak, O.; Mechrez, G.; Domb, A.J. Effect of Ethylene Oxide and Gamma (γ-) Sterilization on the Properties of a PLCL Polymer Material in Balloon Implants. *ACS Omega* **2019**, *4*, 21319–21326. [CrossRef]
20. Pietrzak, W.S. Effects of Ethylene Oxide Sterilization on 82: 18 PLLA/PGA Copolymer Craniofacial Fixation Plates. *J. Craniofac. Surg.* **2010**, *21*, 177–181. [CrossRef]
21. Phillip, E.; Murthy, N.S.; Bolikal, D.; Narayanan, P.; Kohn, J.; Lavelle, L.; Bodnar, S.; Pricer, K. Ethylene Oxide's Role as a Reactive Agent during Sterilization: Effects of Polymer Composition and Device Architecture. *J. Biomed. Mater. Res. Part B Appl. Biomater.* **2013**, *101B*, 532–540. [CrossRef]
22. Ethylene Oxide Sterilization of Medical Devices: A Review. Available online: https://www.sciencedirect.com/science/article/abs/pii/S0196655307000521 (accessed on 18 February 2020).
23. Ethylene Oxide Potential Toxicity: Expert Review of Medical Devices: Vol 5, No 3. Available online: https://www.tandfonline.com/doi/abs/10.1586/17434440.5.3.323 (accessed on 18 February 2020).
24. Lucas, A.D.; Merritt, K.; Hitchins, V.M.; Woods, T.O.; McNamee, S.G.; Lyle, D.B.; Brown, S.A. Residual Ethylene Oxide in Medical Devices and Device Material. *J. Biomed. Mater. Res. Part B Appl. Biomater.* **2013**, *66*, 548–552. Available online: https://onlinelibrary.wiley.com/doi/abs/10.1002/jbm.b.10036 (accessed on 18 February 2020).
25. Kinetics of the Aeration of Ethylene-Oxide Sterilized Plastics. Available online: https://www.sciencedirect.com/science/article/pii/0142961280900381 (accessed on 18 February 2020).
26. Aeration of Ethylene Oxide-Sterilized Polymers. Available online: https://www.sciencedirect.com/science/article/pii/0142961286901080 (accessed on 18 February 2020).
27. FDA. Ethylene Oxide Sterilization for Medical Devices. Available online: https://www.fda.gov/medical-devices/general-hospital-devices-and-supplies/ethylene-oxide-sterilization-medical-devices (accessed on 18 February 2020).
28. Loo, S.C.J.; Ooi, C.P.; Boey, Y.C.F. Radiation Effects on Poly (Lactide-Co-Glycolide) (PLGA) and Poly(l-Lactide) (PLLA). *Polym. Degrad. Stab.* **2004**, *83*, 259–265. [CrossRef]
29. Davison, L.; Themistou, E.; Buchanan, F.; Cunningham, E. Low Temperature Gamma Sterilization of a Bioresorbable Polymer, PLGA. *Radiat. Phys. Chem.* **2018**, *143*, 27–32. [CrossRef]
30. Montanari, L.; Cilurzo, F.; Selmin, F.; Conti, B.; Genta, I.; Poletti, G.; Orsini, F.; Valvo, L. Poly (Lactide-Co-Glycolide) Microspheres Containing Bupivacaine: Comparison between Gamma and Beta Irradiation Effects. *J. Control. Release* **2003**, *90*, 281–290. [CrossRef]

31. Cingolani, A.; Casalini, T.; Caimi, S.; Klaue, A.; Sponchioni, M.; Rossi, F.; Perale, G. A Methodologic Approach for the Selection of Bio-Resorbable Polymers in the Development of Medical Devices: The Case of Poly(l-Lactide-Co-ε-Caprolactone). *Polymers* **2018**, *10*, 851. [CrossRef]
32. Tsuji, H.; Mizuno, A.; Ikada, Y. Properties and Morphology of Poly(L-lactide). III. Effects of Initial Crystallinity on Long-term in Vitro Hydrolysis of High Molecular Weight Poly(L-lactide) Film in Phosphate-buffered Solution. *J. Appl. Polym. Sci.* **2000**, *77*, 1452–1464. Available online: https://onlinelibrary.wiley.com/doi/abs/10.1002/1097-4628%2820000815%2977%3A7%3C1452%3A%3AAID-APP7%3E3.0.CO%3B2-S (accessed on 12 May 2020).
33. Widmer, M.S.; Gupta, P.K.; Lu, L.; Meszlenyi, R.K.; Evans, G.R.D.; Brandt, K.; Savel, T.; Gurlek, A.; Patrick, C.W.; Mikos, A.G. Manufacture of Porous Biodegradable Polymer Conduits by an Extrusion Process for Guided Tissue Regeneration. *Biomaterials* **1998**, *19*, 1945–1955. [CrossRef]
34. Shim, J.-H.; Kim, J.Y.; Park, J.K.; Hahn, S.K.; Rhie, J.-W.; Kang, S.-W.; Lee, S.-H.; Cho, D.-W. Effect of Thermal Degradation of SFF-Based PLGA Scaffolds Fabricated Using a Multi-Head Deposition System Followed by Change of Cell Growth Rate. *J. Biomater. Sci. Polym. Ed.* **2010**, *21*, 1069–1080. [CrossRef]
35. Ragaert, K.; Cardon, L.; Dekeyser, A.; Degrieck, J. Machine Design and Processing Considerations for the 3D Plotting of Thermoplastic Scaffolds. *Biofabrication* **2010**, *2*, 014107. [CrossRef]
36. Woodruff, M.A.; Hutmacher, D.W. The Return of a Forgotten Polymer—Polycaprolactone in the 21st Century. *Prog. Polym. Sci.* **2010**, *35*, 1217–1256. [CrossRef]
37. Soufivand, A.A.; Abolfathi, N.; Hashemi, A.; Lee, S.J. The Effect of 3D Printing on the Morphological and Mechanical Properties of Polycaprolactone Filament and Scaffold. *Polym. Adv. Technol.* **2020**, *31*, 1038–1046. [CrossRef]
38. Peltola, S.M.; Melchels, F.P.W.; Grijpma, D.W.; Kellomäki, M. A Review of Rapid Prototyping Techniques for Tissue Engineering Purposes. *Ann. Med.* **2008**, *40*, 268–280. [CrossRef]
39. Faisant, N.; Siepmann, J.; Richard, J.; Benoit, J.P. Mathematical Modeling of Drug Release from Bioerodible Microparticles: Effect of Gamma-Irradiation. *Eur. J. Pharm. Biopharm.* **2003**, *56*, 271–279. [CrossRef]

Article

Characterization of Bio-Inspired Electro-Conductive Soy Protein Films

Pedro Guerrero [1,2,*], Tania Garrido [1], Itxaso Garcia-Orue [3,4], Edorta Santos-Vizcaino [3,4], Manoli Igartua [3,4], Rosa Maria Hernandez [3,4] and Koro de la Caba [1,2,*]

[1] BIOMAT Research Group, University of the Basque Country (UPV/EHU), Escuela de Ingeniería de Gipuzkoa, Plaza de Europa 1, 20018 Donostia-San Sebastián, Spain; tania.garrido@ehu.eus
[2] BCMaterials, Basque Center for Materials, Applications and Nanostructures, UPV/EHU Science Park, 48940 Leioa, Spain
[3] NanoBioCel Group, Laboratory of Pharmaceutics, School of Pharmacy, University of the Basque Country (UPV/EHU), Paseo de la Universidad 7, 01006 Vitoria-Gasteiz, Spain; itxaso.garcia@ehu.eus (I.G.-O.); edorta.santos@ehu.eus (E.S.-V.); manoli.igartua@ehu.eus (M.I.); rosa.hernandez@ehu.es (R.M.H.)
[4] Biomedical Research Networking Centre in Bioengineering, Biomaterials and Nanomedicine (CIBER-BBN), 01006 Vitoria-Gasteiz, Spain
* Correspondence: pedromanuel.guerrero@ehu.eus (P.G.); koro.delacaba@ehu.eus (K.d.l.C.); Tel.: +34-943-018-535 (P.G.); +34-943-017-188 (K.d.l.C.)

Citation: Guerrero, P.; Garrido, T.; Garcia-Orue, I.; Santos-Vizcaino, E.; Igartua, M.; Hernandez, R.M.; de la Caba, K. Characterization of Bio-Inspired Electro-Conductive Soy Protein Films. *Polymers* **2021**, *13*, 416. https://doi.org/10.3390/polym13030416

Academic Editor: Luis García-Fernández

Received: 20 December 2020
Accepted: 25 January 2021
Published: 28 January 2021

Publisher's Note: MDPI stays neutral with regard to jurisdictional claims in published maps and institutional affiliations.

Copyright: © 2021 by the authors. Licensee MDPI, Basel, Switzerland. This article is an open access article distributed under the terms and conditions of the Creative Commons Attribution (CC BY) license (https://creativecommons.org/licenses/by/4.0/).

Abstract: Protein-based conductive materials are gaining attention as alternative components of electronic devices for value-added applications. In this regard, soy protein isolate (SPI) was processed by extrusion in order to obtain SPI pellets, subsequently molded into SPI films by hot pressing, resulting in homogeneous and transparent films, as shown by scanning electron microscopy and UV-vis spectroscopy analyses, respectively. During processing, SPI denatured and refolded through intermolecular interactions with glycerol, causing a major exposition of tryptophan residues and fluorescence emission, affecting charge distribution and electron transport properties. Regarding electrical conductivity, the value found (9.889×10^{-4} S/m) is characteristic of electrical semiconductors, such as silicon, and higher than that found for other natural polymers. Additionally, the behavior of the films in contact with water was analyzed, indicating a controlled swelling and a hydrolytic surface, which is of great relevance for cell adhesion and spreading. In fact, cytotoxicity studies showed that the developed SPI films were biocompatible, according to the guidelines for the biological evaluation of medical devices. Therefore, these SPI films are uniquely suited as bio-electronics because they conduct both ionic and electronic currents, which is not accessible for the traditional metallic conductors.

Keywords: soy protein; film; semiconductor; biomaterial

1. Introduction

Conventional electronic devices include various organic/inorganic materials, which are non-biodegradable, or even toxic, and lead to the accumulation of electronic wastes, causing serious ecological pollution [1]. In this context, the development of functional materials has shifted from petrochemical feedstock to natural, renewable and more sustainable resources [2,3]. Owing to their biodegradability and biocompatibility, biopolymer-based conductive materials are arising significant interest in researchers, from both academic and industrial fields, and have emerged as promising candidates for the production of multifunctional electro-conductive components for advanced applications, such as strain sensors, wearable devices, and portable electronic equipment (electrical devices) [4]. In this regard, energy harvesting by nature-driven biodegradable and biocompatible materials, which respond to biomechanical activities, is receiving great attention to develop an alternative energy source for next-generation portable biomedical devices [5–7]. Some bio-based materials, such as fish scales [8], cellulose [9], or spider silk [10], have been studied for green energy harvesting. However, the products developed from these raw materials were not completely bio-based materials, since they were chemically treated or/and mixed with other non-bio-based materials.

In order to overcome the above-mentioned drawbacks, the use of proteins can become an attractive alternative for eco-friendly devices. In particular, soy protein isolate (SPI) is a natural, abundant, and available protein, constituted of 18 amino acid residues. This amino acid composition affects the protein structure and, thus, the properties of the products developed [11,12]. In this regard, Las Heras and co-workers [13] determined that SPI was rich in glutamic and aspartic acids, leucine, serine, glycine, alanine and proline amino acids, with cysteine and methionine as the principal sulphur-containing amino acids. In fact, cysteine presents the ability to form inter- and intra-chain disulfide bonds, which play a crucial role in protein-folding pathways and, thus, in protein structure. Furthermore, SPI exhibits a variety of peptides that promote migration and cell proliferation, key factors for tissue regeneration applications [14]; among them, lunasin owns RGD-like sequences required to promote stable cell adhesion [15]. In this regard, the use of SPI films as electronic components can avoid biocompatibility concerns that are present with conventional electronic components based on synthetic polymers, and may allow easier integration into biological systems. Some recent researches [16,17] have shown that natural and nature-inspired materials can be used not only to create organic devices with state-of-the-art performance but also to provide conceptual clues about the molecular design of organic semiconductors. Therefore, understanding interactions in SPI films and their effect on protein structure is essential.

Processing methods also affect the protein structure and, thus, determine the properties of the final product. It is well known that thermomechanical processes, such as extrusion and compression, induce dissociation, denaturation, and aggregation of soy protein subunits [18–20]. The protein is disassembled and then, reassembled by disulfide bonds and non-covalent interactions, such as hydrogen bonding, leading to the formation of stable structures. Breaking intermolecular linkages that stabilize the protein in their native state and the orientation and reconstructing the chains with the formation of new intermolecular linkages stabilize the three-dimensional network formed. Additionally, extrusion and compression are versatile and well-established processes used to manufacture polymeric materials, and these thermomechanical processes are relatively inexpensive and amenable to industrial scale-up. Therefore, in this work extrusion was employed to obtain SPI pellets, converted into films by compression molding. Afterward, the aim of this work was focused on the study of the relationship among processing, structure, and properties of SPI films, including optical, mechanical, electrical and biological properties.

2. Materials and Methods

2.1. Materials

Soy protein isolate (SPI), PROFAM 974, with 90% protein on a dry basis, was supplied by ADM Protein Specialties Division (Amsterdam, The Netherlands). SPI has 5% moisture, 4% fat, and 5% ash and its isoelectric point is 4.6. Glycerol, with a purity of 99.01% (Panreac, Barcelona, Spain), was used as plasticizer.

2.2. Pellets Preparation

SPI and glycerol were mixed in a Stephan UMC 5 mixer for 5 min at 1500 rpm in order to obtain a good blend. Blends were prepared with 80 wt% SPI and 20 wt% glycerol on total mass. These blends were added into the feed hopper of a twin-screw extruder and mixed with water in the extruder barrel. The MPF 19/25 APV Baker extruder used in this study had 19 mm diameter-barrel and a length/barrel diameter ratio of 25:1. Barrel temperatures were set at 70 °C, 80 °C, 95 °C, and 100 °C for the four zones from input to output, and die temperature was set at 100 °C, based on previous work [21]. Water was pumped directly into the extruder barrel at a constant speed of 250 rpm using a peristaltic 504U MK pump (Watson Marlow Ltd., Rommerskirchen, Germany). All trials were carried out using a water speed of 3.0 g/min (0.18 kg/h). The feed rate of the extruder was adjusted to 1 kg/h and a single die of 3 mm diameter was used, giving a throughput per unit area of 0.141 kg/h mm^2. Extrusion process variables were measured; in particular, the specific mechanical energy (SME), the amount of work input from the driver motor into the extruded material.

2.3. Film Preparation

The pellets obtained by extrusion were placed between two aluminum sheets using a caver laboratory press (Specac, Madrid, Spain), previously heated at 120 °C. The pellets were pressed at 80 bar for 2 min to obtain compression-molded films. All samples were conditioned in an ACS Sunrise 700 V bio-chamber at 25 °C and 50% relative humidity for 48 h before testing.

2.4. Film Characterization

2.4.1. Differential Scanning Calorimetry (DSC) and Thermo-Gravimetric Analysis (TGA)

DSC experiments were performed using a Mettler Toledo DSC 822 (Mettler Toledo, Barcelona, Spain). Around 4 mg of sample was heated from −50 °C to 250 °C at a rate of 10 °C/min under nitrogen atmosphere to avoid oxidation reactions. Sealed aluminum pans were used to prevent mass loss during the experiment.

TGA was carried out using a Mettler Toledo SDTA 851 equipment (Mettler Toledo, Barcelona, Spain). All specimens were heated from room temperature to 800 °C at a heating rate of 10 °C/min under nitrogen atmosphere (10 mL/min) to avoid thermo-oxidative reactions.

2.4.2. Fourier Transform Infrared (FTIR) Spectroscopy

Attenuated total reflectance Fourier transform infrared (ATR-FTIR) spectroscopy was used to identify the characteristic functional groups of SPI films. Measurements were performed with a Nicolet Nexus FTIR spectrometer (ThermoFisher, Madrid, Spain) equipped with an MKII Golden Gate accessory (Specac) with a diamond crystal as ATR element at a nominal incidence angle of 45° with a ZnSe lens. Measurements were recorded in the 4000–750 cm^{-1} region, using 32 scans at a resolution of 4 cm^{-1}. All spectra were smoothed using the Savitzky–Golay function. Second-derivative spectra of the amide region were used at peak position guides for the curve fitting procedure, using OriginPro 2019b software.

2.4.3. Ultraviolet-Visible (UV-vis) and Fluorescence Spectroscopies

The spectrophotometer UV-vis-Jasco (Model V-630) (Jasco, Madrid, Spain), coupled to solid sample support, was used to determine light barrier properties of films. Light absorption was measured at wavelengths from 200 nm to 800 nm.

Fluorescence emission measurements were performed with the spectrophotometer (Horiba, Madrid, Spain) of Photon Technology International (PTI) coupled to a solid sample support. SPI films were excited at 370 nm and emission spectra were recorded from 385 nm to 600 nm, using FeliX32 software.

2.4.4. X-ray Diffraction (XRD) and X-ray Photoelectron Spectroscopy (XPS)

XRD was performed with a diffraction unit PANalytical Xpert PRO (Malvern, Madrid, Spain), operating at 40 kV and 40 mA. The radiation was generated from Cu-Kα (λ = 1.5418 Å) source and the diffraction data were collected from 2θ values from 2° to 90°, where θ is the angle of incidence of the X-ray beam on the sample. All spectra were smoothed using the Savitzky–Golay function. Second-derivative of XRD spectra were used at peak position guides for the curve fitting procedure, using OriginPro 2019b software.

XPS was performed in a SPECS spectrometer (SPECS, Barcelona, Spain) using a monochromatic radiation equipped with Al Kα (1486.6 eV). The binding energy was calibrated by Ag 3d5/2 peak at 368.28 eV. All spectra were recorded at 90° take-off angle.

Survey spectra were recorded with 1.0 eV step and 40.0 eV analyzer pass energy and the high-resolution regions with 0.1 eV step and 20 eV pass energy. All core level spectra were referred to the C 1s peak at 284.6 eV. Spectra were analyzed using the CasaXPS 2.3.19PR 1.0 software, and peak areas were quantified with a Gaussian–Lorentzian fitting procedure.

2.4.5. Scanning Electron Microscopy (SEM)

Morphology of the film cross-section was visualized using a scanning electron microscope S-4800 (Hitachi, Madrid, Spain). The samples were mounted on a metal stub with double-side adhesive tape and coated under vacuum with gold (JFC-1100) in an argon atmosphere prior to observation. All samples were examined using an accelerating voltage of 15 kV.

2.4.6. Mechanical Testing

Tensile strength (TS), and elongation at break (EB) were measured according to ASTM D638-03 [22], using Instron 5967 electromechanical testing system with a tensile load cell of 500 N. Five specimens for each film were cut into bone-shaped samples of 4.75 mm × 22.25 mm, the film thickness was 0.626 mm, and the crosshead rate was 5 mm/min.

2.4.7. Electrical Conductivity

Electrical properties of films were measured by a Keithley 4200-SCS (Mouser, Barcelona, Spain) equipment for semiconductors analysis in a Faraday cage at room temperature with low-noise triax cables and results were analyzed with KITE Software (Keithley Interactive Test Environment 9.0). Two point measurements were carried performing linear scans from −20 V to 20 V in order to obtain intensity vs. voltage curves. The conductivity (resistivity) of SPI films was measured by using a four-point collinear probe (Figure 1). Four electrodes were used in order to minimize any measurement error due to the contact resistance. The films were placed in contact with copper sheets, adhered in turn to polycarbonate plates. The distance between the electrodes was 3 mm and the dimensions of the samples were 2×2 cm^2 section and 0.626 mm height.

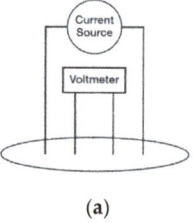
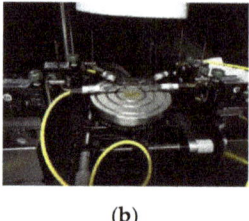

(a)   (b)

Figure 1. (a) Four-point collinear probe resistivity configuration and (b) photograph of the home-built dispositive used for the conductivity measurement.

2.4.8. Water Uptake

First, dry samples of the films were cut into discs of 8 mm in diameter. Then, they were weighed and immersed in 1 mL of phosphate buffer solution (PBS) at 37 °C for determined periods of time. Samples were collected at specific times, the excess water was wiped out with a filter paper and the wet discs weighed. Water uptake (WU) was calculated as:

$$WU(\%) = \frac{W - W_0}{W_0} \times 100 \qquad (1)$$

where W_0 is the weight of the dry discs at the beginning of the study and W is the weight of the wet discs at every time point.

2.4.9. Water Contact Angle

Contact angle measurements were carried out by the sessile drop technique using a Dataphysics OCA 20 system (Aries, Madrid, Spain). Samples were laid on a movable sample stage and about 3 µL of distilled water was placed onto the film surface to estimate the hydrophobic character. The contact angle images were recorded on the sample, using SCA20 software. Five measurements were made at room temperature.

2.4.10. Degradation Analyses

To assess the degradation of the films in an aqueous environment, samples were cut into discs of 8 mm in diameter, weighed and immersed into PBS at 37 °C. At determined time points, films were collected, washed with milliQ water and freeze-dried. Dry samples were weighed and the percentage of the remaining weight (RM) was calculated using the following equation:

$$RM\ (\%) = \frac{final\ weight}{initial\ weight} \times 100 \tag{2}$$

Enzymatic degradation was assessed by immersing discs of 8 mm in diameter into 1 mL of a 0.5 mg/mL collagenase H solution (Sigma-Aldrich, Saint Louise, MO, USA). Samples were incubated at 37 °C for 1, 4 and 7 days. At those time points, discs were collected, washed with milliQ water, freeze-dried and weighed. RM values were calculated as expressed by Equation (2).

Cellular degradation was evaluated using L-929 fibroblasts (ATCC, Manassas, VA, USA). Cells were cultured in Eagle's Minimum Essential Medium (EMEM, ATCC, Manassas, VA, USA) supplemented with 1% (v/v) penicillin-streptomycin and 10% (v/v) inactivated Horse Serum at 37 °C in a humidified incubator with a 5% CO_2 atmosphere. Cell passages were performed every 2–3 days depending on the confluence of cells. First, film samples were cut into discs of 8 mm in diameter and pre-treated to avoid cell toxicity. Briefly, discs were dialyzed in 1 L of MilliQ water for 72 h into a dialyzing bag and then, freeze-dried. Dry discs were weighed and sterilized by UV radiation for 20 min. Subsequently, cells were seeded on top of them at a density of 5000 cells/well, and incubated at 37 °C for 4 and 7 days. Finally, discs were collected, washed with MilliQ, freeze-dried, and weighed. RM values were calculated as expressed by Equation (2). Three independent experiments, each of them with three replicates, were performed.

2.4.11. Cytotoxicity Studies

Cytotoxicity studies were performed according to the ISO 10993-4:2009 [23] guideline for biological evaluation of medical devices. Firstly, film discs of 8 mm in diameter were pre-treated following the same procedure described in cellular degradation. Discs were sterilized exposing them to UV light for 20 min and dialyzed in 1 L of MilliQ water for 72 h. Then, they were transferred into culture medium for 24 h to reach osmotic equilibrium. Finally, pre-treated discs were divided into two groups to conduct direct or indirect cytotoxicity assays.

For direct cytotoxicity assay, L-929 fibroblasts were seeded onto a 24 well plate at a density of 35,000 cells/well and incubated overnight in order to allow cell attachment. Then, pre-treated discs were carefully placed into the wells. Cells incubated without any sample were used as positive control and cells incubated with 10% of DMSO (ATCC, Manassas, VA, USA) as negative control. After 48 h of incubation, discs were removed from the wells and viability was assessed. The different incubation time of direct and indirect assays was decided due to previous studies [24], where direct incubation needed more time to show the real cytotoxicity of the samples.

For indirect cytotoxicity assay, each disc was incubated with 0.5 mL of culture medium for 24 h to obtain the extracted medium. Meanwhile, cells were seeded on a 96 well plate at a density of 5000 cells/well and incubated overnight to allow cell attachment. Then, the medium was replaced with the extracted medium of the discs. Fresh medium was used as positive control and medium with 10% of DMSO as negative control. Cells were incubated for 24 h and then viability was evaluated.

In both cases, viability was assessed using the CCK-8 colorimetric assay (Cell Counting Kit-8, Sigma-Aldrich, Saint Louise, MO, USA). Briefly, 50 µL of the CCK8 reagent in the case of direct cytotoxicity and 10 µL in the case of indirect cytotoxicity were added to the cells, in each case 10% (v/v) of the volume of the wells. Cells were incubated for 4 h and then the absorbance of the wells was read at 450 nm, using 625 nm as a reference wavelength Plate Reader Inifinite M200 (Tecan, Barcelona, Spain). The absorbance values were directly proportional to the number of living cells in each well. Results were expressed as the percentage of living cells regarding the positive control. Three independent experiments, each of them with three replicates, were performed.

2.5. Statistical Analysis

For biological experiments, means were compared through one-way ANOVA. Based on the results of the Levene test for the homogeneity of variances, Bonferroni or Tamhane post-hoc analysis was applied. For non-normally distributed data, Mann–Whitney non-parametric analysis was applied. All the statistical computations were assessed using SPSS 25.0.

3. Results and Discussion

3.1. Physicochemical Properties

The extrusion process used to mix, homogenize, and shape the SPI blend by forcing it through a specifically designed opening led to a continuous filament with no bubble at the extruder die. The process variable measured for this system was the specific mechanical energy (SME) and its value was 648 kJ/kg, with a stable torque of 18%. This SME value indicated the extent of molecular breakdown the material undergoes during the extrusion process [25]. This extrusion process led to the decrease in electrostatic repulsions and allowed intermolecular interactions and a new SPI network formation to occur. In particular, globular proteins, such as SPI, are triggered by thermal treatment and, thus, water and glycerol can enter the protein network and interact with protein chains by hydrogen bonding with easily accessible polar amino acids side chains, preventing protein–protein interactions and, thereby, leading to plasticization. Therefore, water and glycerol acted as a dispersion medium, influencing the extrudate viscosity and flowability. In order to assess the effect of the extrusion process, the thermal properties of SPI films were analyzed by TGA and DSC. As shown in Figure 2, the thermal behavior of SPI was not found to be affected by the extrusion process. The DTG curve showed three main stages (Figure 2a): the first stage around 90 °C was related to the moisture evaporation; the second stage was associated with glycerol evaporation [26] and appeared around 220 °C, a higher temperature than the boiling point of glycerol (182 °C), indicating the interactions between SPI and glycerol; and the third stage above 325 °C was attributed to the thermal degradation of SPI [27]. Therefore, the thermal stability of SPI was not affected by extrusion. Additionally, DSC analysis was carried out and a pronounced endothermic peak was observed around 100 °C (Figure 2b), corresponding to the denaturation of 7S globulins present in SPI, and a second one around 240 °C, related to the high molecular fraction of 11S globulins [28].

This thermal denaturation of the protein involved the disruption of intramolecular bonding and the unfolding and aggregation of protein molecules. Since the formation of a blend occurred upon application of denaturing conditions, the denatured protein may undergo partial refolding, thus regaining some secondary structure during the blending process. The extent of such refolding may affect the number of functional groups available for intermolecular interactions and, thus, the stability of the network formed. Regarding the processing method used, it is generally accepted that proteins are denatured during extrusion, so that the reactive unfolded protein chains can interact with each other and with glycerol, leading to the formation of a new network [18]. Since these intermolecular interactions formed in the SPI network offer great opportunities for the design of specific structures, FTIR analysis was carried out to assess them. FTIR spectrum of SPI films is shown in Figure 3a. The spectrum exhibited three characteristic bands common to proteins: amide I band at 1630 cm^{-1}, associated with the carbonyl group; amide II band at 1530 cm^{-1}, corresponding to N-H bending; and amide III band at 1230 cm^{-1}, related to C-N stretching and N-H bending [29]. The broad band observed in the 3500–3000 cm^{-1} range is related to free and bound O-H and N-H groups, which are able to form hydrogen bonding. The band corresponding to amide I depends on the secondary structure of the protein backbone and it is the most commonly used band for the quantitative analysis of secondary structures. Therefore, a curve fitting treatment was carried out to estimate quantitatively the relative proportion of each component representing a type of secondary structure (Figure 3b–d). The derivative function was calculated to determine the number of components in the amide I region for the curve-fitting process. The areas of assigned amide I bands in the second derivative spectra correspond linearly to the number of different types of secondary structures present in the protein.

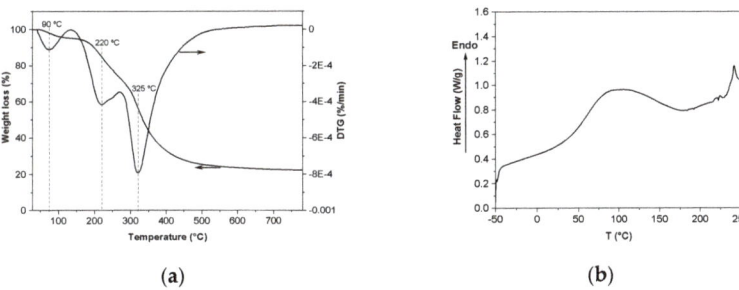

Figure 2. (a) Thermo-gravimetric analysis (TGA) and DTG curves, and (b) differential scanning calorimetry (DSC) curve of soy protein isolate (SPI) films.

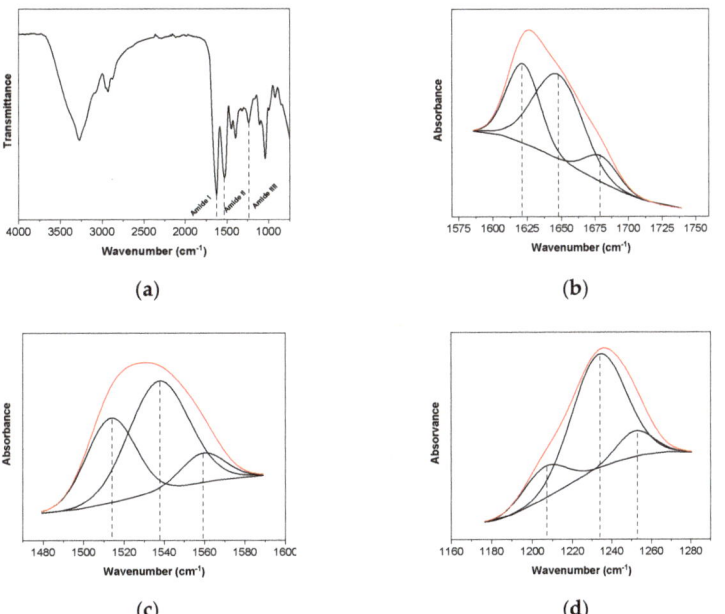

Figure 3. (a) FTIR spectrum and curve fitting of (b) amide I, (c) amide II, and (d) amide III of SPI films.

Table 1 shows the percentage of different types of secondary structures present in SPI films. Regarding amide I region, three bands can be distinguished. The band at 1649 cm^{-1} is observed in the FTIR spectra of most proteins, and it is assigned to the random coil, helical conformation, or both [30]. Additionally, the two bands at 1622 and 1678 cm^{-1} are assigned to β-sheet conformations [31]. On the one hand, the hydrogen bonding frequency dependence of the amide I vibration predominantly derives from the C=O group and, considering the central role hydrogen bonding plays in protein folding, results showed that interpeptide hydrogen bonding stabilized both α-helix/random coils and β-sheet secondary structures, which act as junction zones in the cohesion of film. On the other hand, the hydrogen bonding frequency dependence of the amide II vibration derives from N–H hydrogen bonding, which alters both the geometry and electronic density distribution. In relation to the band corresponding to amide II, the band at 1513 cm^{-1} was attributed to the vibrational modes of tyrosine side chains [32], which contribute at the interfaces of the β sheet regions, while the band at 1537 cm^{-1} was attributed to lysine side chains [32]. Concerning the band at 1559 cm^{-1}, this was related to the carboxyl side groups of aspartic and glutamic acids present in SPI [33]. Finally, the amide III band is known to reveal some information on local stress and coupling effects of C–N stretching and H–N–C in plane bending modes and, thus, the deconvoluted amide III band region showed a band at 1206 cm^{-1} assigned to tyrosine/phenylalanine, a second band at 1233 cm^{-1} assigned to the random coil/helical conformation, and a third one related to β-sheets at 1252 cm^{-1} [34]. Due to these secondary structures, such as β-sheet and α-helix, in addition to the existence of peptide linkage, hydroxyl and carbonyl groups, SPI films show a structure of amino acids strongly interconnected through intra- and intermolecular hydrogen bonding that can provide films with electric properties.

Table 1. Secondary structure determination for amide I, II and III in SPI films.

FTIR Band	Amide I			Amide II			Amide III		
Wavenumber (cm^{-1})	1622	1649	1678	1513	1537	1559	1206	1233	1252
Area (%)	33.90	54.50	11.60	34.25	55.23	10.52	15.83	73.31	10.86

3.2. Optical Properties

Ultraviolet-visible (UV-vis) spectroscopy was used to measure barrier properties against UV-vis light. Figure 4a shows light absorbance values of SPI films. As can be seen, films showed excellent barrier properties against UV light in the range from 200 nm to 300 nm. The absorbance from 200 nm to 240 nm was associated with C=O, COOH, CONH$_2$ groups in protein chains [35], and the absorbance at 250–280 nm was associated with the presence of sensitive chromophores in soy protein, such as tyrosine, phenylalanine and tryptophan [36]. The low absorbance value at 600 nm is associated with the transparency of SPI films and this is an indicator of the compatibility of the components used in the film-forming formulation [37], leading to homogeneous structures at the macroscopic scale.

In order to explore and elucidate the effect of the chromophores in SPI films, the fluorescence steady-state emission was measured and shown in Figure 4b. The fluorescence intensity of the SPI film was measured according to an excitation wavelength at 370 nm and emission wavelengths from 380 nm to 600 nm. The intensity of the peak at 413 nm corresponds to tryptophan, while the small shoulder at 437 nm corresponds to tyrosine and phenylalanine present in SPI [38]. It is known that pure tryptophan presents a fluorescence emission at 350 nm [39]; therefore, the red shift to 413 nm indicated that the tryptophan fluorescence emission was highly influenced by hydrogen bonding and other non-covalent interactions between soy protein and glycerol, as shown by FTIR results, causing a major exposition of tryptophan residues. Based on these results, the photo-induced conduction indicated an alteration of protein structure, conformation, and charge distribution, which directly affected electron transport properties.

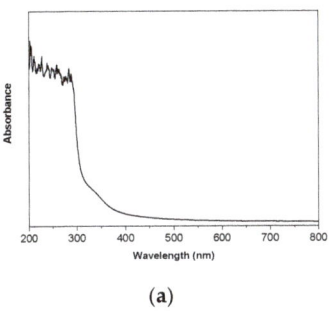
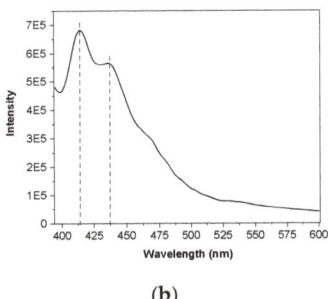

Figure 4. (a) UV-vis spectra and (b) fluorescence emission spectra (excitation at 370 nm) of SPI films.

3.3. X-ray Diffraction (XRD) and X-ray Photoelectron Spectroscopy (XPS)

The crystallinity degree of SPI films was determined by deconvolution from the X-ray diffraction (XRD) pattern. As can be seen by the broad peak in Figure 5a, SPI films are predominantly amorphous. The deconvolution of the XRD pattern showed a broad peak centered at 2θ of ~32.3°, associated with the amorphous structure of SPI, and the other two narrower peaks at 2θ of ~10.4° and ~21.4°, associated with the crystalline character of the films. The areas calculated for each of these peaks were 74.4%, 21.0% and 4.6% for the peaks centered at 2θ of ~32.3°,~21.4°, and ~10.4°, respectively, indicating a crystallinity degree of ~ 26%. The peak at 21.4° is associated with S-S bonds of cysteine in SPI films; these S-S bonds can tightly bind highly aligned β-sheets chains together, playing a relevant role in the electric dipole formation [40].

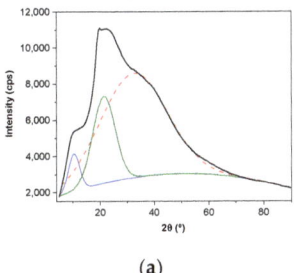
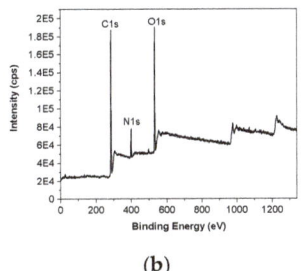
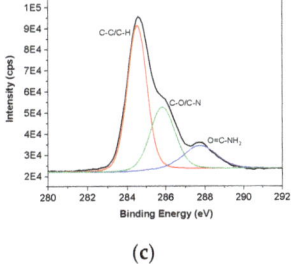

Figure 5. (a) X-ray diffraction (XRD) pattern, (b) X-ray photoelectron spectroscopy (XPS) survey data of SPI films, and (c) deconvolution of C1s spectrum.

The XPS survey data analysis of SPI films (Figure 5b) showed the presence of three peaks, corresponding to C, N, and O. Additionally, the deconvolution of C1s spectrum showed other three peaks (Figure 5c): a dominant peak at 284.6 eV, attributed to C-C and C-H bonds; a peak at 285.8 eV, assigned to C-O/C-N; and the peak at 287.7 eV, assigned to O=C-NH$_2$ bonds [41]. Furthermore, the high-resolution O1s spectra presented only one peak at 531.9 eV, attributed to O-C=O/O=C-N, whereas the high-resolution N1s spectra showed the presence of C=N/C-N at 399.9 eV. The small intensity of the peak related to N1s, compared to the ones corresponding to C1s and O1s, indicated the low exposure of amino groups towards the surface.

3.4. Film Microstructure and Mechanical and Electrical Properties

In order to confirm the good compatibility among the components, previously suggested by optical results, scanning electron microscopy (SEM) analysis was carried out. The surface and cross-sections of SPI films are shown in Figure 6. As can be seen, films exhibited a homogeneous and compact structure (Figure 6a) with a rough surface (Figure 6b).

Therefore, it can be assumed that during extrusion and film formation soy protein and glycerol were mixed without the formation of bubbles or holes. This evidenced significant molecular rearrangements within the polymer matrix, in which intermolecular interactions by hydrogen bonds occurred. This homogeneity of the film microstructure led to easy to handle films, as can be seen in the image in Figure 7a, with an elongation at a break value of 21.5 ± 1.7% and a tensile strength of 9.2 ± 0.5 MPa. Since mechanical properties are largely associated with the distribution and density of intramolecular and intermolecular interactions, which determined spatial structures, the flexibility of the films can also be related to the replacement of intramolecular interactions by intermolecular interactions by hydrogen bonds among SPI and glycerol, as shown above by FTIR results.

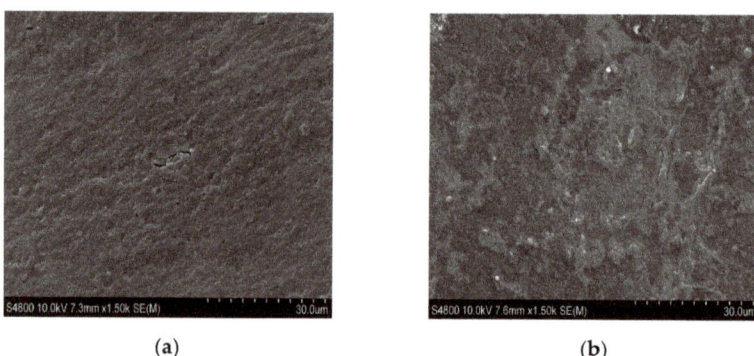

Figure 6. SEM images of (**a**) the cross-section and (**b**) the surface of SPI films at ×1.50 magnification.

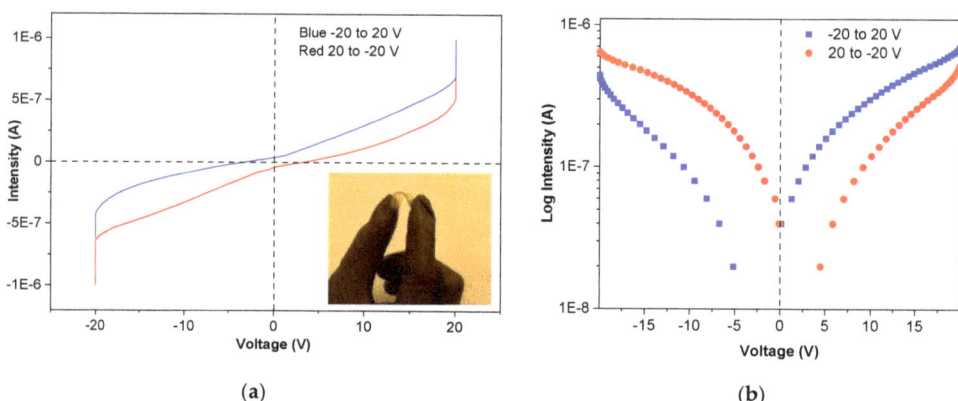

Figure 7. (**a**) Current intensity–voltage (I-V) curves (inset: SPI film flexibility) and (**b**) log I-V curves of SPI films.

In order to analyze the effect of film structure on electrical conductivity, electron transport across SPI monolayer films was studied using solid-state protein-based molecular junctions with a current intensity-voltage (I-V) method, and results are shown in Figure 7a. As can be seen, SPI films showed a pinched hysteretic shape and, thus, behaved like memristors since I-V dependence in the range from −20 V to +20 V was nonlinear, the basis of functional electronic devices, as shown in Figure 7b. In relation to the electrical resistance and conductivity, the values obtained at room temperature were 1.011×10^3 Ω m and 9.889×10^{-4} S/m, respectively, characteristic values of electrical semiconductor materials, such as silicon, and above the values of other well-known natural materials with satisfactory conductive properties [42].

The fundamental mechanism of electron transport (electron motion) via proteins is less understood than electron transfer (electron flow) due to the fact that electron transport measurements are not carried out in solution and, thus, there is no ionic charge in the medium surrounding the protein to screen charging as the electron moves across the protein. In this work, SPI films were measured between electronically conducting electrodes. This type of junction can be thought of as a donor–bridge–acceptor junction; the driving force of the transport is the electrical potential difference. The potential that the electron encounters when transported through the protein is affected by amino acid residues and their spatial arrangement. In this regard, protein structure can be understood at several levels: the primary structure, the linear sequence of amino acids held together by covalent peptide bonds, forming a polypeptide chain; the secondary structure, where the polypeptide chains form highly ordered three-dimensional features (α-helices and β-sheets) based on hydrogen bonding between peptide bonds; the tertiary structure, formed by additional secondary structure elements that undergo the necessary folding by specific interactions, including the formation of hydrogen bonds, and disulfide bonds giving rise to a compact structure [16]. The effect of secondary structure on electronic transport has been experimentally demonstrated with molecular junctions [43]. Therefore, the electron transport confirmed by I-V curves is in accordance with FTIR results, where the most prominent bands were related to the vibrations of peptide bonds at 1630 cm^{-1} and 1530 cm^{-1}, corresponding to amide I and II vibrational modes, with other relevant bands assigned to vibrational modes of side groups, such as NH_2 in-plane bending and C=C or C-N bonds, present in aromatic amino acid residues, such as tyrosine, phenylalanine, and tryptophan. These findings confirm that the side groups of amino acid residues play a relevant role in the inelastic part of the transport across the protein.

3.5. Water Uptake and Contact Angle

Results on water uptake showed that films were able to absorb the 72.1 ± 3.1% of their dry weight from 45 min onwards. No difference in water uptake was observed among time points, and the maximum value of 74.8 ± 2.6% was achieved at 120 min, as can be seen in Figure 8, where experimental values fit well (R = 0.999) with the empirical equation. The moderate swelling of SPI films is in accordance with the compact structure found by SEM analysis, in agreement with the intermolecular interactions between protein and glycerol indicated by FTIR results. This behavior suggests an enhancement of the conductivity properties since films could conduct both ionic and electronic currents [44], performance not provided by traditional metallic conductors.

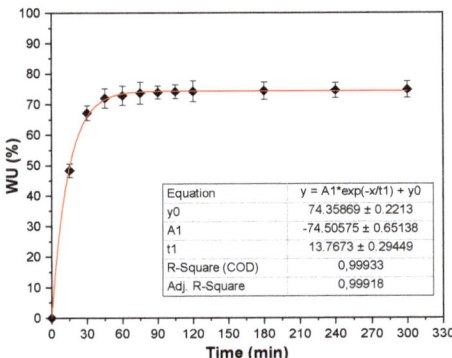

Figure 8. Water uptake of SPI films.

Regarding water contact angles, the values measured were 30.23 ± 0.76°, indicating that the SPI film surface was hydrophilic and, thus, those polar groups were orientated towards the surface. As can be seen in Figure 9a–c, SPI films absorbed the water droplet after 5 min. Since proteins are self-organized macromolecules with environment-sensitive secondary (α-helix and β-sheet), and tertiary (folding) structures, these results indicated that protein chains have the ability to rearrange when subjected to wet environments. Therefore, considering that a large number of functional interactions happen at surfaces, such as cell adhesion and growth, and that high hydrophilicity and water-retaining properties are vital for removing wound exudates and providing a moist environment for cell growth [45], degradation and biocompatibility studies were carried out.

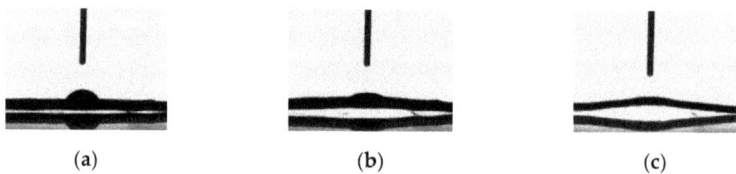

(a)   (b)   (c)

Figure 9. Contact angle of SPI films as a function of time: (a) t = 0 min, (b) t = 3 min, and (c) t = 5 min.

3.6. Degradation and Biocompatibility Studies

SPI films showed a two-stage weight loss in an aqueous environment, as can be seen in Figure 10a. First, a very quick weight loss was observed, since RM values were $78.9 \pm 1.7\%$ in the first hour of immersion. Thereafter, weight loss was slower, since no significant differences were found until day 7, when RM values decreased to $74.5 \pm 1.2\%$. From day 7 until the end of the study on day 28 (RM = $72.3 \pm 1.9\%$), no significant difference was observed. Accordingly, only 28% of weight loss was observed, most of it due to glycerol loss.

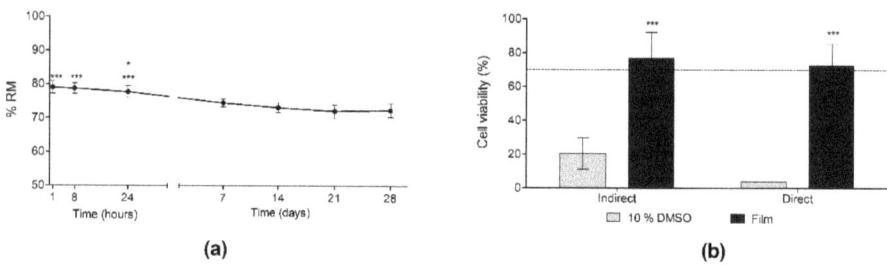

(a)   (b)

Figure 10. (a) Hydrolytic degradation. *** $p < 0.001$ comparing results at 1 h, 8 h and 24 h time points with the remaining time points, except from the comparison between 24 h and 7 days, when the statistical difference was lower (* $p < 0.05$), and (b) cell biocompatibility. *** $p > 0.001$ comparing the films with the negative control (10% DMSO).

Regarding enzymatic degradation, RM values were very similar at the three time points tested, about 72% of the dry weight. Comparing to the hydrolytic degradation, a significantly (** $p < 0.001$) faster weight loss was observed at 24 h in the samples incubated with the enzymatic solution. However, there was no difference in the assays at 4 and 7 days. Therefore, we can conclude that in both cases 28% of the weight was lost throughout the study; however, the weight loss occurred faster in an enzymatic media than in an aqueous media, highlighting that enzymes participate in film degradation.

In relation to cellular degradation, the RM values of SPI films incubated with cells at days 4 and 7 were $95.2 \pm 1.4\%$ and $94.2 \pm 1.2\%$, respectively. The films were dialyzed in an aqueous media for 72 h prior to the study, in order to avoid possible cytotoxicity due to changes in pH and osmolarity and, thus, the initial weight loss of glycerol occurred during the pre-treatment and not during the cell degradation study. In this case, the initial weight of the films was lower than in the other degradation studies, since films were weighed after the pre-treatment, which explains the lower weight loss observed. At any rate, taking into account the weight loss that occurred during the pre-treatment, the real RM was

72.46 ± 9.6% and 73.24 ± 8.4% at days 4 and 7, respectively. This matches with the results obtained in the enzymatic degradation study, where the final weight loss was similar to the hydrolytic degradation study, but occurred faster.

Biocompatibility was evaluated to confirm the lack of cytotoxicity of SPI films. Before conducting the assays, films were dialyzed for 72 h, according to the results obtained in a preliminary study, where changes in pH and osmolarity were observed after the first 72 h of incubation with culture medium (data not shown). As can be seen in Figure 10b, the results obtained from both direct and indirect cytotoxicity assays showed that cell viability was above 70%. This confirmed that the type of assay did not affect cell viability results, which indicate the biocompatibility of SPI films according to the ISO 10993-5:2009 guidelines for the biological evaluation of medical devices [23].

4. Conclusions

In this work, SPI films for molecular electronic devices were modelled as one-dimensional solid-state conductors because it is known that the biological functions of proteins are dependent on their structure in a fundamental way. The SPI films obtained in this work by an industrial thermo-mechanical process, such as extrusion and compression, were transparent and homogeneous, with a compact structure and a rough surface. The intermolecular interactions between SPI and glycerol led to flexible films. Besides flexibility, SPI films showed a pinched hysteretic shaped intensity-voltage curve, indicating their performance as semiconductors. In contrast to the usually hard and dry electronic devices, SPI films were flexible and hydrophilic. This behavior of SPI films is of great relevance when films are intended to be put in close contact with the human body, where soft and flexible electronic devices are more suitable to fit the human body environment. In sum, easy to handle films were developed from natural raw materials, resulting in flexible and biocompatible films, which provide a promising strategy towards biocompatible organic electronics.

Author Contributions: Conceptualization, P.G.; methodology, P.G.; E.S.-V.; formal analysis, P.G., T.G., I.G.-O.; investigation, P.G., T.G., I.G.-O.; resources, K.d.l.C., R.M.H.; data curation, P.G., T.G., I.G.-O., E.S.-V.; writing—original draft preparation, P.G., T.G., I.G.-O, E.S.-V.; writing—review and editing, K.d.l.C., M.I., R.M.H.; supervision, K.d.l.C., M.I., R.M.H.; project administration, K.d.l.C.; funding acquisition, K.d.l.C. All authors have read and agreed to the published version of the manuscript.

Funding: This research was funded by MCI/AEI/FEDER, UE, grant number RTI2018-097100-B-C22.

Institutional Review Board Statement: Not applicable.

Informed Consent Statement: Not applicable.

Data Availability Statement: The data presented in this study are available on request from the corresponding author.

Acknowledgments: Authors acknowledge Advanced Research Facilities (SGIker) from the University of the Basque Country.

Conflicts of Interest: Authors declare no conflict of interest.

References

1. Karan, S.K.; Mandal, D.; Khatua, B.B. Self-powered flexible Fe-doped RGO/PVDF nanocomposite: An excellent material for a piezoelectric energy harvester. *Nanoscale* **2015**, *7*, 10655–10666. [CrossRef] [PubMed]
2. Chen, J.; Xu, J.; Wang, K.; Qian, X.; Sun, R. Highly thermostable, flexible, and conductive films prepared from cellulose, graphite, and polypyrrole nanoparticles. *ACS Appl. Mater. Interfaces* **2015**, *7*, 15641–15648. [CrossRef] [PubMed]
3. Jin, S.; Li, K.; Gao, Q.; Zhang, W.; Chen, H.; Li, J. Development of conductive protein-based film reinforced by cellulose nanofibril template-directed hyperbranched copolymer. *Carbohydr. Polym.* **2020**, *237*, 116141. [CrossRef] [PubMed]
4. Lin, F.; Zheng, R.; Chen, J.; Su, W.; Dong, B.; Lin, C.; Huang, B.; Lua, B. Microfibrillated cellulose enhancement to mechanical and conductive properties of biocompatible hydrogels. *Carbohydr. Polym.* **2019**, *205*, 244–254. [CrossRef]
5. Tollefson, J. Power from the oceans: Blue energy. *Nature* **2014**, *508*, 302. [CrossRef]
6. Zheng, Q.; Shi, B.; Fan, F.; Wang, X.; Yan, L.; Yuan, W.; Wang, S.; Liu, H.; Li, Z.; Wang, Z.L. In vivo powering of pacemaker by breathing-driven implanted triboelectric nanogenerator. *Adv. Mater.* **2014**, *26*, 5851–5856. [CrossRef]

7. Li, Z.; Zhu, G.; Yang, R.; Wang, A.C.; Wang, Z.L. Muscle-driven in vivo nanogenerator. *Adv. Mater.* **2010**, *22*, 2534–2537. [CrossRef]
8. Ghosh, S.K.; Mandal, D. High-performance bio-piezoelectric nanogenerator made with fish scale. *Appl. Phys. Lett.* **2016**, *109*, 103701. [CrossRef]
9. Rajala, S.; Siponkoski, T.; Sarlin, E.; Mettänen, M.; Vuoriluoto, M.; Pammo, A.; Juuti, J.; Rojas, O.J.; Franssila, S.; Tuukkanen, S. Cellulose nanofibril film as a piezoelectric sensor material. *ACS Appl. Mater. Interfaces* **2016**, *8*, 15607–15614. [CrossRef] [PubMed]
10. Lee, S.M.; Pippel, E.; Moutanabbir, O.; Kim, J.-H.; Lee, H.-J.; Knez, M. In situ raman spectroscopic study of al-infiltrated spider dragline silk under tensile deformation. *ACS Appl. Mater. Interfaces* **2014**, *6*, 16827–16834. [CrossRef]
11. Zubeldía, F.; Ansorena, M.R.; Marcovich, N.E. Wheat gluten films obtained by compression molding. *Polym. Test.* **2015**, *43*, 68–77. [CrossRef]
12. Shi, W.; Dumont, M.J. Review: Bio-based films from zein, keratin, pea, and rapeseed protein feedstocks. *J. Mater. Sci.* **2014**, *49*, 1915–1930. [CrossRef]
13. Las Heras, K.; Santos-Vizcaíno, E.; Garrido, T.; Gutiérrez, F.B.; Aguirre, J.J.; de la Caba, K.; Guerrero, P.; Igartua, M.; Hernandez, R.M. Soy protein and chitin sponge-like scaffolds: From natural by-products to cell delivery systems for biomedical applications. *Green Chem.* **2020**, *22*, 3445–3460. [CrossRef]
14. Ahn, S.; Chantre, C.O.; Gannon, A.R.; Lind, J.U.; Campbell, P.H.; Grevesse, T.; O'Connor, B.B.; Parker, K.K. Soy protein/cellulose nanofiber scaffolds mimicking skin extracellular matrix for enhanced wound healing. *Adv. Healthc. Mater.* **2018**, *7*, 1701175. [CrossRef]
15. Chatterjee, C.; Gleddie, S.; Xiao, C.-W. Soybean Bioactive Peptides and Their Functional Properties. *Nutrients* **2018**, *10*, 1211. [CrossRef] [PubMed]
16. Du, N.; Yang, Z.; Liu, X.-Y.; Li, Y.; Xu, H.-Y. Structural origin of the strain-hardening of spider silk. *Adv. Funct. Mater.* **2011**, *21*, 772–778. [CrossRef]
17. Andonegi, M.; las Heras, K.; Santos-Vizcaíno, E.; Igartua, M.; Hernández, R.M.; de la Caba, K.; Guerrero, P. Structure-properties relationship of chitosan/collagen films with potential for biomedical applications. *Carbohydr. Polym.* **2020**, *237*, 11615. [CrossRef]
18. Guerrero, P.; Kerry, J.P.; de la Caba, K. FTIR characterization of protein-polysaccharide interactions in extruded blends. *Carbohydr. Polym.* **2014**, *111*, 598–605. [CrossRef]
19. Sorgentini, D.A.; Wagner, J.R.; Añon, M.C. Effects of thermal treatment of soy protein isolate on the characteristics and structure-function relationship of soluble and insoluble fractions. *J. Agric. Food Chem.* **1995**, *43*, 2471–2479. [CrossRef]
20. Yamauchi, F.; Yamagishi, T.; Iwabuchi, S. Molecular understanding of heat-induced phenomena of soybean protein. *Food Rev. Int.* **1991**, *7*, 721–729. [CrossRef]
21. Guerrero, P.; Beatty, E.; Kerry, J.P.; de la Caba, K. Extrusion of soy protein with gelatin and sugars at low moisture content. *J. Food Eng.* **2012**, *110*, 53–59. [CrossRef]
22. ASTM D 638-03. Standard test method for tensile properties of plastics. In *Annual Book of ASTM Standards*; American Society of Testing and Materials: Philadelphia, PA, USA, 2003.
23. ISO 10993-5. *Biological Evaluation of Medical Devices-Part 5: Tests for In-Vitro Cytotoxicity*; ISO: Geneve, Switzerland, 2009.
24. Etxabide, A.; Vairo, C.; Santos-Vizcaino, E.; Guerrero, P.; Pedraz, J.L.; Igartua, M.; de la Caba, K.; Hernandez, R.M. Ultra thin hydro-films based on lactose-crosslinked fish gelatin for wound healing applications. *Int. J. Pharm.* **2017**, *530*, 455–467. [CrossRef] [PubMed]
25. Zhang, B.; Zhang, Y.; Dreisoerner, J.; Wei, Y. The effects of screw configuration on the screw fill degree and special mechanical energy in twin-screw extruder for high-moisture texturised defatted soybean meal. *J. Food Eng.* **2015**, *157*, 77–83. [CrossRef]
26. Rhim, J.W.; Wang, L.F.; Hong, S.I. Preparation and characterization of agar/silver nanoparticles composite films with antimicrobial activity. *Food Hydrocolloids* **2013**, *33*, 327–335. [CrossRef]
27. Wang, W.; Guo, Y.; Otaigbe, J.U. Synthesis, characterization and degradation of biodegradable thermoplastic elastomers from poly(ester urethane)s and renewable soy protein isolate biopolymer. *Polymer* **2010**, *51*, 5448–5455. [CrossRef]
28. Kumar, R.; Choudhary, V.; Mishra, S.; Varma, I.K.; Mattiason, B. Adhesives and plastics based on soy protein products. *Ind. Crop. Prod.* **2002**, *16*, 155–172. [CrossRef]
29. Steven, E.; Saleh, W.R.; Lebedev, V.; Acquah, S.F.A.; Laukhin, V.; Alamo, R.G.; Brooks, J.S. Carbon nanotubes on a spider silk scaffold. *Nat. Commun.* **2013**, *4*, 2435. [CrossRef] [PubMed]
30. Nguyen, J.; Baldwin, M.A.; Cohen, F.E.; Prusine, S.B. Protein peptides induce α-helix to β-sheet conformational transitions. *Biochemistry* **1995**, *34*, 4186–4192. [CrossRef]
31. Bonwell, E.S.; Wetzel, D.L. Innovative FTIR imaging of protein film secondary structure before and after heat treatment. *J. Agric. Food Chem.* **2009**, *57*, 10067–10072. [CrossRef]
32. Hu, X.; Kaplan, D.L.; Cebe, P. Dynamic protein–water relationships during β-sheet formation. *Macromolecules* **2008**, *41*, 3939–3948. [CrossRef]
33. Barth, A. The infrared absorption of amino acid side chains. *Prog. Biophys. Mol. Biol.* **2000**, *74*, 141–173. [CrossRef]
34. Ling, S.; Qi, Z.; Knight, D.P.; Shao, Z.; Chen, X. Synchrotron FTIR microspectroscopy of single natural silk fibers. *Biomacromolecules* **2011**, *12*, 3344–3349. [CrossRef] [PubMed]
35. Pal, G.K.; Nidheesh, T.; Suresh, P.V. Comparative study on characteristics and in vitro fibril formation ability of acid and pepsin soluble collagen from the skom pf catla (*Catla catla*) and rohu (*Labeo rohita*). *Food Res. Int.* **2015**, *76*, 804–812. [CrossRef] [PubMed]
36. Li, H.; Liu, B.L.; Gao, L.Z.; Chen, H.L. Studies on bullfrog skin collagen. *Food Chem.* **2004**, *84*, 65–69. [CrossRef]

37. Liu, D.; Zhang, L. Structure and properties of soy protein plastics plasticized with acetamide. *Macromol. Mater. Eng.* **2006**, *291*, 820–828. [CrossRef]
38. Eftink, M.R. The use of fluorescence methods to monitor unfolding transitions in proteins. *Biophys. J.* **1991**, *66*, 482–501. [CrossRef]
39. Xiang, B.Y.; Ngadi, M.O.; Simpson, B.K.; Simpson, M.V. Pulsed electric field induced structural modification of soy protein isolate as studied by fluorescence spectroscopy. *J. Food Process Preserv.* **2011**, *35*, 563–570. [CrossRef]
40. Karan, S.K.; Maiti, S.; Kwon, O.; Paria, V.; Maitra, A.; Si, S.K.; Kim, Y.; Kim, J.K.; Khatua, B.B. Nature driven spider silk as high energy conversion efficient biopiezoelectric nanogenerator. *Nano Energy* **2018**, *49*, 655–666. [CrossRef]
41. Zhou, L.; Fu, P.; Cai, X.; Zhou, S.; Yuan, Y. Naturally derived carbon nanofibers as sustainable electrocatalysts for microbial energy harvesting: A new application of spider silk. *Appl. Catal. B* **2016**, *188*, 31–38. [CrossRef]
42. Irimia-Vladu, M.; Głowacki, E.D.; Voss, G.; Bauer, S.; Sariciftci, N.S. Green and biodegradable electronics. *Mater. Today* **2012**, *15*, 340–346. [CrossRef]
43. Aradhya, S.V.; Venkataraman, L. Single-molecule junctions beyond electronic transport. *Nat. Nanotechnol.* **2013**, *8*, 399–410. [CrossRef] [PubMed]
44. Rosenberg, B. Electrical conductivity of proteins. *Nature* **1962**, *193*, 364–365. [CrossRef] [PubMed]
45. Jin, G.; Prabhakaran, M.P.; Kai, D.; Annamalai, S.K.; Arunachalam, K.D.; Ramakrishna, S. Tissue engineered plant extracts as nanofibrous wound dressing. *Biomaterials* **2013**, *34*, 724–734. [CrossRef] [PubMed]

Article

A Full Set of In Vitro Assays in Chitosan/Tween 80 Microspheres Loaded with Magnetite Nanoparticles

Jorge A Roacho-Pérez [1,†], Kassandra O Rodríguez-Aguillón [1,†], Hugo L Gallardo-Blanco [2], María R Velazco-Campos [2], Karla V Sosa-Cruz [3], Perla E García-Casillas [3], Luz Rojas-Patlán [2], Margarita Sánchez-Domínguez [4], Ana M Rivas-Estilla [1], Víctor Gómez-Flores [3], Christian Chapa-Gonzalez [3,*] and Celia N Sánchez-Domínguez [1,*]

1. Departamento de Bioquímica y Medicina Molecular, Facultad de Medicina, Universidad Autónoma de Nuevo León, Monterrey 64460, Mexico; jorge.roacho@uacj.mx (J.A.R.-P.); kora_g416@hotmail.com (K.O.R.-A.); ana.rivasst@uanl.edu.mx (A.M.R.-E.)
2. Departamento de Genética, Facultad de Medicina, Universidad Autónoma de Nuevo León, Monterrey 64460, Mexico; hugoleonid2011@icloud.com (H.L.G.-B.); roble.velazco@gmail.com (M.R.V.-C.); qcb.luz.rojas@gmail.com (L.R.-P.)
3. Instituto de Ingeniería y Tecnología, Universidad Autónoma de Ciudad Juárez, Ciudad Juárez 32310, Mexico; ksosacruz@yahoo.com (K.V.S.-C.); pegarcia@uacj.mx (P.E.G.-C.); victor.gomez@uacj.mx (V.G.-F.)
4. Centro de Investigación en Materiales Avanzados, S.C. (CIMAV, S.C.), Unidad Monterrey, Apodaca 66628, Mexico; margarita.sanchez@cimav.edu.mx
* Correspondence: christian.chapa@uacj.mx (C.C.-G.); celia.sanchezdm@uanl.edu.mx (C.N.S.-D.)
† These authors contributed equally to this work.

Abstract: Microspheres have been proposed for different medical applications, such as the delivery of therapeutic proteins. The first step, before evaluating the functionality of a protein delivery system, is to evaluate their biological safety. In this work, we developed chitosan/Tween 80 microspheres loaded with magnetite nanoparticles and evaluated cell damage. The formation and physical–chemical properties of the microspheres were determined by FT-IR, Raman, thermogravimetric analysis (TGA), energy-dispersive X-ray spectroscopy (EDS), dynamic light scattering (DLS), and SEM. Cell damage was evaluated by a full set of in vitro assays using a non-cancerous cell line, human erythrocytes, and human lymphocytes. At the same time, to know if these microspheres can load proteins over their surface, bovine serum albumin (BSA) immobilization was measured. Results showed 7 nm magnetite nanoparticles loaded into chitosan/Tween 80 microspheres with average sizes of 1.431 µm. At concentrations from 1 to 100 µg/mL, there was no evidence of changes in mitochondrial metabolism, cell morphology, membrane rupture, cell cycle, nor sister chromatid exchange formation. For each microgram of microspheres 1.8 µg of BSA was immobilized. The result provides the fundamental understanding of the in vitro biological behavior, and safety, of developed microspheres. Additionally, this set of assays can be helpful for researchers to evaluate different nano and microparticles.

Keywords: polymers; magnetite nanoparticles; chitosan; Tween 80; synthesis; nanotoxicology; genotoxicity; hemotoxicity

1. Introduction

Conventional methods for drug administration involve the use of tablets, capsules, and injections [1]. Although these methods are functional for the treatment of several health conditions, the complexity of human diseases has led scientists to develop new strategies for drug administration. One of the tissues with difficult access is the brain. Some disorders of the central nervous system such as stroke, brain tumors, and neurodegenerative diseases can be treated with therapeutic proteins. The delivery of these proteins is complicated because blood–brain barriers protect the brain from the entrance of outside substances [2].

Microspheres have been studied as a vehicle for the delivery of therapeutic agents [3]. Polymeric microspheres increase the therapeutic agents' useful lifetime because they protect the therapeutic agent from enzymatic degradation [4]. Chitosan microspheres have been studied and developed by several research groups as a hemostatic [5], antibacterial [6], and anti-inflammatory agent [7], and a vehicle for drug [8] and vaccine delivery [9]. Chitosan (CS), obtained after alkaline deacetylation of chitin, is a pH-sensitive natural polycationic polysaccharide. Some attractive properties of chitosan are its low cost, cationic charge, biocompatibility, and chemical stability [10]. Chitosan combined with magnetite nanoparticles have a broad field of application, but mainly they have been the subject of research in the supply of drugs, proteins, peptides, genes, DNA, and others [11–13]. Magnetic nanoparticles can target the microsphere to a specific tissue, using an external magnetic field [14]. Some magnetic nanomaterials have reached clinical trials, and others are available in the market for their use in humans [15–19]. The most used magnetic nanoparticles are iron oxides, such as magnetite (Fe_3O_4), because of its high degree of biocompatibility and unique magnetic properties that can be controlled by the synthesis methods [20]. Another material that can be used for the formation of the microspheres is the non-ionic surfactant polysorbate 80 (Tween® 80). The novelty of Tween 80 is attributed to its ability to be targeted into the brain after intravenous injection. Tween 80 binds to the plasma lipoprotein Apo-E, which attaches to LDL receptors on brain microvascular endothelial cells. Therefore, Tween 80 can deliver pharmaceutical substances into the brain [21,22].

Before testing the functionality of microspheres, it is essential to develop and apply in vitro tests that can predict the potential toxicity of particles. It is fundamental to evaluate the biological safety of particles to ensure their biocompatibility, especially because different nanoparticle synthesis methodologies generate different physical properties that lead to distinct biological behaviors. The toxicity of nanoparticles occurs at molecular and cellular levels. When particles move through the body, they interact with different biological environments such as blood, extracellular matrix, cytoplasm, cell organelles, and nucleus. Consequently, this interaction can affect some biomolecules and cellular components [23]. The exposed biomolecules can suffer structural and functional alterations [24,25]. Nanoparticles can also cause cytotoxicity by breaking several membranes outside and inside the cell. When the membrane integrity is compromised, the content of the membrane compartments can leak. This leak can cause cell stress and interfere with the normal function of the cell and organelles [26,27].

Although several papers evaluate the cytotoxicity of chitosan microspheres, different synthesis methodologies generate different physical properties that interfere with the biological behavior of microspheres. The main objective of this work was to determine if chitosan/Tween 80 microspheres loaded with magnetite nanoparticles can cause cell damage. We evaluated cell metabolic activity (MTT assay), morphological cell changes (H&E staining), cell lysis of membrane (hemolysis test), chromosome changes (SCE test), and evaluation of the cell cycle (lymphocyte culture).

2. Materials and Methods

The Ethics in the Research Committee of the School of Medicine and Dr. José Eleuterio Gonzalez University Hospital of the Universidad Autónoma de Nuevo León reviewed and approved this methodology in September 2017, with the project identification code BI17-00001. All human donors were treated according to ethical standards.

2.1. Chitosan/Tween 80 Microspheres Loaded with Magnetite Nanoparticles Preparation

Magnetite nanoparticles were synthetized by chemical coprecipitation based on the methodology used in this research previously reported [28]. The preparation of the microspheres is shown in Figure 1. A sample of 90 mg of magnetite nanoparticles was dispersed in 45 mL of a 7% Tween 80 solution (Sigma-Aldrich, St. Louis, MO, USA) using an ultrasonic processor (Fisher Scientific, Pittsburgh, PA, USA) amplitude 80%. A total of

90 mg of medium molecular weight chitosan (Sigma-Aldrich, St. Louis, MO, USA) was dissolved until homogenization in 45 mL of a 2% glacial acetic acid solution (Thermo Fisher Scientific, Waltham, MA, USA). Chitosan solution was added drop by drop into the magnetite nanoparticles solution and mixed with an ultrasonic processor. In order to induce the protonation of the chitosan, the pH of the mix was adjusted to 5.5 with ammonium hydroxide (NH_4OH, CTR Scientific, Monterrey, Mexico). Magnetite-CS/Tween 80 microspheres were isolated by centrifugation at 14,000 rpm for 10 min and washed with distilled water repeatedly until magnetite-CS/Tween 80 microspheres reached pH 7. Magnetite-CS/Tween 80 microspheres were dried at 50 °C for 24 h and pulverized in an agate mortar and stored.

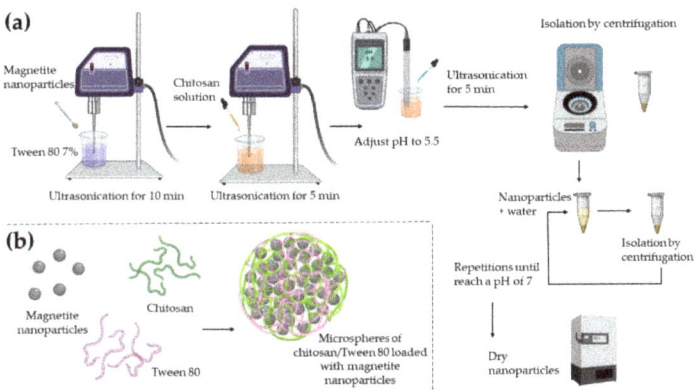

Figure 1. A scheme cartoon of the preparation of the microspheres. (**a**) A complete diagram of the methodology; (**b**) The interaction of chitosan and Tween 80 generate microspheres, which load the magnetite nanoparticles into a polymer network. Created with BioRender.com.

2.2. Nanoparticle Characterization

Functional group analysis was evaluated by Fourier transform infrared spectroscopy (FT-IR NICOLET 6700, Thermo Fisher Scientific, Waltham, MA, USA), and Raman spectra were obtained in an instrument equipped with a 785 nm laser source (Alpha 300 RA, WITec, Ulm, Germany). While, thermogravimetric analysis from room temperature to 700 °C was accomplished on a SDT 2960 Simultaneous DSC-TGA (TA Instrument, New Castle, DE, USA). The samples' morphology was examined with scanning electron microscopy (SEM JSM-7000F, JEOL, Akrishima, Japan). For SEM analysis, the different samples were dispersed in distilled water and then spread on a carbon tape slide. Once the sample had dried, a secondary electron image (SEI) and energy-dispersive X-ray spectroscopy (EDS) study were performed. The samples' distribution size was measured with dynamic light scattering (DLS) equipment (Nanotrack Wave II, Microtrac, Haan, Germany). In order to avoid nanoparticle agglomeration, nanoparticle samples were prepared immediately before their analysis [29]. For DLS analysis, nanoparticles and microspheres were dispersed (1 mg/mL) into previously filtered distilled water.

2.3. Cytotoxicity: MTT Assay and H&E Staining

For the measurement of the microspheres' capacity to cause cell death a 3-(4,5-dimethylthiazol-2-yl)-2,5-diphenyltetrazolium bromide (MTT) assay and a hematoxylin–eosin (H&E) staining was developed using the embryo mouse fibroblast cell line 3T3L1 (ATCC, Manassas, VA, USA). A total of 10,000 cells were seeded per well, in a 96-well microplate, and cultured with supplemented DMEM. The supplemented DMEM medium contained Dulbecco's modified Eagle's medium (Thermo Fisher Scientific, Waltham, MA, USA), 10% of fetal bovine serum (Thermo Fisher Scientific, Waltham, MA, USA), and

1% of penicillin-streptomycin (Thermo Fisher Scientific, Waltham, MA, USA). After 24 h, cells were exposed to the microspheres previously sterilized by 15 min of UV light exposure. The microspheres were evaluated at different concentrations by triplicate from 1 to 10,000 µg/mL. Cells without exposition to any particle were taken as a negative control. Cell death was measured after 24, 48, and 72 h of exposure.

For the MTT assay, cells were processed with the Cell Proliferation Kit I (Roche, Basilea, Switzerland) following the instructions. Absorbance was read by UV–vis spectroscopy (Nanodrop, Thermo Fisher Scientific, Waltham, MA, USA) at 570 nm. For the results analysis, the negative control absorbance was adjusted as 100% of cell viability. The calculation of cell viability of the treated cells was calculated according to the negative control. Thereafter H&E staining was performed. The medium was removed, and cells were washed with PBS (Thermo Fisher Scientific, Waltham, MA, USA) three times. For the fixation of the cells, each well was incubated 10 min at −20 °C with 50 µL of cold methanol. Cells were rewashed with PBS three times. For the stain, cells were incubated 5 min in hematoxylin, tap water wash, HCl (diluted at 0.5% in ethanol) wash, tap water wash, distilled water wash, 5 min in eosin, and a tap water wash. Morphology of the cells was analyzed in an optical microscope (Inverted microscopeCKX41, Olympus, Shinjuku, Japan).

2.4. Hemolysis Test

A hemolysis test was developed to determine the effect of microspheres on erythrocyte lysis. A heparinized tube (Becton Dickinson, Franklin Lakes, NJ, USA) with blood from a healthy donor was used. Erythrocytes were isolated by centrifugation (3000 rpm for 4 min) and washed three times with Alsever's solution (dextrose 0.116 M, sodium chloride 0.071 M, sodium citrate 0.027 M, and citric acid 0.002 M, pH 6.4). Microspheres were dispersed in Alsever's solution at different concentrations (1–10,000 µg/m). Quintupled samples of microspheres were incubated with the erythrocytes in a relation of 1:99 erythrocytes: microspheres v/v. The incubation conditions were 37 °C in agitation at 400 rpm for 30 min. The positive control was a suspension of erythrocytes in distilled water. Negative control was a suspension of erythrocytes in Alsever's solution. After incubation, samples were centrifuged (3000 rpm for 4 min), and the hemoglobin released in the supernatant was measured by UV–Vis spectroscopy at 415 nm. The absorbance of the positive control was adjusted as 100% of hemolysis, and each of the samples were calculated according to the positive control.

2.5. Sister Chromatid Exchange Assay

For this analysis, a primary culture of human lymphocytes was used. Blood from a healthy donor was collected in a heparinized tube. An aliquot of 500 µL of blood was cultured in 5 mL of RPMI-1640 culture medium (Thermo Fisher Scientific, Waltham, MA, USA), 100 µL of phytohemagglutinin for mitosis induction (Thermo Fisher Scientific, Waltham, MA, USA), and microspheres at different concentrations (from 1 to 100 µg/mL). For the positive control, mitomycin C 40 ng/mL (Sigma-Aldrich, St. Louis, MO, USA) was added. After 24 h of culture, bromodeoxyuridine, a synthetic thymidine analog nucleoside, (BrdU, Sigma-Aldrich, St. Louis, MO, USA) was added to the medium at a final concentration of 23.1 µM. In order to stop mitosis in metaphase, after 48 h of incubation with BrdU, KaryoMax Colcedim were added to the medium at a final concentration of 10 µg/mL (Thermo Fisher Scientific, Waltham, MA, USA). After 40 min of incubation with colcedim, cells were centrifuged for 10 min at 1200 rpm. Supernatant was eliminated and cells were incubated with a hypotonic solution of KCl 0.075 M for 20 min, and were washed with a solution of acetic acid and methanol (1:3, v/v) until the precipitate turned white. The cell suspension was dropped into a clean microscope slide for fixation. Cells over the slides were incubated for 40 min with bisbenzimide (Sigma-Aldrich, St. Louis, MO, USA) to stain A-T rich regions of the chromosome. After this incubation, samples were exposed to UV light for 105 min and covered with a saline-sodium citrate buffer $0.5\times$ for 105 min. Finally, samples were covered with Giemsa stain at 2% in PBS, pH 6.8. Samples were dried and analyzed in an optical microscope. Segments of genetic material exchanged from one chromatid to another were counted in 50 metaphases per sample.

2.6. Surface Protein Load

Bovine serum albumin (BSA) was used as a model molecule for measuring protein load over the microspheres surface. With this experiment it can be demonstrated that the synthesized microspheres can immobilize a protein over their surface. The loaded protein can be replaced with other different protein with therapeutic properties for brain diseases. Triplicated samples of 500 µg of microspheres were incubated in agitation for 30 min in 300 mL of a PBS solution (pH 7.4, room temperature) with BSA (10 different amounts from 879 to 8799 µg). After incubation, microspheres were precipitated using a magnetic plate. The non-immobilized proteins were taken from the supernatant and measured by UV–Vis spectroscopy at 280 nm.

2.7. Statistical Analysis

Biological assays data were studied by an analysis of variance (ANOVA) and a Tukey's HSD (honestly significant difference) test, with a confidence interval of 95% to determine significant differences between the control group and the test samples.

3. Results and Discussion

3.1. Microsphere Characterization: FT-IR Spectroscopy

FTIR spectra of bare magnetite and chitosan/Tween 80 microspheres loaded with magnetite nanoparticles are shown in Figure 2. This characterization is based on the vibrations of the chemical bonds of the analyzed samples. It means that FTIR evidences the functional groups present in the sample. Bare magnetite nanoparticles show the main band at 547 cm^{-1}, also reported by other authors, corresponding to Fe-O stretching in the octahedral site of the crystal structure of magnetite [30–32]. The microspheres spectrum shows more bands due to the different functional groups present in the chitosan and Tween 80. The spectrum shows bands at 3453 cm^{-1} and 1637 cm^{-1} correspond to vibrational stretching and bending modes of the hydroxyl group. The bands at 2921 cm^{-1} and 2852 cm^{-1} correspond to C-H asymmetrical and symmetrical, respectively, stretching vibrations. The band at 1267 cm^{-1} was related to the C-H bending mode. The band at 1186 cm^{-1} was related to the stretching vibration of the C-O link. The band observed at 1083 cm^{-1} corresponds to the asymmetric stretch vibration of the C-O-C bond. The band at 818 cm^{-1} corresponds to a C-H deformation. The band that corresponds only to Tween 80 is presented at 1741 cm^{-1}, which indicates a C=O stretching vibration. The bands that correspond only to the chitosan are the vibratory stretching mode of C-N observed at 1468 cm^{-1} and the bending mode of N-H at 1508 cm^{-1} [33–37].

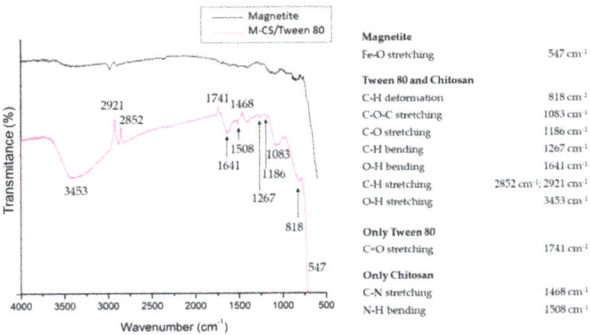

Figure 2. FTIR spectra of bare magnetite and chitosan/Tween 80 microspheres loaded with magnetite nanoparticles (M-CS/Tween 80).

3.2. Microsphere Characterization: Raman Spectroscopy

The Raman signal obtained from the microspheres is shown in Figure 3. The strongest bands showed in the spectrum correspond to the inorganic compound. According with the Bio-Red's HORIBA library Edition KnowItAll, the bands shown at 213.5, 277.6, 389.7, 584.4, and 1249 cm^{-1} correspond to the spectrum RMX #265 Magnetite (heated). This spectrum is the result of the changes produced in the magnetite by the heat of the power laser [38,39]. The intensity of the polymeric organic compounds is lower than the inorganic part, but some bands were identified for the bending vibrations of C-NH-C (277.6 cm^{-1}), COC (469 cm^{-1}), CO-NH (497 cm^{-1}), C-CH$_3$ (497 cm^{-1}), and CH (1249 cm^{-1}). Additionally, the band at 1249 cm^{-1} can be ascribed to the stretching of C-C and C-O [40,41].

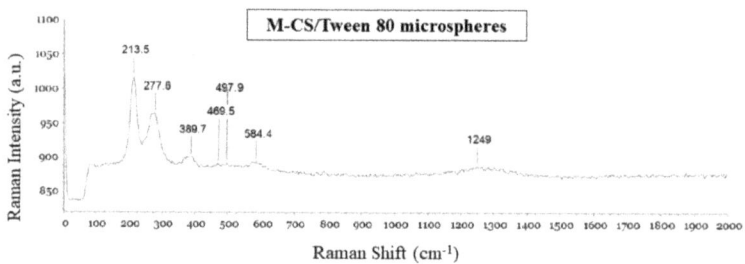

Figure 3. Raman spectrum of chitosan/Tween 80 microspheres loaded with magnetite nanoparticles (M-CS/Tween 80).

3.3. Microsphere Characterization: TGA

Thermal analysis provides information about the polymeric composition of the sample. The results obtained from magnetite nanoparticles and microspheres of chitosan/Tween 80 loaded with magnetite nanoparticles are shown in Figure 4. Magnetite nanoparticles showed a loss of 4% of their weight, possibly because of a residual water loss. Microspheres showed a 16.7% of their weight loss at two different temperatures. Tween 80 weight loss temperature is shown at two different temperatures near 250 and 350 °C respectively [42]. The weight loss at a temperature of 300 °C is given by the degradation of the chitosan [43]. The spectra confirm the presence of both components in the sample.

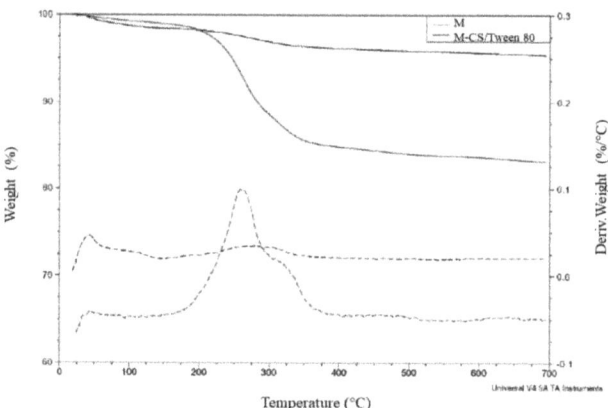

Figure 4. TGA spectrum. The weight loss of microspheres, from 250 to 350 °C, is given by the presence of chitosan and Tween 80.

3.4. Microsphere Characterization: SEM-EDS

To ensure that M-CS/Tween 80 microspheres have adopted the desired spherical morphology, they were observed through SEM. It is observed in Figure 5a that M-CS/Tween 80 microspheres adopt the expected spherical morphology. Images show spherical nanoparticles inside the chitosan microspheres. An EDS study was developed to ensure that the nanoparticles inside the chitosan microsphere correspond to magnetite. Figure 5b shows the distribution of Fe inside the microspheres. Fe distribution corresponds with the nanoparticles inside the microspheres, proving that magnetite nanoparticles are inside the chitosan microspheres.

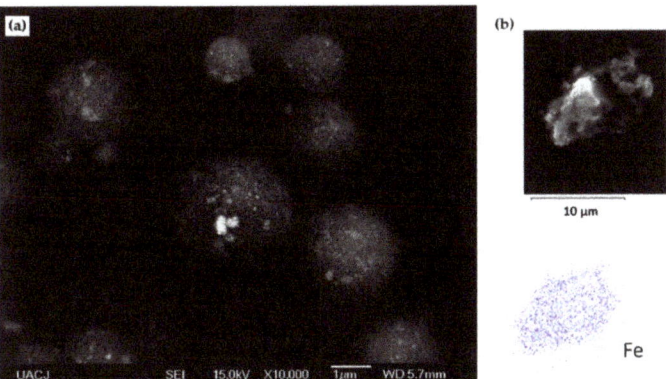

Figure 5. (a) SEM images of M-CS/Tween 80 microspheres at ×10,000; (b) EDS shows the distribution of Fe in the microsphere.

3.5. Microspheres Characterization: DLS

The size distribution of magnetic nanoparticles (Figure 6a) and M-CS/Tween 80 microspheres (Figure 6b) was measured using the DLS technique. Both structures show sizes with a normal distribution. The average size of the magnetite nanoparticles obtained was 7 nm. This size is consistent with that reported by authors for chemical coprecipitation by the rapid injection synthesis method [28]. The average size of 98.8% of M-CS/Tween 80 microspheres was 1.431 µm, with a range size that goes from 1.431 to 1.756 µm. Only 1.2% of synthesized microspheres show average sizes of 6 µm. The size could correspond to agglomerates due to the polydispersity index (PDI) being 0.77 for magnetite nanoparticles and 0.22 for M-CS/Tween 80 microspheres.

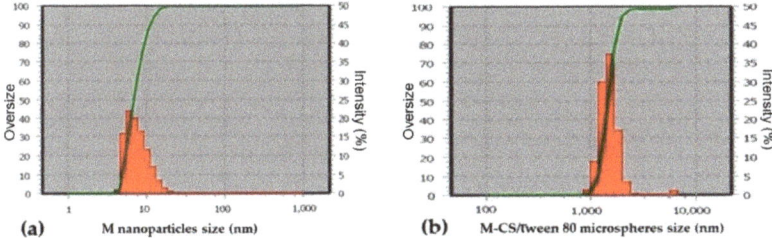

Figure 6. (a) Magnetite nanoparticle and (b) M-CS/Tween 80 microsphere size distribution. Polydispersity index (PDI) 0.77 and 0.22, respectively.

3.6. Cytotoxicity of M-CS/Tween 80 Microspheres: MTT Assay

Cytotoxicity tests evaluate cell damage, growth, and specific aspects of cell metabolism. The MTT assay is based on measuring the metabolic activity of cells. MTT is a yellow and water-soluble agent; the viable cells reduce it into a blue-violet insoluble compound called formazan. The number of viable cells is correlated with the presence of formazan in the sample [44]. Evaluation of the cytotoxicity of magnetite nanoparticles and M-CS/Tween 80 microspheres was measured by exposure of 3T3L1 cells, followed by the MTT assay. The evaluation was performed after 24, 48, and 72 h of exposure. Figure 7 shows a comparison between the cytotoxic effect of magnetite nanoparticles vs. M-CS/Tween 80 microspheres. An increment in particle concentration increments cytotoxicity. An increment in time exposure also increments particle cytotoxicity. Statistical analysis ($p = 0.05$) showed that M-CS/Tween 80 microspheres are less cytotoxic in comparison with bare magnetite nanoparticles. This result demonstrates the non-toxic properties of chitosan and Tween 80. At the first 24 h of exposure, M-CS/Tween 80 microspheres showed a non-cytotoxic behavior at concentrations from 1 to 100 µg/mL. In summary, this experiment showed that the use of chitosan and Tween 80 improved the magnetite biological behavior but was only safe to use at concentrations lower than 100 µg/mL, and preferably, in short exposure times.

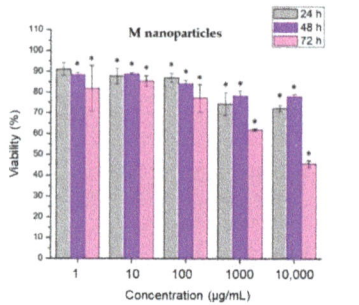 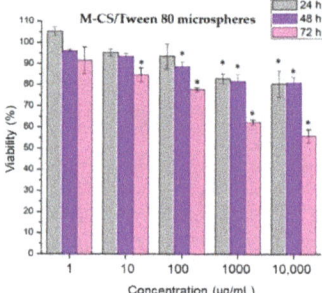

Figure 7. MTT assay of bare magnetite nanoparticles and chitosan/tween 80 microspheres loaded with magnetite nanoparticles. There is a significant difference * of particles cytotoxicity concerning negative controls.

The results obtained by measuring the viability of 3T3L1 cells after exposure with magnetite nanoparticles and M-CS/Tween 80 microspheres at concentrations of 1–10,000 µg/mL were compared by those obtained by Lotfi et al. In their study, they used concentrations of 10, 25, 50, 75, and 100 µg/mL of magnetite nanoparticles coated with chitosan for 24 h in MCF7 cells and fibroblasts. They obtained 78% cell viability for bare magnetite nanoparticles and a cell viability of 80% for magnetite coated with chitosan in the MCF7 cell line [35]. Viabilities obtained by our research group at concentrations of 100 µg/mL at the first 24 h are 87% for magnetite nanoparticles and 93% for M-CS/Tween 80 microspheres. It is essential to evaluate the cytotoxicity of the particles. Changes in the properties of the nanoparticles are due to batch-to-batch variations, which are reflected in their interaction with the cells. Additionally, it must be considered that the results reflect the specific response of each cell line. Additionally, results obtained in this work give us a screening of particles' interaction in 48 and 72 h. The cytotoxicity of these particles increased with an increment in time exposure. Therefore, it is recommended not to use them in applications when a long-time exposure is needed.

3.7. Cytotoxicity of M-CS Microspheres: H&E Staining

To observe morphological changes in the 3T3L1 cells due to their exposure with particles, H&E staining was performed. Figure 8 shows images from an optical microscope at 40× after 24 h of exposure to the particles. Cell morphology changes and a decrease in cell confluence were observed at concentrations of 1000 and 10,000 µg/mL with both types of particles. At concentrations of 1, 10, and 100 μg/mL, there were no evident changes in cytoplasm or nucleus morphology. H&E results are comparable with the results of the MTT assay. Figures 7 and 8 show the changes generated by the microspheres and nanoparticles in 2D cell culture models. However, a possible improvement is the in vitro evaluation of cytotoxicity in 3D models. The 3D models mimic the tissue microenvironment. Authors suggest that microspheres proposed for antitumoral applications should also be tested in a tumor spheroid to elucidate the intracellular and intercellular signaling of cancer [45,46].

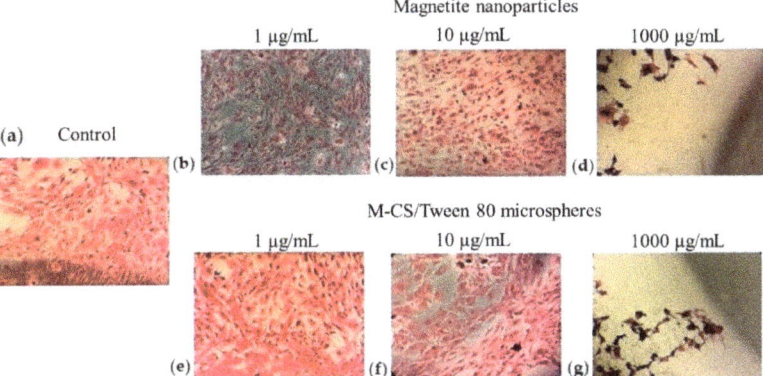

Figure 8. Optical microscope images of 40× of H&E staining at 24 h. (**a**) Negative control with no particles; (**b–d**) Magnetite nanoparticles and (**e–g**) M-CS/Tween 80 microspheres at different concentrations.

3.8. Hemolysis Assay

When the outer membrane of the erythrocytes is destroyed, hemoglobin is released. The lysis produces the release of the intracellular content of the red cells. The number of erythrocytes destroyed is estimated by measuring the amount of hemoglobin released in a sample. Hemolysis assays gives a reference to know the degree of toxicity of the nanoparticles that will be in contact with human blood. Concentrations from 1 to 10,000 μg/mL of bare magnetite nanoparticles and M-CS/Tween 80 microspheres were evaluated. The results were analyzed in accordance with the Standard Practice for Assessment of Hemolytic Properties of Materials ASTM F 756-08, which indicates that any material with a hemolysis rate of less than 2% is considered non-hemolytic [47]. On one hand, at a concentration of 10,000 μg/mL hemolysis produced by bare magnetite nanoparticles is higher than 2%, at this concentration it is considered a hemolytic material (Figure 9). Concentrations from 1 to 1000 μg/mL have a hemolysis rate lower than 2%, at those concentrations magnetite nanoparticles are considered a non-hemolytic material. On the other hand, M-CS/Tween 80 microspheres are considered non-hemolytic material at all concentrations tested. This experiment indicates that chitosan and Tween 80 improve the biological properties of magnetite nanoparticles.

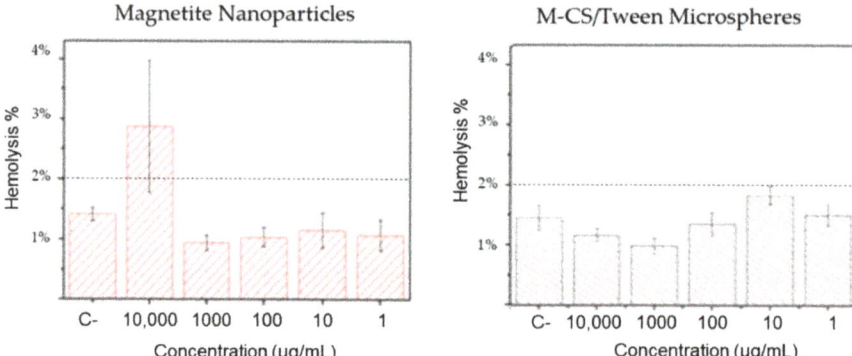

Figure 9. Hemolysis test results of bare magnetite nanoparticles and chitosan/tween 80 microScheme 2016. tested concentrations of 250, 500, and 3000 µg/mL of bare magnetite nanoparticles, reporting hemolysis rates below 2% at all concentrations [47]. In the present study, concentrations Figure 1 to 10,000 µg/mL of bare magnetite nanoparticles and M-CS/Tween 80 nanospheres were tested. Hemolysis rates obtained from magnetite nanoparticles are similar to the results of Macias et al.

3.9. Sister Chromatid Exchange

The formation of sister chromatid exchange (SCE) originated by DNA lesions has been a research subject for a long time. SCE is an exchange between segments of DNA in homologous loci from a chromosome. SCE formation is a normal cellular event, with a constant low frequency. An abnormal frequency of SCE is correlated with the production of chromosomal aberrations. The most accepted SCE formation mechanism is based on a double-strand break that occurs during DNA replication. Some environmental agents can cause a pathologic homologous recombination between the sister chromatids of chromosomes [48]. It is essential to evaluate if magnetite nanoparticles or M-CS/Tween 80 microspheres can cause this kind of exchange between genetic material because some changes in the chromosome sequence can cause several pathologies such as cancer. Turkez et al. evaluated the formation of SCE in cells exposed with nanoparticles of hydroxyapatite. They reported an average of 6.6 ± 1.4 SCE at concentrations of 100 µg/mL, and an average of 5.9 ± 1 SCE at concentrations of 10 µg/mL. They established that below 10 SCE, there is no significant damage in DNA or DNA-repair enzymes [49]. In this project, concentrations of 1, 10, and 100 µg/mL of magnetite nanoparticles and M-CS/Tween 80 microspheres were tested. Figure 10a shows an image of the chromosomes from one lymphocyte exposed to the positive control mitomycin C. Figure 10b shows the results of the number of SCEs generated in 50 samples for each different particle concentration tested. All the lymphocytes exposed with magnetite nanoparticles and with M-CS/Tween 80 microspheres shown less than 5 SCE per sample. In the positive control, the number of SCE was 23.38 ± 5.02, significantly different ($p = 0.05$) to all the tested samples. Results indicate that bare magnetite nanoparticles and M-CS/Tween 80 microspheres did not produce SCE chromosomal changes.

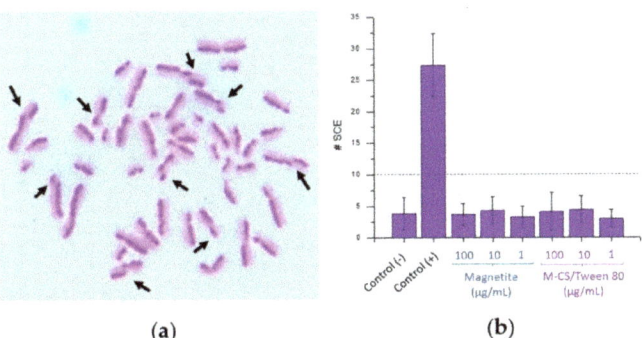

Figure 10. Sister chromatid exchange (SCE) analysis of lymphocytes exposed to microspheres. (**a**) An SCE positive sample (mitomycin C). The exchanges between one sister chromatid to another are witnessed because of the color. (**b**) The number of SCE found in all the samples.

3.10. Damage in The Proliferation Mechanism

The proliferation index (PI) of the cultivated lymphocytes was analyzed using Equation (1)

$$PI = [M1 + 2(M2) + 3(M3)]/n \tag{1}$$

where M1, M2, and M3 means the number of lymphocytes found in the first, second, and third cycle of mitosis, respectively; and n is the total of scored cells. Mitomycin C was used as a positive control of proliferation damage. Data was analyzed based of previous reports from different authors [50,51]. As Figure 11 shows, there was a significative difference ($p = 0.05$) between the positive control versus the negative control. None of the tested samples showed a significant difference ($p = 0.05$) against the negative control. Thus, results show that any particle tested caused changes in the normal cell cycle.

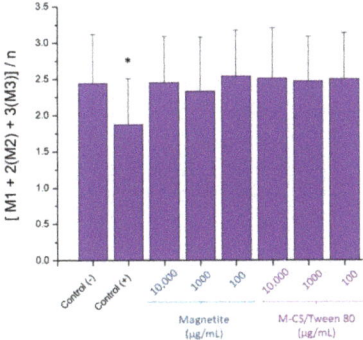

Figure 11. Proliferation index results of bare magnetite nanoparticles and chitosan/tween 80 microspheres loaded with magnetite nanoparticles. There was a significative difference ($p = 0.05$) between the positive control * versus the negative control.

3.11. Protein Load

Microspheres were incubated with the BSA protein at different concentrations. The data analysis and interpretation were done based on other authors' work [52,53]. Figure 12 shows that 500 μg of M-CS/Tween 80 microspheres could load onto their surface 900 μg of BSA protein. This result means that 1 μg of microspheres could load an average of 1.8 μg of BSA protein on their surface. These microspheres could be potentially redesigned to load different therapeutic proteins with similar physical properties than BSA.

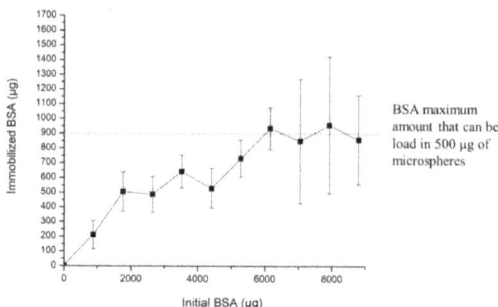

Figure 12. Different amounts of immobilized BSA are shown for different initial amounts of BSA.

4. Conclusions

In this study, we developed chitosan/Tween 80 microspheres loaded with magnetite nanoparticles, and these were characterized by FTIR, Raman, TGA, SEM, EDS, and DLS. The main point of this study was to determine the cell damage caused by the developed microspheres. The results confirm the safety of the use of chitosan/Tween 80 microspheres loaded with magnetite nanoparticles in vitro. There was no evidence of significant changes in mitochondrial metabolic activity, cell morphology, membrane lysis, sister chromatid exchanges, or the cell cycle. The authors recommend using a full set of assays to measure the cytotoxicity of nanoparticles because one single assay does not determine damage in different cell components. Although, these microspheres could be loaded with proteins with similar physical proprieties to BSA. For all the above, these materials represent a viable option in protein transport and release. The present study provides a fundamental understanding of the biological in vitro behavior of developed microspheres. This knowledge will open different research lines for the use of this microspheres as a target protein delivery system. The presented set of assays in this paper can help researchers to evaluate different nano and microparticles.

Author Contributions: Conceptualization, J.A.R.-P., K.O.R.-A., H.L.G.-B., C.C.-G. and C.N.S.-D.; methodology, J.A.R.-P., K.O.R.-A., M.R.V.-C., K.V.S.-C., L.R.-P. and V.G.-F.; investigation, J.A.R.-P., K.O.R.-A. and K.V.S.-C.; resources, M.R.V.-C., P.E.G.-C., A.M.R.-E., C.C.-G. and C.N.S.-D.; writing—original draft preparation, J.A.R.-P.; writing—review and editing, K.O.R.-A., H.L.G.-B., M.R.V.-C., L.R.-P., M.S.-D., C.C.-G. and C.N.S.-D.; visualization, C.C.-G. and C.N.S.-D.; supervision, H.L.G.-B., P.E.G.-C., C.C.-G. and C.N.S.-D.; funding acquisition, H.L.G.-B. All authors have read and agreed to the published version of the manuscript.

Funding: CONACYT funded this research, Call for Basic Research 2017–2018, grant number A1-S-9859.

Institutional Review Board Statement: The Ethics in the Research Committee of the School of Medicine and José Eleuterio Gonzalez University Hospital of the Universidad Autónoma de Nuevo León reviewed and approved this methodology in September 2017, with the project identification code BI17-00001.

Informed Consent Statement: All human donors were treated according to ethical standards.

Data Availability Statement: The data presented in this study are available on request from the corresponding author.

Acknowledgments: The authors want to acknowledge the support given by student Fernando G Ruiz-Hernández (Medical student, UANL), Sergio Lozano-Rodríguez (Scientific publications support coordinator, UANL), Tomás Galván-Alférez (Engr. student, UACJ), Itzel M Morales-Rojas (Engr. student, UACJ), Lorena Rivera-Ríos (Materials laboratory responsible, UACJ), Jorge I Betance-Salcido (SEM technician, UACJ) and Oswaldo Sánchez-Dena (PostDoc researcher, UACJ). Thank to NANOMED-UACJ research group for the feedback received during the research.

Conflicts of Interest: The authors declare no conflict of interest.

References

1. Dhamecha, D.; Movsas, R.; Sano, U.; Menon, J.U. Applications of alginate microspheres in therapeutics delivery and cell culture: Past, present and future. *Int. J. Pharm.* **2019**, *569*, 118627. [CrossRef] [PubMed]
2. Calias, P.; Banks, W.A.; Begley, D.; Scarpa, M.; Dickson, P. Intrathecal delivery of pretein therapeutics to the brain: A critical reassessment. *Pharmacol. Ther.* **2014**, *144*, 114–122. [CrossRef] [PubMed]
3. Ding, C.; Bi, H.; Wang, D.; Kang, M.; Tian, Z.; Zhang, Y.; Wang, H.; Zhu, T.; Ma, J. Preparation of Chitosan/Alginate-ellagic Acid Sustained-release Micro-spheres and their Inhibition of Preadipocyte Adipogenic Differentiation. *Curr. Pharm. Biotechnol.* **2019**, *20*, 1213–1222. [CrossRef]
4. Sinha, V.R.; Singla, A.K.; Wadhawan, S.; Kaushik, R.; Kumria, R.; Bansal, K.; Dhawan, S. Chitosan microspheres as a potential carrier for drugs. *Int. J. Pharm.* **2004**, *274*, 1–33. [CrossRef] [PubMed]
5. Liu, L.; Lv, Q.; Zhang, Q.; Zhu, H.; Liu, W.; Deng, G.; Wu, Y.; Shi, C.; Li, H.; Li, L. Preparation of carboxymethyl chitosan microspheres and their application in hemostasis. *Disaster Med. Public Health Prep.* **2017**, *11*, 660–667. [CrossRef]
6. Shen, J.; Jin, B.; Qi, Y.C.; Jiang, Q.Y.; Gao, X.F. Carboxylated chitosan/silver-hydroxyapatite hybrid microspheres with improved antibacterial activity and cytocompatibility. *Mater. Sci. Eng. C* **2017**, *78*, 589–597. [CrossRef]
7. Cosco, D.; Failla, P.; Costa, N.; Pullano, S.; Fiorillo, A.; Mollace, V.; Fresta, M.; Paolino, D. Rutin-loaded chitosan microspheres: Characterization and evaluation of the anti-inflammatory activity. *Carbohydr. Polym.* **2016**, *152*, 583–591. [CrossRef]
8. Yu, H.L.; Feng, Z.Q.; Zhang, J.J.; Wang, Y.H.; Ding, D.J.; Gao, Y.Y.; Zhang, W.F. The evaluation of proanthocyanidins/chitosan/lecithin microspheres as sustained drug delivery system. *Biomed. Res. Int.* **2018**, *2018*. [CrossRef]
9. Arthanari, S.; Mani, G.; Peng, M.M.; Jang, H.T. Chitosan-HPMC-blended microspheres as a vaccine carrier for the delivery of tetanus toxoid. *Artif. Cells Nanomed. Biotechnol.* **2016**, *44*, 517–523. [CrossRef]
10. Vaghari, H.; Jafarizadeh-Malmiri, H.; Berenjian, A.; Anarjan, N. Recent advances in application of chitosan in fuel cells. *Sustain. Chem. Process.* **2013**, *1*, 16. [CrossRef]
11. Shaterabadi, Z.; Nabiyouni, G.; Soleymani, M. High impact of in situ dextran coating on biocompatibility, stability and magnetic properties of iron oxide nanoparticles. *Mater. Sci. Eng. C* **2017**, *75*, 947–956. [CrossRef] [PubMed]
12. Nirala, N.R.; Harel, Y.; Lellouche, J.P.; Shtenberg, G. Ultrasensitive haptoglobin biomarker detection based on amplified chemiluminescence of magnetite nanoparticles. *J. Nanobiotechnol.* **2020**, *18*, 6. [CrossRef] [PubMed]
13. Taghizadeh, S.; Ebrahiminezhad, A.; Ghoshoon, M.B. Magnetic immobilization of *Pichia pastoris* cells for the production of recombinant human serum albumin. *Nanomaterials* **2020**, *10*, 111. [CrossRef] [PubMed]
14. Frounchi, M.; Shamshiri, S. Magnetic nanoparticles-loaded PLA/PEG microspheres as drug carriers. *J. Biomed. Mater. Res. Part A* **2015**, *103*, 1893–1898. [CrossRef] [PubMed]
15. Caster, J.M.; Patel, A.N.; Zhang, T.; Wang, A. Investigational nanomedicines in 2016: A review of nanotherapeutics currently undergoing clinical trials. *Wiley Interdiscip. Rev. Nanomed. Nanobiotechnol.* **2017**, *9*. [CrossRef] [PubMed]
16. Grootendorst, D.J.; Jose, J.; Fratila, R.M.; Visscher, M.; Velders, A.H.; Ten, B. Evaluation of superparamagnetic iron oxide nanoparticles (Endorem®) as a photoacoustic contrast agent for intra-operative nodal staging. *Contrast Media Mol. Imaging.* **2013**, *8*, 83–91. [CrossRef]
17. Johannsen, M.; Gneveckow, U.; Thiesen, B.; Taymoorian, K.; Cho, C.H.; Waldofner, N.; Scholz, R.; Jordan, A.; Loening, S.A.; Wust, P. Thermotherapy of prostate cancer using magnetic nanoparticles: Feasibility, imaging, and three-dimensional temperature distribution. *Eur. Urol.* **2007**, *52*, 1653–1661. [CrossRef]
18. Karakatsanis, A.; Christiansen, P.M.; Fischer, L.; Hedin, C.; Pistioli, L.; Sund, M.; Rasmussen, N.R.; Jornsgard, H.; Tegnelius, D.; Erisson, S.; et al. The Nordic SentiMag trial: A comparison of super paramagnetic iron oxide (SPIO) nanoparticles versus Tc(99) and patent blue in the detection of sentinel node (SN) in patients with breast cancer and a meta-analysis of earlier studies. *Breast Cancer Res. Treat.* **2016**, *157*, 281–294. [CrossRef]
19. Maier-Hauff, K.; Rothe, R.; Scholz, R.; Gneveckow, U.; Wust, B.; Thiesen, B.; Feussener, A.; Von Deimling, A.; Waldoefner, N.; Felix, R.; et al. Intracranial thermotherapy using magnetic nanoparticles combined with external beam radiotherapy: Results of a feasibility study on patients with glioblastoma multiforme. *J. Neurooncol.* **2007**, *81*, 53–60. [CrossRef]
20. Chomoucka, J.; Drbohlavova, J.; Huska, D.; Adam, V.; Kizek, R.; Hubalek, J. Magnetic nanoparticles and targeted drug delivering. *Pharmacol. Res.* **2010**, *62*, 144–149. [CrossRef]
21. Yadav, M.; Parle, M.; Sharma, N.; Dhingra, S.; Raina, N. Brain targeted oral delivery of doxycycline hydrochloride encapsulated Tween 80 coated chitosan nanoparticles against ketamine induced psychosis: Behavioral, biochemical, neurochemical and histological alterations in mice. *Drug Deliv.* **2017**, *24*, 1429–1440. [CrossRef] [PubMed]
22. Ma, Y.; Zheng, Y.; Zeng, X.; Jiang, L.; Chen, H.; Liu, R.; Huang, L.; Mei, L. Novel docetaxel-loaded nanoparticles based on PCL-Tween 80 copolymer for cancer treatment. *Int. J. Nanomed.* **2011**, *6*, 2679–2688. [CrossRef]
23. Wolfram, J.; Zhu, M.; Yang, Y.; Shen, J.; Gentile, E.; Paolino, D.; Fresta, M.; Nie, G.; Chen, C.; Shen, H.; et al. Safety of nanoparticles in medicine. *Curr. Drug Targets* **2015**, *16*, 1671–1681. [CrossRef] [PubMed]
24. Pelaz, B.; Charron, G.; Pfeiffer, C.; Zhao, Y.; De la Fuente, J.M.; Liang, X.J.; Parak, W.J.; Del Pino, P. Interfacing engineered nanoparticles with biological systems: Anticipating adverse nano-bio interactions. *Small* **2013**, *9*, 1573–1584. [CrossRef] [PubMed]

25. Zuo, G.; Kang, S.G.; Xiu, P.; Zhao, Y.; Zhou, R. Interactions between proteins and carbon-based nanoparticles: Exploring the origin of nanotoxicity at the molecular level. *Small* **2013**, *9*, 1546–1556. [CrossRef]
26. Hamilton, R.F.; Wu, N.; Porter, D.; Buford, M.; Wolfarth, M.; Holian, A. Particle length-dependent titanium dioxide nanomaterials toxicity and bioactivity. *Part. Fibre Toxicol.* **2009**, *6*, 35. [CrossRef]
27. Cho, W.S.; Duffin, R.; Howie, S.E.; Scotton, C.J.; Wallace, W.A.; Macnee, W.; Bradley, M.; Megson, I.L.; Donaldson, K. Progressive severe lung injury by zinc oxide nanoparticles; the role of Zn2+dissolution inside lysosomes. *Part. Fibre Toxicol.* **2011**, *8*, 27. [CrossRef]
28. Roacho-Pérez, J.A.; Ruiz-Hernandez, F.G.; Chapa-Gonzalez, C.; Martínez-Rodríguez, H.G.; Flores-Urquizo, I.A.; Pedroza-Montoya, F.E.; Garza-Treviño, E.N.; Bautista-Villareal, M.; García-Casillas, P.E.; Sánchez-Domínguez, C.N. Magnetite nanoparticles coated with PEG 3350-Tween 80: In vitro characterization using primary cell cultures. *Polymers* **2020**, *12*, 300. [CrossRef]
29. Bhattacharjee, S. DLS and zeta potential—What they are and what they are not? *J. Control. Release* **2016**, *235*, 337–351. [CrossRef]
30. Sarkar, T.; Rawat, K.; Bohidar, H.B.; Solanki, P.R. Electrochemical immunosensor based on PEG capped iron oxide nanoparticles. *J. Electroanal. Chem.* **2016**, *783*, 208–216. [CrossRef]
31. Anbarasu, M.; Anandan, M.; Chinnasamy, E.; Gopinath, V.; Balamurugan, K. Synthesis and characterization of polyethylene glycol (PEG) coated Fe_3O_4 nanoparticles by chemical co-precipitation method for biomedical applications. *Spectrochim. Acta Part A Mol. Biomol. Spectrosc.* **2015**, *135*, 536–539. [CrossRef] [PubMed]
32. Khalil, M.I. Co-precipitation in aqueous solution synthesis of magnetite nanoparticles using iron(III) salts as precursors. *Arab. J. Chem.* **2015**, *8*, 279–284. [CrossRef]
33. Khan, Y.; Durrani, S.K.; Siddique, M.; Mehmood, M. Hydrothermal synthesis of alpha Fe_2O_3 nanoparticles capped by Tween-80. *Mater. Lett.* **2011**, *65*, 2224–2227. [CrossRef]
34. Fu, X.; Kong, W.; Zhang, Y.; Jiang, L. Novel solid–solid phase change materials with biodegradable trihydroxy surfactants for thermal energy storage. *RSC Adv.* **2015**, *5*, 68881–68889. [CrossRef]
35. Lotfi, S.; Ghaderi, F.; Bahari, A.; Mahjoub, S. Preparation and characterization of magnetite–chitosan nanoparticles and evaluation of their cytotoxicity effects on MCF7 and fibroblast cells. *J. Supercond. Nov. Magn.* **2017**, *30*, 3431–3438. [CrossRef]
36. Anand, M.; Sathyapriya, P.; Maruthupandy, M.; Beevi, A.H. Synthesis of chitosan nanoparticles by TPP and their potential mosquito larvicidal application. *Front. Lab. Med.* **2018**, *2*, 72–78. [CrossRef]
37. Jaramillo-Martínez, S.; Vargas-Requena, C.; Rodríguez-González, C.; Hernández-Santoyo, A.; Olivas-Armendáriz, I. Effect of extrapallial protein of *Mytilus californianus* on the process of in vitro biomineralization of chitosan scaffolds. *Heliyon* **2019**, *5*. [CrossRef]
38. Slavov, L.; Abrashev, M.V.; Merodiiska, T.; Gelev, C.; Vandenberghe, R.E.; Markova-Deneva, I.; Nedkov, I. Raman spectroscopy investigation of magnetite nanoparticles in ferrofluids. *J. Magn. Magn. Mater.* **2010**, *322*, 1904–1911. [CrossRef]
39. Maruthupandy, M.; Muneeswaran, T.; Anand, M.; Quero, F. Highly efficient multifunctional graphene/chitosan/magnetite nanocomposites for photocatalytic degradation of important dye molecules. *Int. J. Biol. Macromol.* **2020**, *153*, 736–746. [CrossRef]
40. Zajac, A.; Hanuza, J.; Wandas, M.; Dyminska, L. Determination of N-acetylation degree in chitosan using Raman spectroscopy. *Spectrochim Acta A Mol. Biomol. Spectrosc.* **2015**, *134*, 114–120. [CrossRef]
41. Souza, N.L.G.D.; Salles, T.F.; Brandão, H.M.; Edwards, H.G.M.; Oliveira, L.F.C. Synthesis, vibrational spectroscopic and thermal properties of oxocarbon cross-linked chitosan. *J. Braz. Chem. Soc.* **2015**, *26*, 1247–1256. [CrossRef]
42. Kishore, R.S.K.; Pappenberger, A.; Bauer-Dauphin, I.; Ross, A.; Buergi, B.; Staempfli, A.; Mahler, H.C. Degradation of polysorbates 20 and 80: Studies on thermal autoxidation and hydrolysis. *J. Pharm. Sci.* **2011**, *100*, 721–731. [CrossRef] [PubMed]
43. Mohamed, M.H.; Udoetok, I.A.; Wilson, L.D.; Headley, J.V. Fractionation of carboxylate anions from aqueous solution using chitosan cross-linked sorbent materials. *J. Name* **2013**, 1–3. [CrossRef]
44. Gomez-Perez, M.; Fourcade, L.; Mateescu, M.A.; Paquin, J. Neutral Red versus MTT assay of cell viability in the presence of copper compounds. *Anal. Biochem.* **2017**, *535*, 43–46. [CrossRef] [PubMed]
45. Katrina, A.; Voliani, V. Three-dimensional tumor models: Promoting breakthroughs in nanotheranostics translational research. *Appl. Mater. Today* **2020**, *19*, 100552. [CrossRef]
46. Correia, S.I.; Pereira, H. Silva-Correia, J. VanDijk, C.N.; Espregueira-Mendes, J.; Oliveira, J.M.; Reis, R.L. Current concepts: Tissue engineering and regenerative medicine applications in the ankle joint. *J. R. Soc. Interface* **2013**, *11*, 20130784. [CrossRef] [PubMed]
47. Macías-Martínez, B.I.; Cortés-Hernández, D.A.; Zugasti-Cruz, A.; Cruz-Ortíz, B.R.; Múzquiz-Ramos, E.M. Heating ability and hemolysis test of magnetite nanoparticles obtained by a simple co-precipitation method. *J. Appl. Res. Technol.* **2016**, *14*, 239–244. [CrossRef]
48. Raga, O.; Rimantas, P. Genotoxic properties of *Betonica officinalis*, *Gratiola officinalis*, *Vincetoxicum luteum* and *Vincetoxicum hirundinaria* extracts. *Food Chem. Toxicol.* **2019**, *134*, 110815. [CrossRef]
49. Turkez, H.; Yousef, M.I.; Sönmez, E.; Togar, B.; Bakan, F.; Soizo, P.; DiStefano, A. Evaluation of cytotoxic, oxidative stress and genotoxic responses of hydroxyapatite nanoparticles on human blood cells. *J. Appl. Toxicol.* **2014**, *34*, 373–379. [CrossRef]
50. Buyukleyla, M.G.; Tuylu, B.A.; Sinan, H.; Sivas, H. The Genotoxic and Antigenotoxic Effects of Tannic Acid in Human Lymphocytes. *Drug Chem. Toxicol.* **2012**, *35*, 11–19. [CrossRef]
51. Atlı-Şekeroğlu, Z.; Güneş, B.; Kontaş-Yedier, S.; Şekeroğlu, V.; Aydın, B. Effects of tartrazine on proliferation and genetic damage in human lymphocytes. *Toxicol. Mech. Methods* **2017**, *27*, 370–375. [CrossRef] [PubMed]

52. Chung-Lun, L.; Cheng-Huang, L.; Huan-Cheng, C.; Meng-Chih, S. Protein attachment on nanodiamonds. *J. Phys. Chem. A* **2015**, *119*, 7704–7711. [CrossRef]
53. Sotnikov, D.V.; Berlina, A.N.; Ivanov, V.S.; Zherdev, A.V.; Dzantiev, B.B. Adsorption of proteins on gold nanoparticles: One or more layers? *Colloids Surf. B Biointerfaces.* **2019**, *173*, 557–563. [CrossRef] [PubMed]

Article

Evaluation of Glycerylphytate Crosslinked Semi- and Interpenetrated Polymer Membranes of Hyaluronic Acid and Chitosan for Tissue Engineering

Ana Mora-Boza [1,2,†], Elena López-Ruiz [3,4,5,6,†], María Luisa López-Donaire [1,2,*], Gema Jiménez [3,4,5,6], María Rosa Aguilar [1,2], Juan Antonio Marchal [3,4,6,7], José Luis Pedraz [2,8], Blanca Vázquez-Lasa [1,2,*], Julio San Román [1,2] and Patricia Gálvez-Martín [9]

1. Institute of Polymer Science and Technology, ICTP-CSIC, C/Juan de la Cierva 3, 28006 Madrid, Spain; amorboz@gmail.com (A.M.-B.); mraguilar@ictp.csic.es (M.R.A.); jsroman@ictp.csic.es (J.S.R.)
2. CIBER-BBN, Health Institute Carlos III, C/Monforte de Lemos 3-5, Pabellón 11, 28029 Madrid, Spain; joseluis.pedraz@ehu.eus
3. Biopathology and Regenerative Medicine Institute (IBIMER), Centre for Biomedical Research, University of Granada, E-18100 Granada, Spain; elenalopru@gmail.com (E.L.-R.); gemajg@ugr.es (G.J.); jmarchal@go.ugr.es (J.A.M.)
4. Instituto de Investigación Biosanitaria de Granada (ibs.GRANADA), University Hospitals of Granada University of Granada, E-18071 Granada, Spain
5. Department of Health Sciences, University of Jaén, 23071 Jaén, Spain
6. Excellence Research Unit "Modeling Nature" (MNat), University of Granada, E-18016 Granada, Spain
7. Department of Human Anatomy and Embryology, Faculty of Medicine, University of Granada, E-18016 Granada, Spain
8. NanoBioCel Group, Laboratory of Pharmaceutics, University of the Basque Country (UPV/EHU), School of Pharmacy, Paseo de la Universidad 7, 01006 Vitoria-Gasteiz, Spain
9. R&D Human Health, Bioibérica S.A.U., 08950 Barcelona, Spain; pgalvez@bioiberica.com
* Correspondence: marisalop@ictp.csic.es (M.L.L.-D.); bvazquez@ictp.csic.es (B.V.-L.)
† These authors contributed equally.

Received: 19 October 2020; Accepted: 7 November 2020; Published: 11 November 2020

Abstract: In the present study, semi- and interpenetrated polymer network (IPN) systems based on hyaluronic acid (HA) and chitosan using ionic crosslinking of chitosan with a bioactive crosslinker, glycerylphytate (G_1Phy), and UV irradiation of methacrylate were developed, characterized and evaluated as potential supports for tissue engineering. Semi- and IPN systems showed significant differences between them regarding composition, morphology, and mechanical properties after physicochemical characterization. Dual crosslinking process of IPN systems enhanced HA retention and mechanical properties, providing also flatter and denser surfaces in comparison to semi-IPN membranes. The biological performance was evaluated on primary human mesenchymal stem cells (hMSCs) and the systems revealed no cytotoxic effect. The excellent biocompatibility of the systems was demonstrated by large spreading areas of hMSCs on hydrogel membrane surfaces. Cell proliferation increased over time for all the systems, being significantly enhanced in the semi-IPN, which suggested that these polymeric membranes could be proposed as an effective promoter system of tissue repair. In this sense, the developed crosslinked biomimetic and biodegradable membranes can provide a stable and amenable environment for hMSCs support and growth with potential applications in the biomedical field.

Keywords: interpenetrated polymer network; semi-IPN; methacrylated hyaluronic acid; chitosan; glycerylphytate; mesenchymal stem cell

1. Introduction

Hydrogels derived from natural polymers exhibit potential for tissue engineering (TE) applications as they closely mimic the extracellular matrix (ECM) of native tissues. They also provide a suitable environment for supporting cell adhesion and growth compared to other materials due to their biocompatibility, swelling ability, and possibility of diffusion for nutrients and waste exchange [1,2]. Thus, hydrogel membranes that present mechanical and physicochemical properties similar to those of native tissues have gained much attention in the latest years [3].

Polysaccharides-based hydrogels are promising candidates to fulfil the diversified demands in a variety of biomedical applications [4]. Chitosan (Ch) and hyaluronic acid (HA) are two polysaccharides that can form hydrogels and are widely exploited for its use as scaffolds for TE [2]. Ch is a linear polysaccharide widely applied in the biomedical field due to its structural similarity to the naturally occurring glycosaminoglycans and its susceptibility to degradation by enzymes in humans [2,5]. Ch shows also antimicrobial and hemostatic properties attributed to its cationic nature of amino groups [4–6]. HA is a glycosaminoglycan found in ECM that plays a key role as an environmental cue to regulate cell behavior during embryonic development, healing processes, inflammation [4,7,8]. HA participates in important cell signaling pathways due to the presence of cell surface receptors like CD44 and RHAMM, which is a receptor for hyaluronan-mediated motility [9]. Moreover, HA demonstrated to play powerful multifunctional activity in homeostasis and tissue remodeling processes [6,10]. Ch and HA have been combined to fabricate different matrices for several TE applications [1,2,4,11–17], such as polyelectrolyte complexes for dental pulp regeneration [12] or injectable hydrogels [1,2,14,17,18] for cartilage repair [14,18], peripheral nerve regeneration [17], and adipose tissue regeneration [19], among others. Ch and HA combination has been particularly attractive for osteochondral regeneration applications due to their physicochemical and compositional similarities with native cartilage [7,11,20–22]. For example, Mohan et al. [11] performed a profound study about the regeneration capacity of Ch/HA gels on critical osteochondral defects in knee joints of New Zealand white rabbits, claiming the potential regenerative capacities of their systems. In other work carried out by Erickson et al. [7], HA and Ch were used to fabricate a bilayer scaffold to repair osteochondral defects, showing excellent cellular proliferation results.

Among all available types of polymeric-based matrices, interpenetrating networks (IPNs) and semi-IPNs membranes provide highly tunable platforms regarding composition and physicochemical properties by the combination of different polymers and crosslinking processes. These systems have showed attractive features in terms of enhanced stability and mechanical properties, mainly due to the molecular reinforcement resulted from the network/s of different polymers [4,9,23]. As it is known, an IPN consists of a combination of two (or more) polymer networks which are physically or chemically crosslinked and entangled within each other. For its part, in a semi-IPN, only one of the polymers is crosslinked and the linear polymer is entangled within the network [9]. The present approach provides a promising candidate system for TE applications in the form of natural-occurring polysaccharides semi- and IPN systems using a novel recently developed crosslinker glycerylphytate (G_1Phy). G_1Phy is a natural derived crosslinker that possesses reduced cytotoxicity and antioxidant properties [24], and showed enhanced cellular adhesion and proliferation in comparison to other traditionally used phosphate-based crosslinkers like tripolyphosphate [25]. Specifically, we develop and evaluate semi- and IPN systems formed by Ch/HA and Ch/methacrylated HA (HAMA), respectively, ionically crosslinked with G_1Phy [25]. Although Ch [4] and HA [9,20,26,27] have been combined with other polymers for the preparation of semi- and IPN systems, the reported polymeric composition and applied crosslinking strategies in this work has not been explored before. The obtained materials were characterized by a set of techniques in terms of composition, physicochemical, morphological and mechanical properties as well as in vitro behavior, observing clear differences that are expected to influence its efficacy on their biological performance. Biological assays regarding viability, cell adhesion and proliferation were assessed on human mesenchymal stromal cells (hMSCs). The excellent biocompatibility of our systems was demonstrated by large spreading areas of hMSCs on the hydrogel surfaces. Moreover, semi-IPN system showed a significantly enhancement of hMSCs proliferation over time in comparison to the other systems. Our findings suggested that surface properties and composition, can play a key role in the final application of semi- and IPN membranes as effective matrices for TE, tentatively to guided bone regeneration applications.

2. Materials and Methods

2.1. Materials

HA (Ophthalmic grade, 800–1000 kDa, Bioiberica, Barcelona, Spain) and Ch with a degree of deacetylation of 90% (Medical grade, M_w: 200–500 kDa, Altakitin SA, Lisboa, Portugal) were used as received. Methacrylic anhydride (MA), poly (ethylene glycol) dimethacrylate (PEGDMA, M_n: 8000 Da) and the photoinitiator Irgacure 2959 were purchased from Sigma Aldrich (St. Louis, MO, USA) and used as received. G_1Phy was prepared as previously described by Ana Mora-Boza et al. [24], using phytic acid and glycerol from Sigma Aldrich. Solvents as isopropanol (Scharlau, Barcelona, Spain) and ethanol (BDH Chemicals, Philadelphia, PA, USA) were used as received. Dialysis membranes (3500 Da cut off) were purchased from Spectrum® (Columbia, MO, USA). Additional reagents such as phosphate buffered saline (PBS), calcium chloride, nitric acid 65% (v/v), acetic acid (AA) and sodium hydroxide were purchased from Thermo Fisher Scientific Corporation (Waltham, MA, USA). Tris hydrochloride, 1 M solution (pH 7.5/Mol. Biol.) was purchase from Fisher BioReagents (Waltham, MA, USA).

2.2. Synthesis of Methacrylated Hyaluronic Acid

HAMA was synthesized through an esterification reaction in alkaline conditions following the protocol described by Khunmaneeet et al. [28]. HA (1 g) was dissolved in 100 mL of Milli-Q water in a two necked glass flask for 24 h. MA was added to the HA solution at a MA:HA ratio of 1:1. The mixture was kept at 0 °C using an ice bath and the pH was controlled at 8.5 by adding NaOH (5 M) with the help of an automatic titrator (Metrohm, Switzerland) for 24 h. The final product was purified by precipitation in cold ethanol, subsequently centrifuged (Eppendorf centrifuge 5810 R model, Madrid, Spain), dissolved in double distilled water (ddH_2O), and dialyzed for 4 days. After freeze drying, a white powder was finally obtained. HAMA was characterized by proton nuclear magnetic resonance (^{1}H-NMR, Bruker AVANCE IIIHD-400, MA, USA) and attenuated total reflection–Fourier transform infra-red (ATR-FTIR, Perkin-Elmer (Spectrum One), Waltham, MA, USA) spectroscopies. HAMA methacrylation degree was determined by its ^{1}H-NMR spectrum giving a value of 4.5% (Figure S1).

2.3. Preparation of Ch Membranes

Dried Ch was dissolved at a concentration of 2 wt.% in 1% AA water solution containing 13 wt.% $CaCl_2$ respect to Ch. Once it was dissolved, it was poured into a glass petri dish (internal diameter: 49 mm) and dried under moister conditions at room temperature until constant weight. Then, membranes were detached from the petri dishes after 5 min of incubation in NaOH and subsequently rinsed with Milli-Q water until neutral pH was reached. Finally, membranes were ionically crosslinked by their immersion into a G_1Phy water solution at a concentration of 15 mg/mL (30 wt.% respect to chitosan) for 24 h and room temperature. The uncoupled G_1Phy was removed by rinsing twice the membranes with Milli-Q water.

2.4. Preparation of Ch/HA and Ch/HAMA Membranes

Either type of membranes, Ch/HA or Ch/HAMA, were prepared with a content of 75% of Ch and 25% of HA or HAMA, respectively. Either HA or HAMA solution in 1% AA with $CaCl_2$ (%) was added to the Ch solution together with additional drops of 2 M HCl to achieve the total dissolution of both polymers. For Ch/HAMA membranes, HAMA solution was supplemented with 5% PEGDMA crosslinker and 2% of photoinitiator Irgacure 2959, both respect to HAMA content, in order to trigger photopolymerization by UV-light irradiation. Therefore, the corresponding solution was poured and irradiated at 365 nm for 15 min using a UVP chamber photoreactor (CL-1000, Thermo Fisher Scientific Corporation, MA, US), equipped with 5 bulbs of 365 nm working at an intensity of 2.9 mW/cm^2. Finally, membranes were submitted to ionic crosslinking with G_1Phy following the same protocol as described in Section 2.3. All membranes were prepared in the form of a circle of 5 cm diameter using a petri dish and with thicknesses between 0.22 ± 0.03 to 0.42 ± 0.06 mm. Those membranes were punched with diameters of 12 mm for the in vitro experiments and biological assays. Figure 1 shows the polymer and crosslinker compositions used for the fabrication of each system along with the digital images of the as-obtained membranes.

A

Membrane Sample	Ch (wt.%)	HA (wt.%)	HAMA (wt.%)	G_1Phy (wt.%) [a]	PEGDMA (wt.%) [b]
Ch (Network)	2.0	0	0	30	0
Ch/HA (Semi-IPN)	1.5	0.5	0	30	0
Ch/HAMA (IPN)	1.5	0	0.5	30	5

[a] Percentages calculated respect to Ch content.
[b] Percentage calculated respect to HAMA content

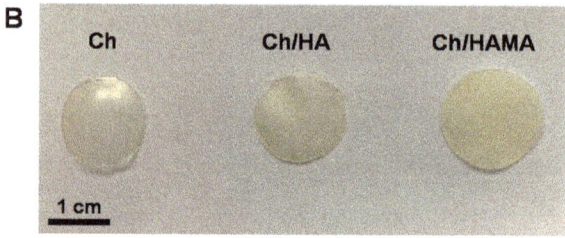

Figure 1. Polymer and crosslinker concentrations (wt.%) used for the fabrication of Ch membranes, semi-, and IPNs developed in this work (**A**); Digital images of the developed systems (**B**).

2.5. Characterisation Techniques

^{1}H-NMR spectra were recorded with a Varian Mercury 400 MHz (Agilent, Santa Clara, CA, USA). The spectra were carried out at 25 °C in D_2O (10% w/v) and referenced to the residual proton absorption of the solvent, D_2O [4.7 ppm].

ATR-FTIR of samples were carried out on a Perkin-Elmer Spectrum BX spectrophotometer (MA, USA). All spectra were recorded from 600 to 4000 cm^{-1} with a resolution of 4 cm^{-1} and 32 scans.

Elemental analysis (EA) was performed with an elemental LECO model CHNS-932 microanalyzer (MI, USA). The determination of C and H was carried out with CO_2 and H_2O specific infrared detectors, while N (N_2) was determined by thermic conductivity. The measurements were conducted at 990 °C using He as transporter gas.

Inductively coupled plasma optical emission spectrometry (ICP-OES) measurements were carried out in a 4300 DV Perkin-Elmer plasma emission spectrometer (MA, USA) under dynamic argon flow at 16 L/min using a Gemcone (Perkin-Elmer, MA, USA) nebulizer under dynamic argon flow at 0.8 L/min, and 1300 W of plasma power.

Scanning electron microscopy (SEM) images were taken in a Hitachi S-8000 instrument (Tokyo, Japan) operating in transmission mode at 100 kV on dry samples.

Atomic force microscopy (AFM) analysis was performed with an apparatus PicoLE (Molecular Imaging) operating in the acoustically driven, intermittent contact ("tapping") mode, using standard silicon AFM probes (NSC11/Cr-Au, Mikromasch, Tallinn, Estonia) having a cantilever spring constant of 48 N/m and a resonance frequency of 330 kHz. 10 × 10 mm^2 AFM images were taken on dry samples.

Topography was examined by AFM using the WSxM 5.0 Develop 9.1 software. Three acquisitions were made with roughness parameters analysis for each sample. Data were expressed as mean ± standard deviation (SD).

Water contact angle (WCA) measurements were performed at 25 °C on dried membranes, by the sessile drop technique using a KSV instruments LTD CAM 200 Tensiometer (Hertfordshire, UK) and employing Milli-Q water as a liquid with known surface tension. A minimum of 10 measurements were taken and averaged for each sample. Data were expressed as mean ± SD.

Rheological measurements were determined using an advanced rheometer from TA instruments, model AR-G2 (DE, US), equipped with a Peltier and a solvent trap. The last one allows leading the measurement in a water-saturated atmosphere by avoiding water evaporation from the membrane. Samples were previously stabilized by their immersion for 24 h in 7.4 PBS at 37 °C. All tests were carried out using a 25 mm diameter steel sand blasted parallel plate. Oscillatory shear tests with strain sweep step were performed at a frequency of 0.5 Hz and a strain ranging from 0.01 to 100% in order to determine the linear viscoelastic region (LVR) of the different membranes. Finally, frequency sweeping tests of membranes were conducted with a frequency scanning from 0.01 to 10 Hz at 0.1% strain and 37 °C to determine the elastic (G') and viscous (G") moduli. Three replicates of each sample were evaluated.

The average mesh size ξ was calculated from G' based on the rubber elasticity theory (RET) using the following Equation (1) [29]:

$$\xi = \left(\frac{G'N_A}{RT}\right)^{-\frac{1}{3}} \quad (1)$$

where G' is the storage modulus, N_A is the Avogadro constant, R is the molar gas constant, and T is the temperature. Three replicates of each sample were evaluated. Data were expressed as mean ± SD.

2.6. G_1Phy Quantification

The amount of G_1Phy ionically crosslinked in the membranes was quantified measuring the P content using ICP-OES (Perkin-Elmer, MA, US). Polymeric membranes were dried at 60 °C after their incubation in G_1Phy solution until constant weight. Afterwards, membranes were digested in nitric acid, 65% (v/v) at 60 °C for 24 h. Then, samples were diluted with Milli-Q water to 5% nitric acid concentration in order to be compatible for ICP-OES analysis. A blank solution was prepared with only nitric acid at 5% in Milli-Q water. All measurements were performed in triplicate. A standard calibration curve of P with concentration from 0–1000 mg/L was used. Three measurements were made for each sample. Data were expressed as mean ± SD.

2.7. Swelling

Dry membranes with a diameter of 12 mm were soaked in 10 mL of PBS (pH = 7.4) at 37 °C. The swollen membranes were removed at different periods of time. After removing the attached excess of water on the surface with filter paper, the membranes were weighed. The water uptake was calculated as described in Equation (2):

$$\text{Swelling (\%)} = [(W_W - W_0)/W_0] \times 100 \quad (2)$$

where W_w and W_0 are the weight of the swollen membrane at time t, and the initial dried weight of the membrane, respectively. For each period of time and sample, a minimum of four replicates were measured and averaged. Data were expressed as mean ± SD.

2.8. Degradation

Dry membranes (12 mm diameter and 0.2 mm thickness) were weighed and then placed in 10 mL of PBS (pH 7.4) at 37 °C. The membranes were retrieved at predetermined time point and washed with

Milli-Q water to remove any remaining salts. The membranes were then dried at 60 °C until constant weight. The percentage of weight loss was calculated following the Equation (3):

$$\text{Weight loss (\%)} = [(W_W - W_t)/W_0] \times 100 \quad (3)$$

where W_0 and W_t are the weight of the initial dry membrane and the dried membrane at time t after incubation in PBS, respectively. For each period of time and sample, a minimum of four replicates were measured and averaged. Data were expressed as mean ± SD.

2.9. Release of G_1Phy

The corresponding membrane (12 mm diameter) was immersed into 10 mL of Tris-HCl 0.1 M buffer (pH 7.4) and incubated at 37 °C. Aliquots of 2 mL were taken at different periods of time (1,2,4,7 and 14 days) and replaced with fresh media. The different aliquots were diluted to 5 mL with Milli-Q water and measured by ICP-OES. For each period of time and sample, a minimum of four replicates were measured and averaged. Data were expressed as mean ± SD.

2.10. Cell Studies

2.10.1. hMSCs Isolation and Culture from Adipose Tissue

hMSCs used in this study were isolated from human abdominal fat obtained from healthy donors undergoing liposuction plastic surgery. Ethical approval for the study was obtained from the Ethics Committee (number: 02/022010) of the Clinical University Hospital of Málaga, Spain. Informed patient consent was obtained for all samples used in this study. hMSCs were isolated from human adipose tissue and characterized as previously reported [30,31]. Cells were cultured in high-glucose Dulbecco's modified Eagle's medium (DMEM; Sigma-Aldrich, St Louis, MO, USA) supplemented with 10% fetal bovine serum (FBS; Sigma-Aldrich), 100 U/mL penicillin and 100 mg/mL streptomycin (Invitrogen Inc., Grand Island, NY, USA) at 37 °C in a humidified atmosphere containing 5% CO_2. Medium was regularly changed every 3 days. At 80% of confluence, cells were subcultured. For all the experiments cells were used between passages 4 and 6.

2.10.2. hMSCs Culture in Hydrogel Membranes

hMSCs isolation and culture from adipose tissue were performed as a described in Section 2.10.1. Hydrogel membranes (12 mm diameter) were sterilized by immersion in 70% ethanol aqueous solution for 1 h, washed several times in PBS and then subjected to UV light (Philips, Pila, Poland) for 20 min on both sizes. Then, hydrogel membranes were incubated in 24 well plates with complete medium overnight before cells were seeded. hMSCs suspension containing 30,000 cells in 200 µL of medium was slowly dropped onto the surface of each membrane and incubated for 2 h at 37 °C. After that, 1 mL of fresh medium was added to each well plate. All samples were incubated under a 5% CO_2 atmosphere at 37 °C. The culture medium was replaced every 2 days and the hydrogel membranes were processed for subsequent analysis.

2.10.3. Cell Viability Assay

Cell viability was determined on days 1, 7 and 21 using Live/Dead™ Viability/Cytotoxicity Kit (Invitrogen Inc., Grand Island, NY, USA). The hydrogel membranes were incubated in PBS containing Calcein AM (2 µM) and ethidium homodimer (4 µM) at 37 °C for 30 min to stain live and dead cells, respectively. Membranes were imaged by confocal microscopy (Nikon Eclipse Ti-E A1, Amsterdam, The Netherlands) and analyzed using NIS-Elements software (Amsterdam, The Netherlands).

2.10.4. Cell Proliferation Assay

Cell proliferation was analyzed using AlamarBlue® assay (Bio-Rad Laboratories, Inc., manufactured by Trek Diagnostic System., Hercules, CA, USA) after 1, 5, 7, 14 and 21 days. The hydrogel membranes were incubated with AlamarBlue® solution at 37 °C for 3 h. Fluorescence of reduced AlamarBlue® was determined at 530/590 nm excitation/emission wavelengths (Synergy HT, BIO-TEK, Winooski, VT, USA).

2.10.5. Environmental Scanning Electron Microscopy (ESEM)

The hydrogel membranes were analyzed using a variable-pressure equipment FEI, mod. Quanta 400 (OR, USA). The analysis was performed to characterize the surface structure of the membranes and cell growth after 21 days in culture. Samples were fixed with 2% glutaraldehyde and, then, were rinsed in 0.1 M cacodylate buffer and incubated overnight at 4 °C. For critical point the samples were then maintained with osmium tetroxide 1% at room temperature during 1 h and dehydrated in a series of ethanol solutions (50%, 70%, 90% and 100%) by soaking the samples in each solution for 15 min. Subsequently, samples were critical point dried (Anderson, 1951) in a desiccator (Leica EMCPD300, Wetzlar, Germany) and covered by evaporating them in a carbon evaporator (Emitech K975X).

2.10.6. Statistical Analysis

All graphed data represent the mean ± SD from at least three experiments. Two-tailed Student T test analysis were performed for Ch/HA and Ch/HAMA samples with respect to Ch ones at each time point at significance level of ** $p < 0.01$, and for Ch/HA samples with respect to Ch/HAMA samples at each time point at significance level of (## $p < 0.01$).

3. Results and Discussion

3.1. Physicochemical Characterization and Viscoelastic Properties of Membranes

Elemental composition of the membranes was determined by elemental analysis. Table 1 shows the theoretical and experimental elemental compositions (C, H, and N) for the different membranes. Experimental compositions correlated very well with those calculated theoretically, which revealed the absence of impurities. Table 1 also shows a decrease of the experimental C/N value (5.25 ± 0.07) respect to the theoretical one (6.1) for Ch/HA membranes. In case of Ch/HAMA membranes, the experimental C/N ratio (5.83 ± 0.15) approached the theoretically expected (6.1). The decrease of C/N ratio in semi-IPN systems could indicate a possible release of HA from these membranes after washing steps since this phenomenon was not observed in IPN systems. This means that the covalent crosslinking of HAMA mediated by UV-light seemed to retain the HA polysaccharide in the IPN. In addition, this fact was confirmed by a decrease in the yield percentage of Ch/HA membranes. Nevertheless, both semi- and IPN systems can be considered promising candidates for tissue regeneration since both of them contain HA, an essential component of the native ECM [7]. Likewise, the presence of HA in the membranes will help to maintain the functionality and characteristic structure of regenerated tissues, which is essential for the final success of these scaffolds [6,11]. It is important to consider that the concentration range in which HA can provide these beneficial properties is quite wide [32].

Table 1. Theoretical (Theo) and experimental (Exp) elemental compositions, crosslinker content, and yield percentage for Ch, Ch/HA and Ch/HAMA membranes.

Membrane Sample	C [a]		H [a]		N [a]		C/N [a]		G_1Phy Content (%) [b,c]	Yield (%)
	Theo	Exp	Theo	Exp	Theo	Exp	Theo	Exp		
Ch	44.9	43.03 ± 0.33	6.80	6.86 ± 0.01	8.60	8.24 ± 0.10	5.2	5.22 ± 0.02	5.4 ± 0.1	95.3 ± 1.1
Ch/HA	44.7	42.72 ± 0.11	6.53	6.77 ± 0.09	7.36	8.12 ± 0.11	6.1	5.25 ± 0.07	4.9 ± 0.5	75.0 ± 0.6
Ch/HAMA	44.7	42.11 ± 0.18	6.53	6.58 ± 0.15	7.36	7.21 ± 0.19	6.1	5.83 ± 0.15	2.5 ± 0.4	93.5 ± 2.5

[a] Determined by elemental analysis. [b] Determined by ICP. [c] Gram of G_1Phy per gram of membrane × 100.

The content of G_1Phy incorporated in the membranes resulted from the ionic crosslinking between amino and phosphate groups was analyzed by ICP-OES (Table 1). The amount of the phytate crosslinker for Ch/HA membrane decreased somewhat compared to that of Ch sample what was attributed to the lower chitosan content in the former membranes (Ch:HA ratio = 75:25). However, the G_1Phy crosslinked in the Ch/HAMA membranes decreased nearly to the half. This less crosslinker content in the IPN membranes in comparison to semi-IPNs maybe due to the fact that the covalently crosslinked HA could hindrance the availability of amine groups of chitosan for ionic interactions with the phosphate groups of G_1Phy.

The FTIR spectra of the membranes are represented in Figure S2. The FTIR spectra of semi- and IPN samples showed the characteristic bands of Ch and HA polysaccharides. The main bands appeared between 3600 and 3200 cm^{-1} (υ O-H and N–H associated); at 2924 and 2854 cm^{-1} (υ C–H); at 1720 cm^{-1} (υ C = O in carboxylic and ester groups); at 1643/1634 cm^{-1} (υ C = O of amide, amide I); at 1579 cm^{-1} (δ N–H); at 1420 cm^{-1} (υ COO$^-$ and δ C–H); at 1373 cm^{-1} (δ –CH$_3$ symmetrical); at 1333 cm^{-1} (υ C–N, amide III); at 1258/1262 cm^{-1} (X P = O); at 1150 cm^{-1} (υ C–O–C asymmetric); at 1066, 1029, 995 and 984 cm^{-1} (υ C–O alcohols, υ P–O and P–O–C, υ C–O glycosidic linkages and vibration of pyranose structure); at 893 and 721 cm^{-1} (υ P–O and P–O–C) [33,34].

Surface morphology is a critical factor for the development of biomaterials that effectively promote cell adhesion and proliferation [35]. Figure 2 shows a detailed examination of surface topography for Ch (A), Ch/HA (B) and Ch/HAMA (C) by SEM and AFM, as well as the calculated roughness parameters (R_a, roughness average, and RMS, root mean square) for the different systems. Qualitative topographic differences among the membranes can be observed in the SEM micrographs. Particularly, Ch system showed the flattest surface followed by the semi-IPN sample, while a much more granular surface was observed in semi-IPN system. This result illustrated a topographic change due to the incorporation of HA that can be a consequence of electrostatic interactions between carboxylic groups of HA and amino groups of Ch, which leads to polyelectrolyte complex formation [32]. The Ch-HA interactions could be hindered in IPN systems by UV curing process, resulting in flatter surfaces as it is observed in their Ch/HAMA micrographs. AFM 3D images show the nano features and the roughness parameters of representative areas of the membranes. R_a and RMS values correlated very well with the topography observed by SEM. As it was expected, R_a and RMS values were significantly higher for Ch/HA membranes in comparison to those of Ch and Ch/HAMA that were very similar to each other. Therefore, we can conclude that covalent crosslinking of HAMA with UV-light resulted in a more compact framework in IPN systems compared to semi-IPNs, which leaded to a decrease of roughness at the nanoscale [9].

Surface parameters such as wettability are important properties that must be studied since hydrophilic-hydrophobic balance greatly determines cell adhesion and proliferation properties of the scaffolds [36]. Surface-wetting characterization is currently carried out by WCA measurements the most common method being the sessile drop goniometry [37], which was used in this work. Measured WCA values for Ch, Ch/HA, Ch/HAMA membranes were 48.18 ± 2.71°, 40.97 ± 3.26°, and 47.73 ± 4.96°, respectively. All the systems showed hydrophilic surfaces (WCA < 90°) as it was expected because of the characteristic water absorption nature of these polysaccharides [38]. Different WCA for Ch samples are reported in literature. For instance, Tamer et al. found higher WCA values (89 ± 0.6°) for Ch surfaces [39] than those obtained in our work. However, it has been reported that Ch polarity highly depends on the type and concentration of the used neutralization solution as well as the time of washing steps. Noriega et al. [40] performed a profound study where they reported a wide range of WCA for Ch surfaces, from 45 to 65°, that highly depended on neutralization parameters. As expected, hydrophilicity increased as neutralization base concentration and incubation time increased [40]. In our samples, the relatively low WCA values observed for Ch could also be due to the contribution of available phosphate groups coming from the G_1Phy crosslinker on the membrane surface, which exhibits high affinity to polar liquids [41]. A decrease of WCA was observed for semi-IPN due to the higher content of G_1Phy in this sample, along with the presence of HA and its polyanionic character [42]. For its part, IPNs showed WCA values similar to those of Ch membranes, which could be due to the reduction of carboxylic groups of HA after methacrylation reaction and further covalent crosslinking. Since membrane surfaces showed different WCA values, we can conclude that wettability properties seem to be an easily tunable parameter in function of composition and applied crosslinking processes in our systems.

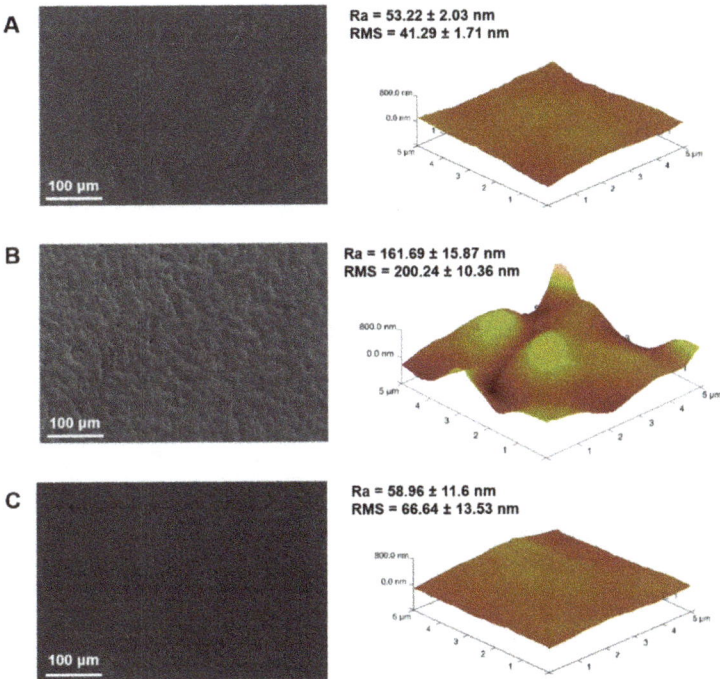

Figure 2. SEM micrographs (left) and AFM 3D perspective images with their respective calculated roughness parameters (right) for Ch (**A**), Ch/HA (**B**), and Ch/HAMA (**C**) polymeric membranes.

Rheological measurements were carried out to study the viscoelastic properties of our systems. The evolution of the elastic and viscous moduli of the membranes was studied in their LVR at a constant strain of 0.1%, and it is represented in Figure 3A. All the systems exhibited a plateau in the studied frequency range, which indicated the stability of the crosslinked network. This plateau also showed a solid-like behavior of the membranes, since elastic moduli was independent on the applied frequency [43]. IPN showed higher G' values in comparison to Ch and semi-IPN systems due to the macromolecular reinforcement of the polymeric network after covalent crosslinking. IPNs have previously demonstrated to improve mechanical properties regarding semi-IPNs, because of double crosslinking mechanisms [4,9]. Finally, loss tangent (tan δ = viscous modulus/elastic modulus), which is an index of the viscoelasticity of the systems, was calculated by taking the ratio between G″ and G' at a frequency of 1 Hz. Values of 0.27, 0.17 and 0.24, were obtained for Ch, Ch/HA and Ch/HAMA membranes, respectively.

The mesh size is defined as the distance between crosslinking points of the membrane, which is related to the mechanical strength of the hydrogel [29,44]. As expected, the IPN membrane exhibited a lower mesh size value due to dual crosslinking, resulting in a more compact structure in comparison to Ch and semi-IPN membranes, which showed similar mesh size values. (Figure 3B) These results corroborate surface and morphology characterization illustrated in SEM and AFM images (Figure 2).

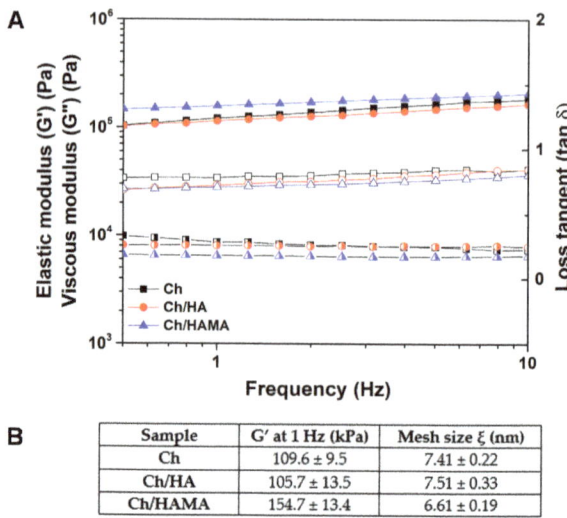

Figure 3. Evolution of elastic (G', filled) and viscous (G", unfilled) moduli, and loss tangent (tan δ, half-filled) as a function of applied frequency at constant strain of 0.1% of Ch, Ch/HA, and Ch/HAMA polymeric membranes (**A**). The average mesh size values ξ of Ch, Ch/HA and Ch/HAMA polymeric membranes determined at a frequency of 1 Hz (**B**).

Collectively, the results described in this subsection showed some differences regarding composition, surface topography, wettability and mechanical properties for the semi- and IPN systems that are expected to also exert relevant differences on their swelling and degradation properties, as well as biological performance.

3.2. In Vitro Swelling and Degradation Studies

Swelling of the developed membranes was studied under physiological conditions and the results are represented in Figure 4A. For all membranes, a fast water uptake during the first hour was observed, reaching a stable value after 3 h, when equilibrium was attained. Ch and Ch/HA membranes showed rather similar swelling profiles giving final swelling values ~100% (Figure 4B). Due to pKa value of Ch at 6.4, its amino groups are not positively charged under physiological conditions and the repulsive forces in the polymeric backbone are not induced, not taking place this increase of the network water uptake [36]. For its part, Ch/HAMA membranes showed the highest equilibrium water absorption (up to 140%) which could be attributed to the formation of a dual crosslinked network able to locate a higher amount of water molecules in their interstices, and to retain a higher HA content in comparison to semi-IPN (Table 1). Nevertheless, all membranes showed moderate swelling that will contribute to maintain their structural stability. If necessary, swelling could be adjusted by varying the content of HAMA in the membrane [45].

In vitro degradation of all hydrogel membranes immersed in a PBS solution at 37 °C was below 10% over 14 days (Figure 4B) and it slightly increased after 28 days. However, for longer incubation time (~2 months), degradation of Ch and Ch/HAMA was maintained while Ch/HA membranes suffered further degradation (~16%). The initial membrane weight loss could be attributed to the progressive breaking of the ionic bonds formed between the phosphate and amino groups, what produces release of the G_1Phy crosslinker and consequently, dissolution of HA and Ch polymeric chains. Similar results were reported for Ch/HA tissue engineering porous scaffolds where it was suggested than the degradation in PBS is only because of polymeric dissolution [46]. The higher weight loss of Ch/HA versus Ch membranes could be associated to the presence of entangled HA within the Ch network, favoring its dissolution. Accordingly, Ch/HAMA membranes, displayed the highest stability. This fact could be explained because the covalently crosslinked HA prevents its dissolution and the hydrolytically degradable ester bonds in the HAMA are sterically hindered [47].

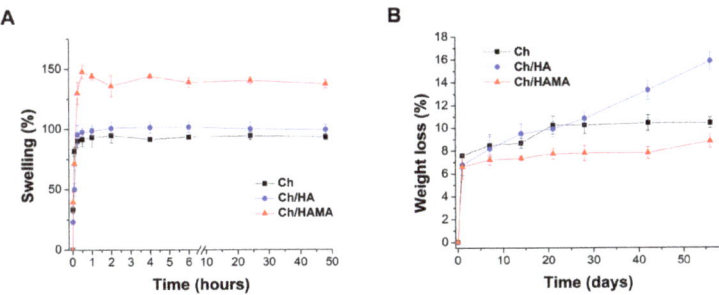

Figure 4. Effect of hydrogel membrane composition on swelling after incubation in PBS 7.4 at 37 °C for different periods of time (**A**). Weight loss of Ch, Ch/HA and Ch/HAMA membranes at different time points after soaking in PBS 7.4 at 37 °C under static conditions (**B**). Data represented the mean ± SD.

3.3. G_1Phy Release

The release profile of G_1Phy from the different hydrogel membranes is showed in Figure 5. All membranes showed a fast release (~72%) during the first 24 h which correspond to a G_1Phy concentration of 0.1 ± 0.003, 0.09 ± 0.005 and 0.04 ± 0.001 mg/mL for Ch, Ch/HA and Ch/HAMA membranes, respectively. It is worth mentioning that at 24 h the G_1Phy concentration of Ch/HAMA membrane is nearly half to that observed for Ch and Ch/HA membranes and this fact is in agreement with the lower initial content of G_1Phy incorporated in the Ch/HAMA membranes, data previously described in Table 1. A plateau in the release profile is observed for the three systems after 7 days, where Ch and Ch/HA membranes reached an 85% release of the initial G_1Phy membrane content, giving a final G_1Phy concentration of 0.11 ± 0.004 and 0.10 ± 0.0005 mg/mL, respectively. On the other hand, Ch/HAMA membranes showed a G_1Phy release of ~90%, corresponding to a final concentration of 0.05 ± 0.005 mg/mL. The slightly higher G_1Phy release in the latter membrane could be due to the higher water uptake of Ch/HAMA membranes what can favor ion diffusion and subsequently ion exchange between G_1Phy anions and negative anions present in the Tris buffer. Electrostatic interaction between G_1Phy ions (PO_4^{2-} or HPO_4^{-}) and the amino groups of Ch and breaking of links with incubation time could account for these release profiles. However, no complete release of initial G_1Phy content was achieved after 14 days. Physical mixture of Ch with phytic acid has been reported by Barahuie et al. [48] showing a complete release after 60 s. Finally, it is worth mentioning that the release pattern of our systems is in agreement with that of Ch/phytic acid systems reported in the literature [48,49].

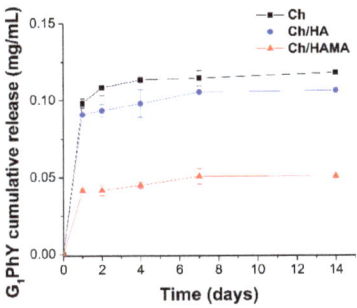

Figure 5. Release profiles of G_1Phy from the Ch, Ch/HA and Ch/HAMA membranes in 0.1 M Tris buffer (pH 7.4) at 37 °C. Data represented the mean ± SD.

3.4. Biological Evaluation

3.4.1. ESEM Microscopy of Hydrogel Membranes

To characterize the microstructural architecture of Ch, Ch/HA and Ch/HAMA membranes and observe the morphology of hMSCs cultured on them, an ESEM analysis was carried out on day 21. It is known that the surface roughness is an important factor in promoting cell attachment [50]. ESEM images of Ch, Ch/HA and Ch/HAMA membranes (Figure 6A) revealed a rougher surface for the Ch/HA membrane compared to those of Ch/HAMA and Ch, corroborating the SEM observations for dried samples (Figure 2) but surfaces of all membranes were able to support cell growth (Figure 6B). ESEM images evidenced the cells covering the surface of Ch, Ch/HA and Ch/HAMA membranes, with good adhesion, spreading, and a homogenous distribution throughout the entire surface. Moreover, ESEM images showed an interconnected cell community that attached to the scaffold which also confirmed the biocompatibility of the membranes [51,52].

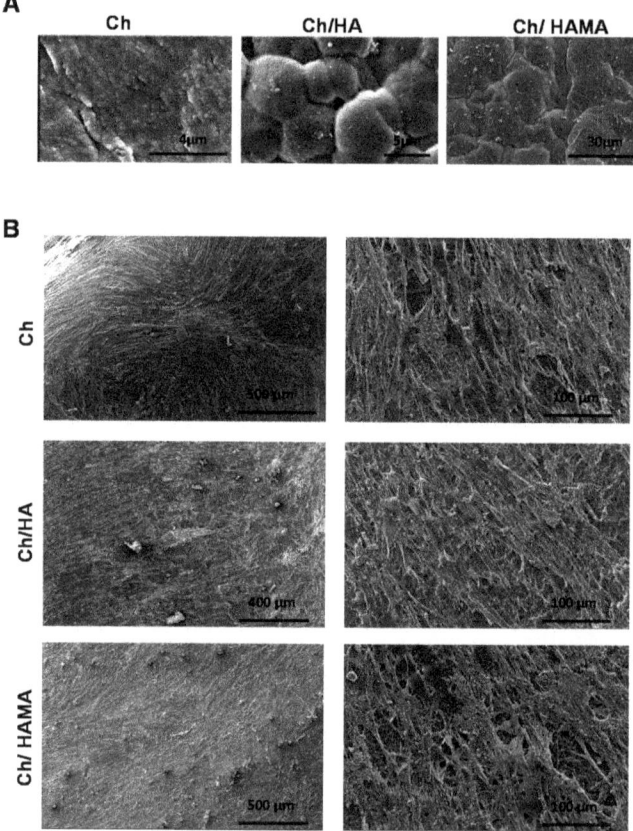

Figure 6. ESEM images of Ch, Ch/HA and Ch/HAMA hydrogel membranes. Representative ESEM images of hydrogel membranes before hMSCs culture (**A**). Representative ESEM images of hMSCs growing on hydrogel membranes after 21 days (**B**).

3.4.2. Cell Viability and Proliferation Assays

In order to evaluate the feasibility of Ch, Ch/HA and Ch/HAMA hydrogel membranes as an adequate support for cell survival, the viability of the seeded hMSCs on the top of the membranes was evaluated. The live/dead assay was employed to visualize the presence of living and dead cells after 1, 7 and 21 days in the hydrogel membranes (Figure 7A). Confocal images showed hMSCs growing on all the membrane surfaces at days 1 and 7. The number of living cells was much higher at day 21 and cells appeared covering the hydrogel membranes with few dead cells. These results indicated that Ch, Ch/HA and Ch/HAMA hydrogel membranes can provide an amenable environment that supports hMSCs growth and confirmed the cell viability with no cytotoxic effects.

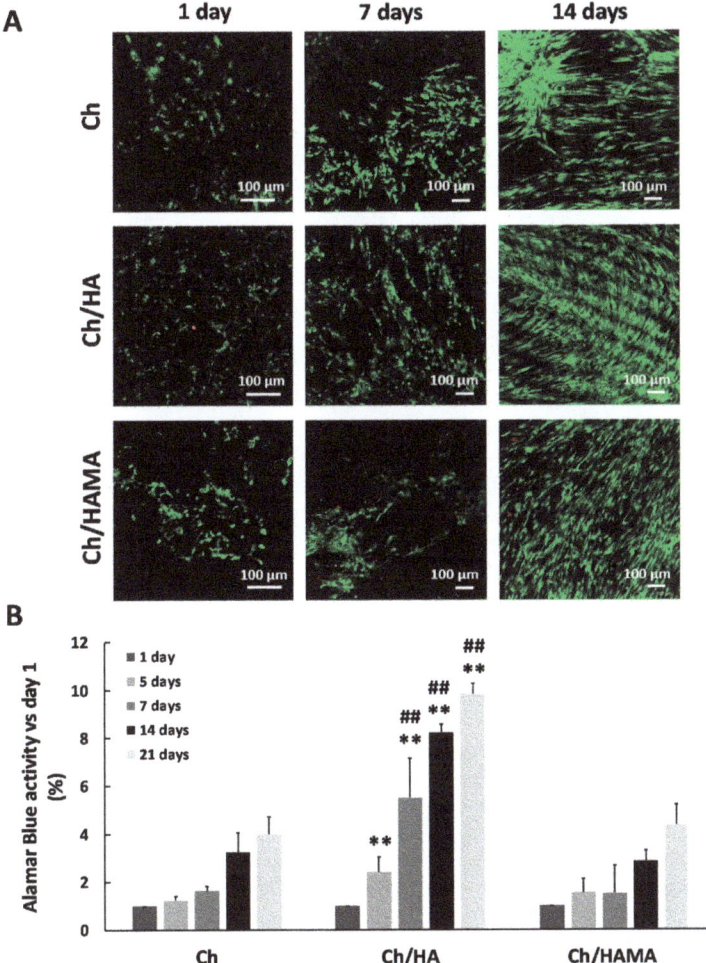

Figure 7. Cytocompatibility of Ch, Ch/HA and Ch/HAMA hydrogel membranes with hMSCs. Representative confocal images of hMSCs stained with Calcein AM (living cells in green) and ethidium homodimer (dead cell in red) at days 1, 7 and 21 using the Live/Dead®® assay (**A**). Cell proliferation at the hydrogel membranes after 1, 5, 7, 14 and 21 days (**B**). Values are represented as mean ± SD (n = 3) and normalized respect to day 1 values. Two-tailed Student T test analysis were performed for Ch/HA and Ch/HAMA samples with respect to Ch samples at each time at significance level of ** $p < 0.01$, and for Ch/HA samples with respect to Ch/HAMA samples at each time point at significance level of (## $p < 0.01$).

Proliferation of hMSCs cultured in Ch, Ch/HA and Ch/HAMA hydrogel membranes was evaluated with AlamarBlue®® assay at 1, 5, 7, 14 and 21 days of cell culture (Figure 7B). Results demonstrated that cell proliferation increased from day 1 until 21 days in all the systems. No significant differences were found between Ch/HAMA and Ch membranes regarding cell proliferation at any time. For its part, Ch/HA demonstrated a significantly enhanced cell proliferation in comparison to Ch at 5, 7, 14 and 21 days, and in comparison to Ch/HAMA system at 7, 14 and 21 days. This result may be due to the supportive microenvironment of semi-IPNs system for cell attachment and proliferation coming from entangled HA and a higher content of G_1Phy in the semi-IPN system leading to a higher released concentration as it was observed in the G_1Phy release profile (Figure 5), which after being released at short times of incubation could be assimilated by hMSCs, exerting a positive effect on cell adhesion and proliferation as it has been previously observed for similar polymeric systems [25,53]. Correira et al. [46] demonstrated that the biological performance of polysaccharides based systems was affected by their physicochemical factors, which could be tuned by the incorporation of HA to Ch at different proportions. In fact, they found that the addition of HA to Ch scaffolds up to 5% improved both physicochemical and biological properties of Ch scaffolds [46]. Additionally, previous works of the authors demonstrated the benefits of the glycerylphytate crosslinker on cell adhesion and proliferation of 3D scaffolds [25]. In overall, the better biological performance of the semi-IPN system can be attributed on the one hand, to its physicochemical features regarding surface roughness, mechanical properties, approaching those of cartilage, and wettability. On the other hand, the enhancement of cell viability and proliferation of Ch/HA sample in comparison to Ch/HAMA system can derived from the presence of linear HA embedded in the semi-IPN and its higher ability to be interchanged with the medium respect to crosslinked HAMA, which is longer retained in the IPN membrane [36,46], along with the higher content of the bioactive G_1Phy for this membrane. Thus, in our study, the semi-IPN highlights as the best candidate to mimic the native tissue ECM. Some authors have claimed the benefits of semi-IPN systems containing HA for tissue regeneration due to its similarities to ECM composition [27,54]. For example, Pescosolido et al. [54] combined HA with photocrosslinkable dextran to overcome instability problems of HA derived from its high hydrophilicity. The presence of the bioactive HA provided excellent biological properties to their systems. For its part, Skaalure et al. [27] developed a semi-IPN consisting on poly(ethylene glycol) and entrapped HA, whose incorporation clearly led to an enhanced cell adhesion and proliferation.

4. Conclusions

Semi-IPN and IPN systems based on HA and Ch crosslinked with G_1Phy were developed as biomimetic and degradable membranes with potential application in TE. Significant differences between semi-IPNs and IPNs were observed in terms of surface topography, mechanical, swelling and degradability properties. IPNs demonstrated to enhance HA retention, as well as mechanical properties of the polymeric network thanks to covalent crosslinking mediated by UV-light irradiation. Dual crosslinking processes of IPNs, consisting of ionic crosslinking of Ch and photopolymerization of HAMA, provided membranes with long-term stability and increased swelling. Moreover, the IPN framework led to flatter surfaces in comparison to semi-IPN. All the studied systems demonstrated high biocompatibility, supporting hMSCs adhesion and proliferation on their surfaces. However, the semi-IPN significantly increased cell proliferation over time respect to IPN, arising as the best candidate of the studied systems. This behavior could be due to the surface features of the semi-IPN (i.e., hydrophilic nature, granular topography and mechanical properties mimicking those of native cartilage), the higher content of the bioactive crosslinker and the entangled HA what seems to be key properties to favor hMSCs performance. These finding suggest that Ch/HA semi-IPNs ionically crosslinked with G_1Phy have potential to be proposed as an effective promoter system of tissue repair. Further studies will be carried out to evaluate both in vitro and in vivo differentiation abilities of hMSCs seeded on these biomimetic ECM membranes and their potential application for guided bone regeneration.

Supplementary Materials: The following are available online at http://www.mdpi.com/2073-4360/12/11/2661/s1, Figure S1: (A) ^{1}H-NMR spectrum of HAMA with 4.5% methacrylation degree in D_2O, and (B) ATR-FTIR spectra of HA and HAMA, Figure S2: ATR-FTIR spectra of Ch, Ch/HA, and Ch/HAMA membranes.

Author Contributions: Conceptualization and methodology, M.L.L.-D., A.M.-B., E.L.-R.; physicochemical analysis, M.L.L.-D., A.M.-B.; Biological characterization, E.L.-R., G.J.; resources, and data interpretation, M.L.L.-D., A.M.-B., E.L.-R.; writing—original draft preparation, M.L.L.-D., A.M.-B.; writing—review and editing, A.M.-B., M.L.L.-D., B.V.-L.; supervision, B.V.-L., P.G.-M.; funding acquisition, P.G.-M., J.S.R., J.A.M., J.L.P., M.R.A. All authors have read and agreed to the published version of the manuscript.

Funding: The authors thanks to "La Caixa" Foundation (ID 100010434), which supported Ana Mora-Boza (scholarship code LCF/BQ/ES16/11570018) and to the Spanish Ministry of Economy and Competitiveness for financial support (project RTC-2016-5451-1) and the Fundación Mutua Madrileña (project FMM-AP17196-2019). M. R. Aguilar and B. Vázquez-Lasa are members of the *SusPlast platform (Interdisciplinary Platform for Sustainable Plastics towards a Circular Economy)* from the Spanish National Research Council (CSIC).

Conflicts of Interest: The authors declare no conflict of interest.

References

1. Gilarska, A.; Lewandowska-Lancucka, J.; Horak, W.; Nowakowska, M. Collagen/chitosan/hyaluronic acid-based injectable hydrogels for tissue engineering applications—Design, physicochemical and biological characterization. *Colloids Surf. B Biointerfaces* **2018**, *170*, 152–162. [CrossRef] [PubMed]
2. Nair, S.; Remya, N.S.; Remya, S.; Nair, P.D. A biodegradable in situ injectable hydrogel based on chitosan and oxidized hyaluronic acid for tissue engineering applications. *Carbohydr. Polym.* **2011**, *85*, 838–844. [CrossRef]
3. Yazdi, M.K.; Vatanpour, V.; Taghizadeh, A.; Taghizadeh, M.; Ganjali, M.R.; Munir, M.T.; Habibzadeh, S.; Saeb, M.R.; Ghaedi, M. Hydrogel membranes: A review. *Mater. Sci. Eng. C* **2020**, 114. [CrossRef]
4. Suo, H.; Zhang, D.; Yin, J.; Qian, J.; Wu, Z.L.; Fu, J. Interpenetrating polymer network hydrogels composed of chitosan and photocrosslinkable gelatin with enhanced mechanical properties for tissue engineering. *Mater. Sci. Eng. C Mater. Biol. Appl.* **2018**, *92*, 612–620. [CrossRef]
5. Abarrategi, A.; Lopiz-Morales, Y.; Ramos, V.; Civantos, A.; Lopez-Duran, L.; Marco, F.; Lopez-Lacomba, J.L. Chitosan scaffolds for osteochondral tissue regeneration. *J. Biomed. Mater. Res. A* **2010**, *95*, 1132–1141. [CrossRef]
6. Muzzarelli, R.A.; Greco, F.; Busilacchi, A.; Sollazzo, V.; Gigante, A. Chitosan, hyaluronan and chondroitin sulfate in tissue engineering for cartilage regeneration: A review. *Carbohydr. Polym.* **2012**, *89*, 723–739. [CrossRef]
7. Erickson, A.E.; Sun, J.; Lan Levengood, S.K.; Swanson, S.; Chang, F.C.; Tsao, C.T.; Zhang, M. Chitosan-based composite bilayer scaffold as an in vitro osteochondral defect regeneration model. *Biomed. Microdevices* **2019**, *21*, 34. [CrossRef]
8. Knudson, C.B. Hyaluronan and CD44: Strategic players for cell-matrix interactions during chondrogenesis and matrix assembly. *Birth Defects Res. Part C Embryo Today Rev.* **2003**, *69*, 174–196. [CrossRef]
9. Suri, S.; Schmidt, C.E. Photopatterned collagen-hyaluronic acid interpenetrating polymer network hydrogels. *Acta Biomater.* **2009**, *5*, 2385–2397. [CrossRef]
10. Tognana, E.; Borrione, A.; De Luca, C.; Pavesio, A. Hyalograft C: Hyaluronan-based scaffolds in tissue-engineered cartilage. *Cells Tissues Organs* **2007**, *186*, 97–103. [CrossRef]
11. Mohan, N.; Mohanan, P.V.; Sabareeswaran, A.; Nair, P. Chitosan-hyaluronic acid hydrogel for cartilage repair. *Int. J. Biol. Macromol.* **2017**, *104*, 1936–1945. [CrossRef] [PubMed]
12. Coimbra, P.; Alves, P.; Valente, T.A.; Santos, R.; Correia, I.J.; Ferreira, P. Sodium hyaluronate/chitosan polyelectrolyte complex scaffolds for dental pulp regeneration: Synthesis and characterization. *Int. J. Biol. Macromol.* **2011**, *49*, 573–579. [CrossRef] [PubMed]
13. Weinstein-Oppenheimer, C.R.; Brown, D.I.; Coloma, R.; Morales, P.; Reyna-Jeldes, M.; Diaz, M.J.; Sanchez, E.; Acevedo, C.A. Design of a hybrid biomaterial for tissue engineering: Biopolymer-scaffold integrated with an autologous hydrogel carrying mesenchymal stem-cells. *Mater. Sci. Eng. C Mater. Biol. Appl.* **2017**, *79*, 821–830. [CrossRef] [PubMed]
14. Park, H.; Choi, B.; Hu, J.; Lee, M. Injectable chitosan hyaluronic acid hydrogels for cartilage tissue engineering. *Acta Biomater.* **2013**, *9*, 4779–4786. [CrossRef] [PubMed]

15. Liu, C.; Liu, D.; Wang, Y.; Li, Y.; Li, T.; Zhou, Z.; Yang, Z.; Wang, J.; Zhang, Q. Glycol chitosan/oxidized hyaluronic acid hydrogels functionalized with cartilage extracellular matrix particles and incorporating BMSCs for cartilage repair. *Artif. Cells Nanomed. Biotechnol.* **2018**, *46*, 721–732. [CrossRef]
16. Smith, L.J.; Gorth, D.J.; Showalter, B.L.; Chiaro, J.A.; Beattie, E.E.; Elliott, D.M.; Mauck, R.L.; Chen, W.; Malhotra, N.R. In vitro characterization of a stem-cell-seeded triple-interpenetrating-network hydrogel for functional regeneration of the nucleus pulposus. *Tissue Eng. Part A* **2014**, *20*, 1841–1849. [CrossRef]
17. Zhang, L.; Chen, Y.; Xu, H.; Bao, Y.; Yan, X.; Li, Y.; Li, Y.; Yin, Y.; Wang, X.; Qiu, T.; et al. Preparation and evaluation of an injectable chitosan-hyaluronic acid hydrogel for peripheral nerve regeneration. *J. Wuhan Univ. Technol.-Mater. Sci. Ed.* **2016**, *31*, 1401–1407. [CrossRef]
18. Tan, H.; Chu, C.R.; Payne, K.A.; Marra, K.G. Injectable in situ forming biodegradable chitosan-hyaluronic acid based hydrogels for cartilage tissue engineering. *Biomaterials* **2009**, *30*, 2499–2506. [CrossRef]
19. Tan, H.; Rubin, J.P.; Marra, K.G. Injectable in situ forming biodegradable chitosan-hyaluronic acid based hydrogels for adipose tissue regeneration. *Organogenesis* **2010**, *6*, 173–180. [CrossRef]
20. Chen, P.; Xia, C.; Mo, J.; Mei, S.; Lin, X.; Fan, S. Interpenetrating polymer network scaffold of sodium hyaluronate and sodium alginate combined with berberine for osteochondral defect regeneration. *Mater. Sci. Eng. C Mater. Biol. Appl.* **2018**, *91*, 190–200. [CrossRef]
21. Deng, C.; Chang, J.; Wu, C. Bioactive scaffolds for osteochondral regeneration. *J. Orthop. Transl.* **2019**, *17*, 15–25. [CrossRef] [PubMed]
22. Hu, X.; Man, Y.; Li, W.; Li, L.; Xu, J.; Parungao, R.; Wang, Y.; Zheng, S.; Nie, Y.; Liu, T.; et al. 3D Bio-Printing of CS/Gel/HA/Gr Hybrid Osteochondral Scaffolds. *Polymers* **2019**, *11*, 1601. [CrossRef] [PubMed]
23. Dragan, E.S. Advances in interpenetrating polymer network hydrogels and their applications. *Pure Appl. Chem.* **2014**, *86*, 1707–1721. [CrossRef]
24. Mora-Boza, A.; Włodarczyk-Biegun, M.K.; del Campo, A.; Vázquez-Lasa, B.; Román, J.S. Glycerylphytate as an ionic crosslinker for 3D printing of multi-layered scaffolds with improved shape fidelity and biological features. *Biomater. Sci.* **2020**, *8*, 506–516. [CrossRef] [PubMed]
25. Mora-Boza, A.; López-Donaire, M.L.; Saldaña, L.; Vilaboa, N.; Vázquez-Lasa, B.; San Román, J. Glycerylphytate compounds with tunable ion affinity and osteogenic properties. *Sci. Rep.* **2019**, *9*, 11491. [CrossRef] [PubMed]
26. Chung, C.W.; Kang, J.Y.; Yoon, I.S.; Hwang, H.D.; Balakrishnan, P.; Cho, H.J.; Chung, K.D.; Kang, D.H.; Kim, D.D. Interpenetrating polymer network (IPN) scaffolds of sodium hyaluronate and sodium alginate for chondrocyte culture. *Colloids Surf. B Biointerfaces* **2011**, *88*, 711–716. [CrossRef]
27. Skaalure, S.C.; Dimson, S.O.; Pennington, A.M.; Bryant, S.J. Semi-interpenetrating networks of hyaluronic acid in degradable PEG hydrogels for cartilage tissue engineering. *Acta Biomater.* **2014**, *10*, 3409–3420. [CrossRef]
28. Khunmanee, S.; Jeong, Y.; Park, H. Crosslinking method of hyaluronic-based hydrogel for biomedical applications. *J. Tissue Eng.* **2017**, *8*. [CrossRef]
29. Gan, Y.; Li, P.; Wang, L.; Mo, X.; Song, L.; Xu, Y.; Zhao, C.; Ouyang, B.; Tu, B.; Luo, L.; et al. An interpenetrating network-strengthened and toughened hydrogel that supports cell-based nucleus pulposus regeneration. *Biomaterials* **2017**, *136*, 12–28. [CrossRef]
30. Galvez, P.; Martin, M.J.; Calpena, A.C.; Tamayo, J.A.; Ruiz, M.A.; Clares, B. Enhancing effect of glucose microspheres in the viability of human mesenchymal stem cell suspensions for clinical administration. *Pharm. Res.* **2014**, *31*, 3515–3528. [CrossRef]
31. Lopez-Ruiz, E.; Jimenez, G.; Kwiatkowski, W.; Montanez, E.; Arrebola, F.; Carrillo, E.; Choe, S.; Marchal, J.A.; Peran, M. Impact of TGF-beta family-related growth factors on chondrogenic differentiation of adipose-derived stem cells isolated from lipoaspirates and infrapatellar fat pads of osteoarthritic patients. *Eur. Cells Mater.* **2018**, *35*, 209–224. [CrossRef] [PubMed]
32. Liu, H.; Yin, Y.; Yao, K.; Ma, D.; Cui, L.; Cao, Y. Influence of the concentrations of hyaluronic acid on the properties and biocompatibility of Cs-Gel-HA membranes. *Biomaterials* **2004**, *25*, 3523–3530. [CrossRef] [PubMed]
33. De Oliveira, S.A.; da Silva, B.C.; Riegel-Vidotti, I.C.; Urbano, A.; de Sousa Faria-Tischer, P.C.; Tischer, C.A. Production and characterization of bacterial cellulose membranes with hyaluronic acid from chicken comb. *Int. J. Biol. Macromol.* **2017**, *97*, 642–653. [CrossRef] [PubMed]

34. Yao, K.D.; Peng, T.; Goosen, M.F.A.; Min, J.M.; He, Y.Y. pH-sensitivity of hydrogels based on complex forming chitosan: Polyether interpenetrating polymer network. *J. Appl. Polym. Sci.* **1993**, *48*, 343–354. [CrossRef]
35. Rocha Neto, J.B.M.; Taketa, T.B.; Bataglioli, R.A.; Pimentel, S.B.; Santos, D.M.; Fiamingo, A.; Costa, C.A.R.; Campana-Filho, S.P.; Carvalho, H.F.; Beppu, M.M. Tailored chitosan/hyaluronan coatings for tumor cell adhesion: Effects of topography, charge density and surface composition. *Appl. Surf. Sci.* **2019**, *486*, 508–518. [CrossRef]
36. Iacob, A.T.; Dragan, M.; Ghetu, N.; Pieptu, D.; Vasile, C.; Buron, F.; Routier, S.; Giusca, S.E.; Caruntu, I.D.; Profire, L. Preparation, Characterization and Wound Healing Effects of New Membranes Based on Chitosan, Hyaluronic Acid and Arginine Derivatives. *Polymers* **2018**, *10*, 607. [CrossRef]
37. Huhtamäki, T.; Tian, X.; Korhonen, J.T.; Ras, R.H.A. Surface-wetting characterization using contact-angle measurements. *Nat. Protoc.* **2018**, *13*, 1521–1538. [CrossRef]
38. Hoffman, A.S. Hydrogels for biomedical applications. *Adv. Drug Deliv. Rev.* **2012**, *64*, 18–23. [CrossRef]
39. Tamer, T.M.; Collins, M.N.; Valachova, K.; Hassan, M.A.; Omer, A.M.; Mohy-Eldin, M.S.; Svik, K.; Jurcik, R.; Ondruska, L.; Biro, C.; et al. MitoQ Loaded Chitosan-Hyaluronan Composite Membranes for Wound Healing. *Materials* **2018**, *11*, 569. [CrossRef]
40. Noriega, S.E.; Subramanian, A. Consequences of Neutralization on the Proliferation and Cytoskeletal Organization of Chondrocytes on Chitosan-Based Matrices. *Int. J. Carbohydr. Chem.* **2011**, *2011*. [CrossRef]
41. Amaral, I.F.; Granja, P.L.; Melo, L.V.; Saramago, B.; Barbosa, M.A. Functionalization of chitosan membranes through phosphorylation: Atomic force microscopy, wettability, and cytotoxicity studies. *J. Appl. Polym. Sci.* **2006**, *102*, 276–284. [CrossRef]
42. Yamanlar, S.; Sant, S.; Boudou, T.; Picart, C.; Khademhosseini, A. Surface functionalization of hyaluronic acid hydrogels by polyelectrolyte multilayer films. *Biomaterials* **2011**, *32*, 5590–5599. [CrossRef] [PubMed]
43. He, Y.; Liu, C.; Xia, X.; Liu, L. Conformal microcapsules encapsulating microcarrier-L02 cell complexes for treatment of acetaminophen-induced liver injury in rats. *J. Mater. Chem. B* **2017**, *5*, 1962–1970. [CrossRef] [PubMed]
44. Welzel, P.B.; Prokoph, S.; Zieris, A.; Grimmer, M.; Zschoche, S.; Freudenberg, U.; Werner, C. Modulating Biofunctional starPEG Heparin Hydrogels by Varying Size and Ratio of the Constituents. *Polymers* **2011**, *3*, 602–620. [CrossRef]
45. Bean, J.E.; Alves, D.R.; Laabei, M.; Esteban, P.P.; Thet, N.T.; Enright, M.C.; Jenkins, A.T.A. Triggered Release of Bacteriophage K from Agarose/Hyaluronan Hydrogel Matrixes by Staphylococcus aureus Virulence Factors. *Chem. Mater.* **2014**, *26*, 7201–7208. [CrossRef]
46. Correia, C.R.; Moreira-Teixeira, L.S.; Moroni, L.; Reis, R.L.; van Blitterswijk, C.A.; Karperien, M.; Mano, J.F. Chitosan scaffolds containing hyaluronic acid for cartilage tissue engineering. *Tissue Eng. Part C Methods* **2011**, *17*, 717–730. [CrossRef]
47. Tous, E.; Ifkovits, J.L.; Koomalsingh, K.J.; Shuto, T.; Soeda, T.; Kondo, N.; Gorman, J.H., 3rd; Gorman, R.C.; Burdick, J.A. Influence of injectable hyaluronic acid hydrogel degradation behavior on infarction-induced ventricular remodeling. *Biomacromolecules* **2011**, *12*, 4127–4135. [CrossRef]
48. Barahuie, F.; Dorniani, D.; Saifullah, B.; Gothai, S.; Hussein, M.Z.; Pandurangan, A.K.; Arulselvan, P.; Norhaizan, M.E. Sustained release of anticancer agent phytic acid from its chitosan-coated magnetic nanoparticles for drug-delivery system. *Int. J. Nanomed.* **2017**, *12*, 2361–2372. [CrossRef]
49. Yang, C.-Y.; Hsu, C.-H.; Tsai, M.-L. Effect of crosslinked condition on characteristics of chitosan/tripolyphosphate/genipin beads and their application in the selective adsorption of phytic acid from soybean whey. *Carbohydr. Polym.* **2011**, *86*, 659–665. [CrossRef]
50. Ayala, R.; Zhang, C.; Yang, D.; Hwang, Y.; Aung, A.; Shroff, S.S.; Arce, F.T.; Lal, R.; Arya, G.; Varghese, S. Engineering the cell-material interface for controlling stem cell adhesion, migration, and differentiation. *Biomaterials* **2011**, *32*, 3700–3711. [CrossRef]
51. Barlian, A.; Judawisastra, H.; Alfarafisa, N.M.; Wibowo, U.A.; Rosadi, I. Chondrogenic differentiation of adipose-derived mesenchymal stem cells induced by L-ascorbic acid and platelet rich plasma on silk fibroin scaffold. *PeerJ* **2018**, *6*, e5809. [CrossRef] [PubMed]
52. Yang, Z.; Wu, Y.; Li, C.; Zhang, T.; Zou, Y.; Hui, J.H.; Ge, Z.; Lee, E.H. Improved mesenchymal stem cells attachment and in vitro cartilage tissue formation on chitosan-modified poly(L-lactide-co-epsilon-caprolactone) scaffold. *Tissue Eng. Part A* **2012**, *18*, 242–251. [CrossRef] [PubMed]

53. Mora-Boza, A.; García-Fernández, L.; Barbosa, F.A.; Oliveira, A.L.; Vázquez-Lasa, B.; San Román, J. Glycerylphytate crosslinker as a potential osteoinductor of chitosan-based systems for guided bone regeneration. *Carbohydr. Polym.* **2020**, *241*, 116269. [CrossRef] [PubMed]
54. Pescosolido, L.; Schuurman, W.; Malda, J.; Matricardi, P.; Alhaique, F.; Coviello, T.; van Weeren, P.R.; Dhert, W.J.; Hennink, W.E.; Vermonden, T. Hyaluronic acid and dextran-based semi-IPN hydrogels as biomaterials for bioprinting. *Biomacromolecules* **2011**, *12*, 1831–1838. [CrossRef]

Publisher's Note: MDPI stays neutral with regard to jurisdictional claims in published maps and institutional affiliations.

© 2020 by the authors. Licensee MDPI, Basel, Switzerland. This article is an open access article distributed under the terms and conditions of the Creative Commons Attribution (CC BY) license (http://creativecommons.org/licenses/by/4.0/).

Article

Formulation and Evaluation of Microwave-Modified Chitosan-Curcumin Nanoparticles—A Promising Nanomaterials Platform for Skin Tissue Regeneration Applications Following Burn Wounds

Hafiz Muhammad Basit [1,2], Mohd Cairul Iqbal Mohd Amin [3], Shiow-Fern Ng [3], Haliza Katas [3], Shefaat Ullah Shah [1] and Nauman Rahim Khan [1,2,*]

[1] Department of Pharmaceutics, Faculty of Pharmacy, Gomal University, DIKhan 29050, KPK, Pakistan; basitkhan053@gmail.com (H.M.B.); shefaatbu@gmail.com (S.U.S.)
[2] Gomal Centre for Skin/Regenerative Medicine and Drug Delivery Research (GCSRDDR), Faculty of Pharmacy, Gomal University, DIKhan 29050, KPK, Pakistan
[3] Centre for Drug Delivery Technology, Faculty of Pharmacy, Universiti Kebangsaan Malaysia, Kuala Lumpur 50300, Malaysia; mciamin@ukm.edu.my (M.C.I.M.A.); nsfern@ukm.edu.my (S.-F.N.); haliza.katas@ukm.edu.my (H.K.)
* Correspondence: naumanpharma@gmail.com

Received: 8 September 2020; Accepted: 5 October 2020; Published: 6 November 2020

Abstract: Improved physicochemical properties of chitosan-curcumin nanoparticulate carriers using microwave technology for skin burn wound application are reported. The microwave modified low molecular weight chitosan variant was used for nanoparticle formulation by ionic gelation method nanoparticles analyzed for their physicochemical properties. The antimicrobial activity against *Staphylococcus aureus* and *Pseudomonas aeruginosa* cultures, cytotoxicity and cell migration using human dermal fibroblasts—an adult cell line—were studied. The microwave modified chitosan variant had significantly reduced molecular weight, increased degree of deacetylation and decreased specific viscosity. The nanoparticles were nano-sized with high positive charge and good dispersibility with entrapment efficiency and drug content in between 99% and 100%, demonstrating almost no drug loss. Drug release was found to be sustained following Fickian the diffusion mechanism for drug release with higher cumulative drug release observed for formulation (F)2. The microwave treatment does not render a destructive effect on the chitosan molecule with the drug embedded in the core of nanoparticles. The optimized formulation precluded selected bacterial strain colonization, exerted no cytotoxic effect, and promoted cell migration within 24 h post application in comparison to blank and/or control application. Microwave modified low molecular weight chitosan-curcumin nanoparticles hold potential in delivery of curcumin into the skin to effectively treat skin manifestations.

Keywords: modified chitosan; curcumin; microwave; nanoparticles

1. Introduction

Being the largest human organ, skin is often easily susceptible to defects sometimes being congenital defect and mostly by burn, trauma or diseases which have long lasting effect on the wellbeing and psychology of the patient. According to WHO statistics, globally 180,000 people die annually due to skin burn injuries only [1]. To date, autografts, allografts and xenografts has been used for the purpose. Although they are able to promote wound healing but they have inherent demerits of limited availability, high expenses, immune incompatibility, risk of infection and shortage of donor sites [2–4]. These demerits gave birth to developing skin scaffolds able to hasten the skin regeneration process by engineering skin substitutes made of different biodegradable polymers and their composites with aided medicaments. An artificial extracellular matrix is thus developed impregnated with skin healing hastening agents [5,6]. The process of tissue engineering enables self-healing potential of human body for regeneration of lost, damaged or injured tissues/organs which is enabled by creating suitable cell environment for its survival and functional achievements [5,6]. Various attempts has been made to formulate/develop skin substitutes to hasten the normal wound healing process including hydrogels [7], sponges [8], polymeric nanoparticles [9], nanofibers [10] and polymeric films [11] etc.

Chitosan has been extensively studied and finds extensive applications in the field of biomedical engineering and drug delivery to construct tissue scaffolds [12,13], as an excipients in drug delivery carriers [14,15], in gene delivery [16], and wound healing [13] and can easily be processed into gels [17], membranes [18], nanofibers [19], beads [20], microparticles [21], nanoparticles [22], scaffolds [23] and sponges [24]. Chitosan promotes cell proliferation, activates macrophages, stimulates tissue reorganization, promotes fibroblast proliferation, promotes collagen deposition, enhances increased production of hyaluronic acid at wound site and also acts as hemostatic agent, thus helps in wound healing with minimal scar formation [25,26]. The physicochemical properties of chitosan i.e., molecular weight and degree of deacetylation play pivotal role in skin tissue regeneration applications where a direct relation is established between these properties and rate of wound healing [27,28].

Curcumin, a natural drug with potent antioxidant and antibacterial attributes has been found to promote tissue regeneration by tissue remodeling, granulation, new tissue formation and collagen deposition [29]. Various carrier mediated drug delivery systems has been developed and tested for their potential in wound healing studies including polymer-curcumin nanoparticles [30,31], polymer-curcumin nanoemulsion gel [32], gel [33], self-assemble nanogel [34] and hydrogels [35,36]. Various attempts have also been made to develop such scaffolds in the form of sponges, polymeric films and nanofibers [37–39]. The curcumin has also been tested as a composite formulation for the purpose of accelerated wound healing applications [18,40]. Curcumin being photosensitive with poor bioavailability necessitates development of a suitable carrier enabling efficient penetration into skin with good bioavailability.

Various chemical, physical, and biological methods are employed for the purpose of reducing the molecular weight of chitosan. Though, chemical method is a low cost method but possess the demerits of difficult process control and high cost of associate steps [41]. Similarly, physical methods like X-rays, gamma radiations and UV radiations though can easily be scaled up but they presents challenges like low quality of resultant product, unaffected degree of deacetylation and additionally special equipment's are also required [42,43], while the enzyme based methods have high production cost [44]. In contrast, microwave is a cheap, nondestructive technique where with efficiently modifying the molecular weight chitosan without damaging the main chitosan chain and significantly reducing the degree of deacetylation of chitosan [45]. The main purpose of this study was to reduce chitosan molecular weight using microwave technique followed by physicochemical analysis of modified chitosan and later its use in the fabrication of curcumin nanoparticles as low molecular weight chitosan with low viscosity and high degree of deacetylation is envisaged to improve curcumin delivery into skin to hasten skin tissue regeneration following 2nd degree burn wound.

2. Materials and Methods

2.1. Chemicals

Chitosan (high molecular weight (HMW), molecular weight: 310 kDa to 375 kDa, degree of deacetylation: ~75 %, Sigma Aldrich, St. Louis, MO, USA)) was used as a starting material with curcumin (~95% purity, Zhejiang metals and minerals Import and Export Corporation, Hangzhou, Zhejiang Province, China) as a model drug for nanoparticle formulation. Ethanol (≥95% Sigma Aldrich, USA), monobasic potassium phosphate (Sigma Aldrich, St. Louis, MO, USA), disodium hydrogen orthophosphate (Merck, Darmstadt, Germany), hydrochloric acid (Merck, Germany), tripolyphosphate (TPP, Sigma Aldrich, USA), acetic acid (Sigma Aldrich, St. Louis, MO, USA), sodium chloride (Sigma Aldrich, St. Louis, MO, USA) and sodium hydroxide (Sigma Aldrich, St. Louis, MO, USA) were used in nanoparticle fabrication as adjuvant chemicals. All chemicals were used without any further purification.

2.2. Chitosan Molecular Weight Modification

The molecular weight of chitosan was reduced using already reported method [45]. Briefly, one accurately weighed gram of chitosan was added into accurately weighed amount of 49 g of 2% (v/v) acetic acid solution under ambient conditions with continuous magnetic stirring until complete dissolution which was then followed by the addition of 2 mL of 0.9% (w/w) sodium chloride solution for the purpose of increasing solution conductivity in order to hasten rapid chain breakage in a shorter time [46]. The mixture was then placed on the turntable of a microwave oven (LG, Model Number MS2022D, Beijing, China) in an off-center position. It was then subjected to microwave treatment at an irradiation power of 800 W for 5 min and 9 min intervals. Following microwave treatment, the solution was properly covered and left to cool at room temperature. The pH of the solution was then adjusted to 7.5 using 2N sodium hydroxide solution and then the polymer was precipitated by adding 40 mL of ethanol (96%). The precipitate was filtered, washed several times with distilled water and incubated in a hot air oven at 40 °C ± 1 °C for 24 h or until completely dry. The dried chitosan was then conditioned in a desiccator under ambient conditions for 24 h, triturated to form powder and then subjected to various characterization tests listed below.

2.3. Specific Viscosity (SV)

The solution viscosity of the modified chitosan was determined using an already reported method [47]. Briefly, a 0.5% (w/w) solution of each chitosan sample was prepared in 2% (v/v) acetic acid and filled into a U-tube viscometer (Ostwalds Viscometer, Zhejiang, China). The solution was sucked to an already marked distance and left to travel under gravity till the lower mark under ambient conditions. The time for solutions to travel from upper mark to lower mark were recorded. The same procedure was used for 2% (v/v) acetic acid solution as a reference. The relative solution viscosity of each chitosan sample was calculated using following relation:

$$SV = \frac{F - F0}{F0} \qquad (1)$$

where F is flow time of test solution, $F0$ is flow time of 2% (v/v) acetic acid solution.

Experiments were conducted at least in triplicate, and the results averaged.

2.4. Degree of Deacetylation (DD)

The DD of chitosan was determined using the vibrational spectroscopic technique. The respective chitosan samples were spread on a diamond crystal for attenuated total reflectance-fourier transform infrared spectroscopy (ATR-FTIR, UATR TWO, Perkin Elmer, Bukinghamshire, UK) and clamped to ensure close contact and high sensitivity. Each sample was scanned at a resolution of 4 cm^{-1} over a wavenumber region of 450 to 4000 cm^{-1}. The DD of chitosan was calculated using the following equation:

$$DD = 100 - \left[\frac{A1655}{A3450} \times \frac{100}{1.33}\right] \qquad (2)$$

where A denoted ATR-FTIR absorbance value at 1655 and 3450 cm^{-1} wavenumbers ascribing amide-I of N-acetyl moiety and hydroxyl/amine moieties of the chitosan. The factor 1.33 denoted the value of A1655/A3450 for a fully N-acetylated chitosan [48]. Experiments were carried out at least in triplicate for each batch of samples and the results averaged.

2.5. Molecular Weight Analysis of Chitosan

The molecular weight analysis of HMW and low molecular weight (LMW) chitosan was done using gel permeation chromatography (Waters 1515 Isocratic HPLC pump, Waters Corporation, Hertfordshire, UK) equipped with dual a wavelength absorbance detector (2487, Waters Corporation, Hertfordshire, UK) and an injector (Waters 717 Plus Autosampler, Waters Corporation, Hertfordshire, UK) by means of a refractive index detector (Waters 2414 refractive index detector) as previously reported [49]. A styragel HR 5 tetrahydrofuran (THF) column (5 µm, 10–5 Å, 7.8 mm × 300 mm, Waters, Hertfordshire, UK) was used with tetrahydrofuran (THF) as the mobile phase. Samples were prepared by dissolving 5 mg of the polymer in 5 mL of 0.001% acetic acid (v/v) and filtered through nylon membrane (5 µm, Mountain Safety Research (MSR), Nylon Syringe Filter, Seattle, WA, USA) and 50 µl of it was then injected into the system. The flow rate of the mobile phase and the column temperature were kept at 1 mL/min and 40 °C, respectively. Experiments were carried out at least in triplicate and the results averaged.

2.6. LMW Chitosan-Curcumin Nanoparticles Preparation

LMW chitosan-curcumin nanoparticles were prepared by4 ionic gelation method, previously described [50] with some modification. Briefly, curcumin was separately dissolved in ethanol (96%) in a concentration of 1% (w/v) and chitosan was dissolved in 2% (v/v) acetic acid solution at pH 5.0 separately at concentrations of 0.3 and 0.4% (w/v), respectively. Two nanoparticles formulations were prepared viz. in one beaker, 3 mL of 0.3% chitosan solution was taken and added to 0.3 mL of curcumin solution, while, in another beaker, 3 mL of chitosan 0.4% was taken and added to 0.4 mL curcumin stock solution. Nanoparticles were spontaneously formed by adding 3 mL of TPP solution (0.1% w/v, in deionized distilled water) into the above solutions under a constant magnetic stirring (MS MP8 Wise Stir, Wertheim, Germany) at 700 rpm for 1 h at room temperature. The resultant modified chitosan-curcumin nanoparticles were harvested by ultracentrifugation (11,200× g) using an Optima L-100 XP Ultracentrifuge (Beckman-Coulter, Brea, CA, USA) at 18 °C for 1 h. The resultant pelleted nanoparticles were re-suspended in deionized distilled water, freeze dried at −48 °C for 3 to 4 days or until completely dry (Freezone loor top freeze dryer, Labconco, Kansas City, MO, USA) and subjected to further characterization tests.

2.7. Size and Zeta Potential

The particle size and zeta potential of the LMW chitosan-curcumin nanoparticles were measured by dispersing the nanoparticles in aqueous ethanolic solution through brief sonication and measured by method of photon correlation spectroscopy using Malvern Zetasizer Nano ZS 90 (Malvern Instruments Ltd., Malvern, UK) at 25 °C in a quartz cell and zeta potential cell with a detect angle of 90°, respectively. Experiments were conducted in triplicate and the results averaged.

2.8. Entrapment Efficiency and Drug Content

The entrapment efficiency and drug content of LMW chitosan-curcumin nanoparticles were determined by employing an indirect method of determining the free drug and the difference in total drug added and free drug determined. Briefly, the freeze-dried nanoparticles were resuspended in deionized water and filled into ultracentrifuge tubes and subjected to ultracentrifugation at 11,200× g using an Optima L-100 XP Ultracentrifuge (Beckman-Coulter, Brea, CA, USA) at 18 °C for 1 h. The obtained pellets were separated and washed with methanol twice to remove any unentrapped curcumin. The methanol rinse and the supernatant were combined and analyzed by HPLC (Waters 1515 Isocratic HPLC pump, Waters Corporation, Hertfordshire, UK) equipped with dual wavelength absorbance detector (2487, Waters Corporation, Hertfordshire, UK) and an injector (Waters 717 Plus Autosampler, Waters Corporation). A reversed-phase column (ODS Hypersil, 250 × 4.6 mm, 5 µm, Thermo Scientific, Loughborough, UK) fitted with a refillable C18 guard column

(Upchurch Scientific, Oak Harbor, WA, USA) was used as the stationary phase for estimation of unentrapped drug. The mobile phase comprised of acetonitrile and trifluoro acetic acid (TFA, 0.2% v/v) in a 1:1 ratio at a flow rate of 1.5 mL/min with a detection wavelength of 430 nm. The experiment was repeated three times and the results averaged. The percentage of entrapped curcumin was calculated using the following relation:

$$Percent\ entrappment\ efficiency = \frac{Total\ Cur\ added - unloaded\ Cur}{Total\ Cur\ added} \times 100 \qquad (3)$$

The total drug content of the nanoparticles was calculated using following relation:

$$Percent\ Drug\ content = \frac{Loaded\ Cur}{Total\ Cur\ added} \times 100 \qquad (4)$$

The experiment was repeated three times and the results averaged.

2.9. Morphology

The Morphology of LMW chitosan-curcumin nanoparticles was analyzed using transmission electron microscopy (Tecnai Spirit, FEI, Eindhoven, The Netherlands) operated at an acceleration voltage of 80 kV. A total of 10 µL of samples (concentration 1 mg/mL) was placed on a 400-mesh copper grid coated with carbon and stained with uranyl acetate (2% w/v). The imaging was then performed, and respective sections photographed.

2.10. In Vitro Drug Release

The in vitro drug release pattern of the curcumin from nanoparticles was studied on vertical glass Franz diffusion cells (PermeGear Inc., Hellertown, PN, USA). Briefly, the receiving compartment of the Franz diffusion cell internal volume (12 mL) was filled with bubble free phosphate buffered saline (PBS) pH 7.4 to mimic wound pH with a surface area 1.76 cm^2 enabling maximal drug solubility in the solvent maintained at 37 °C ± 1.0 °C with continuous magnetic stirring on a magnetic stirrer at 400 rpm. The Tuffryn® membrane (0.2 µm pore size, HT Tuffryn® membrane, Sigma Aldrich, St. Louis, MO, USA) was used as a barrier between the receiving and donor compartments. An accurately weighed 120 mg of nanoparticle was placed on membrane and the experiment run for 24 h. The sample aliquots of 1 mL were withdrawn at regular time intervals of 0, 0.5, 1, 2, 4, 6, 8, 12, 24 and 48 h and analyzed by HPLC (as discussed earlier) for the extent of drug release. Solvent equal in volume to volume withdrawn was added at each sampling point to maintain sink conditions. The cumulative amount of curcumin released and hence which permeated through per surface area (cm^2) of membrane was calculated and plotted against time (h). The experiment was run in triplicate and the results averaged.

2.11. Drug Release Kinetics

The mechanism of drug release from the films was assessed by fitting drug release data into the Korsmeyer–Peppas equation as expressed by:

$$Mt/M\infty = Kt^n \qquad (5)$$

where $Mt/M\infty$ is a fraction of drug released at time t, k is the release rate constant and n is the release exponent. The $0.45 \leq n$ corresponds to a Fickian diffusion mechanism, $0.45 < n < 0.89$ to non-Fickian transport, $n = 0.89$ to Case II (relaxational) transport, and $n > 0.89$ to super case II transport [51].

2.12. Vibrational Spectroscopic and Thermal Analysis of Nanoparticles

The characteristics peaks of all samples (HMW chitosan, LMW chitosan, curcumin and LMW chitosan-curcumin nanoparticles) were recorded by an ATR-FTIR spectrophotometer (UATR TWO, Perkin Elmer, Bukinghamshire, UK). Each sample was placed on to the surface of the diamond crystal and clamped to ensure close contact and high sensitivity. All the samples were scanned over a wavenumber range of 450 to 4000 cm^{-1} with an acquisition time of 2 min. Each sample was analyzed three times and the results averaged.

Thermal analysis was also employed as an additional tool to the assess changes induced in the polymer by microwave treatment and to evaluate effective entrapment of curcumin inside the core of nanoparticles. Briefly, the changes in the transition temperature and enthalpies of all samples (HMW chitosan, LMW chitosan, curcumin and LMW chitosan-curcumin nanoparticles) were recorded via differential scanning calorimetry (PerkinElmer Thermal Analysis, Boston, MA, USA). An accurately weighed 2 to 3 mg of each sample was sealed in standard aluminum pan and heated from 0 °C to 300 °C under a continuous flow of nitrogen gas at rate of 40 mL/min. The characteristic peak temperature and enthalpy of the system were recorded. Each sample was analyzed three times and the results averaged.

2.13. Antimicrobial Activity

The antibacterial activity of LMW chitosan-curcumin nanoparticles was assessed using two selected bacterial strains i.e., *Staphylococcus aureus* (American Type Culture Collection (ATCC) 6538™)) and *Pseudomonas aeruginosa* (ATCC 9027™) under aseptic conditions (class II biological safety cabinet, Clyde Apac, Minto, Australia) via a well diffusion test. Briefly, a sterile swab was dipped in bacterial inoculum suspension and streaked on the surface of Mueller–Hinton agar following which multiple wells (diameter 6 mm) were made in agar plate carefully to ensure similar depth and diameter. Nanoparticles containing different drug concentrations (1000, 500, 250 µg/mL, pre-suspended in sterile deionized water) as well as blank nanoparticles were applied into wells using Gentamicin (20 µg/mL) as a positive control, distilled water as a negative control and plates incubated at 37 °C for 24 h (Memmert, Schwabach, Germany). The following day, zones of inhibition were measured. The results are presented as the mean of three experiments.

The minimum inhibitory concentration (MIC) and minimum bactericidal concentration (MBC) were determined using a broth microdilution assay [52,53]. Briefly, both bacterial inoculums were prepared and adjusted to cell density of 1–2 × 10^8 CFU/mL. A total of 100 µL of Mueller–Hinton broth was dispensed into each target well of a 96-well plate followed by application of various nanoparticles formulations pre-suspended in sterile deionized water containing different concentrations of curcumin (1000, 500, 250, 125, 62.5 and 31.2 µg/mL) in triplicate with Gentamicin (20µg/mL) used as a positive control, distilled water as a negative control and broth without bacteria used as an environment control. The contents of wells were thoroughly mixed with a micropipette and incubated at 37 °C for 24 h. Following incubation, all the wells were visually inspected for the appearance where clear wells signify that bacterial growth has been inhibited. The minimum concentration of curcumin at which the wells remained colorless will be the MIC of the applied formulation. Following day, the contents of wells were then streaked onto agar plates and incubated at 37 °C for 24 h and observed for any bacterial growth. The minimum curcumin concentration at which no growth appears will be the MBC.

2.14. Cell Viability

Human Dermal Fibroblast-Adult (HDFa) cells were purchased from the Tissue Engineering Centre (Kuala Lumpur, Malaysia) and were cultured using Dulbecco's Modified Eagle Medium (DMEM) supplemented with 1% Penicillin/Streptomycin, 5% fetal bovine serum in a T-75 flask and incubated at 37 °C under a 5% CO_2 purge till confluence. Following incubation, cells were counted and 1×10^4 cells per well in a 96 well culture plate were seeded and incubated overnight to ensure cell adhesion. On the following day, nanoparticles (1000, 500, 250, 125, 62.5 and 31.2 µg/mL, premixed with 100 µL of media) were applied and plates were incubated again for 24 h. The next day, media containing formulations were decanted off aseptically and 10 µL of MTT reagent was added into each well and the plates incubated again for 4 h where the live cells converted the MTT reagent to needle like formazan crystals which transit to the cell surface through the process of exocytosis from the endosomal/lysosomal compartment. The formed formazan crystals were dissolved by adding 100 µL of DMSO into each well, shaking for 20 min at 300 rpm followed by analysis on a UV microplate reader (Infinite® 200 Pro NanoQuant, Tecan, Männedorf, Switzerland) at wavelength ah of 570 nm. The cell viability percentage was calculated using the following equation [54]:

$$Cell\ viability = \frac{A1}{A2} \times 100 \quad (6)$$

where A1 = Absorbance of treated cells; A2 = Absorbance of control cells.

The results are presented as a mean of six experiments.

2.15. Cell Migration Analysis

HDFa cells at a density of 1×10^4 cells/well were seeded into a 6-well plate and incubated for 72 h in a humidified incubator maintained at 37 °C with 5% CO_2 (ESCO CCL-170B-8 Singapore). Small linear scratches were created along the confluent monolayer of cells by gently scraping the surface using a sterile pipette tip. Cells were gently rinsed with sterile phosphate buffer saline (PBS) to remove cellular debris. Each well was applied with 1000 µg/mL formulation and blank nanoparticles applied separately and incubated again. The photographs of plates were taken at intervals of time 0, 24, 48 and 72 h using a digital camera connected to the inverted microscope (CK30-F200 Olympus, Tokyo, Japan). Plates without any treatment were used as a control. The time required for cells to fill the scratch completely was recorded for each application. All images were analyzed for the area left unfilled by the cells in the scratch after different time intervals via Image J using the unfilled area of scratch in photographs taken after no time as a standard scale. The experiment was repeated six times and the results compared. The percentage of wound closure was calculated as follows:

$$Wound\ closure\ (\%) = \frac{w(0) - w(t)}{w(0)} \times 100 \quad (7)$$

where $w(0)$ is wound area at 0 time and $w(t)$ is wound area at a specific time.

2.16. Statistical Analysis

All values are expressed as a mean of three readings with respective standard deviation. Student's t-test or analysis of variance (ANOVA) followed by post-hoc analysis was used with the level of significance set at $p < 0.05$.

3. Results

3.1. Physicochemical Analysis of Parent and Modified Chitosan

Under the influence of microwaves, the chitosan chain scission took place at the main chain and at the amide linkage between the amino functional group of the parent chain and acetyl moiety. The HMW chitosan (molecular weight = 30,665 ± 245 Da; degree of deacetylation = 81.29 ± 6.9%) became shorter with 9 min microwave treatment (15,223 ± 182 Da, Student's t-test, $p = 0.002$) and a higher degree of deacetylation (91.41 ± 8.2; Students t-test: $p = 0.06$). The LMW chitosan also had significantly lower solution viscosity than the HMW chitosan (parent chitosan = 13.466 ± 2.690 and 9 min = 2.009 ± 0.0164; $p < 0.05$). Similarly, the particle size of the powdered polymer also significantly reduced (Student t-test, $p < 0.05$) where for HMW chitosan the size of 4123.7 ± 421.4 nm reduced to 667.7 ± 145.7 nm and surface charge significantly increased (Student t-test, $p < 0.05$) from 21.4 ± 2.3 mV to 67.4 ± 6.2 mV with microwave scission of HMW chitosan due to reduced acetyl content (Table 1).

Table 1. Physicochemical analysis of polymer samples and low molecular weight (LMW) chitosan nanoparticles.

Formulation	Chitosan (%)	Curcumin (µg)	Size (nm)	Zeta Potential (mV)	PDI	Drug Content (%)	Entrapment Efficiency (%)
Parent chitosan	-	-	4123.7 ± 421.4	21.4 ± 2.3	1.0 ± 0.0	-	-
LMW Chitosan	-	-	667.7 ± 145.7	67.4 ± 6.2	0.72 ± 0.1	-	-
F1 blank	0.3	-	170.7 ± 2.9	46.4 ± 2.6	0.27 ± 0.05	-	-
F2 blank	0.4	-	223.7 ± 9.3	46.7 ± 1.6	0.5 ± 0.01	-	-
F1	0.3	300	279.7 ± 20.3	52.4 ± 1.50	0.67 ± 0.05	99.99 ± 1.39	99.93 ± 3.43
F2	0.4	400	259.2 ± 19.4	42.7 ± 1.53	0.54 ± 0.05	99.99 ± 0.34	99.96 ± 2.12

3.2. Physicochemical Analysis of Modified Chitosan Curcumin Nanoparticles

The size, polydispersability index (PDI), zeta potential of HMW chitosan, the LMW chitosan variant, drug content and entrapment efficiency of LMW chitosan-curcumin nanoparticles results with varying drug concentrations are given in Table 1. Pearson correlation analysis revealed that the concentration of chitosan is negatively related to size of the nanoparticles showing a strong indirect relation between the two (R = 0.959) while, in case of zeta potential, an indirect but weak relationship between the chitosan concentration and surface charge of the nanoparticles (R = 0.700) was detected.

3.3. Morphology

The morphological analysis of the optimized LMW chitosan-curcumin nanoparticles results are given in Figure 1. The TEM analysis showed nanoparticles formed were spherical to square and pentagonal in shape with even geometry throughout their structure. The increase in curcumin quantity from 300 µg to 400 µg did not significantly alter the surface morphology where both the formulations showed smooth surface appearance. TEM results confirmed our data from dynamic light scattering where an increase in chitosan concentration did not significantly alter the nanoparticle structure and the overall morphology of square to rectangular and pentagonal remained independent of chitosan concentration [55].

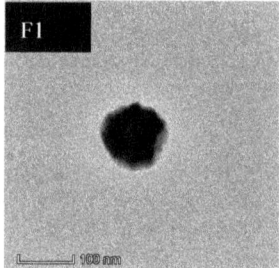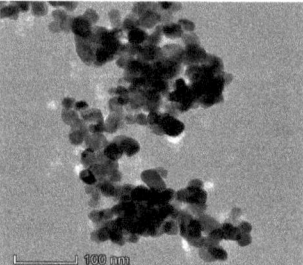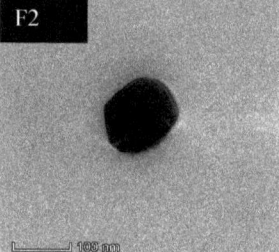

Figure 1. TEM pictographs of different LMW chitosan-curcumin nanoparticles.

3.4. In Vitro Drug Release

The in vitro drug release behavior of the curcumin from both optimized formulations (F1 and F2) is given in Figure 2. Both the formulations showed a sustained drug release pattern with no significant difference between the two (ANOVA, $p > 0.05$) irrespective of the quantity of drug added into the formulation. The F2 sample containing a higher drug quantity showed a higher cumulative drug release pattern compared to the F1 formulation. The release was consistent from the very beginning of the release experiment for both formulations where F2 showed significantly higher cumulative drug released at the end of experiment i.e., 20.67% ± 1.84 µg/cm^2·h compared to F1 where the same was found to be 18.12% ± 0.40 µg/cm^2·h (Student's t-test, $p < 0.05$).

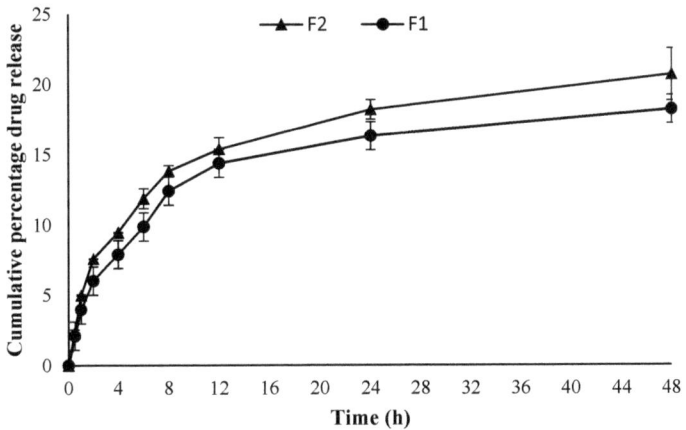

Figure 2. Cumulative percentage drug release ($n = 3$, ±SD).

3.5. Vibrational Spectroscopic Analysis

The ATR-FTIR spectra of HMW chitosan, LMW chitosan, TPP, pure curcumin and F2 nanoparticles are given in Figure 3. The principle peaks of chitosan were observed at 3358–3355 cm^{-1} which is mainly due to OH/NH stretching vibration, weaker bands between 2873–2871 cm^{-1} and 1587–1560 cm^{-1} corresponding to CH stretching and NH$_2$ groups, respectively, and at 1027 cm^{-1} which arises due to C–O–C stretching (Figure 3a–c) [49]. The curcumin principle peaks were observed at 3505–3504 cm^{-1} and 1625 cm^{-1} and 1507 cm^{-1} which corresponds to OH stretching and aromatic (C=C) stretching (Figure 3d) [56]. The spectrum of TPP demonstrated principle signature peaks at 1210.3, 1165.4 cm^{-1} and 1095.3 cm^{-1} which demonstrate phosphate stretching (P–O–C) and phosphate vibration (P=O) and symmetric and asymmetric stretching vibration of the PO$_3$ groups (Figure 3e) [57].

3.6. Thermal Analysis

The thermal analysis results of HMW chitosan, LMW chitosan and the formulation are given in Figure 4. The endothermic peak of parent chitosan moiety at 171.91 °C ± 8.3 °C significantly reduced to 146.85 °C ± 7.4 °C depicting that microwave treatment could successfully scissor the chitosan chain into a small molecular weight moiety, but interestingly, the enthalpy of the system, instead of being reduced, significantly increased with microwave treatment (Student's t-test, $p < 0.05$).

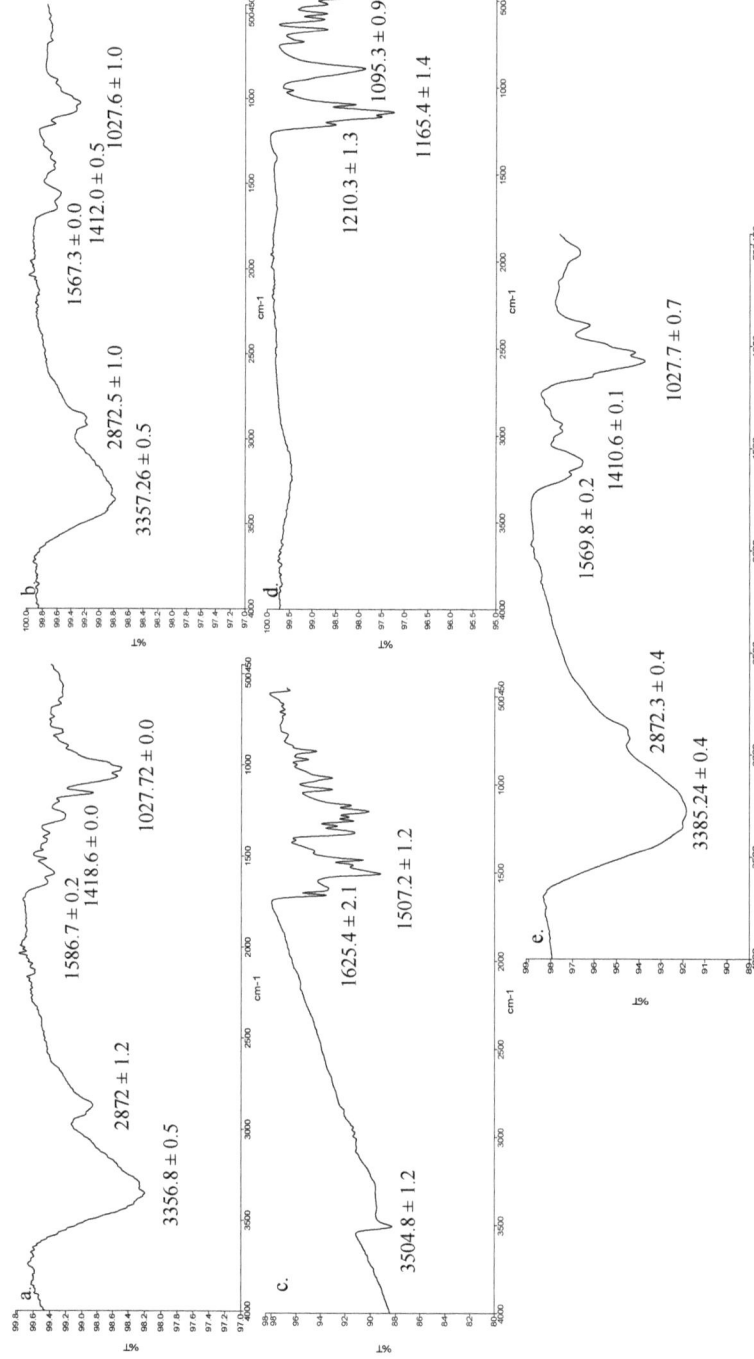

Figure 3. ATR-FTIR spectra of (**a**) high molecular weight (HMW) chitosan, (**b**) LMW chitosan, (**c**) pure curcumin, (**d**) tripolyphosphate (TPP), (**e**) formulation (F2) nanoparticles.

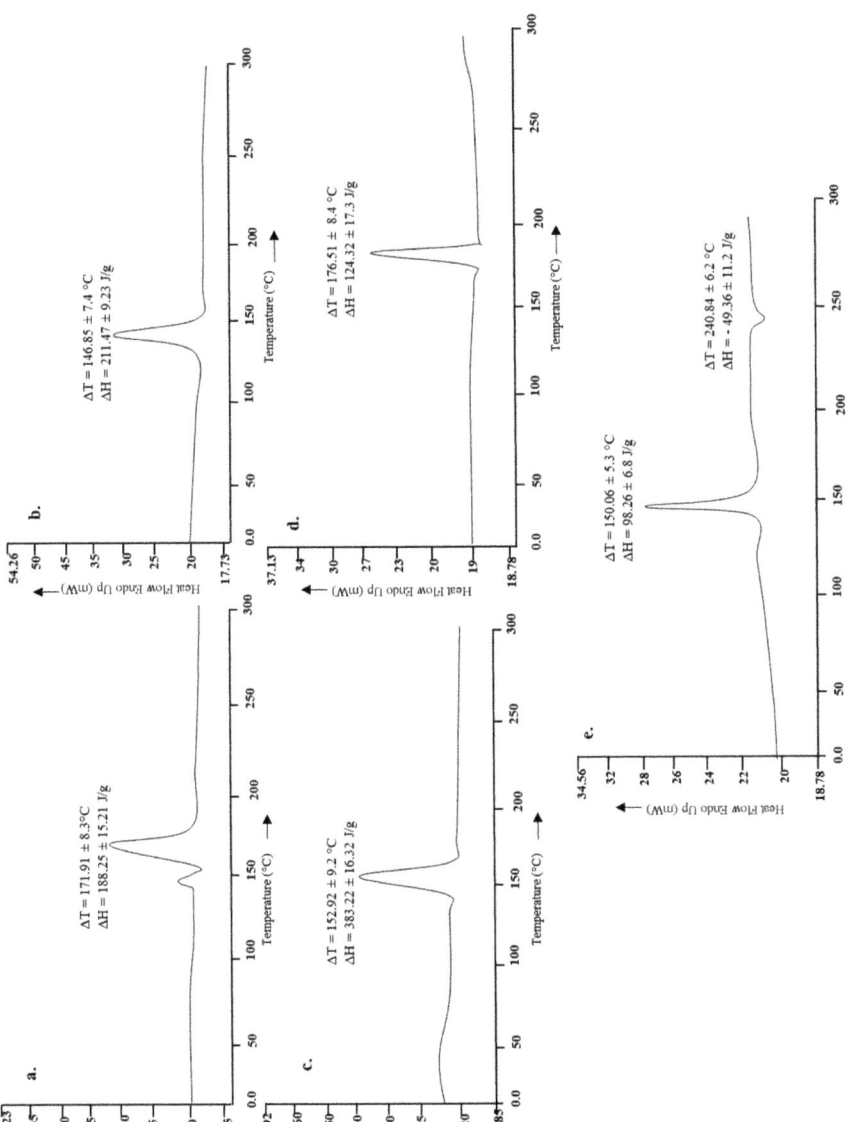

Figure 4. Thermal analysis of (**a**) parent chitosan, (**b**) microwave treated chitosan, (**c**) blank F2, (**d**) curcumin, (**e**) F2.

3.7. Antimicrobial Analysis

The antimicrobial activities of drug loaded, and blank nanoparticles are given in Figure 2. Both the drug loaded formulations (F1 and F2) showed dose dependent antimicrobial activity against selected bacterial strains with significant effect (Student's t-test, $p < 0.05$) observed for the F2 formulation containing higher quantity of chitosan used in nanoparticle fabrication. The formulated chitosan and curcumin nanoparticles were found to synergistically exert antimicrobial activity (Figure 5) where both the F1 and F2 formulations showed significant (Student's t-test, $p < 0.05$) antimicrobial activity against *Staphylococcus aureus* and *Pseudomonas aeruginosa* compared to blank nanoparticles and comparable to the positive control. The MIC for drug loaded nanoparticles was found to be 250 µg for both formulations irrespective of chitosan quantity used while the MBC was found to be 1000 µg for F1 and 500 µg for F2 formulation demonstrating that, since F2 contained a higher concentration of chitosan, chitosan-curcumin exerted more synergistic bactericidal activity compared to F1.

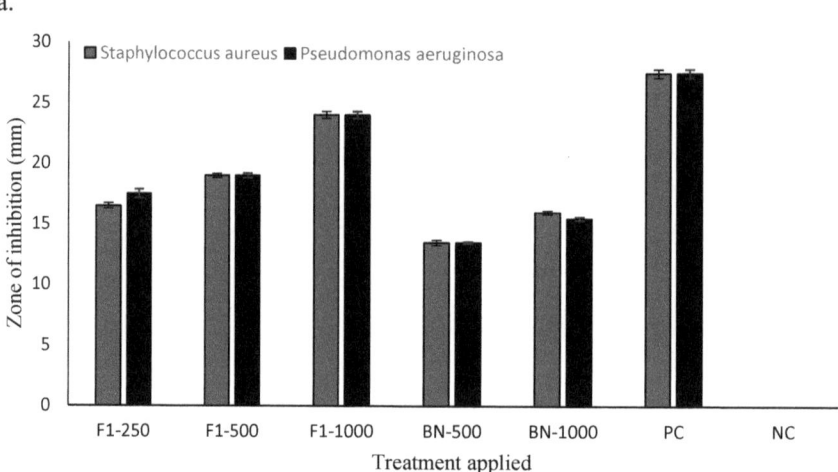

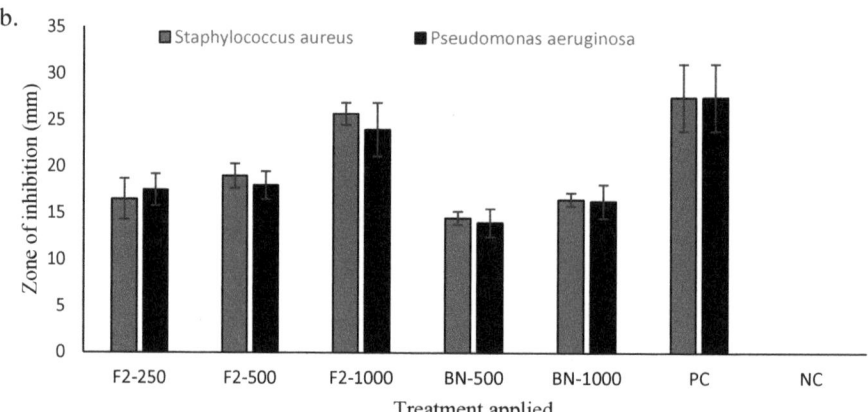

Figure 5. Antimicrobial activities of (**a**) F1 and (**b**) F2 different treatments applied: 250 µg, 500 µg, 1000 µg, BN = Blank nanoparticles, PC = positive control (Gentamicin), NC = negative control (distilled water).

3.8. Cell Viability and Cell Migration Analysis

The drug loaded nanoparticles' effect was evaluated using an adult human dermal fibroblast cell line using vehicle as a control. Both the formulations did not significantly affect the cell growth as the cell viability remained above 80% after 24 h post application due to low cytotoxicity ascribed to modified physicochemical attributes of chitosan (Figure 6). To assess the ability of chitosan and curcumin to promote cell proliferation, a cell migration test was carried out using human dermal fibroblasts with 1000 µg nanoparticle application up to 72 h using drug loaded as well as blank nanoparticles (Figure 7). The scratch area for F2 achieved full confluence within 48 h while F1 and the control within full confluence 72 h. The blank exhibited the slowest cell migration as the scratch area did not achieved full confluence within 72 h. The percentage wound closure was found to be 69.39 % (Student's t-test, $p < 0.05$) post 24 h incubation compared to control and/or blank nanoparticles treatment where it was found to be 40.02% and 19.86%, respectively.

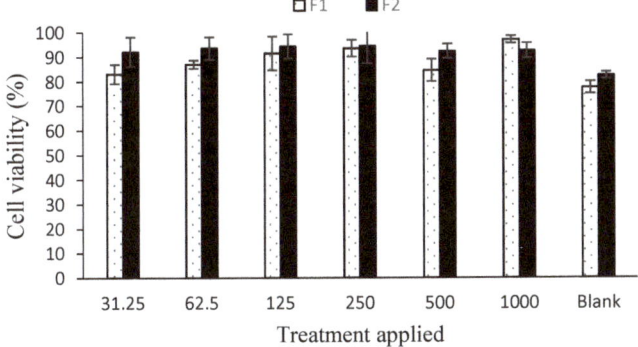

Figure 6. Cell viability analysis results ($n = 6$, ± SD).

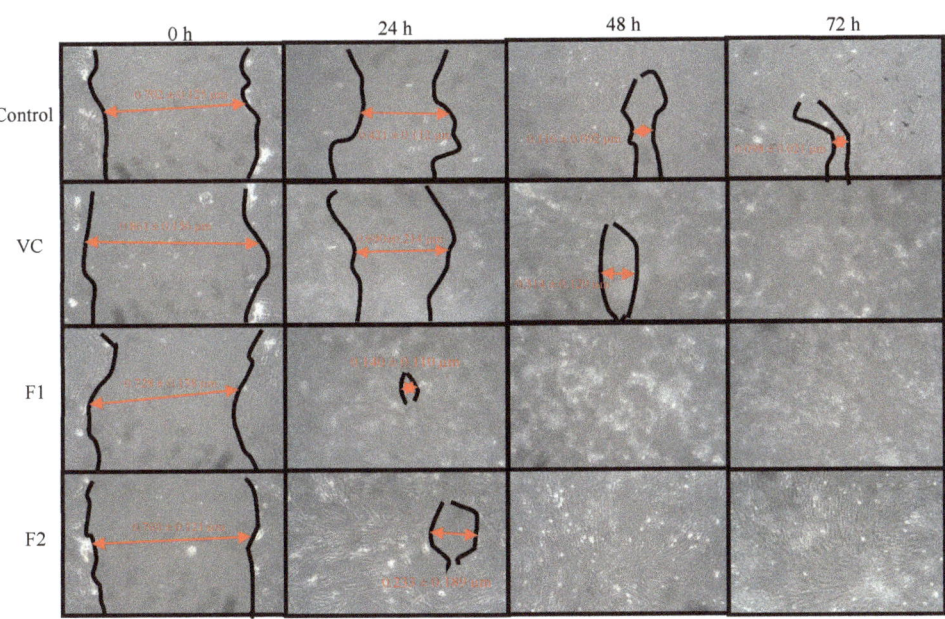

Figure 7. Cell migration analysis.

4. Discussion

Under the influence of microwaves, chitosan chain scission took place at the main chain and at the amide linkage between the amino functional group of the parent chain and acetyl moiety which significantly reduced the particle size and increased the surface charge with microwave scission of HMW chitosan due to reduced acetyl content. The molecular weight of the polymer is directly related to the size of the formulated nanoparticles where higher the molecular weight, the larger the nanoparticles achieved and vice versa. The skin drug and/or nanoparticle penetration is size dependent. Low molecular weight, low viscosity, smaller particle size and higher surface charge chitosan is envisaged to help produce smaller sized nanoparticles with efficient penetration into the skin and/or burn site and hence deliver the drug more effectively. Similarly, increase in the degree of deacetylation leads to formation of more free NH_2 functional groups on the chitosan chain with reduced acetyl content, which is envisaged to impart more positive surface charge onto the chitosan chain due to protonation of NH_2 by the free protons available in the acidic media, leading to a higher magnitude of the positive surface charge which is deemed favorable for the stability of nanoparticles due to mutual repulsion and hence prevents aggregation. Furthermore, skin is inherently negatively charged due to polar lipids, predominantly phospholipids, in the skin structure [58]. The negative charge on skin is deemed favorable to attract positively charged chitosan polymer moieties enabling adhesion and/or drug/nanoparticle penetration into the skin. Though nanoparticle size is inversely related to the chitosan concentration used in their fabrication, a situation where more free NH_3^+ groups are available to react with TPP and having a chitosan concentration beyond 0.3% w/v are reported to increase the nanoparticle size [49,59,60]. Our findings were contradictory to this, where with the use of 0.4% w/v chitosan concentration, a slight decrease in the nanoparticle size (Student's t-test, $p = 0.260$) was observed which can be attributed to the chitosan of low molecular weight used in the nanoparticle fabrication [61] (Table 1). Similarly, drug loading is also expected to further increase nanoparticles size, being an additional formulation moiety which is either adsorbed at the surface or entrapped in the nanoparticle core, and hence the particle size of the drug loaded nanoparticles was found to be significantly increased (Student's t-test, $p = 0.015$) compared to blank nanoparticles [49,62,63] as shown in Table 1. Pearson correlation analysis revealed that the concentration of chitosan is negatively related to size of the nanoparticles, showing a strong indirect relation between the two ($R = 0.959$). The zeta potential analysis showed all nanoparticles bear high positive surface charge irrespective of the drug concentration added but significantly increase with increasing concentrations of chitosan (Student's t-test, $p < 0.05$). The Pearson correlation analysis revealed an indirect but week relationship between the chitosan concentration and surface charge of the nanoparticles ($R = 0.700$). High surface positive charge not only ensures the stability of nanoparticles due to mutual repulsion but also guarantees enhanced antimicrobial activity through interaction with the bacterial cell membrane which is negatively charged. Moreover, blank and curcumin nanoparticles with high concentrations of chitosan (0.3% to 0.4% w/v) showed PDI values in the range of 0.5 to 0.6, which indicated narrow particle size distribution. The drug content of both formulations was found to be in the range of 99% to 100%, irrespective of the quantity of chitosan used in fabrication of nanoparticles which reflected no damage or spillage during experimentation with excellent entrapment efficiency of higher than 99% in both formulations.

Both the formulations (F1 and F2) showed a sustained drug release pattern (Figure 2). Curcumin is a poorly water-soluble and photosensitive drug. Encapsulation of curcumin in chitosan nanoparticles is envisaged to not only improve its higher cellular uptake but also improve stability and protection from photodegradation [64,65]. Sustained release of the drug with comparative good solubility in water (i.e., wound bed) is envisaged to promote skin regeneration with low frequency of dressing changing. The drug release data of F1 and F2 were fitted into the Korsmeyer–Peppas equation where both formulations showed a Fickian diffusion mechanism of drug release (F1 $n = 0.350$, $R^2 = 0.9514$ and for F2 $n = 0.336$, $R^2 = 0.9634$).

The vibrational analysis (Figure 3) indicated that microwave induced scission of chitosan does not significantly alter the main chain structure, confirming the nondestructive effect of microwave on parent chitosan while the nanoparticle spectra was missing the signature peaks observed for curcumin (i.e., aromatic stretching C=C) which demonstrated that the curcumin was entrapped in the core of nanoparticles instead of being adsorbed on the surface. The thermal analysis results (Figure 4) indicated that the glass transition temperature of chitosan and its oligomers are greatly affected by the number of amine groups available as they contribute towards hydrogen bonding between the chitosan chain and water [66]. Reduction of molecular weight leads to reduction of acetyl content and hence more NH_2 functional groups available for hydrogen bonding between the chitosan chains and/or with the water; hence, this led to a significant increase in the energy required to induce a state transition (ΔH = 211.47 ± 9.23 J/g, Figure 4b). Thermal analysis of blank nanoparticles (Figure 4c) revealed a slight increase in the transition temperature and a significant increase in the enthalpy values, which is attributed to the formation of a cross linked structure by the use of tripolyphosphate [67]. Incorporation of curcumin into LMW chitosan nanoparticles insignificantly reduced the transition temperature of the chitosan moiety (Figure 4e), suggesting that that the drug is located between the polymeric chains, and increased the free volume and mobility of the system [68]. The transition temperature corresponding to curcumin moiety significantly increased from 176.51 °C ± 8.4 °C to 240.84 °C ± 6.2 °C (Figure 4d). Increase in the transition melting temperature of curcumin suggests an effective cross-linked polymer network was formed with curcumin entrapped inside the nanoparticle core and shift of curcumin melting temperature is envisaged due to the formation of additional interactive forces between the chitosan and curcumin moieties. Curcumin binds to the new polymeric network via hydrogen bonds with chitosan surface OH and/or NH_2 functional groups [69] and/or electrostatic interaction with the former being the prominent binding force between chitosan and curcumin [70], thus promoting greater drug dispersion, stability and it may play pivotal role in increased drug loading into the nanoparticles.

The degree of deacetylation and molecular weight of the chitosan moiety are important parameters where a higher degree of deacetylation and low molecular weight are deemed favorable to interact with bacterial cell membrane and enable easy penetration into the bacterial cell, thus inserting a bactericidal activity. The antibacterial activity of chitosan is envisaged to be due to presence of charged groups in the polymer backbone which ionically interact with the bacterial membrane constituents resulting in hydrolysis of peptidoglycan, inciting leakage of intracellular electrolytes [71]. Burn wounds often become complicated by secondary bacterial infection where 61% mortality has been reported to be associated with *Staphylococcus aureus*, *Pseudomonas aeruginosa*, *Klebsiella sp.* and *Escherichia coli* based bacterial infections [72,73]. Curcumin is reported to possess significant antimicrobial activity against a wide range of Gram-positive and Gram-negative bacteria [74] while the chitosan moiety is also reported to possess antimicrobial activity against various bacterial strains [75]. The formulated chitosan and curcumin nanoparticles synergistically exerted antimicrobial activity (Figure 5) where both the F1 and F2 formulations showed significant (Student's t-test, $p < 0.05$) antimicrobial activity against *Staphylococcus aureus* and *Pseudomonas aeruginosa* compared to blank nanoparticles and comparable to the positive control. The synergistic antimicrobial activity of chitosan and curcumin is envisaged to prevent infiltration of bacteria into the wound and hence prevent it from becoming complicated which will ultimately translate into better patient compliance and lower treatment cost.

The LMW chitosan-curcumin nanoparticles did not exert any significant cytotoxic effect, which is thought to be due to high degree of deacetylation of chitosan where it exhibits a stimulating effect on cells proliferation compared to counterparts with low degrees of deacetylation [76]. The degree of deacetylation of modified chitosan was significantly higher than parent chitosan (91.41 ± 8.2) which is hence deemed favorable in promoting skin tissue regeneration by facilitating dermal fibroblast proliferation. Curcumin exhibited cytotoxic effects in a dose dependent manner at concentrations higher than 5 µM [77]. Since our formulations contain curcumin in a significantly lower quantity compared to the reported concentration, hence no significant cytotoxic effect was observed in different nanoparticle applications. Both the tested formulations filled the scratch at significantly faster rates compared to the control and blank applications post 24 h incubation where the control and F1 formulations achieved full cell confluence after 72 h while F2 achieved full cell confluence within 24 h which can be attributed to the higher curcumin content embedded in core of the F2 formulation.

5. Conclusions

The molecular weight of chitosan was successfully reduced using microwave treatment. LMW chitosan generated small sized nanoparticles with positive surface charge and good dispersibility with the drug embedded inside the core of nanoparticles, thus facilitating sustained drug release in vitro through the mechanism of Fickian diffusion. The vibrational spectroscopic and thermal analysis results revealed that the chitosan main structure remained unaffected with microwave treatment but physical properties like transition temperature were significantly reduced. LMW chitosan-curcumin nanoparticles (F2) exerted synergistic antibacterial activity, with no significant cytotoxicity and enabled significantly faster scratch filling as well as cell confluence within 24 h post application comparable to control and/or blank treatment. LMW chitosan-curcumin nanoparticles with added synergistic antimicrobial activity and negligible cytotoxicity may prove beneficial in efficient delivery of curcumin into skin to hasten skin tissue regeneration following burn wounds.

Author Contributions: Conceptualization, N.R.K., H.K. and S.-F.N.; methodology, N.R.K. and H.K.; software, H.M.B., N.R.K. and S.-F.N.; validation, H.M.B., N.R.K., and H.K.; formal analysis, H.M.B., N.R.K., M.C.I.M.A., H.K., S.-F.N.; investigation, H.M.B. and H.K.; resources, N.R.K., M.C.I.M.A. and S.-F.N.; data curation, N.R.K., writing—original draft preparation, H.M.B., N.R.K.; writing—review and editing, H.M.B., N.R.K. and S.-F.N.; visualization, N.R.K., M.C.I.M.A., H.K. and S.-F.N.; supervision, S.U.S., N.R.K., M.C.I.M.A. and S.-F.N.; project administration, N.R.K., M.C.I.M.A. and S.-F.N.; funding acquisition, S.U.S., H.M.B. and N.R.K. All authors have read and agreed to the published version of the manuscript.

Funding: This research was funded by Higher Education Commission of Pakistan, Grant Number 7493 and APC was funded partly by UKM Malaysia and partly by Higher Education Commission of Pakistan.

Acknowledgments: The authors acknowledge the Gomal University Pakistan, Higher Education Commission of Pakistan for facility and financial support. The authors also acknowledge the Universiti Kebangsaan Malaysia for facility support.

Conflicts of Interest: The authors report no possible conflict of interest in this work.

References

1. World Health Organization. Burn. Available online: https://www.who.int/en/news-room/fact-sheets/detail/burns (accessed on 27 November 2019).
2. Dumas, A.; Gaudin-Audrain, C.; Mabilleau, G.; Massin, P.; Hubert, L.; Baslé, M.F.; Chappard, D. The influence of processes for the purification of human bone allografts on the matrix surface and cytocompatibility. *Biomaterials* **2006**, *27*, 4204–4211. [CrossRef] [PubMed]
3. Katayama, Y.; Montenegro, R.; Freier, T.; Midha, R.; Belkas, J.S.; Shoichet, M.S. Coil-reinforced hydrogel tubes promote nerve regeneration equivalent to that of nerve autografts. *Biomaterials* **2006**, *27*, 505–518. [CrossRef] [PubMed]
4. Hoffman, M.D.; Xie, C.; Zhang, X.; Benoit, D.S.W. The effect of mesenchymal stem cells delivered via hydrogel-based tissue engineered periosteum on bone allograft healing. *Biomaterials* **2013**, *34*, 8887–8898. [CrossRef] [PubMed]
5. Martínez-Santamaría, L.; Guerrero-Aspizua, S.; Del Río, M. Skin bioengineering: Preclinical and clinical applications. *Actas Dermosifiliogr.* **2012**, *103*, 5–11. [CrossRef]
6. Tabata, Y. Tissue Regeneration Based on Drug Delivery Technology. In *Topics in Tissue Engineering*; Ashammakhi, N., Ferretti, P., Eds.; Oulu University: Oulu, Finland, 2003; pp. 1–32.
7. Noori, S.; Kokabi, M.; Hassan, Z.M. Nanoclay Enhanced the Mechanical Properties of Poly (Vinyl Alcohol)/Chitosan/Montmorillonite Nanocomposite Hydrogel as Wound Dressing. *Procedia Mater. Sci.* **2015**, *11*, 152–156. [CrossRef]
8. Huang, N.; Lin, J.; Li, S.; Deng, Y.; Kong, S.; Hong, P.; Yang, P.; Liao, M.; Hu, Z. Preparation and evaluation of squid ink polysaccharide-chitosan as a wound-healing sponge. *Mater. Sci. Eng. C* **2018**, *82*, 354–362. [CrossRef]

9. Chen, S.A.; Chen, H.M.; Yao, Y.D.; Hung, C.F.; Tu, C.S.; Liang, Y.J. Topical treatment with anti-oxidants and Au nanoparticles promote healing of diabetic wound through receptor for advance glycation end-products. *Eur. J. Pharm. Sci.* **2012**, *47*, 875–883. [CrossRef]
10. Xu, F.; Weng, B.; Gilkerson, R.; Materon, L.A.; Lozano, K. Development of tannic acid/chitosan/pullulan composite nanofibers from aqueous solution for potential applications as wound dressing. *Carbohydr. Polym.* **2015**, *115*, 16–24. [CrossRef]
11. Wong, T.W.; Ramli, N.A. Carboxymethylcellulose film for bacterial wound infection control and healing. *Carbohydr. Polym.* **2014**, *112*, 367–375. [CrossRef]
12. Jayakumar, R.; Prabaharan, M.; Reis, R.L.; Mano, J.F. Graft copolymerized chitosan—Present status and applications. *Carbohydr. Polym.* **2005**, *62*, 142–158. [CrossRef]
13. Madhumathi, K.; Kumar, P.T.S.; Abhilash, S.; Sreeja, V.; Tamura, H.; Manzoor, K.; Nair, S.V.; Jayakumar, R. Development of novel chitin / nanosilver composite scaffolds for wound dressing applications. *J. Mater. Sci. Mater. Med.* **2010**, *21*, 807–813. [CrossRef] [PubMed]
14. Jayakumar, R.; Nwe, N.; Tokura, S.; Tamura, H. Sulfated chitin and chitosan as novel biomaterials. *Int. J. Biol. Macromol.* **2007**, *40*, 175–181. [CrossRef]
15. Wang, J.J.; Zeng, Z.W.; Xiao, R.Z.; Xie, T.; Zhou, G.L.; Zhan, X.R.; Wang, S.L. Recent advances of chitosan nanoparticles as drug carriers. *Int. J. Nanomed.* **2011**, *6*, 765–774.
16. Jayakumar, R.; Chennazhi, K.P.; Muzzarelli, R.A.A.; Tamura, H.; Nair, S.V.; Selvamurugan, N. Chitosan conjugated DNA nanoparticles in gene therapy. *Carbohydr. Polym.* **2010**, *79*, 1–8. [CrossRef]
17. Nagahama, H.; Nwe, N.; Jayakumar, R.; Koiwa, S.; Furuike, T.; Tamura, H. Novel biodegradable chitin membranes for tissue engineering applications. *Carbohydr. Polym.* **2008**, *73*, 295–302. [CrossRef]
18. Jayakumar, R.; Rajkumar, M.; Freitas, H.; Selvamurugan, N.; Nair, S.V.; Furuike, T.; Tamura, H. Preparation, characterization, bioactive and metal uptake studies of alginate/phosphorylated chitin blend films. *Int. J. Biol. Macromol.* **2009**, *44*, 107–111. [CrossRef] [PubMed]
19. Shalumon, K.T.; Binulal, N.S.; Selvamurugan, N.; Nair, S.V.; Menon, D.; Furuike, T.; Tamura, H.; Jayakumar, R. Electrospinning of carboxymethyl chitin/poly(vinyl alcohol) nanofibrous scaffolds for tissue engineering applications. *Carbohydr. Polym.* **2009**, *77*, 863–869. [CrossRef]
20. Jayakumar, R.; Reis, R.L.; Mano, J.F. Phosphorous containing chitosan beads for controlled oral drug delivery. *J. Bioact. Compat. Polym.* **2006**, *21*, 327–340. [CrossRef]
21. Prabaharan, M.; Mano, J.F. Chitosan-based particles as controlled drug delivery systems. *Drug Deliv.* **2005**, *12*, 41–57. [CrossRef]
22. Anitha, A.; Divya Rani, V.V.; Krishna, R.; Sreeja, V.; Selvamurugan, N.; Nair, S.V.; Tamura, H.; Jayakumar, R. Synthesis, characterization, cytotoxicity and antibacterial studies of chitosan, O-carboxymethyl, N,O-carboxymethyl chitosan nanoparticles. *Carbohydr. Polym.* **2009**, *78*, 672–677. [CrossRef]
23. Peter, M.; Binulol, N.S.; Soumya, S.; Nair, S.V.; Tamura, H.; Jayakumar, R. Nanocomposite scaffolds of bioactive glass ceramic nanoparticles disseminated chitosan matrix for tissue engineering applications. *Carbohydr. Polym.* **2010**, *20*, 284–289. [CrossRef]
24. Kazuaki, M.; Shingo, M.; Yusuke, Y.; Akira, F. In vitro degradation behavior of freeze-dried carboxymethyl-chitin sponges processed by vacuumheating and gamma irradiation. *Polym. Degrad. Stab.* **2003**, *81*, 327–332.
25. Baldrick, P. The safety of chitosan as a pharmaceutical excipient. *Regul. Toxicol. Pharmacol.* **2010**, *56*, 290–299. [CrossRef]
26. Paul, W.; Sharma, C.P. Chitin and alginates wound dressings: A short review. *Trends Biomater. Artif. Organs* **2004**, *18*, 18–23.
27. Alsarra, I.A. Chitosan topical gel formulation in the management of burn wounds. *Int. J. Biol. Macromol.* **2009**, *45*, 16–21. [CrossRef] [PubMed]
28. Choi, Y.S.; Lee, S.B.; Hong, S.R.; Lee, Y.M.; Song, K.W.; Park, M.H. Studies on gelatin-based sponges. Part III: A comparative study of cross-linked gelatin/alginate, gelatin/hyaluronate and chitosan/hyaluronate sponges and their application as a wound dressing in full-thickness skin defect of rat. *J. Mater. Sci. Mater. Med.* **2001**, *12*, 67–73. [CrossRef]
29. Joe, B.; Vijaykumar, M.; Lokesh, B.R. Biological Properties of Curcumin-Cellular and Molecular Mechanisms of Action. *Crit. Rev. Food Sci. Nutr.* **2004**, *44*, 97–111. [CrossRef]

30. Krausz, A.E.; Adler, B.L.; Cabral, V.; Navati, M.; Doerner, J.; Charafeddine, R.A.; Chandra, D.; Liang, H.; Gunther, L.; Clendaniel, A.; et al. Curcumin-encapsulated nanoparticles as innovative antimicrobial and wound healing agent. *Nanomed. Nanotechnol. Biol. Med.* **2015**, *11*, 195–206. [CrossRef]
31. Chereddy, K.K.; Coco, R.; Memvanga, P.B.; Ucakar, B.; Rieux, A.D.; Vandermeulen, G.; Préat, V. Combined effect of PLGA and curcumin on wound healing activity. *J. Control. Release* **2013**, *171*, 208–215. [CrossRef]
32. Thomasa, L.; Zakira, F.; Aamir Mirzab, M.; Anwer, K.; Ahmad, F.J.; Iqbal, Z. Development of Curcumin loaded chitosan polymer based nanoemulsion gel: In vitro, ex vivo evaluation and in vivo wound healing studies. *Int. J. Biol. Macromol.* **2017**, *101*, 569–579. [CrossRef]
33. Kanta, V.; Gopala, A.; Pathaka, N.N.; Kumarb, P.; Tandana, S.K.; Kumar, D. Antioxidant and anti-inflammatory potential of curcumin accelerated the cutaneous wound healing in streptozotocin-induced diabetic rats. *Int. Immunopharmacol.* **2014**, *20*, 322–330. [CrossRef]
34. El-Refaie, W.M.; Elnaggar, Y.S.R.; El-Massik, M.A.; Abdallah, O.Y. Novel curcumin-loaded gel-core hyaluosomes with promising burn-wound healing potential: Development, in-vitro appraisal and in-vivo studies. *Int. J. Pharm.* **2015**, *486*, 88–98. [CrossRef] [PubMed]
35. Guo, G.; Li, X.; Ye, X.; Qi, J.; Fan, R.; Gao, X.; Wu, Y.; Zhou, L.; Tong, A. EGF and curcumin co-encapsulated nanoparticle/hydrogel system as potent skin regeneration agent. *Int. J. Nanomed.* **2016**, *11*, 3993–4009. [CrossRef] [PubMed]
36. Gong, C.Y.; Wu, Q.J.; Wang, Y.J.; Zhang, D.D.; Luo, F.; Zhao, X.; Wei, Y.Q.; Qian, Z.Y. A biodegradable hydrogel system containing curcumin encapsulated in micelles for cutaneous wound healing. *Biomaterials* **2013**, *34*, 6377–6387. [CrossRef] [PubMed]
37. Croisier, F.; Jérôme, C. Chitosan-based biomaterials for tissue engineering. *Eur. Polym. J.* **2013**, *49*, 780–792. [CrossRef]
38. Shi, D.; Wang, F.; Lan, T.; Zhang, Y.; Shao, Z. Convenient fabrication of carboxymethyl cellulose electrospun nanofibers functionalized with silver nanoparticles. *Cellulose* **2016**, *23*, 1899–1909. [CrossRef]
39. Arockianathan, P.M.; Sekar, S.; Kumaran, B.; Sastry, T.P. Preparation, characterization and evaluation of biocomposite films containing chitosan and sago starch impregnated with silver nanoparticles. *Int. J. Biol. Macromol.* **2012**, *50*, 939–946. [CrossRef] [PubMed]
40. Cheng, F.; Gao, J.; Wang, L.; Hu, X.Y. Composite chitosan/poly(ethylene oxide) electrospun nanofibrous mats as novel wound dressing matrixes for the controlled release of drugs. *J. Appl. Polym. Sci.* **2015**, *132*, 1–8. [CrossRef]
41. Tømmeraas, K.; Vårum, K.M.; Christensen, B.E.; Smidsrød, O. Preparation and characterisation of oligosaccharides produced by nitrous acid depolymerisation of chitosans. *Carbohydr. Res.* **2001**, *333*, 137–144. [CrossRef]
42. Sato, K.; Kishimoto, T.; Morimoto, M.; Saimoto, H.; Shigemasa, Y. Hydrolysis of acetals in water under hydrothermal conditions. *Tetrahedron Lett.* **2003**, *44*, 8623–8625. [CrossRef]
43. Cravotto, G.; Tagliapietra, S.; Robaldo, B.; Trotta, M. Chemical modification of chitosan under high-intensity ultrasound. *Ultrason. Sonochem.* **2005**, *12*, 95–98. [CrossRef]
44. Naveed, M.; Phil, L.; Sohail, M.; Hasnat, M.; Baig, M.M.F.A.; Ihsan, A.U.; Shumzaid, M.; Kakar, M.U.; Mehmood Khan, T.; Akabar, M.D.; et al. Chitosan oligosaccharide (COS): An overview. *Int. J. Biol. Macromol.* **2019**, *129*, 827–843. [CrossRef]
45. Nawaz, A.; Wong, T.W. Microwave as skin permeation enhancer for transdermal drug delivery of chitosan-5-fluorouracil nanoparticles. *Carbohydr. Polym.* **2017**, *157*, 906–919. [CrossRef]
46. Xia, W.; Liu, P.; Zhang, J.; Chen, J. Biological activities of chitosan and chitooligosaccharides. *Food Hydrocoll.* **2011**, *25*, 170–179. [CrossRef]
47. Wong, T.W.; Nurulaini, H. Sustained-release alginate-chitosan pellets prepared by melt pelletization technique. *Drug Dev. Ind. Pharm.* **2012**, *38*, 1417–1427. [CrossRef] [PubMed]
48. Bagheri-Khoulenjani, S.; Taghizadeh, S.M.; Mirzadeh, H. An investigation on the short-term biodegradability of chitosan with various molecular weights and degrees of deacetylation. *Carbohydr. Polym.* **2009**, *78*, 773–778. [CrossRef]
49. Katas, H.; Wen, C.Y.; Siddique, M.I.; Hussain, Z.; Fadhil, F.H.M. Thermoresponsive curcumin/DsiRNA nanoparticle gels for the treatment of diabetic wounds: Synthesis and drug release. *Ther. Deliv.* **2017**, *8*, 137–150. [CrossRef] [PubMed]

50. Chuah, L.H.; Billa, N.; Roberts, C.J.; Burley, J.C.; Manickam, S. Curcumin-containing chitosan nanoparticles as a potential mucoadhesive delivery system to the colon. *Pharm. Dev. Technol.* **2013**, *18*, 591–599. [CrossRef]
51. Dash, S.; Murthy, P.N.; Nath, L.; Chowdhury, P. Kinetic modeling on drug release from controlled drug delivery systems. *Acta Pol. Pharm.-Drug Res.* **2010**, *67*, 217–223.
52. Lai, J.C.Y.; Lai, H.Y.; Rao, N.K.; Ng, S.F. Treatment for diabetic ulcer wounds using a fern tannin optimized hydrogel formulation with antibacterial and antioxidative properties. *J. Ethnopharmacol.* **2016**, *189*, 277–289. [CrossRef]
53. Andrews, J.M. Determination of minimum inhibitory concentrations. *J. Antimicrob. Chemother.* **2002**, *49*, 1049. [CrossRef]
54. Khan, N.R.; Wong, T.W. 5-Fluorouracil ethosomes–skin deposition and melanoma permeation synergism with microwave. *Artif. Cells, Nanomed. Biotechnol.* **2018**, *46*, 568–577. [CrossRef] [PubMed]
55. Sreekumar, S.; Goycoolea, F.M.; Moerschbacher, B.M.; Rivera-Rodriguez, G.R. Parameters influencing the size of chitosan-TPP nano- and microparticles. *Sci. Rep.* **2018**, *8*, 1–11.
56. Nair, R.S.; Morris, A.; Billa, N.; Leong, C.-O. An Evaluation of Curcumin-Encapsulated Chitosan Nanoparticles for Transdermal Delivery. *AAPS PharmSciTech* **2019**, *20*, 68–69. [CrossRef] [PubMed]
57. Martins, A.F.; de Oliveira, D.M.; Pereira, A.G.B.; Rubira, A.F.; Muniz, E.C. Chitosan/TPP microparticles obtained by microemulsion method applied in controlled release of heparin. *Int. J. Biol. Macromol.* **2012**, *51*, 1127–1133. [CrossRef] [PubMed]
58. Stillwell, W. *An Introduction to Biological Membranes from Bilayers to Rafts*, 1st ed.; Elsevier B.V.: Oxford, UK, 2013; ISBN 9780444521538.
59. Raja, M.A.G.; Katas, H.; Hamid, Z.A.; Razali, N.A. Physicochemical properties and in vitro cytotoxicity studies of chitosan as a potential carrier for dicer-substrate siRNA. *J. Nanomater.* **2013**, *2013*, 1–10. [CrossRef]
60. Shu, X.Z.; Zhu, K.J. The influence of multivalent phosphate structure on the properties of ionically cross-linked chitosan films for controlled drug release. *Eur. J. Pharm. Biopharm.* **2002**, *54*, 235–243. [CrossRef]
61. Villegas-Peralta, Y.; López-Cervantes, J.; Madera Santana, T.J.; Sánchez-Duarte, R.G.; Sánchez-Machado, D.I.; Martínez-Macías, M. del R.; Correa-Murrieta, M.A. Impact of the molecular weight on the size of chitosan nanoparticles: Characterization and its solid-state application. *Polym. Bull.* **2020**. [CrossRef]
62. Chen, Y.; Mohanraj, V.J.; Parkin, J.E. Chitosan-dextran sulfate nanoparticles for delivery of an anti-angiogenesis peptide. *Lett. Pept. Sci.* **2003**, *10*, 621–629. [CrossRef]
63. Gan, Q.; Wang, T. Chitosan nanoparticle as protein delivery carrier-Systematic examination of fabrication conditions for efficient loading and release. *Colloids Surf. B Biointerfaces* **2007**, *59*, 24–34. [CrossRef]
64. Mora-Huertas, C.E.; Fessi, H.; Elaissari, A. Polymer-based nanocapsules for drug delivery. *Int. J. Pharm.* **2010**, *385*, 113–142. [CrossRef] [PubMed]
65. Preetz, C.; Rübe, A.; Reiche, I.; Hause, G.; Mäder, K. Preparation and characterization of biocompatible oil-loaded polyelectrolyte nanocapsules. *Nanomed. Nanotechnol. Biol. Med.* **2008**, *4*, 106–114. [CrossRef] [PubMed]
66. Dhawade, P.P.; Jagtap, R.N. Characterization of the glass transition temperature of chitosan and its oligomers by temperature modulated differential scanning calorimetry. *Pelagia Res. Libr. Adv. Appl. Sci. Res.* **2012**, *3*, 1372–1382.
67. Rampino, A.; Borgogna, M.; Blasi, P.; Bellich, B.; Cesàro, A. Chitosan nanoparticles: Preparation, size evolution and stability. *Int. J. Pharm.* **2013**, *455*, 219–228. [CrossRef] [PubMed]
68. Parize, A.L.; Stulzer, H.K.; Laranjeira, M.C.M.; Brighente, I.M.D.C.; De Souza, T.C.R. Evaluation of chitosan microparticles containing curcumin and crosslinked with sodium tripolyphosphate produced by spray drying. *Quim. Nova* **2012**, *35*, 1127–1132. [CrossRef]
69. Kittur, F.S.; Prashanth, K.V.H.; Sankar, K.U.; Tharanathan, R.N. Characterization of chitin, chitosan and their carboxymethyl derivatives by differential scanning calorimetry. *Carbohydr. Polym.* **2002**, *49*, 185–193. [CrossRef]
70. Khezri, A.; Karimi, A.; Yazdian, F.; Jokar, M.; Mofradnia, S.R.; Rashedi, H.; Tavakoli, Z. Molecular dynamic of curcumin/chitosan interaction using a computational molecular approach: Emphasis on biofilm reduction. *Int. J. Biol. Macromol.* **2018**, *114*, 972–978. [CrossRef]
71. Goy, R.C.; Morais, S.T.B.; Assis, O.B.G. Evaluation of the antimicrobial activity of chitosan and its quaternized derivative on E. Coli and S. aureus growth. *Braz. J. Pharmacogn.* **2016**, *26*, 122–127. [CrossRef]

72. Kehinde, A.O.; Ademola, S.A.; Okeshola, O.A.; Oluwatosin, O.M.; Bakare, R.A. Pattern of Bacterial Pathogens in Burn Wound Infections in Ibadan, Nigeria. *Ann. Burn. Fire Disasters* **2004**, *17*, 12–15.
73. Motayo, B.O.; Akinbo, J.A.; Ogiogwa, I.J.; Idowu, A.A.; Nwanze, J.C.; Onoh, C.C.; Okerentugba, P.O.; Innocent-Adiele, H.C.; Okonko, I.O. Bacteria Colonisation and Antibiotic Susceptibility Pattern of Wound Infections in a Hospital in Abeokuta. *Front. Sci.* **2013**, *3*, 43–48.
74. Da Silva, A.C.; Santos, P.D.D.F.; Silva, J.T.D.P.; Leimann, F.V.; Bracht, L.; Gonçalves, O.H. Impact of curcumin nanoformulation on its antimicrobial activity. *Trends Food Sci. Technol.* **2018**, *72*, 74–82. [CrossRef]
75. Benhabiles, M.S.; Salah, R.; Lounici, H.; Drouiche, N.; Goosen, M.F.A.; Mameri, N. Antibacterial activity of chitin, chitosan and its oligomers prepared from shrimp shell waste. *Food Hydrocoll.* **2012**, *29*, 48–56. [CrossRef]
76. Howling, G.I.; Dettmar, P.W.; Goddard, P.A.; Hampson, F.C.; Dornish, M.; Wood, E.J. The effect of chitin and chitosan on the proliferation of human skin fibroblasts and keratinocytes in vitro. *Biomaterials* **2001**, *22*, 2959–2966. [CrossRef]
77. Cianfruglia, L.; Minnelli, C.; Laudadio, E.; Scirè, A.; Armeni, T. Side effects of curcumin: Epigenetic and antiproliferative implications for normal dermal fibroblast and breast cancer cells. *Antioxidants* **2019**, *8*, 382. [CrossRef]

Publisher's Note: MDPI stays neutral with regard to jurisdictional claims in published maps and institutional affiliations.

© 2020 by the authors. Licensee MDPI, Basel, Switzerland. This article is an open access article distributed under the terms and conditions of the Creative Commons Attribution (CC BY) license (http://creativecommons.org/licenses/by/4.0/).

Article

3D Printing of a Reactive Hydrogel Bio-Ink Using a Static Mixing Tool

María Puertas-Bartolomé [1,2], Małgorzata K. Włodarczyk-Biegun [3], Aránzazu del Campo [3,4], Blanca Vázquez-Lasa [1,2,*] and Julio San Román [1,2]

1. Institute of Polymer Science and Technology, ICTP-CSIC, Juan de la Cierva 3, 28006 Madrid, Spain; mpuertas@ictp.csic.es (M.P.-B.); jsroman@ictp.csic.es (J.S.R.)
2. CIBER's Bioengineering, Biomaterials and Nanomedicine, CIBER-BBN, Health Institute Carlos III, Monforte de Lemos 3-5, 28029 Madrid, Spain
3. INM—Leibniz Institute for New Materials, Campus D2 2, 66123 Saarbrücken, Germany; m.k.wlodarczyk@rug.nl (M.K.W.-B.); aranzazu.delcampo@leibniz-inm.de (A.d.C.)
4. Chemistry Department, Saarland University, 66123 Saarbrücken, Germany
* Correspondence: bvazquez@ictp.csic.es; Tel.: +34-915-618-806 (ext. 921522)

Received: 29 July 2020; Accepted: 27 August 2020; Published: 31 August 2020

Abstract: Hydrogel-based bio-inks have recently attracted more attention for 3D printing applications in tissue engineering due to their remarkable intrinsic properties, such as a cell supporting environment. However, their usually weak mechanical properties lead to poor printability and low stability of the obtained structures. To obtain good shape fidelity, current approaches based on extrusion printing use high viscosity solutions, which can compromise cell viability. This paper presents a novel bio-printing methodology based on a dual-syringe system with a static mixing tool that allows in situ crosslinking of a two-component hydrogel-based ink in the presence of living cells. The reactive hydrogel system consists of carboxymethyl chitosan (CMCh) and partially oxidized hyaluronic acid (HAox) that undergo fast self-covalent crosslinking via Schiff base formation. This new approach allows us to use low viscosity solutions since in situ gelation provides the appropriate structural integrity to maintain the printed shape. The proposed bio-ink formulation was optimized to match crosslinking kinetics with the printing process and multi-layered 3D bio-printed scaffolds were successfully obtained. Printed scaffolds showed moderate swelling, good biocompatibility with embedded cells, and were mechanically stable after 14 days of the cell culture. We envision that this straightforward, powerful, and generalizable printing approach can be used for a wide range of materials, growth factors, or cell types, to be employed for soft tissue regeneration.

Keywords: 3D-bioprinting; static mixer; reactive hydrogel; chitosan; hyaluronic acid

1. Introduction

3D bioprinting is a booming additive manufacturing technology that allows the layer-by-layer deposition of a cell-laden material to fabricate 3D constructs with spatial control over scaffold design. This technology has been widely used in the last few years for tissue engineering and regenerative medicine applications as it allows the artificial reconstruction of the complexity of native tissues or organs [1–3]. To date, great efforts have been made to develop suitable bio-inks to provide cell-laden scaffolds with good mechanical properties as well as high cell viability. Hydrogels are often used as supporting material in the bio-ink due to their favorable intrinsic properties for supporting cellular growth [4–13]. Their unique inherent properties similar to the extracellular matrix (EMC), such as porosity that allows nutrient and gaseous exchange, high water content, biodegradability, and biocompatibility make them attractive for cell therapy applications [14–16]. Specifically, bio-inks based on natural hydrogels, such as alginate, agarose, gelatin, collagen, chitosan, or hyaluronic acid are regularly used [4,17–23].

Extrusion is the most frequently used technique for 3D bio-printing with hydrogel precursors [24,25]. The minimal requirements that the hydrogel bio-ink has to fulfil for successful extrusion include: (1) bio-ink must easily flow through the needle during printing but retain the shape after extrusion, (2) printed strands should have a good structural integrity to provide self-supporting structures with good adhesion between layers, and (3) bio-ink must ensure cell survival and proliferation within the printed scaffold [26,27]. Naturally derived hydrogels are still challenging to print due to their weak mechanical properties that lead to poor printing accuracy and low stability of the printed structures [9,13,28–31]. Traditional approaches based on increasing the polymer content and viscosity or the crosslinking density have been attempted to improve printability of naturally derived hydrogel bio-inks and the mechanical performance of their printed scaffolds [5,27]. However, bio-inks with high polymer contents or viscosities can compromise cell viability due to the high shear forces and lower nutrients transport through the printed constructs [5,27,32]. Thus, the development of low viscosity bio-inks and suitable printing extrusion processes for bio-fabrication are still in demand.

In this paper, we present a methodology for printing homogeneous strands from a reactive hydrogel using a dual syringe system with a static mixing tool coupled to an extrusion bio-printer. The two reactive hydrogel precursors are loaded into separate syringes, simultaneously extruded by mechanical displacement and transported to the static mixer in a 1:1 ratio. They are homogeneously mixed during the short residence time in the static mixer and the crosslinking reaction is initiated prior to extrusion. The partially crosslinked hydrogel is then extruded from the printhead. This approach has several advantages for 3D extrusion printing: (1) it uses low viscosity starting solutions of the hydrogel precursors and avoids high shear stress during extrusion, (2) the in situ crosslinking provides appropriate structural integrity to the printed thread to maintain the printed shape, (3) it avoids post-printing cell seeding (cells are embedded in the ink), washing steps, or additional physical factors that may complicate the fabrication process [28].

Two component systems have been used for 3D printing. Skardal et al. [33] used methacrylated gelatin and methacrylated hyaluronic acid [33] to print scaffolds with gradient material properties, but exchange of the syringes during the printing process was required. Bakarich et al. [24] and O'Connell et al. [34] presented a two component extrusion printing system (alginate/polyacrylamide and an acrylated urethane [24], or gelatin–methacrylamide and hyaluronic acid–methacrylate [34]) where materials were mixed prior to printing in a static nozzle or in the needle, and UV irradiation was required for stabilization of the printed structures. Reactive hydrogels have been used in 3D printing. Gregor et al. [35] and Zimmermann et al. [36] prepared 3D scaffolds by fusing individual droplets of two precursors (fibrinogen and thrombin [35] or thiol-terminated starPEG and maleimide-functionalized heparin [36]), and Lozano et al. [37] used a hand-held system with a coaxial syringe tip to extrude precursors (gellan gum-RGD and $CaCl_2$). However, mixing and crosslinking take place during droplet deposition, and scaffolds with spatially graded material properties were obtained [36,37]. Maiullari et al. [38] used a microfluidic 3D printing approach where alginate/PEG-fibrinogen and $CaCl_2$ precursors were mixed after extrusion from a coaxial needle and an ultraviolet (UV) crosslinking step was needed to increase stability of the printed structure. Bootsma et al. [39] used a mixing head in order to combine hydrogel precursors (alginate/acrylamide/N,N-methylenebisacrylamide/D-glucono-δ-lactone and alginate/$CaCO_3$/Irgacure 1173) even though additional UV crosslinking was still necessary to induce covalent crosslinking. On the other hand, the static mixing tool has been used for different reactive hydrogel systems in several reported works. Deepthi et al. prepared an injectable fibrin hydrogel containing alginate nanobeads using a double syringe connected to the static mixer [40]. Hozumi et al. studied the gelation process through a static mixer of alginate hydrogels with Ca^{2+} [41]. In addition, different studies have reported the use of static mixers in other hydrogel processing technologies, like in injectable [42–45] or moulding [46,47] formulations, but, in all these examples, shape fidelity has not been addressed. To our best knowledge, static mixers have not been explored for 3D bioprinting of reactive hydrogels with good shape fidelity and resolution.

In order to take advantage of a static mixer for 3D printing of a reactive two-components ink, the gelation kinetics has to be carefully adjusted to the residence time in the tool. In this work, we used the naturally derived polysaccharides chitosan [28,48–50] and hyaluronic acid [21,51–53] modified with reactive groups and formulated in two separate precursor solutions. Chitosan is a great candidate for tissue engineering applications since it exhibits notable biological features such as great cytocompatibility and biodegradability, antibacterial, hemostatic, or muco-adhesive properties [54,55]. Specifically, a carboxymethyl chitosan derivative (CMCh) was selected because of its good solubility at physiological pH, which allows straightforward encapsulation of cells in the bio-ink [56,57] and avoids any neutralization or washing steps, commonly used for pure chitosan-based printing [28,49,58,59]. Hyaluronic acid (HA) is an important component of the extracellular matrix (ECM), which favors cell affinity and proliferation [60,61]. HA is not suitable by itself for 3D printing, but it can improve printability of the bio-ink due to its shear-thinning behavior [52], and, when it is in combination with chitosan, it can counteract the brittle mechanical properties of the former [54]. Thus, when using a CMCh/HA ink, the free amines of CMCh can react with the aldehyde groups of partially oxidized HAox [62] after mixing via Schiff base formation [56,63–65], which gives rise to a hydrogel structure. This system has been demonstrated to allow viable cell encapsulation [66]. In this paper, this printing methodology has been optimized in order to fulfil the requirements for successful bioprinting to lead to cell laden 3D hydrogel constructs with good resolution and shape fidelity.

2. Materials and Methods

2.1. Materials

Carboxymethyl chitosan (CMCh) (degree of deacetylation 85–90%, viscosity = 5–300 mPas, Chitoscience, Halle (Saale, Germany), sodium hyaluronan (HA) (~1.5–1.8 × 10^6 Da, Sigma-Aldrich, St. Louis, MO, USA), sodium periodate ($NaIO_4$) (Alfa Aesar, Haverhill, MA, USA), ethylene glycol (Sigma), hydroxylamine (Sigma-Aldrich), iron chloride (III) (Sigma-Aldrich), and phosphate buffered saline solution (PBS) 10 mM pH 7.4 (Gibco, Thermo Fisher, Waltham, MA, USA) were used as received. Sodium hyaluronan of low molecular weight (M_w ~200 kDa, Bioiberica, Barcelona, Spain) was oxidized (HAox) prior to use, as reported elsewhere [62], with a final oxidation degree of 48 ± 3.2% [67,68].

2.2. Bio-Ink Formulation

Hydrogel inks were formulated in two separate solutions. CMCh was dissolved in PBS (pH = 7.4) unless otherwise noted. HAox or HAox with HA mixture was dissolved in 0.1 M NaCl. The final ink formulation is named CMChn/HAoxn-HAn, where the number (n) that follows the polysaccharide abbreviation means the weight percentage of the precursor solution. Initially, different CMCh/HAox and CMCh/HAox-HA compositions were tested for optimization of the printing process. Lastly, 3D printed scaffolds were prepared with the optimized hydrogel formulation: CMCh2/HAox4-HA0.4. A 1:1 volume ratio of the two solutions was used for printing. To increase in vitro stability of the printed scaffolds for longer term cell studies, a post-printing stabilization step was carried by immersion in a 20 mM $FeCl_3$ aqueous solution for 7 min.

2.3. 3D Printing with a Static Mixing Tool

The 3D Discovery printer (RegenHu, Villaz-Saint-Pierre, Switzerland) was modified to accommodate a static mixing tool. The mixing tool system consists of two 1-mL disposable syringes coupled to a single disposable static mixer (2.5 mL length, helical screw inside) provided by RegenHu Company, and a printing needle (Figure 1A). To employ the mixing system, the original high precision plunger dispenser of the printer was adjusted with a custom-made holder for the static mixing tool. The obtained dual extrusion printing head employs simultaneous mechanically-driven movement of two syringe plungers using a single motor (Figure 1B). This leads to a 1:1 mixing ratio of the liquids from the connected syringes that have the same dimensions. The plungers' speed is controlled by the software, and accordingly modified by RegenHu. The solutions are transported to the static mixer, and then extruded through the connected needle (Figure 1C).

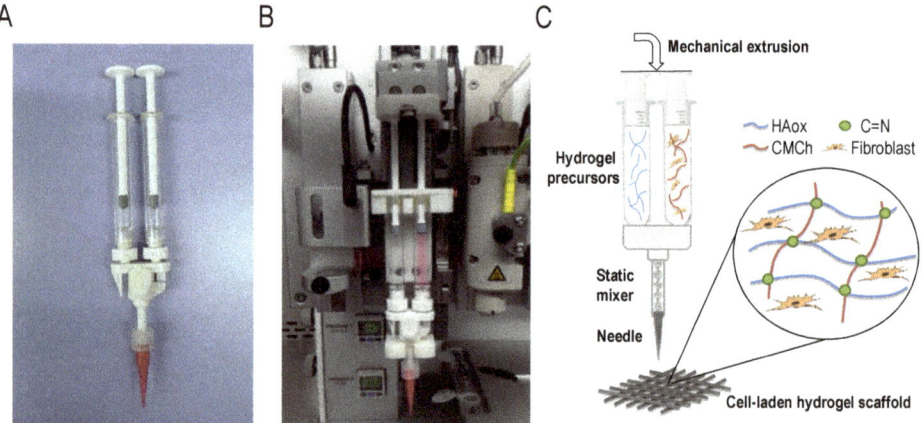

Figure 1. (**A**) Picture of the static mixing tool. (**B**) Static mixing tool coupled to the 3D printer. (**C**) Scheme of the bioprinting process using the mixing device coupled to the 3D printer.

Printing and plunger speeds were optimized for each tested ink formulation. Values in the range of 5 to 25 mm/s (print head movement speed) and 0.04 to 0.1mm/s (plunger speed) were tested, using a design of 4-cm long parallel lines. A conical polyethylene needle with an inner diameter of 200 µm was used. 3D scaffolds (2 or 4 layer grid square: 12×12 mm^2, 1.5 mm separation between strands) were printed with 15 mm/s printing speed and 0.06 mm/s plunger speed. Prior to start of the designed architecture printing, one sacrificial 4-cm long line was printed to allow material homogenization in the mixer. Scaffolds of formulations without encapsulated cells were printed onto granulated paper (hp laserjet transparency film) and cell-laden scaffolds of the optimized formulation CMCh2/HAox4-HA0.4 were printed onto glass coverslips.

2.4. Bio-Ink and 3D Printed Scaffolds Characterization

2.4.1. Rheological Analysis

Rheology experiments were performed at a controlled temperature of 25 °C, using a rheometer (ARG2, TA Instruments, New Castle, DE, USA) equipped with a parallel plate sand-blasted geometry (25-mm diameter).

Gelation times were measured in oscillation mode in time sweep experiments. Storage modulus (G') and loss modulus (G'') were recorded at a frequency of 1 Hz and 1% strain over 5 min. Furthermore, 75 µL of CMCh solution was deposited on the lower plate. Then, the same volume of the corresponding HAox or HAox-HA solution was deposited on top, shortly after being mixed by pipetting, and the mixture was immediately compressed between measuring plates (300 µm measuring gap). Different compositions were tested: CMCh1/HAox1, CMCh1/HAox2, CMCh2/HAox2, CMCh2/HAox4, CMCh3/HAox3, CMCh3/HAox6, and CMCh2/HAox4-HA0.4. Gelation time was defined as the time when G' crossed-over G''. Each sample was measured three times and the average gelation time value was given. Additionally, to determine final viscoelastic properties, a frequency sweep experiment was performed for the optimized bio-ink formulation CMCh2/HAox4-HA0.4. Storage modulus (G') and loss modulus (G'') were recorded at 1% strain and increasing frequencies from 1 to 200 Hz.

Viscosities of 2 wt % CMCh, 4 wt % HAox, and HAox-HA blends (4 wt % HAox with 0.2, 0.4, or 0.6 wt % HA amounts) solutions were determined by rotational shear measurements at an increasing shear rate from 1 to 500 s^{-1}. Final viscosity of CMCh2/HAox4-HA0.4 hydrogel was measured at shear rates increasing from 0.1 to 150 s^{-1}.

Each sample was measured three times and the average and standard deviation were given.

2.4.2. Attenuated Total Reflection Fourier Transform Infra-Red (ATR-FTIR) Spectroscopy

Attenuated Total Reflection Fourier Transform Infra-Red spectra of CMCh, HAox-HA, and CMCh2/HAox4-HA0.4 samples were measured for structural characterization (ATR-FTIR, Perkin-Elmer Spectrum One, Madrid, Spain).

2.4.3. Scaffolds Morphology

Light microscopy characterization of printed formulations was performed with a stereomicroscope SMZ800N (Nikon, Tokyo, Japan) equipped with home-made bottom illumination and camera (13MPx, Samsung, Seoul, Korea) for imaging.

2.4.4. In Vitro Swelling and Degradation Studies

Swelling and degradation in vitro were carried out in physiological (PBS pH = 7.4 at 37 °C) conditions to evaluate stability of just-printed scaffolds, and scaffolds with additional post-printing stabilization. The additional post-printing stabilization step was performed by immersion in a 20 mM FeCl$_3$ aqueous solution for 7 min. For the in vitro swelling experiments, printed samples (two layers of square-based scaffolds) were incubated in PBS for 4 h and, after gentle removal of excess of PBS, imaged using a stereomicroscope SMZ800N (Nikon, Dusseldorf, Germany). Swelling was evaluated by measuring strand widths in the scaffold with imageJ software before and after incubation. A minimum of four replicates was analyzed and results were given as mean ±SD. In vitro degradation was qualitatively analyzed by microscope pictures after 1, 4, 7, 14, and 21 days of incubation in PBS. Images were taken using a microscope Nikon Eclipse TE2000-S with camera NikonDS-Ri2.

2.5. CMCh2/HAox4-HA0.4 Based Bioprinting

2.5.1. Cell Culture

L929 Fibroblasts (ATCC, Manassas, VA, USA) were cultured in RPMI 1640 phenol red free medium (Gibco, 61870-010) supplemented with 20% fetal bovine serum (FBS, Gibco, 10270), 200 mM L-glutamine, and 1% penicillin/streptomycin (Invitrogen, Thermo Fisher). Incubation was carried out at 37 °C, 95% humidity, and 5% CO$_2$. Cell culture media was changed every two days.

2.5.2. Bio-Ink Preparation

L929 Fibroblasts (ATCC) were loaded in the 2 wt % CMCh solution (prepared in RPMI 1640) at a 2×10^6 cells/mL concentration. Solutions of both hydrogel precursors (2 wt % CMCh loaded with L929, and 4 wt % HAox-0.4 wt % HA) were transferred into the printing syringes and two layers of square-based (120×120 mm^2) scaffolds were printed at r.t onto glass coverslips. The final cell concentration in the scaffold was 1×10^6 cells/mL. An additional post-printing stabilization step was carried by immersion in a 20 mM FeCl$_3$ aqueous solution for 7 min. After that time, solution was removed and samples were incubated in cell culture media.

2.5.3. Cell Assays and Staining

In order to evaluate cell viability within the hydrogel scaffold over a 14-day period, staining with fluorescein diacetate (FDA) and propidium iodide (PI) was carried out to detect live and dead cells, respectively. In brief, scaffolds were washed with PBS at different time points (1, 4, 7 and 14 days) and incubated with 20 μg/mL FDA and 6 μg/mL PI for 10 min at r.t. Then, samples were washed with

PBS 3× and fluorescence images were taken using Nikon Ti-Ecllipse microscope (Nikon Instruments Europe B.V., Amsterdam, The Netherlands). Quantification was performed by image analysis using the Image-J 1.52p software counting both green and red cells with the function "find maxima." The cell viability percentage was calculated by quantifying the live cells between the total amount of cells in at least five images for three independent samples. As a control, cell viability studies of cells encapsulated in bulk hydrogels without using the static mixer were performed on the first day. Analysis of variance (ANOVA) using the Tukey grouping method of the results for the printed samples was performed at each time with respect to the first day at a significance level of *** $p < 0.05$ with respect to non-printed samples at a significance level of ### $p < 0.05$.

Fluorescence staining of nuclei was carried out to quantify cell proliferation within the 3D constructs over a 14-day period. Cells were fixed with PFA (paraformaldehyde) 3.7% w/v for 15 min at different time points (1, 4, 7, and 14 days), which is followed by permeabilization with 0.5% Triton -X 100 in PBS for 15 min and incubated with 1:1000 DAPI (4′,6-diamidino-2-fenilindol) dillution in PBS for 20 min. Lastly, samples were imaged using a LSM 880 confocal microscope (Zeiss, Jena, Germany). Image analysis was performed using the Image-J 1.52p software using the functions "Z project" and "find maxima" to count the number of nuclei observed in all the z levels analyzed by confocal. Cell quantification with ImageJ was performed in three images per sample in a 425.1 × 425.1 μm tested area, and quantification obtained on day 1 was normalized to 100%. Analysis of variance (ANOVA) using the Tukey grouping method of the results for the printed samples was performed at each time point at a significance level of * $p < 0.05$.

3. Results and Discussion

3.1. 3D Printing with a Static Mixing Tool

The printing conditions to obtain stable threads from the two-component hydrogel system using the static mixer tool were evaluated. Once the two components enter the static mixer, diffusion and a covalent reaction between the amine groups of CMCh and aldehyde groups of HAox is started. Non-covalent interactions such as hydrogen bonds or ionic interactions might further stabilize the gel [69–71] and can be beneficial for the extrusion process. For high fidelity printing without clogging the nozzle, the crosslinking kinetics must be adjusted to ensure adequate mixing and good printing quality. The polymer concentration and components ratio, the gelation kinetics, and the viscosity of the inks are relevant parameters to adjust.

The matching between hydrogel crosslinking kinetics and extrusion speed is essential to obtain an adequate crosslinking degree in the static mixer to allow flow while providing smooth and stable strands [39,41]. The crosslinking kinetics for different CMCh and HAox weight concentrations (CMCh1/HAox1, CMCh1/HAox2, CMCh2/HAox2, CMCh2/HAox4, CMCh3/HAox3, and CMCh3/HAox6) were studied in rheological experiments. Figure 2A presents the variation of the shear modulus G′ and the loss modulus G″ as a function of time for the CMCh/HAox formulations that gave a measurable gelation point. As the system began to crosslink through the formation of Schiff base linkages, G′ increased at a faster speed than G″, which indicates a change in the viscoelastic behavior of the system to a more solid-like state. These differential growth speeds led to crossover point of G′ and G″, defined as a gelation point, which indicates that the 3D hydrogel network was formed [66,72]. The corresponding gelation time ranged from 0.90 ± 0.06 to 4.68 ± 0.10 min for the formulations studied (Figure 2B). Regarding ink composition, gelation time decreased with a dropping CMCh/HAox ratio and with an increasing CMCh concentration. The printability of the ink formulations was evaluated by image analysis of printed threads (Figure 2C). Printed threads with 1 wt % CMCh were liquid, which are in agreement with the rheology data that did not show gel formation (undergelation). Broken lines with small gel blocks were visible for 3 wt % CMCh formulations. In these cases, gel formation was faster than the residence time of the solution in the static mixer (over-gelation), and the shear force needed to extrude the ink caused gel fracture. For the intermediate CMCh compositions

(2 wt %), semi-solid printed strands were observed, so 2 wt % CMCh was considered a minimum concentration threshold for gel formation. The CMCh2/HAox2 formulation yielded broad lines with low shape fidelity. A feasible region was found for CMCh2/HAox4 formulation, which rendered smooth lines with shape fidelity. In this formulation, the crosslinking degree achieved in the mixing head provided adequate viscosity for extrusion with enough mechanical stability for high fidelity printing. CMCh2/HAox4, with a gelation time of 3.64 ± 0.43 min, was selected as the most appropriate ink for the subsequent experiments.

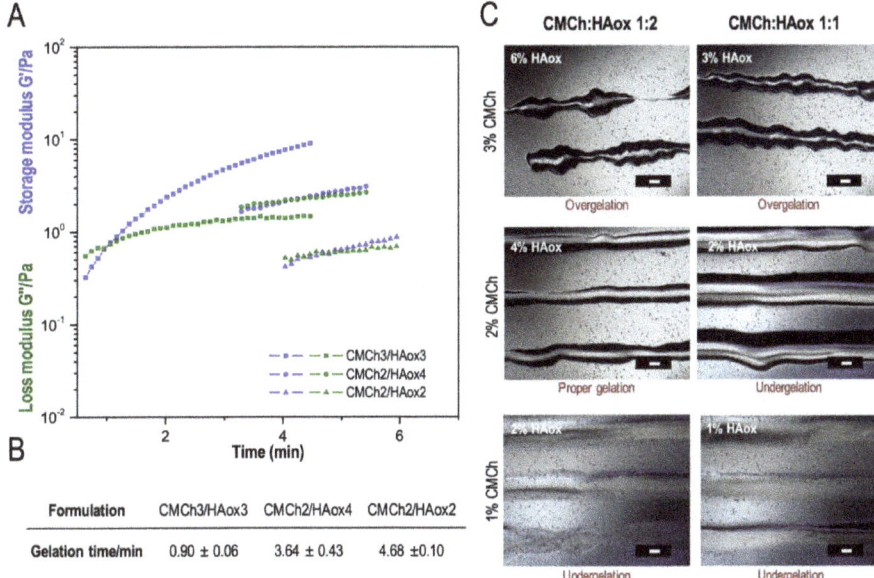

Figure 2. (**A**) Storage (G′) and loss (G″) moduli obtained in time sweep rheological experiments, (**B**) gelation times, defined as G′ and G″ crossover points, and (**C**) light microscope pictures of printed samples of CMCh/HAox formulations with different weight concentrations of CMCh and HAox solutions. Scale bars in a white color correspond to 500 µm.

While printing the CMCh2/HAox4 mixture, bubbles were observed in the needle (Figure 3A, black arrows). In addition, the printed lines had an irregular shape (Figure 3A). We hypothesized that the different viscosities of the precursor solutions due to the different molecular weights of the polymers [64,73] would be the reason for these features. Figure 3B shows that the viscosity of 2 wt % CMCh is 2 orders of magnitude higher than the viscosity of 4 wt % HAox. A different viscosity of the precursor's solutions is reported to lead to non-homogeneous mixtures due to their different flow through the mixer during extrusion [39,41]. Different strategies have been used in order to adjust viscosities of precursor solutions when using static mixers. For example, Hozumi et al. [41] used carboxymethyl cellulose as a thickening agent, and Bootsma et al. [39] distributed the solution with the largest impact on viscosity in the two syringes. In order to increase the viscosity of the HAox solution, we supplemented it with non-oxidized HA. The viscosity of different HAox-HA blends is also plotted in Figure 3B. All tested solutions presented a shear thinning behavior that facilitates extrusion and shape fidelity [5]. The addition of increasing amounts of HA to the HAox solution lead to a higher viscosity of the mixture. Based on the obtained results, the addition of 0.4 wt % of HA to the 4 wt % HAox solution resulted in a similar viscosity to the 2 wt % CMCh solution. This addition did not influence crosslinking kinetics of the formulation (Figure S1). The printing test with CMCh2/HAox4-HA0.4 formulation (Figure 3C) showed regular and smooth lines without broken parts and no bubbles were formed during the printing process. The CMCh2/HAox4-HA0.4 formulation provided stable filaments with low deviance from the needle geometry and minimized collapsing between the superposed layers visible in the cross-points (Figure 3C). These observations indicate that static mixing of solutions with comparable viscosities improves mixing performance, printing quality, and the resolution.

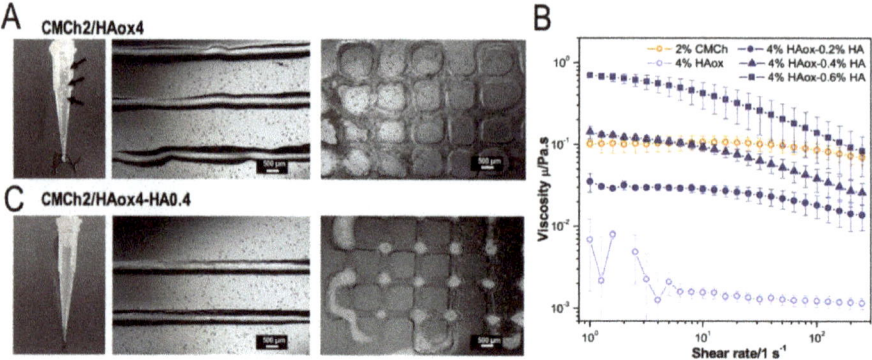

Figure 3. (A) Image of the needle during printing. Black arrows highlight bubbles inside the bio-ink. Light microscopy pictures of printed strands and 3D printed scaffolds using CMCh2/HAox4 formulation. (B) Viscosity measurements of 2 wt % CMCh, 4 wt % HAox, and different HAox/HA blends. (C) Image of the needle during printing and light microscopy pictures of printed strands and 3D printed scaffolds using CMCh2/HAox4-HA0.4 formulation.

The printing protocol described in this case allows high fidelity printing of hydrogel structures with low-viscosity ink solutions, which is favorable for cell laden scaffolds [5,27,32]. The hydrogel viscosity, flow rate, and gelation kinetics of the components as they pass through the static mixer affect the mixing performance, homogeneity, and self-support capacity of the bio-ink.

3.2. Characterization of the Optimized Bio-Ink

In order to confirm the formation of covalent crosslinks between the CMCh and HAox components of the printing mixture, the CMCh2/HAox4-HA0.4 formulation was characterized by FTIR spectroscopy (see Figure S2). The characteristic peaks corresponding to the functional groups of the CMCh and HAox/HA precursors were observed in the mixture [56,57,62], together with a band at 1653 cm^{-1}, which can be attributed to the stretching vibration of the C=N bond of the Schiff base formed by a reaction of amine and aldehyde groups. This indicates that covalent crosslinking was successfully achieved [63,64,74]. Furthermore, a peak was observed at 885 cm^{-1}, corresponding to the hemiacetal structure obtained due to the unreacted aldehyde groups of HAox after crosslinking [62]. Intensity of this peak is lower than in the HAox spectrum, which indicates that the rest of the aldehyde groups had participated in the crosslinking reaction.

The viscoelastic properties of the crosslinked CMCh2/HAox4-HA0.4 hydrogel were studied by rheology in frequency sweep experiments. Hydrogel formation was corroborated since storage modulus was always higher than loss modulus. Additionally, a slight frequency-dependent viscoelastic behavior was observed. Presumably, the shear modulus values were mainly due to the covalent crosslinking of the CMCh and HAox functional groups, and HA did not influence crosslinking kinetics or final modulus (Figure S1). Gels were soft with a shear modulus in the range of 50–100 Pa (Figure 4A). This value indicates that these hydrogel scaffolds are promising candidates for regeneration of soft tissues [39], and is comparable to reported chitosan/hyaluronic acid injectable hydrogels with encapsulated cells for abdominal reparation and adhesion prevention [39,63,74,75]. The viscosity of the crosslinked CMCh2/HAox4-HA0.4 bio-ink vs. shear rate is plotted in Figure 4B. The ink viscosity found was relatively low, especially when compared to air pressure-based extruded inks (in the range of

30–6 × 10^7 mPa) [1,2], which is a desirable feature since low-viscosity bio-inks usually allow higher cell viability [5,27,32]. Solution behaved as a non-Newtonian fluid, where viscosity decreased linearly with an increasing shear rate. This shear thinning behavior is a favorable property for printing. It implies a decrease in the viscosity when the shear stress increases inside the needle under applied pressure, which is followed by a sharp increase of viscosity after extrusion. This facilitates both extrusion and shape fidelity [5].

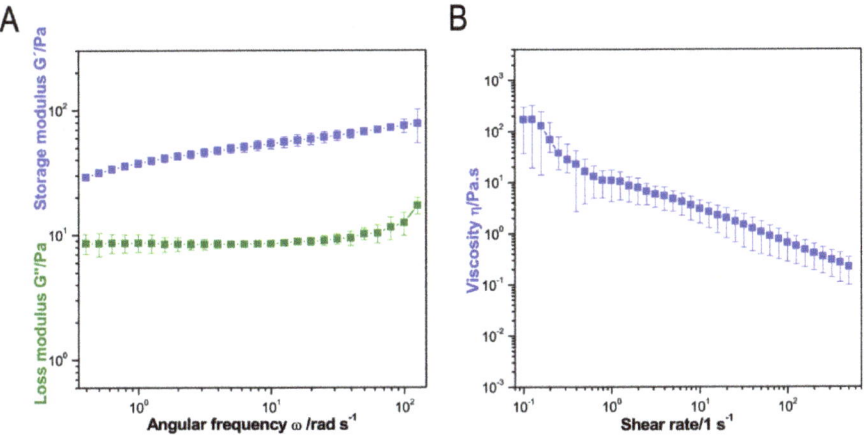

Figure 4. (**A**) Frequency sweep experiment and (**B**) viscosity analysis of CMCh2/HAox4-HA0.4 bio-ink formulation.

3.3. Characterization of the 3D Printed Hydrogel Scaffolds

2- and 4-layered grid square scaffolds (12 × 12 mm^2 printed area) were printed using the CMCh2/HAox4-HA0.4 formulation (Figure 5). Good printing accuracy and resolution was observed and stable scaffolds with filaments of uniform dimensions (diameter 357 ± 58 μm) were obtained.

Swelling and degradation rates are relevant parameters when using hydrogels' bio-inks since they affect the fidelity and stability of the bio-printed scaffolds, as well as allow cellular ingrowth and tissue regeneration [27,76,77]. In this study, swelling and degradability of the printed scaffolds were analyzed by imaging the scaffolds after incubation in PBS for given times and by quantification of the width of the strands (Figure 6A–C). Figure 6A shows the microscopy images of 2-layer printed scaffolds after 4 h of immersion in PBS. A 49 ± 9% swelling was observed under these conditions (Figure 6A). A progressive and notable decrease in scaffold volume was observed with incubation time, up to nearly complete degradation after 7 days (Figure 6B). This degradation rate is slightly faster than previously reported for CMCh/HAox injectable hydrogels (10–14 days) [63,74,75], which can be assigned to the higher surface area and open structure of the grid scaffold that makes them more sensitive for degradation. Schiff's base crosslinked hydrogels have low stability due to the dynamic nature of the bond [57,78,79]. Thus, to increase the long-term stability of the scaffolds, a new crosslinking approach was proposed since stability is directly related to the crosslinking degree of the hydrogel [76,80,81]. A post-printing crosslinking step was adopted by immersing the scaffold in 20 mM FeCl$_3$ for 7 min. Fe (III) forms coordination complexes with hyaluronic acid units at a physiological pH [82], which are expected to act as additional crosslinking points in the printed scaffold. Figure 6A shows lower swelling (19 ± 8%) of the printed scaffold after the second crosslinking step and 4 h after swelling. Additionally, slower degradation of the scaffold was observed after post-printing stabilization (Figure 6C). The scaffold maintained its structural integrity up to 28 days of incubation, although signs of erosion were appreciated in the last stage. In conclusion, the treatment with iron

(III) leads to 3D scaffolds with higher structural integrity and long-term stability. This reinforces the stability of Schiff's base crosslinked hydrogels, which has remained a challenging issue [45,78].

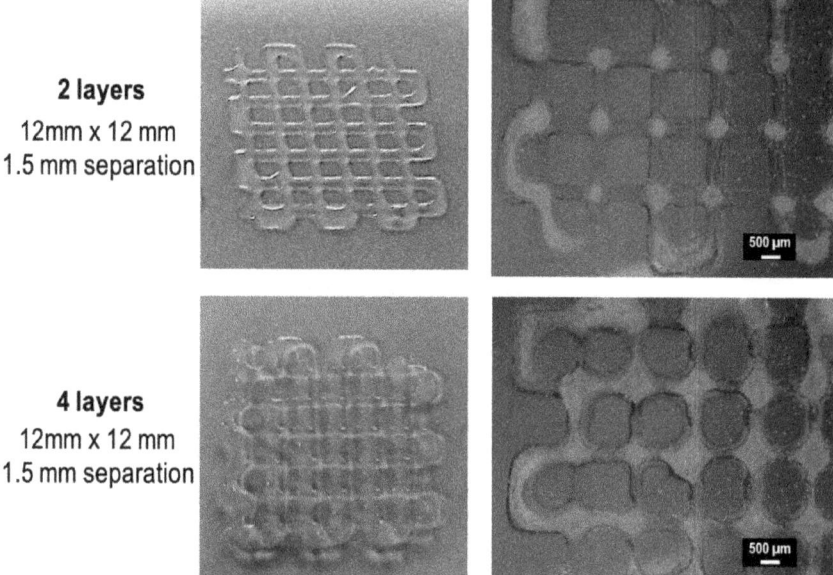

Figure 5. Camera pictures (**left**) and light microscopy pictures (**right**) of 2-layer and 4-layer square scaffolds printed using CMCh2/HAox4-HA0.4 formulation.

3.4. CMCh2/HAox4-HA0.4 Based Bio-Printing

In general, 3D printed hydrogels have been demonstrated to protect cells from mechanical damage induced during the extrusion process, while providing an appropriate environment for the encapsulated cells after printing (by mimicking the ECM) [27,30]. Nevertheless, it is a critical aspect in bio-fabrication to evaluate whether viscosity, in situ crosslinking, and printing process are compatible with encapsulated living cells [4,10]. Thus, the ability of CMCh2/HAox4-HA0.4 formulation to be used as a bio-ink was tested by printing scaffolds with encapsulated L929 fibroblasts. Cell viability in the bio-printed scaffolds was studied during a 14-day period. Live/dead staining allowed imaging of the cells in 2-layer printed scaffolds (Figure 7A). Abundance of cells homogeneously distributed in the scaffold were observed, which reflects the good mixing performance during the printing process. Additionally, cells were released from the hydrogel after 7 and 14 days of culture, which can be due to the degradation rate of the scaffolds. This is a desirable characteristic for potential regenerative approaches for wound healing, where delivered cells would migrate out of the scaffold to heal the injured site [4]. Quantification of cell viability is displayed in Figure 7B, together with data from 3D cultures in non-printed hydrogels of the same formulation as the control. Cells in the printed or non-printed materials showed viability around 60–65% after 1 day of incubation. There are no significant differences between printed and non-printed formulation, which indicates that the printing process did not affect the cells short-term viability. Cell viability in the printed scaffold increased at longer culture times and reached 96% and 95% 7 and 14 days of culture, respectively, which are both significantly different from printed and non-printed formulations after 1 day of incubation. These data suggest that neither the covalent reaction responsible of gelation nor the shear stress produced by the printing process or the stabilization process with iron (III) cause adverse long-term effects on the cells. The proliferation rate of the cells in the scaffolds was analyzed after DAPI staining. Cell proliferation increased over the 14-day period in the printed scaffolds (Figure 7C), and values reached after 14 days of incubation were significantly different from those found after 1 and 4 days of culturing. Lastly, the scaffolds maintained their structural integrity during the whole culture processes (Figure 7D), which indicates that the optimized printed formula and the subsequent stabilization step with iron produced mechanically robust scaffolds with good biocompatibility.

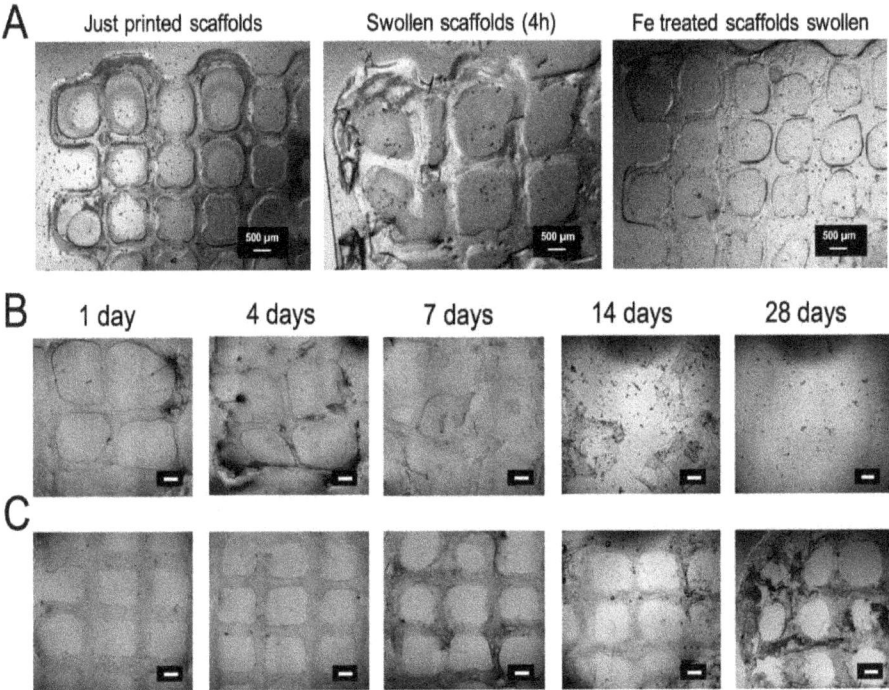

Figure 6. (A) Microscopy images of CMCh2/HAox4-HA0.4 printed scaffolds just after printing, after swelling in PBS for 4 h, and after iron treatment and swelling in PBS for 4 h. (B) Degradation study of CMCh2/HAox4-HA0.4 scaffolds with no additional treatment after incubation in PBS and (C) after iron treatment and incubation in PBS at different time points. Scale bars correspond to 500 μm.

These observations were consistent with other studies based on naturally derived hydrogels bioprinting [17,51,83]. For example, Akkineni et al. studied the encapsulation of endothelial cells in a low viscosity hydrogel core (1% gelatin and 3% alginate) by obtaining cell viability values around 65% one day after printing, which is comparable to our results. A high viscosity shell composed of 10% alginate and 1% gelatin and a secondary crosslinking with $CaCl_2$ provided the structural integrity to the scaffold [17]. In addition, Gu et al. presented the direct-write printing of stem cells within a polysaccharide-based bio-ink comprising alginate, carboxymethylchitosan, and agarose. The time course of dead cells content within the optimized bio-ink (containing 5% w/v of carboxymethylchitosan) demonstrated a relatively high (around 25%) cell death after printing. Subsequently, this decreased to around 10% by day 7, following a trend very similar to that of our work [83]. On the other hand, some reactive hydrogels have been reported for cell encapsulation such as: injectable hydrogels with proliferation trends similar to that found in our work [42], layered platforms with constant cell viability values around 70% until 5 days [37], or gradient formulations [36] where cell viability values slightly decrease with time until around 80% after 7 days. Based on the in-vitro studies, we conclude that the proposed printing technology and bio-ink formulation of this work are suitable as a 3D printing platform for potential biomedical applications as cell carriers in the tissue engineering field.

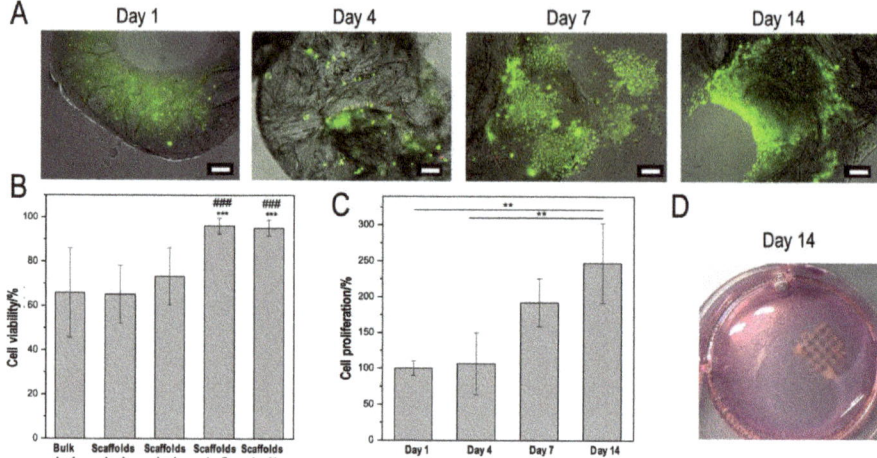

Figure 7. Biological results of CMCh2/HAox4-HA0.4 printed scaffolds loaded with L929 fibroblasts and treated with Fe over a 14-day period: (**A**) Fluorescence imaging of live/dead stained scaffolds at different culture days. (**B**) Quantification of live/dead results including bulk hydrogels at 1 day as a control. Analysis of variance (ANOVA) of the results for the printed samples was performed at each time point with respect to day 1 at a significance level of *** $p < 0.05$, and with respect to non-printed samples at a significance level of ### $p < 0.05$. (**C**) Proliferation assay by quantification of nuclei after DAPI staining. Analysis of variance (ANOVA) of the results for the printed samples was performed at each time point at a significance level of ** $p < 0.01$. (**D**) Picture of a stable printed scaffold after 14 days of incubation. Scale bars correspond to 200 µm.

4. Conclusions

The present study describes the development of a reactive hydrogel bio-ink with an extrusion printing methodology based on a dual-syringe system with a static mixing tool. This method shows multiple advantages for 3D extrusion bio-printing. (1) Gelation during the extrusion process provides enough viscosity for printing with good shape fidelity while using low viscosity precursor solutions, (2) the crosslinking during extrusion provides enough structural integrity to retain the printed shape, and (3) the stability of the scaffold, if required for long-term culturing, can be increased in a simple incubation step. Bio-printed scaffolds obtained with our approach showed good biocompatibility, moderate swelling, and shape stability during 14 days of culturing. Since precursors' concentrations and printing conditions can be easily varied, this printing approach offers high versatility and we envision that it can be adaptable to a wide range of reactive systems with appropriate crosslinking kinetics to be employed in the future for broad applications in regenerative medicine and tissue-engineering.

Supplementary Materials: The following are available online at http://www.mdpi.com/2073-4360/12/9/1986/s1. Figure S1. (A) Storage (G′) and loss (G″) moduli obtained in time sweep rheological experiments and (B) gelation times, defined as G′ and G″ crossover points, obtained for CMCh2/HAox4 and CMCh2/HAox4-HA0.4 formulations. Figure S2. FTIR spectra of CMCh, HAox, and CMCh/HAox reactive hydrogel.

Author Contributions: M.P.-B. extensively contributed to the investigation and writing-original draft preparation. M.K.W.-B. contributed to methodology and draft preparation. B.V.-L., J.S.R., and A.d.C. contributed to supervision. All the authors contributed to analysis, discussion, and redaction of the paper. All authors have read and agreed to the published version of the manuscript.

Funding: This research was funded by CIBER-BBN, Spain and the Spanish Ministry of Economy and Competitivity (project MAT2017-84277-R and M. Puertas-Bartolomé scholarship BES-2015-075161). B. Vázquez-Lasa and J. San Román are members of the *SusPlast platform (Interdisciplinary Platform for Sustainable Plastics towards a Circular Economy)* from the Spanish National Research Council (CSIC).

Acknowledgments: The authors acknowledge RegenHu Company, and, particularly, Sandro Figi, Dominic Ernst, Michael Kuster, and Andreas Scheidegger for the fruitful collaboration, development, and providing the mixing tool. The authors thank Emmanuel Terriac from INM, Germany, for assistance in the confocal imaging.

Conflicts of Interest: The authors declare no conflict of interest.

References

1. Derakhshanfar, S.; Mbeleck, R.; Xu, K.; Zhang, X.; Zhong, W.; Xing, M. 3D bioprinting for biomedical devices and tissue engineering: A review of recent trends and advances. *Bioact. Mater.* **2018**, *3*, 144–156. [CrossRef] [PubMed]
2. Cui, H.; Nowicki, M.; Fisher, J.P.; Zhang, L.G. 3D bioprinting for organ regeneration. *Adv. Healthc. Mater.* **2017**, *6*, 1601118. [CrossRef] [PubMed]
3. Włodarczyk-Biegun, M.K.; del Campo, A. 3D bioprinting of structural proteins. *Biomaterials* **2017**, *134*, 180–201. [CrossRef] [PubMed]
4. Murphy, S.V.; Skardal, A.; Atala, A. Evaluation of hydrogels for bio-printing applications. *J. Biomed. Mater. Res. Part A* **2013**, *101*, 272–284. [CrossRef]
5. Malda, J.; Visser, J.; Melchels, F.P.; Jüngst, T.; Hennink, W.E.; Dhert, W.J.; Groll, J.; Hutmacher, D.W. 25th anniversary article: Engineering hydrogels for biofabrication. *Adv. Mater.* **2013**, *25*, 5011–5028. [CrossRef]
6. Hospodiuk, M.; Dey, M.; Sosnoski, D.; Ozbolat, I.T. The bioink: A comprehensive review on bioprintable materials. *Biotechnol. Adv.* **2017**, *35*, 217–239. [CrossRef]
7. Chaudhari, A.; Vig, K.; Baganizi, D.; Sahu, R.; Dixit, S.; Dennis, V.; Singh, S.; Pillai, S. Future prospects for scaffolding methods and biomaterials in skin tissue engineering: A review. *Int. J. Mol. Sci.* **2016**, *17*, 1974. [CrossRef]
8. Mora-Boza, A.; Puertas-Bartolomé, M.; Vázquez-Lasa, B.; Román, J.S.; Pérez-Caballer, A.; Olmeda-Lozano, M. Contribution of bioactive hyaluronic acid and gelatin to regenerative medicine. Methodologies of gels preparation and advanced applications. *Eur. Polym. J.* **2017**, *95*, 11–26. [CrossRef]
9. Zhang, J.; Allardyce, B.J.; Rajkhowa, R.; Zhao, Y.; Dilley, R.J.; Redmond, S.L.; Wang, X.; Liu, X. 3D printing of silk particle-reinforced chitosan hydrogel structures and their properties. *ACS Biomater. Sci. Eng.* **2018**, *4*, 3036–3046. [CrossRef]
10. Aydogdu, M.O.; Oner, E.T.; Ekren, N.; Erdemir, G.; Kuruca, S.E.; Yuca, E.; Bostan, M.S.; Eroglu, M.S.; Ikram, F.; Uzun, M. Comparative characterization of the hydrogel added PLA/β-TCP scaffolds produced by 3D bioprinting. *Bioprinting* **2019**, *13*, e00046. [CrossRef]
11. Yan, J.; Wang, Y.; Zhang, X.; Zhao, X.; Ma, J.; Pu, X.; Wang, Y.; Ran, F.; Wang, Y.; Leng, F. Snakegourd root/Astragalus polysaccharide hydrogel preparation and application in 3D printing. *Int. J. Biol. Macromol.* **2019**, *121*, 309–316. [CrossRef] [PubMed]
12. Chen, S.; Jang, T.-S.; Pan, H.M.; Jung, H.-D.; Sia, M.W.; Xie, S.; Hang, Y.; Chong, S.K.M.; Wang, D.; Song, J.J.I.J.o.B. 3D Freeform Printing of Nanocomposite Hydrogels Through in situ Precipitation in Reactive Viscous Fluid. *Int. J. Bioprinting* **2020**, *6*, 258.
13. Jang, T.-S.; Jung, H.-D.; Pan, M.H.; Han, W.T.; Chen, S.; Song, J. 3D printing of hydrogel composite systems: Recent advances in technology for tissue engineering. *Int. J. Bioprinting* **2018**. [CrossRef]
14. Annabi, N.; Tamayol, A.; Uquillas, J.A.; Akbari, M.; Bertassoni, L.E.; Cha, C.; Camci-Unal, G.; Dokmeci, M.R.; Peppas, N.A.; Khademhosseini, A. 25th anniversary article: Rational design and applications of hydrogels in regenerative medicine. *Adv. Mater.* **2014**, *26*, 85–124. [CrossRef] [PubMed]
15. Ratner, B.D.; Bryant, S.J. Biomaterials: Where we have been and where we are going. *Annu. Rev. Biomed. Eng.* **2004**, *6*, 41–75. [CrossRef]
16. Caló, E.; Khutoryanskiy, V.V. Biomedical applications of hydrogels: A review of patents and commercial products. *Eur. Polym. J.* **2015**, *65*, 252–267. [CrossRef]
17. Akkineni, A.R.; Ahlfeld, T.; Lode, A.; Gelinsky, M. A versatile method for combining different biopolymers in a core/shell fashion by 3D plotting to achieve mechanically robust constructs. *Biofabrication* **2016**, *8*, 045001. [CrossRef]
18. Roehm, K.D.; Madihally, S.V. Bioprinted chitosan-gelatin thermosensitive hydrogels using an inexpensive 3D printer. *Biofabrication* **2017**, *10*, 015002. [CrossRef]

19. Huang, J.; Fu, H.; Wang, Z.; Meng, Q.; Liu, S.; Wang, H.; Zheng, X.; Dai, J.; Zhang, Z. BMSCs-laden gelatin/sodium alginate/carboxymethyl chitosan hydrogel for 3D bioprinting. *RSC Adv.* **2016**, *6*, 108423–108430. [CrossRef]
20. Park, J.Y.; Choi, J.-C.; Shim, J.-H.; Lee, J.-S.; Park, H.; Kim, S.W.; Doh, J.; Cho, D.-W. A comparative study on collagen type I and hyaluronic acid dependent cell behavior for osteochondral tissue bioprinting. *Biofabrication* **2014**, *6*, 035004. [CrossRef]
21. Haring, A.P.; Thompson, E.G.; Tong, Y.; Laheri, S.; Cesewski, E.; Sontheimer, H.; Johnson, B.N. Process-and bio-inspired hydrogels for 3D bioprinting of soft free-standing neural and glial tissues. *Biofabrication* **2019**, *11*, 025009. [CrossRef]
22. Li, C.; Wang, K.; Zhou, X.; Li, T.; Xu, Y.; Qiang, L.; Peng, M.; Xu, Y.; Xie, L.; He, C. Controllable fabrication of hydroxybutyl chitosan/oxidized chondroitin sulfate hydrogels by 3D bioprinting technique for cartilage tissue engineering. *Biomed. Mater.* **2018**, *14*, 025006. [CrossRef] [PubMed]
23. Mora-Boza, A.; Włodarczyk-Biegun, M.K.; del Campo, A.; Vázquez-Lasa, B.; Román, J.S. Glycerylphytate as an ionic crosslinker for 3D printing of multi-layered scaffolds with improved shape fidelity and biological features. *Biomater. Sci.* **2020**. [CrossRef] [PubMed]
24. Bakarich, S.E.; Gorkin, R., III; Gately, R.; Naficy, S.; in het Panhuis, M.; Spinks, G.M. 3D printing of tough hydrogel composites with spatially varying materials properties. *Addit. Manuf.* **2017**, *14*, 24–30. [CrossRef]
25. Kirchmajer, D.M.; Gorkin Iii, R. An overview of the suitability of hydrogel-forming polymers for extrusion-based 3D-printing. *J. Mater. Chem. B* **2015**, *3*, 4105–4117. [CrossRef]
26. Kiyotake, E.A.; Douglas, A.W.; Thomas, E.E.; Nimmo, S.L.; Detamore, M.S. Development and quantitative characterization of the precursor rheology of hyaluronic acid hydrogels for bioprinting. *Acta Biomater.* **2019**, *95*, 176–187. [CrossRef]
27. Hölzl, K.; Lin, S.; Tytgat, L.; Van Vlierberghe, S.; Gu, L.; Ovsianikov, A. Bioink properties before, during and after 3D bioprinting. *Biofabrication* **2016**, *8*, 032002. [CrossRef]
28. Wu, Q.; Therriault, D.; Heuzey, M.-C. Processing and properties of chitosan Inks for 3D printing of hydrogel microstructures. *ACS Biomater. Sci. Eng.* **2018**, *4*, 2643–2652. [CrossRef]
29. Li, H.; Tan, Y.J.; Liu, S.; Li, L. Three-dimensional bioprinting of oppositely charged hydrogels with super strong Interface bonding. *ACS Appl. Mater. Interfaces* **2018**, *10*, 11164–11174. [CrossRef]
30. Highley, C.B.; Song, K.H.; Daly, A.C.; Burdick, J.A. Jammed microgel inks for 3D printing applications. *Adv. Sci.* **2019**, *6*, 1801076. [CrossRef]
31. Hinton, T.J.; Jallerat, Q.; Palchesko, R.N.; Park, J.H.; Grodzicki, M.S.; Shue, H.-J.; Ramadan, M.H.; Hudson, A.R.; Feinberg, A.W. Three-dimensional printing of complex biological structures by freeform reversible embedding of suspended hydrogels. *Sci. Adv.* **2015**, *1*, e1500758. [CrossRef] [PubMed]
32. Aguado, B.A.; Mulyasasmita, W.; Su, J.; Lampe, K.J.; Heilshorn, S.C. Improving viability of stem cells during syringe needle flow through the design of hydrogel cell carriers. *Tissue Eng. Part A* **2011**, *18*, 806–815. [CrossRef] [PubMed]
33. Skardal, A.; Zhang, J.; McCoard, L.; Xu, X.; Oottamasathien, S.; Prestwich, G.D. Photocrosslinkable hyaluronan-gelatin hydrogels for two-step bioprinting. *Tissue Eng. Part A* **2010**, *16*, 2675–2685. [CrossRef] [PubMed]
34. O'Connell, C.D.; Di Bella, C.; Thompson, F.; Augustine, C.; Beirne, S.; Cornock, R.; Richards, C.J.; Chung, J.; Gambhir, S.; Yue, Z. Development of the Biopen: A handheld device for surgical printing of adipose stem cells at a chondral wound site. *Biofabrication* **2016**, *8*, 015019. [CrossRef]
35. Gregor, A.; Hošek, J. 3D printing methods of biological scaffolds used in tissue engineering. *Rom Rev Precis Mech Opt Mechatron.* **2011**, *3*, 143–148.
36. Zimmermann, R.; Hentschel, C.; Schrön, F.; Moedder, D.; Büttner, T.; Atallah, P.; Wegener, T.; Gehring, T.; Howitz, S.; Freudenberg, U. High resolution bioprinting of multi-component hydrogels. *Biofabrication* **2019**, *11*, 045008. [CrossRef]
37. Lozano, R.; Stevens, L.; Thompson, B.C.; Gilmore, K.J.; Gorkin, R., III; Stewart, E.M.; in het Panhuis, M.; Romero-Ortega, M.; Wallace, G.G. 3D printing of layered brain-like structures using peptide modified gellan gum substrates. *Biomaterials* **2015**, *67*, 264–273. [CrossRef]
38. Maiullari, F.; Costantini, M.; Milan, M.; Pace, V.; Chirivì, M.; Maiullari, S.; Rainer, A.; Baci, D.; Marei, H.E.-S.; Seliktar, D. A multi-cellular 3D bioprinting approach for vascularized heart tissue engineering based on HUVECs and iPSC-derived cardiomyocytes. *Sci. Rep.* **2018**, *8*, 13532. [CrossRef]

39. Bootsma, K.; Fitzgerald, M.M.; Free, B.; Dimbath, E.; Conjerti, J.; Reese, G.; Konkolewicz, D.; Berberich, J.A.; Sparks, J.L. 3D printing of an interpenetrating network hydrogel material with tunable viscoelastic properties. *J. Mech. Behav. Biomed. Mater.* **2017**, *70*, 84–94. [CrossRef]
40. Deepthi, S.; Jayakumar, R. Alginate nanobeads interspersed fibrin network as in situ forming hydrogel for soft tissue engineering. *Bioact. Mater.* **2018**, *3*, 194–200. [CrossRef]
41. Hozumi, T.; Ohta, S.; Ito, T. Analysis of the Calcium Alginate Gelation Process Using a Kenics Static Mixer. *Ind. Eng. Chem. Res.* **2015**, *54*, 2099–2107. [CrossRef]
42. Zhang, Y.; Chen, H.; Zhang, T.; Zan, Y.; Ni, T.; Liu, M.; Pei, R. Fast-forming BMSC-encapsulating hydrogels through bioorthogonal reaction for osteogenic differentiation. *Biomater. Sci.* **2018**, *6*, 2578–2581. [CrossRef] [PubMed]
43. Whitely, M.; Cereceres, S.; Dhavalikar, P.; Salhadar, K.; Wilems, T.; Smith, B.; Mikos, A.; Cosgriff-Hernandez, E. Improved in situ seeding of 3D printed scaffolds using cell-releasing hydrogels. *Biomaterials* **2018**, *185*, 194–204. [CrossRef] [PubMed]
44. Moglia, R.S.; Whitely, M.; Dhavalikar, P.; Robinson, J.; Pearce, H.; Brooks, M.; Stuebben, M.; Cordner, N.; Cosgriff-Hernandez, E. Injectable polymerized high internal phase emulsions with rapid in situ curing. *Biomacromolecules* **2014**, *15*, 2870–2878. [CrossRef] [PubMed]
45. Hozumi, T.; Kageyama, T.; Ohta, S.; Fukuda, J.; Ito, T. Injectable hydrogel with slow degradability composed of gelatin and hyaluronic acid cross-linked by Schiff's base formation. *Biomacromolecules* **2018**, *19*, 288–297. [CrossRef]
46. Wu, H.-D.; Yang, J.-C.; Tsai, T.; Ji, D.-Y.; Chang, W.-J.; Chen, C.-C.; Lee, S.-Y. Development of a chitosan–polyglutamate based injectable polyelectrolyte complex scaffold. *Carbohydr. Polym.* **2011**, *85*, 318–324. [CrossRef]
47. Kohn, C.; Klemens, J.; Kascholke, C.; Murthy, N.; Kohn, J.; Brandenburger, M.; Hacker, M. Dual–component collagenous peptide/reactive oligomer hydrogels as potential nerve guidance materials–from characterization to functionalization. *Biomater. Sci.* **2016**, *4*, 1605–1621. [CrossRef]
48. Caballero, S.S.R.; Saiz, E.; Montembault, A.; Tadier, S.; Maire, E.; David, L.; Delair, T.; Grémillard, L. 3-D printing of chitosan-calcium phosphate inks: Rheology, interactions and characterization. *J. Mater. Sci. Mater. Med.* **2019**, *30*, 6. [CrossRef]
49. Intini, C.; Elviri, L.; Cabral, J.; Mros, S.; Bergonzi, C.; Bianchera, A.; Flammini, L.; Govoni, P.; Barocelli, E.; Bettini, R. 3D-printed chitosan-based scaffolds: An in vitro study of human skin cell growth and an in-vivo wound healing evaluation in experimental diabetes in rats. *Carbohydr. Polym.* **2018**, *199*, 593–602. [CrossRef]
50. Demirtaş, T.T.; Irmak, G.; Gümüşderelioğlu, M. A bioprintable form of chitosan hydrogel for bone tissue engineering. *Biofabrication* **2017**, *9*, 035003. [CrossRef]
51. Noh, I.; Kim, N.; Tran, H.N.; Lee, J.; Lee, C. 3D printable hyaluronic acid-based hydrogel for its potential application as a bioink in tissue engineering. *Biomater. Res.* **2019**, *23*, 3. [CrossRef] [PubMed]
52. Ouyang, L.; Highley, C.B.; Rodell, C.B.; Sun, W.; Burdick, J.A. 3D printing of shear-thinning hyaluronic acid hydrogels with secondary cross-linking. *ACS Biomater. Sci. Eng.* **2016**, *2*, 1743–1751. [CrossRef]
53. Mazzocchi, A.; Devarasetty, M.; Huntwork, R.; Soker, S.; Skardal, A. Optimization of collagen type I-hyaluronan hybrid bioink for 3D bioprinted liver microenvironments. *Biofabrication* **2018**, *11*, 015003. [CrossRef] [PubMed]
54. Muxika, A.; Etxabide, A.; Uranga, J.; Guerrero, P.; de la Caba, K. Chitosan as a bioactive polymer: Processing, properties and applications. *Int. J. Biol. Macromol.* **2017**, *105*, 1358–1368. [CrossRef] [PubMed]
55. Pellá, M.C.G.; Lima-Tenório, M.K.; Tenório-Neto, E.T.; Guilherme, M.R.; Muniz, E.C.; Rubira, A.F. Chitosan-based hydrogels: From preparation to biomedical applications. *Carbohydr. Polym.* **2018**, *196*, 233–245. [CrossRef]
56. Nguyen, N.T.-P.; Nguyen, L.V.-H.; Tran, N.M.-P.; Nguyen, T.-H.; Huynh, C.-K.; Van, T.V. Synthesis of cross-linking chitosan-hyaluronic acid based hydrogels for tissue engineering applications. In Proceedings of the International Conference on the Development of Biomedical Engineering in Vietnam; Springer: Singapore, 2017; 63, pp. 671–675. [CrossRef]
57. Qian, C.; Zhang, T.; Gravesande, J.; Baysah, C.; Song, X.; Xing, J. Injectable and self-healing polysaccharide-based hydrogel for pH-responsive drug release. *Int. J. Biol. Macromol.* **2019**, *123*, 140–148. [CrossRef]

58. Ye, K.; Felimban, R.; Traianedes, K.; Moulton, S.E.; Wallace, G.G.; Chung, J.; Quigley, A.; Choong, P.F.; Myers, D.E. Chondrogenesis of infrapatellar fat pad derived adipose stem cells in 3D printed chitosan scaffold. *PLoS ONE* **2014**, *9*, e99410. [CrossRef]
59. Lee, J.-Y.; Choi, B.; Wu, B.; Lee, M. Customized biomimetic scaffolds created by indirect three-dimensional printing for tissue engineering. *Biofabrication* **2013**, *5*, 045003. [CrossRef]
60. Catoira, M.C.; Fusaro, L.; Di Francesco, D.; Ramella, M.; Boccafoschi, F. Overview of natural hydrogels for regenerative medicine applications. *J. Mater. Sci. Mater. Med.* **2019**, *30*, 115. [CrossRef]
61. Song, R.; Murphy, M.; Li, C.; Ting, K.; Soo, C.; Zheng, Z. Current development of biodegradable polymeric materials for biomedical applications. *Drug Des. Dev. Ther.* **2018**, *12*, 3117–3145. [CrossRef]
62. Tan, H.; Chu, C.R.; Payne, K.A.; Marra, K.G. Injectable in situ forming biodegradable chitosan–hyaluronic acid based hydrogels for cartilage tissue engineering. *Biomaterials* **2009**, *30*, 2499–2506. [CrossRef] [PubMed]
63. Song, L.; Li, L.; He, T.; Wang, N.; Yang, S.; Yang, X.; Zeng, Y.; Zhang, W.; Yang, L.; Wu, Q.; et al. Peritoneal adhesion prevention with a biodegradable and injectable N,O-carboxymethyl chitosan-aldehyde hyaluronic acid hydrogel in a rat repeated-injury model. *Sci. Rep.* **2016**, *6*, 37600. [CrossRef] [PubMed]
64. Khorshidi, S.; Karkhaneh, A. A self-crosslinking tri-component hydrogel based on functionalized polysaccharides and gelatin for tissue engineering applications. *Mater. Lett.* **2016**, *164*, 468–471. [CrossRef]
65. Puertas-Bartolomé, M.; Benito-Garzón, L.; Fung, S.; Kohn, J.; Vázquez-Lasa, B.; San Román, J. Bioadhesive functional hydrogels: Controlled release of catechol species with antioxidant and antiinflammatory behavior. *Mater. Sci. Eng. C* **2019**, *105*, 110040. [CrossRef] [PubMed]
66. Nguyen, N.T.-P.; Nguyen, L.V.-H.; Tran, N.M.-P.; Nguyen, D.T.; Nguyen, T.N.-T.; Tran, H.A.; Dang, N.N.-T.; Van Vo, T.; Nguyen, T.-H. The effect of oxidation degree and volume ratio of components on properties and applications of in situ cross-linking hydrogels based on chitosan and hyaluronic acid. *Mater. Sci. Eng. C* **2019**, *103*, 109670. [CrossRef] [PubMed]
67. Zhao, H.; Heindel, N.D. Determination of degree of substitution of formyl groups in polyaldehyde dextran by the hydroxylamine hydrochloride method. *Pharm. Res.* **1991**, *8*, 400–402. [CrossRef]
68. Yan, S.; Wang, T.; Feng, L.; Zhu, J.; Zhang, K.; Chen, X.; Cui, L.; Yin, J. Injectable in situ self-cross-linking hydrogels based on poly (L-glutamic acid) and alginate for cartilage tissue engineering. *Biomacromolecules* **2014**, *15*, 4495–4508. [CrossRef]
69. Florczyk, S.J.; Wang, K.; Jana, S.; Wood, D.L.; Sytsma, S.K.; Sham, J.G.; Kievit, F.M.; Zhang, M. Porous chitosan-hyaluronic acid scaffolds as a mimic of glioblastoma microenvironment ECM. *Biomaterials* **2013**, *34*, 10143–10150. [CrossRef]
70. Kim, Y.; Larkin, A.L.; Davis, R.M.; Rajagopalan, P. The design of in vitro liver sinusoid mimics using chitosan–hyaluronic acid polyelectrolyte multilayers. *Tissue Eng. Part A* **2010**, *16*, 2731–2741. [CrossRef]
71. Wu, J.; Wang, X.; Keum, J.K.; Zhou, H.; Gelfer, M.; Avila-Orta, C.A.; Pan, H.; Chen, W.; Chiao, S.M.; Hsiao, B.S. Water soluble complexes of chitosan-g-MPEG and hyaluronic acid. *J. Biomed. Mater. Res. Part A* **2007**, *80*, 800–812. [CrossRef]
72. Weng, L.; Chen, X.; Chen, W. Rheological characterization of in situ crosslinkable hydrogels formulated from oxidized dextran and N-carboxyethyl chitosan. *Biomacromolecules* **2007**, *8*, 1109–1115. [CrossRef] [PubMed]
73. Purcell, B.P.; Lobb, D.; Charati, M.B.; Dorsey, S.M.; Wade, R.J.; Zellars, K.N.; Doviak, H.; Pettaway, S.; Logdon, C.B.; Shuman, J.A. Injectable and bioresponsive hydrogels for on-demand matrix metalloproteinase inhibition. *Nat. Mater.* **2014**, *13*, 653. [CrossRef] [PubMed]
74. Deng, Y.; Ren, J.; Chen, G.; Li, G.; Wu, X.; Wang, G.; Gu, G.; Li, J. Injectable in situ cross-linking chitosan-hyaluronic acid based hydrogels for abdominal tissue regeneration. *Sci. Rep.* **2017**, *7*, 2699. [CrossRef] [PubMed]
75. Li, L.; Wang, N.; Jin, X.; Deng, R.; Nie, S.; Sun, L.; Wu, Q.; Wei, Y.; Gong, C. Biodegradable and injectable in situ cross-linking chitosan-hyaluronic acid based hydrogels for postoperative adhesion prevention. *Biomaterials* **2014**, *35*, 3903–3917. [CrossRef]
76. Huber, D.; Grzelak, A.; Baumann, M.; Borth, N.; Schleining, G.; Nyanhongo, G.S.; Guebitz, G.M. Anti-inflammatory and anti-oxidant properties of laccase-synthesized phenolic-O-carboxymethyl chitosan hydrogels. *New Biotechnol.* **2018**, *40*, 236–244. [CrossRef]
77. Caliari, S.R.; Burdick, J.A. A practical guide to hydrogels for cell culture. *Nat. Methods* **2016**, *13*, 405. [CrossRef]

78. Xin, Y.; Yuan, J. Schiff's base as a stimuli-responsive linker in polymer chemistry. *Polym. Chem.* **2012**, *3*, 3045–3055. [CrossRef]
79. Zhang, Z.P.; Rong, M.Z.; Zhang, M.Q. Polymer engineering based on reversible covalent chemistry: A promising innovative pathway towards new materials and new functionalities. *Prog. Polym. Sci.* **2018**, *80*, 39–93. [CrossRef]
80. Peng, X.; Peng, Y.; Han, B.; Liu, W.; Zhang, F.; Linhardt, R.J. Stimulated crosslinking of catechol-conjugated hydroxyethyl chitosan as a tissue adhesive. *J. Biomed. Mater. Res. Part B Appl. Biomater.* **2018**, *107*, 582–593. [CrossRef]
81. Barbucci, R.; Magnani, A.; Consumi, M. Swelling behavior of carboxymethylcellulose hydrogels in relation to cross-linking, pH, and charge density. *Macromolecules* **2000**, *33*, 7475–7480. [CrossRef]
82. Mercê, A.L.R.; Carrera, L.C.M.; Romanholi, L.K.S.; Recio, M.a.Á.L. Aqueous and solid complexes of iron (III) with hyaluronic acid: Potentiometric titrations and infrared spectroscopy studies. *J. Inorg. Biochem.* **2002**, *89*, 212–218. [CrossRef]
83. Gu, Q.; Tomaskovic-Crook, E.; Lozano, R.; Chen, Y.; Kapsa, R.M.; Zhou, Q.; Wallace, G.G.; Crook, J.M. Functional 3D neural mini-tissues from printed gel-based bioink and human neural stem cells. *Adv. Healthc. Mater.* **2016**, *5*, 1429–1438. [CrossRef] [PubMed]

 © 2020 by the authors. Licensee MDPI, Basel, Switzerland. This article is an open access article distributed under the terms and conditions of the Creative Commons Attribution (CC BY) license (http://creativecommons.org/licenses/by/4.0/).

Article

Development of Biocomposite Polymeric Systems Loaded with Antibacterial Nanoparticles for the Coating of Polypropylene Biomaterials

Mar Fernández-Gutiérrez [1,2], Bárbara Pérez-Köhler [2,3,4,*], Selma Benito-Martínez [2,4,5], Francisca García-Moreno [2,4,5], Gemma Pascual [2,3,4,*], Luis García-Fernández [1,2], María Rosa Aguilar [1,2], Blanca Vázquez-Lasa [1,2] and Juan Manuel Bellón [2,4,5]

[1] Institute of Polymer Science and Technology, Spanish National Research Council (ICTP-CSIC), 28006 Madrid, Spain; mar.fernandez.gutierrez@csic.es (M.F.-G.); luis.garcia@csic.es (L.G.-F.); mraguilar@ictp.csic.es (M.R.A.); bvazquez@ictp.csic.es (B.V.-L.)
[2] Biomedical Networking Research Centre on Bioengineering, Biomaterials and Nanomedicine (CIBER-BBN), 28029 Madrid, Spain; selma.benito@uah.es (S.B.-M.); francisca.garciam@uah.es (F.G.-M.); juanm.bellon@uah.es (J.M.B.)
[3] Department of Medicine and Medical Specialties, University of Alcalá, 28805 Madrid, Spain
[4] Ramón y Cajal Health Research Institute (IRYCIS), 28034 Madrid, Spain
[5] Department of Surgery, Medical and Social Sciences, University of Alcalá, 28805 Madrid, Spain
* Correspondence: barbara.perez@uah.es (B.P.-K.); gemma.pascual@uah.es (G.P.)

Received: 15 July 2020; Accepted: 13 August 2020; Published: 15 August 2020

Abstract: The development of a biocomposite polymeric system for the antibacterial coating of polypropylene mesh materials for hernia repair is reported. Coatings were constituted by a film of chitosan containing randomly dispersed poly(D,L-lactide-co-glycolide) (PLGA) nanoparticles loaded with chlorhexidine or rifampicin. The chlorhexidine-loaded system exhibited a burst release during the first day reaching the release of the loaded drug in three or four days, whereas rifampicin was gradually released for at least 11 days. Both antibacterial coated meshes were highly active against *Staphylococcus aureus* and *Staphylococcus epidermidis* (10^6 CFU/mL), displaying zones of inhibition that lasted for 7 days (chlorhexidine) or 14 days (rifampicin). Apparently, both systems inhibited bacterial growth in the surrounding environment, as well as avoided bacterial adhesion to the mesh surface. These polymeric coatings loaded with biodegradable nanoparticles containing antimicrobials effectively precluded bacterial colonization of the biomaterial. Both biocomposites showed adequate performance and thus could have potential application in the design of antimicrobial coatings for the prophylactic coating of polypropylene materials for hernia repair.

Keywords: biocomposite; chitosan; chlorhexidine; coating; hernia; mesh infection; nanoparticles; PLGA; polypropylene; rifampicin

1. Introduction

In hernia repair, the implantation of biomaterials constitutes the main surgical strategy aimed at reinforcing the abdominal wall, providing a tension-free repair that restores the biological, biomechanical, and functional properties of the damaged tissue [1]. Among the numerous advantages afforded by these devices, the use of hernia mesh materials allows for the reduction of recurrence rates and recovery period [2]. Nevertheless, mesh implantation is not exempt from drawbacks, being that infection is one of the most severe postoperative complications having a big impact on patients and the healthcare system [3]. Many factors influence the incidence of mesh-related infection, such as the type of hernia pathology, the surgical procedure carried out, the prostheses implanted, and the patient co-morbidities among the most relevant [4].

Typically, biomaterial-related infections are triggered when bacteria invade the wound and adhere to the surface of the implanted mesh and surrounding tissues. Once adhered, bacteria proliferate and interact with others, forming clusters that colonize the implant [5]. Many of these bacteria secrete an extracellular polysaccharide matrix that covers the cluster and strengthen their adhesion to the surface, establishing a biofilm [6]. Protected by this barrier, bacteria are strongly attached to the substrate and exert high resistance to the action of antibiotics and host defense cells [7], events that make biofilm-based infections extremely difficult to treat. Most of these infections, either involving biofilms or not, are caused by *Staphylococcus aureus* and *S. epidermidis* [8].

Considering that bacterial adhesion to the implant surface plays a key role in the pathogenesis of mesh-related infections, there is a growing interest in developing novel surgical meshes endowed with antibacterial properties. In this sense, coating meshes with drug-loaded polymeric systems is currently one of the main approaches being evaluated by researchers [1,9]. In recent years, numerous laboratory and preclinical studies demonstrated that antimicrobial coatings could reduce mesh-related infections by inhibiting bacterial adhesion, implant colonization, and biofilm formation [3,10,11].

Different antimicrobials have been used to endow hernia mesh materials with antibacterial properties. Silver metal [12] or antibiotics such as gentamicin [13], vancomycin [14], and rifampicin–minocycline combinations [15] are among the most widely used drugs for hernia mesh coating. Although exerting a potent antibacterial activity, it is widely known that antibiotics carry the risk of provoking bacterial resistances, thus, the prophylactic use of these biocides should be reduced to a minimum. Antiseptic agents emerge as a plausible alternative to antibiotics for these purposes. From the wide variety of antiseptics available, some have been used to coat mesh materials with promising results, being a potential tool for developing antimicrobial devices. Such is the case of broad-spectrum antiseptics like triclosan [16] and chlorhexidine [17] both of which are currently being applied in the manufacturing of antimicrobial surgical devices for clinical use [11].

The simplest way to coat a mesh is via dipping or soaking the device in aqueous solutions containing the target antimicrobial immediately before its implantation. Although this strategy is often applied in human clinics [18], it fails to provide controlled levels of the drug at the surgical site. This limitation can be alleviated if the mesh is coated with a polymeric compound loaded with antimicrobials. Theoretically, antibacterial mesh polymeric coatings must provide a local, sustained, and slow drug release at the implant site, neither producing local nor systemic detrimental effects due to the diffusion of the antimicrobial [14]. Furthermore, coatings must be non-toxic and biocompatible, ensuring adequate mesh integration into the host tissue without compromising the biomechanical properties of the implant [19]. Over the last decade, many polymeric systems to be used as drug carriers were developed and applied for the prophylactic coating of hernia repair materials, with promising outcomes [11,20]. From this plethora of candidates, the polymeric systems loaded with drug releasing nanoparticles represent one of the most effective approaches to endow these devices with antibacterial activity [5].

In the present study, we developed a novel antimicrobial polymeric coating loaded with biodegradable nanoparticles containing chlorhexidine or rifampicin for hernia mesh materials. The nanoparticles were loaded with two different antimicrobials, chlorhexidine or rifampicin, as candidates for antiseptic and antibiotic agents, respectively. The former ($C_{22}H_{30}Cl_2N_{10}$) is a cationic biguanide antiseptic widely used as a preoperative skin disinfectant [21]. The latter ($C_{43}H_{58}N_4O_{12}$) is considered one of the most potent semisynthetic antibiotics to treat bacterial infections [22]. Both drugs were selected due to their antimicrobial activity as well as their promising application with biomedical devices. The morphology, drug release profile, cytocompatibility, and antimicrobial performance of coated polypropylene meshes was evaluated in vitro using a model of bacterial mesh infection caused by *S. aureus* and *S. epidermidis*.

2. Materials and Methods

2.1. Chemicals

Commercial reagents and antimicrobials used for the coating preparation were: chitosan (medium molecular weight, deacetylation degree 75%; Sigma-Aldrich, St. Louis, MO, USA); dichloromethane, DCM (Sigma-Aldrich); poly(D,L-lactide-*co*-glycolide), PLGA, Resomer RG503 (50/50; Evonik, Essen, Germany); chlorhexidine, CHX (Sigma-Aldrich); and rifampicin, RIF (Sigma-Aldrich).

2.2. Coating Preparation

The elaboration of the polymeric coatings was conducted at room temperature. For the establishment of the unloaded polymer, chitosan was dissolved in 1% acetic acid with continuous magnetic stirring for 24 h at a concentration of 1.0% w/v. Then, PLGA was dissolved in DCM (25% w/w vs. chitosan) and subsequently loaded into a 1 mL syringe. This PLGA solution was added to the chitosan solution drop by drop using a 23-gauge needle, and the obtained chitosan–PLGA dispersion was stirred for 12 h for evaporation of the organic solvent DCM. In the case of the drug-loaded polymers, CHX (2% w/w vs. chitosan) or RIF (5% w/w vs. chitosan) were dissolved in DCM and the mixture with the solution of PLGA was added drop by drop slowly, to the chitosan aqueous solution under stirring for 12 h. In this time period, the organic solvent DCM evaporated with the subsequent formation of solid nanoparticles containing the biodegradable polymer and the corresponding antiseptic or antibiotic drug as a nanodispersion in the original aqueous solution of chitosan, which was applied for the homogeneous coating of the commercial mesh. According to the amount ratio of PLGA and drug, CHX nanoparticles contained 10% (w/w) of the drug, whereas those prepared with RIF contained 15% (w/w) of the bioactive drug.

2.3. Mesh Coating and Study Groups

A lightweight, monofilament, polypropylene (PP) mesh material was used (Optilene Mesh Elastic; B. Braun, Melsungen, Germany). Under sterile conditions, the mesh was cut into 1 cm^2 squares that were subsequently coated with the different polymeric compounds, by means of the casting technique. Using a Pasteur pipette, 300 µL of the corresponding compound were dropped onto each mesh, and coated samples were dried at room temperature for 24 h. The coated meshes were sterilized with ethylene oxide at 40 °C for 20 h. Four study groups were established:

- Control (uncoated PP).
- Pol (PP coated with unloaded polymer).
- Pol-CHX (PP coated with CHX-loaded polymer).
- Pol-RIF (PP coated with RIF-loaded polymer).

2.4. Visualization of the PLGA-Nanoparticles Loaded Biocomposites

The size and distribution of the nanoparticles from the original dispersion of nanoparticles in the solution of chitosan was analyzed under scanning electron microscopy (SEM), via the deposition of a few drops on the sample holder of the microscope. Moreover, samples from the different study groups ($n = 3$ each) were used to evaluate the homogeneity and distribution of the polymeric coating. The meshes were dehydrated in an increasing graded ethanol series (30%, 50%, 70%, 90%, and 100%, incubation for 15 min each), desiccated in a Polaron CPD7501 (Fisons Instruments, Ipswich, UK), gold-palladium coated, and visualized in a Zeiss DSM950 SEM (Carl Zeiss, Oberkochen, Germany).

2.5. Drug Release

Samples from the different study groups ($n = 3$ each) were transferred to vials containing 5 mL of phosphate buffer saline (PBS, pH 7.4) at 37 °C. At different time intervals in a 14-day period, aliquots were taken for further analysis, and the vials were refilled with the same volume of fresh PBS.

The collected aliquots were mixed with the mobile phase (1:1), consisting of a solution of acetonitrile: water (20:80), and filtered. The drug release was analyzed by means of high-performance liquid chromatography (HPLC) using a Shimadzu SL-200 apparatus (Shimadzu Scientific Instruments Inc., Columbia, SC, USA) with a Kromaphase C18 column (250 mm × 4.6 mm) and eluting rate of 1 mL/min, using a UV detector at 254 and 280 nm for CHX and RIF, respectively. Calibration curves using a free drug were previously obtained with good correlation values (r = 0.9887 for CHX and r = 0.9668 for RIF). Three measurements were taken for each sample. Results were plotted in a graph representing the cumulative percentage drug release over time.

2.6. Cell Viability

A cell viability assay was conducted to determine the in vitro biocompatibility of the designed mesh coatings, by means of the alamarBlue test. The alamarBlue is a colorimetric reagent (active ingredient: resazurin) used to measure cell proliferation and cytotoxicity of eukaryotic cells or bacteria, following exposure to chemicals or drugs; cells exerting a proliferative status will reduce this reagent (converting resazurin to resorufin), which provokes a colorimetric change in the culture medium measurable by spectrophotometry. For this assay, rabbit skin fibroblasts were used, which were isolated by the explant method, as described elsewhere [17]. Biopsies were harvested from the dermis tissue of 3 male New Zealand White rabbits belonging to another study, (experimental protocol approved by the Committee on the Ethics of Animal Experiments of the University of Alcalá, reference PROEX 045/18). Cells were cultured in a controlled humid atmosphere (37 °C, 5% CO_2) in Dulbecco's modified Eagle medium (DMEM) containing 10% fetal bovine serum (FBS) plus 1% antibiotic–antimycotic cocktail for cell culture (Life Technologies, Carlsbad, CA, USA), and visualized in a Zeiss Axiovert 40C phase-contrast inverted microscope (Carl Zeiss, Oberkochen, Germany). Cells were grown in 6-well plates (2.5×10^5 cells per well in 2 mL of DMEM) and incubated overnight under the conditions described above. Next, the medium was renewed, and a mesh fragment of the corresponding study group was transferred to each well (*n* = 9 per group). Following a 24 h incubation period under the same conditions, the medium was discarded, wells were washed in Hank's balanced salt solution (Life Technologies) and 2 mL of FBS-free DMEM supplemented with 10% alamarBlue viability reagent (Bio-Rad, Hercules, CA, USA) was added. After 5 h of incubation at 37 °C, several 100 µL aliquots were collected from each well to read absorbance (OD570 and OD600) in an iMark microplate absorbance reader (Bio-Rad, Hercules, CA, USA). Data were analyzed using software provided by the manufacturer. Results were expressed as the mean percentage cell viability.

2.7. Antibacterial Performance of the Biocomposites

To determine the performance of the antibacterial coatings, fragments from the different study groups were inoculated with *S. aureus* (Sa) ATCC25923 and *S. epidermidis* (Se) ATCC35984, purchased from the Spanish Type Culture Collection (Valencia, Spain). In all the experiments carried out, the inoculating dose was 10^6 CFU/mL. Bacterial suspensions were prepared immediately before starting the procedures. To avoid cross-contamination, all the assays described below were developed in an independent manner for Sa and Se.

2.7.1. Elaboration of the Bacterial Suspensions

Bacteria were thawed, spread onto lysogeny broth (LB) agar plates (Biomérieux, Marcy l'Etoile, France) and incubated for 24 h at 37 °C. A colony was transferred to 25 mL of LB broth. Following overnight incubation at 37 °C, the culture was diluted in sterile 0.9% saline to prepare the target inoculum ($1.25–1.5 \times 10^6$ CFU/mL), by spectrophotometry (OD600). The number of viable CFU in each inoculum was determined by the spot plaque method.

2.7.2. Sequential Agar Well Diffusion Test

A variant of the agar well diffusion test was carried out to determine the capacity of the drug-loaded mesh coatings to sequentially create zones of inhibition (ZOIs) over a 14-day period of study. Briefly, several LB agar plates were inoculated with 1 mL of the corresponding bacterial suspension. Then, samples from the different study groups ($n = 6$ per group and strain) were individually placed onto the agar and incubated at 37 °C for 24 h. Following incubation, all the plates were photographed for further ZOIs measurement. Those samples exhibiting ZOIs were transferred to freshly inoculated LB plates and incubated again under the same conditions. For each sample, this process was repeated every 24 h until no ZOI was observed in the agar, or up to day 14, if the ZOIs were still being developed. Once the assay was finished, the pictures previously taken were used to quantify the ZOIs amplitude at every 24-h interval. For this quantification, two perpendicular diameters of the ZOIs were measured, using the ImageJ software (National Institutes of Health, Bethesda, MD, USA). Results were plotted in a graph representing the mean ZOI (mm) of the different study groups over time.

2.7.3. Bacterial Adhesion to the Meshes

The ability of the different coatings to avoid bacterial adhesion to the mesh surface was assessed by sonication. Samples ($n = 6$ per group per strain) were individually immersed in 3 mL of LB broth, in 6-well plates. Then, each well was inoculated with 1 mL of the corresponding inoculum, and plates were incubated at 37 °C for 24 h. Following incubation, the meshes were carefully washed in 1 mL of sterile 0.9% saline to remove the non-adhered bacteria and subsequently transferred to Falcon tubes containing 10 mL of peptone water. The tubes were sonicated for 10 min at 40 KHz in a Bransonic 3800-CPXH ultrasonic bath (Branson Ultrasonics, Danbury, CT, USA). The supernatant in each tube was vortexed for 1 min, serially diluted in peptone water (6 tenfold dilutions), and 100 µL of the supernatant plus serial dilutions were spread on LB agar plates. Following a 24-h incubation at 37 °C, the plates were counted to quantify the bacterial adhesion to the surface of the meshes. Results were expressed as the mean bacterial yields per study group.

2.7.4. Turbidimetric Determination of the Bacterial Growth

Spectrophotometric assays were carried out to determine whether the presence of the coated meshes inhibited the bacterial growth in culture. As previously described, samples from the different groups ($n = 6$ per group per strain) were immersed in 3 mL of LB broth, inoculated with 1 mL of the corresponding bacteria and incubated 37 °C for 24 h. As blank, mesh samples immersed in 3 mL of LB broth plus 1 mL of sterile 0.9% saline were kept under the same conditions. Following incubation, the liquid was aspirated, vortexed, and measured by spectrophotometry (OD600) using an Ultrospec 3100 Pro spectrophotometer (Amersham Biosciences, Little Chalfont, UK). Results were expressed as the mean absorbance recorded for the different study groups.

2.8. Statistical Analysis

The data collected in the different experiments were expressed as the mean ± standard error of the mean. Data were compared by a one-way analysis of variance (ANOVA) and Bonferroni as the post hoc test. All statistical tests were performed using GraphPad Prism 5.0 (GraphPad Software Inc., La Jolla, San Diego, CA, USA). Significance was set at $p < 0.05$.

3. Results

3.1. Bioactive Polymeric Film

The immiscibility of the organic solvent DCM with the diluted acetic acid solution in water provided an excellent tool for the preparation of a biocomposite system, formed by nanoparticles of PLGA loaded with either CHX or RIF dispersed in the diluted aqueous solution of chitosan.

The chitosan solution containing the nanoparticles was a homogeneous dispersion of the particles formed after the evaporation of the volatile organic solvent DCM under the stirring of the dispersion of chitosan, thus avoiding the need of isolating such dispersion before the application of the biocomposite system to the mesh materials. Figure 1 shows a micrograph of a dry film of chitosan containing nanoparticles and it is easy to distinguish that the nanoparticles were homogenously and randomly distributed in all the films analyzed by SEM, with a PLGA-CHX nanoparticle size ranging from 300 to 400 nm on average. A similar result with the PLGA-RIF system was obtained. We observed that after the evaporation of the water and sterilization at 40 °C, the amount of residual DCM in the dry film was practically zero. The architecture of the different coatings with pure chitosan or chitosan charged with nanoparticles was visualized in SEM after the application to the commercial meshes (Figure 2). The uncoated mesh exhibited its characteristic architecture, consisting of a knitted PP monofilament that created large pores with a honeycomb-like structure. Once coated, the surface of this biomaterial turned different. The polymeric compound formed a thin layer that covered the mesh filaments and the pores in a continuous fashion.

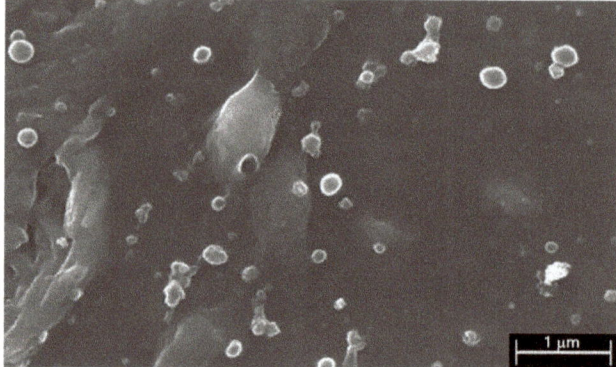

Figure 1. Scanning electron microscopy (SEM) micrograph of PLGA nanoparticles loaded with chlorhexidine, randomly distributed in a chitosan film (scale: 1 μm). The nanoparticles' size ranges from 300 to 400 nm. Similar observations were recorded for the rifampicin-loaded polymer.

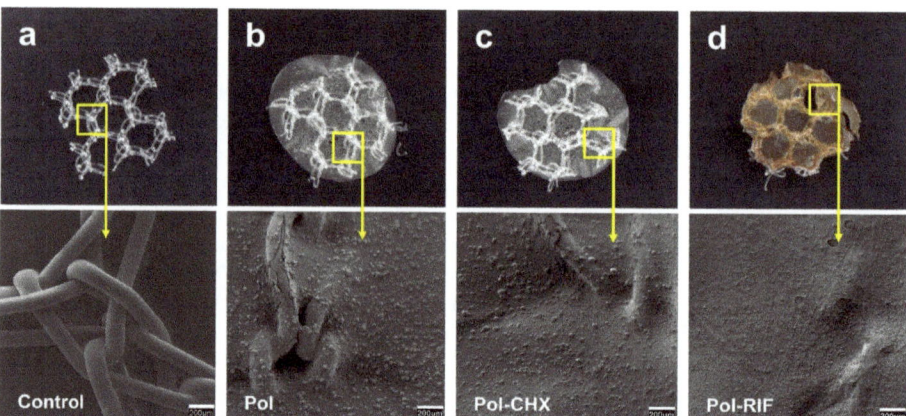

Figure 2. Macroscopic pictures and SEM micrographs (scales: 200 μm) of (**a**) nude control meshes, (**b**) meshes coated with the unloaded biocomposite (Pol), (**c**) meshes coated with chlorhexidine-loaded biocomposite (Pol-CHX), and (**d**) meshes coated with the rifampicin-loaded biocomposite (Pol-RIF).

3.2. Release Kinetics of CHX and RIF

Results from the drug release studies revealed a different response between the two antimicrobials (Figure 3). The antiseptic drug CHX exhibited a moderate burst release during the first 12 h, followed by a gradual release that was extended for 14 days. However, the antibiotic RIF displayed a more gradual release profile, reaching the total release by day 11. No release was detected in the Pol group, revealing the absence of products of degradation generated by the polymeric coating.

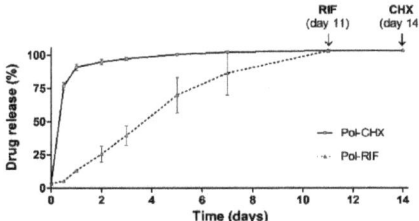

Figure 3. Release profile of CHX and RIF from the chitosan–PLGA nanoparticles polymeric coating over time, following its hydration in PBS ($n = 3$ each). Maximum cumulative release was observed after 11 and 14 days for RIF and CHX, respectively.

3.3. Cell Compatibility of the Biocomposite

Cultured fibroblasts exhibited a typical elongated shape and multilayered growth pattern, showing numerous cells undergoing mitosis. This morphology remained unchanged following exposition to the control, Pol and Pol-RIF meshes (Figure 4), keeping fibroblasts their proliferative status. However, cells influenced by Pol-CHX meshes turned stellate and a diminished proportion of mitotic figures was evidenced. Results from the alamarBlue assays revealed equivalent viability among cells cultured in the presence of either the uncoated (99.51% ± 1.10%), Pol (93.21% ± 1.13%), and Pol-RIF meshes (94.65% ± 1.21%), while fibroblasts exposed to Pol-CHX (70.74% ± 2.69%) showed a significantly lower percentage viability ($p < 0.001$).

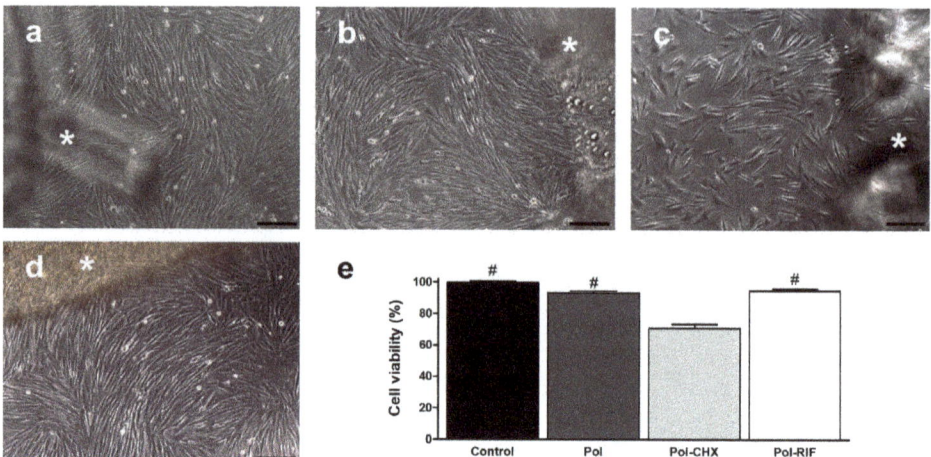

Figure 4. Light-microscopy micrographs (scales: 100 µm) of rabbit skin fibroblast cultured in the presence of a (**a**) control, (**b**) Pol, (**c**) Pol-CHX, or (**d**) Pol-RIF material. In each picture, the corresponding mesh is depicted (asterisks). The cells influenced by Pol-CHX coated meshes (**c**) exhibited slight morphological changes. (**e**) Analysis of cell viability after a 24-h exposure to the different materials ($n = 9$ each). Viability of fibroblasts decreased when cultured with Pol-CHX, compared to the rest of the study groups (#: *** $p < 0.001$).

3.4. Antibacterial Performance of the Biocomposites

3.4.1. Control of the Bacterial Load

The quantification of all the bacterial suspensions used yielded viable CFU values within the expected range. On average, bacterial load from the different inocula were 1.43×10^6 CFU/mL and 1.41×10^6 CFU/mL for Sa and Se, respectively.

3.4.2. Antibacterial Activity over Time

The sequential agar well diffusion test was aimed at determining the ability of the polymeric coatings to exert antibacterial activity over time (Figure 5). As expected, no ZOIs were developed by the control and Pol meshes, while a different response was observed with the antiseptic and the antibiotic-loaded coatings. Regardless of the bacteria inoculated, Pol-CHX meshes developed stable ZOIs that gradually decreased until full depletion following 7 days of the initial inoculation. By contrast, the ZOIs displayed by Pol-RIF were still stable by day 14. The statistical evaluation of the ZOIs amplitude was performed individually for each bacterial strain. When inoculated with Sa, ZOIs from the Pol-RIF meshes were wider than those from the Pol-CHX group, especially at days 1, 2 ($p < 0.001$), 6 ($p < 0.01$), and 7 ($p < 0.01$). Similar findings were collected in the Se-inoculated samples, with statistical relevance at days 1, 2, 3, 6, and 7 ($p < 0.001$).

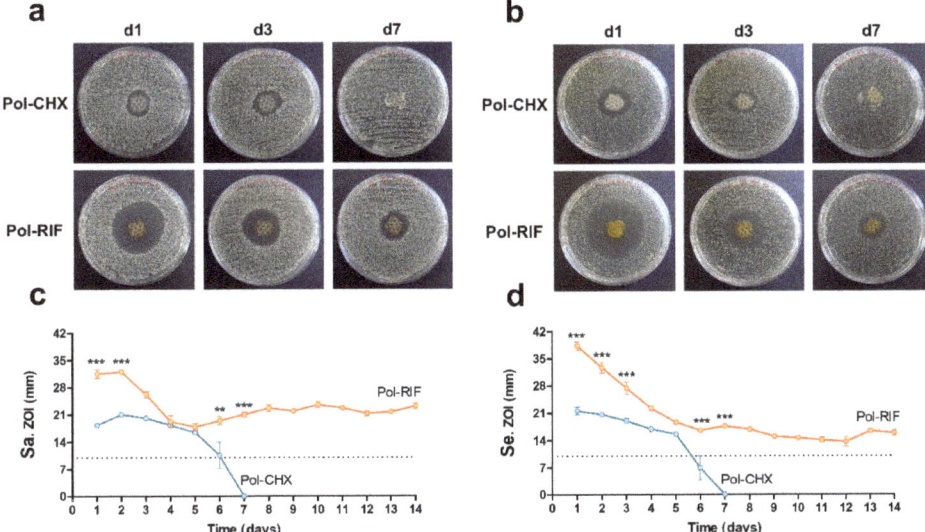

Figure 5. Antibacterial activity over time. Upper panel: Macroscopic pictures of the zones of inhibition (ZOIs) developed by Pol-CHX and Pol-RIF coated meshes at days 1, 3, and 7 of the initial inoculation with (**a**) Sa and (**b**) Se. Lower panel: Representation of the 24-h variation of the ZOIs amplitude of Pol-CHX and Pol-RIF coated meshes challenged with (**c**) Sa and (**d**) Se (** $p < 0.01$; *** $p < 0.001$). The length of the mesh fragments is depicted (dashed lines).

3.4.3. Prevention of Bacterial Adhesion

The quantification of bacteria yielded from the different meshes following sonication (Table 1) revealed a strong adhesion to the surface of the control and Pol materials. For both Sa and Se, loads collected from the uncoated meshes were about 2-log smaller than the yields from the Pol materials ($p < 0.01$). Contrary to this, the antimicrobial coatings were highly effective; out of the total Pol-CHX and Pol-RIF meshes inoculated with either Sa or Se, only one sample belonging to the Pol-CHX group exhibited positive Se counts (1.10×10^3 CFU), while the surface of the rest of fragments was free of bacteria. Furthermore, the bacterial load found in this sample was significantly lower than the yields from the control and Pol groups ($p < 0.01$).

Table 1. Bacterial adhesion to the surface of the different materials. Results were expressed as mean, median, minimum, and maximum bacterial loads (recorded as colony forming units, CFU) yielded per study group and bacteria ($n = 6$ each). The percentage of samples yielding positive counts (%) is provided.

Bacteria	Value	Control	Pol	Pol-CHX	Pol-RIF
Sa	Mean (CFU)	1.09×10^7	3.41×10^9	0	0
	Median (CFU)	9.15×10^6	2.54×10^9	0	0
	Min. (CFU)	5.00×10^6	2.24×10^8	0	0
	Max. (CFU)	2.01×10^7	1.23×10^{10}	0	0
	Positive (%)	100 (6/6)	100 (6/6)	0 (0/6)	0 (0/6)
Se	Mean (CFU)	1.12×10^5	7.08×10^7	1.83×10^2	0
	Median (CFU)	6.00×10^4	6.70×10^7	0	0
	Min. (CFU)	2.00×10^4	1.98×10^7	0	0
	Max. (CFU)	3.60×10^5	1.49×10^8	1.10×10^3	0
	Positive (%)	100 (6/6)	100 (6/6)	16.67 (1/6)	0 (0/6)

3.4.4. Inhibition of Bacterial Growth

Bacterial suspensions containing the different meshes were incubated for 24 h to determine whether the antimicrobial biocomposites did influence the growth pattern of Sa and Se, by means of a turbidimetric assay (Figure 6). Data collected from the absorbance recordings revealed a similar response of Sa and Se to the mesh coatings. For both strains, the absorbance slightly increased in those cultures containing Pol meshes compared to the control ones, being this increase statistically relevant for Sa ($p < 0.05$). In turn, and regardless of the bacterial strain, cultures containing Pol-CHX and Pol-RIF exhibited no turbidity and the values of absorbance significantly decreased in comparison with the rest of the groups ($p < 0.001$).

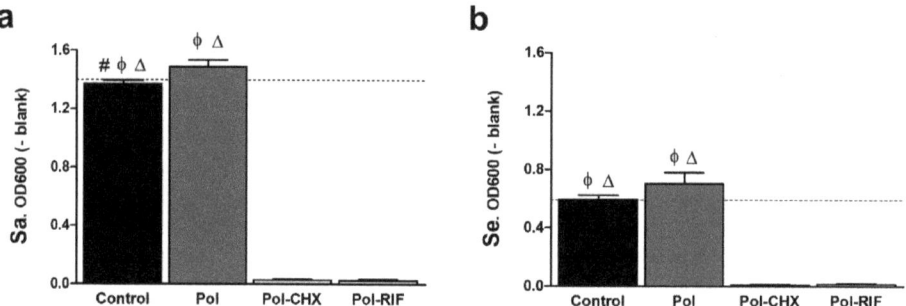

Figure 6. Absorbance (OD600) of (a) Sa and (b) Se cultures influenced by the different materials ($n = 6$ each). The absorbance of bacteria cultured without mesh is depicted (dashed lines). Regardless of the strain, no growth was recorded when bacteria were exposed to the antimicrobial biocomposites. Statistical significance was as follows: #: vs. Pol (* $p < 0.05$); ϕ: vs. Pol-CHX (*** $p < 0.001$); Δ: vs. Pol-RIF (*** $p < 0.001$).

4. Discussion

In general surgery, abdominal wall hernia repair constitutes one of the most frequently developed procedures. According to a recent Cochrane review, approximately 20 million surgeries are being performed per year worldwide [23]. In those patients, bacterial mesh infections increase the morbidity and mortality rates, lengthen the hospital stay and raise the sanitary costs [24]. Hence, the clinical and social consequences of mesh infections are of major concern.

One of the biggest issues associated with this postoperative complication is that currently there are no gold-standard therapies for preventing mesh-related infections [25]. Preoperative surgical antibiotic prophylaxis is common practice, although it fails to avoid infections in many cases [26] and there is no clear consensus among surgeons regarding its administration [27]. In this milieu, the implantation of antimicrobial meshes constitutes a promising option to reduce the incidence of infection.

Antimicrobial prostheses can be developed following different procedures; the most attractive alternative consists of coating the mesh with bioactive polymeric systems. Ideally, these coatings must cover the surface uniformly, providing an adequate drug concentration at the implant site [28]. Moreover, the coating should be biodegradable, facilitating both infiltration of neoformed connective tissue across the mesh pores and mesh integration into the host tissue [14]. Finally, the drug diffusion from the polymer should be gradual and maintained for about 3–4 days [29,30] to keep bactericidal activity over time. The innovative chitosan–PLGA nanoparticles system designed in this study combines all these requisites, especially when loaded with the antibiotic; firstly, observations from the different experiments carried out revealed signs of polymer degradation, suggesting adequate biodegradability of the biocomposite. Secondly, not only is this coating homogeneously distributed throughout the mesh surface, but also allows for a sustained release of rifampicin for up to 11 days. Nevertheless, the meshes coated with the antiseptic-loaded polymer exhibited an initial burst release during the first 24 h. These findings can be related to the chemical nature of chlorhexidine; this molecule is positively charged and has high binding affinity, features that often hinder the dilution of this antiseptic in polymeric systems and drug carriers [31,32].

Together with the chemical composition, the biocide activity of the selected drug is of great importance when developing antibacterial coatings for biomedical devices. The two antimicrobials included in this study were chosen because of their broad-spectrum activity against Gram-positive and Gram-negative bacteria, as well as their potential usefulness to prevent biomaterial-related infection following hernia repair. Although both drugs are effective biocides, the antiseptic and the antibiotic used have different mechanisms of action. On one side, chlorhexidine interacts with the proteins of the bacterial cell wall, provoking the disruption of the cytoplasmic membrane and the leakage of several cytoplasmic components [33]. By comparison, rifampicin provokes genetic damage caused by a strong, specific inhibition of the bacterial DNA-dependent RNA polymerase that hinders the RNA transcription processes [34]. Observations from our study revealed that polymeric films containing these drugs killed bacteria during the first 24 h of exposure, suggesting a fast, almost immediate biocide performance of these biocomposites, regardless of their mode of action.

Given that mesh infection can arise days, weeks, or even months after surgery [25], the development of antibacterial coatings exerting sustained biocide activity over time would be highly advantageous. It has been suggested that implanted antimicrobial devices should keep activity for 2–3 weeks [30]. Consistent with this statement, our rifampicin-loaded biocomposite exhibits biocide activity for at least 2 weeks. Notwithstanding this, the activity of the coating containing chlorhexidine lasted for at about 1 week. Likewise, rifampicin-loaded materials fully inhibited bacterial adhesion to the mesh surface, while the chlorhexidine-loaded ones allowed mild adhesion of *S. epidermidis* in just one sample. Together, these results reveal a stronger effect and better long-term activity of the antibiotic compared to the antiseptic bioactive system. To the best of our knowledge, no previous data were reported comparing the effect of these two drugs relative to hernia mesh materials, except for another study recently carried out by us [35]. In alignment with this, a clinical trial on the use of antibacterial central venous catheters to reduce bloodstream infections revealed better performance of those devices containing rifampicin compared to others provided with chlorhexidine [36].

Biocompatibility of the polymeric system is another key feature to be considered when designing antimicrobial coatings. In a setting of infection, mesh integration can be hampered if bacteria win the so-called "race for the surface" and colonize the implant before the host cells [37]. Biocompatible devices enhance the interaction of host cells with the implant promoting tissue integration [30,38], thereby indirectly reducing the establishment of infections. Recently, PLGA nanoparticles modified with chitosan were reported as effective drug carriers given their great biocompatibility and sustained release [39]. Our findings align with these observations and show that the biocomposites, especially those provided with rifampicin, exert no detrimental effects on cultured cells.

Given that rifampicin specifically targets the bacterial DNA-dependent RNA polymerase, we confirmed that, at the concentration tested, this drug does not provoke any side effect on fibroblasts in terms of cell viability. However, cytotoxicity of chlorhexidine to eukaryotic cells is well known [33,40]. It was demonstrated that this antiseptic could reduce the growth pattern of cultured cells, as well as provoke morphological modifications on fibroblasts [41], which corroborates our cell viability results. The main reason for this effect is that chlorhexidine can also disrupt the cytoplasmic membrane of eukaryotic cells, even when used at very low concentrations [42]. Together, these findings suggest a potential limitation of this antiseptic for further biomedical applications. In previous studies, our group managed to reduce the toxicity of this antiseptic via incorporation into a quaternary-ammonium-based polymeric compound [17]. With the development of this chitosan–PLGA polymeric system we have optimized the cell response to chlorhexidine, producing a more efficient antimicrobial mesh coating that exerts reduced toxicity while keeping a proper antibacterial activity.

In summary, the results described in the present study demonstrate that a chitosan–PLGA polymeric system can be loaded with antimicrobials to develop a bioactive coating for hernia repair mesh materials. Although the antibiotic rifampicin exhibits an optimal effect, the incorporation of an antiseptic agent such as chlorhexidine also shows an adequate response. Both coatings exert a desirable drug release profile, cell compatibility, or antimicrobial activity. The development of in vivo preclinical studies would provide further data regarding the potential application of these antimicrobial biocomposites to reduce postoperative infections following hernia repair. These steps will be carried out in the next future.

5. Conclusions

Polypropylene mesh materials for hernia repair can be coated with a bioactive chitosan–PLGA nanoparticle polymeric system loaded with either an antibiotic (rifampicin) or an antiseptic (chlorhexidine). Both biocomposite coatings effectively inhibited the bacterial colonization of the mesh without compromising the viability of eukaryotic cells. Rifampicin exerted slightly better performance than chlorhexidine in terms of drug release kinetics, cell compatibility, and antibacterial activity over time. The prophylactic application of these coatings offers a potential strategy to protect meshes from bacterial adhesion following implantation.

Author Contributions: Conceptualization, B.P.-K., G.P., J.M.B. and M.R.A.; methodology, B.P.-K., G.P., M.F.-G. and S.B.-M.; software, B.V.-L.; F.G.-M. and L.G.-F.; validation, B.V.-L., G.P., J.M.B. and M.F.-G.; formal analysis, B.P.-K., B.V.-L., M.F.-G. and S.B.-M.; investigation, B.V.-L., F.G.-M. and L.G.-F.; resources, F.G.-M., G.P. and L.G.-F.; data curation, B.P.-K., F.G.-M., M.F.-G. and S.B.-M., writing—original draft preparation, B.P.-K., G.P., J.M.B. and M.F.-G.; writing—review and editing, B.P.K., G.P., L.G.-F. and M.F.-G.; visualization, B.P.-K., L.G.-F., M.F.-G. and S.B.-M.; supervision, G.P., J.M.B. and M.R.A.; project administration, B.P.-K.; G.P., J.M.B. and M.R.A.; funding acquisition, G.P., J.M.B. and M.R.A. All authors have read and agreed to the published version of the manuscript.

Funding: The study was supported by Grants SAF2017-89481-P and MAT2017-84277-R from the Spanish Ministry of Science, Innovation and Universities. Financial support from the CIBER-BBN is acknowledged.

Acknowledgments: The study was supported by Grants SAF2017-89481-P and MAT2017-84277-R from the Spanish Ministry of Science, Innovation and Universities. Financial support from the CIBER-BBN is acknowledged (CIBER-BBN Internal Collaborations 2018-2021: "Functionalization of polymer coatings using antimicrobial agents for prevention of mesh infection following hernia surgery" and "Polymeric nanoparticles for the treatment of bacterial diseases").

Conflicts of Interest: The authors declare no conflict of interest.

References

1. Kalaba, S.; Gerhard, E.; Winder, J.S.; Pauli, E.M.; Haluck, R.S.; Yang, J. Design strategies and applications of biomaterials and devices for hernia repair. *Bioact. Mater.* **2016**, *1*, 2–17. [CrossRef] [PubMed]
2. Rastegarpour, A.; Cheung, M.; Vardhan, M.; Ibrahim, M.M.; Butler, C.E.; Levinson, H. Surgical mesh for ventral incisional hernia repairs: Understanding mesh design. *Plast. Surg. (Oakv)* **2016**, *24*, 41–50. [CrossRef] [PubMed]
3. Falagas, M.E.; Kasiakou, S.K. Mesh-related infections after hernia repair surgery. *Clin. Microbiol. Infect.* **2005**, *11*, 3–8. [CrossRef] [PubMed]
4. Montgomery, A.; Kallinowski, F.; Köckerling, F. Evidence for replacement of an infected synthetic by a biological mesh in abdominal wall hernia repair. *Front. Surg.* **2016**, *2*, 67. [CrossRef] [PubMed]
5. De Miguel, I.; Prieto, I.; Albornoz, A.; Sanz, V.; Weis, C.; Turon, P.; Quidant, R. Plasmon-based biofilm inhibition on surgical implants. *Nano. Lett.* **2019**, *19*, 2524–2529. [CrossRef] [PubMed]
6. Arciola, C.R.; Campoccia, D.; Montanaro, L. Implant infections: Adhesion, biofilm formation and immune evasion. *Nat. Rev. Microbiol.* **2018**, *16*, 397–409. [CrossRef] [PubMed]
7. Veerachamy, S.; Yarlagadda, T.; Manivasagam, G.; Yarlagadda, P.K. Bacterial adherence and biofilm formation on medical implants: A review. *Proc. Inst. Mech. Eng. H* **2014**, *228*, 1083–1099. [CrossRef]
8. Jacombs, A.S.W.; Karatassas, A.; Klosterhalfen, B.; Richter, K.; Patiniott, P.; Hensman, C. Biofilms and effective porosity of hernia mesh: Are they silent assassins? *Hernia* **2020**, *24*, 197–204. [CrossRef]
9. Blatnik, J.A.; Thatiparti, T.R.; Krpata, D.M.; Zuckerman, S.T.; Rosen, M.J.; von Recum, H.A. Infection prevention using affinity polymer-coated, synthetic meshes in a pig hernia model. *J. Surg. Res.* **2017**, *219*, 5–10. [CrossRef]
10. Pérez-Köhler, B.; Bayon, Y.; Bellón, J.M. Mesh infection and hernia repair: A review. *Surg. Infect.* **2016**, *17*, 124–137. [CrossRef]
11. Guillaume, O.; Pérez-Tanoira, R.; Fortelny, R.; Redl, H.; Moriarty, T.F.; Richards, R.G.; Eglin, D.; Petter Puchner, A. Infections associated with mesh repairs of abdominal wall hernias: Are antimicrobial biomaterials the longed-for solution? *Biomaterials* **2018**, *167*, 15–31. [CrossRef] [PubMed]
12. Nergiz Adıgüzel, E.; Esen, E.; Aylaz, G.; Keskinkılıç Yağız, B.; Kıyan, M.; DoğanÜnal, A.E. Do nano-crystalline silver-coated hernia grafts reduce infection? *World J. Surg.* **2018**, *42*, 3537–3542. [CrossRef] [PubMed]
13. Binnebösel, M.; von Trotha, K.T.; Ricken, C.; Klink, C.D.; Junge, K.; Conze, J.; Jansen, M.; Neumann, U.P.; Lynen Jansen, P. Gentamicin supplemented polyvinylidenfluoride mesh materials enhance tissue integration due to a transcriptionally reduced MMP-2 protein expression. *BMC Surg.* **2012**, *12*, 1. [CrossRef] [PubMed]
14. Fernández-Gutiérrez, M.; Olivares, E.; Pascual, G.; Bellón, J.M.; San Román, J. Low-density polypropylene meshes coated with resorbable and biocompatible hydrophilic polymers as controlled release agents of antibiotics. *Acta Biomater.* **2013**, *9*, 6006–6018. [CrossRef] [PubMed]
15. Majumder, A.; Scott, J.R.; Novitsky, Y.W. Evaluation of the antimicrobial efficacy of a novel rifampin/minocycline-coated, noncrosslinked porcine acellular dermal matrix compared with uncoated scaffolds for soft tissue repair. *Surg. Innov.* **2016**, *23*, 442–455. [CrossRef] [PubMed]
16. Cakmak, A.; Cirpanli, Y.; Bilensoy, E.; Yorganci, K.; Caliş, S.; Saribaş, Z.; Kaynaroğlu, V. Antibacterial activity of triclosan chitosan coated graft on hernia graft infection model. *Int. J. Pharm.* **2009**, *381*, 214–219. [CrossRef]
17. Pérez-Köhler, B.; Fernández-Gutiérrez, M.; Pascual, G.; García-Moreno, F.; San Román, J.; Bellón, J.M. In vitro assessment of an antibacterial quaternary ammonium-based polymer loaded with chlorhexidine for the coating of polypropylene prosthetic meshes. *Hernia* **2016**, *20*, 869–878. [CrossRef]
18. Yabanoğlu, H.; Arer, İ.M.; Çalıskan, K. The effect of the use of synthetic mesh soaked in antibiotic solution on the rate of graft infection in ventral hernias: A prospective randomized study. *Int. Surg.* **2015**, *100*, 1040–1047. [CrossRef]
19. Harth, K.C.; Rosen, M.J.; Thatiparti, T.R.; Jacobs, M.R.; Halaweish, I.; Bajaksouzian, S.; Furlan, J.; von Recum, H.A. Antibiotic-releasing mesh coating to reduce prosthetic sepsis: An in vivo study. *J. Surg. Res.* **2010**, *163*, 337–343. [CrossRef]
20. Majumder, A.; Neupane, R.; Novitsky, Y.W. Antibiotic coating of hernia meshes: The next step toward preventing mesh infection. *Surg. Technol. Int.* **2015**, *27*, 147–153.

21. Chen, S.; Chen, J.W.; Guo, B.; Xu, C.C. Preoperative antisepsis with chlorhexidine versus povidone-iodine for the prevention of surgical site infection: A systematic review and meta-analysis. *World J. Surg.* **2020**, *44*, 1412–1424. [CrossRef] [PubMed]
22. Chai, W.M.; Lin, M.Z.; Song, F.J.; Wang, Y.X.; Xu, K.L.; Huang, J.X.; Fu, J.P.; Peng, Y.Y. Rifampicin as a novel tyrosinase inhibitor: Inhibitory activity and mechanism. *Int. J. Biol. Macromol.* **2017**, *102*, 425–430. [CrossRef] [PubMed]
23. Orelio, C.C.; van Hessen, C.; Sánchez-Manuel, F.J.; Aufenacker, T.J.; Scholten, R.J. Antibiotic prophylaxis for prevention of postoperative wound infection in adults undergoing open elective inguinal or femoral hernia repair. *Cochrane Database Syst. Rev.* **2020**, *4*, CD003769. [CrossRef] [PubMed]
24. Engelsman, A.F.; van der Mei, H.C.; Ploeg, R.J.; Busscher, H.J. The phenomenon of infection with abdominal wall reconstruction. *Biomaterials* **2007**, *28*, 2314–2327. [CrossRef] [PubMed]
25. Grafmiller, K.T.; Zuckerman, S.T.; Petro, C.; Liu, L.; von Recum, H.A.; Rosen, M.J.; Korley, J.N. Antibiotic-releasing microspheres prevent mesh infection in vivo. *J. Surg. Res.* **2016**, *206*, 41–47. [CrossRef] [PubMed]
26. Binnebösel, M.; von Trotha, K.T.; Jansen, P.L.; Conze, J.; Neumann, U.P.; Junge, K. Biocompatibility of prosthetic meshes in abdominal surgery. *Semin. Immunopathol.* **2011**, *33*, 235–243. [CrossRef] [PubMed]
27. MacCormick, A.P.; Akoh, J.A. Survey of surgeons regarding prophylactic antibiotic use in inguinal hernia repair. *Scand. J. Surg.* **2018**, *107*, 208–211. [CrossRef]
28. Labay, C.; Canal, J.M.; Modic, M.; Cvelbar, U.; Quiles, M.; Armengol, M.; Arbos, M.A.; Gil, F.J.; Canal, C. Antibiotic-loaded polypropylene surgical meshes with suitable biological behaviour by plasma functionalization and polymerization. *Biomaterials* **2015**, *71*, 132–144. [CrossRef]
29. Guillaume, O.; Lavigne, J.P.; Lefranc, O.; Nottelet, B.; Coudane, J.; Garric, X. New antibiotic-eluting mesh used for soft tissue reinforcement. *Acta Biomater.* **2011**, *7*, 3390–3397. [CrossRef]
30. Busscher, H.J.; van der Mei, H.C.; Subbiahdoss, G.; Jutte, P.C.; van den Dungen, J.J.; Zaat, S.A.; Schultz, M.J.; Grainger, D.W. Biomaterial-associated infection: Locating the finish line in the race for the surface. *Sci. Transl. Med.* **2012**, *4*, 153rv10. [CrossRef]
31. Lee, D.Y.; Spångberg, L.S.; Bok, Y.B.; Lee, C.Y.; Kum, K.Y. The sustaining effect of three polymers on the release of chlorhexidine from a controlled release drug device for root canal disinfection. *Oral. Surg. Oral. Med. Oral. Pathol. Oral. Radiol. Endod.* **2005**, *100*, 105–111. [CrossRef] [PubMed]
32. Greenhalgh, R.; Dempsey-Hibbert, N.C.; Whitehead, K.A. Antimicrobial strategies to reduce polymer biomaterial infections and their economic implications and considerations. *Int. Biodeter. Biodegr.* **2019**, *136*, 1–14. [CrossRef]
33. Lim, K.S.; Kam, P.C. Chlorhexidine: Pharmacology and clinical applications. *Anaesth. Intensive Care* **2008**, *36*, 502–512. [CrossRef] [PubMed]
34. Artsimovitch, I.; Vassylyeva, M.N.; Svetlov, D.; Svetlov, V.; Perederina, A.; Igarashi, N.; Matsugaki, N.; Wakatsuki, S.; Tahirov, T.H.; Vassylyev, D.G. Allosteric modulation of the RNA polymerase catalytic reaction is an essential component of transcription control by rifamycins. *Cell* **2005**, *122*, 351–363. [CrossRef] [PubMed]
35. Pérez-Köhler, B.; Linardi, F.; Pascual, G.; Bellón, J.M.; Eglin, D.; Guillaume, O. Efficacy of antimicrobial agents delivered to hernia meshes using an adaptable thermo-responsive hyaluronic acid-based coating. *Hernia* **2019**, in press. [CrossRef]
36. Bonne, S.; Mazuski, J.E.; Sona, C.; Schallom, M.; Boyle, W.; Buchman, T.G.; Bochicchio, G.V.; Coopersmith, C.M.; Schuerer, D.J. Effectiveness of minocycline and rifampin vs. chlorhexidine and silver sulfadiazine-impregnated central venous catheters in preventing central line-associated bloodstream infection in a high-volume academic intensive care unit: A before and after trial. *J. Am. Coll. Surg.* **2015**, *221*, 739–747. [CrossRef]
37. Pham, V.T.; Truong, V.K.; Orlowska, A.; Ghanaati, S.; Barbeck, M.; Booms, P.; Fulcher, A.J.; Bhadra, C.M.; Buividas, R.; Baulin, V.; et al. "Race for the surface": Eukaryotic cells can win. *ACS Appl. Mater. Interfaces* **2016**, *8*, 22025–22031. [CrossRef]
38. Baylón, K.; Rodríguez-Camarillo, P.; Elías-Zúñiga, A.; Díaz-Elizondo, J.A.; Gilkerson, R.; Lozano, K. Past, present and future of surgical meshes: A review. *Membranes (Basel)* **2017**, *7*, 47. [CrossRef]
39. Lu, B.; Lv, X.; Le, Y. Chitosan-modified PLGA nanoparticles for control-released drug delivery. *Polymers (Basel)* **2019**, *11*, 304. [CrossRef]

40. Burroughs, L.; Ashraf, W.; Singh, S.; Martínez-Pomares, L.; Bayston, R.; Hook, A.L. Development of dual anti-biofilm and anti-bacterial medical devices. *Biomater. Sci.* **2020**, *8*, 3926–3934. [CrossRef]
41. Wyganowska-Swiatkowska, M.; Kotwicka, M.; Urbaniak, P.; Nowak, A.; Skrzypczak-Jankun, E.; Jankun, J. Clinical implications of the growth-suppressive effects of chlorhexidine at low and high concentrations on human gingival fibroblasts and changes in morphology. *Int. J. Mol. Med.* **2016**, *37*, 1594–1600. [CrossRef] [PubMed]
42. Liu, J.X.; Werner, J.; Kirsch, T.; Zuckerman, J.D.; Virk, M.S. Cytotoxicity evaluation of chlorhexidine gluconate on human fibroblasts, myoblasts, and osteoblasts. *J. Bone Jt. Infect.* **2018**, *3*, 165–172. [CrossRef] [PubMed]

© 2020 by the authors. Licensee MDPI, Basel, Switzerland. This article is an open access article distributed under the terms and conditions of the Creative Commons Attribution (CC BY) license (http://creativecommons.org/licenses/by/4.0/).

MDPI
St. Alban-Anlage 66
4052 Basel
Switzerland
Tel. +41 61 683 77 34
Fax +41 61 302 89 18
www.mdpi.com

Polymers Editorial Office
E-mail: polymers@mdpi.com
www.mdpi.com/journal/polymers